3D Virtual Reconstruction and Visualization of Complex Architectures (3D-ARCH 2015)

ISPRS Archives Volume XL-5/W4, 2015

Avila, Spain
25-27 February 2015

Editors:

D. Gonzalez-Aguilera
F. Remondino
J. Boehm
T. Kersten
T. Fuse

ISBN: 978-1-5108-2366-2

Printed from e-media with permission by:

Curran Associates, Inc.
57 Morehouse Lane
Red Hook, NY 12571

Some format issues inherent in the e-media version may also appear in this print version.

Printed by Curran Associates, Inc. (2016)

International Society for Photogrammetry and Remote Sensing (ISPRS)
Attn: Chen Jun
National Geomatics Centre of China
28 Lianhuachixi Road Haidian
Beijing 100830 China

Phone: 86 10 63 88 1102
Fax: 86 10 68 42 4101

chenjun@nsdi.gov.cn

Additional copies of this publication are available from:

Curran Associates, Inc.
57 Morehouse Lane
Red Hook, NY 12571 USA
Phone: 845-758-0400
Fax: 845-758-2633
Email: curran@proceedings.com
Web: www.proceedings.com

TABLE OF CONTENTS

3D MODEL OF AL ZUBARAH FORTRESS IN QATAR – TERRESTRIAL LASER SCANNING VS. DENSE IMAGE MATCHING

T. KERSTEN [a, *], K. MECHELKE [b] and L. MAZIULL [a]

[a] HafenCity University Hamburg, Photogrammetry & Laserscanning Lab, Überseeallee 16, 20457 Hamburg
(thomas.kersten, klaus.mechelke)@hcu-hamburg.de, Lena-Maziull@web.de
[b] HafenCity University Hamburg, Geodetic Lab, Überseeallee 16, 20457 Hamburg

Commission V – WG V/4

KEY WORDS: 3D, archaeology, cultural heritage, meshing, reconstruction, virtual

ABSTRACT:

In September 2011 the fortress Al Zubarah, built in 1938 as a typical Arabic fortress and restored in 1987 as a museum, was recorded by the HafenCity University Hamburg using terrestrial laser scanning with the IMAGER 5006h and digital photogrammetry for the Qatar Museum Authority within the framework of the Qatar Islamic Archaeology and Heritage Project. One goal of the object recording was to provide detailed 2D/3D documentation of the fortress. This was used to complete specific detailed restoration work in the recent years. From the registered laser scanning point clouds several cuttings and 2D plans were generated as well as a 3D surface model by triangle meshing. Additionally, point clouds and surface models were automatically generated from digital imagery from a Nikon D70 using the open-source software Bundler/PMVS2, free software VisualSFM, Autodesk Web Service 123D Catch beta, and low-cost software Agisoft PhotoScan. These outputs were compared with the results from terrestrial laser scanning. The point clouds and surface models derived from imagery could not achieve the same quality of geometrical accuracy as laser scanning (i.e. 1-2 cm).

1. INTRODUCTION

In order to be able to fulfil the current high geometrical precision requirements of precise 3D data recording of large and complex objects geodetic measuring methods, photogrammetric multi-image procedures and terrestrial laser scanning, are used. Nowadays, low-cost systems, which make automatic object reconstruction possible using digital cameras and evaluation procedures from photogrammetry and computer vision, are increasingly available on the market. These budget-friendly procedures use open-source algorithms, such as Structure from Motion or Dense Image Matching, in order to automatically derive 3D models from imagery. However, the accuracy potential of such low-cost systems must still be examined in different applications and, if necessary, must be optimized in relation to metrology aspects (accuracy and reliability).

Practical examples of image-based modelling for the documentation of cultural monuments have been shown by Remondino & Menna (2008). Remondino et al. (2008) demonstrated that the same results can be obtained with both image-based and range-based procedures for the recording and modelling of monuments .Barazzetti et al. (2009) presented the combined use of photogrammetric and computer vision procedures for the automatic and exact 3D modelling of terrestrial objects. They also showed that similar results can be achieved with image-based and range-based recording systems. Kersten & Lindstaedt (2012c) showed that image-based modelling supplies geometrically similar results for archaeological finds when compared to a reference structured light system.

In this contribution, such low-cost systems (using the same image data set with open-source software, free software, web services, and low-cost software) are compared with reference data from terrestrial laser scanning that was collected for the huge and complex Al Zubarah fortress in Qatar.

2. THE AL ZUBARAH FORTRESS, QATAR

The Al Zubarah fort is situated in the historically-important town of Zubarah, which in turn is located on the north-western coast of the Qatar peninsula in the Madinat ash Shamal municipality. It is located 105kms from the Qatari capital of Doha (Figure 1). The impressive fortress is one of the most well-known sights and tourist attractions in Qatar with an area of 34m × 34m and a height of 9m. Near the fortress is the former city Al Zubarah, which was the largest settlement of the country with up to 6.000 inhabitants. It was the centre of the pearl divers from the middle of the 18th up to the beginning of the 20th Century. Today's fortress Al Zubarah was completely rebuilt in 1938 on behalf of Sheikh Abdullah Qasim Al-Thani (1871-1957) approximately 2 km inland of the former city. This military structure attained a strategic meaning in this location because of the constant conflict with the neighbouring state of Bahrain. Thus, the fortress still served as an important military base even after the independence of Qatar in the year 1971 and up until the middle of 1980s. In 1987 it was renovated into a museum to display diverse exhibits and artworks, particularly contemporary, topical archaeological findings. The square fort consists of solid loam walls, in whose corners three round and one rectangular military tower in the typical Arab architectural style rule are available (Figure 1).

Today the area of the former city, which was abandoned after lengthy military conflicts approx. 100 years ago, is an archaeological excavation site (Richter et al. 2011) for which the status of UNESCO world cultural heritage was requested together with the fortress (Thuesen & Kinzel 2011). In June 2013 UNESCO added the abandoned historical coastal town Al Zubarah Archaeological Site (including the Al Zubarah fortress) to the World Heritage List. Zubarah is one of the largest and best-preserved examples of an 18th-19th century traditional pearl fishing and merchant town in the Gulf. It is the first site in Qatar to be entered on to the UNESCO World Heritage List. In the

* Corresponding author

This contribution has been peer-reviewed.
doi:10.5194/isprsarchives-XL-5-W4-1-2015

Figure 1: Southern and western facade of the Al Zubarah fortress in Qatar (left), map of Qatar (right)

future a friendship bridge (Qatar Bahrain Causeway) is to be built in this area as a fast connection by car between the two States of Qatar and Bahrain.

3. INSTRUMENTS AND SOFTWARE USED

For 3D data acquisition of the fort the terrestrial laser scanning system IMAGER 5006h, manufactured by the company Zoller+Fröhlich in Wangen/Allgäu (Germany), was used. The scans were registered with the software Z+F LaserControl and the processed in the software Geomagic Studio. The image data was acquired with a digital reflex camera Nikon D70 (3008 × 2000 pixels), equipped with a Nikkor zoom lens with a focal length of 35mm. The automatic generation of 3D point clouds and 3D surface models from the image data was carried out with the following software packages: the open-source software Bundler/PMVS2 (Patch based Multi View Stereo Software; Snavely et al. 2008, Furukawa & Ponce 2010), the free software VisualSFM (Wu 2007, Wu 2011), both developed by the University of Washington (USA), the low-cost software Agisoft PhotoScan from St. Petersburg, Russia, and the Autodesk Web service 123D Catch beta (Abate et al. 2011). The basis of 123D Catch beta is the software smart3Dcapture from the French company acute3D in Nice, whose basic algorithms are described in Courchay et al. (2010). For automatic operational use of Bundler/PMVS2 the HafenCity University Hamburg (HCU) has developed a graphical user interface (Kersten et al. 2012). VisualSFM and PhotoScan ran their main algorithms on the GPU (graphics processing unit) to significantly speed up the data processing. These above-mentioned software solutions are briefly described in Kersten & Lindstaedt (2012a, 2012b, and 2012c). Furthermore, they were already used in different projects at HCU Hamburg, in which good results were obtained and good experiences were made for smaller objects (Kersten et al. 2012).

4. DATA EVALUATION - TERRESTRIAL LASER SCANNING

The 3D laser scanning of the Al Zubarah Fort was conducted by two people in September 2011 using the scanner IMAGER 5006h. 19 scans around the outside of the object were acquired in approx. 3.5 hrs, while 48 scans were acquired inside in 6.5 hrs, both using the High scan resolution which corresponds to a grid spacing of 6mm at 10m distance. As well as the inner court all interiors were also scanned and additional scans were acquired from all four towers in order to achieve better registration between exterior and interior scans. For the registration of the scans only a few targets were fixed inside the building. However, in the area close to the exterior of the fortress paper targets (DIN A4) were fixed on concrete piers while rotatable and swivelling targets with a diameter of 6cm were anchored in the ground (Figure 2).

The local registration of 62 scans was carried out with LaserControl using 169 targets. An average deviation of 3.8mm was obtained for the targets. Five scans of interiors were not used. For the subsequent RGB colouring of the scans nine photographs were taken after scanning at each scan station in order to compute a full-spherical panorama. Therefore, a Nikon D90 (4288 × 2848 pixels) with a 10.5mm Nikkor fisheye lens was used on an optical pivot point adapter (nodal point) at each scan station. The entire point cloud was sampled down to a point spacing of 1cm using the software Geomagic Studio. Thus, the data volume could be reduced to 11.6 million points, which were used for further processing of the fortress (Figure 3). After filtering of this point cloud in Geomagic Studio, a 3D mesh of 10 million triangles was generated (Figure 3 right). The processing of the laser scanning data is summarised as follows: ten hours for the acquisition of 67 laser scans, 31 hours for the registration and RGB colouring of all scans in LaserControl, four hours for 3D meshing in Geomagic Studio, resulting in a total of 45 hours for the whole workflow.

Figure 2: Terrestrial laser scanner IMAGER 5006h in use at fort Al Zubarah

This contribution has been peer-reviewed.
doi:10.5194/isprsarchives-XL-5-W4-1-2015

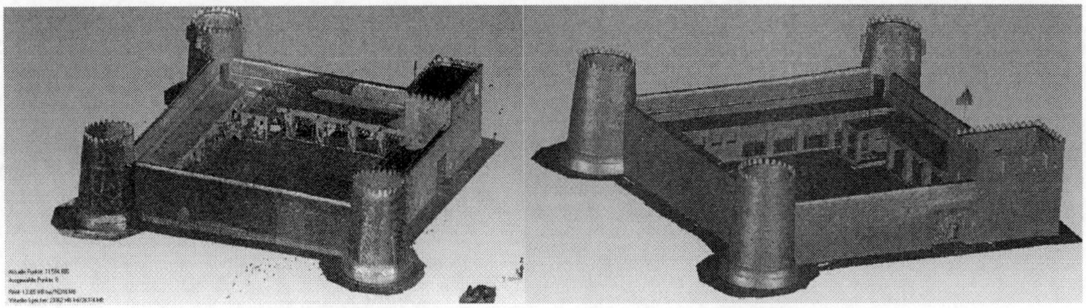

Figure 3: Al Zubarah fortress in Qatar as 3D point cloud (left) and as meshed 3D model (right)

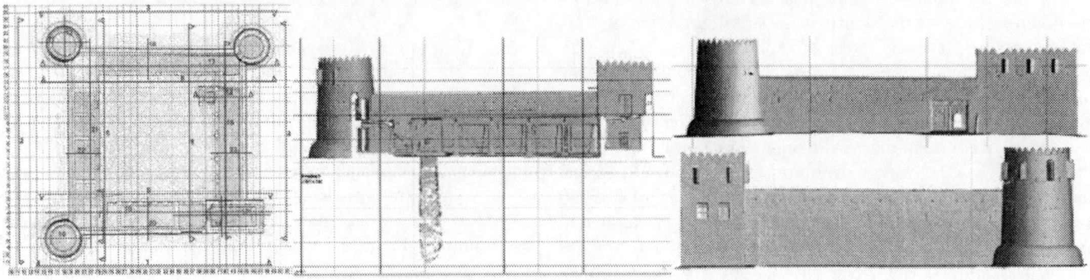

Figure 4: Overview sketch of the exported sections (left), three selected sections of Al Zubarah fortress – interior court yard with water well (centre) and two exterior walls (right)

As basis for the planning of the restoration work, 23 sections (profiles) were exported as DXFs from the point cloud (Figure 4 left). Figure 4 (centre) shows the eastern front of the inner court with water well (section 9) and two sections of the exterior walls as examples of the typical output – section 1 (top right) of the southern exterior facade and section 4 (bottom right) of the exterior eastern front.

5. DATA EVALUATION - IMAGERY

5.1 Imagery for Panorama Generation

The image photography was accomplished for two tasks – computation of RGB panorama images for the RGB colouring of the scans and image-based 3D modelling from RGB point clouds. First, for the computation of a full-spherical panorama, a set of

Figure 5: Camera Nikon D90 at the nodal point adapter in use (left) and eight of nine panorama images (right)

Figure 6: Virtual tour of Al Zubarah fortress using a set of 16 full-spherical panorama images

This contribution has been peer-reviewed.
doi:10.5194/isprsarchives-XL-5-W4-1-2015

Figure 7: Example for camera positions at Al Zubarah fortress – western exterior wall (left) and southern interior façade (right)

nine photographs were taken on the nodal point adapter, which was fastened to the tripod of the scanner at each scan station – one pointing towards the sky, two towards the ground and six photos for the 360° circle in the horizontal position (Figure 5). The 62 panoramas were automatically computed with the software PTGui V9.04. These panoramas were computed with a low resolution of approx. 6300 × 3100 pixels and were afterwards used for the RGB colouring of the point clouds in LaserControl (see Figure 3 left, western front of the fortress). Spherical panoramas with the full resolution of approx. 11700 × 5850 pixels (data volume 43 MB) were also automatically generated to provide a virtual tour of the fortress as a visualisation tool for architects, restorers, and archaeologists and for other project partners. These panorama images were converted into a set of six cube images (in total ca. 5 MB), while a panorama viewer program simultaneously generated one interactive virtual tour for all available panorama images. Hotspots, an overview map, photographs, text, music, videos and many more custom actions can be integrated using XML-based programming. A connection between different panoramas is created over so-called Hotspots, enabling the viewer to navigate by themselves between these different panoramas. Figure 6 shows the graphical user interface for the interactive virtual tour of the Al Zubarah fortress (www.hcu-laserscanning.de/vt/zubarah/fort_110917/fort.html), in which all spherical panoramas are linked to each other via hotspots or via the overview map (right corner). Thus the user can zoom in and out, rotate the panorama in all directions, and navigate through all panoramas by clicking the hotspots or the marked positions in the map.

5.2 Imagery for Generation of Points Clouds and 3D Models

Secondly photographs of the exterior walls and the interior area of the fortress were taken with the Nikon D70 in order to generate 3D point clouds and/or surface models using different software packages. The images were captured without adherence to multi-photo photogrammetric criteria with many photographs taken at eye level. The focal length of the uncalibrated camera was fixed to 35mm. In Figure 7 an image configuration for an exterior wall and a subrange of the inner court is represented. Altogether 176 photographs were taken for the exterior area and 219 for the interior area of the fortress.

5.3 Generation of Points Clouds with Bundler/PMVS2 and VisualSFM

The generation of 3D point clouds with Bundler/PMVS2 and VisualSFM was conducted on a high performance computer (Intel Xeon CPU E5540 processor with 2.53 GHz, 24 GB RAM, using the operating system Windows 7 Enterprise 64 bits; status 2011/2012). Due to the high number of photos only the exterior walls of the forts were computed first with both software packages. In Table 1 statistics on the generation of an entire point cloud for the four exterior walls of Al Zubarah fortress are

Figure 8: 3D point cloud of the exterior walls (right) and of the court yard (centre) from Bundler/PMVS2, and the entire point cloud of Al Zubarah fortress from VisualSFM (right)

Figure 9: Entire 3D point cloud of the fortress from Bundler/PMVS2 and VisualSFM (left) and 3D model meshed in Geomagic (right)

Figure 10: 3D surface model of the exterior walls (left) and the interior area (right) of the fortress Al Zubarah with and without texture generated with Autodesk web service 123D Catch beta

summarised. For the 176 photos the computation time for VisualSFM using the GPU was a factor of 30 faster than Bundler/PMVS2. However, 450,000 fewer points (~10%) were counted. The reason for the slight smaller number of points in VisualSFM compared to Bundler/PMVS2 could not be discerned. The point clouds for the exterior walls and interior area from Bundler/PMVS2 are represented in Figure 8 (right and centre).

	Visual SFM	Bundler/PMVS2	Difference
# images	176	176	0
# points	4.147.129	4.591.875	444.746
Time	01 hrs 31 min	46 hrs 08 min	Factor 1:30

Table 1. Statistics on the point cloud generation of the four exterior walls using different software

In VisualSFM a 3D point cloud (7,089,119 points, see Figure 8 right) for the entire fortress could be computed in 3 hrs 31 min with an entire image data set of 393 photos, while the task with the same image data had be cancelled after more than 14 days computing time in Bundler/PMVS2 with the computer specified above due to problems with the internal memory (RAM).

All generated point clouds were scaled using different, well distributed reference distances from the laser scanning data. The precision of the reference distances was significantly influenced by the point density within the point cloud, since it was difficult to measure distinctive points in the point clouds depending upon the zoom level used. Conclusively, the entire point cloud from VisualSFM was registered with the dense point cloud of the interior area from Bundler/PMVS2 using a best-fit algorithm in Geomagic Studio with an average deviation of 1.6cm. In Figure 9 the result of this registration (10.3 million points) and the triangle meshing of the point cloud (15.5 million triangles) is illustrated.

5.4 Generation of Surface Models with Autodesk 123D Catch Beta

Using 123D Catch beta, a web service of Autodesk, a 3D surface model was computed within 10-20 minutes depending upon data set and/or number of images after the upload of the image data on an external computer server of Autodesk. The user could download the model from the Autodesk server after notification by email. After further data processing the model was sent to the server again for a renewed calculation with maximum resolution offered by the web service. In different tests with the service it was recognized that a data set with more than 100 images had led rather to bad, partial incomplete and incorrect results. Therefore surface models were computed in the maximum resolution for the four exterior and interior fronts using up to 70-80 images (see also Figure 7). The single models of each facade were transferred to Geomagic Studio using the data format OBJ. In Geomagic the individual fronts were manually corrected for blunders. Especially redundant triangles, which particularly represented the sky at the edges of the walls (see Figure 7), were manually marked and deleted. Using measured distances from the laser scanning data the individual exterior and interior fronts were scaled and registered afterwards as separated models for the exterior walls (1.1 million triangles) and the interior area (2.3 million triangles) using ICP algorithms in Geomagic.

Figure 10 shows the 3D models of the exterior walls (average deviation of 1.4cm after registration) and of the interior court (deviation 2-3cm) of the Al Zubarah fortress with and without texture. A data fusion of the exterior and interior models could not be realized due to unacceptable geometrical differences and gaps. It can be assumed that these differences are a result of a too small overlap between the models and the insufficient determination of the reference distances by point measurements both within the scan data and the 3D model, although the point identification was much better in the highly resolved textures of the 3D models than in the point clouds. Furthermore, an inhomogeneous scale within the models could be possible.

Figure 11: 3D model of the Al Zubarah fortress generated with Agisoft PhotoScan 1.0.4

This contribution has been peer-reviewed.
doi:10.5194/isprsarchives-XL-5-W4-1-2015

5.5 Generation of Point Clouds and Surface Models with Agisoft PhotoScan

A third 3D surface model of the fortress was generated with the software Agisoft PhotoScan version 1.0.4 (professional edition). PhotoScan is a multi-view image-based 3D modelling solution for 3D reconstruction of textured mesh models. The general workflow of PhotoScan is as following: add photos (data import), align photos (generation of a sparse point cloud for image orientation using SIFT algorithm and bundle block adjustment), build dense point cloud (dense image matching), build mesh (3D triangulation), and build texture (texture mapping). The advantage of this software is the capability to measure control points in the high resolution images, to use control points in the bundle adjustment, and to use a sophisticated camera calibration model including the calibration parameter focal length, principle point, radial lens distortion (k1-k4), decentring (tangential) distortion (p1, p2), and skew (shearing, defining the angle between the x and y pixel axes). For the data processing with PhotoScan the image data set was reduced to 379 images, which covers the exterior walls and the interior area of the fortress. However, the full image data set of 379 images was used as one image block (group/chunk) for the computation in PhotoScan. The computer used for this data processing was different to the computer used with the other software packages: PC Fujitsu Celsius, Intel® Xeon® CPU E5540, 64 Bit processor with four kernels, NVIDIA GeForce GTX 690 VRAM graphic card, 12 GB RAM and Windows 7 as operating system. The following parameters were set for the batch processing workflow: align photos (high accuracy with a point limit of 5 million points), build dense point cloud (high quality, aggressive depth filtering), build mesh (arbitrary surface type, dense point cloud as source data, interpolation enabled, custom face count 6 million), and build texture (generic mapping mode, texture from all cameras, blending mode Mosaic, texture size 3008, texture count 1, and no colour correction). Five signalised targets were measured in the images, while the 3D coordinates of these control points were determined in LaserControl. These five ground control points were well-distributed in object space and fixed at the walls of the interior area of the fortress. However, the control point configuration was not optimal, since signalised control points were not available at the exterior walls. Furthermore, the blue sky was manually masked in the images before data processing with PhotoScan. This was necessary to reduce the number of erroneous measurements in the transition between object and sky significantly. The following results were achieved: sparse point cloud for image orientation and camera calibration with 724,495 points, dense point cloud with 36.6 million points and a mesh with 6 million triangles (faces). The residuals of the control points were between 1-2mm in the XYZ coordinate. The meshed 3D model of the fortress is illustrated in Figure 11. As a first quality control the horizontal section of the fortress were compared to the reference data (laser scanning). The maximum difference between the image-based and range-based section was 2cm (Figure 14 right).

6. 3D MODELS IN COMPARISON

The 3D models of the Al Zubarah fortress, which were generated with VisualSFM & Bundler/PMVS2, 123D Catch beta, and with Agisoft PhotoScan, were compared to reference data from laser scanning in Geomagic Studio, in order to evaluate the geometrical quality of the 3D models derived from imagery. For this reason the 3D point clouds and models derived from images were manually registered to the 3D reference point cloud and model (from laser scanning data) using 3-4 common points in a first step. In the second step they were automatically registered using an ICP (iterative closest point) algorithm.

The colour-coded deviations between the compared 3D data are illustrated in Figure 12, 13 and 14. The average deviations to the reference data is between 5cm and 10cm with maximum deviations of approx. 60cm for the models generated with scaling by distances. Thus, the quality of the 3D data (point cloud and models) from Bundler/PMVS2 in combination with VisualSFM and 123D Catch beta does not correspond to the high geometrical quality of the laser scanning data for this large object.

Figure 12 shows the 3D point cloud (left), generated with Bundler/PMVS2 and VisualSFM from image data, and the 3D model (right) compared with the reference model from laser scanning data. The green areas in Figure 12 represent the deviations between point cloud and model (from laser scanning), which are smaller than 5cm (left), and between the two 3D models, which are smaller than 10cm (right). However, it is evident to see in Figure 12 that many areas have also higher deviations, which indicates systematic characteristics from the partial imprecise scaling. The light blue and yellow colours illustrate deviations, which are between 5cm and 15cm (left) and between 10cm and 20cm (right).

Slightly smaller deviations are computed in the 3D comparison between the model from the interior court of the fort (derived with 123D Catch beta) and the reference model from laser scanning (Figure 13 left). Most of the areas are green representing deviations of less than 5cm. Contrary, the result for the exterior walls is the worst of all 3D data, since differences between 5cm and 15cm (light blue) are predominantly visible (Figure 13 right).

Only the results derived by Agisoft PhotoScan with an average deviation of 2cm to the reference were significantly better (Figure 14), since signalised control points were used for the registration of this point cloud, while for the other image-based solution only reference distances were used for the scaling of the dense point clouds. It was not possible to measure the signalised control points in these data sets, since they were fixed on walls of the interior courtyard of the fortress. These results of the 3D comparison show the potential of the image-based software solution for large architectural objects. It could be shown, that instead of object scaling using different reference distances a 3D transformation of the different objects into the reference coordinate system using well-distributed (signalised) control points in object space supplied geometrically better and more reliable results. Therefore, it is necessary to use precise and well-distributed (signalised) control points for the generation of dense point clouds from imagery for large and complex architectural objects. Furthermore, the software used for dense image matching must be capable to use the information of control points (XYZ coordinates and its weights) in a rigorous bundle block adjustment to derive reliable results for image orientation and camera calibration.

7. CONCLUSION AND OUTLOOK

In this contribution the investigations showed the fact that beside terrestrial laser scanners used as standard technology for 3D recording also camera-based low-cost systems are able to record and to generate 3D data of large objects such as the Al Zubarah fortress. The data acquisition with cameras is fast, flexible and economical in comparison to laser scanning. However, the geometrical quality of the laser scanning data could not be achieved for all tested software packages using dense image matching, since average deviations between 5cm and 10cm to the reference model were evaluated. Nevertheless, the results of PhotoScan with average deviations of 2cm are geometrically very

This contribution has been peer-reviewed.
doi:10.5194/isprsarchives-XL-5-W4-1-2015

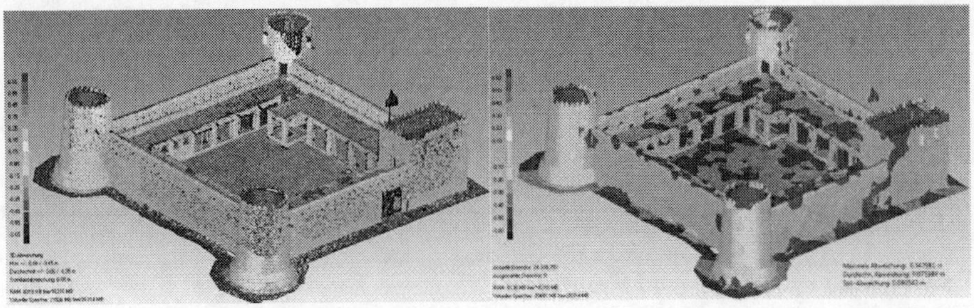

Figure 12: Bundler/PMVS2 & VisualSFM vs. reference – deviations of the 3D point cloud (left, green = ±5cm) and the meshed 3D model (right, green = ±10cm)

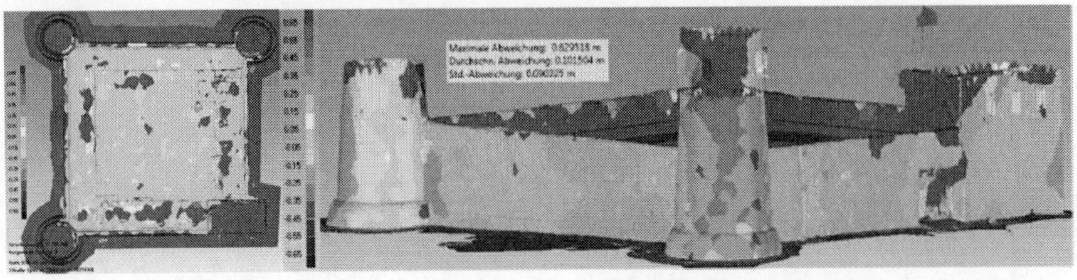

Figure 13: 3D data in comparison – 3D model from laser scanning (IMAGER 5006h) vs. 3D model of the interior area (left) & 3D model of the exterior walls (right) from 123D Catch Beta (green = ±5cm)

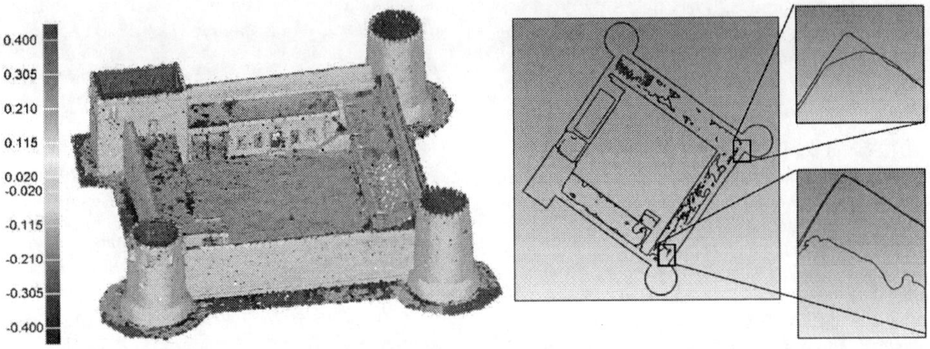

Figure 14: Agisoft PhotoScan - 3D comparison to reference (left) generated in Geomagic (green = ±2cm) and comparison of horizontal sections (black = image-based, red = laser scanning data) with maximum difference of 2cm (right)

close to the laser scanning data. Regarding quality and reliability, the limiting factors for the photogrammetric procedure are in general and especially for large objects the photo scale, the lighting conditions, the image block configuration, and the resolution of camera/lens. Also the quality of the measuring procedures as well as the control point determination for scaling or for 3D transformation is of great significance. The comparison of the measuring procedures showed that also image measuring procedures used in Bundler/PMVS2, VisualSFM, 123D Catch beta, and PhotoScan produced results with different quality. To use a camera with higher resolution and qualitatively better lens, could improve the results slightly. However, significantly improved results would have been supplied using a geometrical optimised image configuration by taking photography from elevated positions (e.g. from an Octocopter). The image recording strategy must be changed from classical manual photogrammetric evaluation procedures to automatic point measuring and point transfer methods, which need a higher overlap of the images during the object recording to adapt to the structure-from-motion method. Furthermore, instead of scaling the objects using different reference distances a 3D transformation using well distributed control points will presumably provide reliable results. Nevertheless, the problem of point identification and measurement in 3D data with different resolution will still exist. Kersten & Lindstaedt (2012c) showed that comparable geometrical quality from terrestrial laser scanners and structure light systems can be achieved for smaller objects (e.g. archaeological finds) using low-cost systems including their measuring procedures in combination with the use of scale bars for object scaling. The performance ability of computers is crucial for the minimisation of data processing time of image data evaluation. Especially for larger objects, which were recorded by a huge number of photographs, the evaluation software must be able to use the computing power of the graphics processing unit (GPU), in order to obtain acceptable computing time. Using the available web services relieve own computer capacities, but data security after the upload of the data is still an open question. Meanwhile there is a large number of algorithms for pixel-based matching in stereo or multi-view images available from the computer vision community providing different

This contribution has been peer-reviewed.
doi:10.5194/isprsarchives-XL-5-W4-1-2015

achievement potential (Scharstein & Szeliski 2009). Many so-called open-source and/or low-cost software packages for image-based evaluation will enter the market, in which combined methods from computer vision and photogrammetry are used as new measuring and evaluation procedures. Thus, they constitute a reasonable alternative or addition for range-based measuring systems such as laser scanning or structured light systems (fringe projection), which could herald the start of the beginning of a new development in optical measurement technique. However, further system investigations are necessary to compare the geometrical quality with reference data. Care has to be taken that these procedures fulfil the geodetic and photogrammetric requirements for geometrical quality analysis (metric methodology), which could make further adaptation essential for these procedures.

ACKNOWLEDGEMENTS

The support of the Qatar Islamic Archaeology and Heritage Project (QIAH), which was launched jointly by the Qatar Museum Authority and the University of Copenhagen (Carsten Niebuhr Centre for Multicultural Heritage, Department of Cross-Cultural and Regional Studies - ToRS), is gratefully acknowledged.

REFERENCES

Abate, D., Furini, G., Migliori, S., Pierattini, S., 2011. Project Photofly: New 3D Modelling Online WEB Service. *International Archives of Photogrammetry, Remote Sensing and Spatial Information Sciences* 38(5/W16), http://www.isprs.org/proceedings/XXXVIII/5-W16/pdf/abate_etal.pdf.

Barazzetti, L. Remondino, F., Scaioni, M., 2009. Combined use of photogrammetric and computer vision techniques for fully automated and accurate 3D modeling of terrestrial objects. *Proc. of SPIE Optics+Photonics*, Vol. 7447.

Courchay, J., Pons, J.-P., Monasse P., Keriven, R., 2010. Dense and Accurate Spatio-temporal Multi-view Stereovision. Computer Vision – ACCV 2009, *Lecture Notes in Computer Science, Volume 5995*, pp. 11-22.

Furukawa, Y., Ponce, J., 2010. Accurate, Dense, and Robust Multi-View Stereopsis. *IEEE Transactions on Pattern Analysis and Machine Intelligence*, 32(8), 2010, 1362-1376.

Kersten, Th., 2007. Virtual Reality Model of the Northern Sluice of the Ancient Dam in Marib/Yemen by Combination of Digital Photogrammetry and Terrestrial Laser Scanning for Archaeological Applications. *International Journal of Architectural Computing,* Special Focus on Cultural Heritage, 02 (05), Multi-Science, pp. 339-354.

Kersten, Th., Lindstaedt, M., 2012a. Automatic 3D Object Reconstruction from Multiple Images for Architectural, Cultural Heritage and Archaeological Applications Using Open-Source Software and Web Services. *Photogrammetrie – Fernerkundung – Geoinformation*, Heft 6, pp. 727-740.

Kersten, Th., Lindstaedt, M., 2012b. Potential of Automatic 3D Object Reconstruction from Multiple Images for Applications in Architecture, Cultural Heritage and Archaeology. *International Journal of Heritage in the Digital Era*, Volume 1, No. 3, a Multi-Science Publication, Brentwood, UK, pp. 399-420.

Kersten, Th., Lindstaedt, M., 2012c. Image-Based Low Cost Systems for Automatic 3D Recording and Modelling of Archaeological Finds and Objects. EuroMed 2012 - Int. Conference on Cultural Heritage, Ioannides, M.; Fritsch, D.; Leissner, J.; Davies, R.; Remondino, F.; Caffo, R. (Eds.), Lecture Notes in Computer Science (LNCS), Volume 7616, Springer-Verlag Berlin Heidelberg, pp. 1-10.

Kersten, Th., Lindstaedt, M., Mechelke, K., Zobel, K., 2012. Automatische 3D-Objektrekonstruktion aus unstrukturierten digitalen Bilddaten für Anwendungen in Architektur, Denkmalpflege und Archäologie. 32. Wissenschaftlich-Technische Jahrestagung der DGPF, 14.-17. März 2012 in Potsdam, *Publikationen der Deutschen Gesellschaft für Photogrammetrie, Fernerkundung und Geoinformation e. V.*, E. Seyfert (Hrsg.), Vol. 21, Proceedings on CD-ROM.

Remondino, F., Menna, F., 2008 Image-based surface measurement for close-range heritage documentation. *International Archives of Photogrammetry, Remote Sensing and Spatial Information Sciences*, 37(B5-1), pp. 199-206.

Remondino, F., El-Hakim, S.F., Gruen, A., Zhang, L., 2008. Turning images into 3-D models. *IEEE Signal Processing Magazine*, 25(4), pp. 55-65.

Richter, T., Wordsworth, P. D., Walmsley, A. G., 2011. Pearlfishers, townsfolk, Bedouin and Shaykhs: economic and social relations in Islamic Al-Zubarah. *Proceedings of the Seminar for Arabian Studies*, 41, 2011, 317-332.

Scharstein, D., Szeliski, R., 2009. Middlebury Stereo Vision Page. http://vision.middlebury.edu/stereo/, visited at 7.1.2012.

Snavely, N., Seitz, S.M., Szeliski, R., 2008. Modeling the World from Internet Photo Collections. *International Journal of Computer Vision*, 80(2), pp. 189-210.

Thuesen, I. Kinzel, M., 2011. Al-Zubarah Archaeological Park as a UNESCO World Cultural. Heritage Site - a master plan for its site management, preservation, and presentation (poster). *Proceedings of the Seminar for Arabian Studies*, 41, pp. 371-376.

Wu, C., 2007. SiftGPU: A GPU implementation of Scale Invariant Feature Transform (SIFT). http://cs.unc.edu/~ccwu/siftgpu.

Wu, C., 2011. VisualSFM: A Visual Structure from Motion System. http://www.cs.washington.edu/homes/ccwu/vsfm/.

ROBOTIC MAPPING OF CULTURAL HERITAGE SITES

Dorit Borrmann, Robin Heß, HamidReza Houshiar, Daniel Eck, Klaus Schilling, and Andreas Nüchter

Informatics VII – Robotics and Telematics, Julius-Maximilians University Würzburg, Germany
borrmann@informatik.uni-wuerzburg.de

Commission V, WG V/4

KEY WORDS: laser scanning, 3D modeling, multi-sensors

ABSTRACT:

In archaeological studies the use of new technologies has moved into focus in the past years creating new challenges such as the processing of the massive amounts of data. In this paper we present steps and processes for smart 3D modelling of environments by use of the mobile robot Irma3D. A robot that is equipped with multiple sensors, most importantly a photo camera and a laser scanner, enables the automation of most of the processes, including data acquisition and registration. The robot was tested in two scenarios, Ostia Antica and the Würzburg Residence. The paper describes the steps for creating 3D color reconstructions of these renown cultural heritage sites.

1 INTRODUCTION

Archaeology is a historical science of high social interest. It studies the human being and its legacy, such as buildings, tools and art. Cultural heritage sites can be found all over the world and they tell us the story of humanity in different areas of the world. Remote sensing has become state of the art in modeling archaeological sites. This way of digitization of entire buildings or areas gives as a unique opportunity to preserve the current state of prehistoric buildings and to join forces of experts all over the world. Collecting the data is tedious work. It includes finding the best position for a laser scan, moving the equipment to the position and georeferencing of the scanning position. Letting a robotic system take over this work reduces the time spent in the field by 75 % and decreases the impact to the sites. We present the robot Irma3D, that was designed to create in a tele-operated fashion digital 3D models of environments. This paper describes the setup and the capabilities of the robot and the steps to create these 3D models automatically from multiple sensor sources. The entire process is demonstrated by means of experiments carried out at cultural heritage sites.

The robot was tested in two scenarios, Ostia Antica and the Würzburg Residence Palace. In this paper we describe the data collection with the robot Irma3D in these two renowned historic sites, the post-processing needed to create a full 3D color model and present the resulting models from a garden house in Ostia Antica (http://youtu.be/sf-gq5xlaIc), the White Hall (http://youtu.be/_wPug_So_iE) and the Imperial Hall (http://youtu.be/jKVxlLvu7Pk) at the Würzburg Residence. The results can interactively be viewed in a 3D viewer or with an Oculus Rift (cf. Bruder et al. (2014)).

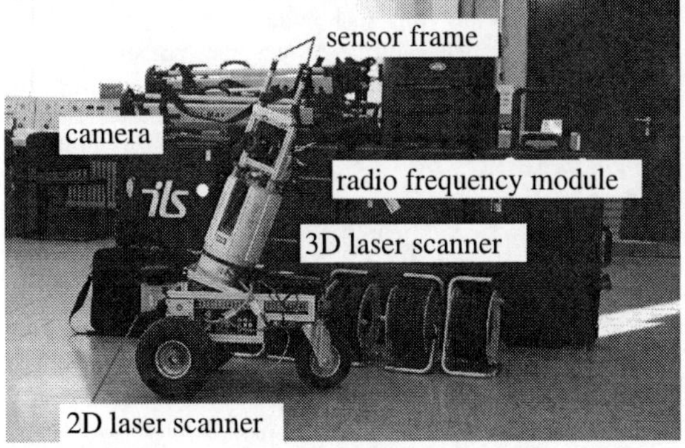

Figure 1: The Irma3D robot with the setup used in the Residence.

2 EXPERIMENTAL SETUP

In this section we explain the setup for robotic mapping of cultural heritage sites. First, we describe the environments where measurements were to be taken. Second, the hardware used in both scenarios is presented.

2.1 Cultural Heritage Sites

The robot was tested in two scenarios, Ostia Antica and the Würzburg Residence. Ostia Antica is a large archeological site, close to the modern suburb of Ostia (Rome). Due to the exceptionally well preserved state of the city, Ostia is of immense interest for the study of the Roman empire. According to archeologically unverified tradition Ostia was founded during the second half of the 7th century B.C. at the mouth of the Tiber river. This region was said to be strategically important for Rome due to the there existing salt marshes. Supported by evidence is the settlement at the mouth of the Tiber as early as the beginning of the 4th century B.C. Initially only a military camp meant to defeat Rome towards the sea Ostia grew to be an autonomous harbor city of about 50 hectares with a population of approximately 50,000 in the second half of the 2nd century A.D. In the following began the decline of the city. Large parts of the town were abandoned following historical events. The last inhabitants left the city after raids of the Saracen pirates from the sea around the 9th century A.D. (Ostia Antica, 2014).

The experiments took place in one of the garden houses in Ostia Antica. Historians suspect that these garden houses were a large building project consisting of many serial prepared houses that were rented out afterwards. Of special interest is the context of architecture and accouterment, i.e., the wall paintings and the floor mosaics. The garden house in question contains of one large semi-open area and several smaller rooms that are connected with narrow hallways. Scanning the entire area completely without holes requires a large number of scan positions.

This contribution has been peer-reviewed.
doi:10.5194/isprsarchives-XL-5-W4-9-2015

The Residence Palace in Würzburg Germany is a baroque palace in the city center of Würzburg, Germany and was labeled a UNESCO World Cultural Heritage site in 1981. Being built from 1720 to 1744 with the interior finished in 1780 it is now one of Europe's most renowned baroque castles. It was laboriously reconstructed after being heavily damaged during World War II. Not destroyed during the war remained the large unsupported trough vault above the main stair-case designed by architect Balthasar Neumann, the Garden hall with ceiling paintings by Johann Zick, the white hall with the impressive stucco work by Antonio Bossi and the Imperial hall with frescos by Giovanni Battista Tiepolo. With its large colorful paintings by the Venetian painter Giovanni Battista Tiepolo and fine stucco work by stuccoist Antonio Giuseppe Bossi in many of the almost 400 rooms the Würzburg Residence is a unique example of baroque style (Würzburg Residence, 2014).

Experiments were carried out in both the White hall and the Imperial hall, two large halls with impressive 3D structure. Together with the colorful paintings in the Imperial hall the environment can only be captured by the combination of two technologies, e.g., laser scanning and photography.

2.2 HARDWARE

The data was acquired with the mobile robot Irma3D (Intelligent Robot for Mapping Applications in 3D). Irma3D is a small, battery-powered, light weight three wheeled vehicle. It consists of a modified Volksbot RT 3 chassis with two front wheels. Each is actuated by an individual 150 W DC Maxon motor. The motors are powerful enough to move the robot at a maximum velocity of 2.2 m/s. The third wheel is in the back of the chassis and is swivel-mounted and thus completely passive as it follows the directions of the front wheels. The high-powered electrical two-wheel drive is equipped with rotary encoders to measure wheel rotations. This information is used to provide pose estimates of the robot via odometry. The pose estimates are improved using data from the Xsens MTi IMU device that is also attached to the robotic platform. For obstacle avoidance when moving autonomously a Sick LMS 100 2D laser scanner is added to the front of the robot. This sensor can also be used to improve the localization of the robot. The central sensor of Irma3D is the 3D laser scanner VZ-400 by RIEGL Measurement GmbH. The scanner is mounted on top of the Volksbot RT 3 chassis. Attached to the top of the scanner is a Canon 1000D DSLR camera. After a 3D scan has been acquired the camera is used to acquire color information for the point cloud.

In the Würzburg Residence an iSpace sensor frame is also mounted on top of the laser scanner. iSpace is a high-precision position and tracking system from Nikon Metrology (2014). The optical laser based system consists of several transmitters. These are mounted on a wall or on tripods to cover the experimental area both indoors and outdoors. The rotating head of each transmitter emits two perpendicular fan-shaped laser beams at a unique distinguishable frequency near 40 Hz. The vertical opening angle of the laser beams is limited to 40 degrees and the detectable range lies between 2 to 55 meters. Several sensor frames can be located within the system. A sensor frame consists of at least one detector, a photo diode with a horizontal opening angle of 360 degrees and a vertical opening angle of 90 degrees. A small radio frequency module transmits the sensor data wirelessly to the base station of the iSpace system, a PC running the iSpace control software. A sensor frame with one detector is sufficient to acquire 3D position information. To measure also the rotation and to increase the accuracy of the position data the sensor frame used on the robot has a total of four detectors. The iSpace system differs from other position and tracking systems as the transmitters do not actively observe the position of the sensor frames. Instead, each sensor frame receives the laser data from the transmitters and sends the information on to the control PC. The control PC calculates the elevation and azimuth angles between all detectors for a sensor frame and each visible transmitter based on the received data defining a straight line between transmitter and detector. Given the relative transformation between the transmitters the length of the lines is calculated using triangulation. To determine the position of the transmitters a calibration procedure using a few hundred points from a special sensor frame is applied. An optimization process calculates the position of all transmitters in a self-defined coordinate system. In typical environments the iSpace system is able to perform measurements at a sampling rate of 40 Hz with a maximum error of $[\pm 0.25]$mm. In practice environmental factors such as size, reflection of the surface and occlusions of the transmitters have to be taken into consideration.

3 ROBOTIC MAPPING

Scanning a large interconnected area consisting of many rooms requires a large number of scanning positions if one wants to cover the area completely without holes. In terrestrial laser scanning this is done by manually moving the laser scanner. To determine the exact position of the laser scanner for each position special targets are commonly placed in the environment whose positions are measured with a total station. Afterwards, the position of the scanner is either determined by placing another target on the tripod at the position where the scan is to be taken or by identifying the targets directly in the laser scan. This procedure is time-consuming and has to be repeated for each scanning position. The idea of using a robot for scan acquisition is to eliminate this time and thus reduce the data acquisition time. In this section we describe the methods used for creating precise 3D models of the collected data, i.e., the registration and calibration methods for bringing all the data from the various sensors into one common coordinate system automatically.

3.1 Sequential scan matching

To collect data at several locations the robot is moved to a scanning location and stops there for data collection. In robotics research methods have evolved over the past years that are specifically designed to register point clouds collected by a mobile robot automatically. Commonly used for this task is the Iterative Closest Point (ICP) algorithm (Besl and McKay, 1992). The algorithm takes advantage of the fact that robots usually have a rough estimate of their current pose (position and orientation). For the robot used in this paper this pose estimate is calculated from the movements measured with the rotary wheel encoders and the IMU device. Starting from such initial pose estimates the algorithm calculates effectively the correct transformation between two point clouds by means of minimizing distances between corresponding points. Corresponding points are chosen based on Euclidean distances. The algorithm is described in Algorithm 1.

Given a registered point cloud M and a point cloud D with an initial pose estimate the ICP first tries to find for each point $\mathbf{d}_i$ from D the point $\mathbf{m}_i$ in M that is closest to d_i. Then one needs to solve for a transformation $(\mathbf{R}, \mathbf{t})$ (orientation $\mathbf{R}$ and translation $\mathbf{t}$) that minimizes the error function E_{ICP}. Nüchter et al. (2010) presents several minimization methods. These two steps are iterated to find the best transformation between the two point clouds. For best results a threshold t_{dist} is introduced and all point pairs with a distance larger than t_{dist} are discarded from the calculations. N is the number of remaining point pairs.

This contribution has been peer-reviewed.
doi:10.5194/isprsarchives-XL-5-W4-9-2015

Algorithm 1 The ICP algorithm

Require: point clouds M and D

1: find point correspondences

2: minimize for rotation $\mathbf{R}$ and translation $\mathbf{t}$

$$E_{\text{ICP}}(\mathbf{R}, \mathbf{t}) = \frac{1}{N} \sum_{i-1}^{N} ||\mathbf{m}_i - (\mathbf{R}\mathbf{d}_i + \mathbf{t})||^2 \quad (1)$$

3: **return** pose (R, t)

4: iterate 1 and 2

In practise this procedure is adopted as follows. Using the first scanning position as reference the nth scan is always registered against the $(n-1)$th scan. This way all scans are sequentially transformed into the same coordinate system.

3.2 Global optimization

When applying the sequential procedure small local errors for the registration of each pair add up, possibly leading to larger errors for long sequences. To overcome this issue a global optimization was proposed by Borrmann et al. (2008). Given a graph of corresponding scans point pairs are determined for each scan pair (j, k) and the new error function E_{opt} requires finding transformations for all scans simultaneously:

$$E_{\text{opt}} = \sum_{j \to k} \sum_{i} ||\mathbf{R}_j \mathbf{m}_i + \mathbf{t}_j - (\mathbf{R}_k \mathbf{d}_i + \mathbf{t}_k)||^2 \quad (2)$$

Methods to minimize E_{opt} are presented in (Nüchter et al., 2010). The implementation used in this publication is freely available from *The 3D Toolkit* (3DTK) (Andreas Nüchter et al., 2014).

3.3 Acquiring initial pose estimates

ICP and the global optimization rely on initial pose estimates to determine the correct transformation between laser scans. Different methods to acquire pose estimates are described in the following.

3.3.1 Odometry On mobile robots pose estimates are commonly attained from odometry. Odometry for wheeled robots such as Irma3D is based on calculating the distance traveled by the robot based on wheel rotations. For this purpose the relation between the count c of the wheel encoders and the wheel rotations are related to each other using a factor f. Knowing the diameter d of the tires the distance traveled by one wheel is calculated as $\Delta s = \pi \cdot d \cdot f \cdot c$. Considering the distance B between the two wheels and the distances traveled by each wheel $\Delta s_l, \Delta s_r$ the pose at time step n (x, y, θ) of the robot is calculated as:

$$\theta_n = \theta_{n-1} + (\Delta s_r - \Delta s_l)/B \quad (3)$$

$$x_n = x_{n-1} + 0.5 \cdot (\Delta s_r + \Delta s_l) \cdot \cos -\theta_n \quad (4)$$

$$y_n = y_{n-1} - 0.5 \cdot (\Delta s_r + \Delta s_l) \cdot \sin -\theta_n \quad (5)$$

The quality of these measurements depends highly on the behavior of the robot on the ground. If the floor is slippery and the wheels spin uncontrolled the estimates lack precision. To increase the precision of the position estimates the xSens IMU is attached to the robot and thus the measured accelerations are integrated into the position estimates. This is especially helpful for rotations.

Odometry works sufficiently when short distances are covered and the robot follows smooth trajectories. With increasing path lengths, many rotations and frequent stops along the trajectory errors sum up.

Figure 2: Illustration of a scan pair with Mercator projection and matched features from the garden house in Ostia Antica.

3.3.2 Feature-based registration If no pose estimates were acquired during data collection the remaining option is to determine them directly from the data. Apart from range information modern laser scanners often capture the amount of light that is returned to the sensor. This information, known as reflectance value, can be used to detect features in the data. In terrestrial laser scanning it is common practice to attach high reflective markers to the environment and detect them in the data to register the laser scans. This procedure takes a lot of time to place the markers in good positions, find them in the data and register the data. Often times the exact positions of the markers are also measured using a tachymeter extending the time needed in the field even further. As an additional disadvantage the markers will be visible in the data thus adulterating the scene and the mapping. Alternative methods use natural features rather than manually attached markers. Conditioned by the buildup of the laser scanning platform the points are captured as range, reflectance and spherical coordinates. This facilitates the generation of a panorama image using the spherical coordinates and the reflectance information. The 3D data is thus projected onto an image. Different projection methods are evaluated by Houshiar et al. (2012).

The panorama generation enables the use of image based feature matching methods. These methods analyze the image and create a description of areas with high changes in intensity. The most common features are SIFT (Scale invariant feature transform) features. They also show superior performance for feature-based point cloud registration (Houshiar et al., 2012). As the SIFT feature detector works in gray-scale images the panorama images from reflectance values of laser scans are ideally suited for feature matching. For the automatic registration of point clouds using these panorama images corresponding features are detected in the panorama images of scan pairs. Feature correspondences found in two reflectance panoramas are used to calculate pairwise transformation matrices of the point clouds with a RANSAC-like (Fischler and Bolles, 1981) approach. For this purpose the algorithm identifies the feature in one image that is the closest to the sampled feature from the other image based on a comparison of their descriptors (see Figure 2). Several algorithms such as the $k-$nearest neighbor (KNN) search and the radius KNN search are possible solutions to this problem. The ratio nearest search as presented by Lowe (2004) has shown the most promising results. The registration proceeds by testing a subset of 3 point pair matches and examining the two triangles that are defined by these points. The algorithm translates the triangles so that their centroids lie at the center of the common reference frame. The orientation that minimizes the error between the points is then computed by the closed form solution proposed by Horn (1987).

Depending on the projection method used for the panorama generation the optimal image size varies. The larger the image, the

This contribution has been peer-reviewed.
doi:10.5194/isprsarchives-XL-5-W4-9-2015

Figure 3: The calibration board seen from Irma3D in the Würzburg Residence as photo (left) and point cloud (right).

more detail is depicted. At the same time, however, the processing time for the registration increases. Therefore image size optimization can be carried out by calculating the optimized size of the image based on the aspect ratio of the projection to reduce the extra distortion introduced by fitting a panorama image to a fixed size image (Houshiar et al., 2013).

3.4 Camera and laser scanner co-calibration

The sensors on Irma3D used for reconstruction of the environment are the Riegl VZ-400 laser scanner from terrestrial laser scanning and the Canon EOS-1000 DSLR camera. The laser scanner acquires data with a field of view of $360° \times 100°$. The camera is equipped with a Canon EF-S 18-55 mm IS II zoom lens. Thus it has a much smaller opening angle. To achieve the full horizontal field of view the scanner head rotates around the vertical scanner axis when acquiring the data. We take advantage of this feature when acquiring image data. Since the camera is mounted on top of the scanner, it is also rotated. We acquire several images during one scanning process to cover the full $360°$. To avoid blurring and the problems that come from the necessity of synchronization we refrain from taking the images while scanning. Instead we perform a full $360°$ rotation for scanning and rotate back with stops at the image positions. A further advantage of this strategy is that the camera can be connected with regular USB cables because the cable is unwound after each rotation.

To join the data from the laser scanner and the camera automatically the two sensors have to be co-calibrated. This process is described in detail by Borrmann et al. (2012). Each sensor perceives the world in its own local coordinate system and to join the perceived information we need the specific parameters of these coordinate systems. Laser scanners are calibrated to acquire precise geometric coordinates. A camera has unique parameters that define how a point (x, y, z) in world coordinates is projected onto the image plane. These intrinsic camera parameters are calculated through a process known as geometric camera calibration. Intrinsic camera calibration is typically done using a chessboard pattern because the corners are reliably detected in the images. For determining the transformation between the camera and the scanner coordinate system, i.e., extrinsic calibration, we attach the calibration pattern onto a board that is mounted on a tripod (cf. 3). This way the board can easily be positioned at different locations and hangs almost freely in the air facilitating the the detection of the calibration board in the point cloud data. The RANSAC algorithm (Fischler and Bolles, 1981) or the Randomized Hough Transform (Borrmann et al., 2011) are common methods for plane detection in point clouds. The search area is easily reduced with a simple thresholding technique leaving the board as the most prominent plane in the data. In the Residence, the scanner was tilted when mounted onto the robot. On the one hand this enables free view of the ceiling. On the other hand, it makes removal of the floor more difficult as the scanner coordinate system is not aligned to the floor anymore and the distance between the board and the floor is small. This is easily overcome by searching for the two most prominent planes and removing the floor from the search results. To calculate the exact pose (position and orientation) of the board a plane model is generated by subsampling points on a plane with the dimensions of the calibration board. This model is transformed towards the center of the detected plane facing the same direction as the plane. The model is then fitted perfectly to the data using the ICP algorithm (see Section 3.1). Since the positions of the chessboard corners on the board are known, their exact positions in 3D space can be calculated from the pose of the board.

The precise transformation between scanner and camera is determined by acquiring several pairs of data and evaluating the reprojection error of each individual transformation. The thus calculated transformation is directly used to project the scanned points onto the camera image and color the points according to the pixel color. Due to the different fields of view of the sensors, they each perceive slightly different parts of the world. A region that is visible from one sensor might be occluded for the other sensor. When mapping the color information to the point cloud this causes wrong correspondences and therefore faulty assigned values. To solve this problem a ray tracing procedure was implemented that checks whether a point in the point cloud can be seen by the camera. We connect the point $\mathbf{P}$ and the camera position $\mathbf{C}$ with a straight line PC and select all points with a distance less than a threshold t to PC, i.e., all points $\mathbf{O}_i$ for which

$$|\mathbf{P} - \mathbf{O}_i|^2 - \frac{|(\mathbf{P} - \mathbf{O}_i) \cdot (\mathbf{P} - \mathbf{C})|^2}{|\mathbf{P} - \mathbf{C}|^2} < t^2 \quad (6)$$

holds true. If any point $\mathbf{O}_i$ lies between $\mathbf{P}$ and $\mathbf{C}$, $\mathbf{P}$ is not visible from the camera and is therefore discarded. The threshold t accounts for small inaccuracies in the calibration. To speed up the checking procedure the points are organized in a kD-tree data structure. With a quick check those voxels are immediately discarded that are not traversed by the ray and therefore all the points within are ignored.

3.5 Calibration of sensors for localization

To achieve precise pose estimates for the 3D laser scanner the localization sensors of the robot have to be calibrated to the laser scanner. The method used for odometry, IMU and the 2D laser scanner is explained in Elseberg et al. (2012). In this paper an iSpace sensor frame is attached to the VZ-400 laser scanner to localize the robot in the iSpace coordinate system. After setting up the transmitters of the iSpace system several reflective markers were attached to objects in the environment. The centers of the markers are measured with the iSpace handvector bar, thus determining their position in the iSpace coordinate system. These markers show up nicely in the reflectance data of the laser scanner. To measure their precise position first a full scan of the environment is carried out. The RiScan Pro Software is used to detect the markers in the environment. The correct markers are manually chosen from a list of automatically detected candidates. In a second step, fine scans of the markers are performed. The software controls the scanner automatically to scan the area around the selected markers with a very high resolution to determine the precise position in the local coordinate system. Third, the coordinates of the markers in the coordinate system defined by iSpace are imported as control points and the scans registered to these control points based on the marker position. This yields the position and orientation of the laser scanner in the iSpace coordinate

This contribution has been peer-reviewed.
doi:10.5194/isprsarchives-XL-5-W4-9-2015

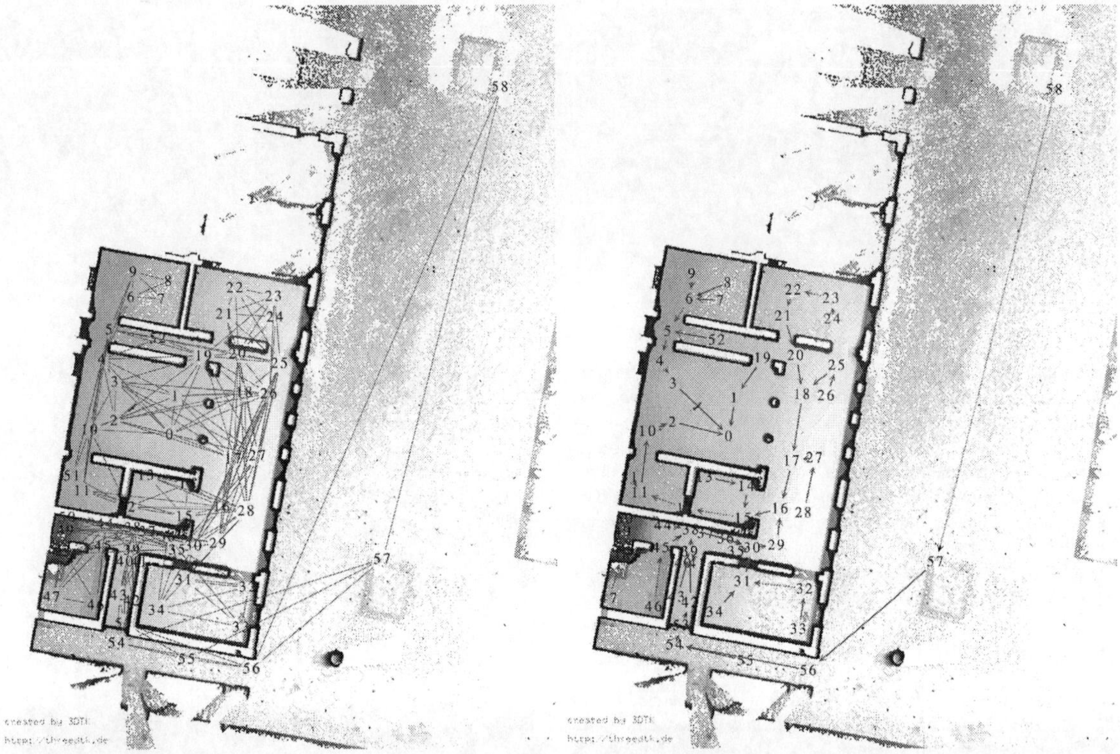

Figure 4: Floor plan of the garden house in Ostia Antica generated as an orthogonal projection of the 3D model without the protection roof. The scan positions are marked. Left: The red lines show the graph used during the global optimization of the model reconstruction. Right: The arrows indicate how the scans were registered with the feature-based registration. Green arrows with solid lines indicate the scans that were registered with the standard settings. Red dashed lines mark those pairs where a modification of the parameters was necessary. The outdoor scan pair connected with a black dotted line was not registrable at all.

system at the time the scan was taken. Additionally, the pose of the sensor frame is also recorded. In the following poses will be treated as transformation matrices $\mathbf{T}$, consisting of the rotation $\mathbf{R}$ and the translation $\mathbf{t}$. Repeating this procedure for n scans gives n pairs of poses for the Riegl laser scanner $\mathbf{T}_{r,i}$ and the sensor frame $\mathbf{T}_{m,i}$. From these poses the transformation $\mathbf{T}_{m \to r}$ between the coordinate systems is calculated as:

$$\mathbf{T}_{m \to r,i} = \mathbf{T}_{r,i} \mathbf{T}_{m,i}^{-1}. \tag{7}$$

To reduce noise the average $\mathbf{T}_{m \to r}$ over all transformation matrices $\mathbf{T}_{m \to r,i}$ is calculated. Afterwards, for each new scan position the position of the laser scanner in the iSpace coordinate system is calculated as:

$$\mathbf{T}_{r,i} = \mathbf{T}_{m \to r} \mathbf{T}_{m,i}. \tag{8}$$

4 EXPERIMENTAL RESULTS

4.1 Ostia Antica

Scanning took place over the course of 5 days. A total of 59 scans were acquired. Every morning scanner and camera were recalibrated to avoid inaccuracies caused by the transport. Also, the exposure time of the camera had to be adjusted according to the lighting conditions. The number of scans acquired per day are listed in Table 1. The first day was mainly used to prepare the environment for data collection. Especially the floor had to be cleaned from dust to allow for perceiving the floor mosaics. The first scans in the afternoon of day 1 were taken to check the functionality of all components. On day 2 the full day was reserved

Table 1: Work progress in Ostia Antica

	Day 1	Day 2	Day 3	Day 4	Day 5
# scans	6	16	16	15	6

for scanning and the number of scans could be increased. On day 3 a technical defect caused a break in the data acquisition phase but the defect could be resolved and the scanning resumed. At the end of day 4 the planned parts of the garden house had been captured. Day 5 was left to take some additional scans in areas that were hard to access and from the outside to complete the model.

A floor plan was created from a 2D orthogonal projection of the final model without the protection roof and is depicted in Figure 4. The scan positions are marked in the floor plan. The robot coordinate system is defined by the position of the robot at the start of the robot control program. Thus, for every run the coordinate system is different. We tried to overcome this issue by starting the robot at the same position every day. However, the inaccurate starting position in combination with the long and curvy robot path to the first scan of the run often lead to large errors in the pose estimates causing the odometry based mapping algorithms to fail. The discontinuity of the path also caused problems for the sequential ICP algorithm as subsequent scans often have no overlap. Therefore, the poses were manually improved and scan pairs for matching with sufficient overlap manually chosen to create a coarse registration. Afterwards, the global optimization was applied to create the final model using the graph shown in Figure 4.

After the problems with initial pose estimates from odometry the feature-based algorithm was implemented to automate the reg-

This contribution has been peer-reviewed.
doi:10.5194/isprsarchives-XL-5-W4-9-2015

Figure 5: Irma3D at the garden house in Ostia Antica.

istration process further. The floor plan was used to determine which scans should have correspondences. In Figure 4 the arrows point towards the scan each scan was registered to. Green arrows with solid lines mark the scan pairs that were successfully registered with the standard parameters. Red dashed lines mark those pairs where a modification of the parameters was necessary. The registration failed for only one scan pair, the two outdoor scans which had a distance of approximately 50 m between them and therefore very little resemblance in the panorama images. Images showing the final reconstruction are shown in Figure 6

4.2 Würzburg Residence

To evaluate the methods further experiments were carried out in the Würzburg Residence. The iSpace system was used to measure the robot position with high precision. The localization system was set up using six transmitters that were placed in the Imperial Hall as depicted in Figure 7. The robot was manually driven to 11 scanning positions. At each scanning position the position of the robot was recorded using the iSpace system. Figure 8 shows a top view of the resulting models. The scanning positions are connected by black lines. The points are color-coded based on the scanning position. Several observations can be made. First, it is obvious that the position estimates from odometry contain very large errors. Thanks to the fact that the environments consists of a single large hall it is still possible to correct the model using point-based registration. In more complex environments such as the garden house, however, the approach is likely to fail. Second, feature-based registration leads to really good results. In fact, when comparing the maps generated with iSpace measurements and FBR, the walls appear thinner for FBR, suggesting that the positions are more accurate. Third, point-based registration further improves the map quality. This can be seen by having a close look at the color distribution in the maps. Close to the individual scanning positions the density of the color of that scan becomes higher, while in the other maps the distribution is more random. This effect is the same, when comparing the results with start estimates from different methods.

Figure 10 explains the inaccuracies in the position estimates from the iSpace system. Shown are the uncertainties in estimating the position of the robot's sensor frame for each scanning position as calculated by the system. The high uncertainties come most likely from the reflections caused by the windows and the chandeliers. This makes clear, that registration based solely on the iSpace system does not yield the required accuracy. A further disadvantage of the method are the transmitters that act as obstacles during the data acquisition. In combination with the time needed for setup this makes the system infeasible for the purpose.

(a)

(b)

(c)

Figure 6: The model of the garden house in Ostia Antica. (a) two views of the data acquired on the first day. (b) The complete model with data from all days. The white lines mark the scanning sequence. (c) Outside view of the complete model.

The effectiveness of the ray tracing procedure is demonstrated with an examplary scan of the Imperial Hall in Figure 7(b)+(c). Especially behind the tripods and the chandeliers a lot of errors are removed after raytracing. In the White Hall the experiences from the Imperial Hall were repeated. The reconstruction generated from 9 scans is shown in Figure 11.

(a) Irma3D (b) No raytracing (c) Raytracing

Figure 7: (a) Irma3D in the Imperial Hall of the Würzburg Residence during calibration. In the background is one of the transmitters for the iSpace localization system and a box with two reflective markers for the calibration procedure. (b) + (c) One scan from the Imperial Hall colored with the information from the photos without (a) and with (b) correction from using raytracing. It is obvious that the raytracing methods removes wrongly colored points.

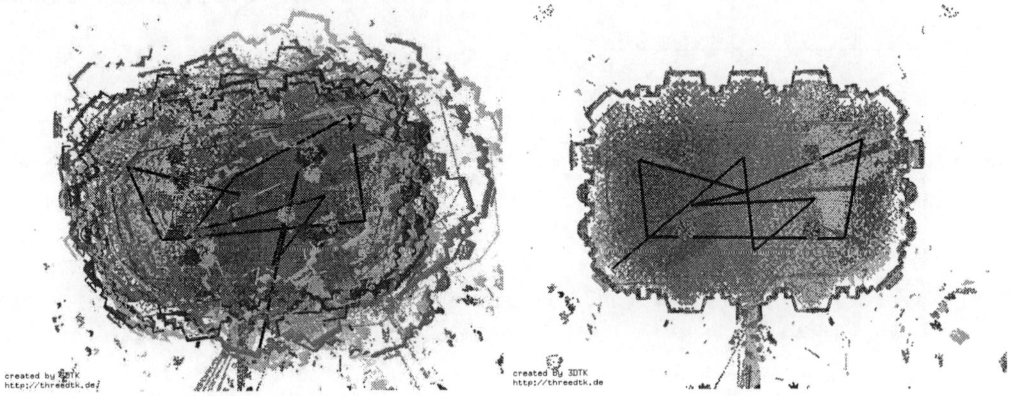

(a) Odometry + IMU (b) iSpace

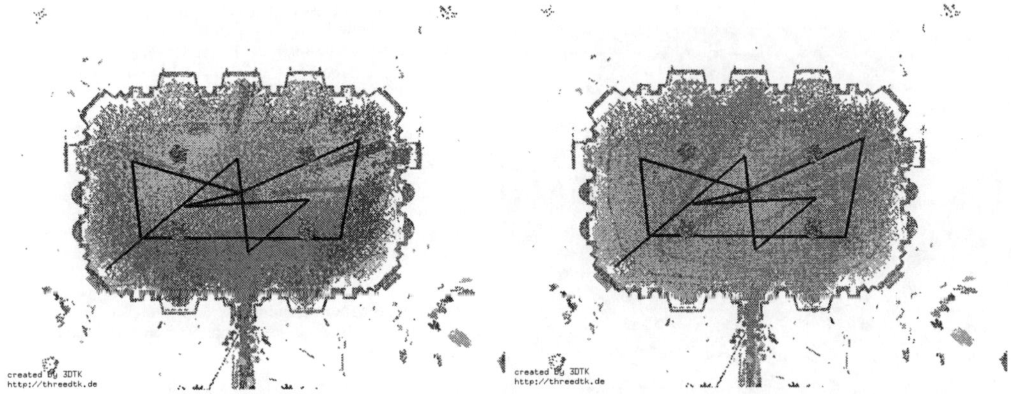

(c) FBR (d) ICP

Figure 8: Floor plan of the Imperial Hall at the Würzburg Residence. The reconstruction of the garden house in Ostia Antica. The black lines connect the scan positions. Poses from (a) Odometry + IMU, (b) the iSpace system, (c) feature-based registration, (d) the registration with the global ICP using FBR as initial pose estimates.

5 CONCLUSIONS AND OUTLOOK

The 3D reconstruction of cultural heritage sites is tedious work. The acquisition and registration of data from multiple data sources requires skilled personal and a lot of time to create a complete virtual 3D environment. In this paper we outlined steps for 3D modelling of environments by use of the mobile robot Irma3D. The robot collects simultaneously color images and 3D laser scans. Several processes are described that help automate not only the data acquisition but also registration step. These processes are evaluated in two examplary environments to demonstrate their effectiveness. These experiments show the limitations of the approaches but also outline the ability to overcome them. Future work focusses further on the evaluation of the different methods to evaluate the accuracies of the presented approach. The use of the 2D laser scanner to acquire initial pose estimates will also be evaluated in comparison to the methods presented here.

This contribution has been peer-reviewed.
doi:10.5194/isprsarchives-XL-5-W4-9-2015

Figure 9: Reconstruction of the Imperial Hall in the Würzburg Residence. Visible in the hall are four of the six transmitters from the iSpace localization system.

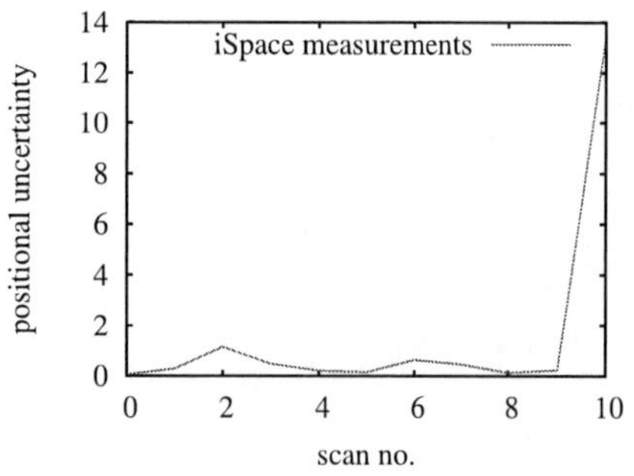

Figure 10: Uncertainties of the pose estimates from the iSpace localization system.

Figure 11: Reconstruction of the White Hall in the Würzburg Residence. In the close-up view the impressive stucco work is clearly visible.

6 ACKNOWLEDGEMENTS

We are grateful to the Soprintendenza Speciale per i Beni Archeologici di Roma - Sede di Ostia (A. Pellegrino) and Università Sapienza di Roma (S. Falzone) for the work permit and the coordination at Ostia Antica and to the Bayerische Verwaltung der staatlichen Schlösser, Gärten und Seen for access to the Würzburg Residence. We would like to thank Jan Elseberg, Norbert Zimmermann and Irmengard Mayer for their help with the experiments in Ostia Antica.

References

Andreas Nüchter et al., 2014. 3DTK – The 3D Toolkit. Webpage. http://www.threedtk.de.

Besl, P. J. and McKay, N. D., 1992. A Method for Registration of 3-D Shapes. IEEE Transactions on Pattern Analysis and Machine Intelligence 14(2), pp. 239 – 256.

Borrmann, D., Afzal, H., Elseberg, J. and Nüchter, A., 2012. Mutual Calibration for 3D Thermal Mapping. In: Proceedings of the 10th Symposium on Robot Control (SYROCO), Dubrovnik, Croatia.

Borrmann, D., Elseberg, J., Lingemann, K. and Nüchter, A., 2011. The 3D Hough Transform for Plane Detection in Point Clouds – A Review and a new Accumulator Design. 3D Research.

Borrmann, D., Elseberg, J., Lingemann, K., Nüchter, A. and Hertzberg, J., 2008. Globally consistent 3d mapping with scan matching. Robotics and Autonomous Systems 56(2), pp. 130 – 142.

Bruder, G., Steinicke, F. and Nüchter, A., 2014. Immersive Point Cloud Virtual Environments. In: Proceedings of IEEE Symposium on 3D User Interfaces 3DUI Proceedings of IEEE Symposium on 3D User Interfaces (3DUI '14), pp. 161–162. Poster.

Elseberg, J., Borrmann, D. and Nüchter, A., 2012. Automatic and Full Calibration of Mobile Laser Scanning Systems. In: Proceedings of the 13th International Symposium of Experimental Robotics (ISER '12), Vol. 79, Springer Tracts in Advanced Robotics, Quebec City, Canada.

Fischler, M. A. and Bolles, R. C., 1981. Random Sample Consensus: A Paradigm for Model Fitting with Applications to Image Analysis and Automated Cartography. Communications of the ACM 24, pp. 381 – 395.

Horn, B. K. P., 1987. Closed–form Solution of Absolute Orientation using Unit Quaternions. Journal of the Optical Society of America A 4(4), pp. 629 – 642.

Houshiar, H., Elseberg, J., Borrmann, D. and Nüchter, A., 2012. A Study of Projections for Key Point Based Registration of Panoramic Terrestrial 3D Laser Scans. Journal of Geo-spatial Information Science.

Houshiar, H., Elseberg, J., Borrmann, D. and Nüchter, A., 2013. Panorama based point cloud reduction and registration. In: Proceedings of the 16th IEEE International Conference on Advanced Robotics (ICAR '13), Montevideo, Urugauy.

Lowe, D. G., 2004. Distinctive Image Features from Scale-invariant Keypoints. International Journal of Computer Vision.

Nikon Metrology, 2014. iSpace – Portable Metrology System User Manual and Startup Guide. Webpage. http://www.nikonmetrology.com.

Nüchter, A., Elseberg, J., Schneider, P. and Paulus, D., 2010. Study of Parameterizations for the Rigid Body Transformations of The Scan Registration Problem. Journal Computer Vision and Image Understanding (CVIU) 114(8), pp. 963–980.

Ostia Antica, 2014. SOPRINTENDENZA SPECIALE PER I BENI ARCHEOLOGICI DI ROMA – SEDE DI OSTIA. Webpage. http://www.ostiaantica.beniculturali.it.

Würzburg Residence, 2014. Bayerische Verwaltung der staatlichen Schlösser, Gärten und Seen. Webpage. http://www.residenz-wuerzburg.de.

3D SURVEYING & MODELING OF UNDERGROUND PASSAGES IN WWI FORTIFICATIONS

P. Rodríguez-Gonzálvez [a], E. Nocerino [b], F. Menna [b], S. Minto [b], F. Remondino [b]

[a] Department of Cartographic and Land Engineering, High School of Ávila, University of Salamanca, Ávila, Spain
Email: pablorgsf@usal.es
[b] 3D Optical Metrology (3DOM) unit, Bruno Kessler Foundation (FBK), Trento, Italy
Email: (nocerino, fmenna, minto, remondino)@fbk.eu

Commission V, WG V/4

KEY WORDS: Digitization, Reconstruction, Modelling, Photogrammetry, Laser scanning, Cultural Heritage, Visualization

ABSTRACT:

The virtual reconstruction of subterranean structures is a suitable scenario for the integration of different geomatics techniques although narrow passages, lack of light and irregular surface can arise various problems in the data acquisition as well as processing procedures. Generally the final product is a dense and detailed 3D model, whose number of triangles increases quickly according to the complexity of the object. This complexity reduces the efficient use and dissemination of the produced information therefore innovative solutions are sought. The article presents the 3D surveying and modelling of underground passages of World War I (WWI) fortifications. After the acquisition of dense point clouds by means of terrestrial scanning (TLS), a simplification and optimization workflow is performed with the aim of generating a lightweight product that keeps the maximum amount of significant information. A continuous scene representation with a 87% triangle reduction is generated, while the final precision is preserved according to a tolerance predefined by the final user. Such 3D product can be employed as basis for reconstruction, consolidation, preservation and valorisation of the WWI tunnels.

Figure 1. a) The entire complex of WWI fortifications in Lusern (Trento, Italy) with the surveyed tunnel, ca 100 m long with a section of 2 x 2 m (after [http://de.wikipedia.org/wiki/Werk_Lusern]). b) *Werk* Lusern from the air (source: www.trentinograndeguerra.it). c) The entrance of a tunnel.

1. INTRODUCTION

The paper presents the on-going interdisciplinary project VAST (VAlorizzazione Storia e Territorio – Valorization of History and Landscape, http://vast.fbk.eu/; Nocerino et al., 2014) which aim to develop a multi-technique approach for the 3D modelling of Austro – Hungarian World War I (WWI) fortifications located in the Trentino region (nowadays part of Italy). The work is part of cultural heritage valorisation activities related to the celebrations for the centenary of the WWI, promoted by local authorities to valorise military buildings and areas that were the scene of important and crude battles. Before the end of WWI, the Trentino – Alto Adige region was part of the Austro–Hungarian Empire and constituted a fundamental strategic area, one of the most dangerous frontiers of the Empire. To protect and reinforce the territory, more than 50 fortifications (*werk* in German) were built, many of them constructed on plateaus and mountain tops for monitoring and protecting the border against the Italian army. The fortifications, built in concrete and hand-carved local stone, were made up of a main body, secondary casemates and several underground passages. Nowadays, the great majority of these buildings are ruined and the main idea behind the VAST project is to virtually reconstruct them to valorise these heritage monuments, allow virtual sightseeing and produce communication material.

The multi-technique methodology so far developed for the 3D reconstruction of WWI fortifications (Nocerino et al., 2014) is based on reality-based techniques (photogrammetry, laser scanning, topographic surveying) integrated with Computer Aided Architectural Design (CAAD) methods. While the previous work was focused on the 3D modelling of the main building of the fortifications, in this contribution the virtual reconstruction of subterranean passages is presented.

2. RELATED WORKS AND WORK AIMS

Such tunnels were excavated to connect adjacent fortifications or as evacuation passages. Most of them are now blocked due to collapses and need to be restored to ensure their proper preservation and allow accesses to visitors. Therefore a proper and detailed 3D survey is mandatory. In underground environments, several geomatics techniques have been applied and described in literature: –3D active sensors such as terrestrial laser scanning – TLS (Beraldin et al., 2011; Roncat et al., 2011; González-Aguilera et al., 2011; Remondino et al., 2011, photogrammetry (Roncella et al., 2012; Arles et al., 2013), mapping mobile systems (Ziparo et al., 2013; Zlot and Bosse, 2014) or 3D SLAM (Ferreira et al., 2012).

This contribution has been peer-reviewed.
doi:10.5194/isprsarchives-XL-5-W4-17-2015

Each of these methods has its own advantages and disadvantages. TLS is normally the most suited technology for tunnel documentation, but when the space is limited and the shape very elongated and linear (such as in the case of underground passages of fortifications) the number of TLS stations increases exponentially, the network configuration becomes weak for standard alignment procedure with an ICP algorithm and the precision of 3D points is reduced due to the slant incidence angles (Soudarissanane et al., 2009). On the other hand, photogrammetry can directly provide both geometric and high resolution RGB information, although underground tunnels are unfavourable environments for passive sensors with limited field of view and poor or completely absent illumination.

This paper deals with the 3D surveying and modelling of underground passages (tunnels) of WWI fortifications located in the Trentino plateaus and highlands of Luserna, Lavarone and Folgaria, in Italy (Fig. 1). For some of these tunnels (Fig. 2), original maps are stored in national archives. Such drawings are usually mainly sketches that contain minimal information about the predesigned shape, generally different from the one actually built (time restrictions for building adaptation to the geology, etc.). Since the actual shape differs significantly from the original design plans, TLS and photogrammetric surveys have been carried out to record both the real geometry and colour information. This approach provides a high dense and resolute 3D model, with a huge number of points and/or triangles, not directly suitable for dissemination and virtual interaction due to limits of currently available hardware and software. To this end, a final lightweight product is required and a simplification process should be carried out. In this work, in order to obtain an optimal mesh model with the minimum loss of significant information different approaches are tested. These approaches are designed and implemented to comply with the project requirements (i.e. tolerance and level of detail to be represented) taking into consideration the characteristics of different kinds of tunnels.

As result, the final simplified 3D model could be easily integrated in a visualization tool (Jiménez Fernández-Palacios et al., 2013) for cultural dissemination. Furthermore, as the produced 3D model is precisely geo-referenced (by GNSS) with the rest of the fortifications structures, it could also be used for further analyses and interpretations by experts (historians, conservators, etc.).

3. AREA OF STUDY

The analysed tunnel (Fig. 2) belongs to the *werk* Lusern (Fig. 1), connects two outposts of the military fortification named Viaz and Oberwiesen. It is more than 100 m long with a section of about 2 x 2 m. Some lateral rooms, not reported in the original military drawings, are also present and were probably used to store material or as resting place. The tunnel walls and ceiling surfaces are very rough, due to the construction techniques moreover in some areas the ceiling has collapsed. The irregular surfaces and narrow passages, often adapted to the rock stratum, complicated the data acquisition, due to obstacles not considered in the planning of the survey.

4. DATA COLLECTION AND PROCESSING

4.1 Equipment

The surveying of the tunnel was performed with a terrestrial laser scanner - TLS (Table 1, Fig. 2 right), a total station - TS (Table 2) as well as a digital camera (Table 3).

The total station was used for the accuracy assessment of the final 3D model and for geo-referencing reasons. Along the tunnel, several high contrast targets were placed every 25 m on the walls and ceiling to serve as GCPs (Fig. 2 left and middle). The targets were designed for laser and surveying, with binary code and a circle diameter of 35 mm. These targets were surveyed with a topographic network while two exterior bases were observed to link the interior survey with a global coordinate system.

	Leica HDS7000
Type	Phase shift
Wavelength	1500 nm
Angular precision (H & V)	125 µrad
Beam divergence	< 0.3 mrad
Range noise (black 14%)	0.5 mm rms @ 10 m
Spot size	6.8 mm @ 10 m

Table 1. Terrestrial Laser Scanner technical specifications.

	Topcon GPT7001
Range measurement accuracy (non-prism)	±5 mm
Range (non-prism)	1.5 to 250 m
Angle measurement accuracy (non-prism)	1"
Tilt correction	Dual axis
Compensating range	±4"

Table 2. Total Station technical specifications.

Figure 2. The tunnel - connecting the Viaz and Oberwiesen outposts of the Lusern fortification - with the topographic targets (left and middle) and a TLS station (right).

This contribution has been peer-reviewed.
doi:10.5194/isprsarchives-XL-5-W4-17-2015

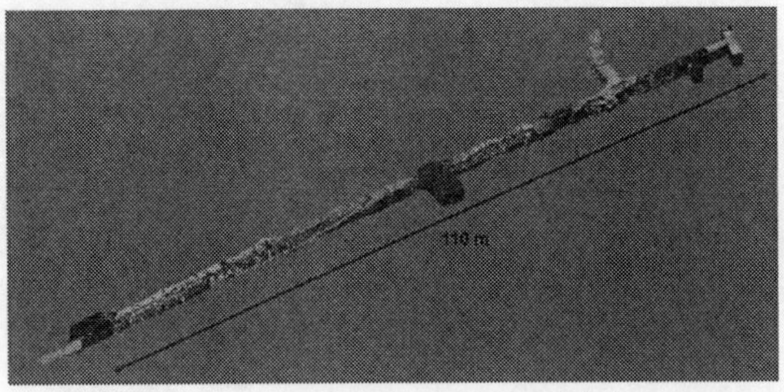

Figure 3. TLS 3D data after the alignment of all the scans.

The TLS survey involved a total number of 19 stations and ca 285 million points (average spatial resolution of ca 4 mm). The TLS stations were distributed uniformly in the central passage and in the discovered lateral rooms, with a mean distance between stations of ca 8 m. This distance was chosen according to the average tunnel section, which affected the point density distribution, the point precision (due to the high incidence angle) and occlusions caused by the rough surface.

The photogrammetric survey was performed mainly for texturing purposes (Table 3). The images were acquired using an external flash attached to the camera due to the absence of illumination inside the tunnel. About 400 photos were acquired with a mean ground sample distance (GSD) of 2 mm.

	Nikon d3x
Sensor type	FX CMOS
Sensor size	35.9 × 24.0 mm
Resolution (effective pixels)	24.5 Mpx
Image size	6048 × 4032 px
Pixel density	2.8 MP/cm^2
Pixel size	5.95 μm
Employed focal length	14 mm

Table 3. Digital camera technical specifications.

4.2 Data processing

The complete range-based 3D point cloud of the tunnel was obtained employing a combination of classical range data processing procedures. Firstly, five scans ("key stations") with an optimal view of the GCPs were aligned to the local coordinate system with a solid-rigid transformation (6 parameters). Secondly, the adjacent TLS stations were aligned to the key stations through a standard ICP algorithm (Bels and McKay, 1992). Finally, all stations, with the exception of the central one, were freed to refine the alignment using an implementation of the ICP algorithm. The final ICP error was 3.6 mm, employing ca 420,000 points in every pair of adjacent stations.

The accuracy assessment of the registered TLS data inside the entire tunnel (Fig. 3) was performed using the available GCPs distributed on the wall and ceiling of the tunnel.

A spatial similarity transformation was used as mathematical model.

A total of 54 GCPs were employed and a blunder rejection test was applied (Pope, 1976) according to the standardized residuals (w):

$$w_i = \frac{(\mathbf{P}\cdot v)_i}{\hat{\sigma}_0\sqrt{(\mathbf{P}\cdot\mathbf{Q}_{vv}\cdot\mathbf{P})_{ii}}} \tag{1}$$

where v_i is the i-th residual of the transformation, $\mathbf{P}$ is the weight matrix, $\mathbf{Q}_{vv}$ the cofactor matrix of residuals and σ_o is the a-posteriori standard deviation. Since the residuals are computed using the a posteriori variance, they will not be normally distributed and will rather follow a Tau distribution. The critical value of the Tau distribution may be computed from the tables of the t-student distribution (Heck, 1981) according to:

$$\tau_{r,\alpha_0/2,} = \sqrt{\frac{r\cdot\left(t_{r-1,\alpha_0/2}\right)^2}{r-1+\left(t_{r-1,\alpha_0/2}\right)^2}} \tag{2}$$

where r are the degree of freedom of the adjustment and α_0 the significance level for a single observation. The value of α_0 is computed from the total significance level (α) and the number of observations (n):

$$\alpha_0 = 1-(1-\alpha)^{\frac{1}{n}} \approx \frac{\alpha}{n} \tag{3}$$

According to Eq. 1, none of the GCPs employed were found to be blunders at 5% of confidence level, since the maximum standardized residual was 2.89 (the threshold established by Eq. 2 was 3.19). Since the Pope test could keep more observations that would normally be discarded as outliers by the normal distribution, the Baarda test (Baarda, 1968) was also computed to assure the data quality and given the high number of control points. This test is computed in a similar way as Eq. 1, but employing the a-priori standard deviation. As result a maximal normalized residual value of 2.76 was obtained, while the critical value of Baarda test for the same confidence level as Pope - and a power of the test (β) of 20% - established the

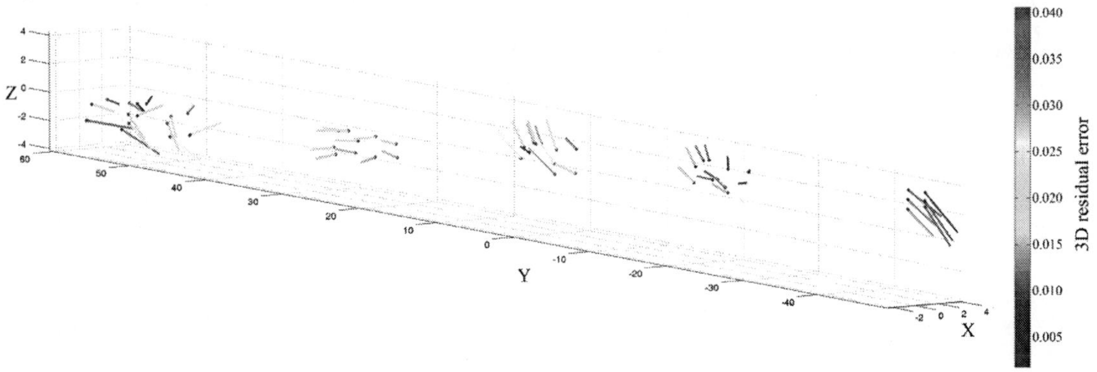

Figure 4. 3D residual error of the targets used to co-register all the TLS scans with a 6 parameters transformation. The standard deviations resulted: $std_X = 1.55$ cm, $std_Y = 0.6$ cm, $std_Z = 1.37$ cm, respectively.

This contribution has been peer-reviewed.
doi:10.5194/isprsarchives-XL-5-W4-17-2015

threshold at 2.8, thus confirming the absence of blunders. The final accuracy assessment of the survey is showed in Figure 4, along with the spatial distribution of the errors (note that the Y axis is aligned with the tunnel main axis). The standard deviation along the tunnel axis is 2.2-2.6 times smaller with respect to the other two directions. This can be explained by the fact that the targets were located in the lateral walls but facing always the Y-axis for a proper surveying and to avoid the precision degradation due to incident angle (Fig. 2). The uncertainty related to centre marking, which implies an error in the target plane, is related to the target definition, the target's colour and the tunnel environment (mainly the humidity).
The obtained a-posteriori standard deviation from the solid-rigid transformation was 12.5 mm. This value complies with the a-priori expected precision as the precision from the adjustment of the total station survey was in average 1.3 mm. On the other hand, the a-priori error of 3.0 mm (at 10 metres) was estimated for the TLS (Table 1) according to the error propagation of Reshetyuk (2009):

$$\sigma=\sqrt{\sigma_{\rho}^{2}+\left(\sigma_{\varphi}^{2}+\sigma_{beam}^{2}\right)+\left(\sigma_{\theta}^{2}+\sigma_{beam}^{2}\right)} \qquad (4)$$

where σ_ρ, σ_φ and σ_θ are the standard deviation of the range, horizontal angle and vertical angle measurement, respectively. The term σ_{beam} is computed on the basis of the beam divergence (γ), according to (Lichti and Gordon, 2004):

$$\sigma_{beam}=\pm\frac{\gamma}{4} \qquad (5)$$

The final error propagation taken into account both error sources (total station and TLS) and the total number of stations (19) reaches up to 13.1 mm, for the worst scenario of only two consecutive overlapping clouds. This result is statistical compatible with the obtained a-posteriori standard deviation.

Before the successive steps, the TLS point cloud (285 million of points) was reduced to fulfil the VAST project requirements, producing a 1 cm spatial resolution 'integrated 3D point cloud'.

5. POST-PROCESSING

Most common surface reconstruction algorithms implemented in reverse modelling commercial packages try to wrap a surface over a set of points while keeping their original coordinates (Ramos and Sadri, 2007). These algorithms aim at the most faithful representation of the produced geometric model on the basis of the available data. In case of irregular point sampling steps, the final geometric model may present surface holes, worsening the final look of the model. Therefore, further time-consuming post-processing operations, using for example hole filling algorithms, will be mandatory to achieve a proper visual appearance of the final 3D model.
After the creation of a complete and topologically correct 3D model, a decimation process is normally needed to adapt it to the user requirements that are mostly based on spatial resolution and/or number of triangles. However, these parameters do not take into account the visual perceived representation of the information, being merely based on sampling theory.
In order to achieve a more convenient visualization of large and complex 3D models - such as the WWI tunnels - while still keeping low the total number of triangles, an optimization process was designed. The simplification process has been carried out with free software, namely CloudCompare (2015) and MeshLab (Cignoni et al., 2008), developing a systematic procedure that can allow the traceability of the whole process and avoiding the human subjective interpretation in the modelling process (Fig. 5).
The principle that drives the simplification procedure is that an end user should be able to define the total number of triangles or the required spatial resolution. Ideally, considering equilateral triangular meshes, the target number of triangle (N) could be derived applying simple geometrical considerations:

$$N=\frac{4\cdot A}{\sqrt{3}\cdot L^{2}} \qquad (6)$$

where A is total lateral area of the meshed object, L is the face edge length (equal to the user-defined spatial resolution) and N the number of triangles.
In the following subsections the different simplification approaches and results on the Lusern tunnel are described.

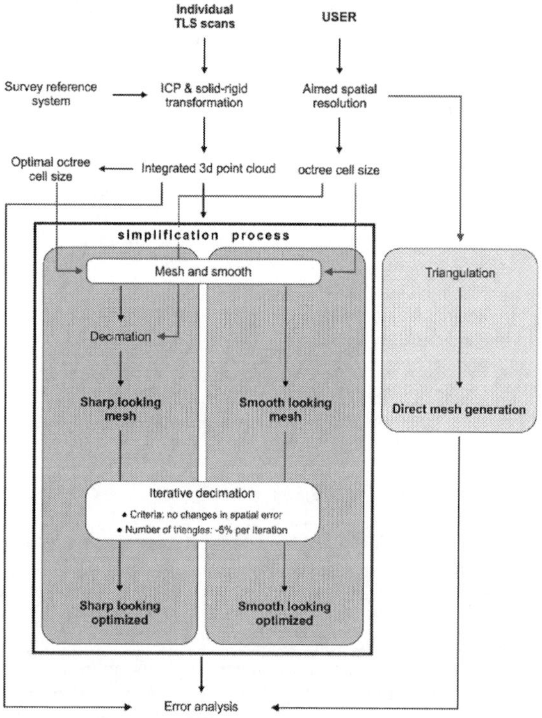

Figure 5. The proposed simplification pipeline (bolded area) for the post-processing of the surveyed data.

5.1 Smooth-looking

The first simplification approach here proposed consists in a straightforward mesh generation procedure where the Poisson algorithm (Kazhdan et al., 2006) implemented in CloudCompare (CloudCompare, 2015) is applied to the input point cloud. The mesh resolution is determined by the octree level, chosen according to the user defined spatial resolution.
That means that the target spatial resolution, or number of triangles, is achieved directly in one single step. Unlike a direct mesh generation process, which usually requires mesh editing operations such as filtering and refinement (Attene, 2010; Varnuška et al., 2005), the Poisson algorithm directly encloses smoothing step and provides a continuous geometry. These characteristics are very important when further simplification

This contribution has been peer-reviewed.
doi:10.5194/isprsarchives-XL-5-W4-17-2015

procedures, such as reverse modelling (Herrero-Huerta et al., 2015), are required.

5.2 Sharp-looking

For a sharp-looking mesh, the proposed method involves a two-step procedure: (i) a high spatial resolution mesh generated with Poisson algorithm from the input point cloud and (ii) a successive geometric optimization in Meshlab.
Unlike the one-step method presented in section 5.1, in this case the first action produces a polygonal mesh (named "intermediate mesh") with a spatial resolution as high as the resolution of the original point cloud. In other words, the octree level is chosen to fit the input point cloud sampling step.
In order to reduce the number of polygons of the intermediate mesh to accomplish the user-defined final spatial resolution, the quadratic edge collapse algorithm (Garland and Heckbert, 1997) implemented in Meshlab is applied. In this way the final result preserves a high level of perceived detail while the number of triangles is reduced for visualization purposes.

5.3 Further mesh optimization

Both methods described in the previous sections take into account only the spatial resolution as procedural criteria. An improvement is here proposed consisting in further reducing the final number of triangles through the "collapse" of areas without relevant information. The idea behind this additional step is that the final number of triangles can be further reduced beyond the number of triangles originally defined by the user, without losing significant level of detail according to a predefined tolerance. For this purpose, an iterative process is proposed where the mesh derived from the previous procedures (both sharp and smooth-looking) is "collapsed" of 5% of total number of triangles using the quadratic edge collapse algorithm. If the resulted mesh error against the original input point cloud remains unchanged (with respect to the previous iteration), the collapse process continues. An iterative procedure is required since the quadratic edge collapse algorithm implementation does not allow decimation at fixed spatial resolution.
Between the two methods in sections 5.1 and 5.2, the smooth looking approach is more susceptible to achieve high level of reduction, since there is a high percentage of areas lacking relevant information.

5.4 Results of mesh optimization for the Lusern fortification's tunnel

For the studied area, different geometric resolution of the final 3D tunnel were investigated keeping in mind the future reconstruction, consolidation, preservation and valorisation purposes. Starting from the integrated 3D point cloud (1 cm spatial resolution), the resolution of the final 3D geometric model was set to 10 cm for two reasons:

1. Such resolution is a good compromise for an easy manipulation and fruition of the 3D tunnel.
2. The 3D model will be integrated with the other 3D products of the VAST project. Most of these outcomes (Nocerino et al., 2014) are CAAD-based models, produced from original drawings at various scale (1:50 up to 1:800). Considering a common drawing error of 0.25 mm, an average resolution - and accuracy - of 5 cm should be chosen. But additional error sources (paper deteriorations, digitization error, draw idealizations, etc.) must be considered thus leading to a deterioration of the results.

This reference value implies, that for an ideal meshing of equilateral triangles, a total number N of 247,350 faces should be obtained (Eq. 6), where the lateral initial area A is computed from a coarse approximate mesh based on alpha-shape computation (Edelsbrunner and Mücke, 1994).
The three different scenarios described in 5.1, 5.2 and 5.3 were tested. Firstly, for a **smooth looking** mesh case, the original point cloud was directly meshed and smoothed by the Poisson algorithm, for the 10 cm average spatial resolution objective (172,876 triangles).
Secondly, for a **sharp looking** mesh case, the intermediate mesh resulted in 12 million triangles, which was simplified, in the second step, by quadratic edge collapse algorithm for the 10 cm average final spatial resolution goal. A final number of 168,936 triangles (4^3 times reduction, where the exponent is related to the octree level) was obtained.
Thirdly, the sharp **optimized** method was tested. In this case, the number of triangles of both the meshes is optimized, collapsing the areas without relevant information. As final result, the number of triangles is reduced of a 14% (which corresponds to 144,842 triangles) and 34% (114,690 triangles) for the sharp and the smooth respectively. Against the direct mesh generation, the reduction percentage reaches up to 10% (15,598 triangles less) and 29% (45,750 triangles less).

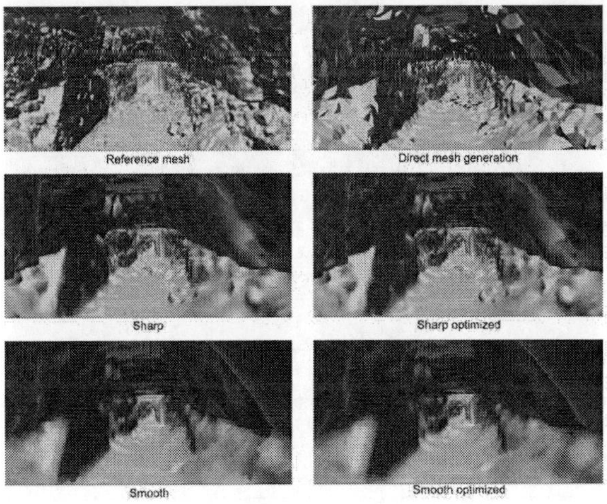

Figure 6. Example of the simplification approach in a high-detail area of the underground passage (10 cm mesh).

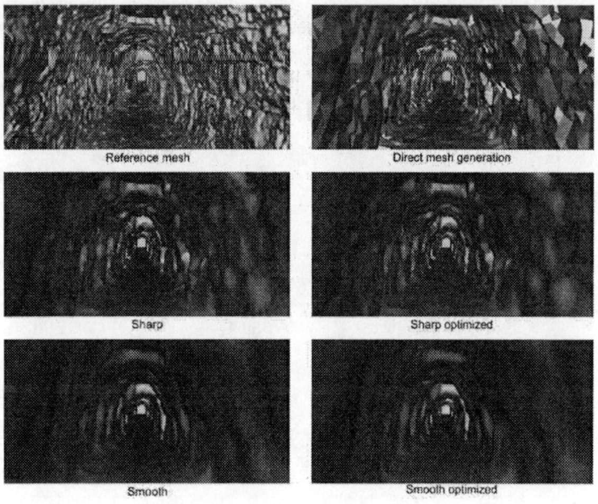

Figure 7. Example of the simplification approach in a mid-detail area of the underground passage (10 cm mesh).

This contribution has been peer-reviewed.
doi:10.5194/isprsarchives-XL-5-W4-17-2015

Finally, the result of a **direct mesh generation** without smoothing for the objective resolution is compared. In this case, the number of triangles, is similar to the previous one (160,440 triangles), but it is affected by the presence of several holes (approx. 1800), which worsen the final model appearance.
In Figure 6 and 7, the four different simplification procedures (named accordingly to the previous section) are showed for two different areas of the tunnel, together with the maximum resolution meshing (named reference mesh), and the direct mesh generation result (named direct triangulation).

	Sharp looking	*Sharp optimized*
Number of triangles	168,936	144,842
Signed error against input point cloud	+0.1 ± 7.9 mm	+0.1 ± 7.9 mm
Average spatial resolution	105.9 mm	114.0 mm
Spatial resolution C.I. (1 σ)	[91.1 - 131.8] mm	[97.5 - 143.1] mm

Table 4. Result of the sharp simplification method.

	Smooth looking	*Smooth optimized*
Number of triangles	172,876	114,690
Signed error against input point cloud	-0.6 ± 22.5 mm	-0.6 ± 22.5 mm
Average spatial resolution	112.8 mm	136.0 mm
Spatial resolution C.I. (1 σ)	[104.2 - 123.9] mm	[122.7 - 154.8] mm

Table 5. Result of the smooth simplification method.

Tables 4 and 5 summarises the results achieved by the presented methodologies, the final spatial resolution (and its confidence intervals) and the error distance of the 3D model from the original unprocessed point cloud. The average spatial resolutions, obtained in the basis on ideal equilateral triangles for the non-optimized methods, are very similar to the reference values, as well as to the direct mesh generation (mean resolution of 104 mm).
Regarding the model error, the sharp looking method achieved the best precision with a standard deviation of 8 mm, while the direct mesh generation reached up to 18 mm. As far as the smooth method is concerned, it is aimed to an extra geometric simplification, therefore it shows the highest final error (23 mm). In none of the models a significant systematic bias was detected. As example of the graphical error distribution for the sharp looking approach is showed in Figure 8, where no bias is showed nor patterns in error distribution.

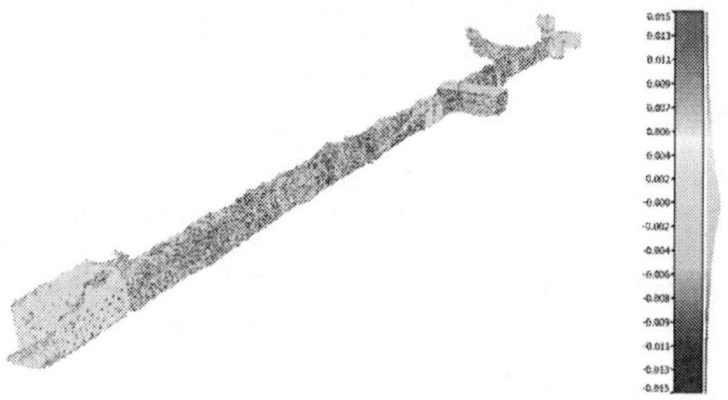

Figure 8. Signed distance error for the sharp looking methodology against the original TLS point cloud.

In table 6 the result for the spatial resolution of 5 cm are showed. The reference number of triangles is 989409, but with a presence of more than 7500 holes. The improvement methods allowed a triangle reduction of 9% and 64% (sharp and smooth looking) against the direct triangulation. The model errors between the different approaches are distributed in a similar way than the 10 cm test. For a coarser resolution (20 cm) the reduction achieved is the 47% and 88% (sharp and smooth looking) against the direct mesh generation.
All the tests so far presented were done on the basis of the Poisson octree level. However, there are additional parameters that could be affected the quality of the meshing process. The first is the octree cell size that was evaluated in the experimental results. The second more relevant is the minimum number of sample points that should fall within an octree node as the octree construction is adapted to sampling density. According to Kazhdan et al., (2006), this parameter ranges from 1 to 5 for noise free data, and from 15 to 20 for noisy data. Since the precision of the data acquisition is higher than point cloud used for the evaluation, the employment of one sample per node is possible. However, the same cases evaluated previously were computed with 5 samples per node which delivers a smoother surface, without significant changes. The precision of the smooth approach was slight reduced (up to 19 mm for the 10 cm case), while in the sharp one remains without changes. The only significant effect was that the optimization methods efficiency was reduced, so the parameters should be adapted to geomatics acquisition techniques.

5.5 Texturing of the optimised mesh of the Lusern fortification's tunnel

Finally the optimized sharp looking model at 10 cm resolution, as best compromise between geometric reduction, level of detail and discrepancy with the original data, was selected for the final texturing procedure. The acquired images were firstly relatively oriented using an automatic procedure. In order to mitigate the error accumulation when orienting large elongated

	Sharp looking	*Sharp looking optimized*	*Smooth looking*	*Smooth looking optimized*	*Direct mesh generation*
Numbe of triangles	723,644	593,390	725,06	234,548	651,343
Signed error against input point cloud	+0.1 ± 5.7 mm	+0.1 ± 5.7 mm	-0.2 ± 11.9 mm	-0.2 ± 12.1 mm	-0.0 ± 7.4 mm
Average spatial resolution	51.9 mm	57.5 mm	54.4 mm	91.3 mm	51.4 mm
Spatial resolution C.I. (1 σ)	[45.4 - 62.1] mm	[50.4 - 68] mm	[49.8 - 60.7] m	[79.4 - 110.8] mm	[45.3 - 60.7] mm

Table 6. Results of the simplification methods for a 5 cm target resolution.

This contribution has been peer-reviewed.
doi:10.5194/isprsarchives-XL-5-W4-17-2015

blocks of images (Nocerino et al, 2014), the 3D coordinates of some topographic targets (4 at the main entrance 3 in the middle and 4 at the exit) were used as GCPs to define the absolute orientation and scale (final spatial RMSe of 8 mm on check points) and bring the photogrammetric output into the TLS reference system. Finally the images were projected onto the 3D geometric model using also a blending process to reduce the texture jumps caused by the light source (Fig. 9).

Figure 9. The final photorealistic 3D model of the tunnel, with the texture applied over the sharp optimized 10 cm mesh.

6. CONCLUSIONS AND FUTURE PERSPECTIVES

The paper presented the 3D surveying and modelling of the underground passages on WWI fortifications. These complex scenarios were surveyed with TLS to record millions of points which needed massive post-processing in order to be easily used by non-experts. A methodology was proposed showing that a final model >85 times lighter than the original (144,842 triangles vs 12 millions triangles at 10 cm resolution) can be derived, still preserving significant geometric information.

The degree of improvement in the number of triangles is higher if the final geometrical resolution decreases. Also, the smooth error is related to the Poisson process, and adds an extra component in the final error, explaining the higher values that exceed the user spatial tolerance. In the case of the sharp looking approach the signed error against the point cloud are lower than the desired spatial resolution.

In the evaluated cases, a small bias is detected for the smoothed meshes, since in this process the data experiences a little shrink. Although the sharp looking approach has the lower performance (from 9% to 47% reduction), the significant information conserved is higher, as is shown visually (qualitative) and in the precision parameters (quantitative).

The proposed workflow requires further analysis, since for the 10 cm resolution the results obtained in the optimized smooth mesh were inferior to the corresponding 5 and 20 cm cases. The reduction in a 29% of the triangles is away from the 64% and 88% achieved in the other cases.

As future work of the VAST project, we aim to test a photogrammetric approach to survey and model further underground tunnels, exploiting its low-cost approach and it additional advantage of the texture acquisition.

ACKNOWLEDGEMENTS

The VAST project is funded by *Fondazione Cassa di Risparmio di Trento e Rovereto* – CARITRO. The authors are really thankful to all the partners (*Centro Documentazione Luserna, Istituto Cimbro Luserna, Centro Internazionale Studi di Archeologia di Superficie, Fondazione Forte Belvedere*) involved in the VAST project (http://vast.fbk.eu/) as well as *Archivio Storico di Trento* and the *Azienda per il Turismo a Folgaria, Lavarone e Luserna* for their fundamental contribution and valuable support in the bibliographic and historical research.

REFERENCES

References from Journals:

Attene, M., 2010. A lightweight approach to repairing digitized polygon meshes. *The visual computer*, 26 (11), pp. 1393-1406.

Besl, P., McKay, N. A., 1992. Method for registration of 3-D shapes. *IEEE transactions on pattern analysis and machine intelligence*, 14, pp. 239-256.

Edelsbrunner, H., Mücke, E. P., 1994. Three-dimensional alpha shapes. *ACM Transactions on Graphics (TOG)*, *13*(1), pp. 43-72.

González-Aguilera, D., Muñoz-Nieto, Á., Rodríguez-Gonzálvez, P., Menéndez, M., 2011. New tools for rock art modelling: automated sensor integration in Pindal Cave. *Journal of archaeological science*, 38 (1), pp. 120-128.

Heck, B., 1981. Der einfluß einzelner beobachtungen auf das ergebnis einer ausgleichung und die suche nach ausreißern in den beobachtungen (Examining the effects of observations to the adjustment results and analysis of outliers). *Allgemeine vermessungs-nachrichten*, 88, pp.17-34.

This contribution has been peer-reviewed.
doi:10.5194/isprsarchives-XL-5-W4-17-2015

Herrero-Huerta, M., González-Aguilera, D., Rodriguez-Gonzalvez, P., Hernández-López, D., 2015. Vineyard yield estimation by automatic 3D bunch modelling in field conditions. *Computers and electronics in agriculture*, 110, pp. 17-26.

Remondino, F., Rizzi, A., Jimenez Fernandez-Palacios, B., Agugiaro, G., Baratti, G., De Amicis, R., 2011. The Etruscan in 3D: from space to underground. *Geoinformatics FCE CTU journal*, 6, pp. 283-290.

Varnuška, M., Parus, J., Kolingerová, I., 2005. Simple holes triangulation in surface reconstruction. In: *Algoritmy*, pp. 280-289.

Zlot, R., Bosse, M., 2014. Efficient large-scale 3D mobile mapping and surface reconstruction of an underground mine. *Field and service robotics*, Springer Berlin, Heidelberg, pp. 479-493.

References from Other Literature:

Arles, A., Clerc, P., Sarah, G., Téreygeol, F., Bonnamour, G., Heckes, J., Klein, A., 2013. 3D Reconstruction and modeling of subterranean landscapes in collaborative mining archaeology projects: techniques, applications and experiences In: *XXIV international CIPA symposium*, Strasbourg, France, pp. 61-66.

Baarda, W., 1968. A testing procedure for use in geodetic networks. *Netherlands geodetic commission*, Vol. 2 (5), Delft.

Beraldin, J.A., Picard, M., Bandiera, A., Valzano, V., Negro, F., 2011. Best practices for the 3D documentation of the Grotta dei Cervi of porto Badisco, Italy. In: *S&T/SPIE electronic imaging*, San Francisco, USA, Vol. 7864, pp. 78640J-78640J-7.

Cignoni, P., Corsini, M., & Ranzuglia, G., 2008. Meshlab: an open-source 3D mesh processing system. *Ercim news*, 73, pp. 45-46.

Ferreira, A., Almeida, J.M., Silva, E., 2012. Application of visual-inertial slam for 3D mapping of underground environments. In: *IROS12 4th international workshop on planning, perception and navigation for intelligent vehicles*, Algarve, Portugal, pp. 61-66.

Garland, M., Heckbert, P., 1997. Surface simplification using quadric error metrics. In: *SIGGRAPH 97*, pp. 209-216.

Jiménez Fernández-Palacios, B., Remondino, F., Stefani, C., Lombardo, J., De Luca, L., 2013. Web visualization of complex reality-based 3d models with Nubes. In: *IEEE conference digital heritage 2013*, 1, pp. 701-704.

Kazhdan, M., Bolitho, M., Hoppe, H., 2006. Poisson surface reconstruction. In: *4th eurographics symposium on geometry processing*.

Lichti, D., Gordon, S., 2004. Error propagation in directly georeferenced terrestrial laser scanner point clouds for cultural heritage recording. In: *FIG working week*, Athens, Greece.

Lowe, D. 1999. Object recognition from local scale-invariant features, In: *International conference on computer vision*, Corfu, Greece.

Nocerino, E., Fiorillo, F., Minto, S., Menna, F., Remondino, F., 2014. A non-conventional procedure for the 3D modeling of WWI forts. In: *The international archives of the photogrammetry, remote sensing and spatial information sciences*, Riva del Garda, Italy, Vol. XL-5, pp. 457-464.

Nocerino, E., Menna, F., & Remondino, F., 2014. Accuracy of typical photogrammetric networks in cultural heritage 3D modeling projects. *ISPRS-International Archives of the Photogrammetry, Remote Sensing and Spatial Information Sciences*, *1*, 465-472.

Pope, A. J. 1976. The statistics of residuals and the detection of outliers. NOAA technical report NOS 65 NGS 1, National ocean service, national geodetic survey, US department of commerce. Rockville, MD, Washington, 133pp.

Ramos, E.A., Sadri, B., 2007. Geometric and topological guarantees for the wrap reconstruction algorithm. In: *18th annual ACM-SIAM symposium on discrete algorithms*, pp. 1086-1095.

Reshetyuk, Y., 2009. Self-calibration and direct georeferencing in terrestrial laser scanning. PhD thesis, KTH - Royal institute of technology, Stockholm, Sweden.

Roncat, A., Dublayansky, Y., Spötl, C., Dorninger, P., 2011. Full-3D surveying of caves: A case study of Mär-Chenhöhle (Austria). In: *IAMG conference*, Salzburg, Austria.

Roncella, R., Umili, G., Forlani, G., 2012. A novel image acquisition and processing procedure for fast tunnel DSM production. In: *The international archives of the photogrammetry, remote sensing and spatial information Sciences*, Melbourne, Australia, Vol. XXXIX-B5, pp. 297-302.

Soudarissanane, S., Lindenbergh, R., Menenti, M., Teunissen, P., 2009. Incidence angle influence on the quality of terrestrial laser scanning points. In: *The international archives of the photogrammetry, remote sensing and spatial information sciences*, Paris, France, Vol. XXXVIII-3/W8, pp. 183-188.

Ziparo, V.A., Zaratti, M., Grisetti, G., Bonanni, T.M., Serafin, J., Di Cicco, M., Proesmans, M., Van Gool, L., Vysotska, O., Bogoslavskyi, I., Stachniss, C., 2013. Exploration and mapping of catacombs with mobile robots. In: *IEEE international symposium on safety, security, and rescue robotics (SSRR)*, pp. 21-26.

References from Websites (Last accessed on 2015, Jan. 24):

CloudCompare, 2015. EDF R&D, Telecom ParisTech. http://www.cloudcompare.org/.

AN APPROACH TO 3D DIGITAL MODELING OF SURFACES WITH POOR TEXTURE BY RANGE IMAGING TECHNIQUES. 'SHAPE FROM STEREO' VS. 'SHAPE FROM SILHOUETTE' IN DIGITIZING JORGE OTEIZA'S SCULPTURES

J. García Fernández [a], A. Álvaro Tordesillas [b], S. Barba [c]

[a] Architectural Photogrammetry Lab., College of Architecture, University of Valladolid - jgarciaf@mtu.edu
[b] Dept. of Graphical Representation, College of Architecture, University of Valladolid - tordesillas@arq.uva.es
[c] Dep. of Civil Engineering, Faculty of Engineering, University of Salerno - sbarba@unisa.it

Commission V, WG V/4

KEY WORDS: Cultural Heritage, Modelling, Radiometric-geometric comparison, Surface

ABSTRACT:

Despite eminent development of digital range imaging techniques, difficulties persist in the virtualization of objects with poor radiometric information, in other words, objects consisting of homogeneous colours (totally white, black, etc.), repetitive patterns, translucence, or materials with specular reflection. This is the case for much of the Jorge Oteiza's works, particularly in the sculpture collection of the *Museo Fundación Jorge Oteiza (Navarra, Spain)*.
The present study intend to analyse and asses the performance of two digital 3D-modeling methods based on imaging techniques, facing cultural heritage in singular cases, determined by radiometric characteristics as mentioned: *Shape from Silhouette* and *Shape from Stereo*. On the other hand, the text proposes the definition of a documentation workflow and presents the results of its application in the collection of sculptures created by Oteiza.

1. INTRODUCTION

The transitions from analogue to digital photogrammetry, and the rapid development of personal computers processing capacity, have allowed a theoretical-practical framework to develop. This guarantees digital modelling of three-dimensional shapes with high levels of accuracy and reliability in terms of geometry and radiometry.
The 3D digital modelling based on digital photogrammetry has become one of the most important tools in solving actual needs for *heritage analyses*, *procedural modelling* and *accuracy requirement in 3D documentation*. Connected with this goal has been conducted the R&D project: HAR201[1] *Low cost digital survey: sculpture pieces with poor texture by Jorge Oteiza.* During the documentation stage, singular pieces of poor textures surfaces presented special difficulties for an accurate 3D digital reconstruction. These were the cases for much of the Jorge Oteiza's sculptures in the collection of the *Museo Fundación Jorge Oteiza (Navarra, Spain)*. These works are a set of small pieces sculpted in white chalk, white marble, and welded sheet metal that was painted black (Fig. 1). The collection, which Oteiza never gave graphic definition prior to their creation, has not been subjected to analytic processes or metric documentation.

2. THE MODELLING METHODS

In order to generate a reliable documentation strategy, the study begins with the theoretical analysis of two methods of digital image-based survey. While having the premise of carried on procedures focused on low cost technologies (based on image information), the selected methods have radically different performances in terms of dependence on the object texture: Shape from Silhouette - SfSi, based on the detection of the item shape; and Shape from Stereo - SfS ,based on detection of homologous pixels in stereo/multiple image pairs.

2.1 Reconstruction based on geometric properties: the method Shape from Silhouette (SfSi)

One of the major difficulties for 3D reconstruction in cultural heritage is the poverty of information on images to reconstruct the scene. This occurs because, at present, the implemented algorithms in the most common software are based on the massive correspondence of pixels or set of pixels, from a theoretically stereoscopic positioning. The SfSi method, however, bases reconstruction processes on 3D shape detection from different viewpoints. The idea of using silhouettes for this purpose was first introduced in 1974 by Baumgart and formalized by Laurenti (Laurentini, 1994). This computes intersections of the visual pyramids formed by the silhouettes (generatrix) and the director beam (irregular pyramid axis) from a well-known viewpoint (Haro, 2012). Therefore, this method obtains the surface and the volume contained within the intersection of both visual pyramids, Visual Hull (Fig 2). The visual pyramid can follow two paths, either by an approach based on volume (voxels) or by an approach based on surfaces:

- Approach based on voxels: The space segmentation occurs into a large number of cubes (volumetric elements), called voxels, and the shape is represented on a number of these three-dimensional measurement units. Therefore, the Visual Hull resolution is defined by the size of the voxels and the discrimination ability of full and empty voxels (Forbes, 2007).

[1] The project is financed by the MICINN (Spanish Ministry of Science and Innovation).

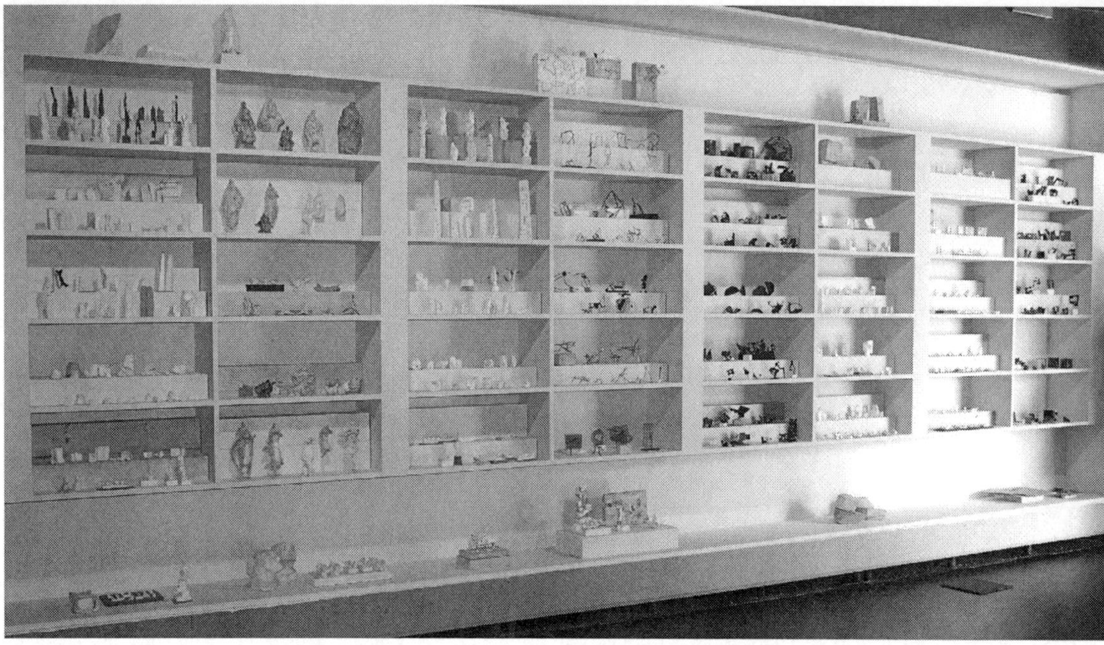

Figure 1. Collection of sculptures by Jorge Oteiza, *Museo Fundación Jorge Oteiza,* Navarra, Spain

- Approach from surfaces: The second path is defined by the surface analysis. The process begins with the specific geometry analysis of the visual pyramids, and comprises the intersection calculus of polygons. This intersection is defined by the spatial projection of the silhouettes and it is computed by various procedures that prioritize calculation speed (real time processing) or the quality of geometric result (Michoud et al., 2006).

2.2 Reconstruction based on radiometric properties: the method Shape from Stereo (SfS)

In the other case, SfS bases acquisition process and 3D reconstruction in the encounter and detection of homologous points (RGB pixels) between stereo and multiple images. Although dense stereo correspondence has been a widely studied problem, the evolution of correspondence and densification algorithms continue to rise. Today, analysing correspondence structures just based on local measurements, such as image intensity (or colour) and phase (Ogale and Aloimonos, 2005), is not enough. We must also consider solutions based on external geometric constrains, symbolic descriptions or patterns (Zhang et al., 2014). Although the study is focused on the SfS method, the results can be extended to the more general Structure from Motion (SfM) method, where internal camera parameters do not need to be known. This project's preference is to conduct digital reconstruction based on the accurate camera calibration (internal orientation), made by the calibration module on PhotoModeler Scanner. However, current developments based on reverse engineering make possible the simultaneously reconstruction of 3D points obtaining (recovering) projection matrices, and maintaining similar accuracies in the correspondence and massification of points.

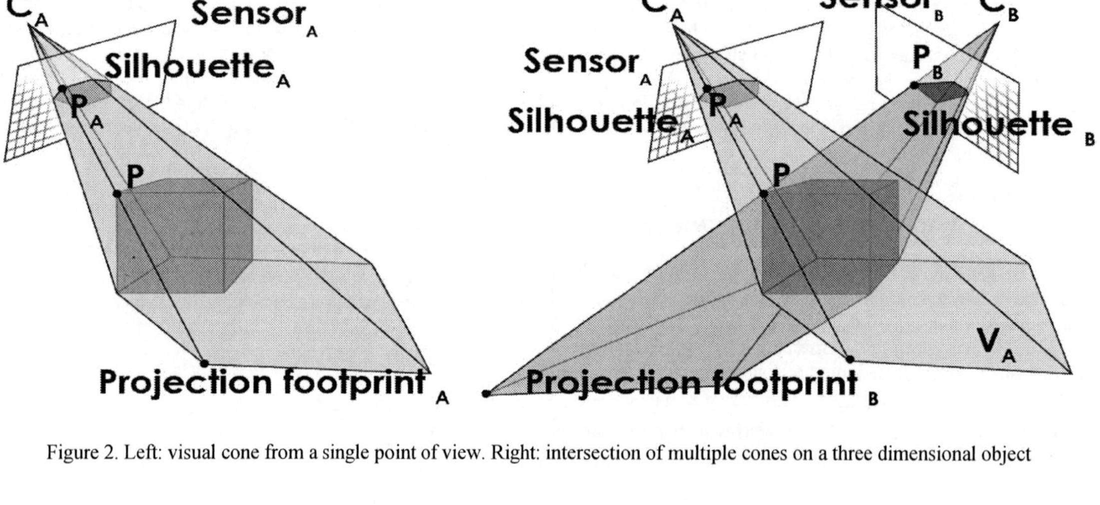

Figure 2. Left: visual cone from a single point of view. Right: intersection of multiple cones on a three dimensional object

This contribution has been peer-reviewed.
doi:10.5194/isprsarchives-XL-5-W4-25-2015

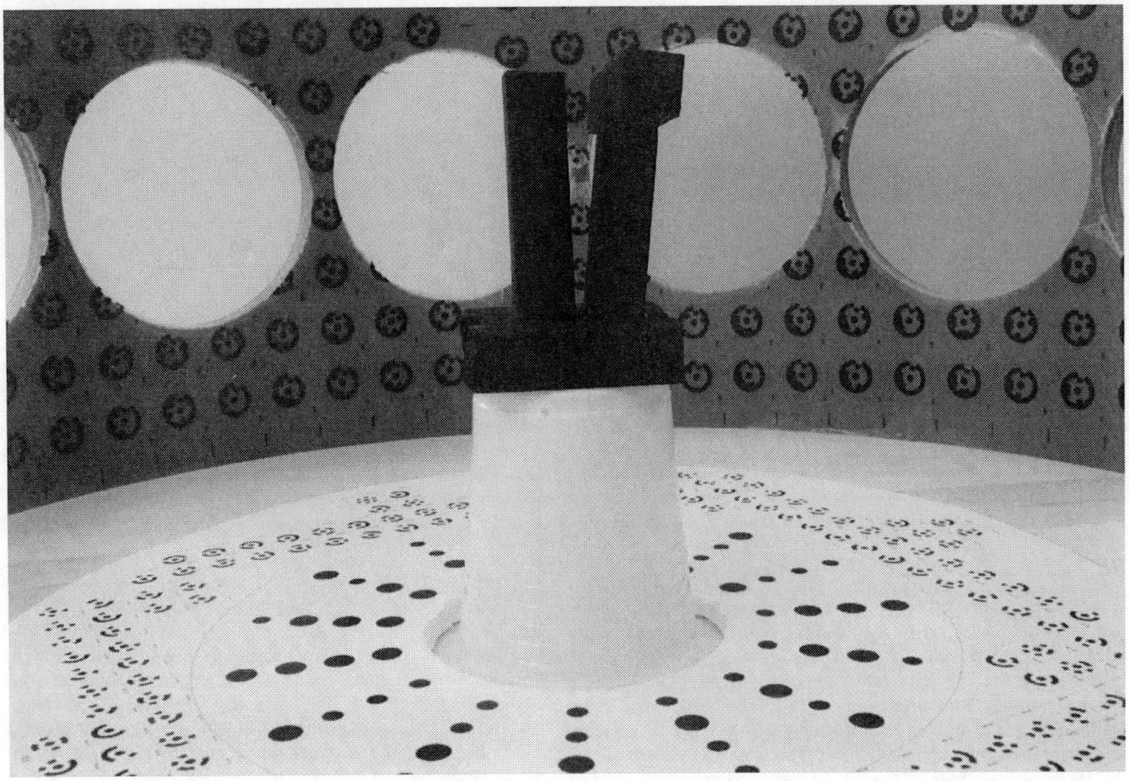

Figure 3. *Barrel of Capture*, designed for conducting simultaneous documentation process based on SfSi and SfS

2.3 Documentation process

The need for capturing images while maintaining optimal (and similar) external conditions, forced us to establish a capture flow in parallel. Once the strengths and limitations of the two methods become known, it was designed a controlled space named the *Barrel of Capture* (Fig. 3). This *Barrel*, consisting in a rotary mechanized structure, was designed to supply the optimal capture condition of both methods. To accomplish this, were first established criteria for ideal 3D reconstruction under the typical conditions for each technique. That includes conditions for image capture, taking into account that the data processing was conducted with PhotoModeler Scanner 1.0.4 (for SfS Method) and CX2 Strata 3D Photo v2.1 (for SfSi method, Fig. 1). Therefore, the barrel had to meet the following conditions:

(for SfS method)

- Permissibility of photo capture in a parallel or close to parallel orientation. Cameras with relation base/height of ¼. Images with an overlap close to 40%;
- space in the scene for identifying code targets of the reconstruction software (PhotoModeler Scanner);
- scene structure capable to document the piece sample while occupying the entire frame of the sensor;
- scenario that allows the entry of uniform light, without shadows.

(for SfSi method)

- Permissibility steps capture every 24 degrees to the lower ring and 40 degrees to the upper ring;
- space in the scene for identifying templates of the reconstruction software (Strata 3D Photo);
- scenario with the possibility of swapping templates. Templates with different colours were used in order to avoid colour contained in the scene.

For the metric comparison was selected a chalk sculpture with homogeneous colour and dimensions of 120mm x 8mm x 62mm. The documentation process, applied to the selected sample, led us to the creation of two digital 3D models (one model for each method). Figure 4 shows the relative orientation of the cameras and the mesh obtained from SfSi method.

3. DATA COMPARISON

Once established the theoretical framework and the procedures used for a parallel documentation, this study focuses on the metric comparison of digital models obtained from both techniques in order to assess the potentials in the accurate modelling, as well as the limitations. For an accurate comparison, a geometric reference with a rigorous reliability was necessary. In the case of studied sculptural pieces, the lack of earlier dependable documentation led to the surface modelling by using High Dynamic Range Scanning (HDRS) Technique (Zhang and Yau, 2009). For this purpose, this study used the scanner *Perceptron ScanWorks V5* with an articulated arm *Romer*. Because the starting position of the laser beam (beam pattern) and the arrival through two CCD sensors (in the case of the Perceptron V5) are known, it is possible to calculate the incidence angle of the beam plane relative to the documented surface.

This contribution has been peer-reviewed.
doi:10.5194/isprsarchives-XL-5-W4-25-2015

Figure 5: Left: laser beam plane projected onto a piece for three-dimensional reconstruction. Right: mesh obtained from 3D point cloud.

Using two sensors, each capturing half of the projected plane, enhances the accuracy of the feature being documented and maintains small dimensions of the device base.
The scans are referenced to a unique (and portable) coordinate system through the articulated arm. The HDRS enables calibration of the emitted bean intensity with respect to surface properties of the documented object. This allows reconstruction of surfaces and materials with different properties – semi-translucent materials (high beam penetration), dark materials (high light absorption), or highly reflective materials (specular) – fitted to those required in this project (Fig. 5).
To perform the metric comparison, the proposed project used *Innovmetric Polyworks* software and two modules. First, IMAlign creates surface based on the laser scanner point cloud, without application of smoothing, tolerance or reduction. Second IMInspect aligns data and calculates the metric differences between the reference and the surfaces obtained from both SfSt and SfSi methods (Fig. 6, A-B). The conducted comparison used the criteria of shortest distance perpendicular to the plane. Therefore, each point was correlated with an error vector (Fig. 6, C). The result of the comparison are shown in Table 1.

4. COMPARISSON ANALYSIS AND APPLICATION

The behaviour of the main parameters (Tab. 1) reflects how insofar as homologous points and its reduced neighbourhood cannot be identified, the method SfSi will have better performances compared to the SfS method. As described in the data comparison, the SfSi method directed the 3D reconstruction of the entire surface without jumping or discontinuities (Gig. 6 B, C), while keeping the error in range below 3mm.
The metric comparison, together with the previous performance analysis, allows us to define SfSi the more efficient method for the lower variance and for the richness of the radiometric information: in this line was defined a systematic survey of digital 3D sculptures with poor radiometric information, based on the method SfSi.
Its use will have better results while documentation criteria mentioned in section 2.3 are fulfilled and the expected accuracy is within the range. The application of the SfSi methods in the Jorge Oteiza collection (*Museo Fundación*, Navarra) able possible the creation of an interactive and online digital database of the not documented pieces (Fig. 7).

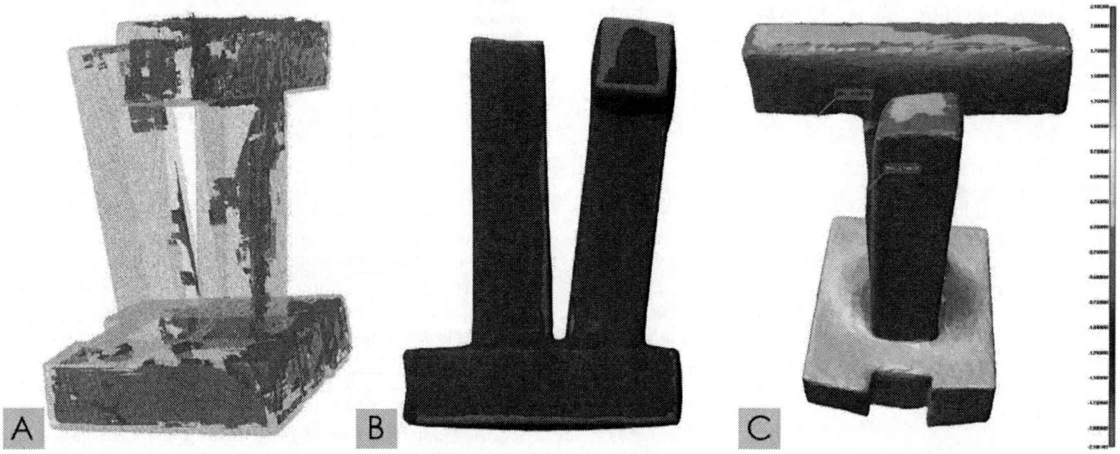

Figure 6. A: overlapping point clouds obtained from Shape from Stereo method (darker) versus laser scanner data (lighter).
B: overlapping meshes obtained from Shape from Silhouette (in red) versus laser scanner data mesh (in green). C: error scale map

This contribution has been peer-reviewed.
doi:10.5194/isprsarchives-XL-5-W4-25-2015

Table 1. Main parameters in the comparison process

Parameter	SfSi (vs Reference)	SfS (vs Reference)
Point total	10344	9516
Points in the range of ± 1.5mm	7100 (68.64%)	9174 (96.40%)
Points in the range of ± 2.5mm	9852 (95.24%)	9432 (99.12%)
Point distribution	Homogeneous	Non-Homogeneous
Error average	1.3474mm	0.438mm
Standard Deviation	0.5174mm	0.3278mm

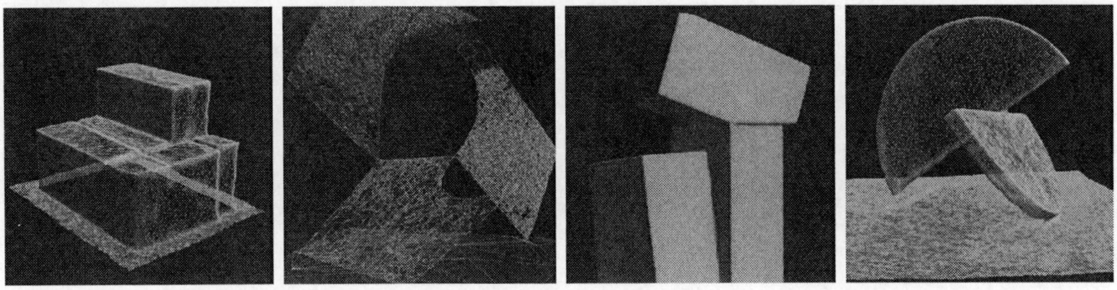

Figure 7. Database of digitized Oteiza's sculptures based on the proposed workflow

5. CONCLUSIONS AND FUTURE WORKS

The present text, allows the estimation of real potentialities and limitations of two range imaging techniques: SfSi and SfS, applied in the field of cultural heritage. The study bases this estimation in the analysis of geometric modelling behaviour and the geometric comparison of 3D models. The obtained result helps to improve the selection and use of these methods in the cultural interpretation, dissemination and preservation.
In particular, the study of the Shape from Silhouette method allows us to retake math implementations over 20 years ago, which can be reused now thanks to the new technologies and new cultural needs. Related to the documentation of sculptural pieces with of poor texture (dark colours, repetitive patterns and materials with little response to light stimuli); this method obtains significant advantages over others based on range information or massive correspondence of pixels. Its gain will be higher while the texture of the piece is more homogeneous.
The need to improve accuracy while is efficiently reconstructed the geometry of shapes with poor texture compels us to continue the study in two new research lines. On one hand, it is the analysis of the influence and behaviour of visual hull calculus in samples characterized by both curves and straight edges; in the other hand, it is the analysis of geometric reconstruction based on a unified system that comprises both SfSi and SfS.

ACKNOWLEDGE

We gratefully acknowledge support from the Laboratory of Architectural Photogrammetry (University of Valladolid, Spain), and from the Museo-Fundación Jorge Oteiza in Pamplona (Spain).

REFERENCES

Forbes, K., 2007, *Calibration, Recognition, and Shape from Silhouettes of Stones*, PhD Thesis, University of Cape Town.

Haro, G., 2012, Shape from Silhouette Consensus. *Pattern Recognition,* 45(9), pp. 3231-3244.

Laurentini, A., 1994, The visual hull concept for silhouette-based image understanding. *IEEE Trans, Pattern Analysis and Machine Intelligence*, 16(2), pp. 150-162.

Michoud, B., Guillou, E. and Bouakaz, S., 2006, Shape From Silhouette: Towards a Solution for Partial Visibility Problem. *Eurographics*, pp. 13-16.

Ogale, A. and Aloimonos, Y., 2005, Shape and the stereo correspondence problem. *International Journal of Computer Vision*, 65(3), pp. 147-162.

Zhang, S. and Yau, S., 2009, High Dynamic Range Scanning Technique. *Optical Engineering Journal*, 48(3), 033604.

Zhang, D., Lee, D. and Chang Y., 2014, A New Profile Shape Matching Stereovision Algorithm for Real-time Human Pose and Hand Gesture Recognition. *International Journal of Advanced Robotic Systems*, 11(16), pp. 1-12.

This contribution has been peer-reviewed.
doi:10.5194/isprsarchives-XL-5-W4-25-2015

A HAND-FREE SOLUTION FOR THE INTERACTION IN AN IMMERSIVE VIRTUAL ENVIRONMENT: THE CASE OF THE *AGORA* OF SEGESTA

Riccardo Olivito [1, *], Emanuele Taccola [2], Niccolò Albertini [1]

[1] Scuola Normale Superiore, Italy – (riccardo.olivito, niccolo.albertini)@sns.it
[2] University of Pisa, Italy – emanuele.taccola@unipi.it

Commission V, WG V/4

KEY WORDS: Interaction, Immersive Archaeology, Cyber-Archaeology, Virtual environments, CAVE, Image-based modeling.

ABSTRACT: The paper illustrates the project of an interdisciplinary team composed of archaeologists and researchers of the Scuola Normale Superiore and the University of Pisa. The synergy between these Centres has recently allowed for a more articulated 3D simulation of the *agora* of Segesta. Here, the archaeological excavations have brought to light the remains of a huge public building (*stoa*) of the Late-Hellenistic Period. Computer graphics and image-based modeling have been used to monitor, document and record the different phases of the excavation activity (layers, findings, wall structures) and to create a 3D model of the whole site.
In order to increase as much as possible the level of interaction, all the models can be managed by an application specially designed for an immersive virtual environment (CAVE-like system). By using hands tracking sensor (Leap) in a non-standard way, the application allows for a completely hand-free interaction with the simulation of the *agora* of Segesta and the different phases of the fieldwork activities. More specifically, the operator can use simple hand gestures to activate a natural interface, scroll and visualize the perfectly overlapped models of the archaeological layers, pop up the models of single meaningful objects discovered during the excavation, and obtain all the relative metadata (stored in a dedicated server) which are visualizable on external devices (e.g. tablets or monitors) without further wearable devices. All these functions are contextualized within the whole simulation of the *agora*, so that it is possible to verify old interpretations and enhance new ones in real-time, simulating within the CAVE the whole archaeological investigation, going over the different phases of the excavation in a more rapid way, getting information which could have been ignored during the fieldwork, and verifying, even *ex-post*, issues not correctly documented during the fieldwork.
The opportunity to physically interact with the 3D model of the *agora* increases drastically the level of affordance between operator and 3D simulation, and to verify with a completely new approach, issues which can be hardly investigated by using traditional 2D documentation.

1 INTRODUCTION

Since 2001 the archaeological investigations in the *agora* of Segesta, in northwestern Sicily, have brought to light a large part of the buildings which bordered the ancient public square (Ampolo, Parra 2012). In particular, the most interesting acquisitions of planimetric, architectural, and monumental nature are related to a large *stoa* with two projecting wings, which was built in the Late-Hellenistic period (end of the 2nd century BCE).
The accurate architectural study of the *stoa* (Abate, Cannistraci 2012) led to the formulation of a hypothetical reconstruction of the vertical development of the building.
Within this articulated monumental complex, during the very last years it has been possible to use new digital techniques of documentation, which have allowed for a significant improvement in the data acquisition phase, in the graphic documentation, and in the following elaboration processes.

2 WORKFLOW

The 3D models of the *agora* were created with computer graphics (fig. 1) and aerial and terrestrial close range photogrammetry techniques, using software like AutoCAD, 3ds Max, Agisoft PhotoScan and Geomagic (Taccola 2012, Olivito, Taccola forthcoming).

Figure 1. 3D model of the *agora* of Segesta

The distinction between these two methods is obvious: in the first case, although we rely on measured data and consolidated studies of the elevations, we reconstruct the buildings as we presume they had to appear in a given historical period. In the second case, we use reality-based procedures in order to record objects and structures in their current condition.
In so doing, it is possible to enhance the quality of the scientific research on the Segestan public square, improving the level of interaction between user and digital models to be visualized within the virtual environment suitably created for the project.

2.1 A Virtual-Archaeological approach

In a first experimental project (Albertini *et al.* 2014), the reconstruction of the *agora* of Segesta was imported in Unity 3D Game Engine, using techniques of polygons decimation, but

This contribution has been peer-reviewed.
doi:10.5194/isprsarchives-XL-5-W4-31-2015

preserving the high resolution textures without compromising the real-time visualization.
The model was implemented in a dedicated application for a CAVE-like System. Within this virtual environment, a "walk" interaction metaphor was developed to make the exploration as natural as possible. The interaction metaphor, limited to small movements, solved the conflict between a very large virtual environment (100 meters ca.) and the actual environment of the CAVE, that is much smaller (3 meters) (fig. 2).

Figure 2. 3D model of the *agora* within the CAVE

Overall, the interface ensures excellent functionality and immediacy of use.
Obviously, this 3D reconstruction, realized with computer graphics, has positive and negative features. As to the positive ones, the possibility to visualize and walk within the virtual environment allows us to think about issues that are hardly investigable by using traditional 2D tools. For instance: possible solutions adopted in the carpentry; physical and spatial relations between the *stoa*, its internal space, and the other monuments of the public square; relations between architecture, decorations, and documentary apparatus (e.g. statues and inscriptions); peculiar functions of the different sectors of the *agora*; the role played by natural and artificial lighting in different moments of the day.
On the other hand, this model was firstly conceived as a reconstruction more than a simulation (Clark 2010), with a limited level of interaction.
As a result, at present, it neither allows us to extrapolate or draw in real time information and data from the visualized elements, nor can be continuously modified.

2.2 A Cyber-Archaeological approach

With the most recent techniques used during the fieldwork activity and in the labs, a new approach was proposed, mainly oriented towards simulation and Cyber-Archaeology, as suggested by the recent literature (Forte 2010, Forte 2012). In this sense, the 3D model of the *agora* could represent a basis for further studies or, quoting Jeffrey Clark, a Simulation Slice.
The starting point of this new research methodology is constituted by the introduction of techniques of close range photogrammetry, both aerial and terrestrial, used for the real-time documentation of the fieldwork activities and the digital drawings of the elements discovered during the excavation (Dell'Unto 2014).
As to the aerial photogrammetry, in the case of the *agora* of Segesta a MikroKopter Hexacopter UAV was employed, for whom an automatic flight plan was set up (fig. 3). The survey of the whole area of the *stoa* has required 3 flights, with a total of 664 shoots.

Figure 3. Flight plan for the aerial photogrammetry

The images acquired were turned into a 3D model of the entire *agora*, which constitutes the necessary basis for the positioning and alignment of all the surveys created with SfM terrestrial procedures. In fact, an appropriate number of GCPs was positioned and measured with total station to obtain the coordinates relative to the reference system used within the area of excavation.
During the last campaign of excavation on the site of Segesta, 6 image-based models were created (fig. 4), following the progressive development of field activities. The excavation levels include the most superficial layers, up to the traces of medieval re-occupation of the area and those relating to the collapse and abandonment of the Late-Hellenistic *stoa*.

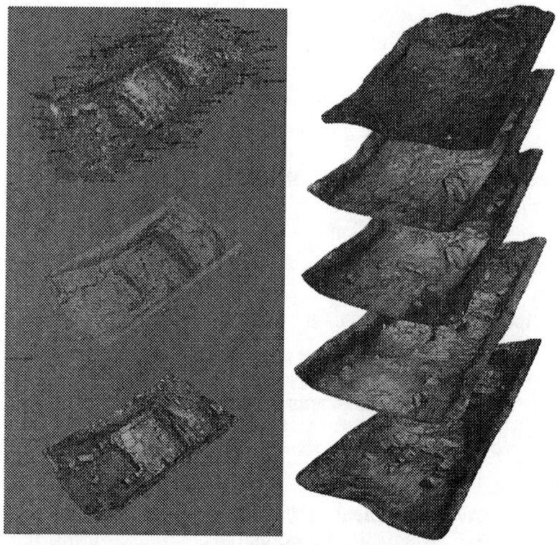

Figure 4. Some of the image-based models realized during the fieldwork activities

Each 3D model consists of several stratigraphic layers. In this first phase of the fieldwork, the nature of the stratigraphic deposit, simplified and uniform, did not required a work of semantic segmentation within the single model. However a first segmentation process has been started and will be enhanced in the next campaigns.
At this stage, the selection of the isolated elements has been entirely subjective and based on the needs of the research activity but also on the simplified stratigraphic features of the archaeological context. As a result, in the case of the new application, individual and separated models of two meaningful elements were realized: a limestone ionic *geison* (or dripstone) of the second floor of the portico, that was discovered during

This contribution has been peer-reviewed.
doi:10.5194/isprsarchives-XL-5-W4-31-2015

the excavation of the collapse layers of the *stoa*; and a medieval well-preserved oven of the Svevian period (fig. 5).

Figure 5. Image-based model of a medieval oven

The time required on the field to achieve a single model was about 20 minutes, including the acquisition of approximately 60 images for each level and the measuring of Ground control points with total station.

2.3 The application

Similarly to the 3D model of the *agora* realized with computer graphics, a virtual environment was created, in which the image-based models were loaded with Unity 3D Game Engine.
Although still in progress, the application used to display and interact with models has already positive aspects, both from the scientific and technical point of view.
In the first phase the issues related to hardware were addressed, trying to figure out how to best use the tools of tracking. A support for glasses was modeled and printed in 3D, allowing for the use of an ultrasonic sensor and hand position sensor at the same time (fig. 6).

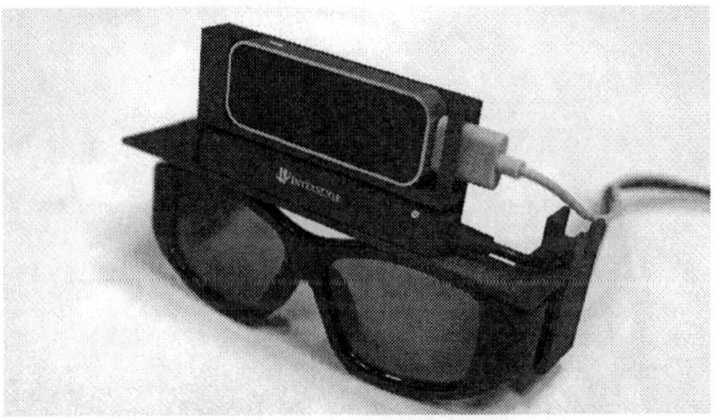

Figure 6. 3D glasses with the mounted leap-sensor

In fact, the use of the sensor applied on the user's head is not standard, since it has been designed to remain on a flat surface. Despite this, the device is used in a non-invasive way, thanks to the compactness and light weight of the Leap.
In particular, some of the new features of the 2.0 sdk leap sensor were employed as, for example, "automatic orientation", which allows for the automatic correction of the orientation of the hands, and "image acquisition", which allows for the use of the images that come directly from the infrared cameras in addition to the raw data tracking.
At this point we started the study phase of 3D interface and gesture linked to the exploration of the layers of the excavation.

The process of creating a gesture is structured as follows:

- Creation of base gestures (fig. 7): analysis on raw data input provided by the "leap sensor" and improvement through algorithms for data interpolation. Analysis of the data already processed and programming of "gesture base" (as the rotation of the hand, swipe, grab, etc.).

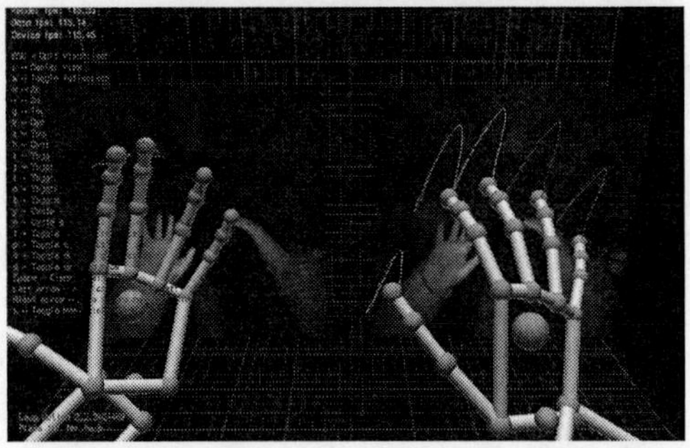

Figure 7. Snapshot of the creation of base gestures

- Creation of Complex gesture (fig. 8): by merging multiple "gesture bases" we created interactions in the environment, such as touch and rotation of objects, scroll and analysis between different archaeological data, etc.

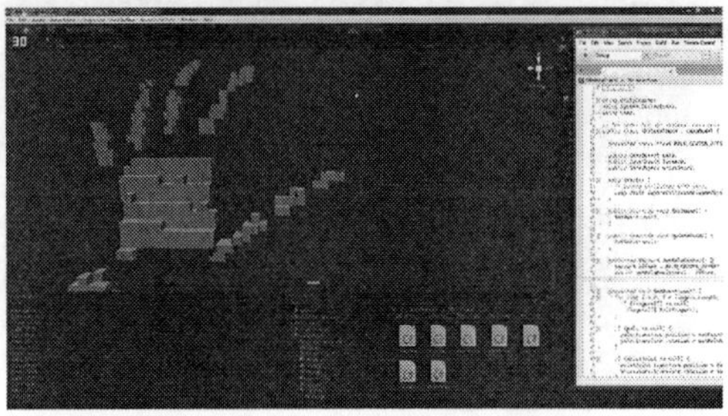

Figure 8. Snapshot of the creation of complex gestures

- Evaluation: study of gesture designed to understand which would allow better interaction with the data at a theoretical level, according to usability principles redesigned for the 3D world (understanding of the interaction, ease of use, effectiveness, etc.).
- Implementation (fig. 9): implementation of gestures valued as realizable in the virtual environment through dedicated script. A plugin for the game engine was developed, that allows us to change the gesture with a graphical editor.

This contribution has been peer-reviewed.
doi:10.5194/isprsarchives-XL-5-W4-31-2015

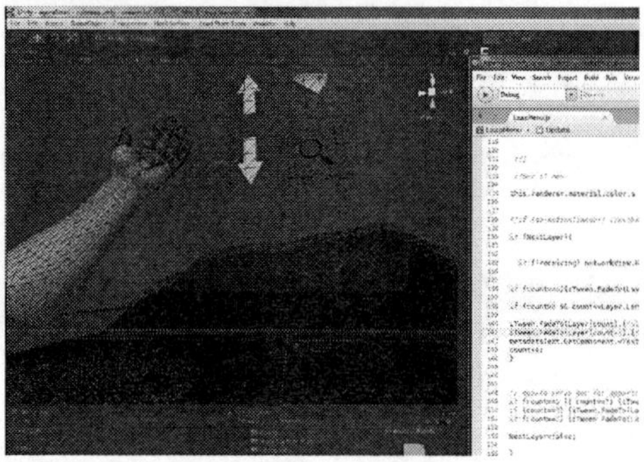

Figure 9. Snapshot of the Implementation phase

- Study of user behaviour (fig 10): testing phase on different users with different skills to analyze the actual effectiveness of the gestures. Through a real-time feedback the gestures were corrected and revised according to the behaviour of the users.

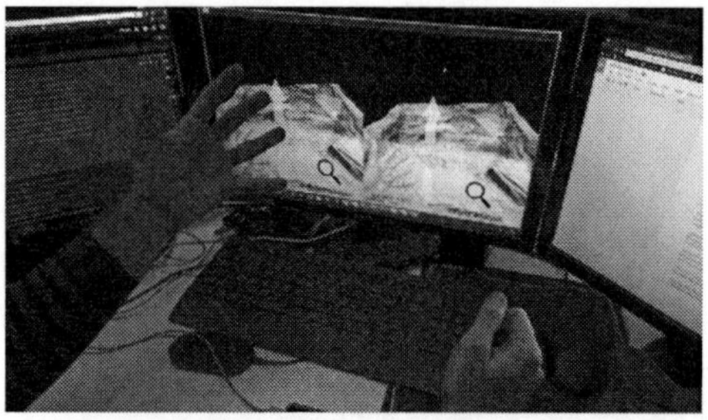

Figure 10. Snapshot of the study of the user behaviour

- Review: revision of gestures based on user experience and new cycle evaluation, implementation and study of user behaviour.

Through this process it was possible to remove all the gestures too complicated for the device (for example pinch to zoom or rotate with the fingers) or too complicated for the user (gestures tiring or unnatural), thus implementing an interface the most natural and efficient as possible.

3 RESULTS

The purpose of the new application is to enhance the number and quality of functions that the operator can use and, as a consequence, the embodiment within the virtual environment. For this reason special attention is paid to the gestures available in the simulation.

In particular, the application aims at improving the level of interaction between users and virtual environment, thanks to the possibility to visualize the perfectly overlapped and aligned digital models of the stratigraphic sequence realized during the fieldwork activity, to scroll between them within the virtual environment, and to query and interrogate the relative metadata only with hand-gestures (fig. 11)

The interface which has been realized is usable in a natural way, by hand movements. This makes the application friendly and easy to use.

At this stage, the control of gestures is limited to one user, whereas the visualization in the virtual environment is multi-user (3 or 4 people depending on the size of the CAVE). As to the leap, the detection of the position of the hand is very accurate, due to a lightweight and non-invasive device. Obviously there are some features still to be developed, such as the possibility to increase the number of users that can interact through gestures. Nevertheless, these problems will be overcome with the introduction of wireless hand trackers that will make simpler to perform gestures in the virtual environment. At the same time, the introduction of the new tracker will increase the level of complexity of gestures, such as the rotation of the models or the measuring of distances, areas, volumes and so on.

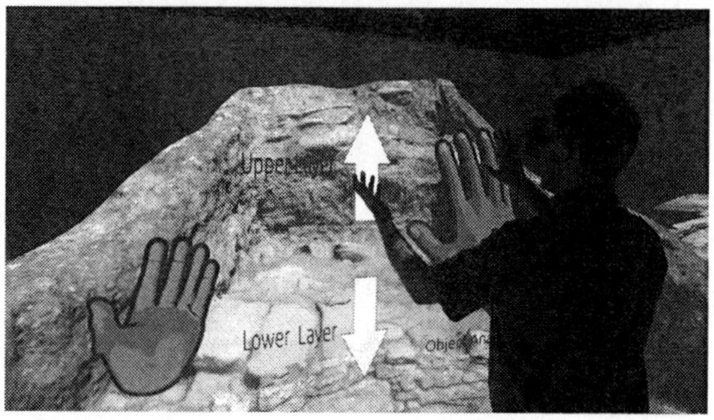

Figure 11. The new application within the CAVE

A further element that has been tried to solve is the difficulty to visualize large amounts of text and metadata in the virtual environment (for instance records of the individual layers, records of architectural elements and artefacts found during the excavation). In fact, only the essential information, such as the layer number, the date of acquisition, or a brief description of the main objects visualized, are displayed within the CAVE.

However, the problem of the analysis of large amounts of data was solved by creating links to the metadata that can be viewed through additional gestures. The links connect to the database used in the phase of data collection, allowing for the examination of layers records, text files, and information on finds and objects visualized inside the CAVE, but on an external device (for instance tablet). In this way, the different users may view the models and at the same time interact directly with them, acquiring information on individual elements or modifying in real time the interpretations of the excavation data.

With regard to the models of the main objects found in the archaeological levels, both these elements were isolated within the single layers and perfectly overlapped into the stratigraphic sequence. Then, thanks to an interactive icon that appears only when the elements are visualizable, it is possible to pop up the models, rotate them, and zoom-in for a closer view of the objects.

Due to the future and obvious increase of the number of models, an extension of the natural interface will be introduced allowing for the rapid selection of single layers and for a quick browsing of the stratigraphic sequence. In this sense, an excessive semantic segmentation would involve the creation of a large amount of data and, consequently, a larger complexity of the database to be managed within the virtual environment. Nevertheless, this is still an open question that will require further and wider reflections, both from the theoretical and the technical point of view.

4 CONCLUSIONS

In studying the *agora* of Segesta we can now rely on a 3D model that allows us to have a more articulated and interactive idea of the ancient square and its buildings at the end of the 2nd

century BCE. As a result, it is possible to shed new light on issues of great importance such as physical and spatial relations between objects and monumental context; internal and external visibility; relations between open and closed spaces; lighting inside the building.
We can now definitively affirm that this 3D model is enhancing a better scientific knowledge of the Segestan *agora* and a wider dissemination of the collected data to both the experts and the non-experts. At the same time, although in the development phase, the new application allows us to reach a higher level of interaction with the digital models, to interrogate them, visualize the metadata available for the single items, and, due to the hand-free nature of the application, to use external devices in order to examine records and texts while staying inside the CAVE. In so doing, it will be possible to virtually reproduce the field-work activity, working in an autonomous or collaborative way.
In the next months, the main goal will be the improvement of the quality and quantity of visualizable models and data. In addition to this, and aware of the limits of devices such as the Oculus Rift, the possibility to use this approach on this kind of device is evaluating, the final aim being a further reduction of the gap between the phases of data collection and elaboration, and that one of interpretation, even during the fieldwork.

AKNOWLEDGEMENTS

The excavation activities in the *agora* of Segesta are conducted by the Laboratorio di Scienze dell'Antichità (LSA) of the Scuola Normale Superiore (Director Prof. Carmine Ampolo) and the Department of Civilization and Forms of Knowledge of the University of Pisa (Prof. Maria Cecilia Parra).
The data acquisition is realized by the Laboratorio di Disegno e Restauro (LADIRE) of the University of Pisa, in cooperation with the Laboratorio di Scienze dell'Antichità.
The application has been developed in the Cave Automatic Virtual Environment of the DreamsLab of the Scuola Normale Superiore (Director Prof. Vincenzo Barone).
A demonstration video of the application is available at the following link: www.youtube.com/watch?v=vBJIpwSHQXA.

REFERENCES

Abate A., Cannistraci O., 2012. La *stoa* tardoellenistica dell'*agora* di Segesta: alcune note preliminari sull'elevato architettonico del lato Ovest. In C. Ampolo (ed.), Agora *greca e* Agorai *di Sicilia,* pp. 305-319.

Albertini N., Barone V., Brogni A., Legnaioli S., Licari D., Taccola E., 2014. The *agora* of Segesta in immersive virtual environments. In *Science and Technology for the Safeguard of Cultural Heritage in the Mediterranean Basin*, Proceedings of the sixth international conference, Athens, 22-25 October 2013, Vol. I, 2014, pp. 299-304.

Ampolo C., Parra M.C., 2012, L'agora di Segesta: uno sguardo d'assieme tra iscrizioni e monumenti, in C. Ampolo (ed.), Agora *greca e* Agorai *di Sicilia,* pp. 271-285.

Clark J.T., 2010. The Fallacy of Reconstruction. In *Cyber-Archaeology*, BAR International Series 2177, pp. 63-73.

Dell'Unto N. 2014. The Use of 3D Models for Intra-Site Investigation in Archaeology. In *3D Recording and Modelling in Archaeology and Cultural Heritage. Theory and best practices*, BAR International Series 2598, pp. 151-158.

Forte M, 2010. Introduction to Cyber-Archaeology. In *Cyber-Archaeology*, BAR International Series 2177, pp. 9-13.

Forte M., 2014. Virtual Reality, Cyberarchaeology, Teleimmersive Archaeology. In *3D Recording and Modelling in Archaeology and Cultural Heritage. Theory and best practices*, BAR International Series 2598, pp. 113-127.

Olivito R., Taccola, E., forthcoming. 3D modelling in the *agora* of Segesta: techniques and data interpretation. In *Archeologia e Calcolatori* 25, pp. 61-74.

Taccola E., 2012. Ricostruzioni 3d per l'*agora* di Segesta. In C. Ampolo (ed.), Agora *greca e* Agorai *di Sicilia,* pp. 287-289.

A MULTI-DATA SOURCE AND MULTI-SENSOR APPROACH FOR THE 3D RECONSTRUCTION AND VISUALIZATION OF A COMPLEX ARCHAELOGICAL SITE: THE CASE STUDY OF TOLMO DE MINATEDA

J.A. Torres-Martínez [a] *, M. Seddaiu [b], P. Rodríguez-Gonzálvez [a], D. Hernández-López [c], D. González-Aguilera [a]

[a] Department of Cartographic and Land Engineering; High School of Ávila, University of Salamanca, Ávila, Spain – (josealberto,pablorgsf, daguilera)@usal.es
[b] Dipartimento di Storia, Scienze dell' Uomo e della Formazione, Università degli Studi di Sassari, Sassari, Italy - marcelloseddaiu@gmail.com
[c] Regional Development Institute-IDR, University of Castilla-La Mancha, Albacete, 02071, Spain - david.hernandez@uclm.es

Commission V, WG V/4

KEY WORDS: Reconstruction, Multisensor, Aerial Photogrammetry, Laser scanning, Computer vision, Archaeology.

ABSTRACT:

The complexity of archaeological sites hinders to get an integral modelling using the actual Geomatic techniques (i.e. aerial, close-range photogrammetry and terrestrial laser scanner) individually, so a multi-sensor approach is proposed as the best solution to provide a 3D reconstruction and visualization of these complex sites. Sensor registration represents a riveting milestone when automation is required and when aerial and terrestrial dataset must be integrated. To this end, several problems must be solved: coordinate system definition, geo-referencing, co-registration of point clouds, geometric and radiometric homogeneity, etc. Last but not least, safeguarding of tangible archaeological heritage and its associated intangible expressions entails a multi-source data approach in which heterogeneous material (historical documents, drawings, archaeological techniques, habit of living, etc.) should be collected and combined with the resulting hybrid 3D of "Tolmo de Minateda" located models. The proposed multi-data source and multi-sensor approach is applied to the study case of "Tolmo de Minateda" archaeological site. A total extension of 9 ha is reconstructed, with an adapted level of detail, by an ultralight aerial platform (paratrike), an unmanned aerial vehicle, a terrestrial laser scanner and terrestrial photogrammetry. In addition, the own defensive nature of the site (i.e. with the presence of three different defensive walls) together with the considerable stratification of the archaeological site (i.e. with different archaeological surfaces and constructive typologies) require that tangible and intangible archaeological heritage expressions can be integrated with the hybrid 3D models obtained, to analyse, understand and exploit the archaeological site by different experts and heritage stakeholders.

1. INTRODUCTION

Several techniques have been applying so far for the 3D reconstruction and visualization of archaeological settlements based on the use of aerial or close-range photogrammetry (Gomez-Lahoz and Gonzalez-Aguilera, 2009), or using terrestrial laser scanner (Gonzalez-Aguilera et al., 2011). However due to the inherent complexity of these sites, a multi-sensor approach is preferred for the 3D reconstruction and visualization of these environments (Torres et al., 2014).

The goal of this study is to propose a multi-data source and multi-sensor workflow in order to obtain high quality archaeological products. To this end, terrestrial scans and images will be registered with aerial images acquired from a paratrike and an unmanned aerial vehicle (UAV).

The application the aforementioned workflow was carried out in the archaeological site in Albacete (Spain). This settlement is characterized by the presence of a monumental defensive system sited in the area denominated as "El Regueròn" that encloses three defensive walls of different chronological and architectural nature. Furthermore, the relevant archaeological stratification of this area with different structures and building types requires of a multi-data source and multi-sensor approach that allows us to properly record and classify archaeological surfaces establishing an integration of topographic data with documents of archaeological excavations.

To carry out studies of characterization, measurement and analysis of the surface and the elements presented in the site, the geometrical information (acquired with geomatic sensors) is combined with other available thematic data such as, photographs, sketches, restorations reports, schedules, etc. in a 3D geographical information system (3DGIS). By this way a description of the site through time is possible.

In order to fulfill the archaeological documentation requirements, the aerial data gathering was performed with a paratrike which allows us to enclose the whole extension of the archaeological site using a full frame reflex camera and assisted by a specific gyrostabilized platform (MUSAS-MUltiSpectral Airborne Sensors) with a ground sample distance (GSD) of 3 cm. Trying to record with higher spatial resolution the area of interest of the site, "El Regueron", an UAV with an ultra-compact camera acquired the fortified walls with a GSD of 1 cm.

Both flight planning (paratrike and UAV) were carried out based on the classical aerial photogrammetric principles (Kraus, 1993). The in-house software MFlip and PFlip (Hernandez-Lopez et al., 2013) were used for flight planning and control. Considering the important topographic relief of the site, special attention was focused on the GSD variation (i.e. scale) during flight, guarantying a level of detail adapted to the archaeological work. In addition, to increase the level of detail in the fortification walls and thus avoid the occlusions due to

* Corresponding author.

This contribution has been peer-reviewed.
doi:10.5194/isprsarchives-XL-5-W4-37-2015

the effects of terrain relief (i.e. areas occluded or without information in the model generated by aerial photogrammetry), a combination of terrestrial photogrammetry and terrestrial laser scanner (TLS) was employed for improving the final 3D hybrid model.
This paper is organized as follows: after this introduction, in section 2, the different sensors employed and their characteristics are detailed. In section 3, the proposed multi-data source and multi-sensor approach is described. Experimental results are shown and discussed in section 4. Conclusions and future directions are given in section 5.

2. MATERIALS

2.1 Paratrike

The main aerial platform employed to the documentation of the archaeological area was a paratrike (Figure 1, up-left). It is a low-cost aircraft with more flexibility than conventional aircrafts, and more autonomy and payload capacity than the UAVs. This last characteristic allows the possibility of boarding better sensors than the UAVs, or even multiple sensors in a stabilized gimbal (MUSAS).

Motor	Rotax 503 two-stroke
Trike	Tandem Trike AIRGES
Tandem paraglide	MAC PARA Pasha 4
Emergency system	Ballistic parachutes
Weight	110 Kg
Weight capability	165-250 Kg or 195-295
Air velocity range	30-60 km/h

Table 1. Technical specifications of the paratrike.

The use of the gyro-stabilized camera platform, MUSAS, guarantees the orientation of the camera according to the flight planning by two servomotors arranged on the *x* and *y* axes, controlled by an Arduino board, which incorporates an IMU with 6 degrees of freedom: 3 accelerometers, a double-shaft gyroscope (for pitch and roll) and an additional gyroscope for yaw.

2.2 Unmanned Aerial Vehicle

For the aerial photogrammetric acquisition in the "El Reguerón" site, the paratrike platform was rejected due to the morphological characteristics of the terrain and the height of flight required to reach the high spatial resolution (GSD of 1 cm). The main drawback was the need to vary the flying height to keep the same scale, due to the presence of high relieves along the strips. As additional disadvantage, the walled constructions were occluded between walls of natural rock. Therefore, in order to complete the archaeological site documentation, an UAV was employed. Concretely, a Microdrone md4-200 was used (Figure 1, down).

UAV weight	900 g
Payload	up to 200 g
Size	54 cm between rotors
Flight time	10 a 20 min
Operating temperature	-10 a 50 °C
Max. height flight	500 m
Max. wind	5 m/s

Table 2. Technical specifications of the Microdrone md4-200 platform.

In spite of the manoeuvrability provided by this UAV, its limited payload required to employ a compact camera for the photogrammetric flight. The flight was planned according to the boarded camera technical specifications (see section 2.3) using the MFlip in-house software.

2.3 Cameras

According to the different methodologies and platforms used, two different cameras were employed:

For the paratrike, a full frame reflex camera, Canon 5D MkII, with a fixed focal length to achieve a GSD of 3 cm, was used (Table 3). This camera was also used for the terrestrial photogrammetry and photorealistic texture mapping due to its better image quality.

Sensor type:	CMOS
Sensor size:	36 x 24 mm
Image size:	5616 x 3744 pixels
Resolution:	21.1 Mp
Radiometric resolution:	14 bits
Pixel size:	6.4 μm
Size	152 × 114 × 75 mm
Weight	810 g (without battery)
Focal length:	50 mm (fixed)

Table 3. Technical specifications of the full-frame camera, Canon 5D MkII.

For the UAV, an ultra-compact camera, Canon IXUS 115 HS (Table 4), was chosen allowing a GSD of 1 cm.

Sensor type:	BSI-CMOS
Sensor size:	6.17 x 4.55 mm
Image size:	4000 x 3000 pixels
Resolution:	12.1 Mp
Radiometric resolution:	JPEG compression
Pixel size:	1.5 μm
Size	93 × 56 × 20 mm
Weight	140 g (battery included)
Focal length:	4.9 – 20 mm (28 – 112 mm equivalent)

Table 4. Technical specifications of the compact camera, Canon IXUS 115 HS.

2.4 Terrestrial laser scanner

In those complex zones a phase shift terrestrial laser scanner, Faro Focus 3D, was employed (Table 5).

Model	Faro Focus 3D
Principle	Phase Shift
Wavelength	905 nm (Near infrared)
Field of view	360 °H x 320 °V
Range std. deviation	2 mm at 25m
Measurement range	0.19 mrad
Beam divergence	8 mm a 50 m
Scanning speed	976,000 points/s

Table 5. Technical specifications of TLS, Faro Focus 3D.

This contribution has been peer-reviewed.
doi:10.5194/isprsarchives-XL-5-W4-37-2015

According to the archaeological settlement characteristics and the TLS technical specifications, laser stations were established in a network that guaranteed an average spatial resolution of 5 mm for the whole scenario. The mean distance acquisition was 15 m.

2.5 Navigation system

In contrast to the UAV platform which encloses an integrated navigation system (GPS, IMU and barometric altimeter), the paratrike requires an external set of sensors in order to provide navigation capabilities and thus fulfill photogrammetric constraints for data acquisition. In particular, the planimetric position is provided by a GPS antenna (Trimble Bullet III), installed in the camera platform close to the optical centre of the camera, connected to a mono-frequency receiver Ublox EVK-6T-0. This system yields an absolute precision of ±9 m on the horizontal axis for 95% of the time (Takasu, 2009). During data acquisition the pilot follows the planned photogrammetric mission in a rugged tabled connected to the GPS system, where the real time track is showed against the planned flight. The final component in the navigation system, the altimetry, which affects the GSD, is controlled by an altimetric barometer (DigiFly VL100) with an absolute precision of ±8 m. This solution is chosen instead of GPS receiver, since the absolute GPS precision is just ±15 m.

2.6 Geo-referencing system

The establishment of the mapping frame in the study area is performed with two GNSS bi-frequency (L1, L2) receivers, Topcon manufacturer. The GNSS observation method was real time kinematic (RTK) getting a precision of 1 cm.

The coordinate reference system was the official coordinate system established by the Spanish law (ETRS89); and the cartographic projection UTM in the 30N zone (EPSG: 25830). This mapping frame was materialized by a GNSS surveying using natural features, which could be clearly identified on aerial images (e.g. corners of well-defined objects, small features with great contrast, etc.).

Figure 1. Paratrike employed (up-left); detail of the stabilization platform "MUSAS" (up-right) and UAV used in those complex areas (down).

3. METHODOLOGY

Given the complexity of the archaeological site and the archaeological documentation requirements, the aerial and terrestrial data acquisition were planned in order to provide an integral and integrated recording of the site.

All the data from the different sensors and plataforms were processed according to the workflow outlined in Figure 2 and explined in the following subsections.

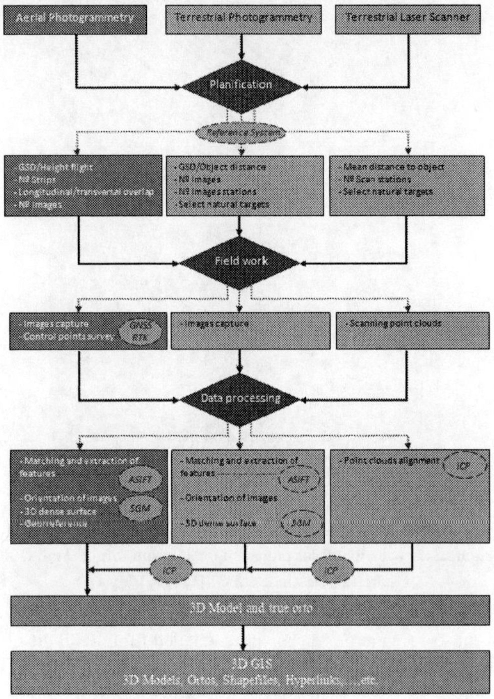

Figure 2. Multi-sensor and multi-data source workflow for the 3D reconstruction of complex archaeological sites.

3.1 Data acquisition

Each technique (i.e. photogrammetry, laser scanning) requires different procedures and protocols for data acquisition.

In the case of aerial data acquisition, image projection centers and camera attitude must be previously defined according to the classical photogrammetric parameters. The flight planning was done using the in-house software; MFlip and PFlip, for the UAV and paratrike flights, respectively (Hernandez-Lopez et al., 2013). The main difference between both flights was the type of flight planning: for the paratrike, a classical stereoscopic photogrammetric flight was done, whereas for the case of UAV, oblique images were also considered.

Since the whole archaeological site (9 ha) must be recorded with a 3 cm resolution, paratrike was used due to the autonomy limitations of the UAV. According to the GSD desired for the whole archaeological site (3 cm) and considering the camera specifications (Table 3), a maximum flight height of 224 m was established (Figure 3 up). As result, a total of 268 images along 7 strips (NW-SE direction) were required for guarantying a side overlap >30% and a forward overlap >70%.

This contribution has been peer-reviewed.
doi:10.5194/isprsarchives-XL-5-W4-37-2015

Figure 3. Images footprints projected over terrain for the 3 cm GSD flight (up) and the 1 cm GSD flight (down).

For the "El Regueròn" area a more detailed flight with the UAV was designed using the ultra-compact camera (Table 4).The terrain relief effect is showed in the Figure 3 (down), where 25 images and a flight height of 32 m guarantee the desired spatial resolution. The overlap parameters were the same.
An important issue for the aerial data acquisition was the image sharpness, which affects the final photogrammetric reconstruction. This issue is affected by the camera parameters (i.e. aperture, shutter time and sensibility), the platform performance (i.e. flight speed and efficiency of the stabilized gimbal to absorb the paratrike vibrations) and the scenario illumination conditions. The flight was executed in a cloudy day to avoid shadows projected, and the camera sensibility and shutter time were setup for these conditions. The aperture and the focal length were fixed constant to avoid variations in the internal camera parameters.

As it was commented previously, the high vertical relieves of the site and the level of detail required in some areas (e.g. those constructive elements that integrate the defensive system of the entrance, details of the walls, etc.) entailed that aerial images were not suitable; so terrestrial images were acquired with the full-frame reflex camera. As a result, a total of 36 oblique images were acquired in order to avoid areas without information, guarantying a sub-centimeter resolution. In addition, terrestrial laser scans were used in those complex areas where photogrammetry could entails problems, requiring a lot of images to enclose the whole geometry or due to the presence of textureless objects or materials. In particular, a network of 12 TLS stations (Figure 4) was designed for an average distance of 15 m with an average spatial resolution of 5 mm.

Figure 4. TLS stations distribution.

3.2 Data processing

3.2.1 TLS alignment: The raw TLS scans were filtered removing noise and undesired information (e.g. vegetation) which could affect the alignment process. Since artificial targets were not used, scans alignment was done by a solid rigid transformation of an iterative closest point (ICP) technique (Besl and McKay, 1992). This iterative process was applied in pairwise stations, so the final a priori error was computed in the basis of the number of stations and the technical specifications of the TLS (Table 5) reaching 1.3 cm for the assumption of two consecutive overlapping point clouds.

3.2.2 Photogrammetric processing: The generation of the dense point cloud, from both aerial and terrestrial images, was automatized through the Photogrammetry Workbench (PW) in-house software, following a three step workflow: image matching, camera orientation and dense model generation.

The aerial images, coming from the paratrike and UAV, were processed to generate a hybrid model, with a total of 293 images and two different cameras. The dataset was checked to assure the sharpness of the images, which could be decreased by the motion blur.

Firstly, the image matching was solved by the ASIFT - Affine Scale Invariant Feature Transform (Morel and Yu, 2009) algorithm, which is robust to the image perspective variations by an affine transformation:

$$A = \begin{bmatrix} a & b \\ c & d \end{bmatrix} = \lambda \begin{bmatrix} \cos k & -\sin k \\ \sin k & \cos k \end{bmatrix} \cdot \begin{bmatrix} t & 0 \\ 0 & 1 \end{bmatrix} \cdot \begin{bmatrix} \cos\omega & -\sin\omega \\ \sin\omega & \cos\omega \end{bmatrix} \quad (1)$$

Where the affine transformation (**A**) is composed by a scale factor (λ), the three angles variation, being κ and ω the roll and yaw angles, while the tilt (φ) is defined as $\varphi = \arccos(1/t)$.

In order to accelerate the process, the overlapped aerial images were identified by their approximate camera orientations provided by the navigation system. In the case of terrestrial images, an all-to-all comparison was applied. This sub-step is a time consuming process which increases exponentially with the number of images (Rieke-Zapp and Nearing, 2005).

This contribution has been peer-reviewed.
doi:10.5194/isprsarchives-XL-5-W4-37-2015

Secondly, camera orientations were computed by Apero module (Pierrot-Deseilligny and Clery, 2011) using the tie-points obtained in the previous step. In this process the external camera parameters (position and attitude) and the internal camera parameters (self-calibration) were obtained. For the internal camera parameters, the corrections to apply in the collinearity equation (Δx, Δy) were defined according to the simplified Fraser model (Fraser et al., 1995) (Pierrot-Deseilligny and Clery, 2011).

$$\Delta x = -x_0 - \frac{x'}{f}\Delta f + x'\left(r^2k_1 + r^4k_2 + r^6k_3\right) + \left(2x'^2 + r^2\right)p_1 + 2p_2x'y' \quad (2)$$
$$\Delta y = -y_0 - \frac{y'}{f}\Delta f + y'\left(r^2k_1 + r^4k_2 + r^6k_3\right) + 2p_1x'y' + \left(2y'^2 + r^2\right)p_2$$

Where the f is the principal distance, (k_1, k_2, k_3) the radial lens distortion coefficients, (p_1, p_2) the tangential ones and (x', y') the image coordinates referred to the principal point.

Finally, a dense 3D model was obtained by means of ray intersection (Kraus, 2007) (Pierrot-Deseilligny and Clery, 2011). To solve this process, a SGM (Semi-Global Matching) technique (Hirschmuller, 2008) was applied. In order to provide metric capabilities to the model, scaling was defined through the manual identification of ground control points (GCPs) in the images and solving a Helmert 3D transformation of seven parameters.

3.2.3 Data fusion: The remaining step is the registration of point clouds derived from the different sensors under the same coordinate system. Firstly, a GNSS campaign of 3 hours based on 3 permanent ERGNSS stations was performed in order to provide precise coordinates to the GNSS base station used in the archaeological settlement. Secondly, GNSS observations were processed guarantying an absolute error of 3 cm. Thirdly, the RTK surveying of the GCPs where corrected to obtain the final coordinates of the 3D model integrated under the same coordinate system, reaching a final relative precision of 1 cm. Finally, the ICP technique was employed to transfer the global coordinate reference system from the geo-referenced aerial point cloud (based on GNSS observations) to the TLS point cloud.

3.2.4 Post-processing: Once the metric and geo-referenced point cloud was generated, a triangulation strategy was applied to generate the digital surface model (DSM). More concretely, the incremental Delaunay algorithm was applied (Bourke, 1989).
To improve the model quality, break lines were incorporated as geometric constraints. Break lines were manually restituted by the operator. Its use is relevant for the accurate representation of significant slope changes, as well as for correctly defining the defensive walls of the site. In those cases where images come from different methodologies or acquisition time, it was necessary a radiometric adjustment, to improve the final model visualization and avoid abrupt radiometric changes in the texture.

Once the DSM was obtained, it was possible to generate an orthoimage from the oriented images. Since multi-data source and multi-senor approach provide different DSM resolutions, it was possible to generate orthoimages with different spatial resolutions. For instance, an orthoimage that depicts the complete archaeological site was generated with a resolution of 10 cm; whereas a 2 cm spatial resolution was employed for the settlement entrance area in order to appreciate the construction features.

4. EXPERIMENTAL RESULTS

4.1 Area of study

The city of "Tolmo de Minateda" (Figure 5) was a strategic settlement of great importance for several centuries, largely because of its peculiar topography and geographical location. It is placed in a plateau hill of about 50 m of height, located at the junction of the route from Complutum to Carthago Nova, one of the principal Roman route that connect the interior of the Plateau with the southeast coast, and a road that connected Castulo with Saetabis.

Figure 5. Aerial view of the study area "Tolmo de Minateda" (from Abad Casal et al., 2012).

The archaeological research developed in the last 30 years has highlighted the importance of this site: from the Middle Bronze Age, through the Iberian era, the Roman period, the Middle Ages to the twentieth century. The Middle Age period provided most of the information by an important Visigoth settlement located in the upper part of the "El Tolmo", where was found between houses and cemeteries, an important Christian basilica.
One of the most interesting areas of the archaeological site is the called "El Reguerón" (Figure 6); an area of natural drainage of 12 m width, which had the main entrance to the top of the hill where the city was located. In the settlement an important fortification system consisting of three walls of different chronology and architectural typology has been documented. The oldest fortification is represented by the so-called "embanked" wall and was built during the final phase of the Iberian period (IV-II B.C.). Currently, only remains a wall of 6 m of height and 10 m width in the top, embanked in the external front and built in irregular masonry work stuck with soil. During the archaeological works in this ancient wall, an earlier phase, which dates back to the middle Bronze Age, was found. The Iberian wall was successively used as a retaining wall for a new fortification building during Roman times when the "Tolmo de Minateda" probably received the title of municipality.
During the period of peninsula occupation by the Visigoth (V-VI A.D.) is referred the last fortification found in "El Reguerón". This wall is presented as solid bulwark with L-shape that closes the valley and flanking the main road access to the city. At this point was located a monumental gate probably formed by two solid towers of blocks of which are now only partial remains are preserved. The wall is formed by a

This contribution has been peer-reviewed.
doi:10.5194/isprsarchives-XL-5-W4-37-2015

lining blocks with inscriptions and architectural elements from older constructions (among which are from Roman period).

Figure 6. Location of the defensive area "El Reguerón".

4.2 Workflow

The application of the multi-data source and multi-sensor fusion workflow was tested in the study area of "Tolmo de Minateda", which includes a total of 9 ha.

The aerial images were processed by the automated photogrammetric approach and the cameras were self-calibrated according to equation (2) as is showed in Table 6.
The raw reconstructed 3D model contains 4,271,354 points, while the TLS integrated point cloud reaches up to 1,314,136 points.

	Full frame reflex Canon 5D	Ultra-compact Canon IXUS 115
f (mm)	52.80	6.70
x_0 (mm)	18.18	3.07
y_0 (mm)	12.18	2.23
k_1	-5.04 x10^{-5}	-2.00 x10^{-3}
k_2	1.79 x10^{-8}	-6.25 x10^{-5}
k_3	9.78 x10^{-12}	3.56 x10^{-6}
p_1	-1.32 x10^{-5}	-5.81 x10^{-4}
p_2	-6.79 x10^{-5}	-1.41 x10^{-4}

Table 6. Internal orientation parameters of the cameras boarded in the aerial platforms.

The geo-referencing of the 3D model was solved by the employment of 28 control points homogeneously distributed in the site (but with higher density in the interest area), while 10 were used as check points.

The final multi-resolution 3D model after triangulation is showed in Figure 7, enclosing the whole study area with a spatial resolution of 3 cm. A total of 10,542,505 triangles were obtained.

For the detailed area which integrates different data sources (i.e. terrestrial laser scanner and photogrammetry), a spatial filtering of 0.5 cm was applied to avoid areas with excessive point density caused by the different overlaps.

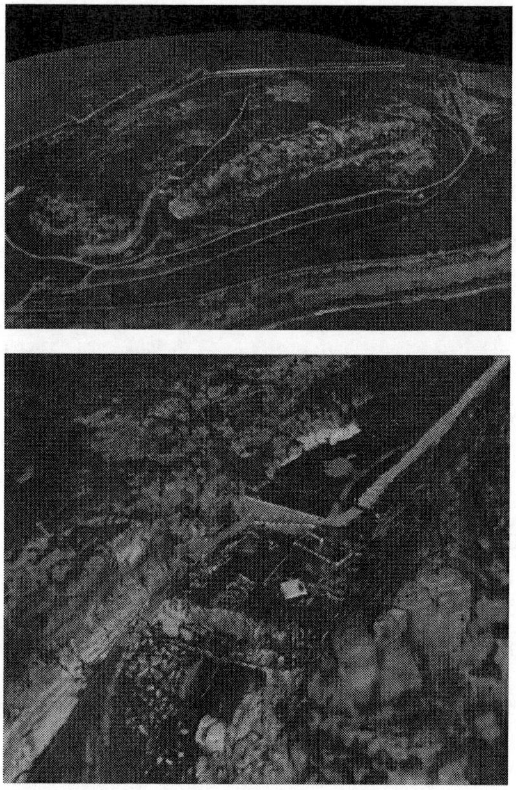

Figure 7. Multi-resolution 3D model of the archaeological settlement (up). Detailed 3D model of the "Reguerón" (down).

By this multi-resolution approach, the final inspection of the archaeological site could be adapted to any area, as is showed in the interest area of "El Reguerón", where the subcentimeter resolution achieved for the walls and construction (Figure 8 down) was integrated with the rest of the archaeological site (Figure 8 top). This multi-resolution capability opens a range of possibilities for a spatial analysis, settlement interpretation, patterns recognition, and the establishing relationships among the elements (sites, artefacts locations, etc.).

Figure 8. Multi-resolution model of "El Regueron" area (up) and detail for the wall (down).

In order to validate the final integration of the different data source, a series of check points were used. The different error components are showed in Table 7. The average altimetric error (1.4 cm) was higher than the planimetric error, as was expected for the GNSS technique. An average precision of 4.1 cm for the 3D vector error was obtained, which is statistical compatible with the expected error of 4.2 cm composed by the model and GNSS check point errors.

Check points			Discrepancies with 3D model coordinates		
X (m)	Y (m)	H (m)	ΔX (m)	ΔY (m)	ΔH (m)
621321.452	4259638.120	448.672	0.035	0.016	-0.013
621375.933	4259646.226	454.775	-0.019	0.044	0.002
621416.258	4259667.686	463.661	0.027	-0.011	0.030
621419.022	4259658.232	464.618	0.001	0.028	0.046
621345.185	4259659.272	454.267	0.022	0.006	0.004
621321.298	4259660.311	452.106	-0.010	-0.033	0.016

Table 7. Geo-referencing errors.

To manage more efficiently the available information of the archaeological settlement, orthoimages were generated as derived products (Figure 9), since they combined the photorealistic texture with the metric properties in an easy to use document for non-experts.

Figure 9. Orthoimages of the whole settlement (up) and defensive area "El Regueron" (down).

However, to allow a more complete interpretation and to make use of the potential of the generated 3D product, an integration with a 3DGIS tool, GeoWeb3D, was done. In particular, the 3D geometry provided by the multi-sensor approach has been integrated with 2D archaeological archives such as sketches, pictures, part details, etc. enhancing the subsequent analysis, and providing problem solving and taking decision capabilities.. In Figure 10 is showed the integration of different historical events over the 3D model reconstructed.

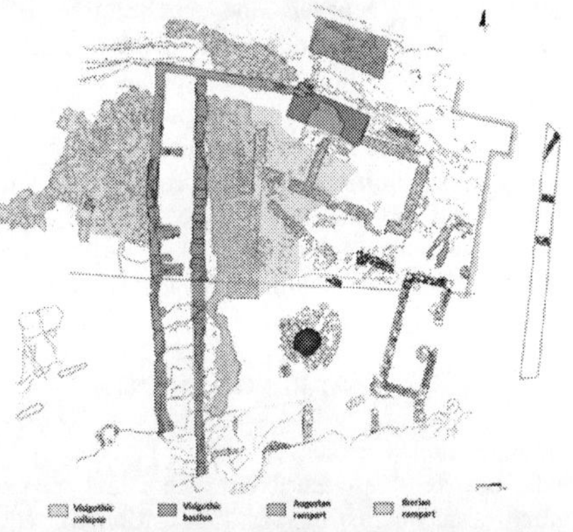

Figure 10. Archaeological sketch (up) overlapped with the 3D model (down).

This contribution has been peer-reviewed.
doi:10.5194/isprsarchives-XL-5-W4-37-2015

5. CONCLUSIONS

This paper presents a methodology based on a combination of multiple sensors, platforms and techniques, which has been tested in a complex archaeological site. As is shown in the experimental results, the automation provided by the photogrammetric and laser techniques, along with the versatility of the aerial platforms (paratrike and UAV), provide the suitability of this methodology for complex archaeological sites.

The potential of ultralight aerial platforms (paratrike) is highlighted, due to its payload and flight autonomy, which overpasses the UAVs capabilities. Regarding the usual archaeological surveys, one of the main differential factors in this work has been the integration of aerial images at different resolutions, terrestrial images and terrestrial laser scans.

The generation of high resolution products based on photogrammetry requires high computational costs. To this end, the presented algorithms, such as ASIFT, have been developed taking advance of GPU capabilities, reducing time operations and consequently improving the workflow efficiency.

The management of the final models through a 3DGIS tool opens new analysis capabilities for the archaeologist (e.g. analysis though time, archaeological investigations, integration of historical and geometric models).

The data acquisition and processing methodologies from Geomatic science broaden the possibilities of sensors, configurations and/or combination as future perspectives. Concretely, the resolution of the external orientation of the images directly, just using the integrated GNSS/IMU of the aerial platform, would allow us to speed up the field work and processing time. Furthermore, in order to obtain realistic textured models and orthoimages, the use of techniques such BRDF (Bidirectional Reflectance Distribution Function) to improve the radiometric parameters will be studied.

ACKNOWLEDGEMENTS

This research has been developed in the framework of the research project "Infraestructura de datos espaciales de patrimonio arqueológico de Castilla-La Mancha" (POII-2014-004-P) of the 2014-2017 Scientific Research Projects cofounded by the European Regional Development Fund.
The authors would like to thank the Ministry of Education, Culture and Sport of Castilla-La Mancha, especially to the Directorate-General for Universities, Research and Innovation, the Directorate-General for Cultural and the Museum of Albacete.

REFERENCES

References from Books:

Kraus, K., 2007. *Photogrammetry: geometry from images and laser scans*. Walter de Gruyter. Berlin, Germany.

References from Journals:

Abad Casal, L., Gutierrez Lloret, S., Gamo Parras, B., Canovas Guillen, P., 2012. El Tolmo de Minateda (Hellín, Albacete, España): un proyecto de investigación y puesta en valor del patrimonio. *Debates de arqueología medieval*, 2, pp. 351-381.

Besl, P., McKay, N., 1992. A method for registration of 3-D Shapes. *IEEE transactions on pattern analysis and machine intelligence*, 14(2), pp. 239-256.

Gomez-Lahoz, J., Gonzalez-Aguilera, D., 2009. Recovering traditions in the digital era: the use of blimps for modelling the archaeological cultural heritage. *Journal of archaeological science*, 36(1), pp. 100-109.

González-Aguilera, D., Muñoz-Nieto, Á., Rodríguez-Gonzálvez, P., Menéndez, M., 2011. New tools for rock art modelling: Automated sensor integration in Pindal cave. *Journal of archaeological science*, 38(1), pp. 120-128.

Hernandez-Lopez, D., Felipe-Garcia, B., Gonzalez-Aguilera, D., Arias-Perez, B., 2013. An automatic approach to UAV flight planning and control for photogrammetric applications: a test case in the Asturias region (Spain), *Photogrammetric engineering and remote sensing*, 79, pp. 87-98.

Hirschmuller, H., 2008. Stereo processing by semiglobal matching and mutual information. *IEEE transactions on pattern analysis and machine intelligence*, 30, pp. 328-341.

Morel, J.M., Yu, G., 2009. ASIFT: A new framework for fully affine invariant image comparison. *SIAM journal on imaging sciences*, 2(2), pp. 438-469.

Rieke-Zapp, D.H., Nearing, M.A., 2005. Digital close range photogrammetry for measurement of soil erosion. *Photogrammetry record*, 20, pp. 69-87.

Torres, J.A., Hernandez-Lopez, D., Gonzalez-Aguilera, D., Moreno Hidalgo, M.A., 2014. A hybrid measurement approach for archaeological site modelling and monitoring: the case study of Mas D'is, Penàguila. *Journal of archaeological science*, 50, pp. 475-483.

References from Other Literature:

Bourke, P., 1989. An algorithm for interpolating irregularly-spaced data with applications in terrain modelling. In: *Pan pacific computer conference*, Beijing, China. 6 pages.

Fraser, C.S., Shortis, M.R., Ganci, G., 1995. Multisensor system self-calibration. In: *SPIE conference 2598, Videometrics IV*, Philadelphia, USA, pp. 2-18.

Pierrot-Deseilligny, M., Clery, I., 2011. Apero, an open source bundle adjusment software for automatic calibration and orientation of set of images. In: *3DARCH'11*, Trento, Italy, pp. 269-276.

Takasu T., 2009. RTKLIB: Open source program package for RTK-GPS, FOSS4G 2009 Tokyo, Japan.

This contribution has been peer-reviewed.
doi:10.5194/isprsarchives-XL-5-W4-37-2015

Restoring Fort Frontenac in 3D: Effective Usage of 3D Technology for Heritage Visualization

M. Yabe[a] , E. Goins[b], C. Jackson[a], D. Halbstein[a], S. Foster[a], S. Bazely[c]

[a] Department of Visual Communication Design, Rochester Institute of Technology, Rochester, New York, USA
<mxy3663><cbjpgd><scffaa><dlhfaa>@rit.edu, http://mxy3663.wix.com/fortfrontenac
[b] Department of Cultural Materials Science and Fine Arts, Rochester Institute of Technology, Rochester, New York, USA
esggsh@rit.edu
[c] Cataraqui Archaeological Research Foundation, Kingston, Ontario, Canada
sue@bazely.ca

Commission V, WG V/4

KEY WORDS: Heritage Visualization, 3D Computer Graphics, Interactive Virtual Design, Web Design, Effective Usage of Learning

ABSTRACT

This paper is composed of three elements: 3D modeling, web design, and heritage visualization. The aim is to use computer graphics design to inform and create an interest in historical visualization by rebuilding Fort Frontenac using 3D modeling and interactive design. The final model will be integrated into an interactive website to learn more about the fort's historic importance. It is apparent that using computer graphics can save time and money when it comes to historical visualization. Visitors do not have to travel to the actual archaeological buildings. They can simply use the Web in their own home to learn about this information virtually. Meticulously following historical records to create a sophisticated restoration of archaeological buildings will draw viewers into visualizations, such as the historical world of Fort Frontenac. As a result, it allows the viewers to effectively understand the fort's social system, habits, and historical events.

1. INTRODCUTION

Fort Frontenac was an important military fortress and fur trading post that was built in Kingston, Ontario, Canada in 1673. Fort Frontenac was embroiled in occasional outbreaks of fighting between the Iroquois and the British, and was destroyed in 1689 but was then later reconstructed in 1695 to strengthen military activity and accommodation. Over the course of 50 years, fur trade gradually became more unprofitable, and the fort was relinquished to British forces during the Seven Year's War in 1758 who devastated it. The fort was never rebuilt, and was abandoned because the fortification was no longer needed. New buildings and streets were then erected above its ruins for more than a hundred years. In 1983, archaeological research finally successfully excavated only a small portion of the fort, the northwest bastion (Fig. 1). The fort was designated as a Canadian National Historic Site and a part of the northwest bastion was reconstructed in 1984 (Cataraqui Archaeological Research Foundation, 2013).

The aim of this paper is to virtually reconstruct all the buildings to look as they did originally. The field that embraces this process is called Heritage Visualization, which is derived from a combination of the use of conventional archaeology and cutting-edge 3D computer graphics design. The field has been used to recreate digital models from antiquity, such as ancient Egypt and Rome, at several research institutes around the world. It represents a great opportunity to rebuild Fort Frontenac using this technique because many forts have been renovated in this manner to show historical reenactment sites.

This research aims at virtually restoring an archaeological building, Fort Frontenac, which in turn will draw viewers into the historical world of Fort Frontenac. This is a difficult task because there is limited information and only a small amount of archaeological remains from the original Fort Frontenac. While the original maps and plans of Fort Frontenac are incomplete, a fertile imagination and computer graphics design skills are essential if the field is to help correct false assumptions regarding historical viewpoints. For example, what were the colors of the buildings? Color may be involved with an overly subjective viewpoint. What were its social system, habits, and historical events at that time? Application of this field will also allow viewers to effectively understand the fort's social system and historical events. The target audience is those who are interested in history and archaeology. For those who are not able to visit Fort Frontenac because it is too far away from their residence, this thesis is designed to expand their knowledge, while saving

This contribution has been peer-reviewed.
doi:10.5194/isprsarchives-XL-5-W4-45-2015

money and time. Historical Visualization is a vital field by means of Internet access to view not only the fort, but also many other forts as well as other historic buildings and their architectural features, such as castles, all over the world. The method used in this process is important to disseminate in this paper for educational and economic purposes.

Figure 1. Aerial imagery of surroundings around the remnants of Fort Frontenac © The City of Kingston Copyright 2011.

2. PROCESS

2.1 Design Approach

Figure 2. The remnants of Fort Frontenac.

First, in order to approach a true picture of Fort Frontenac and its surroundings as they existed in the past, meticulous research was required via historical maps, plans, and architectural styles of the fort at that time using the Internet and libraries. Print and online documentation, such as books, are good resources for this process. However, a physical visit and an archeologist's opinion are more advantageous to developing exceptional research. The dates that this researcher visited Kingston, Canada, were on December 3rd, 2013 and on September 26th, 2014 (Fig. 2) in order to photograph and glean information about the target fort. Furthermore, a consultation with a local archaeologist, Mrs. Susan Bazely, was well worth sharing her knowledge, experience in archaeological research, and insights for research development. Envisioning the fort's restoration was taken advantage of by filtering the collected information and photographs and then sketching out designs on paper. A good guide for the 3D modeling using Maya software was to set up to scan drawings and blueprints.

After the design research was completed, the following practical software was essential to building models: Adobe Photoshop, CrazyBumps, and Autodesk Maya. An initial plan was to model the interior of the buildings, but that plan was cancelled because research was lacking for credible information regarding the 17th century. The exterior of the fortress was constructed in five steps, including the following years: 1673, 1675, 1680, 1685, and 1688. The 3D modeling techniques used included polygon modeling, shading, and lighting. The textures of the building were shaded, and the background of the building and its surrounding environment were added by lighting adjustment. The completed models were rendered to produce video clips and interactive animations through camera movement, which set in motion as a 3D camera was placed to create a bird's eye-view and a 360-degree view from the fort.

After the modeling was complete, interactive sceneries in Adobe Flash were devised to provide an effective way to implement visual communication and learning through the use of a mouse. The interactivity was integrated into the 3D models of the fort with unseen clickable buttons in order to allow the viewer to learn more about the fort's structure and historical importance. The video clips and interactive sceneries were integrated into the website in order to inform the user of the fort's social systems and historical events. Also, the purpose for this is that people prefer to stay at home without visiting a location that is far from their residence, so the uploaded website's goal is to excite the user's curiosity about the fort.

2.2 Modeling Buildings in Maya

2.21 Constructing Buildings and Environment: Maya 2014 is a professional tool that has the means of modeling physical objects and rendering sequences to create animations. By complying with the notes regarding the archaeological and historical consequences of the research, such as photos taken during the visit to the forts, online documentation, and historical plans and print documentation provided by the City of Kingston, as well as Mrs. Bazely's evidence[1], and the extremely restricted

[1] Bazely, Susan M., email message to Mitsuyoshi Yabe, February 26, 2014.

information about Fort Frontenac in the 17th century that was available, the archaeological features of the building construction and their exteriors were required to be meticulously elaborated. The careful operation was performed in Maya in order to fashion the length, height, and size of each building, including the log palisades, curtain walls, and moats.

In addition, Crazybumps is a convenient software program for shading textural materials where photos were simultaneously turned into displacement and normal maps. The software is the most effective, simple and easiest way of creating these types of maps, since the software is beneficial to the enhancement of aesthetics and contrast. Four, six, or eight images became two or four of each image by means of flipping the images vertically and horizontally and then stitching the original whole images together with much repetition in Photoshop. The images were offset, and the clone tool was used to make seamless repeating patterns (Fig. 3). The more images that were put on the one large image in Photoshop, the more realistic the textures of the buildings were when the images were transformed to displacement and normal maps adjusted by Crazybumps according to intensity sharpen, detail, shade, and highlight, which were then imported into Maya. Finally, the normal maps were placed along with the displacement maps onto a Blinn bump map for particular buildings. Most texture images were provided by a website referenced as Cgtextures, which has abundant and wide resources of textures that range from ancient to modern.

Figure 3: A classic sample of stitched textures for the fort's curtain wall done in Photoshop.

For the 1673 imagery, the constructed buildings included the following: two storehouses for provisions and ammunitions storage, log palisades whose north and south sides were forty-six feet and the other sides which were twenty feet in length, and a well (Stewart, 1982). For the restoration of the wooden palisades, as a sequel to researching their detailed textures among old forts in America that were built in the 17th century, their average colors were shabby, drab, and gray with a few knots which could possibly be attributed to a kind of oak, beech, or cedar. For the 1675 imagery, the log palisades had the addition of four limestone square bastions in an area that was 800 yards in circumference (Stewart, 1982). The masonry textures of the bastions were originally made from remnants of Fort Frontenac because Mrs. Bazely mentioned that the surfaces appeared to be yellowish, rough, and blunt [2]. For the 1680 imagery, the constructed buildings included the following: a one-hundred-foot-long quarters, a blacksmith shop, a guardhouse, a cow house, a well, wooden platforms and scaffolds and log palisades on the North, East, and South sides, as well as masonry curtain walls on the West side, with the addition of four limestone square bastions. Around the fort were 15-foot-wide ditches (Stewart, 1982). The masonry curtain walls included many casements and loopholes, and it took a lot of time to meticulously embed the casements and loopholes within the wall (Stewart, 1982). For the 1685 imagery, the fort was enlarged in a square alongside each of four bastions, which measured approximately 96 feet from curtain wall to curtain wall. The South Side had a combination of a 4-foot-high masonry foundation wall and log palisade that stood on the foundation wall. The inside included a masonry quarters or barracks, a guardhouse, a forge, mills, a powder magazine, a bakery, trade house, a well, a sentry-box, and moats that were fifteen feet deep (Stewart, 1982). For the 1688 imagery, the additionally constructed buildings included the following: a large tower and four small towers, wooden platforms and scaffolds, a wooden bridge, a masonry gate, and full limestone curtain walls with many casements and loopholes (Stewart, 1982). The flag with the pole on the big tower was also added. Most of buildings were completely transformed from wooden surfaces to limestone surfaces. However, some buildings still remained with wooden textures. Both textures were replaced because the fort had to be strongly reinforced due to the improvement of the military services and the interior exploration.

The log textures were from Cgtexture, and the limestone textures were from the remnants of Fort Frontenac. The six similar, repeated images were stitched together using offset in Photoshop in order to respectively adjust the surfaces of each pole on the UV texture editor in Maya. For the roofs and wooden walls of the two storehouses, the textures were also found in Cgtexture, and were created by the same means of Photoshop and Crazybumps with the results being placed onto bump maps using Maya. Textures were then put on each building by following these methods (Figs. 4 - 6).

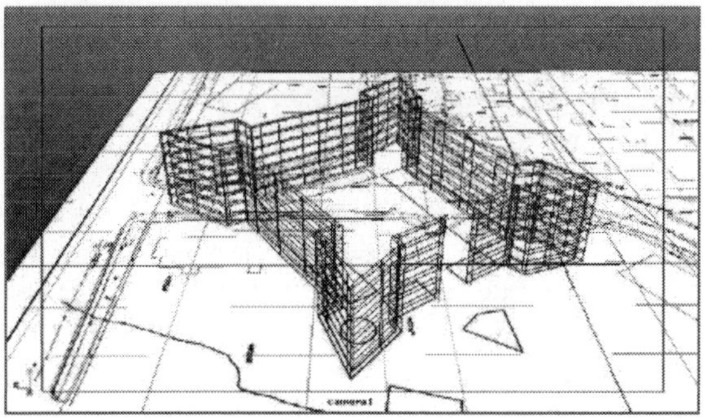

Figure 4: The wireframe model of the 1688 fortress.

[2] Bazely, Susan M., email message to Mitsuyoshi Yabe, February 26, 2014.

This contribution has been peer-reviewed.
doi:10.5194/isprsarchives-XL-5-W4-45-2015

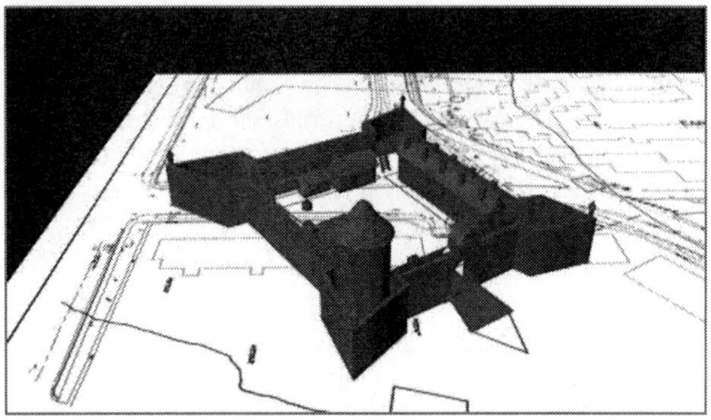

Figure 5: The wireframe model of the 1688 fortress.

Figure 6: The wireframe model of the 1688 fortress.

Geographic landscapes were created in Maya by means of tracing curves over the outline of the plans for 1685 [3] which were provided by the City of Kingston. The outline became a planar surface and that surface was converted into a polygon surface with numerous quarter geometries. As the ensuing landscapes for 1673, 1675, 1680, 1685, and 1688, were created, the geometric maps were shot by the UV Snapshot included in the UV texture editor in Maya and were imported into Photoshop to create color geographic topography. A displacement map and a normal map were created in Crazybumps and were then imported back into Maya to be placed onto a bump map for the landscapes. Moreover, the sea was created by means of using an Ocean Shader and Ocean, and minor corrections were required until the color was very similar to the color of the sea around Kingston that was dully emerald or bluish in color.

Three lights were also set up: Key Light, Fill Light, and Back Light. The key light was like sunlight set as a directional light, turned towards the north, and its color was slightly bright yellowish. The Fill Light was a secondary light set as a spot light facing southwestward, and its color was slightly bluish. The Back Light was the third light set as an additional spot light in a southern direction, and its color was dimly warm orange. Originally, the fort's gate was turned northeast so that the entrance was in shadow in the northeastern direction. The three lights were crucial materials used to heighten an aesthetic sense of natural imagery. In order to create the dynamic moving sky, the sky image was distorted using polar coordinators in Photoshop and the hemisphere modeling was mapped with the image onto the UV texture editor under spherical mapping to eliminate seams in the top sky. The hemisphere modeling was placed so as to entirely hide the forts and geographical landscapes. It was designed for use as a camera application to further an appreciation of the area's natural aesthetics. Lastly, in order to make shadows look realistic, ambient occlusion (AO) can be used to help soften and deepen shadows and to refine a clearly visible separation between objects in a totally white scene. This technique can create a more realistic AO rendering (Fig. 7).

Figure 7. The restoration of the entrance of Fort Frontenac in 3D computer graphics.

2.22 Rendering Sequences to Create Videos and Interactive Design: Regarding the camera animation, rendering PNG sequences featured an overview of the forts for a series of all construction dates and inside views of the forts for 1675, 1680, 1685, and 1688, as well as providing ambient occlusion renderings for all those scenes in video clips. The 1673 version was not included because the inside was too narrow to create a video. A "Two Node" camera was set as the camera body and an aim constraint. This allows the camera to focus on one object while the body is moved. Different circular curves were created to become a motion path attached to each camera in order to make the cameras move very smoothly. Nine hundred frames were required for an overview of the exterior of the 1688 fort and ambient occlusion renderings for the fort in sequence. It took an average of six to nine hours to get this smooth for the videos.

Both of the completed sequences were imported into Adobe After Effects software and then placed as AO sequences under the "multiple" color sequences setting in order to be semitransparent, making the sceneries appear more realistic, and then rendered for videos. The preferable time length for the videos was twenty-four seconds for an overview of the exterior. These setups were needed to make viewers visually comfortable according to the motion and timing.

Two videos were set as a chronological series of the two kinds of entire exteriors of the forts. The images for each were simply

[3] *Fort Frontenac Based on A 1685 Map.* Map. Scale 1:1000. Kingston, Ontario: City of Kingston, 2006.

This contribution has been peer-reviewed.
doi:10.5194/isprsarchives-XL-5-W4-45-2015

taken from Maya, which was a simple and easy way to ensure that each image was ordered chronologically and the corresponding chronological date texts were aesthetically put on each of the images in After Effects. For the aspect of interactive design, only thirty-six frames were required to render an overview of the exterior and interior of the forts, including ambient occlusion renderings for both. It took thirty minutes to set all the frames and install the historical sequences respectively for 1673, 1675, 1680, 1685, and 1688.

2.3 Creating Interactive Imagery in Flash

Flash CC is reliable software designed to play a major role in interactivity. An SWF format generated from Flash was designed for the interactive application for this project and was equipped to become integrated with the website. From a technical view, the interactive animation was set up to play at a rate of only thirty-six frames. The coding was typed and fixed on the action scripts in order to fit the interactive scenery (Fig. 8). The names and descriptions of each building were placed in panels. The descriptions were integrated from the research result derived from trustworthy online and print documentation. These panels were placed around each building using transparent black color with high opacity, and *Palatino Linotype* was used as the font in white text on the basis of its visual effect. Once the SWFs were published, the interactive system enabled users to hover over a certain building with a mouse where the panels including its name and description popped out (Fig. 9). This application could help to explore users' comfort, the effective application of learning about the fort through interactivity, and animation principles. This idea will be a useful educational tool to let users learn about the fort's historical importance and to acquire new knowledge, as well as assisting in the development of educational and study materials.

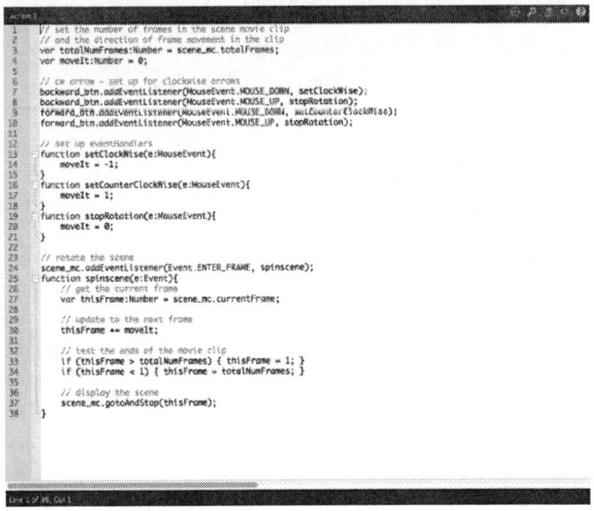

Figure 8: Codes in action scripts.

Figure 9: The interactive scenery of the 1675 interior fortress.

2.4 Designing the Website

Wix, as a website hosting and design tool, provided indispensable web assistance, which is helpful to a beginner in designing attractive websites without coding within a short time frame. Wix is able to convey information more effectively within a limited time than multimedia. Even so, it took a lot of time for a novice to design a website that included numerous entangled codes. In terms of layout design, the color of the top bar was a slight crimson, and the main background was slightly yellowish beige or grey with high opacity under the old map. The reason for the color choices were that the outline and background of the old fort's map plan had colors in common with the roof and wall of the National Defence College that were built on the front of the fort's remnants. The classic font *Palatino Linotype* was chosen again in order to appear similar to old, traditional French fonts. The site reanimates time reenactments of French settlements in Canada (Figs. 10 and 11).

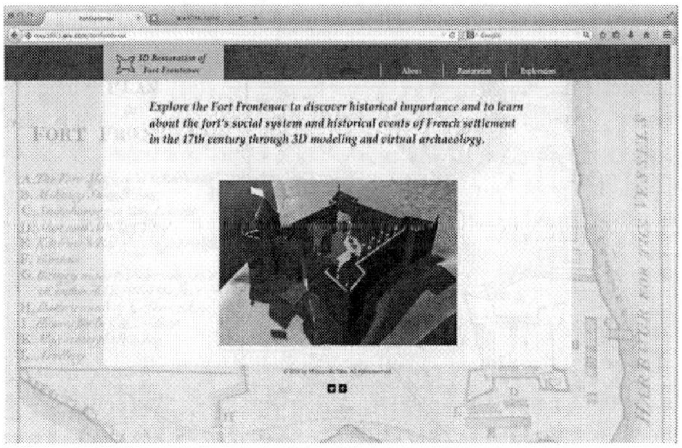

Figure 10: The website of the 3D restoration of Fort Frontenac.

3D Restoration of Fort Frontenac

Figure 11: A sample of the representative logo.

This contribution has been peer-reviewed.
doi:10.5194/isprsarchives-XL-5-W4-45-2015

2.5 Critical Analysis and Troubleshooting

In fact, there were several issues that occurred during this research. Each struggle greatly affected the process of how the research and project was critiqued and improved. The following is a critical analysis and troubleshooting of what went right and/or wrong and how to solve the issues.

The first plan was to create three kinds of videos: an outside overview of the forts, an inside overview of the forts, and a close-up of the gate for each of the years: 1673, 1675, and 1688. Nine hundred frames were essentially required to produce a 36-second video for the outside overview, six hundred frames to produce a 24-second video for the inside overview, and three hundred sixty frames to produce a 12-second video for the close-up of the gate. Rendering nine hundred frames greatly impacted the researcher's lifestyle on a daily basis. It took one day to complete one video. To create the full set of nine videos required a week. On the other hand, to create interactive sceneries for the outside and inside of the forts in Flash CC took only 30 minutes to render all twenty-four frames necessary to build a 360-degree interaction in Flash. At that rate, more than 10 interactions could be created on a day.

However, designing a related website was influenced by the video file formats and interactive sceneries for each page. User-centered design determined the best method to incorporate the videos. Consequently, most videos were removed due to their memory-intensive usage and the time management required. Only the videos for the completed exterior fort in 1688 were uploaded for the webpage. The website featured nine interactive sceneries: the interiors for 1673, 1675, 1680, 1685, and 1688 as well as their exteriors respectively, except for 1673. The appearance of the website was innovative and comfortable for users to deal with and enhanced their knowledge in an effective way. In comparison, watching the videos was not innovative and banal. To sum up, this design could assert that interactive applications are attractive to users and usefully economize the time spent creating them, but featuring videos is relatively conventional and expends considerable effort and time.

3. CONCLUSIONS

Modeling buildings and the related environment created a lot of challenges and many difficulties because there was very little information available about the 17th-century Fort Frontenac. Further information was needed to clarify that Fort Frontenac was compared with other forts that were relatively similar to the target fort, according to their various sizes, such as the height, length, and width of buildings as well as the textures of the buildings and the environment surrounding them. To make the visual motion smooth and slow, nine hundred frames in the rendering sequences were required to create the video clips. Needless to say, the average rendering time was nine hours, and occasionally several defections from the rendering occurred, so the constant supervision was important to the rendering a few times each hour.

The successful website can be made available to public; however, for the past few years there have been further issues with some functions that depend on the consistently upgraded versions of several software programs, such as Flash. In a few years after the website is published, the website may not be useful due to the expiration of a version of a certain software that may not match a particular browser, or a computer that requires a newer version. This would require upgrading to new versions determined by users' needs. In addition, annually hosting the website may prove too costly as well.

Nevertheless, the achievement of a 3D virtual model and an interactive website will enable users to effectively learn about the historical importance of Fort Frontenac. In addition, users can quickly view its exterior from various different perspectives. The development of an interactive website is essential to, and entirely prevalent in, the current world through the development of a many-sided website that enhances digital visualization. The aim was to take the content and make it a more beneficial experience through the use of a vivid video screen rather than by merely studying textbooks. This technique also encourages users to understand the fort's social system and the historical events that occurred at that time of the fort's operation.

By changing the consumer's understanding to an archaeological viewpoint, the field continues to evolve step-by-step. 3D techniques and web design gradually supports this evolution, even though the technology is still developing in areas such as comfort and convenience for modeling, importing colors and sounds, lighting, shading, setting items in motion in 3D, visually communicating a 3D screen, and interactively using a mouse in a website. The more the technology improves, the more heritage visualization can develop. Historical sites which are in repair and abolished cultural methods can be resuscitated and examined through the use of the latest technology.

4. ACKNOWLEDGEMENT

The author would like to thank Dr. Elizabeth Goins, Mr. Chris Jackson, Mr. David Halbstein, Mr. Shaun Foster, and Mrs. Susan Bazely for giving useful advice on this research work.

5. REFERENCES

Bazely, Susan M., Jonas Fernandez, and Helen Sheldon. *Archaeological Assessment Place D'Armes Re-Alignment Kingston, Ontario.* Kingston, Ontario: Cataraqui Archaeological Research Foundation, 2007.

Bazely, Susan M., email message to Mitsuyoshi Yabe, February 26, 2014.

This contribution has been peer-reviewed.
doi:10.5194/isprsarchives-XL-5-W4-45-2015

Canada's historic Place, s.v. "Fort Frontenac National Historic Site of Canada" accessed September 10, 2013, http://www.historicplaces.ca/en/rep-reg/place-lieu.aspx?id=12130

Cataraqui Archaeological Research foundation, s.v. "Fort Frontenc," accessed September 10, 2013, http://www.carf.info/kingston-past/fort-frontenac

Fort Frontenac Based on A 1685 Map. Map. Scale 1:1000. Kingston, Ontario: City of Kingston, 2006.

Frischer, Bernard. "Rome Reborn." Last modified August 1, 2013. http://romereborn.frischerconsulting.com

Preston, Richard Arthur. *Royal Fort Frontenac.* Toronto: The Champlain Society, 1958.

Stewart, Bruce W. *The Kingston Harbourfront Archaeological Project Phase II, Fort Frontenac: Results of the 1982 Test Excavations.* Kingston, Ontario: Cataraqui Archaeological Research Foundation, 1982.

This contribution has been peer-reviewed.
doi:10.5194/isprsarchives-XL-5-W4-45-2015

STRUCTURE FROM MOTION SYSTEMS FOR ARCHITECTURAL HERITAGE. A SURVEY OF THE INTERNAL LOGGIA COURTYARD OF PALAZZO DEI CAPITANI, ASCOLI PICENO, ITALY

E. Ippoliti [a], A. Meschini [b], F. Sicuranza [b]

[a] Dept. of Architectural History, Representation and Conservation (DiSDRA), Sapienza University, Piazza Fontanella Borghese, Roma, Italy – elena.ippoiti@uniroma1.it
[b] School of Architecture and Design (SAD), University of Camerino, Viale della Rimembranza, Ascoli Piceno, Italy – (alessandra.meschini, filippo.sicuranza)@unicam.it

Commission V, WG V/4

KEY WORDS: Structure From Motion, Image-Based Survey, 3d Data Processing, Architectural Survey, Architectural Heritage Representation, Architectural 3d Modeling

ABSTRACT:

We present the results of a point-cloud-based survey deriving from the use of image-based techniques, in particular with multi-image monoscopic digital photogrammetry systems and software, the so-called "structure-from-motion" technique.
The aim is to evaluate the advantages and limitations of such procedures in architectural surveying, particularly in conditions that are "at the limit". A particular case study was chosen: the courtyard of Palazzo dei Capitani del Popolo in Ascoli Piceno, Italy, which can be considered the ideal example due to its notable vertical, rather than horizontal, layout.
In this context, by comparing and evaluating the different results, we present experimentation regarding this single case study with the aim of identifying the best workflow to realise a complex, articulated set of representations—using 3D modelling and 2D processing—necessary to correctly document the particular characteristics of such an architectural object.

1. INTRODUCTION

Continual technological and computational advancements made in a particularly short period of time, estimated at no longer than five years, have oriented the development of multi-image monoscopic digital photogrammetry systems and software towards automatising the process. This is the main motivation for the present paper, which, through a particular structure-from-motion experiment, reflects on the role of the operator/surveyor, a role that would seem marginal today in a field where scientific and technological progress have rendered the procedures ever more automatised. In contrast to this, one of the objectives of the present paper is to underline how, especially in these cases, the role of the operator is still central. The operator is required to have not only a deep understanding of the tools and systems used, but also adequate preparation, which allows for procedures to be directed according to the goals of the survey and therefore to interpret and evaluate the results obtained. For example, even reflecting just on the question of the metric/geometrical and chromatic accuracy of the restitution, one can understand the reasons why the operator is so important.

With regard to the accuracy, the literature regarding structure-from-motion procedures highlights some particular factors, such as the characteristics of the photographic equipment, which greatly influence the quality of the orientation and the subsequent restitution. Using amateur cameras, these can be controlled only by knowing how to evaluate the calibration measures of the camera or the results of the self-calibration (Brown, 1971; Fraser, 1997; Grün & Beyer, 2001). Other factors depend on the correct design and execution of the photograph with respect to the number and position of individual snapshots and the arrangement of the optical axes. With a small number of photos or a photography scheme that does not consider the computational procedures used by the restitution system, the data will not be very reliable. Upon closer inspection, these are factors that essentially depend on the operator, who is certainly required to have the necessary technical abilities to control and correctly manage the procedures. These are skills that allow the operator to exercise control and verify the quality of the results, which is even more necessary the more automatic the processes become.

2. CASE STUDY. THE COURTYARD OF THE PALAZZO DEI CAPITANI IN PIAZZA DEL POPOLO IN ASCOLI PICENO

In order to investigate the use of structure-from-motion systems and place particular attention on the role of the operator, an experiment was conducted on a single case study chosen for its particular characteristics: the Courtyard of the Palazzo dei Capitani in Piazza del Popolo in Ascoli Piceno (Figure 1). Piazza del Popolo, the "square in the form of a palazzo" (Troli, 1984), or the square which gives the impression of being a hall because it is surrounded by porticoes (Piovene, 1956) assumed its current configuration after many changes and renovations beginning in the second half of the thirteenth century. In the thirteenth century the buildings facing the square were all very different; most of them were humble workshops or shops used by craftsmen or merchants. At the time, the square – more square and smaller than it is today - functioned as the venue for the famous Herb Market. During the second half of the thirteenth century, the first nucleus of the Palazzo dei Capitani del Popolo was built in the middle of the long west side. It was only in the fourteenth century that the Palace was enlarged by incorporating some of the nearby buildings; around this time

This contribution has been peer-reviewed.
doi:10.5194/isprsarchives-XL-5-W4-53-2015

these buildings were formally united towards the square by creating a new, unitary façade. Further renovation of the Palazzo dei Capitani del Popolo took place in the first half of the sixteenth century. This included the very unique façade along Via del Trivio by Cola dell'Amatrice, redesigning the main façade with the insertion of the triumphal central door and the monument to Paul III (all between 1519 and 1520), and changes to the interior, with the realisation of the current courtyard.

The formal layout of the courtyard dates from 1549, when the architect Camillo Merli introduced into the already small rectangular space a system of porticos and loggias distributed differently on the different levels. The arrangement of these connective elements was aimed especially at resolving the problem of distribution, which is why the solution identified is not the same on every level. In fact, the porticos on the first two levels, with their overhanging loggias, are present only on two of the four sides. Together with the presence of a deep entryway, the porticos make the courtyard deeper, with a surface area of uncovered space of about 57 m2 and sides that vary in length from 7.2 to 7.9 m. On the third level, on the side facing Piazza del Popolo, the architect instead introduced an additional loggia, thus further reducing the open space, with a smaller side of length 5.2 m and a longer side of length 7.8 m. The fourth and last level, with the same dimensions as the third, is instead equipped with a simple gallery. This is, therefore, an enclosed space whose above-mentioned peculiarities present a particular proportion between footprint and height, with a volume that is relatively tall with respect to the reduced horizontal dimension. In fact, comparing the height with the shorter sides of the portico and loggias on the different floors, the first two levels give a ratio of about 2 to 1, while on the third and fourth levels the ratio is even closer to 3 to 1.

Figure 1. The courtyard of Palazzo dei Capitani del Popolo

3. DATA ACQUISITION AND DATA PROCESS

3.1 The workflow

The combination of these size and morphological characteristics, which render the different phases of the survey more complex, make the case study the perfect example of one of the established objectives, i.e., testing and verifying the possibility of using structure-from-motion systems even in limiting conditions.

With such an intent, the article therefore pays particular attention to the different phases and main issues that qualify the use of such technologies, illustrating how the complexity of the particular operating conditions (or the characteristics of the survey object) guide the entire process, directing all the steps in the different phases of the procedure, such as: the choice of optics, the photography strategies, quality control of the auto-calibration, the accuracy of the point cloud, the means of dealing with the digital models, etc. With the scope of expanding upon the different steps through theoretical/experimental considerations, we illustrate the choices that guided the experimentation, that is, the workflow adopted. Starting from a comparative analysis of the different point clouds, the workflow is aimed at appropriately describing the unique characteristics of this architectural object and documenting the process in detail.

The workflow adopted in the experimentation can be summarised concisely in the three "usual" main steps. The steps are: operation planning, starting from prior knowledge of the survey object within the framework of the survey goals; data acquisition via photography; and the use of the photos, that is, processing the representations. We mention again the importance of correctly designing the survey, which always requires prior knowledge, critiques, and awareness of the object to be surveyed, as well as a clear definition of the objectives. This preliminary but deep approach to the theme is fundamental for the subsequent evolution of the survey since it allows the operations to be carefully planned not only as a function of the survey goals but also as a function of the characteristics of the building and state of the places in which the survey is made. The final phase of the photography project should first guarantee the total photographic coverage of the subject with an adequate superposition between consecutive photos, but also the best photographic quality possible so the photos can be used correctly and profitably.

In using the structure-from-motion systems, it is evident how the accuracy of the rendering is particularly dependent on the characteristics of the photographs. Different elements should be considered, such as the distribution of the snapshots and their orientation, the average object distance, the characteristics and quality of the optics, and especially the camera sensor. The latter in particular contribute to the resolution of the rendering. At this point it is useful to recall that in traditional photogrammetry, control of the restitution scale was derived from the scale of the photo. In turn, this was a function of the average object distance (d) and the focal length of the photographic equipment (f), from which the scale of the photo (sf) could be expressed through the relationship sf = f/d. For example, with a focal length of 50 mm and a distance of 5 m, the frame scale was 50/5000, or 1:100. With digital cameras, the concept of nominal scale was introduced. This derives from the pixel, the smallest indivisible unit, or elementary component, of a digital image.

This contribution has been peer-reviewed.
doi:10.5194/isprsarchives-XL-5-W4-53-2015

The minimum value that can be represented (whatever the scale of representation) is conventionally fixed at 0.3 mm. If the dimension of the object represented in a pixel of a photographic image (mr) is known, it is then possible to obtain the nominal scale of a photo, which is equal to mr/0.3 mm. For example, if the size of the object represented in a pixel is 15 mm, the nominal scale of the photo is equal to 15 mm/0.3 mm, or 1:50. This means that at the nominal scale of 1:50, the pixel represents a segment of 1.5 cm (0.3 mm x 50). With the physical dimension of a sensor pixel (p), the focal length (f), and the object distance (d), it is possible to determine the size of the object represented in one pixel of a photographic image, which is called the "image resolution" (mr), expressed with the relationship $mr = p * d / f$.

For the photography projects adopted for the tests, the horizontal and vertical superposition between consecutive shots was normally 80% but at any rate never less than 60%. The panorama and convergent-axis schemes were favoured because they guarantee better results in the orientation procedures, but the parallel-axis scheme was also tested. The distance between camera and object was assessed from time to time for the different cases in relation to the characteristics of the object, the survey goals, the operating conditions, and the photographic equipment available, with respect to the camera body and the optics. Different photography campaigns were made based on what was established for the photography projects Prior to image processing, the quality of the images was verified and, if necessary, they were either discarded or adjusted to improve the exposure, brightness, chromatism, etc.

The photo-acquisition phase was followed by the utilisation phase. In the first step related to image processing, which can properly be defined as "structure from motion", the goal is to orient the system by determining the camera parameters and parameters to correct distortions. This step ends with the creation of a low-density point cloud.

Once the orientation is resolved, the second step extracts a denser point cloud according to different accuracy criteria. The computational procedure used is called "dense matching" and consists in analysing the details of the photographic images, which have already been oriented, in order to identify the homologous points from which the points in the three coordinates can be derived.

In the first two steps to process the snapshots, the proprietary software Agisoft Photoscan was used (Fassi, Fregonese, Ackermann, De Troia, 2013; Remondino, Spera, Nocerino, Menna, Nex, 2014). In addition, tests were made at the end of the first step in order to verify the results generated by the automatic orientation procedures, i.e., self-calibration. For small groups of snapshots chosen as samples, the success of automatic corrections was compared to those made using software specifically dedicated to calibrating photographic equipment. In particolar, the images corrected with Photoscan were compared to those corrected with PhotoModeler.

The third step in the image-processing phase is aimed at transforming the dense cloud into a polygonal model. In this step, particular attention was placed on the effects of tessellation in dealing with zones characterised by gaps or a low point density. In fact, in these cases, the polygons were created by interpolating the data available in the immediate surrounding area, and it was therefore necessary to be very familiar with the algorithms used by the software, or at least the corresponding effects, and to favour those systems that include options to adjust the degree of interpolation.

Different methods were adopted during the experimentation to transform the dense cloud into a polygonal model. Photoscan continued to be used for clouds deriving from it, making use of the different interpolation options. In particular, in dealing with gaps, the "Enabled" option was always used because it reduces the metrical and geometric disturbance, but intervenes automatically only when reconstructing small gaps.

The fourth and last step in processing the images was aimed at reconstructing the visual appearance, i.e., the material and chromatic appearance, of the surveyed object, resolved by dealing with the surfaces of the model. In all tests, a procedure appropriately developed in 3DS Max Autodesk was used with scripts to automatise the principal steps: importing the oriented snapshots, generating materials to reproject the photos on the polygonal model, and processing high-resolution renderings (Sicuranza, 2013). The decision to use this method arises from the possibility of controlling the different steps and its appropriateness for realising high-resolution geometric and chromatic renderings from which orthographic and elaborate two-dimensional features can be derived.

Following this procedure, the first script, which correctly orients the snapshots, was run in 3DS Max Autodesk. The script also generates fictitious geometries associated with each photo, called "dummies", which cannot be rendered but which are then used to memorise the material connected to the snapshot.

Once the snapshots have been oriented, the polygonal model is imported and the second script is run. For each snapshot, the script processes a material that projects the photo on the geometry, using the direction of the snapshot as the projection direction and the associated photo as the image. The material used is composed of an auto-illuminating shader whose shading is given by an "ambient occlusion" map to which the photo to be projected is added as texture. It is thus possible to render the model without using any light since it is already simulated by the ambient occlusion map together with the chromatic and light information used in the photo providing the texture.

At this point it is possible to impose a view, which is orthographic in our case, and generate as many renderings as there are snapshots, considering that each rendering will present the geometry mapped with a single snapshot. The third script realises the renderings in sequence: the material of the individual snapshot is automatically assigned and the calculation for all the snapshots specified in the script is made automatically.

3.2 Data Acquisition

The particular characteristics of the courtyard, especially the dimensional characteristics, therefore made all the different survey phases exceedingly complex, starting with the photo-acquisition campaign. In establishing the plan of photography, highly recommended optics, i.e., the Nikon Nikkor 15 mm f/3.5 Ai-S, were chosen and mounted on the Nikon D700 equipped with a full-frame DSLR sensor (36 x 23.9 mm) with a maximum resolution of 4256 x 2832 pixels (12.1 million effective pixels) and a physical pixel size of 0.0084 mm (p = 23.9 mm/2832). The panorama scheme was then chosen, due to both the emptiness of the courtyard overall and the internal spaces of the loggias (Figure 2).

This contribution has been peer-reviewed.
doi:10.5194/isprsarchives-XL-5-W4-53-2015

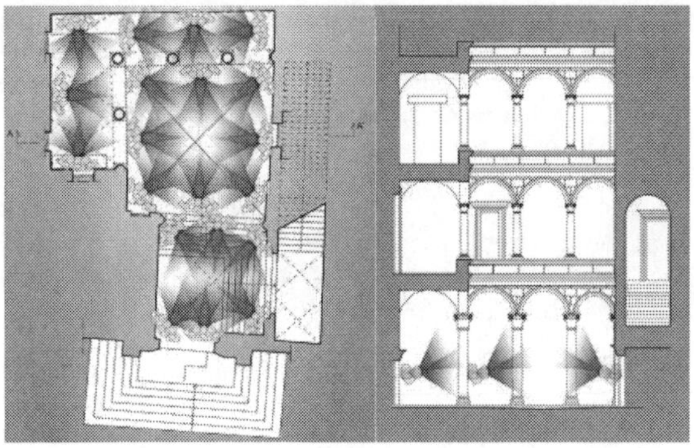

Figure 2. Diagrams of photography campaign in the courtyard

Such a photography strategy consists in taking a series of snapshots for each station in different directions, keeping the camera horizontal (landscape mode). In this way, the groups of images acquired are considerable portions of spherical images, and the horizontal and vertical superposition between the individual snapshots give rise to a virtual extension of the field of view.

The different snapshots were designed to provide ample overlaying areas between successive snapshots, both in the horizontal and in the vertical directions. The overlapping areas were around 80% where possible, but never less than 60%. Specifically, since the panorama scheme was chosen, the means by which the snapshots were taken was different for the different levels due to the different operating conditions. Two series of snapshots were made for the first level. One was made from the greatest possible distance (under the porticos and in the entryway) and the second was made from the centre of the open space (close to the projections from the colonnade) in order to have photographic coverage without shaded areas. Obviously on the upper levels it was possible only to realise the large-distance series of snapshots, with stations in the galleries under the loggias (Figure 3).

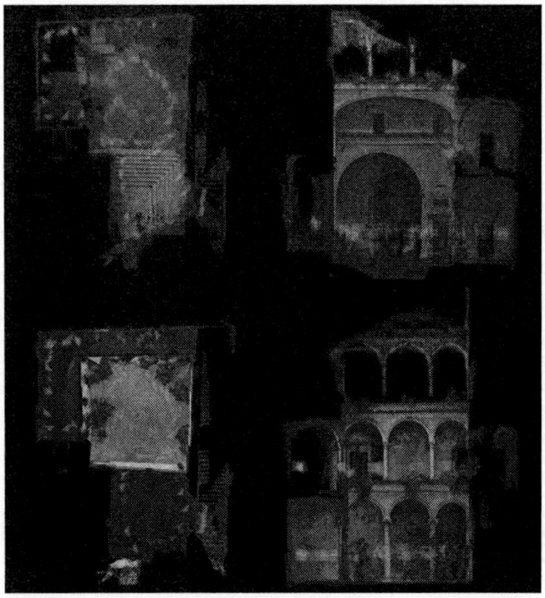

Figure 3. Screenshots of the snapshots on the ground floor (top) and on the first floor (bottom)

The particular volumetric configuration of the courtyard, which gives rise to very different lighting conditions, was also considered during the photography campaign. The main source of light is natural light entering the open space from above. This obviously creates very different exposure conditions between the higher and lower zones, as well as between the façades directly facing the central space and those hidden beneath the porticos and loggias.

The necessity of taking snapshots with very different exposures was already considered when planning the campaign. For this reason, the images were saved in both JPG format and RAW format, which allows eventual exposure defects to be compensated for by taking advantage of its greater colour depth (12 bit) with respect to the JPG format (8 bit). Snapshots were taken contextually, with the correct exposure for the well-lit parts, underexposing those in the shadow. The latter were then corrected by brightening only the areas in shadow, without overexposing the light area, which also avoided the burned effect (Figure 4).

Figure 4. Comparison of the exposures of the snapshots in the courtyard. JPG format (left) and RAW processing (right)

Once the photography project was established, 152 snapshots were made, distributed thus:

- 26 related to the deep entryway, positioned at a height slightly lower than the courtyard and connected to it by a wide stairway, with 18 horizontal snapshots for the faces and 8 for the cross vault of the intrados;
- 84 related to the central space of the courtyard, with 38 horizontal snapshots, 35 looking up, and 11 looking down;
- 42 related to the vaulted spaces of the porticos and loggias, with 22 horizontal snapshots and 20 to capture the cross vaults of the intrados.

For the panorama scheme, 4–6 snapshots were made from each station as follows:

- at least three snapshots with nearly horizontal optical axes, one of which was as perpendicular as possible to the average plane of reference, which passes through the face of the subject of the photo;
- 1 or 2 snapshots with the optical axis pointing upwards or downwards.

This contribution has been peer-reviewed.
doi:10.5194/isprsarchives-XL-5-W4-53-2015

In particular, the snapshots where the optical axis was horizontal but oblique with respect to the average plane of reference were made to obtain a better result in the orientation phase since the spatial reconstruction is made by identifying homologous points distributed evenly throughout space, that is, on many planes at different depths. In contrast, the snapshots where the optical axis was horizontal but perpendicular to the average plane of reference were made in order to extract a better chromatic structure due to the better orientation.

Thirteen photos were discarded on the basis of a qualitative analysis because they presented small movements. Through a test made on a small group of photos chosen as a sample, the results of the photos corrected with PhotoModeler were compared to those automatically corrected with Photoscan, the structure-from-motion system chosen for this experimentation. Once a positive result was verified, the phase aimed at processing the restitutions began (Figure 5).

Figure 5. Upper left: An original photographic image from the campaign. Upper right: Photo corrected by Photomodeler. Lower left: Photo corrected by Photoscan. Lower right: The photo corrected with Photoscan was scaled and aligned with the PhotoModeler correction and superimposed with 50% opacity.

3.3 Data processing

In the first step, which oriented the photos, the 139 snapshots were processed simultaneously in order to obtain a more accurate result, even though this lead to a greater processing time. Once the orientation was made, analysis of the sparse cloud showed that six snapshots had orientation errors that were deemed to be too large. These snapshots were removed and the orientation quality was again verified. Processing then moved on to the next step, which extracted the dense cloud using the high quality setting (equal to ¼ of the area, i.e., of the resolution).

The dense cloud was then recleaned, that is, the points with a position different from what was expected were eliminated. The operation was carried out automatically using the option to filter out anomalous points (outliers), which were selected according to statistical analysis.

Among the three filtering options provided by the program, we chose the one that achieved a more consistent selection because it produced a cloud characterised by greater dimensional accuracy, even if it was less dense. The point cloud produced at the end of the process was therefore composed of more than 30 million points: 30,778,595 to be exact (Figure 6 and Table 1).

Figure 6. The dense point cloud after dense matching. Views

Photo group processed	Total snapshots	Snapshots used	Features extracted (sparse cloud)	Average alignment error (pixels)	Number of points in the dense cloud
All	152	133	508011	0,52	30.778.595

Table 1. Characteristics of the point cloud

In the third step, the problem of the scale of the dense cloud was resolved in an operation called "scaling", for which it is necessary to have reference measurements for the object photographed. This phase was accomplished by selecting three non-aligned points that were also easily recognisable on the cloud extracted by Photoscan. The points were selected from a laser scan made previously for another research project. To carry out the scaling procedure for the cloud in Photoscan, the three points were first identified, highlighted with markers, and assigned with the three-dimensional coordinates obtained from the laser scanner (Figure 7). Based on the coordinates assigned to the markers, the software scaled and oriented the cloud, calculating an average uncertainty of 34 mm, a variance also measured from samples on both clouds (the one from the scanner and the one from structure-from-motion(Figure 8).

Figure 7. The point cloud scaling procedure in Photoscan

Figure 8. Sample measurements to check the scaling and orientation of the point cloud in Photoscan

This contribution has been peer-reviewed.
doi:10.5194/isprsarchives-XL-5-W4-53-2015

Referring to the details in the section "The workflow", the resolution for images in the tests on the courtyard was 4.2 mm. The physical size of a pixel was about 0.0084 mm, the average object distance was 7.5 m, and the focal length was 15 mm (0.0084*7500/15). In other words, each pixel at the centre of the original photo represents a 4.2-mm portion of the subject.

Recalling that the dense cloud was extracted with high quality, given the quantity of photos to process simultaneously, the software therefore worked with images at a quarter of the original resolution, where the pixels at the centre of the photo each represent a portion of the subject of about 4.2 mm * 4, i.e., 16.8 mm. The resolution of the representation expected is therefore 16.8 mm, which is coherent with the visual analysis. Therefore, details are only recognisable if they are about 2 cm or larger. Summarising, the overall uncertainty of the courtyard area is estimated to be around 34 mm, while the dimensional resolution of the restitution is about 16.8 mm.

We then recall that the admissible tolerance, i.e., the uncertainty, in measurements and survey drawings is fixed as the product of the so-called graphing error (+ 0.3 mm) and the denominator of the representation scale adopted. This is, for example, equal to 3 cm for 1:100-scale drawings and 1.5 cm for 1:50-scale graphics. Therefore, in the tests, graphical drawings (3D models and 2D restitutions) can be derived with a scale of detail of 1:50, while the reliability of general measurements will agree with the reliability of a survey at a scale of 1:100.

Figure 9. Snapshots of the capital from different distances and with different focal lengths

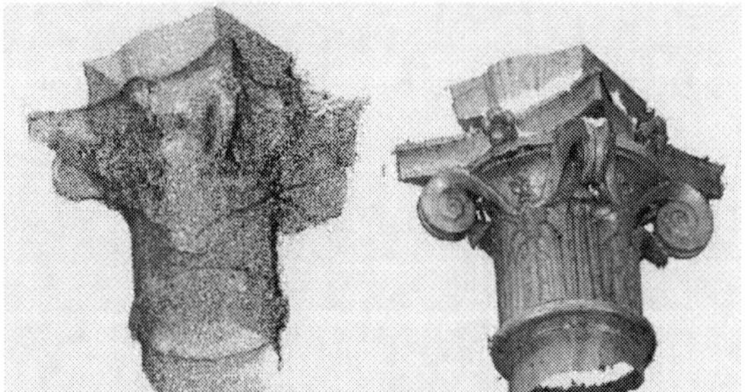

Figure 10. The difference in density between the capital represented by the entire courtyard point cloud (69,000 points) using 15-mm optics and the capital extracted with detail photos (4,156,625) taken with 70-mm optics

It can be assumed that with greater computational power to process photographic images with better resolution, and with high-quality cameras and optics, even more accurate metrical and geometric information could be derived. Figures 9 and 10 show the result in the relief of some details with shooting close-ups.

The next phase was aimed at transforming the point cloud into a polygonal model. The steps are characterised above all by the effects of tessellation and treatment of the gaps. Therefore, the cloud was examined to identify areas with evident "shortages" of the restitution, searching contextually to identify the principal reasons why. Shortages due to excessively reflective surfaces were found in an internal part of the portico on the first floor where there was a glass door, in a recently shined portion of the flooring on the ground floor, and where glass was installed to view the archaeological ruins discovered under the palace. Other missing points or sparse points characterised by low accuracy were found in different areas on the second floor. In these cases, the defects and gaps were most likely due to the particular characteristics of the courtyard area, the lighting, and the tall, narrow spatial proportions, which made the photography conditions difficult, yielding inadequate images.

During the tessellation procedure in Photoscan, treatment of the gaps was done with the "Interpolation" command with three different options: leaving the gaps unaltered ("Disabled"), intervening only on the smallest gaps ("Enabled"), or acting indiscriminately on all the gaps ("Extrapolated"). The "Enabled" option was chosen for this experimentation. It allowed the metrical/geometric noise with the three-dimensional reconstruction to be reduced, even in zones with small gaps. Although it still lacked surface treatments, the polygonal model obtained at the end of this phase already permitted an in-depth understanding of the layout and arrangement of the courtyard space (Figure 11).

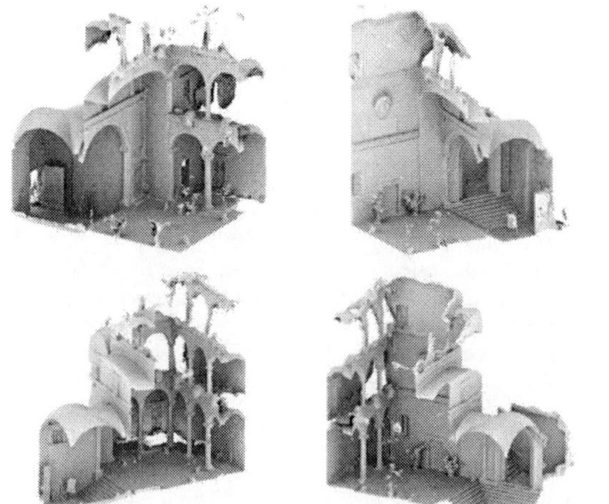

Figure 11. Initial processing of the polygonal model for the courtyard. Views

The last step was dedicated to visualising the appearance, i.e., the materials and colours, which was done by treating the surfaces of the model (texture mapping). This deals with projecting the images made during the photo-acquisition campaign onto the numerical surfaces of the non-decimated model in order to create a high-resolution geometric and

This contribution has been peer-reviewed.
doi:10.5194/isprsarchives-XL-5-W4-53-2015

chromatic static rendering. This was used to create orthographical images and then two-dimensional drawings (Figures 12 and 13). Autodesk 3DS Max was used to reproject the photo according to the procedure illustrated in the section "The workflow". To texturise the complete model for use in visualising combination views (PDF 3D, online 3D player), a mesh with a low number of polygons was used, projected with the photos that were then blended (Figures 14 and 15).

Figure 12. Rendering in orthographic projection of the four faces of the courtyard processed from the textured model

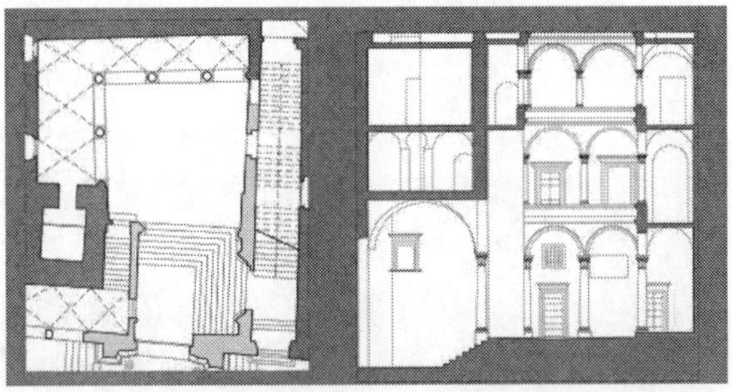

Figure 13. Example of two-dimensional images of the floors and perspective sections obtained

Figure 14. Initial results obtained from texture-mapping

Figure 15. Initial results obtained from texture-mapping

4. CONCLUSION AND FUTURE WORK

In conclusion, the present article demonstrates how the role of the operator is still central in the entire surveying process, even when particularly automatised procedures such as the structure-from-motion technique are adopted. In fact, it is always the operator who is asked to rigorously check the input (photographic campaign, snapshot quality, auto-calibration control) and output (accuracy and density of the point cloud, the presence of gaps and their treatment in converting to a polygonal model, etc.).

But above all, the operator/surveyor is required not only to have extensive knowledge of the instruments and systems used, but also adequate preparation, which will allow him or her to guide the procedures according to the aims of the surveying process and therefore interpret and assess the results obtained.

Developments regarding in the specific tests presented here have been made recently. A second survey campaign was conducted eight months later with new snapshots chosen after assessing the data presented in this essay. In general, the aim was for greater survey coverage, both integrating it with parts of the courtyard that had not been covered (additional data for the second level) and improving the quality of the survey by compensating for areas in the point cloud that showed major gaps in the first campaign.

Data processing is currently moving in two directions. The first deals with aligning the two clouds obtained from two successive surveys (Figures 16 and 17) and the second relates to processing both campaigns together in order to verify the possibility of processing two sets of data with different lighting conditions. Comparing the two procedures will illustrate the possibility of refining the whole body of data not only in terms of greater final point-cloud density, but also in terms of obtaining more accurate results to create the single texturised polygonal model.

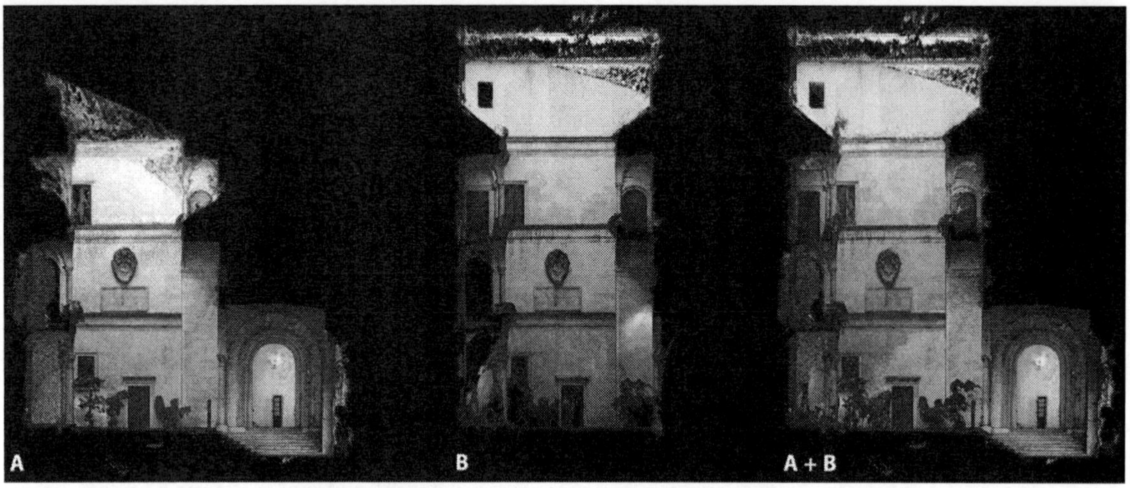

Figure 16. Sections of the point cloud in orthographic projection obtained from the two photography campaigns. A: first campaign, B: second campaign, A+B: overlay of the two campaigns

REFERENCES

Brown, D., 1971. Close-range camera calibration. *Photogrammetric Engineering*, 37(8), pp. 855-866.

Fassi, F., Fregonese, L., Ackermann, S., De Troia, V., 2013. Comparison between laser scanning and automated 3D modelling techniques to reconstruct complex and extensive Cultural Heritage Areas. In: *The International Archives of the Photogrammetry, Remote Sensing and Spatial Information Sciences*, Trento, Italy, XL(5/W1), pp. 73-80.

Fraser, C. S., 1997. Digital self-calibration. In: *ISPRS Journal of Photogrammetry & Remote Sensing*, 52(4), pp. 149-159.

Grün, A., Beyer, H., 2001. System calibration through self-calibration. In: A. Grün, & T. S. Huang (Eds.), *Calibration and orientation of cameras in computer vision*,. Berlin, Germany: Springer, pp. 163-193.

Ippoliti, E. (Ed.), 2013. Valorizzare il patrimonio culturale: esperienze per Ascoli Piceno - Shedding light on the Cultural Heritage: the Ascoli Piceno experience. Roma: Aracne.

Meschini, A., Petrucci, E., Rossi, D., Sicuranza, F., 2014. Point cloud-based survey for cultural heritage – An experience of integrated use of range-based and image-based technology for the San Francesco convent in Monterubbiano. In: *The International Archives of the Photogrammetry, Remote Sensing and Spatial Information Sciences*, *XL*(5), pp. 413-420.

Piovene, G., 1957. *Viaggio in Italia*. Milano: Mondadori.

Remondino, F., Spera, M.G., Nocerino, E., Menna, F., Nex, F., 2014. State of the art in high density image matching. In: *The Photogrammetric Record*, 29(146), pp. 144-166.

Rozzi, R., Sori, E., 1984. Ascoli e il suo territorio: struttura urbana e insediamenti dalle origini ad oggi. Cinisello Balsamo (Italy): Silvana editoriale.

Sicuranza, F., 2013. Sperimentazione di sistemi di structure from motion per la restituzione di apparati decorativi (Doctoral dissertation). La Sapienza Università di Roma, Dipartimento di Storia, Disegno e Restauro dell'Architettura. Dottorato di Ricerca in Scienze della Rappresentazione e del Rilievo, Scuola Nazionale di Dottorato, http://hdl.handle.net/10805/2116.

Figure 17. Sections of the point cloud in orthographic projection obtained from the two photography campaigns. A: first campaign, B: second campaign, A+B: overlay of the two campaigns

This contribution has been peer-reviewed.
doi:10.5194/isprsarchives-XL-5-W4-53-2015

AUGMENTED REALITY TO PRESERVE HIDDEN VESTIGES IN HISTORICAL CITIES. A CASE STUDY

José L. Martínez[a], Sonia Álvarez[a], Jaime Finat[b], Francisco J. Delgado[a]*, Javier Finat[a]

[a] MoBiVAP, Departamento de Informática
Parque Científico de la Universidad de Valladolid
Paseo de Belén 11, Valladolid, Spain, 47010
(joseluis.martinez.jimenez, sonia.alvarez.diaz, franciscojavier.delgado, javier.finat.codes)@uva.es
[b] Rapture Games S.L.
Barbecho 29, Valladolid, Spain, 47014
jaime.finat@rapture-games.com

Commission V, WG V/4

KEY WORDS: Cultural Heritage Dissemination, Archaeological Surveying, Augmented Reality, Serious Games

ABSTRACT:

Mobile devices provide an increasingly sophisticated support to enhanced experiences and understanding the remote past in an interactive way. The use of augmented reality technologies allows to develop mobile applications for indoor exploration of virtually reconstructed archaeological places. In our work we have built a virtual reconstruction of a Roman Villa with data arising from an urgent partial excavation which were performed in order to build a car parking in the historical city of Valladolid (Spain). In its current state, the archaeological site is covered by an urban garden. Localization and tracking are performed using a combination of GPS and inertial sensors of the mobile device. In this work we prove how to perform an interactive navigation around the 3D virtual model showing an interpretation of the way it was. The user experience is enhanced by answering some simple questions, performing minor tasks and puzzles which are presented with multimedia contents linked to key features of the archaeological site.

1. INTRODUCTION

Currently, most conventional approaches to display Cultural Heritage (CH in the successive) contents consist of showing image and video contents which are streamed from multimedia repositories. The design of interactive contents requires more complex interaction patterns. Interaction concerns to the subject and its relation with the environment through a Visualization tool, involving the capability to identify key features to generate smart answers such that users can improve their body of knowledge. The main problem is how to manage the huge amount of knowledge in an attractive way for most people, in order to provide a consciousness of users as citizens. This far-reaching goal concerns to more efficient ways of Edutaintment based in enhanced presentations of multimedia contents.

Digitization of physical resources contained in large public repositories (museums, archives, Foundations) is the first step for the valuation of CH goods. Their reuse can be performed through distributed networks which are organized by means of a specific Ontology which has been previously developed. The lowest level supporting this organization concerns to metadata which are included as a basic layer. Multiple articulations can be developed from specific CH metadata towards EU multimedia data. So, we intend to provide an Internet support to create new digital contents and constructing an European citizenship aware of CH richness as a reference for diversity, tolerance and common values sharing. However, neither of these technological solutions contributes to make more sustainable the CH tourism if there are no visitors.

Edutainment is a challenge for knowledge transmission on multimedia platforms with large implications in education, cultural tourism and Small & Medium Enterprises (SME) activities bound to content production. This area involves a combination of techniques from modeling (interactive or semi-automatic), educational resources (text, images, videos, 3D models) and smart multimodal interfaces (tactile, speech recognition, facial expressions). All these functions are currently available for tablets and smartphones.

Virtual reconstructions in cultural heritage are hypothetical recreations of destroyed sites which are projected towards the present. For this reason, virtual and augmented reality applications are implicitly based in an eye's wink: they do not show the reality in a literal way (obviously impossible). Instead, we must reinterpret facts and sites, taking into account the most meaningful aspects for the current state of our knowledge. The same knowledge gives the opportunity of a dialogue not only between different representations, but also between different possibilities inside the same discourse. In this way, every virtual reconstruction becomes a place to understand the same reality from different viewpoints, where each user reconstructs its own version on a common platform interacting with other public / private involved agents.

The modeling of virtual contents for cultural heritage can be started from documentation obtained from image-and/or range-based devices. A sustainable economic activity requires an industrial development with a well specified methodology for business model, able of providing an economic return; it is necessary to achieve cost-effective solutions according to currently available standards. These goals imply the development of basic models and strategies for information fusion which allow the generation of low-resolution geometric models as a first step; these solutions are refined in successive stages by adding more sophisticated rendering solutions according to technological advances.

This approach is not efficient enough to achieve a wider audience

*Corresponding author: franciscojavier.delgado@uva.es

This contribution has been peer-reviewed.
doi:10.5194/isprsarchives-XL-5-W4-61-2015

from the conception to the content production. It is desirable to develop more advanced models and tools for augmented cognition technologies. Last ones are supported by three-dimensional models which can be obtained from images with a large baseline and other representations arising from sketches, gravures or other sources not always properly georeferenced. Nevertheless recent advances in all these domains, it is necessary to develop more flexible tools for knowledge management in an interactive way. This far-reaching goal implies a more structured approach by specifying the flowchart between different building blocks. An extended version of the proposed storyboard must be able to integrate educational resources, content creation by graphical artists and mobile technologies around an interactive reconstruction of cultural contents.

The rest of the paper has been organized as follows. Section 2. starts with a description of the current location of the urban environment and the archaeological site to provide a contextualization to the reader. Section 3. shows our proposal of reconstruction of a Roman Villa and its environment from archaeological vestiges, with special regard to the *hypocaustum* and some hypothetical interpretation of rooms over this structure. Section 4. is focused on the selection of key features which define the interaction between the user and the virtual environment on mobile devices. Section 5. discusses the results to reinforce those aspects which can be improved. Finally, Section 6. presents the conclusions and some open challenges to be solved in the near future.

2. SURVEYING THE HIDDEN ARCHAEOLOGICAL SITE

Valladolid is a city of Spain. It was built between several rivers (Pisuerga and several branches of Esgueva) which were populated by vaccei before the Roman domination over Hispania. The Roman Domus and its archaeological remains, which have been the object of study in this work, are located very close to a branch of the Esgueva River and away from an original medieval nucleus which was reduced later.

In the 11th century, the old village grew towards this zone and Medieval, Renaissance and more modern constructions destroyed most Roman vestiges in the zone. After some preliminary archaeological excavations performed between the last years of the 20th century and the beginning of the 21st century, the City Hall decided to perform an underground parking in the zone, which is plenty of archaeological vestiges. Figure 1 displays an aerial orthophoto of the surveyed area near of "La Antigua" at the moment of excavations (left), and the same site nowadays with the garden covering the archaeological vestiges (right).

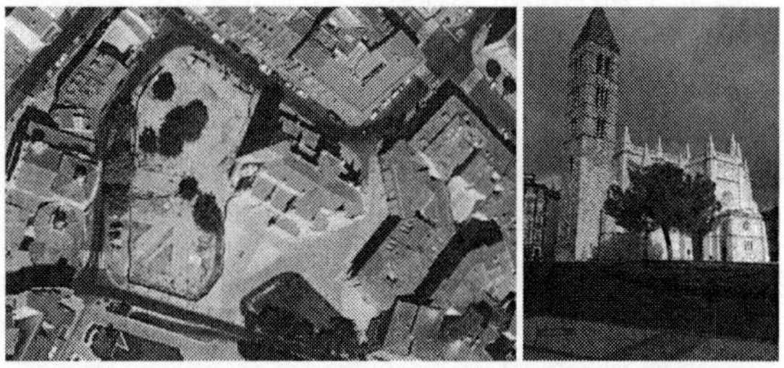

Figure 1: Aerial image taken at the time of performing the archaeological excavations[1](left) and a picture of the current state of the surveyed area (right).

[1]PNOA ©"Instituto Geográfico Nacional de España" - ("Junta de Castilla y León") http://www.ign.es/PNOA.

The opposition and annoyance of citizenship aborted this initiative. The citizen movement requested to declare the zone as archaeological site, but the City Hall decided to cover the remains with an urban garden. Because of this actuation, almost nobody can imagine the historic richness in the underground of the modern city. All those hidden elements are extremely meaningful and rich for self-location, as much as the church of "La Antigua" is. The church has been independently surveyed in its current state and modeled both in high and low polygonization using Blender, an open source 3D modeling software[2]. Figure 2 displays a rendered image of the current state of "La Antigua".

Figure 2: The virtual reconstruction of the current state of "La Antigua" rendered with Blender.

3. AN ESSAY OF REINTERPRETATION

The archaeological excavation was performed in a limited zone according to the available documentation from (STRATO Gabinete de Estudios sobre Patrimonio Histórico y Arqueológico, 2009). This excavation only reveals some elements of the Roman vestiges so it is not possible to obtain a complete picture of Roman constructions, which in the best of the cases are currently hidden. Indeed, most documented vestiges belong to historical periods different of the Roman period. The list of documented vestiges of several periods mainly include, among others:

1. several walls,
2. pipes,
3. pavements,
4. culverts,
5. ceramic vessels,
6. mosaics,
7. two *hypocaustums*,
8. human rests in an ancient necropolis.

In fact, and perhaps due to the urgency of the actuation, the archaeological documentation was generated using only traditional methods; it is incomplete and displays metric inconsistencies between documents which make difficult to track the changes. The main experienced issue was the disagreement between the archaeological strata of the different excavations and the height of the found walls. We have solved these issues by making a coherent adjustment of the 3D position of the elements considering just a single global reference for the 3D model. Figure 3 shows how strata information has been fused in the 3D model.

Blender has been chosen to design the 3D model since it is a multipurpose software covering several stages of the content production pipeline. Using Blender as a common sink for the available

[2]www.blender.org

This contribution has been peer-reviewed.
doi:10.5194/isprsarchives-XL-5-W4-61-2015

Figure 3: The 3D model of the archaeological strata retrieved from documentation which includes vestiges from different historical periods

graphic resources we can collaborate and share elements of the design such as geometry, materials and textures, at the same time we obtain quick renders and previews for a more realistic visualization (see Figure 4).

The virtual reconstruction of the documented vestiges has been performed using the following methodology: 1) the original documentation has been surveyed for each stratigraphic unity, which has been compared with the photographic documentation; 2) this information has been used as a blueprint in Blender to extrude the topology of the site, with special regard to the original dimensions, position and orientation of each element; 3) a collection of tiled textures, normal and specular maps has been designed taking into account the original collection of pictures taken during excavation. Pictures have been manipulated when necessary using the GNU Image Manipulation Program [3]; 4) textures and different maps have been applied over the geometry to define the appearance of the model and to achieve a more realistic visualization in real time.

Figure 4: UV mapping of the roof in Blender (top) and render of the reconstruction of the archaeological vestiges (bottom).

[3]www.gimp.org

The large number of available pictures of the outside of the *hypocaustum* (more than 30) allows us to recover metric information about the dimensions and location of this element at the same time that a high resolution texture using photogrammetry. The reconstruction has been performed using Agisoft PhotoScan to calibrate camera shots, building a sparse point cloud, and converting it into a dense point cloud, which has been finally triangulated and textured. The result is a single dense mesh composed by more than 70.000 faces with a high definition and realistic texture.

Unfortunately the model is too dense to be rendered with conventional mobile devices so we have baked normal and diffuse maps over a retopologized low polygonal 3D model of the *hypocaustum*. Thus, we have obtained a mesh composed only by quad faces with a quite realistic appearance, thanks to the use of conventional shader programs running in the GPU in real time. This mesh can be merged together with the rest of vestiges, improving the realism of the initial reconstruction. A similar strategy has been used previously in the literature such as in (Aparicio Resco et al., 2014) and (López Romero, 2014). Figure 5 displays two screenshots to illustrate this procedure which combines Agisoft PhotoScan with Blender to embed realistic photogrammetric models into virtual reconstructions. Nowadays it is increasingly common to find models reconstructed from multiple views in films and other kind of animations from computer graphics productions where there is no need to pay attention to the weight and complexity of the model. However, this is critical when the model is deployed in mobile devices whether the model is available in the own device or it has to be downloaded from remote repositories.

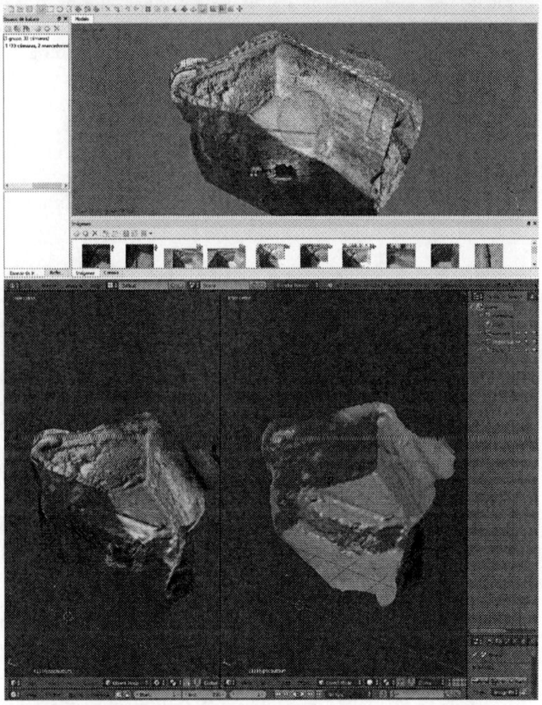

Figure 5: The exterior side of the *hypocaustum* reconstructed in Agisoft PhotoScan (top) and the same 3D model imported in Blender (bottom).

The Villa has been modeled from the position of some evidences of Roman settlements, which were found along excavations that took place between 2006 and 2009. The boundary of the Villa to the west has been determined by a well-preserved *hypocaustum* chamber and its corresponding *atarjea* (see Figure 6). The

This contribution has been peer-reviewed.
doi:10.5194/isprsarchives-XL-5-W4-61-2015

other boundaries of the Villa have been established by portions of walls, an *opus signinum* ground, the interpretation of the path of the river and the current location of the church of "La Antigua". The orientation and position of the second *opus signinum* ground (located at the square H2 of dimensions 220x220 cm) have provided the basis for a hypothetical distribution of the other rooms of the Villa. Besides, the Roman vestiges previously discovered inside of "La Antigua" make coherent the hypothesis that the Villa could be extended reaching the church. Figure 7 displays the hypothetical location of the Villa over the place where the archaeological excavations were performed.

In summary, the archaeological evidences of the Roman period have made us capable of recreating a hypothetical location of the Villa which has been combined with a hypothetical distribution of its rooms taking into account other Roman villas and domus described in the literature(Fernández Vega, 2003). Some of the most relevant examples in Castilla y León which have been previously documented are the Roman Villa of Almenara de Adaja, Puras (Valladolid) in (González Gallero, 2008), and the Roman Villa of Quintanilla de la Cueza (Palencia) in (García Guinea, 2000).

The superimposition of different stratigraphic layers with the current situation makes difficult to interpret properly the original Roman vestiges. In order to solve this problem, Figure 6 includes information from several archaeological excavations performed in the area between the years 2006 and 2009 which are divided in numbered stratigraphic units (U.E.). This figure shows the excavated areas and the architectural vestiges of several historical periods found in them. Please, note that different colours are used to represent different materials instead of different archaeological periods as could be expected. Actually, in this region there have been successive constructions over the vestiges of the previous period without taking care of preserving them, which has created a mess in the plane which makes it difficult to understand. The hypothetical contour of the branch of the River "Esgueva" has been used to limit the Villa to the west while the church limits the Villa to the east, as Figure 7 shows.

Finally, we have designed the textured 3D model of the Roman Villa, including both rooms and ornamental objects, just following the same methodology as we used in the case of the archaeological remains found during the excavations. Results are illustrated by Figure 8 which shows images rendered from different camera locations of representative places of the Villa[4]. The steps to create the complete 3D model of the Roman Villa were the following ones:

1. The 2D designs of the Villa (see Figure 7) were used as blueprint in Blender.
2. The walls of the main lounges, rooms, baths and halls were extruded to create a basic mesh for the model.
3. Some openings were allocated on the walls making loop cuts in the mesh of the rooms to create doors and windows.
4. The different parts of the roof which belongs to different wings of the Villa were connected, forming a special kind of cross hipped roof.
5. Several textures for Roman mosaics were created using resources publicly available with Creative Commons licences[5].

[4]A video preview of a camera travelling through the Roman Villa can be seen at `http://vimeo.com/116531713`. This video does not contain the latest improvements made in the Villa which will be available soon

[5]Textures with Creative Commons licence have been collected from CGTextures`http://www.cgtextures.com`

Figure 6: The archaeological remains and the current location of "La Antigua" church.

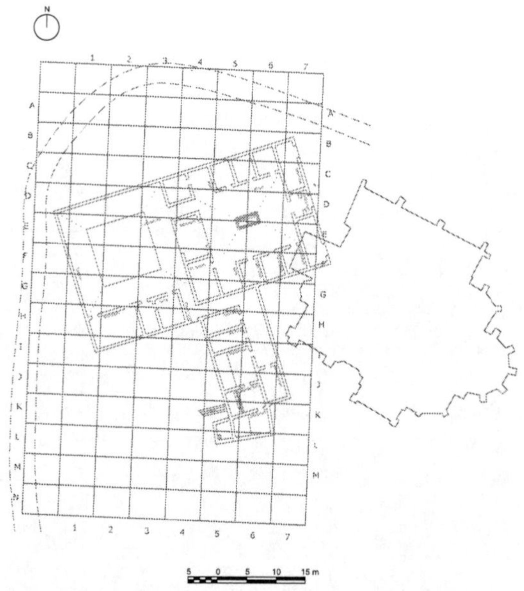

Figure 7: The outline of the Villa located over the archaeological remains which have been hidden for better visualization.

These images have required minor improvements and manipulation for tiling.

6. A complete collection of materials was defined by creating different kind of maps in form of textures including, at least, a diffuse map to give the main color to the surface of the material, a normal map to specify how shadows derived

This contribution has been peer-reviewed.
doi:10.5194/isprsarchives-XL-5-W4-61-2015

from the roughness of the material must be projected without adding extra geometry, and a specular map to determine the brightness level and color reflected at each point of the surface of the material.

7. Ornamental details and furniture such as statues, pictures, bath pools, oil lamps, banks and columns were added to complete the inside of the model.
8. A simplified environment was also created, including a terrain mesh for the ground and the path of the River Esgueva.
9. Several light and a global sun light were added to the scene to create realistic lightning conditions for rendering.
10. Rendering parameters such as ambient occlusion, blurring and anti-aliasing were tweaked to improve the quality of the rendered images.

At each stage we have paid special attention to preserve a coherent topology in the 3D model using only quadrangular faces, avoiding triangular faces and maintaining the simplest mesh for representing each object. For instance, the oil lamps have been modeled using just a couple of planes for the flame and a few more for the chains and the dish. The flame material contains a short animation and alpha transparency. Material maps have been created at the lowest possible resolution with a power of two dimension for better performance at loading textures. Our goal was always to design the most realistic model which can be rendered in real time by modern mobile devices GPUs. This goal involves the use of several maps for materials and different levels of detail to improve the mobile experience of the user in augmented reality applications.

4. AUGMENTED REALITY IN MOBILE APPLICATIONS

The first applications using virtual 3D models for showing cultural heritage contents based on mobile devices were developed as Archaeological guides from the late nineties going from (Vlahakis et al., 2002) to more recent contributions such as (Eggert et al., 2014). The information about the user's current location provides a context to display specific contents. First applications provided only information on laptops extended with head mounted displays (HMD) devices. A HMD consists of an external web camera, a digital compass, a backpack with a GPS receiver, a laptop wireless communication equipment and a battery. This equipment can be incorporated to a wheelchair but is not practical from the user's viewpoint. It is necessary to advance to increase the portability and to improve the performance of low-cost devices able of supporting AR-based applications on mobile devices. Recent advances by Oculus VR are very promising along the same line, but the current cost and the scarce developments are dissuasive.

Some applications to Cultural Tourism can be found in (Fritz et al., 2005), where the authors use a binocular system to visualize enhanced environments to find additional contents by means of a virtual walk providing an immersive experience in an indoor scenario. Unfortunately, this solution is not very wearable, and poses several problems for its extension to outdoor environments.

An extension of methodology based in Archaeoguides to more advanced Mobile Augmented Reality solutions (MAR in the successive) can be found in (Stricker et al., 2010). In this work, the authors develop an application to integrate several tools arising from position-based tracking, computer vision, and interactive visualization. The mobile devices is given by an Ultra Mobile PC with 15 fps. We have developed a similar solution which is based on laptops embedded in a wheelchair which has been proved in the historical center of Segovia (Spain). More recent technological advances for smartphones and tablets have allowed us to extend their approach by incorporating efficient solutions on Android-based mobile devices. A first version was introduced in (Delgado et al., 2013).

Figure 8: Rendered images of two rooms in the Roman Villa: the *atrium* (top) and the *caldarium*, placed over the *hypocaustum* (bottom).

A recent general approach to Augmented Reality on mobile devices can be seen in (Adhani and Awang, 2012). In regard to their applications to Cultural Heritage, authors emphasize their role to visualize the invisible or transparent contents. To warrant an efficient management, it is necessary to combine different aspects which involve content providers and user's needs by means of and appropriate choice of involved technologies. In this work, we are using a markerless approach, having in account only GPS to display the contents in an interactive way.

From the user's viewpoint, some relevant advantages of AR solutions for customers are linked to the wearabililty of portable devices, a friendly management and the capability for experiencing in situ different levels of interaction with very rich objects. This situation must be compatible with very complex technologies which must be transparent for users. Technologies are supplied by different kinds of stakeholders and, by adapting some items of (Adhani and Awang, 2012), they involve to different aspects such as:

- *computational platform* to provide and manage virtual objects in an interactive way;

- *display mechanisms* for showing multimedia contents which can be switched according to user's preferences;
- *registration and tracking* for aligning contents depending on changes in relative location;
- *interaction technologies* depending on the user's profile and enabling him/her the usual tasks in mobile devices;
- *wireless networking* by minimizing adverse effects linked to inertial phenomena and other artifacts:
- *data storage and access technology* to provide customized information on different platforms.

Currently, smartphones and tablets (generally called mobile devices) fit to our purposes as they include sensors to receive GPS position, digital compass for orientation, accelerometer for tracking small motions and wireless communication using different protocols. All of them are integrated in a convenient size and shape allowing to disseminate the cultural heritage value in an interactive way. However, the user interaction is still limited because the user needs both hands to interact with the mobile device. To solve this problem and provide the user with a true free-hand device two authors of this paper have designed and implemented an eye-based tracking for interaction with any low cost camera (like the frontal cameras of most current mobile devices).

The incorporation of a video camera and additional systems to support user tracking allow us to provide meaningful information and a realistic model for position tracking at the same time. An image-based solution for tracking consists of aligning views captured by the mobile device with other views of the environment which are already available (or with projections of a previous available 3D model. This approach extends the traditional vision-based tracking systems (using natural landmarks) and 3D reconstruction from multiple views.

The high resolution model has been designed for a desktop application which can be explored and navigated at home. In this way, the user can search key features to be applied for in situ interactions by using mobile devices. Obviously, models to be displayed in mobile devices requires a lower resolution, specially with respect to the graphic resources (images and the different kind of maps). However, they also support advanced real time rendering effects such as normal mapping and shadow casting thanks to the advances in mobile GPU computing (see Figure 9). A better looking model improves the visual perception of cultural heritage content by the user, at the same time it encourages him to interact with the environment. The application can be tailored for two different kind of users depending on their age, playing capabilities and the level of interaction they want to achieve.

5. EVALUATION AND DISCUSSION

Our software provides a support for different multimedia contents, which can be displayed from an approximate location in an interactive and ludic way. Disseminated remains and the lack of historical information about the archaeological site are two important issues to provide a reconstruction of the current state at the end of the Roman Empire. The interpretation is performed from similar but better preserved Roman villas in the neighborhood nearby. Thus, we are neither intending to provide the true history nor a faithful approach to a reality, which is unknown in most aspects. Our goal is to understand the city as an open museum which is evolving over time as a consequence of the interaction with social agents.

Figure 9: A Nexus 5 viewing the vestiges from different periods discovered in the same area in the three archaeological excavations.

The developed application shows an example of how to integrate implicit and explicit knowledge arising from different sources to reinterpret the reality in connection with some snapshots of historical past from multimedia contents. Currently available data are unstructured, and most data are not accessible. A challenge for the next future is how to discover and integrate the huge amount of information as different layers to be superimposed on an evolving environment. In this way, we intend to contribute to illustrate the changing nature of different relationships between agents from the past to the present. This implies the development of more pleasant tools for the Internet of Things, including tools for content creation which can provide storyboards with different customized paths depending on the profile, e.g.

We have demonstrated how to integrate mobile technologies inside a historical discourse based on a interpretation of hidden archaeological vestiges. For most people, the use of technology must be transparent and in fact they should not appear in an explicit way. However, this goal involves to integration and georeferenciation of data, processing and analysis of models, fusion of information in different LoD in a common framework, rendering capabilities from open-source tools, access and management from mobile devices, remote access to Cultural Heritage repositories and reuse of multimedia contents. All of them require a combination of photogrammetric techniques, information processing and analysis, manual and semi-automatic modeling and rendering, advanced visualization techniques for interactive navigation, and local management from mobile devices.

It is necessary to develop a flexible amenable software architecture and able to integrate services and processes linked to multimedia contents and advanced tools for interactive visualization. Some still as *standing technological issues* concern to a) a more efficient reuse of resources (reactive algorithms, real time performance) under more flexible environments and compatible with several rendering environments (such as HTML5 which provides support for WebGL), but which are not fully integrated with mobile devices; b) the development of more friendly software tools to integrate the above technologies; c) the design and implementation of models to warrant interoperability between tools and data repositories in the Web 3.0 framework; d) the capability of reusing web resources, including cloud computing for custom applications.

Neither of the above challenges can be achieved if there are no economic resources to support them. In view of decreasing resources for Cultural Heritage and the lack of a pan-European data portal, *economic sustainability* of this kind of applications has become a major issue to obtain a return for the performed work. The above integration is focused towards different kinds of agents which are not passive consumers, but active agents wishing to

This contribution has been peer-reviewed.
doi:10.5194/isprsarchives-XL-5-W4-61-2015

participate in the discovery of other realities, linking a reinterpretation of the past to the current reality.

Paths cannot be uniquely determined, but they must be discovered as part of the action performed by each agent; thus, the adopted profile is crucial for edutainment purposes. To achieve this goal it is necessary to develop business models for applications based on mobile devices which can incorporate sponsorship, subscriptions, advertising-support, custom contents to be played at different levels and with different business models (including premiums linked to the achievement of partial goals), and creation of communities of users and content-creators. In absence of a truly interest by the large operators of videogames companies, it is clear that there is not a unique solution, and more structured mash-ups communities involving developers and content creators are necessary.

6. CONCLUSIONS AND FUTURE WORK

In this work, we provide answers to several basic issues concerning to the communication of CH contents by following an interactive strategy based in games with CH contents in historic urban environments. To achieve this goal, we have designed and implemented a software application which includes web services, Mobile Augmented Reality and Serious Games to enhance cultural experiences through immersive interactions performed at smart mobile devices.

The prototype of the developed application is illustrated with some examples corresponding to a currently hidden archaeological zone of Valladolid (Spain). The augmented representation allows to understand the city as a superposition of layers of different periods along the History, where social agents interact between them. In other words, the architectural space is reconstructed from the interaction between different social agents. Therefore, the knowledge of evolution in old times provides some keys to understand their projection towards the future in an active way at the same time users can receive historical education shaped in a game.

The archaeological and historical richness of the zone suggests different extensions of the performed work in the near future, which would be developed depending on additional economic resources. Next steps include the development of a true multiplatform solution and the development of virtual characters for providing user guidance in the area; some more advanced applications could include avatars to ease a more friendly interaction. In the current state the prototype does not include interaction between avatars; in the next version the application would allow the user to select the avatar representing the player and interact with the other players through their avatars.

ACKNOWLEDGEMENTS

Authors are very grateful to the "Servicio Territorial de Cultura de la Junta de Castilla y León" for allowing the access to the archaeological inventory of Valladolid. The documentation of the archaeological excavations was created between 2006 and 2009 by "STRATO Gabinete de Estudios sobre Patrimonio Histórico y Arqueológico S.L." as requested by "CORSÁN-CORVIAM CONSTRUCCIÓN S.A. (GRUPO ISOLUX-CORSÁN)" in a job promoted by the Council of Valladolid ("Área de Urbanismo. Servicio de obras").

REFERENCES

Adhani, N. I. and Awang, R. D. R., 2012. A survey of mobile augmented reality applications. In: 1st International Conference on Future Trends in Computing and Communication Technologies, pp. 89–96.

Aparicio Resco, P., Carmona Barrero, J., Fernández Díaz, M. and Martín Serrano, P., 2014. Involuntary photogrammetry: rescuing 3d geometric information from library pictures. Virtual Archaeology Review 5(10), pp. 11–20.

Delgado, F. J., Martinez, R., Finat, J., Martinez, J., Puche, J. C. and Finat, F. J., 2013. Enhancing the reuse of digital resources for integrated systems to represent, understand and dynamize complex interactions in architectural cultural heritage environments. ISPRS Annals of Photogrammetry, Remote Sensing and Spatial Information Sciences 1(1), pp. 97–102.

Eggert, D., Hücker, D. and Paelke, V., 2014. Augmented reality visualization of archeological data. In: Cartography from Pole to Pole, Springer, pp. 203–216.

Fernández Vega, P. Á., 2003. La casa romana. second edn, Ediciones AKAL.

Fritz, F., Susperregui, A. and Linaza, M., 2005. Enhancing cultural tourism experiences with augmented reality technologies. 6th International Symposium on Virtual Reality, Archaeology and Cultural Heritage (VAST).

García Guinea, M. Á., 2000. La villa romana de Quintanilla de la Cueza (Palencia). Diputacion de Palencia.

González Gallero, I., 2008. Villa Romana Almenara de Adaja Puras. Valladolid: Confuencia Educativa.

López Romero, E., 2014. 'out of the box': exploring the 3d modelling potential of ancient image archives. Virtual Archaeology Review 5(10), pp. 11–20.

STRATO Gabinete de Estudios sobre Patrimonio Histórico y Arqueológico, S. L., 2009. Trabajos arqueológicos integrados en el proyecto de construcción del aparcamiento subterráneo en la plaza de la antigua, en valladolid. Private report allocated in the "Servicio Territorial de Cultura de la Junta de Castilla y León (Valladolid)".

Stricker, D., Pagani, A. and Zoellner, M., 2010. In-situ visualization for cultural heritage sites using novel, augmented reality technologies. Virtual Archaeology Review 1(2), pp. 141–145.

Vlahakis, V., Ioannidis, N., Karigiannis, J., Tsotros, M., Gounaris, M., Stricker, D., Gleue, T., Daehne, P. and Almeida, L., 2002. Archeoguide: an augmented reality guide for archaeological sites. IEEE Computer Graphics and Applications 22(5), pp. 52–60.

This contribution has been peer-reviewed.
doi:10.5194/isprsarchives-XL-5-W4-61-2015

COMPLEX BUILDING DETECTION THROUGH INTEGRATING LIDAR AND AERIAL PHOTOS

R. Zhai

College of Informatics, Huazhong Agricultural University, 430070, Wuhan, China, rfzhai@mail.hzau.edu.cn

Commission V, WG V/4

KEY WORDS: Building Detection, LiDAR, Aerial Imagery, Image Segmentation, Multiple Seed Selection, Region Growing

ABSTRACT:

This paper proposes a new approach on digital building detection through the integration of LiDAR data and aerial imagery. It is known that most building rooftops are represented by different regions from different seed pixels. Considering the principals of image segmentation, this paper employs a new region based technique to segment images, combining both the advantages of LiDAR and aerial images together. First, multiple seed points are selected by taking several constraints into consideration in an automated way. Then, the region growing procedures proceed by combining the elevation attribute from LiDAR data, visibility attribute from DEM (Digital Elevation Model), and radiometric attribute from warped images in the segmentation. Through this combination, the pixels with similar height, visibility, and spectral attributes are merged into one region, which are believed to represent the whole building area. The proposed methodology was implemented on real data and competitive results were achieved.

1. INTRODUCTION

The acquisition of building positions can not only provide basic information for wise earth, and also it provides decision making support for governments. Therefore, there is an increasing demand for digital building model (DBM) generation. Automated building detection in urban areas is one of the difficult problems in photogrammetry even with the advent of LiDAR techniques.

Digital Building Model generation is usually performed in two steps: building detection/recognition and building reconstruction. Building detection is the process of generating the building hypothesis by differentiating buildings from other objects within the data. Building reconstruction, on the other hand, utilizes the hypothesized building regions in the data to derive the building model parameters for its 3D representation (Ma, 2004). A number of authors have proposed the integration of different data sources in 3D building reconstruction. The integration of imagery and 2D GIS data (i.e. building ground plans) has been proposed by Suveg (2004). The use of the ground plans helps in the building hypothesis generation and provides clues about the structure of the building. This is particularly useful in the process of partitioning the building into simple parts when using the CSG model. The integrated use of LiDAR and 2D GIS data for DBM generation has been exploited by Haala (1998) and Suveg (2001). They also employ the CSG model in the reconstruction process. The ground plans are subdivided into rectangular primitives and the selection of the parametric primitive for each 2D rectangle is based on the analysis of the LiDAR DSM. Although existing ground plans can be useful information in the sense that they reduce the search space in the reconstruction procedure, they might be out-of-date, incomplete or not co-aligned with LiDAR or imagery data.

Since the advent of LiDAR data, the integration of LiDAR and photogrammetric data for DBM generation has been the target of research of a number of authors (Rottensteiner, 2002; McIntosh, 2002; Seo, 2003; Ma, 2004). This integration is mainly motivated by their complementary nature (Habib, 2004; Baltasavias, 1999). The main pitfall of DBM generation using only LiDAR is the lack of accuracy of the generated building boundaries. Since photogrammetry provides accurate breakline definition, and its integration with LiDAR can lead to a more accurate and automated DBM generation. Most of the researches proposing the integration of LiDAR and imagery data perform the building hypothesis generation and the building reconstruction using only LiDAR data (Seo, 2003; Ma, 2004), before the data are integrated. The outcome is an initial building model in the sense that its boundaries must be refined. Before the two data are integrated to refine the DBM, the photogrammetric and LiDAR data should be co-aligned relative to the same reference frame. For this purpose, LiDAR features can be used to georeference the imagery relative to the LiDAR reference frame. Since point primitives are not appropriate for LiDAR surfaces due to its irregular nature, other primitives such as lines and planar patches are usually employed (Habib, 2004; Ma, 2004). After the co-registration of the data, the initial LiDAR-derived building boundaries can be projected to the image space. The image edges are usually detected using the Canny edge detector (Canny, 1986). The projected building model is very close to the building boundary edges in the image. In this regard, valuable clues are provided for reducing the search space for matching of conjugate building edges. In addition, the plane parameters for the building roof planes, which are derived from the LiDAR data, provide additional constraints in the photogrammetric reconstruction process increasing its reliability and accuracy. After the imagery edges are correctly detected, they can be used to refine the projected model edges. Finally the refined DBM can be generated. Kwaw(2013) focused on generating building models by using a model-driven method. The focus of the research is generating models with right-angled-corners, which can be described with a collection of rectangles (e.g., L-shape, T-shape, U-shape, gable roofs, and more complex building shapes which are combinations of the aforementioned shapes), under the assumption that the majority of the buildings in urban areas

This contribution has been peer-reviewed.
doi:10.5194/isprsarchives-XL-5-W4-69-2015

belong to this category.

In this research, building rooftop patch can be segmented from large amount of LiDAR data, and rough building rooftop boundaries are generated as well, however, the derived building boundaries are still not exact to some extent because of the irregularity of LiDAR data. It is believed the integration of LiDAR data and aerial imagery will come up with more reliable and high accurate building rooftop boundaries. Line feature detection and matching is the common way employing aerial imagery for building rooftop reconstruction. It is relatively easy to detect linear features from imagery, on the other hand, failed linear feature detection and line matching will also lead to non-reliable building boundaries. In this case, region growing methods are believed to give good segmentations that correspond well to the observed edges (Zhai, 2008). Combining region growing results from left and right images will represent whole rooftop area. This paper will be mainly focusing on the issues related to region oriented building rooftop generation.

The paper is organized as follows. Section 2 introduces and analyzes the proposed methodology used in our research, and the subparts will look at each of these questions. It will be shown that these questions are useful as a suitable base for analyzing the proposed techniques, although some of the questions are not entirely independent. Section 3 will do the experiments by using real datasets and analysis, which demonstrates the feasibility of the proposed method. The conclusion and future work is drawn in Section 4.

2. METHODOLOGIES

The process of extracting and representing information from an image to group pixels together into regions of similarity is commonly called segmentation (Fan, 2004). In 2D imagery, we would like to group pixels together according to the rate of change of their intensity and other properties (attributes) over a region. As for building rooftops, it is known that most building rooftops are represented by different regions from different seed pixels. Considering the purpose of this research and the principals of image segmentation, this paper employs a new region based technique to segment images, combining the advantages of cluster based and region based techniques. In the image segmentation process, we combine the elevation attribute from LiDAR data, visibility attribute from DEM (Digital Elevation Model), and radiometric attribute from warped images in the segmentation. Through this combination, the pixels with similar height, visibility, and spectral attributes are merged into one region. There are several questions to be answered through the process of automated building rooftop reconstruction based on this methodology. Each of the subsections will be explaining all the questions in detail.

2.1 Input Data

As introduced above, the segmentation procedures proceed by taking advantage of several attributes from different data source. Thus, several data are considered as the input data source, building roof patch generated from LiDAR data, building boundaries generated from LiDAR data, left and right warped image, left and right visibility maps, and Colour histogram of the building rooftop area. Each of them will be described in detail as follows.

Building patch generated from LiDAR data. As abovementioned, segmentation technique of LiDAR data has been employed to derive exact building rooftop planar patches and rough building boundaries (Kim, 2007). Planar patch is generated by fitting large set of discrete LiDAR points located at a same plane. Therefore, the segmentation technique can successfully identify the LiDAR points located on a building rooftop, which will be the rough building rooftop area because of the irregularity of LiDAR data.

Building boundaries generated from LiDAR data. The derived boundaries (Kim, 2007) do not represent the actual boundaries of the physical objects being segmented. In other words, these are the boundaries as determined by the LiDAR footprints, and their deviation from the actual physical boundaries depends on the average point density of the LiDAR data. However, they are still the rough boundaries of building rooftops. If these boundaries are projected onto imagery, it is expected that the derived boundaries will be rough estimates of rooftop boundaries.

Left and Right warped images. Distortions caused by different angles of viewing must be rectified. It is necessary to ensure that the imaging distortion does not affect the estimation of building rooftop area. For this purpose, we use a combination of a warping algorithm with assigning pixel values to an output image (matrix) after the transformation. The assigning is performed by resampling algorithms. Thus, in warping algorithms there are two main steps: (i) transformation and (ii) resampling. A warped image is computed by indirect transformation between input image and out image. Here, the mathematical model of the transformation is projective transformation. First, a plane segmented from LiDAR data is extracted and its range is determined from the derived rough boundary. The range of the plane is more extended than the exact boundary of the derived rough boundary while considering the resolution of LiDAR data. Using simple projective transformation, the corresponding colour (or grey) value from the original imagery is registered on the positions of each object point on the segmented plane. Two warped images are generated by projecting the left and right original images of a stereo-pair onto the relevant segmented plane. In addition, the original imagery and LiDAR data are integrated through the common reference frame and the warped images are produced. The utilization of the warped images will provide great benefit in the proposed DBM generation methodologies by handling spectral and positional information simultaneously (Kim, 2009).

Left and Right visibility maps (Habib, 2007). The visibility map is a data structure that describes the projection of the visible scene onto the image plane. The visibility of every cell from the observer cell is computed. Visibility is calculated by measuring the off-nadir angle from the observer's eye to each cell starting from the cells which are the closest to the observer. As long as the off-nadir angle increases in the line of site from the observer, the cell is considered as visible. If the off-nadir angle decreases, the cell is not visible anymore. In this research, the visibility map is generated using the generated DEM from LiDAR data and corresponding image EOP and IOP. The resolution of the visibility image should be consistent with the resolution of warped images. Therefore, interpolation techniques are introduced in the visibility map generation.

Colour histogram of the building rooftop area. A colour histogram is a representation of the distribution of colours in an image, derived by counting the number of pixels of each of a given set of colour ranges in a typically two dimensional or three dimensional colour space. In this work, the colour histogram will be done in 3D colour space, and only related to the image area located inside the projected boundary.

2.2 Multi Seed Pixel Selection

One building rooftop is most likely composed of multiple regions. Therefore, it is not probably that only one region growing from one seed pixel will represent an entire building rooftop. The projected boundaries from LiDAR, as we have mentioned, will be the rough estimates of building boundaries. We are mainly interested in the image area located inside the LiDAR boundary, which is referred to as the Area of Interest (AOI) in this research. As previously mentioned, one building rooftop is probably composed of several different parts, i.e., multiple seed pixels should be selected before region growing proceeds. Therefore, this section will be focusing on this issue and proposing several conditions that one seed pixel should satisfy.

In order to find appropriate seed pixels, the following criteria that a seed pixel should satisfy are: ①Seed pixel should be located inside the area defined by the projected boundary from the LiDAR data, i.e. within the AOI. ②Colour at the seed pixel should correspond to, or be sufficiently close to, a peak in the colour histogram in the AOI. ③Seed pixel should be located within a homogeneous area. That is, the standard deviation of gray values in a window (e.g., 5*5) should be less than a given threshold (e.g., 10.0). ④Height at seed pixel is consistent with the height of the segmented LiDAR patch; ⑤Seed pixel should be located inside the visible area. As we have a visibility map, if one pixel satisfies the first four constraints, but is not visible in the visibility map, this pixel should not be considered as a seed pixel.

The detailed procedures of multiple seed point selection can be illustrated as follows.①Pick up all the pixels inside the initial LiDAR boundary to generate a new image (Image_1), with same size to original image (Image_0), shown as Figure 1(a) and Figure 1(b), respectively. ②Do histogram statistics (Figure 1(c)) to the generated image (Image_1), without considering the black area; ③Pick up the pixel at the first peak in the histogram, check the height constraint, homogeneous characteristic, and the statistics number of this pixel. If all of these conditions are satisfied, this point will be determined as the first seed point, do region growing. ④A new image (Image_2) will be generated through region growing using the first seed point. Check the ratio between the area inside the LiDAR boundary from Image_2 and Image_1. If the ratio is smaller than the given threshold, go to Step ⑤. Area ratio computation is defined as Equation (1). ⑤Generate a new image with all pixels as black. Considering image (I_2), set the gray values of all segmented pixels inside the LiDAR boundary as black, to generate this new image (I_3). ⑥ Do histogram statistics for the new generated image (I_3). ⑦ Repeat Step ③-⑤, till there is no new seed point, or the ratio between the segmented region to the region inside the LiDAR boundary is smaller than one threshold (e.g., 0.95)

$$area\_ratio = \frac{growing\_area\_1}{image\_area\_inside\_LiDAR\_boundary} \quad (1)$$

Where area_ratio defines the ratio between growing regions and the building area

growing_area 1 defines the region area from one seed point

image_area_inside_LiDAR_boundary defines the whole building roof top area inside the initial building boundary generated from LiDAR data

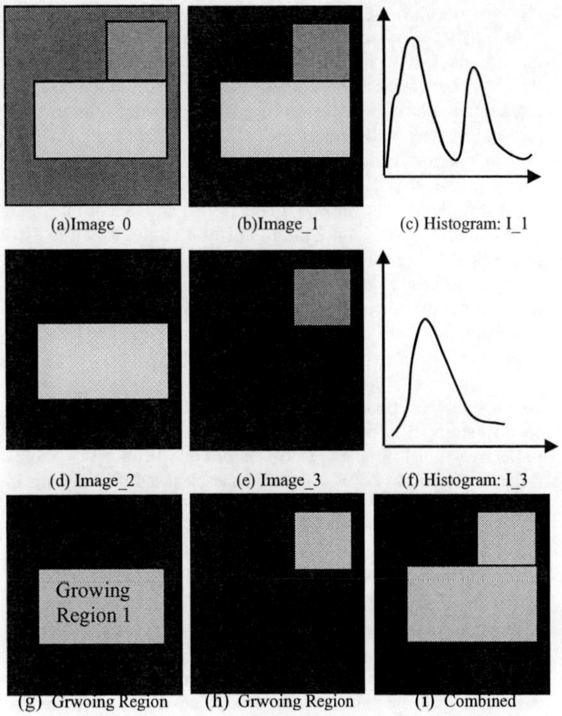

(a)Image_0 (b)Image_1 (c) Histogram: I_1

(d) Image_2 (e) Image_3 (f) Histogram: I_3

(g) Grwoing Region from Seed 1 (h) Grwoing Region from Seed 2 (i) Combined Regions

Figure 1 Multiple seed selection

2.3 Region growing strategy based on multiple seed pixels

After one seed pixel is selected, region growing proceeds increasing the size of the region by adding in neighbouring pixels that are similar with respect to one or more properties, e.g. colour, visibility and height properties.

Colour Consistency in Region Growing. As we have multiple bands(R, G and B), there can be various stopping conditions combining stop conditions in individual bands during the region growing procedures, strict condition and loose condition(Habib, 1999). In particular, strict condition means the region growing procedure stops only when all the three bands, Red, Green, and Blue, satisfy the stop condition. While loose condition means the growing procedure stops one of them satisfies the stop condition. In the research, we are using strict condition, but we have one extra condition as in the 8-connected neighbourhood, number of pixels satisfying this loose condition should be greater than the 4. Two threshold values are defined in region growing procedures by considering colour consistency: Mean threshold value (μ), and range threshold value (σ). Mean threshold value is the mean gray value of segmented area. The initial value is defined by the gray value of seed pixel. In the process of region growing, it is determined by the mean gray value of the segmented image area based on one seed pixel. Range threshold value is the standard deviation of the gray value in growing the region. The initial value is given by the user, or comes from the standard deviation of the check window, which is used for seed pixel selection.

This contribution has been peer-reviewed.
doi:10.5194/isprsarchives-XL-5-W4-69-2015

The segmented building rooftop patch from LiDAR data is more reliable than the LiDAR-derived boundaries. Thus, the height of the image pixels located on the roof area should be consistent with the segmented LiDAR patch, which is introduced in the region growing procedures as height consistency with the segmented LiDAR building rooftop patchA pixel, with the height constraint no more than a given threshold between the 3D point corresponding to this pixel and the LiDAR plane, will be added to the growing region. Two threshold values will be defined in the region growing process, by considering height consistency between the point being checked and the height of the segmented building rooftop: radius threshold value in X, Y plane (r_thr) and height threshold value (h_thr). r_thr defines the radius of a circle, which is used to check whether there is a LiDAR point close to a certain 3D point in this circle., while height threshold value (h_thr); defines the maximum distance between the searched LiDAR point corresponding with the checking pixel and segmented LiDAR planar patch.

Given one seed pixel shown in Figure 2 in image space, the corresponding 3D coordinates (X, Y, Z) can be computed using projective transformation. Define a circle with a given radius r_thr, centered at (X, Y), to search the closest LiDAR point in all the LiDAR data, including ground LiDAR points and non-ground LiDAR points. If there is no LiDAR data available in the certain search area, region growing will stops at that pixel, on the other hand, once some LiDAR points are found in the circle, the distance between the searched LiDAR data and the segmented LiDAR planar patch is computed, only when the distance is less than the given threshold h_thr, the region will keep growing, or it will also stop. Growing pixel 2 shows this case. It avoids the case that some building regions will invade into other building rooftops or ground area, given the color between the building rooftop and other objects are consistent.

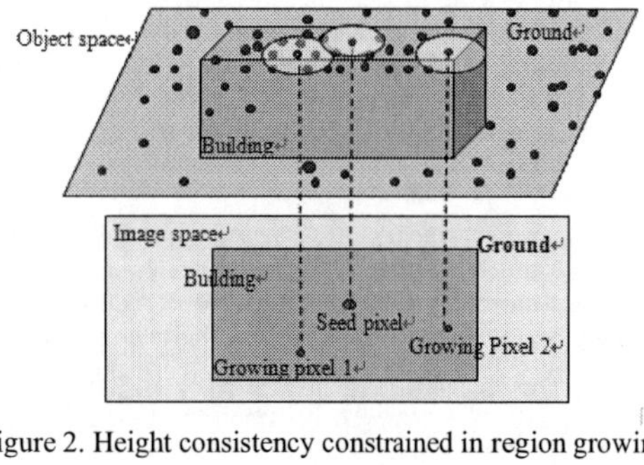

Figure 2. Height consistency constrained in region growing

By considering height consistency in region growing, a radius threshold value is defined when utilizing visibility information. Radius threshold value in visibility map (r_thr_visiblity) defines the radius of a circle so as to check whether the added point is located inside the visible area. When one new pixel (x, y) is added to a region, all the pixels located inside the circle, centered at the checking pixel (x, y) with the given radius (r_thr_visiblity), are checked to find if there are image pixels that are visible in this circle area. If yes, the region growing process will proceed, otherwise it will stop at the pixel (x, y).

Visibility images are generated using a DEM from LiDAR data and available image EOPs. In building areas, especially in urban areas, a visibility map is a useful way to determine whether a certain area is visible in image space. If there are occluded parts on the building rooftop, they will be invisible in the visibility map.

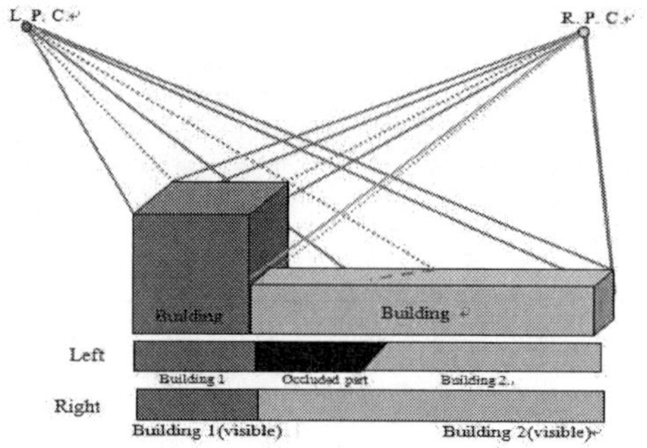

Figure 3. Visibility Consistency constrained in growing

2.4 Region growing procedures

The rough building rooftop boundaries derived from LiDAR data, visibility map generated from DEM, and warped images from different viewpoints, are all considered in order to segment the whole building rooftop area from warped images. The procedures of the region growing for segmentation are as follows. ①Select seed pixels that satisfy the criteria specified above;②Region growing from the first selected seed pixel is performed until there are no other neighbouring pixels to be added to the region;③Remove the segmented region in the AOI, and check whether there is another seed pixel that can be selected;④Region growing from another seed pixel occurs, until there are no other pixels to be added to the current region; ⑤Repeat Step 3)-4). Region growing will stop if there is no other possible seed pixel; ⑥Implement all the above steps in both left warped image and right image. Rooftop regions are represented in the left and right images respectively .The whole rooftop building area will be represented by combining all the regions grown from left and right images, which are illustrated as Figure 4.

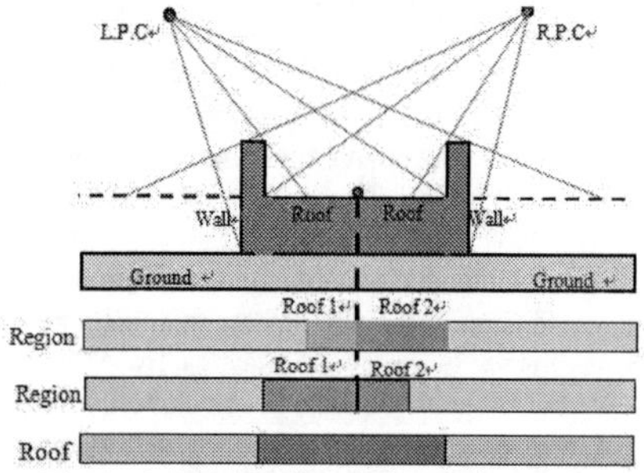

Figure 4. Roof region defined by combining the segmented left and right regions

2.5 How to derive boundaries using segmented regions

In order to generate building boundaries using segmented regions from warped images. As mentioned above, the segmented regions are composed of small regions coming from different seed pixels, also, there are some small holes existing in the combination of images. What we are interested are the outer boundaries of the segmented regions. Thus, inflation and erosion is first used to remove the small holes. Then, **the** generated area is utilized for boundary generation. In this research, LSD (Line Segment Detector) Detector is adapted to generate the building boundary (Rafael, 2012).

This contribution has been peer-reviewed.
doi:10.5194/isprsarchives-XL-5-W4-69-2015

3. EXPERIMENTAL RESULTS AND DISCUSSION

To test the feasibility and evaluate the performance of the proposed methodology, we will present experimental results from real LiDAR and image data. An airborne laser scanning dataset was acquired. The dataset is from an Optec 3100 Sensor model, was acquired at a flying height of 1400 meters, has a ground point spacing of 0.75m, and vertical accuracy of 15cm. Aerial imagery, with an image scale 1:5000, was also acquired. To illustrate the proposed procedures used in this research, the intermediate results for two specific building roofs are presented in Figure 5 and Figure 6, respectively. As shown in Figure 5, the building is a rectangular building with one protruding part which is shown inside the ellipse. The initial building boundary generated from LiDAR data were projected onto the left and right images to generate a small image which only include the building roof top area, respectively. Figure 5(b) shows the segmented results from left and right image based on the proposed methodology in this research. As you can see, the protruding part is not included in the region growing results because of the height consistency. In other words, since the height of the protruding part is higher than that of the roof top, region growing procedures will not invade this area. Figure 5(c) shows the building area combining the left and right image. The building rooftop area after inflation and erosion is shown as Figure 5(d). Figure 5(e) shows the line segments derived from the segmentation results by using LBD method. In contrast with the result by using the edge information, straight line segment extraction and matching were also implemented on the same building, which was shown as Figure 5(f) (Habib, 2010). Figure 6 shows the intermediate results on another building. Part of this building is occluded by another higher neighbouring building as shown in the red ellipse. The boundaries we can see from the image come from higher building, and some of the area in lower rooftop is occluded by the higher building. Since multiple seed selection is taken into consideration, the region outside of the building rooftop is also included shown as the ellipse in Figure 6(c).

During the process of implementing the procedures, several thresholds should be set. The threshold for standard deviation of gray value to check whether the point is in homogeneous area is set as 10.0. Meanwhile, the gray value threshold in the region growing procedure is set as 20.0. Besides, area ratio between the segmented image area inside LiDAR boundary and the original area inside LiDAR boundary is set as 0.95.

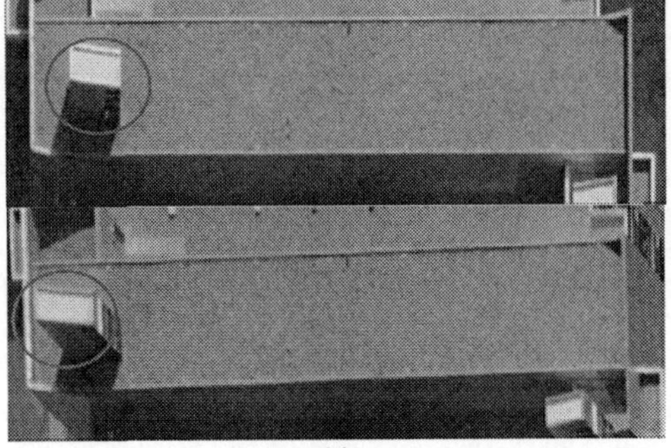

(a)

(b)

(c)

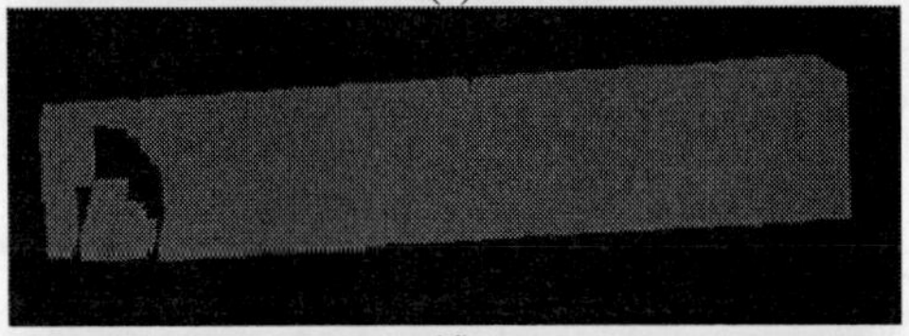

(d)

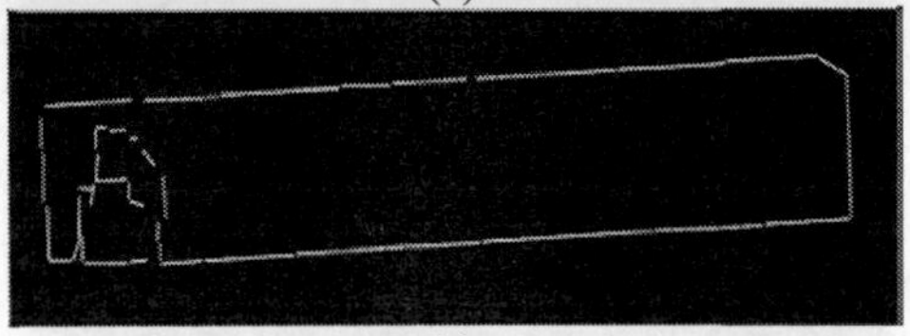

(e)

(f)

Figure 5. Intermediate results on building one (a) Inintial building boundaries from LiDAR data projected onto the left & right images; (b) Segmented results from left & right images; (c) Building image area combining left & right images; (d) Building area after inflation and erosion; (e)the line segments derived from the segmented results; (f) matched line segments from left & right images

This contribution has been peer-reviewed.
doi:10.5194/isprsarchives-XL-5-W4-69-2015

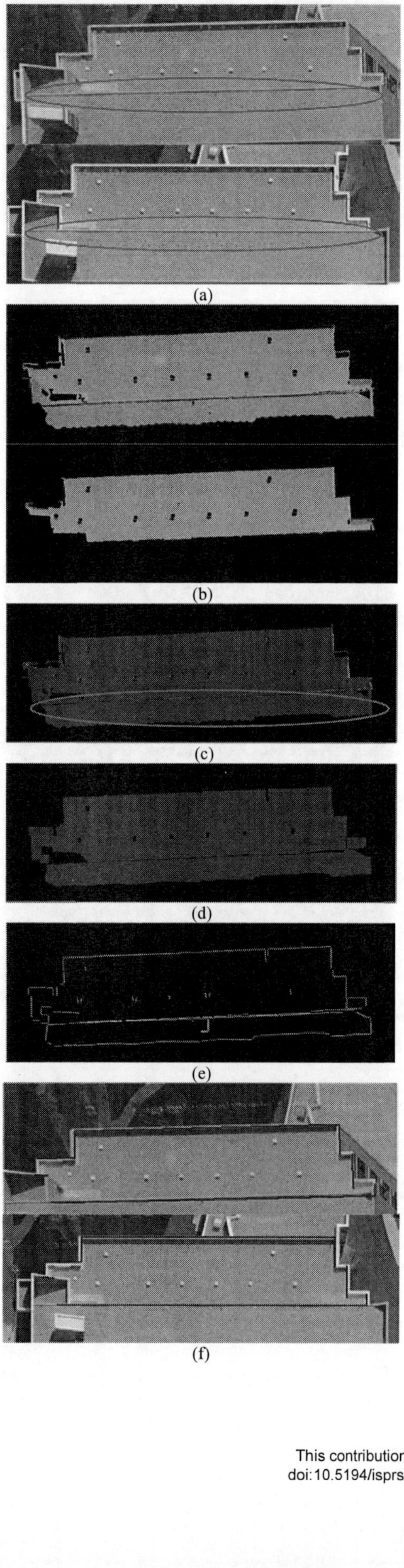

(a) (b) (c) (d) (e) (f)

Figure 6. Intermediate results on building one (a) Inintial building boundaries from LiDAR data projected onto the left & right images; (b) Segmented results from left & right images; (c) Building image area combining left & right images; (d) Building area after inflation and erosion; (e)the line segments derived from the segmented results; (f) matched line segments from left & right images

4. CONCLUSIONS

We have introduced an automated approach to recognize and detect complex buildings from aerial imagery and LiDAR data. Experiments with real data show that the proposed technique could be used to generate building boundaries by combining LiDAR data and aerial imagery. The building areas could be improved by introducing the assistance of the LiDAR data. It confirms that the combination of LiDAR data and imagery data will derive better results than using either of them individually.

Future work will focus on more testing with real datasets to generate building model by using the boundaries generated from this research, and the accuracy assessment will be conducted and compared with other modethodologies. Besides, the combination of region based and edge based methodologies will be investigated in order to generate high accurate building models.

ACKNOWLEDGEMENTS

The author would like to express sincere appreciation to Prof. Ayman F. Habib, who is with Civil Engineering, College of Engineering, Purdue University, for his great contribution to this work. Also, the author would like to thank the support (Program No. 41101409) from the National Natural Science Foundation of China.

REFERENCES

Baltsavias, E. P., Mason, S. O., Stallmann, D., 1995. Use of DTMs/DSMs and orthoimages to support building extraction. Workshop on Automatic Extraction of Man-made Objects from Aerial and Space Images, Switzerland, 1995:199-210.

Canny, J., 1986. A computational approach to edge detection. *IEEE Transactions on Pattern Analysis and Machine Intelligence*, 8, pp. 679–690.

Elaksher F. A., Bethel S. J., and Mikhail M. E., 2003, Roof boundary extraction using multiple images. *Photogrammetric Record*, 18(10)27-40.

Fan, Sophie, Liu Xiao, Zhihong Man, Rashi Samur, 2004, Edge based region growing – a new image segmentation method, *Proceedings of the 2004 ACM Siggraph International Conference on Virtual Reality Continuum and Its Applications in Industry*, 302-305.

Habib A., Morgan, M., Kim, E., and Cheng, R. (2004). Linear Features in Photogrammetric Activities, *ISPRS Congress, Istanbul, Turkey, 2004, PS ICWG II/IV: Automated Geo-Spatial Data Production and Updating*, 610.

Habib, A., Zhai R. and Kim C., 2010. Generation of Complex Building Models by Integrating Stereo-Aerial Imagery and

This contribution has been peer-reviewed.
doi:10.5194/isprsarchives-XL-5-W4-69-2015

LiDAR Data. *Photogrammetric Engineering and Remote Sensing Journal,* 76(5), pp. 609-623.

Habib, A., E. Kim, and C. Kim, 2007, New methodologies for true orthophoto generation, *Photogrammetric Engineering and Remote Sensing*, 73(1), pp. 25-36.

Habib, A., Uebbing, R., and Novak, K., 1999. Automatic Extraction of Road Signs from Color Terrestrial Imagery. *Photogrammetric Engineering and Remote Sensing*, 65(5): 597-601.

Haala, N., Brenner, C., and Anders, K., 1998. 3D urban GIS from laser altimeter and 2D map data, http://www.ifp.uni-stuttgart.de/publications/1998/ohio_laser.pdf (accessed Jan 2015).

Kwak, E., 2013. Automatic 3D Building Model Generation by Integrating LiDAR and Aerial Images Using a Hybrid Approach, ph.D. dissertation, The University of Calgary, Canada.

Kim C., Habib A., and Mrstik P., 2007, New approach for planar patch segmentation using airborne laser scanner. *ASPRS 2007 annual conference*, Tampa, Florida May 7-11.

Ma, R., 2004. Building Model Reconstruction from LIDAR Data and Arial Photographs, Ph.D. dissertation, The Ohio State University.

McIntosh, K. and Krupnik, A., 2002. Integration of laser-derived DSMs and matched image edges for generating an accurate surface model, *ISPRS Journal of Photogrammetry and Remote Sensing*, 56, pp. 167-176, 2002.

Rafael Grompone von Gioi, Jérémie Jakubowicz, Jean-Michel Morel, and Gregory Randall, LSD: a Line Segment Detector, Image Processing On Line, 2 (2012), pp. 35–55.

Rottensteiner, F. and Briese, Ch., 2002. A new method for building extraction in urban areas from high-resolution LIDAR data, *International Archives of Photogrammetry Remote Sensing and Spatial Information Sciences*, 2002, 34(3/A): 295～301.

Seo, S., 2003. Model Based Automatic Building Extraction from LIDAR and Aerial Imagery, Ph.D. dissertation, The Ohio State University.

Suveg, I., Vosselman, G., 2004. Reconstruction of 3D building models from aerial images and maps. *ISPRS Journal of Photogrammetry and Remote Sensing*, 58, pp. 202–224.

Zhai R., Habib F., 2008, Use of LiDAR data to constraint the matching of conjugate features in large-scale imagery over urban areas, *ASPRS 2008 annual conference*, Portland, Oregon, April 28 – May 2.

This contribution has been peer-reviewed.
doi:10.5194/isprsarchives-XL-5-W4-69-2015

Low cost 3D-modelling of a complex archaeological site using aerial photography in the hinterland of Petra, Jordan

R. Emaus [a, b] *, R. Goossens [b]

[a] Udhruh Archaeological Project, Faculty of Archaeology, Leiden University, Netherlands – R.Emaus@Artefact-info.nl
[b] Deptartment of Geography, Ghent University, Krijgslaan 281 S8, 9000 Ghent, Belgium – R.Goossens@UGent.be

Commission V, WG V/4

KEY WORDS: Udhruh, Roman Archaeology, Quarries, Photogrammetry, Aerial Photography, Kite

ABSTRACT:

Individual archaeological sites can sometimes show a complex morphology. One such site is the Roman quarries located one kilometre northwest of the Roman fortress at Udhruh, in the hinterland of Petra. In archaeology there are various platforms such as balloons, kites and unmanned aerial vehicles (uav's) from which low altitude aerial photographs can be taken that have been proven to work. All with their specific advantages above others in different circumstances. By designing a very distinct setup for the kite and optimizing the workflow accordingly, an effective and steady platform for the camera was created. The Quarries were photographed from the air and the images then provided enough material to accurately create a 3D model of the site.

1. INTRODUCTION

Using photogrammetrical software a high resolution DEM or other surface model of a single archaeological site can be recorded accurately. One only needs photographs of the site with ample ground resolution, which in general means that the photographs have to be taken from a low altitude above the ground. In archaeology there are various platforms in use such as balloons, kites and unmanned aerial vehicles (uav's) from which low altitude aerial photographs can be taken that have been proven to work (Verhoeven, 2009). All have their specific advantages above others in different circumstances, mainly dependent on wind conditions and the platforms' logistical needs.

This paper presents a research effort into designing and optimizing a low-cost, effective and accurate way to record archaeological sites in the hinterland of Petra. In particular we will look into the choice of platform and materials that are best suited for this environment whilst still not too expensive and optimize the workflow in the field. Further, we will examine the processing of the data and eventually the accuracy of the model.

Individual archaeological sites can sometimes show a complex morphology that cannot be recorded easily with classical topometric instruments (Verhoeven et al, 2012). One such site is the Roman quarries located one kilometre northwest of the Roman fortress at Udhruh, Jordan. The quarries provide an excellent example of a site that has a very distinct morphology that needs to be recorded with great precision. A high resolution DEM of the site can provide a richer documentation than any total station survey can, and with similar accuracy.

2. BACKGROUND

The ancient city of Petra is located just west of the Jabal ash-Shara ridge, running in a north to south direction east of Wadi 'Araba (Figure 1). As part of the Jordan Rift Valley, Wadi

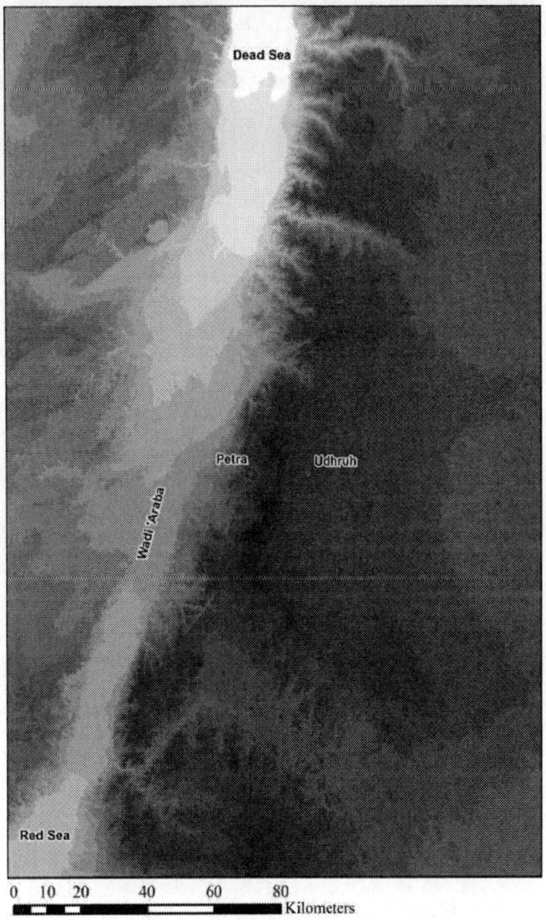

Figure 1. Map showing the topography of Southern Jordan with the locations of Petra and Udhruh.

* Corresponding author.

Figure 2. Part of the large Roman quarries northwest of Udhruh. Photograph by A. Gazenbeek.

'Araba is part of the 6000 km long fault system that runs from Northern Syria to East Africa. The main geological sequence in the Petra region is made up of Cretaceous limestone with impermeable sandstone under it. Because the geology is tilted towards the east, the eastern part of the region is largely made up of limestone while the sandstone is surfacing west of the ash-Shara ridge (Kherfan, 1998). Petra lies within one of the many valleys that have cut deep in the surfacing sandstone stretching west from the ridge. The soils are mostly made up of bare bedrock or stony soils. The landscape east of the ridge consists mainly of rolling hills and highlands that continue into the pre-desert plateau. Wide valleys extend eastward from the ridge into the pre-desert area where they create floodplain environments towards the al-Jafr Basin. The bedrock is mainly made up of limestone and marl. Soils mainly consist of alluvial sand and gravel deposits.

Based on the 6th century Petra Papyri, the hinterland of Petra can at least in easterly direction be estimated to have reached to about 20 kilometres from the city. Both Udhruh and Sadaqa by then still fell under the jurisdiction of Petra, suggesting a political and economic coherent region dominated by the antique city (Kouki, 2012). The transition from a highland environment to the pre-desert plateau can be roughly situated at the town of Udhruh, some 15 km east of Petra. Udhruh, identified as Augustopolis in the Petra Papyri, is mainly known for its second century Roman fortress (Kennedy, 2004). Already in Nabataean times the site probably acted as an important satellite station along the trade route into Petra, providing the desert caravans with water from the local spring and food from the local both dryland and irrigated fieldsystems (Abudanah, 2006; Driessen & Abudanah, 2013). Later, in the Byzantine and Islamic periods, Udhruh was still an important site as the many rural settlements in the hinterland of Petra became more concentrated into larger villages and towns such as Augustopolis since the Byzantine period (Kouki, 2012). The area directly around Udhruh can therefore be seen as an important sub-region in the hinterland of Petra.

While there have been some small excavations within and around the fortress (Kilick, 1983 & 1986), Abudanah has investigated a wider region around Udhruh (Abudanah, 2006) and since 2011 an archaeological survey project has been set up to prospect the region more directly around and in relation to the fortress (Driessen & Abudanah, 2013). Next to the Roman fortress the area has shown to contain various archaeological sites such as water management and field systems, fortlets, watchtowers and the large quarries. The large quarries have yet to be investigated thoroughly but have probably supplied building material mainly for the building of the fortress itself. These quarries therefore have to be examined and documented precisely as part of the research into the large Roman fortress of Udhruh. Modelling the quarries is a first step in this process

3. CHOICE AND DESIGN OF KITE AND CAMERA

3.1 Choice of platform

The soil in the sub-region around Udhruh is mainly covered with loose stones and sometimes razor sharp flint. Together with the abundant presence of small and thorny ground vegetation this makes for a very hostile environment to work with certain platforms such as helium balloons that are easily punctuated. The wind in the region can be steady and of mediate strength when coming from the desert in the east. Coming from the ash-Shara ridge and highland environment in the west, the wind can sometimes be relatively unpredictable and strong. These particular conditions automatically disqualify some of the more commonly used platforms:

3.1.1 Balloon: The variety of the wind-conditions can render a helium balloon very unstable (Verhoeven et al, 2009). Next, balloons are easily damaged or punctuated on the ground by the local vegetation. Also, balloons need a supply of helium, which can be a logistical challenge to acquire locally and handle in the field.

3.1.2 UAV: The sometimes unpredictable winds also create challenges for smaller uav's as they can be suddenly blown a distance away. In the event of a rough landing, the relatively delicate parts of a uav may easily be damaged by the sharp rocks and stones on the surface. Also, a uav can be very costly compared with the other platforms. Though, uav's have been proven to work very well in other parts of the world (Verhoeven, 2009; Hendrickx et al, 2011).

3.1.3 Kite: In combination with the following advantages and considerations the kite was selected as the platform to be used. First, the kite is naturally able to handle various wind conditions while maintaining the needed stability (Aber et al, 2002). Second, all parts of a kite are very easily repaired in the field with only some basic spare parts and a leatherman in case of a rough landing. Third, a kite is very low-cost and maintenance free when compared to other platforms as its parts are easily available all over the world and a kite doesn't need a supply of helium to fly. Fourth, there are no legal constraints on using kites whatsoever. In an increasing number of countries a permit is needed to fly a uav or there can be constraints on importing one into the country.

All of these disadvantages can of course be mitigated but for this region the kite is the best option. Perhaps uav's are feasible in longer running projects in the area or later on in the local Udhruh Archaeological Project.

3.2 Design of the Kite

In order to improve the functionality of the platform, a kite was chosen that can be manoeuvred very well by the use of two steering lines. The operator is therefore able to precisely manoeuvre the kite in virtually every location above the site to get the best coverage from just a few operating positions. This has the advantage that the operator doesn't have to walk across a site in order to cover it from the air as is normally the case with single-line operated kites. This is especially advantageous with sites that have limited accessibility or a difficult terrain such as the Roman quarries with steep cliffs and numerous holes in the ground (Figure 2). From a single operating position, a site can be covered from the whole wind-window of the kite (Figure 6). The wind-window of the kite can further by enlarged by using longer steering lines. But this, of course, also effects the ground resolution of the imagery (see section 5.1).

Kites generally can provide the needed lift from a wind speed of 2 bft. and upwards. The only consideration that has to be made is the size of the kite in function of the average wind speed and the weight of the camera it has to lift; larger kites with lower wind speeds or heavier cameras and smaller kites with larger wind speeds or lighter cameras. Obviously, the kite mustn't be too small or it simply won't fly. But adversely, for safety reasons it better not be too large as well or one will be unable to handle the kite which pulls too hard on the person operating it, leading to disaster. Although a backup kite of 1.6 m^2 was available during fieldwork, a kite of 2.1 m^2 was used to take all the necessary imagery. This kite generally provided just enough lift in the mildest of wind conditions whilst still being controllable in some of the rougher wind conditions we experienced.

The length of the steering lines, and thus the maximum camera height, can be chosen in function of the wanted ground resolution and coverage. Both long (>50 meters) and short (<30 meters) steering lines have their advantages and disadvantages. When using relatively short steering lines the wind-window is of course also relatively small and the operator will have to use multiple operating points when a large area has to be covered. The advantage of shorter steering lines is that there is less of a delay in the steering of the kite, resulting in being able to manoeuver with greater precision. When the exactness of the flight plan is not so much of an issue but the terrain poses some difficulties, longer steering lines will be more advantageous. The operator will be able to cover a larger area not only because the actual wind-window is larger, but also because the kite can take imagery from higher altitudes. The latter advantage will have its detrimental effect on the ground resolution of course.

3.3 Design of the cradle

The camera has to be mounted to the kite in a way that it is stable and pointing in the right direction i.e. down to the earth. The cradle is made of simple iron strips that are bolted together in such a way that a small frame comes into being. The camera can then be mounted inside this frame using a single bolt or thumbscrew that fits in most cameras' tripod screw threads. All space between camera and frame is then filled up with foam (usually an ordinary sponge that has been manually 'customized' to make a perfect fit). The foam provides some extra stability and protection for the camera. The weight of the cradle and camera now amounts to circa 1 kg.

Figure 3. The kite with cradle (side). Photograph by A.K. Helming.

Figure 4. The kite with cradle (front). Photograph by A.K. Helming.

Figure 5. Operation of the kite. Photograph by M.H. Sepers.

The cradle is then attached between the two steering lines of the kite with a rod so the kite won't collapse under the weight of cradle and camera that would otherwise pull the lines together. In case the wind drops completely the rod also makes sure the kite stays open and gently comes down like a parachute (Figure 4).

This construction also provides stability to the camera and prevents it from some unwanted movement. With the camera up in the air, the biggest problem is holding it as motionless as possible not to create any motion-blur in the imagery. With the cradle and camera attached to the two steering lines of the kite, the camera is automatically unable to swing in at least one direction; from left to right in the operators' perspective, or the φ-pivot in classical aerial photography. Because this movement is firmly controlled by the two steering lines.

The movement of the cradle around the carbon rod to which it is attached, creating a swing to and from the operator (or the ω-pivot) now is the only unwanted movement within the whole system. Unfortunately, we have found no structural answer to this problem that doesn't also create any difficulties in e.g. steering the kite. One partial solution however comes from the weight of the cradle; because it is relatively heavy, it automatically dampens all movement, especially sharp movements.

3.4 Choice of the Camera

The camera used was a Panasonic dmc-ft4. This is a standard off the shelf compact camera with zoom lens. Since this research was conducted to establish a workflow and acted as a feasibility study, the choice of camera was mainly driven by its ruggedness and cost. Especially because of the sometimes unpredictable winds we didn't want to put an expensive camera on the line before testing the whole workflow and outcome.

4. OPERATION OF KITE AND CAMERA

4.1 Operation of the Kite

The kite can then be operated by a single person (Figure 5) although a second person is required at the end and start of the flight. Contrary to single line kites, the double or even four lined kites have to be put airborne with their lines completely out; the length of the lines cannot be changed during flight. In order to put the kite airborne, the operator has to hold the steering lines, attached to a bar or other handles, while a second person holds the kite in such a position that it can catch the airflow and already generates some pull on the steering lines. This is important because the kite's intrinsic stability and shape are created by the pull from the steering lines. Obviously, it is also important that the kite is instantly under control of the operator once it is airborne. Naturally, the kite then has to be at a distance from the operator to make sure the steering lines are already generating some pull towards the kite even before it is fully airborne.

Once the kite is airborne, the operator can easily manoeuvre it within its whole wind-window. With the operator standing downwind from the kite, the wind-window is the semi-dome that the kite can describe in function of the length of its steering lines (Figure 6). This has the advantage that a potentially large area can be covered from a single operating point; the full area within the semi-circle that the kite can describe from the operators' standing point. But it also has the potential disadvantage that the kite must take imagery at different altitudes when using the full semi-dome wind-window causing different ground resolutions in the imagery. When the kite is kept at a specific height, the flight plan describes just a semi-circle instead of including the whole

This contribution has been peer-reviewed.
doi:10.5194/isprsarchives-XL-5-W4-77-2015

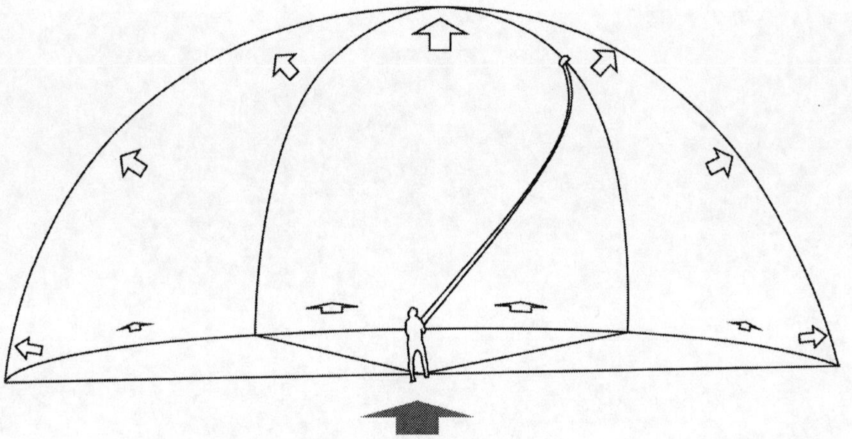

Figure 6. The wind window: grey arrow indicating the wind direction and white arrows all possible kite movement.

area beneath that semi-circle. Of course these flight characteristics may or may not be problematic and have to be considered on a site to site basis.

During fieldwork, relatively long steering lines of 60 meter were used which enabled occupying less operating points and better overall coverage of the site. The disadvantage of these longer steering lines is the increase in drag on these lines, pulling the lines with the wind. Drag is caused by the wind blowing on the surface of the lines and although special lines were used, the sheer length of the used lines still created a significant amount of surface for the wind to create a drag. One of the more obvious signs of the drag is the curve the steering lines show between the operator and the kite. Whereas shorter steering lines (without drag) will show a much more straight line. The actual disadvantage of this drag is that the steering lines follow a longer path between kite and operator, and thus enable the kite to fly lower than theoretically possible. But the biggest drawback of the drag is the dampened and delayed response of the kite to the actions of the operator, leading to somewhat less control over the kite.

4.2 Operation of the camera

In order to take pictures from the kite, one has to remotely operate it or set it to take pictures at a predetermined interval. Although the latter is the easiest and cheapest we found it has some serious disadvantages. The main disadvantage is that the operator can't see which part of the site is photographed exactly. Judging the position of the kite from the ground, the operator only has an approximate idea of the area the camera captures. Also, the exact instance the camera takes a photograph isn't known to the operator, only by approximation. To be sure that enough imagery of good quality was taken in the right direction more pictures were taken than was absolutely necessary. This meant that more flying hours were necessary.

Another disadvantage of the used operation mode of the camera was that only the default automatic mode was available when shooting in interval mode. This meant that image sensitivity, aperture and exposure time could not be set by the operator but where chosen by the camera computer anew for every picture. This of course produced imagery with slightly varying characteristics regarding to aperture and exposure time which the camera computer thought was best for the imagery at the time. In general, though, the variation in the imagery characteristics turned out to be hardly problematic. Only some motion-blur in a few images could have been prevented by setting a better exposure time. The pictures with the motion-blur were excluded from further processing and again more imagery had to be taken with the kite to overcome the problem.

4.3 Flight Plan

The flight plan was designed to fly at the same altitude as much as possible. Thus describing an arc, from left to right in the operators' perspective with the kite. Due to some variety in the wind conditions during flight, the kite had to be flown higher or lower as wind conditions were sometimes more favourable at those altitudes. This brought some variety in the ground resolution of the imagery, but, as unstable winds could be largely omitted, the camera could to some degree be held free of unwanted motion in those stable wind conditions.

Because of the length of the steering lines, only a few operating positions needed to be occupied in order to cover the whole area. As three quarries were recorded, only one operating point per quarry turned out to be necessary. This enabled the operator to completely focus on the kite and its flight plan without having to be aware of his step whilst flying the kite.

5. HIGH RESOLUTION DEM

5.1 Ground Resolution of the Imagery

Operated with two to three people the simple and low-cost setup as described above can acquire good aerial imagery (Figure 7) with a high ground resolution to be further processed to a DEM or other surface model. Because the exact height of the camera is unknown, the exact ground resolution of the imagery cannot be determined precisely. On the basis of the length of the steering lines the theoretical maximum flying altitude of the kite was 60 meters above the ground. In this case the imagery was taken with a 28mm focal length (35mm frame equivalent) which implies that the area

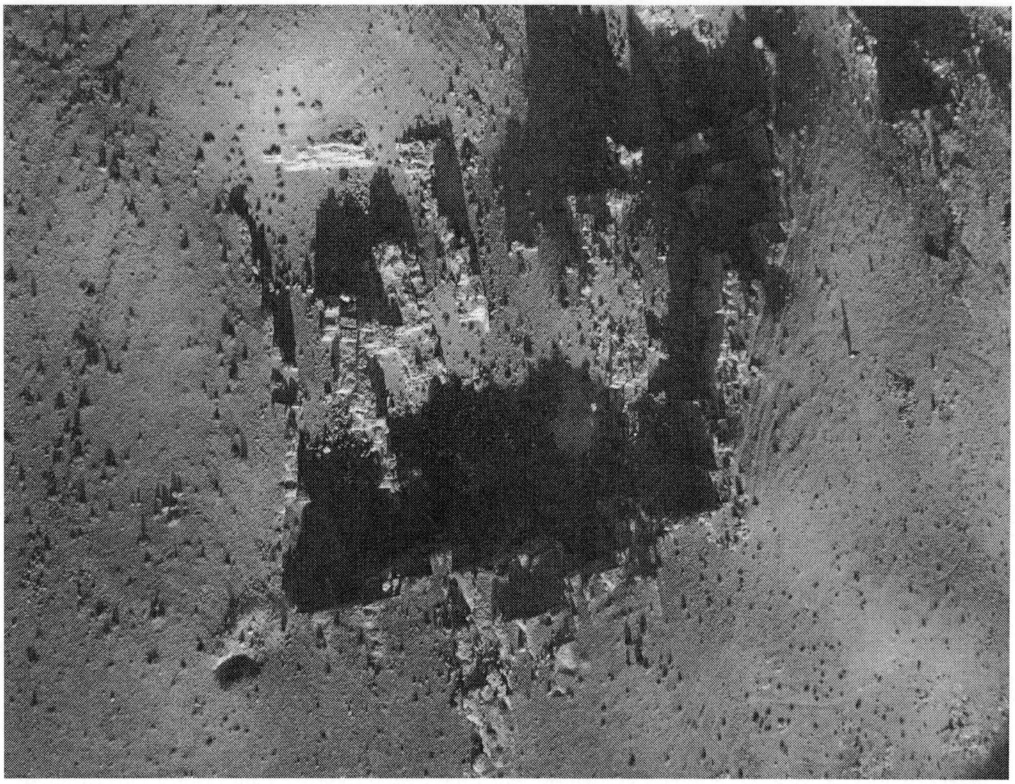

Figure 7. Typical aerial photograph of one part of the quarries.

contained within the image was 0.035m x 60m / 0.028m = 75 meters wide. With an image resolution of 4000 x 3000 pixels, the minimum ground resolution of one pixel can be estimated to be 75m / 4000 = 0.019 meters. Most images however will not have been taken at that altitude but somewhere between 30 to 50 meters. At that altitude the ground resolution will have almost doubled to approximately 1 cm. per pixel.

5.2 Ground Control Points and Accuracy

With the estimated ground resolution of the imagery between 1 and 2 cm. the accuracy of the ground control points (gcp's) optimally will have to be half of that. The topometric instrument available to us at the time of fieldwork was a total station of the type Topcon GTS-601C, with kind permission of the Centre for Nabataean Archaeology at the Al-Hussein bin Talal University of Ma'an. Through a local calibration of the instrument[1] it was found that the used total station setup was able to deliver an accuracy of 1 to 2 cm. depending mainly on wind conditions.

Ideally, the uncertainty of the measurements should be in the range of 0.5 to 1 cm. but this would require either a different setup of the topometric measurements or extensive post-processing of the measurements. Since the main focus of this research was the suitability of the platform and establishing a workflow according to it, it was decided that the current accuracy of the measurements would suffice for the time being. Improving the quality of the gcp's in a later stage of the research.

5.3 Image Processing

All imagery was processed with the Agisoft PhotoScan software to produce a point cloud and surface model of the site. The software utilises a Structure from Motion algorithm and the individual steps of the processing are already described elsewhere (e.g. Verhoeven, 2011). Although, some of the imagery from the kite did need some specific processing steps. Because of the camera shooting in automatic mode and the obliqueness (ω-pivot) of some of the imagery due to the wind, some images could not be used fully. A rather large aperture sometimes caused parts of some images to be out of focus while the main object in the picture was still in focus. The out of focus part of the picture could easily be omitted by the algorithm by using the mask-function within the PhotoScan software. The object in focus will then still be used to render the 3D point cloud and surface while the out of focus part of the image doesn't disturb the processing at all. By using the masking function of PhotoScan, more imagery could be used more easily to eventually create a realistic point cloud and surface model of the quarries.

5.4 Model Accuracy Estimation

Using the accompanying total station measurements PhotoScan not only can georeference the model, but can also calculate an accuracy estimation of the acquired model. The software needs a minimum of three gcp's to do this but the

[1] A calibration following the ISO 17123-5:2005 procedures for Electronic Tacheometers, executed by R. Emaus.

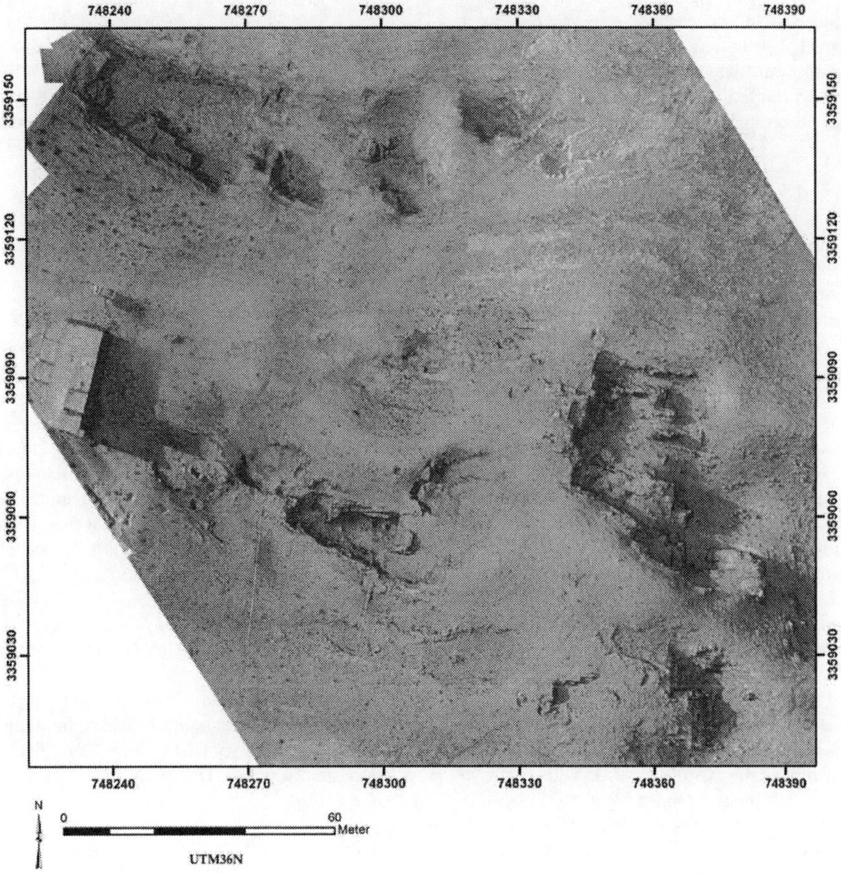

Figure 8. Georeferenced orthophoto of a part of the quarries as created with the surface model and imagery from the kite.

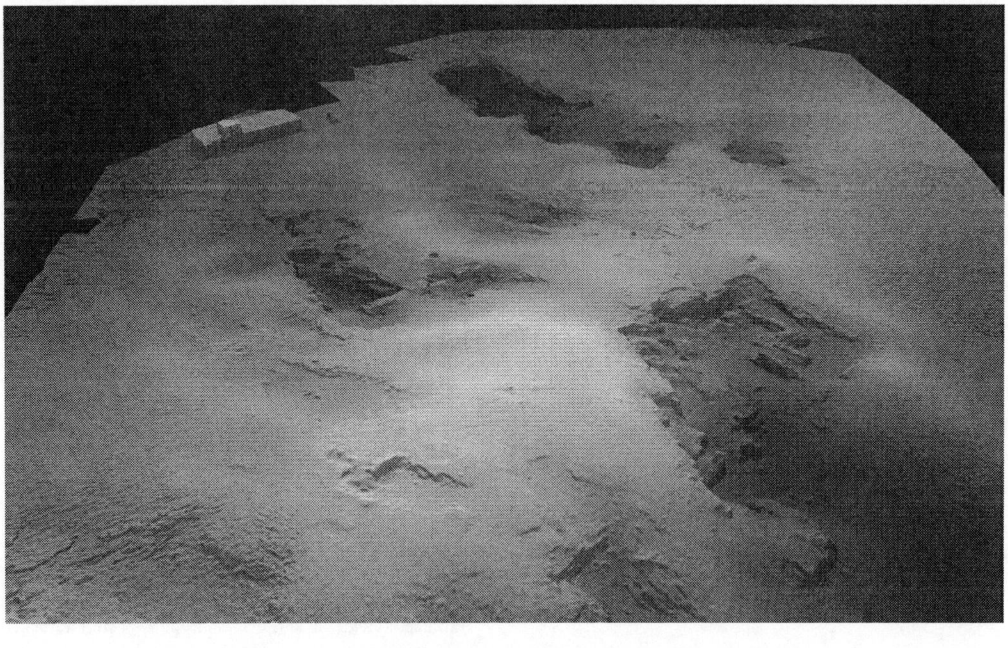

Figure 9. A rendering of the created DEM of a part of the quarries.

This contribution has been peer-reviewed.
doi:10.5194/isprsarchives-XL-5-W4-77-2015

reliability will improve with more gcp's used. When using 4 gcp's with an equal distribution over the model, a model accuracy of 1.1 cm. is calculated. This means that according to the software a high resolution surface model of the Roman quarries can now be created with a more than satisfactory accuracy.

6. CONCLUSIONS AND FUTURE WORK

The equipment and workflow as described above have proven to be able to deliver a good quality and high resolution surface model of an archaeological site showing a complex morphology (Figure 8 & 9). Since this was a feasibility study of the equipment and workflow, there is some room for improvement.

First, the workflow can be improved through the various wi-fi technologies that enable a direct link with the camera from the ground. As stated, this will reduce the time in the air because of the greater efficiency with which the imagery is taken but also because of not having to shoot in automatic mode. A direct link between the camera and the operator is therefore recommendable. Recent advances in radio but certainly low-cost and mobile wi-fi technology that are already integrated in most camera's these days will help create these optimal working conditions and workflow. When operating the camera through a direct link, a second person is probably required to operate the camera.

Second, although it was possible to create good results using the simple Panasonic camera, the results will improve significantly by using a better camera. With a full-frame camera and standard lens the image quality will improve significantly; improving the resulting DEM accordingly. Third, the accompanying total station survey that provided the gcp's will have to be improved. Certainly when having a ground resolution of less than 2 cm. a standard total station survey will no longer suffice in most cases. The accuracy of measurement will have to be improved a good deal by using reflectors on fixed mounts instead of human hold reflectors. Also, numerical post-processing of the measurements can reduce the uncertainty to millimetre levels.

Using low-cost equipment, a complex archaeological site was recorded in 3D with an accuracy of almost 1 cm. The final product couldn't have been created with only a total station survey in the same amount of time and with the same amount of detail that the above described setup has made possible. With some further recommendations made on the camera and workflow not only the eventual surface model of the site will be improved significantly, also the amount of time spent in the air will be reduced as well.

ACKNOWLEDGEMENTS

The research would not have been possible without the support of the Udhruh Archaeological Project and the support of the Faculty of Archaeology and Tourism of the Al-Hussein bin Talal University of Ma'an. We would like to thank the directors of the Udhruh Archaeological Project, Mark Driessen and Fawzi Abudanah and its members for their support in the field.

REFERENCES

Aber, J.S., Aber, S.W., Pavri, F., 2002. Unmanned small-format aerial photography from kites for acquiring large-scale, high-resolution, multiview-angle imagery. In: Morain, S., Budge, A. (ed.) *Proceedings of the XXIVth ISPRS congress: Commission I.* Denver: ISPRS.

Abudanah, F., 2006. *Settlement Patterns and Military Organisation in the Region of Udhruh (southern Jordan) in the Roman and Byzantine Periods*. Newcastle: University of Newcastle.

Driessen, M.J., Abudanah, F., 2013. The Udhruh Lines of Sight: Connectivity in the Hinterland of Petra. *Tijdschrift Mediterrane Archeologie*, 50, 45-52.

Hendrickx, M., Gheyle, W., Bonne, J., Bourgeois, J., De Wulf, A., Goossens, R., 2011. The use of stereoscopic images taken from a microdrone for the documentation of heritage: an example from the Tuekta burial mounds in the Russian Altay. *Journal of Archaeological Science,* 38 (11), 2968-2978.

Kennedy, D.L., 2004. *The Roman army in Jordan*. London: Council for British Research in the Levant.

Kherfan, A.M., 1998. *Geological Map of Bir Khidad 3150 IV.* Hashemite Kingdom of Jordan Natural Resources Authority, Geology Directorate, Geological Mapping Division. Amman: The Royal Jordanian Geographical Centre.

Killick, A.C., 1983. Udhruh. The Frontier of an Empire: the 1980 and 1981 seasons, a preliminary report. *Levant*, 15, 110 – 131.

Killick, A.C., 1986. Udhruh and the Southern Frontier. In: Freeman, P.W.M., Kennedy, D.L., (ed.) *The Defence of the Roman and the Byzantine East*. Oxford: Archaeopress, pp. 431-446.

Kouki, P., 2012. *The Hinterland of a City: Rural settlement and land use in the Petra region from the Nabatean-Roman to the Early Islamic period*. Helsinki: University of Helskinki.

Verhoeven, G.J.J., 2009. Providing an archaeological bird's-eye view: an overall picture of groundbased means to execute low-altitude aerial photography (laap) in archaeology. *Archaeological Prospection*, 16, 233 – 249.

Verhoeven, G.J.J., 2011. Taking Computer Vision Aloft – Archaeological Three-dimensional Reconstructions from Aerial Photographs with PhotoScan. *Archaeological Prospection*, 18, 67-73.

Verhoeven, G.J.J., Loenders, J., Vermeulen, F., Docter, R., 2009. Helikite aerial photography: a versatile means of unmanned, radio controlled, low-altitude aerial archaeology. *Archaeological Prospection*, 16, 125 – 138.

Verhoeven, G.J.J., Taelman, D., Vermeulen, F., 2012. Computer vision-based orthophoto mapping of complex archaeological sites: the ancient quarry of Pitaranha (Portugal – Spain). *Archaeometry*, 54, 1114 – 1129.

EMERGENCY SURVEY OF REMOTE AND ENDANGERED ARCHAEOLOGICAL SITES.

F. Fassi [a, *], C. Rossi [b], A. Mandelli [a]

[a] Politecnico di Milano, ABC Department, 20133 Milano, Italy - (francesco.fassi, alessandro.mandelli)@polimi.it
[b] Centro MUSA, University of Napoli "Federico II", 80055, Portici (NA), Italy – progettooasis@centromusa.it

Commission V, WG V/4

KEY WORDS: Close range photogrammetry, 3D reconstruction, emergency survey, archaeology

ABSTRACT:

The paper describes the survey activities of the late Roman archaeological site of Umm al-Dabadib (Egypt). The interesting case-study can be taken as an example in case of emergency surveys, as this method allows the complete 3D acquisition of a vast and complex area in a very short time and with the aid of simple instruments.

Figure 1. Panorama of Umm el Dabadib (Photo by Fassi F.).

1. INTRODUCTION

1.1 The site.

Umm al-Dabadib lies at the outskirts of the Kharga Oasis, one of the five large oases of Egypt's Western Desert. It is a vast and well-preserved Roman installation consisting of an inhabited area served by an extensive agricultural system, active in the third and fourth centuries AD. The site was abandoned at the beginning of the fifth century, and apart from a short-lived occupation dating to the early twentieth century (Beadnell 1909), it survived until today relatively intact and virtually unknown. Re-discovered in 1998 by one of the co-authors (Rossi 2000), it is now the object of a multi-disciplinary project aiming at studying in parallel both the settlement and its relating agricultural system, jointly carried out by the Politecnico di Milano and the MUSA Centre of the University of Napoli Federico II.

The site consists of three settlements (the Fortified Settlement, the Northern Settlement and the Eastern Settlement), a Fort, a temple, a church, nine cemeteries, a scatter of minor buildings and a vast agricultural installation including seven underground aqueducts and three cultivated areas. It covers a total area of over 10 square km and its remains are extremely well preserved. The historical importance of this site is still awaiting a proper evaluation: it is clearly the result of a specific, huge effort to colonise a harsh environment along what was at that time the southern border of the Roman Empire. Its isolation (it lies at a distance of over 40 km from the oasis' main town) and its adverse climatic conditions (strong winds in winter and spring and stifling heat in the summer), combined with the current lack of available water, had a double effect: on the one hand, they allowed the site to survive in excellent conditions to sixteen centuries of abandonment; on the other hand, they make working there a challenging enterprise, both from a logistic and a financial point of view. Fieldworks must be compact and intense, and not all technology can make it all the way to a place that still lies at the edge of the inhabited world.

1.2 Previous studies and surveys

Until 1998, no survey at all existed of the local antiquities. The first cursory survey was conducted basically by hand and on foot and produced a first general sketch of the existing antiquities (Rossi 2000). The relative position of the main features was marked by GPS and the rest inserted by hand; the main buildings were sketched on the basis of a few basic measurements. This first recognition allowed a rough estimate of the existing remains and represented the basis for the birth of the North Kharga Oasis Survey (NKOS), an archaeological project that run from 2001 to 2007.

Co-directed by Dr. C. Rossi (at that time Cambridge University) and Dr. S. Ikram (American University in Cairo), NKOS produced the first scientific explorations and surveys not only of Umm al-Dabadib, but also of the other sites belonging to the chain of Late Roman military-looking installations that punctuate the Kharga Oasis. The 2003 season, funded by National Geographic, was entirely dedicated to Umm al-

* Corresponding author. This is useful to know for communication with the appropriate person in cases with more than one author.

This contribution has been peer-reviewed.
doi:10.5194/isprsarchives-XL-5-W4-85-2015

Dabadib: NKOS carried out the first scientific survey of the central core of the site, an area of 1x3 km, carried out by means of a theodolite, whereas the farthest features and the long lines of the underground aqueducts were recorded by GPS (Rossi and Ikram 2006). NKOS also experimented aerial photography with the aid of a kite, which yielded significant results in terms of documentation of the nature and extent of this and other sites nearby (Ikram and Rossi 2002).
Looking back at the previous lack of information, these surveys represented a fundamental step towards a better understanding of this still elusive chain of late Roman installations. Looking forward in the direction of a more detailed investigation, they represented a starting point, in need of further refinement: they offer relatively precise information on the location of the various archaeological features, but do not contain the detailed survey of the settlements' subunits (domestic units or groups of houses). For instance, the position of the Fortified Settlement as a whole was fixed thanks to a network of points, but it was impossible to record by theodolite the position of every single wall of every single house. Such an operation would have required weeks of work and would have resulted in a very expensive and tiring expedition. As the site is unlikely to ever change its condition of isolated and remote place, and thus the problems that affect each expedition are likely to remain the same, a different technique had to be adopted to perform a more detailed survey of the local antiquities.

1.3 The current project

Umm al-Dabadib is now the focus of a specific project named OASIS (Old Agricultural Sites and Irrigation Systems), born as a spin-off of NKOS. The scope of OASIS is to study in parallel two sets of contemporary remains: the Fortified Settlement and the agricultural system. Planned, designed and built to work together (one could not exist without the other), they represent a unique chance to investigate the foundation of a self-sufficient installation in a harsh and semi-desert area. Moreover, the study of this site is also likely to yield significant information on the Late Roman policy and strategy of control and defence of Egypt's desert frontiers (Rossi 2013).

Figure 2. The photogrammteric survey of the castle tower.

The presented paper will describe the survey activities conducted during the 2014 archaeological mission, generously funded by National Geographic and Waitt Institute, with the support of MUSA and Politecnico di Milano. A complete photogrammetric terrestrial survey was performed in order to capture the 3D complexity of the place. The best strategy would have been to use a UAV, eventually integrated with terrestrial acquisitions, which would have allowed a fast acquisition from the air thus speeding up the elaboration phase. However, bureaucratic difficulties made this solution unfeasible: the UAV and others instruments were not allowed to the field and so the survey team was left to face an emergency survey, to be conducted on foot on the ground level. The only instruments at our disposal were a compact camera, a reflex camera, a total station and two photographic tripods.

2. THE SURVEY

2.1 Some methodological general considerations

Automation: the key aspect for complex, emergency cases. For the elaboration, we used Agisoft Photoscan. This well-known software (probably one of the best on the market) allows completely autonomous photogrammetric process from the orientation to the construction of a high resolution textured mesh model. It is only thanks to the automation provided by this type of software, that it is possible to tackle a survey like the one presented in this case study.
The past. The use of manual or semi-manual methods have until recently prevented or severely limited the use of an image-based approach to three-dimensional survey applied on large and complex objects. The main reason was the need to acquire a very high number of images and to collect manually a huge number of tie points.
A further problem was the camera calibration. Until recently, it was essential to use semi-metric cameras or pre-calibrate digital consumer camera off-site. This aspect has always limited "a free photographic acquisition" forcing the operator to shot at predetermined distances and using pre-calibrated fix focal values. Some methods of on-site pre-calibration allowed to set and pre-calibrate the camera on the field.
Until recently the high number of images, the manual collection of tie points and the need of camera pre-calibration have been the main reasons why the photogrammetric techniques were extremely time-consuming and challenging, normally reserved to industry experts. Some interesting case of the past, as the survey of Pompei Forum (Guidi G., et al., 2008) and the main Spire of Milan's Cathedral (Fassi et al., 2011), are clear examples that photogrammetric surveys of large heritages were possible, but required a long time, a large group of operators and sensor integration..
Substantially close-range image-based surveys of large and complex areas were virtually impossible, especially in emergency cases where short elaboration time and low costs are inalienable necessities. That has favoured the use of TLS techniques, that are far more expensive but definitely allow a simpler and fast three-dimensional real-time survey of environments, architectures and large dimension objects.
The automatic image based methods. The marriage between photogrammetry and image analysis techniques typical of computer vision have revolutionized in recent years and are still revolutionizing the way to measure objects, especially in the field of Cultural Heritage. (Barazzetti L., et al., 2009) The software born from this match basically work like photogrammetric software following the classical pipeline but using automatic methods of point recognition. This allows a complete automatism in the image orientation phase and the possibility to extract dense 3D point cloud in relative short time. In the extreme case, particularly interesting in case of emergency, the process can be entirely and automatically performed without any intervention by the operator from the beginning to the end. We no longer speak of automation of the process but of autonomous process.
This involves the possibility to orientate in a short time and without "loss of time" a substantial number of images thus

This contribution has been peer-reviewed.
doi:10.5194/isprsarchives-XL-5-W4-85-2015

allowing the acquisition of images for large environments or complex objects.
The identification of a high number of points allows a "very high observation redundancy" that normally leads to stable solutions and a simultaneous camera self-calibration (Gruen, 2001). This allows to: i) avoid pre –calibration steps, ii) to manage different cameras in the same process or iii) change camera settings during the acquisition. It is possible for example to use auto focus and different zoom levels.
Thus the acquisition camera is released by the past photogrammetric constrains retrieving the flexibility and "ease of acquisition" typical of photography. Aware of the possible consequences in terms of achievable measurement accuracy, these simplifications allow freedom of movement around the object, multi-scale acquisition, and make possible the acquisition of objects in difficult position.
Is equally important to consider the economic aspect. Surveying with a photo camera is extremely low cost.
These features make image based method ideal for Cultural Heritage and for complex areas, especially in case of an emergency.
The case of study here presented is the occasion to describe a good practice to survey also with makeshift equipment, showing as some rules of the past and few precautions can provide high quality results in terms of accuracy and completeness of the results.

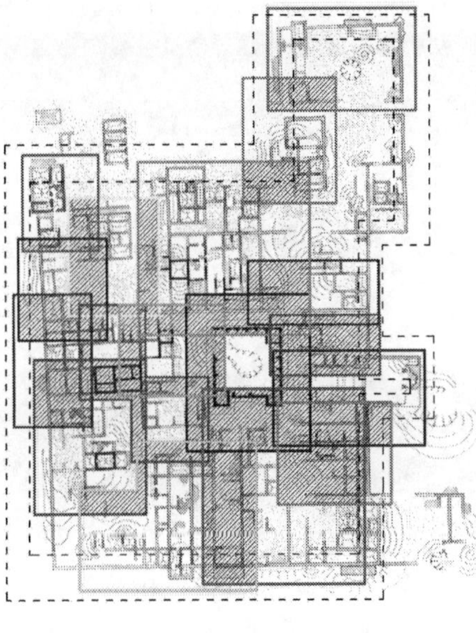

Figure 3: The subdivision of the area in the 18 sub/areas. The hatched parts are the main overlap among the different image blocks.

2.2 The area to be surveyed and its issues

The 2014 season of OASIS focussed on the Fortified Settlement. This is a compact, L-shaped, built-up area, covering an area of about 100 square metres. It is surrounded by a linear wall and crossed by long, parallel walls among which the various domestic units were inserted. No open-air streets have been found (apart from a passage all around the Fort): the settlement was served by long vaulted passages, some running one on top of the other. This means that the settlement was basically solid: this arrangement allowed the best protection from the strong sun and sand-laden winds that batter the area.
Over the centuries, the compact design of the Fortified Settlement generated a solid an undulating mass of ruins: the upper portions of the houses still stand, but parts collapsed and created tall mounds and deep depressions. It is possible to walk among the mounds corresponding to groups of buildings and even to carefully climb on them, but this situation adds to, rather than reduces, the complexity of the object to be surveyed. The uneven ground, the strong and changing light, and the time constrain required the team to subdivide the work into small portions and to engineer *ad hoc* solutions to solve specific problems, as we shall see below.

2.3 The survey phase

2.3.1 The survey design. Due to extension of the archaeological site, the natural choice was to divide all the area in sub/areas. Each area corresponds to a specific building. The subdivision is designed *ad hoc* in order to complete the survey with a sufficient overlap with neighbour areas. Further subdivision in micro-areas is done to complete the "interiors" of the buildings or specific interesting object as arcs, ceilings and pillars. An external, circular take of images around the perimeter was also done in order to materialize a closed external constrain. This created photogrammetric block is used to check the global alignment. (View table 1)
Some practical tricks in order to avoid lack of data or difficulties during the bundle adjustment are:

- to use ups and downs of the terrain to acquire object from progressive raised positions. In this case the natural mounds of rubble were very useful to acquire the upper parts of the buildings.
- to design the acquisition during the hours of noon when the sun creates fewer shadows on the ground and on the structures.
- to speed up the acquisition of the single area in order to avoid dramatic changes of light;
- to take care to framing of every shots, avoiding, if possible, large chunks of sand and sky on the images that could cause problems during image orientation phase.

2.3.2 Topographic survey. For a complex structure, it is be very important to have a good net of GCP useful to georeference all data together, to scale correctly the single models and to have sufficient check points to estimate the accuracy of the reconstruction. (Fassi et al., 2007) In archaeological and architectonical case studies, this continues to be a valid paradigm even if today many procedures are becoming more and more automatic, quite autonomous and user-friendly, and seems to promise perfection without any intervention.
In many emergency cases, one does not necessarily have a topographic instrumentation at one's disposal, and maybe the accuracy and precision of the survey are secondary to the need to take home a result as complete as possible and as quickly as possible. For this reason, one can follow simple tricks which allow to obtain results close to the true without following, by necessity and compulsion, ideals procedures.
We fortunately had a total station, but without additional necessary instrumentation (no tripods, topographic bases, tricuspids and prisms) in order to build a right triangulation network. A good photographic tripods was converted on site for topographic instrumentation and the other one was transformed in "target reference tripods" to materialize all the networks

doi:10.5194/isprsarchives-XL-5-W4-85-2015

points. In order to assure the verticality we use a bubble level tied to the tripod legs with normal tape. A distinctive cross mark on the tripod was used as target point.

The initial idea was to materialize a close network around the central castle but the presence of huge amount of sand and the understandable difficulties allowed us to materialize a network only in the southwest portion of the field thus allowing the detection of at least three sides of the village and of two sides of the main castle. The network was created as a closed triangle using two point on the same ground level and one more point in a higher position. A global 3D accuracy of about 4 mm was obtained. The areas covered by the network are used as reference block to georeference and scale all the other blocks together.

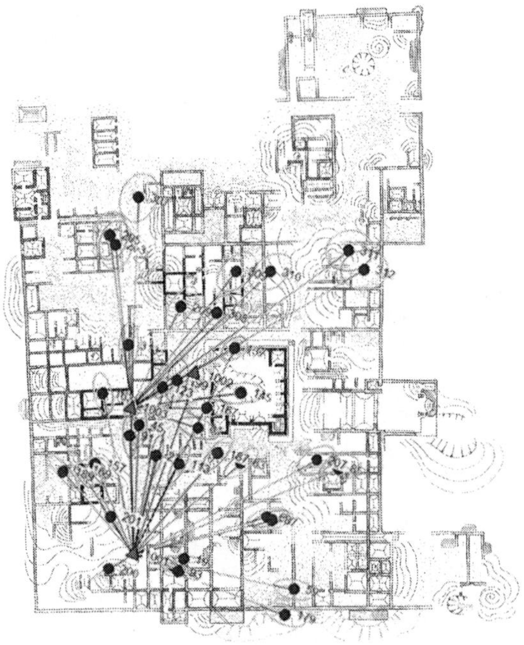

Figure 4. The topographic network. The red triangle are the tree base station.

2.3.3 The photogrammetric acquisition. The best capture geometry to adopt in this case is a closed circular acquisition around every "building". This helps, like a closed traverse, to reduce automatically image orientation errors and false inner orientation parameters calculation without the use of GCP or other geometric external constrains. It is how to create a closed traverse but only made of images.

It is very important, if possible, to use targets because necessary in the photogrammetric elaboration even if you do not have at disposal a topographic instrumentation. You can build or draw and materialize them very quickly and in many ways even using makeshift materials. Where impossible, it is strongly recommended the use of physic architectonical points picked manually on the images as reference points. This happened in particular in the presence of high inaccessible structures

Targets should be well-positioned on the structures. It means that: i) they should be distributed quite uniformly on the structures, ii) they should be seen by a sufficient amount of images iii) in that areas where the geometry is particularly complex and the automatic processes can fail. Targets help a lot in particularly complex areas to tie up discontinued structures and badly illuminated areas. If distributed uniformly in the areas, they act as external constrains to the process of the image orientation and allow a better calculation of lens distortion parameters in the auto-calibration process.

In total, we shot 5100 images using a Canon 5D MarkIII with fix 35mm F2 lens and one Canon G1X. The reflex camera was mainly used for the main buildings and for the largest areas, due to the high resolution (22,3 Mp), the high quality lenses, and the major configuration possibilities. The compact one was used for the smaller building, for small details and when it was necessary to climb some structures or enter in some narrow places.

Every image blocks was pre-oriented in low resolution during the night to verify on site the quality of the image orientation and the completeness of the areas covered by the survey.

Even if the elaboration software provides a result to achieve a high-quality results in terms of accuracy, some photographic practical rules must be followed. (Nocerino et al, 2014) These rules were valid in the past and are even more essential today, because the process is nearly completely automatic and the operator cannot intervene in the process of identification of points. For these reason, even more than in the past, the image quality should be superb.

	N° of images	Total Error (Pix)	Total Error (m)	N° of GCP (Target)	N° of aGCP (architectonical)
Castle	**509**	**0.343**	**0.003**	**10**	
Little House	**197**	**0.336**	**0.001**	**3**	
The Church	121	0.256	/	2	3
Mountain vessels	257	0.290	/	2	4
Building 1	**182**	**0.739**	**0.008**	**7**	
Building 2	**260**	**0.777**	**0.008**	**11**	
Building 3	**412**	**0.252**	**0.004**	**7**	
Building 4	**444**	**0.601**	**0.006**	**17**	
Building 5	**265**	**0.353**	**0.002**	**4**	**1**
Building 6	**176**	**0.308**	**0.005**	**4**	**1**
Building 7	**123**	**0.229**	**0.001**	**3**	
Building 8	**158**	**0.611**	**0.010**	**3**	
Big Building	**1051**	**0.093**	**0.005**	**4**	**4**
North side	126	0.188	/		6
North ruins	226	/	/	7	
West ruins	93	0.184	/	4	4
Church ruins	157	0.483	/	2	4
Perimeter	**341**	**1,214**	**0.017**	/	**10 (common with other blocks)**
Total	5098				

Table 1. Summary table of the diverse image block. In bold characters the blocks covered by total station survey used as reference for the whole area. On the right columns the number of target GCP and manually added architectonical ones (aGCP).

During the acquisition at Umm El Dabadib all images were taken in raw format. This implies to post elaborate the images but allow working on shadows areas, lightning some details and in general increasing contrast and clarity of the images. (Guidi et al., 2014). The best choice would have been to work with fix focal lens, but due to the extension of the area, the extremely tight timetable and the sudden changes of light we decided to speed up the acquisition. In order to avoid dramatic changes of

This contribution has been peer-reviewed.
doi:10.5194/isprsarchives-XL-5-W4-85-2015

focal and achieve a better precision, we tried to maintain the same average distance to the object using always hyper-focal. This is a right way to maintain greater depth of field and limit focal changes
Averagely the capture distance was around 10-15 meters for building acquisition and 5-6 meters for the details. Only for the central castle was used a largest distance, circa 15-20 m. The choice was taken for two reason: i) to facilitate the acquisition in a single block of the high castle walls and ii) to frame in the foreground all the surrounding structures useful for the automatic recording of the blocks. The sensor resolution and the planned capture geometry assured a medium GSD (Ground Sample Distance) of 1mm for front objects (object of the survey) and a max of 5 mm for back objects (useful for registration with others block). The images were acquired with a mean overlap between two consecutive normal views of about 80% to guarantee the automated identification of homologous features and assure a minimum number of intersecting rays for improving reliability. The mean baseline between adjacent images was about 1.5 m with a max of 5 m for more distant object.

Figure 5. Orthoimage of the external façades of the castle. In transparency the internal areas. It is clearly possible to study the level of internals remains and buildings techniques.

The average base/distance ratio was of 0.1. When in the presence of long walls, as for external wall block, convergent images were also acquired assuring a B/D ratio of 0.5. In the presence of sharp edges, the image base was reduced. For more simple building was used a circular convergent capture geometry centred in the hypothetical centre of gravity of the object and moving the camera of circa 10°.

Figure 6. An interesting "Point Cloud view" of the central castle from the actual entrance.

2.4 The automatic elaboration process.

The orientation. As already said we used Agisoft Photoscan for all the process. All the elaboration were conducted using full resolution images for orientation. Each image block was processed separately and optimized using GCP recognized manually in a second phase. This was conducted without assigning the coordinate acquired by total station. These data were added in a third phase to georeference and scale the model. In this way accidental error coming from topography did not affect the orientation. In two cases only, due to a lack of data, some GCPs were added manually before the orientation process in order to compute the incomplete orientation.
In a fourth step all block were aligned together sequentially, starting from the southwest block and gradually adding one by one all the neighbouring blocks. The alignment of the blocks was made using the "point-based" method that uses only image matched points to scale and rigidly align all the block together. The final accuracy was checked step by step using the deviation among common GCPs and, where missing, visually extracting small sections of point cloud to inspect any misalignments.
At the end all, the internal blocks were positioned together with the external view that runs all around the area. This was a sort of final global check of the global alignments. The mean georeferencing errors on the common GCPs was 15mm with a max error of 21 mm.
The model. The dense matching was done in high quality, which means an image size reduction by the factor of 4. Due to the high initial image resolution the dense point cloud have a final resolution of circa 5 mm for a total of circa 650 million of points. It is a sort of very high density LIDAR point cloud from which it is possible to extract 3D models, orthophotos, videos and interesting views and representations.

3. RESULTS AND FUTURE APPLICATIONS

The results are impressive. The feasibility of the survey of each single building was quite clear from previous work (Fassi at al., 2013) and literature (Gonizzi, 2013; Forte 2013).. The general accuracy achieved by the whole process, instead, was not that obvious. Following the described rules of good practice, the general final precision is sub-centimetre and uniform in all areas.
Moreover, the survey is complete and the point model has very high resolution, significantly higher than the classical resolution suitable for architectural representation typical in these cases.
In the specific case of Umm al-Dabadib, a photogrammetric 3D survey represented the quickest and most efficient way to record the huge mass of information yielded by the Fortified Settlement, thus allowing the project to use in a more efficient way the significant financial resources that would have been absorbed by a classic theodolite survey. The same method could be adopted to survey endangered archaeological sites, when a quick and precise documentation of the visible evidence represents the only work that archaeologists might ever be able to perform.
Among future applications, there is the survey of indoor environments. The use of fisheye lens and nets of panorama images are being studied and tested to be used on the next survey mission in Umm el Dabadib (25 January -5 February 2015) . Fisheye survey will results very useful to speed up and simplify the acquisition of particularly narrow areas; panoramas could be very useful for very large environment.
Needless to say that the use of a UAV would have facilitated and speeded up the survey, even if a survey from the ground would still be necessary to integrate the aerial data. The

This contribution has been peer-reviewed.
doi:10.5194/isprsarchives-XL-5-W4-85-2015

impossibility to survey from a highest point of observation is the only negative aspect of this experience. In fact, it was impossible to survey the top of the farthest buildings from the castle.

The collected data may represent the basis for a number of possible operations. First, the data themselves represent a faithful representation of the current state of the archaeological remains in a certain moment; this can be later used to monitor the natural decay of the ruins or, in the worst case, of the damages inflicted by vandals. A comparison between these data with old surveyed data could be also useful to understand how the place has changed in the last 10 years: by comparing the old village map on figure 3 and figure 4 with the orthoimage in the figure 7, it is immediately clear that many areas have disappeared in just a few years.

The survey can be used to study the layout of the settlements, the alignment of the various buildings, as well as the design of the various units. Finally, these data can also be used to prepare a digital reconstruction of the original appearance of the site, to be used both for academic purposes and for popular publications.

Figure 7. Orthophoto of the complex. The images is visualized as a depth map to better understand the different elevation of the buildings.

4. CONCLUSIONS

The survey technique described in this article proved to be efficient and successful. The possibility to use simple instruments and to quickly obtain the raw data make it the ideal survey technique to be used in the case of logistically challenging sites as well as endangered archaeological sites. This technique allows an unparalleled combination of precision and speed, as well as the collection of a large number of raw data that can later be processed and reused for different scopes.

In 2014 we performed the 3D survey of the Fortified Settlement, and we are planning to extend the survey to the Eastern and Northern Settlements as well. The same method, however, could be extended to other archaeological features of the site of completely different nature: the cultivations. The ancient fields are barely visible from the ground, and their layouts can only be spotted for a few minutes at dawn and sunset when the sun is low; surveying them from the ground is out of the question, and would anyway yield a poor result. An aerial survey performed with a drone, instead, is likely to offer precise and detailed results, that would be otherwise impossible to obtain. We hope to be able to experiment on this front and to report our results on a later occasion.

ACKNOWLEDGEMENTS

We wish to express our gratitude to the National Geographic Society and the Waitt Institute for their generous grant in support of the 2014 expedition. We also wish to thank Dr. C. Achille (Politecnico di Milano) and Prof. S. Mazzoleni (Centro MUSA) for their support and advice. Thanks to the Supreme Council for the Antiquities of Egypt for allowing our work; to the Chief Inspector of the Kharga Oasis Mr Ahmed I. Bahgat and to our field inspector Mr Ahmed Moussa for their support on the field; to Mr Magdy Aly for his collaboration as Project's Representative in Egypt; to Mr Ahmed Moussa and Ms Marwa Fawzy (Pan Arab Tours) for their logistic and personal support; and last but not least, to the expedition's supporting team: Hamdy, Walid and Muhammad and Shukri.

REFERENCES

Beadnell, H. J. Ll., 1909. An Egyptian Oasis, London 1909.

Barazzetti L., Remondino F., Scaioni M., 2009. Combined use of photogrammetric and computer vision techniques for fully automated and accurate 3D modeling of terrestrial objects. In Proc. SPIE 7447, Videometrics, Range Imaging, and Applications X, 74470M (September 03, 2009); doi:10.1117/12.825638; http://dx.doi.org/10.1117/12.825638

Fassi F., 2007. 3D modeling of complex architecture integrating different techniques – a critical overview. In: Int. Archives of Photogrammetry, Remote Sensing and Spatial Information Sciences. Zurich, Switzerland, vol. XXXVI-5/W47

Fassi, F., Achille, C., Gaudio, F., and Fregonese, L., 2011. Integrated strategies for the modeling very large and complex architectures, Int. Arch. Photogramm. Remote Sens. Spatial Inf. Sci., XXXVIII-5/W16, 105-112, doi:10.5194/isprsarchives-XXXVIII-5-W16-105-2011.

Fassi F., Fregonese L. Ackermann S., De Troia V., 2013. Comparison between laser scanning and automated 3D modelling techniques to reconstruct complex and extensive cultural heritage areas. ISPRS Workshop 3D-ARCH: 3D virtual reconstruction and visualization of complex architectures, Trento, 25 26 February 2013

Forte, M, Dell'Unto, N., Haddow, S., Lercari, N., 2013. 3D Documentation at Çatalhöyük: New Perspectives for Digital Archaeology. Proceedings of Digital Heritage 2013, Marseille, 2013, IEEE. October, 2013.

Gonizzi Barsanti, S., Remondino, F., and Visintini, D., 2013. 3D surveying and modeling of archaeological sites – some critical issues –, ISPRS Ann. Photogramm. Remote Sens. Spatial Inf. Sci., II-5/W1, 145-150, doi:10.5194/isprsannals-II-5-W1-145-2013.

Gruen, A., Beyer, H. A., 2001: System calibration through selfcalibration. In Calibration and Orientation of Cameras in Computer Vision, Gruen and Huang (Eds), pp. 163-193.

Guidi, G., Remondino, F., Russo, M., Menna, F., Rizzi, A., Ercoli, S., 2009. A Multi-Resolution methodology for the 3D modeling of large and complex archaeological areas. International Journal of Architectural Computing, Vol. 7(1), pp. 39-55

Guidi, G., Gonizzi, S., and Micoli, L., 2014. Image pre-processing for optimizing automated photogrammetry performances, ISPRS Ann. Photogramm. Remote Sens. Spatial Inf. Sci., II-5, 145-152, doi:10.5194/isprsannals-II-5-145-2014.

Guidi, G., Remondino, F., Russo, M., Rizzi, A., Voltolini, F., Menna, F., Fassi, F., Ercoli, S., Masci, M.E., Benedetti, B., 2008. A multi-resolution methodology for archeological survey: the Pompeii Forum. Proc. of 14th Int. Conference on Virtual Systems and MultiMedia (VSMM 2008), pp. 51-59, Limassol, Cyprus

Haukaas, Colleen, 2014. New Opportunities in Digital Archaeology: The Use of Low-Cost Photogrammetry for 3D Documentation of Archaeological Objects from Banks Island, NWT. University of Western Ontario - Electronic Thesis and Dissertation Repository. Paper 2117.

Nocerino, E., Menna, F., and Remondino, F., 2014. Accuracy of typical photogrammetric networks in cultural heritage 3D modeling projects. Int. Arch. Photogramm. Remote Sens. Spatial Inf. Sci., XL-5, 465-472, doi:10.5194/isprsarchives-XL-5-465-2014, 2014.

Remondino F., Del Pizzo S., P. Kersten T., Troisi S., 2012. Low-Cost and Open-Source Solutions for Automated Image Orientation – A Critical Overview. In: Lecture Progress in Cultural Heritage Preservation. Notes in Computer Science Volume 7616, pp 40-54

Rossi, C., 2000. Umm el-Dabadib, Roman Settlement in the Kharga Oasis: Description of the Visible Remains. With a Note on ‘Ayn Amur. MDAIK 56, pp. 235-52.

Rossi C., 2013. Controlling the borders of the empire: the distribution of Late-Roman ‘forts’ in the Kharga oasis, in Bagnall R., Davoli P, and Hope C. (eds.) The Oasis Papers 6, Proceedings of the Sixth Conference of the Dakhla Oasis Project, , Dakhla Oasis Project Monograph 15, Oxford: Oxbow, 2013: 331-6.

Rossi, C. and Ikram, S., 2006. North Kharga Oasis Survey 2003. Preliminary Report: Umm el-Dabadib. MDAIK 62, pp. 279-306.

Ikram S. and Rossi, C., 2002. Surveying the North Kharga Oasis. KMT 13(4), pp. 72-9.

Figure 8. The south view of the settlement extracted from the high-resolution point cloud.

Figure 9. Northern Prospect of the Fortified Settlement (slightly transparent).

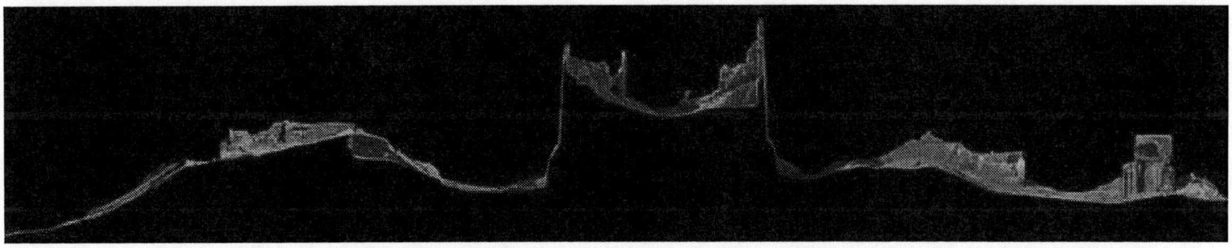

Figure 10. East-West Section of the Fortified Settlement.

This contribution has been peer-reviewed.
doi:10.5194/isprsarchives-XL-5-W4-85-2015

FIRST EXPERIENCES WITH KINECT V2 SENSOR FOR CLOSE RANGE 3D MODELLING

E. Lachat, H. Macher, M.-A. Mittet, T. Landes, P. Grussenmeyer

Photogrammetry and Geomatics Group, ICube Laboratory UMR 7357, INSA Strasbourg, France
(elise.lachat, helene.macher, marie-anne.mittet, tania.landes, pierre.grussenmeyer)@insa-strasbourg.fr

Commission V, WG V/4

KEY WORDS: Kinect v2 Sensor, RGB-D camera, Range Imaging, Calibration, Point Cloud, Assessment

ABSTRACT:

RGB-D cameras, also known as range imaging cameras, are a recent generation of sensors. As they are suitable for measuring distances to objects at high frame rate, such sensors are increasingly used for 3D acquisitions, and more generally for applications in robotics or computer vision. This kind of sensors became popular especially since the Kinect v1 (Microsoft) arrived on the market in November 2010. In July 2014, Windows has released a new sensor, the Kinect for Windows v2 sensor, based on another technology as its first device. However, due to its initial development for video games, the quality assessment of this new device for 3D modelling represents a major investigation axis. In this paper first experiences with Kinect v2 sensor are related, and the ability of close range 3D modelling is investigated. For this purpose, error sources on output data as well as a calibration approach are presented.

1. INTRODUCTION

Capturing three dimensional environments and objects is a necessary task for many applications in different fields. This is nowadays widely made possible thanks to various technologies such as laser scanners, stereo-vision or triangulation-based systems for example. Regarding some aspects like price or computation time, the recent RGB-D cameras offer new possibilities for the modelling of complex structures particularly in cultural heritage documentation.

This type of sensors presents several benefits demanded initially by the gaming and entertainment industry. They are easy to use and some of them are low-cost compared with laser scanners. Regarding these advantages, their use was extended to many other areas such as 3D scene reconstructions or pattern recognition (Kolb et al., 2009) as well as in the robotic field (May et al., 2006).

Kinect v1 and Kinect v2 are based on two different imaging principles. As the accuracy of Kinect v1 limits its use for some engineering measurement tasks, Kinect v2 will probably give better results considering the new technology.

The aim of this paper is first to present the new sensor and its functionalities. In a second part, several tests are carried out and a calibration approach is proposed in order to improve output data. This will highlight not only the errors related to environment and captured object but also errors related to the system itself. Finally, the benefits of the new sensor are listed.

2. KINECT V2 SENSOR PRESENTATION

2.1 Characteristics

Even though Kinect for Windows v2 sensor relies on a different technology than Kinect v1, it also allows the acquisition of three different output streams. It is composed of two cameras, namely a RGB and an infrared (IR) camera, from which just one can be seen on the left side of the sensor (Figure 1). The active illumination of the observed scene is insured by three IR projectors.

Figure 1. Kinect v2 sensor on a photographic tripod

The RGB camera captures color information with a resolution of 1920x1080 pixels, whereas the IR camera is used for the real-time acquisition of depthmaps and also IR data with a 512x424 pixels resolution. The whole acquisitions can be carried out with a framerate up to 30 Hz. A last feature to be mentioned is the field of view for depth sensing of 70 degrees horizontally and 60 degrees vertically.

The technical specifications provided by Microsoft announce an operative measurement range from 0,5 m to 4,5 m. This last characteristic will be verified through different experiments presented in this paper.

To enable the use of the sensor for developers and researchers, the official Microsoft SDK 2.0 (Software Development Kit) is free downloadable. It provides not only the drivers, but also a set of functions or code samples that can be used for own implementations. One should also note the introduction of the

This contribution has been peer-reviewed.
doi:10.5194/isprsarchives-XL-5-W4-93-2015

Kinect Fusion tool in this SDK, which was initially released by Microsoft for the Kinect v1 because of the success of its device. This tool enables thanks to ICP algorithms the direct creation of 3D meshes under different formats (.stl, .obj or .ply) by using the Kinect as a scanning device moved slowly around the object.

2.2 Distance Measurement Principle

As mentioned on Microsoft website and on a Canesta patent, the Kinect v2 sensor is based on time-of-flight principle, whereas the previous Kinect device uses structured light to reconstruct the third dimension. Even though time-of-flight range imaging is a recent technology, many books deal with its principles and its applications (Grzegorzek et al., 2013, Hansard et al., 2012, Kolb and Koch, 2009, Remondino and Stoppa, 2013). Moreover in the last decade, a few companies developed range imaging cameras based on this measurement principle, which can be direct or indirect. The distinction between these two different measurement types is based on the light source used, either light pulses (direct ToF) or amplitude modulated light (indirect ToF). Nevertheless, the basic principle stays unchanged: knowing the speed of light, the distance to be measured is proportional to the time needed by the active illumination source to travel from emitter to target.

As Canesta made use of indirect time-of-flight before Kinect v2 sensor development, we can assume that this methodology also appears in the new sensor investigated here. Indeed, direct time-of-flight requires very costly clocks to accurately determine the runtime of single pulses. Therefore it seems unlikely that such a component be part of a low-cost sensor. For indirect time-of-flight systems based on light modulation, a phase shift between emitted and received signal is measured, replacing the direct measurement of runtime. In that case, the estimated distance between sensor and captured object depends on the determined phase shift $\Delta\varphi$ by equation (1) (Kolb and Koch, 2009).

$$d = \frac{\Delta\varphi}{4\pi f} \cdot c \tag{1}$$

where f = modulation frequency
c = speed of light

It should be noticed that, unlike other time-of-flight sensors, it is impossible to act on the modulation frequency or on the integration time of the input parameters with Kinect v2 sensor.

2.3 Point Cloud Acquisition

Sensors based on the digital imaging technology introduced above deliver a measurement of the distances between an entire scene and the sensor. These distances values are directly stored in a matrix which size corresponds to the depth sensor resolution. Indeed, for each pixel of the 512x424 depth images, the measuring device estimates in real-time a distance value to the corresponding object point. This output data is the so-called depthmap (see an example Figure 2), a two-dimensional image from which the 3D-coordinates of the scene can be calculated. After a few post-processing steps, it is then possible to obtain indirectly point clouds of the captured scene or object.

Assuming that Kinect sensor uses a standard lens, camera intrinsic parameters such as focal length f and principal point coordinates (c_x, c_y) can be determined. The Z-coordinate for a pixel is already known because it corresponds to the value stored in the depthmap. Using these parameters and the perspective projection relationship (see Dal Mutto et al., 2013), a 3D point X in the camera coordinate system can be mapped from the homogenous image coordinates of a pixel $p = [u, v, 1]^T$ through equation (2). As each pixel of the depthmap represents a value for the corresponding point cloud, the calculated point clouds count a number of 217088 points.

$$\lambda p = KX = \begin{bmatrix} f_x & 0 & c_x \\ 0 & f_y & c_y \\ 0 & 0 & 1 \end{bmatrix} \begin{bmatrix} X \\ Y \\ Z \end{bmatrix} \Leftrightarrow \begin{cases} X = \frac{u - c_x}{f_x} \cdot Z \\ Y = \frac{v - c_y}{f_y} \cdot Z \end{cases} \tag{2}$$

Because of the recent release of the Kinect v2 sensor, so far only a little amount of algorithms allows an off-the-shelf use of the sensor. However, the Microsoft SDK (Software Development Kit) provides multiple functions for the recording of the different data types. Respecting the fact that we don't control the mapping functions used by Microsoft to generate point clouds from depthmaps, we developed our own acquisition protocol. Implemented in C# through some code samples available in the SDK, this program enables us to acquire timed acquisitions for each of the different possible output data (RGB images, infrared images, depthmaps, and colorized point clouds).

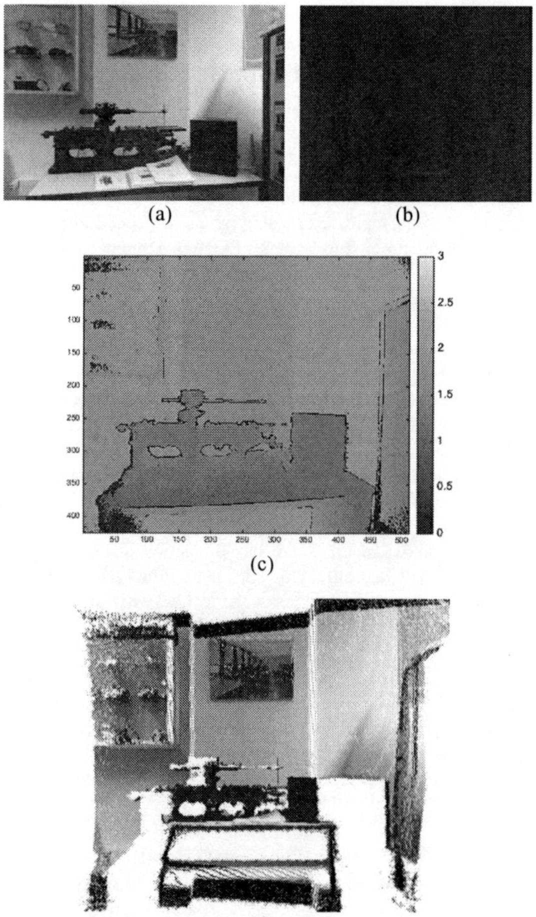

Figure 2. Output data: (a) RGB image; (b) infrared image; (c) colored depthmap (colorbar in meters); (d) colorized point cloud

This contribution has been peer-reviewed.
doi:10.5194/isprsarchives-XL-5-W4-93-2015

3. INVESTIGATION OF SYSTEM PROPERTIES

A good knowledge of sources of errors affecting the measurements of a system is needed to quantify the accuracy of the data provided by it. Even if a large number of these sources are interdependent, we can however try to classify them into four groups, as it is done for terrestrial laser scanners (Reshetyuk, 2009): instrumental errors, errors related to the scanned object, environmental errors and methodological errors. Instrumental errors are usually systematic errors assigned to the system design. They should potentially be removed by system design improvement or by calibration. The environment in which the acquisitions are performed has an influence (temperature, brightness, humidity, etc.), as well as the characteristics of the observed object (e.g. albedo, reflectivity, geometry). Once the camera is on its tripod, the user intervention is limited to choosing settings (e.g. time interval, types of output data), therefore the automation level seems to be relatively high. However, errors can occur according to the distance-to-object chosen for the acquisition, the incidence angle, etc. The point cloud quality depends also strongly on the algorithms used for creating the output data (mapping functions, calibration parameters…). It is essential to determine the influence of these parameters on the quality of the distance measurement. To do that, a few tests are presented in this section.

3.1 Influence of Frames Averaging

Since individual frame acquisitions suffer from noise inherent to the sensor and its technology, averaging successive frames is a possible improvement step to overcome this phenomenon. This part deals with the influence of the number of acquisitions on the precision of the final measurement. Considering a static scene, several repeated measurements with 500 ms interval are carried out from one unique camera location. A statistical analysis over a large sample of 500 acquired depthmaps has been performed. We also analyzed the behavior observed with 20, 50, 100 and 200 successive depthmaps.

First, acquisitions have been made from an office scene which mainly contains objects of squared shape (Figure 3a). Using the acquired depthmaps, for each pixel the mean value and the standard deviation of distance measurements were calculated. For this non planar scene, the computed standard deviations shown in Figure 3b are particularly wide at the objects boundaries, but also on reflective surfaces such as metallic parts. This phenomenon can be explained by the measurement noise which appears in these specific cases. It can also be associated to speckle noise present in infrared data. Moreover, considering the different sizes of samples, one should notice a low impact on the computed standard deviations from the 50 depthmaps. Indeed, for 20 successive frames the result seems to be less smooth than for the other samples sizes. However the effect observed at objects borders does not change a lot between the 50 and 500 samples. While frames are added, the main differences arise at far ranges (from about 3 meters), where depth inhomogeneities are reduced and the visual results smoothed. We will see in section 4 that far range measurements are not reliable. For this reason, they will not be treated and considering 50 successive depthmaps seems to be enough.

Secondly, the same test was repeated by positioning the Kinect device in front of a planar wall, at a distance of approximately 1,15 m. Histograms of the distance measurements performed for a central area of one pixel and its neighbors (3 x 3 pixels) were realized for the different samples mentioned before. Figure 4 shows the result obtained with 100 successive measurements. In that case, the standard deviation is lower than 1 mm (approximately 0,7 mm). It can be noticed that this value is almost constant for all sizes of samples we considered. In conclusion, as for the experiment presented above, a larger sample does not provide a better precision. The future tests will thus be carried out with a smaller amount of acquisitions.

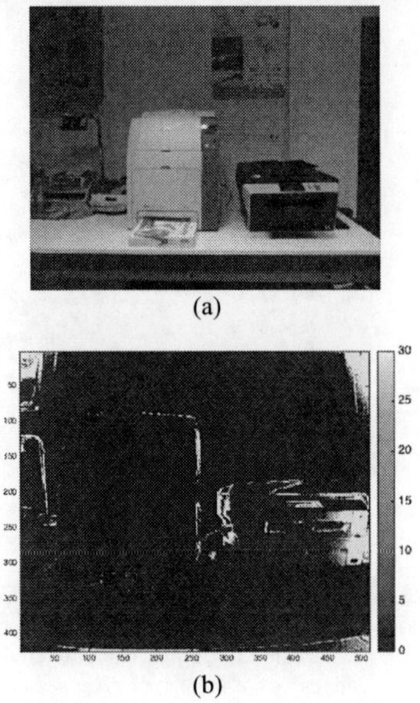

(a)

(b)

Figure 3. (a) Acquired scene; (b) Standard deviations calculated for each pixel of 100 successive depthmaps (colorbar in cm)

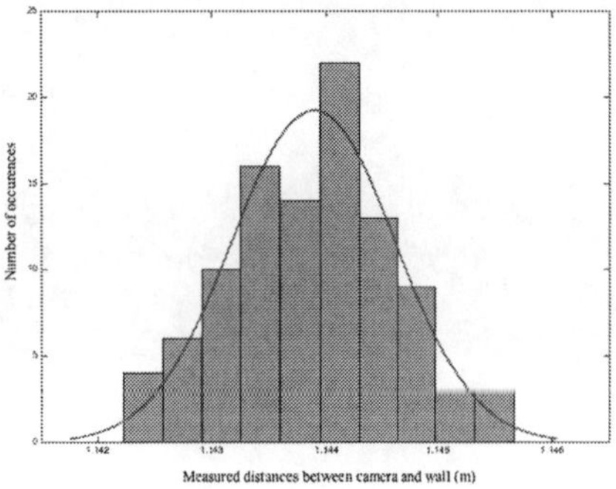

Figure 4. Histogram of 100 distance measurements performed for a 3 x 3 pixels central area

3.2 Pre-heating Time

Previous studies have shown that some RGB-D sensors need a pre-heating time before providing reliable range measurements (Mittet et al., 2013). For checking this phenomenon, the time delay necessary for Kinect v2 to achieve constant measurements has been determined. The sensor was placed parallel to a white planar wall and measurements were carried out each 15 seconds during one and a half hour (360 depthmaps recorded). For each recorded depthmap, the distance from the sensor to the wall was determined in a central area of ten per ten pixels. This area was chosen according to the maximal intensities in the corresponding infrared data.

This contribution has been peer-reviewed.
doi:10.5194/isprsarchives-XL-5-W4-93-2015

This test permits to establish the value of distance measurements as a function of time. Figure 5 shows that the distance varies from 5 mm up to 30 minutes and becomes then almost constant (more or less 1 mm). It can be noticed that the sensor's ventilator starts after 20 minutes. For future tests, the pre-heating time will be respected even if the distance variation is quite low.

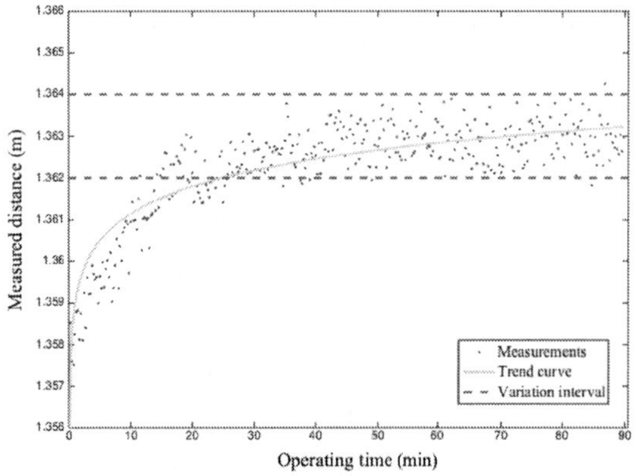

Figure 5. Value of distance measurements as a function of time

3.3 Influence of Materials and Colors

To assess the effects of different materials on intensity as well as on distance measurements, samples characterized by different albedos and roughness were measured. The sensor was placed parallel to the plane regrouping the samples (Figure 6a).

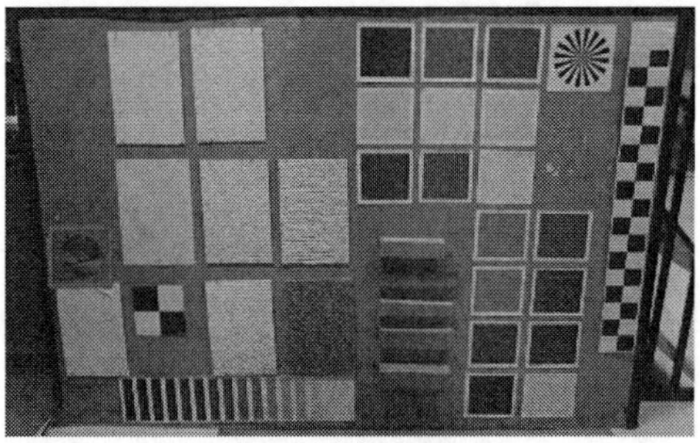

(a)

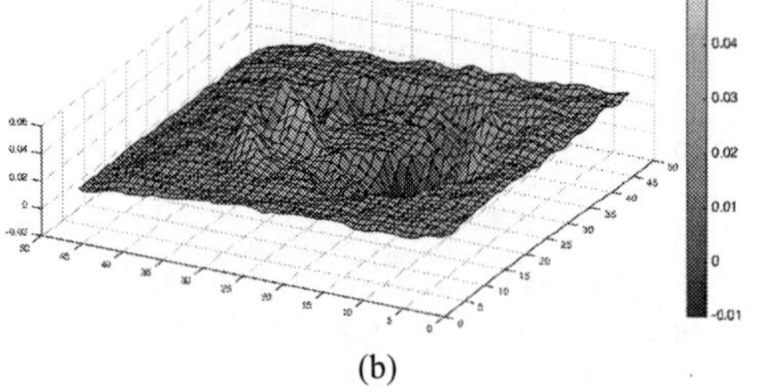

(b)

Figure 6. (a) Samples characterized by different albedos and roughness; (b) zoom on the red frame: depth variations (m) observed on this area including a compact disk

As shown in Figure 6b and Figure 7, reflective and dark materials stand out among all the samples. Indeed, the intensity provided for very reflective as well as very dark surfaces is particularly low. Thus, the corresponding distances in the depthmaps are larger than expected. This effect in the depthmap is mainly important for a compact disk, i.e. high reflective material as illustrated in Figure 6b. For this sample, distances vary up to 6 cm. It is also visible in Figure 7, which presents a checkerboard of grey levels and the corresponding intensity image. The intensity increases with the brightness of the object (except when it is also very reflective or transparent). On the white parts of this checkerboard, a speckle phenomenon appears and illustrates the measurement noise.

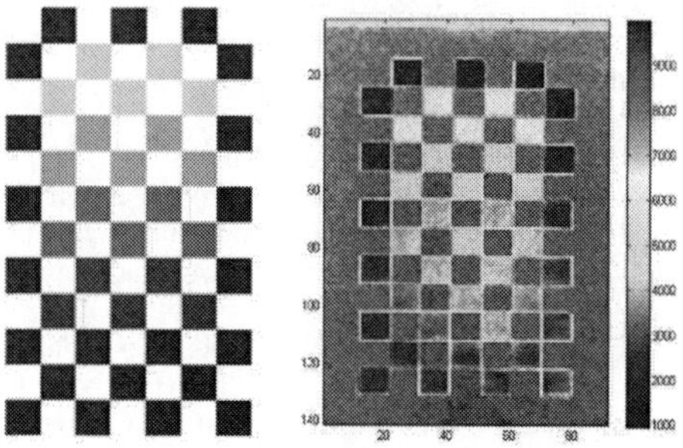

Figure 7. Checkerboard of grey levels (left) and corresponding intensity image (right)

Concerning the level of roughness of the grainy textures on the test panel in Figure 6a, their influence is not clearly visible because of the noise inherent to the point cloud.

3.4 Outdoor Efficiency

As the previous Kinect was not adapted for sunny outdoor acquisitions, the influence of the brightness conditions on the measurements was also studied. Outdoor acquisitions were performed during a sunny day.

This experiment shows that the sensor is able to work during a sunny day provided that the light does not directly illuminate the sensor. Indeed, strong backlighting conditions cause sensor's disconnections from the computer.

Considering the quality of the acquisitions carried out during a sunny day, two phenomena appear: the number of "flying pixels" increases particularly on the edges of the sensor field of view and the number of points decreases with the light intensity. The "flying pixels" effect is shown Figure 8.

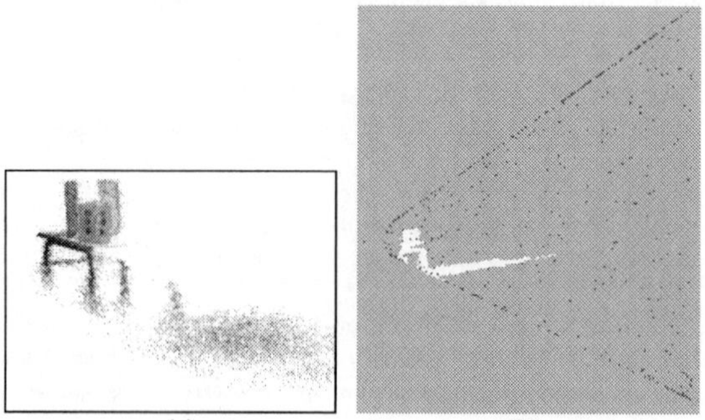

(a) (b)

Figure 8. Point cloud acquired in sunny conditions: (a) scene without "flying pixels"; (b) scene (white) and "flying pixels" (red) in a profile view

This contribution has been peer-reviewed.
doi:10.5194/isprsarchives-XL-5-W4-93-2015

4. GEOMETRIC AND DEPTH CALIBRATION

As the lenses used in time-of-flight cameras are standard, the acquired data suffer from distortion effects that can be estimated and corrected through successive calibration steps. Since we hope to use the Kinect v2 sensor as a measuring device, calibration should enable the acquisition of reliable 3D metric information from the acquired point clouds. Because of the specificity of range imaging cameras, the distinction between two successive approaches has to be made.

4.1 Geometric Calibration

It appears in lots of works e.g. (Hansard et al., 2014) that time-of-flight cameras can be geometrically calibrated with standard methods. For this purpose, as for common 2D sensors, several images of a planar checkerboard were taken under different points of view. It is worth noting that the infrared data were used to handle this geometric calibration, because depth and infrared output streams result from the same sensor.

To determine the necessary intrinsic parameters, our dataset was treated by a Camera Calibration Toolbox proposed by (Bouguet, 2013) under the Matlab software. This algorithm is mainly based on well-known camera models. However, one should underline the fact that changing or removing some of the images makes the computed results vary from a few pixels. This phenomenon is only due to the low sensor resolution. As a matter of fact, best calibration results are considered regarding the lower uncertainties on the parameters.

Note that the Microsoft SDK mentioned earlier also provides a function that returns all these intrinsic parameters (namely *GetDepthCameraIntrinsics()* function). Deviations towards self-computed results are shown in Table 9.

	Self-computed calibration		SDK function	Deviations
	Values *(pixels)*	*Std.*	Values *(pixels)*	*(absolute values)*
Focal length (x)	364,7	*1,6*	366,0	1,3
Focal length (y)	366,1	*1,5*	366,0	0,1
Principal point (x)	255,8	*2,4*	258,6	2,8
Principal point (y)	203,7	*3,1*	206,5	2,8
	Distortion parameters (radial, then tangential)			
K1	0,08708	*0,01976*	0,09357	0,00649
K2	-0,16515	*0,10795*	-0,27394	0,10879
K3	-0,00321	*0,00365*	0,09288	0,09609
P1	-0,00345	*0,00286*	n.a.	
P2	0	*0*	n.a.	

Table 9. Estimated intrinsic parameters using self-calibration, with Microsoft function, and related deviations

The application of these distortion coefficients directly on our point clouds computation algorithm allows the gaining of calibrated three dimensional dataset, which is no more affected by the actual distortions of the initial depthmap. It can also be noticed that the point clouds obtained by applying directly the Microsoft mapping functions seem to be pre-calibrated at least for radial distortions of the lens. Indeed, considering acquisitions of a planar surface, the results are visually similar between a point cloud computed through our method with corrections and a point cloud directly obtained with mapping functions. This phenomenon appears especially on the corners of the captured plane, which are not right angles but rather distorted. As a matter of fact, the intrinsic parameters stored in the device are certainly considered in the point clouds mapping functions.

Moreover, a geometric calibration could also be processed for the RGB sensor, which uses a different lens with a different resolution. However, as we do not need the colorimetric information for the rest of the experiments, this calibration has not yet been performed.

4.2 Depth Calibration

A systematic depth-related deformation, also depicted as wiggling error by some authors (e.g. Lindner et al., 2010) can be observed while working with time-of-flight devices. This deformation is due in part to inhomogeneities in modulation process. Lots of works have been devoted to the understanding of error sources in order to minimize their effects. As a matter of fact, calibration of time-of-flight sensors was a major issue in many studies in the last decade. Most of the time, the introduced approaches are based either on Look-Up Tables (Kahlmann et al., 2006) or on curve approximations with B-splines (Lindner et al., 2010). In both cases, the aim is the storage of depth errors depending on the measured distance.

A common way to assess the distance inhomogeneity consists on positioning the camera parallel to a white planar wall at different well-known distances. In our study, the wall has been surveyed beforehand with a laser scanner (FARO Focus3D) in order to assess his planarity with a device of assumed higher accuracy as the investigated sensor. After that, a line has been implanted by tachometry perpendicularly to the wall. Marks have then been fixed on this line at predetermined distances, from which the acquisitions were made with the Kinect placed on a static tripod, as shown in Figure 10.

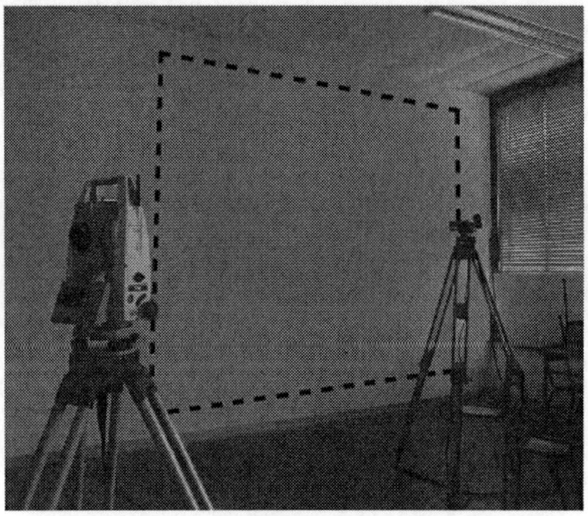

Figure 10. Depth calibration experiment
(observed area in dotted lines)

The marks were implanted at ranges from 1m up to 6m from the reference wall, with 25 cm steps to progressively move the sensor away from this wall. An additional position at 0,8m was also set up, even though Microsoft announces a minimal range of 0,5 m. Indeed, measurements acquired at a range of 0,8 m already present a high standard deviation compared to other ranges, as can be seen on Figure 11. This phenomenon will be more important for measurements at a closer range. Moreover to limit the influence of noise, at each sensor position 50 successive depthmaps of the wall were acquired with intervals

This contribution has been peer-reviewed.
doi:10.5194/isprsarchives-XL-5-W4-93-2015

of one second. Finally, to insure the parallelism between sensor and wall and thus reduce the possible rotation effect, the distances were also surveyed by tachometry at each new position, using little prisms at the two sensor extremities.

Inspired by the work of (Lindner & Kolb, 2006), a two steps distance calibration based on B-splines approximation was carried out. Firstly, the accuracy of distance measurement for a small central matrix of 10x10 pixels has been investigated. Secondly, the deformations extended to the whole sensor array have been assessed, in order to highlight the need of processing a potential per-pixel correction.

4.3 Results

First aim of this experiment was to determine the deviations between real and measured distances as a function of the depth, to produce the result presented in Figure 11. As depicted in this graph, distortions of the averaged central area vary from about 0,5 cm to 2,7 cm. Knowing the curve parameters allows the correction of the measured distances. Another interesting information arises from the observation of the standard deviations calculated over the 50 depthmaps acquired per station. As it can be seen, the standard deviation increases with the range, which means that the measurements become more scattered around the estimated distance when the sensor moves away from the scene. Moreover, at the nearest range of 0,8 m acquired in this test, the standard deviation is higher than for all other sensor positions. As a matter of fact, measurements realized at the minimal announced distance of 0,5 m would probably be still less reliable. Secondly, an important variation of about 1,5 cm at range 4,5 m is observed. This observation confirms the upper value for distance measurement reliability given by Microsoft. The future acquisitions will be limited to this suitable range, observing that in this case the distance deviations are reduced to an interval of about 1 cm.

Looking at this graph, it can also be noticed that despite satisfying results all deviations are positive, meaning that the sensor would systematically measure too short distances. It seems that a systematic offset occurs because of the constant distance (approximately 2 cm) between fixing screw and lens. Indeed, the Kinect was accurately placed on the tripod with respect to its fixing screw, which definitely does not correspond to the optical center of the lens.

In order to eliminate this systematic effect, we suggest subtracting the values of the nearest range (0,8 m) to the values of each respective position. Thus, observing calculated intervals with respect to a first station also affected by the systematic offset provides values no more depending on the unknown misalignment. The realized graph presents the same trend curve as Figure 11, but with values varying between about 5 and -5 mm for ranges from 0,8 to 4,5 m.

Finally, plotting the measured distances over the real distances provides a very high correlation coefficient of r = 0,999. According to the measurement task, it is possible to neglect the depth calibration step.

The second part of depth calibration consists on an extension of the analysis to the whole sensor array. For this purpose, for datasets of the planar wall acquired before at the ten first positions (from 0,80 m to 3 m), fitting planes were computed by means of least-squares adjustment. To fit correct planes in the Kinect point clouds, some outlying points corresponding to null values in the depthmaps and thus very far from the wall were not considered. Then, the Euclidean distances between a point cloud and its corresponding plane are calculated. It provides a residuals matrix which represents for each pixel of the considered depthmap the deviation between approximating plane and actual measured distance. This representation is shown in Figure 12a for a distance of 1,25 meters with deviations in millimeters. In addition, the histogram of the residuals obtained for this range is also presented (Figure 12b). It appears that the values are normally distributed and almost centered on the null value, with a standard deviation of about 4 mm. It can be mentioned that the standard deviations calculated for the fitted planes increase with the range from about 3 mm up to 16 mm for the latest range of 3 meters.

Moreover, as it can be seen in Figure 12a, the deviations with respect to the adjusted plane are more important at image boundaries and especially on the corners. As known from the literature, these radial effects increase with the distance to the wall. In our case, they can reach variations of a few centimeters up to tens of centimeters in the corners at 3 meters range.

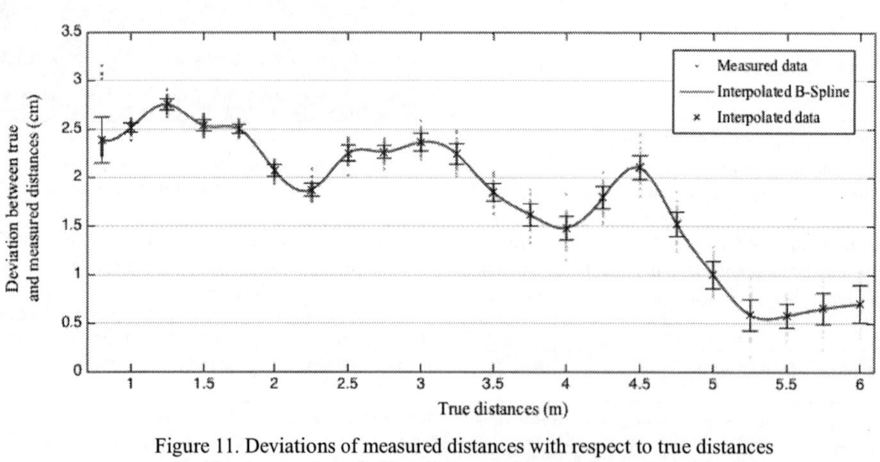

Figure 11. Deviations of measured distances with respect to true distances (without correction of offset distance between fixing crew and lens)

This contribution has been peer-reviewed.
doi:10.5194/isprsarchives-XL-5-W4-93-2015

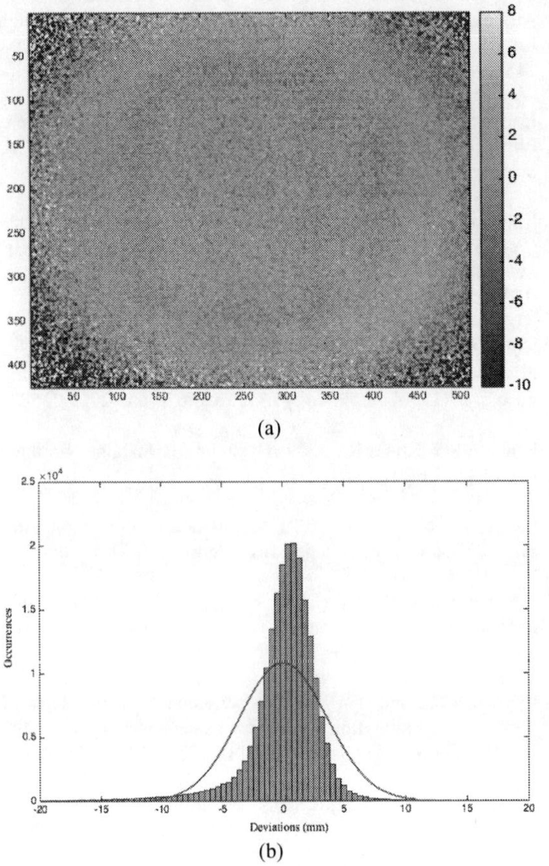

Figure 12. Residual distances between points and a fitted plane at the range of 1,25 m: (a) map with colorbar in millimeters; (b) corresponding histogram

Because of this loss of precision and reliability of the measurements at the boundaries, a first approach to reduce depth errors of entire sensor consists on considering only a restricted central array. Tests were realized with a reduced depthmap size of approximately 360x300 pixels from the 512x424 initial frame. Obviously, the area was chosen very central to remove border effects. The same phenomenon can always be detected at the boundaries of the new considered area but with widely reduced residuals. They reach a few millimeters at the range of 1,25 m. In addition, the standard deviations calculated for the fitted planes on the shorter areas were reduced from about 2 to 5 mm.

Nevertheless, a correction approach for the entire sensor area as proposed in (Pagliari et al., 2014) can be adapted to the device investigated here. It consists on delivering signed corrections interpolated from the residuals matrices computed at different ranges. For each pixel of the array, given the distance value it contains, a correction value will be interpolated considering the nearest ranges for which residuals were computed. The obtained correction matrix then just needs to be added to the initial depthmap in order to apply the per-pixel correction.

A two-way conclusion arises from these observations. On the one hand, using the inspected sensor for high accuracy metrology tasks justifies the need of a per pixel correction managed over the whole sensor. Nevertheless, this correction step is time-consuming due to the local treatment of each pixel. Thus, on the other hand, people who foresee to use the Kinect for Windows v2 for lower-accuracy object reconstruction should locate their object on a central and restricted area of the sensor.

5. DISCUSSION

5.1 Main Advantages of Kinect v2

Thanks to the change of technology carried out by Microsoft for the release of its recent Kinect for Windows v2 sensor, the depth measurements achieved are more accurate compared to the first Kinect device. In addition, the computed point clouds present a better resolution because a distance measurement occurs for each pixel of the captured depthmaps. This allows a more accurate detection of small objects for example. The quality of color images has been also increased. These are the two main technical advantages of Kinect v2 compared to Kinect v1.

Tests about the influence of measurements repetition have shown that the averaging procedure has not necessarily a great influence on the final accuracy of the measurements. Indeed, the noise inherent to the sensor is already reduced in comparison to the first Kinect device, and is essentially present at objects boundaries where artifacts are hardly avoidable. Moreover, it has been shown that outdoor acquisitions can be envisaged. Kinect v2 device seems to be much less sensitive to daylight since point clouds were obtained even on a very sunny day. However, some filters need to be applied to overcome the increased noise effects and to reach exploitable results.

Another advantage is the simultaneous release of Kinect sensor and of its official SDK, which allows development of solutions for many tasks such as algorithms for 3D reconstruction.

Nonetheless, a few material drawbacks related to the use of the Kinect v2 sensor itself can be mentioned, such as the necessity of working under a Windows 8 or 8.1 station equipped with USB3 port. The required additional power input makes it also less mobile for SLAM applications for example.

5.2 Future Works

To complete our current work, it would be interesting to check the potential of the Kinect v2 device for small object modelling purposes. In this context, point clouds of a limestone fragment (approximately 25 x 20 cm) have already been captured. The acquisitions were made under different points of view all around the stone. Visual results look very satisfying, as it can be seen on Figure 13. The multiple point clouds (eight different positions) of the object now need to be registered in order to obtain a complete 3D point cloud. A first approach will consist on a manual segmentation of each individual point cloud in order to remove the artefacts (or flying pixels) that are clearly visible at the edges and boundaries of the object. Once the final mesh of this object will be produced, an accuracy assessment will be carried out thanks to comparisons with reference meshes acquired by other imaging techniques.

This contribution has been peer-reviewed.
doi:10.5194/isprsarchives-XL-5-W4-93-2015

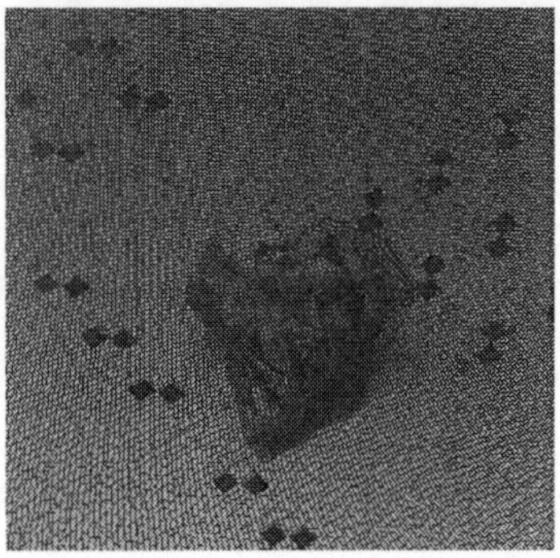

Figure 13. Point cloud of a limestone fragment obtained with Kinect v2

In a second time, tests will be performed using the Kinect Fusion tool mentioned before. It enables to directly acquire a meshed representation of the investigated object. This approach has the advantage to use much less point clouds as the manual approach and is a faster modelling method. However the model accuracy achieved through this method must be assessed. For sure, it depends on the way the Kinect is moved around the object.

6. CONCLUSION

The aim of this paper was to investigate through a set of tests the capacity of the Kinect v2 sensor, a low cost sensor, and to study if it could be an alternative to close range laser scanners for 3D measurements.

Several tests highlighted errors related to environment and to the properties of the captured object but also errors related to the system itself. Moreover, geometric as well as depth calibrations were carried out. Based on the obtained results, the physical limitations of the device could be assessed.

Considering depth measurement precision and outdoor efficiency, the achieved results look promising. Accordingly, the arrival of the second Kinect version can be seen as a real progress for computer vision tasks dealing with range imaging cameras.

In the future, the potential of the Kinect v2 device for small object modelling purposes will be assessed in a qualitative and a quantitative way.

REFERENCES

Bouguet, J.Y., 2013. Camera Calibration Toolbox for Matlab. http://www.vision.caltech.edu/bouguetj/calib_doc (2 Dec. 2013).

Dal Mutto, C., Zanuttigh, P., & Cortelazzo, G.M., 2013. Time-of-Flight Cameras and Microsoft Kinect™. Springer, Berlin, pp. 108.

Grzegorzek, M., Theobalt, C., Koch, R., & Kolb, A., 2013. Time-of-Flight and Depth Imaging - Sensors, Algorithms and Applications. Springer, Berlin, pp. 319.

Hansard, M., Horaud, R., Amat, M., & Evangelidis, G., 2014. Automatic detection of calibration grids in time-of-flight images. Computer Vision and Image Understanding, 121, 108-118.

Hansard, M., Lee, S., Choi, O., & Horaud, R., 2012. Time of Flight Cameras: Principles, Methods, and Applications. Springer, Berlin, pp. 95.

Kahlmann, T., Remondino, F., & Ingensand, H., 2006. Calibration for increased accuracy of the range imaging camera SwissRanger™. Image Engineering and Vision Metrology (IEVM), 36(3), 136-141.

Kolb, A., Barth, E., Koch, R., & Larsen, R., 2009. Time-of-Flight Sensors in Computer Graphics. Computer Graphics Forum, 29(1), 141–159. doi:10.1111/j.1467-8659.2009.01583.x

Kolb, A., & Koch, R., 2009. Dynamic 3D Imaging. Springer, Berlin, pp. 117.

Lange, R., & Seitz, P., 2001. Solid-state time-of-flight range camera. Quantum Electronics, IEEE Journal of, 37(3), 390-397.

Lindner, M., & Kolb, A., 2006. Lateral and depth calibration of PMD-distance sensors. In: *Advances in Visual Computing*. Springer Berlin Heidelberg, pp. 524-533.

Lindner, M., Schiller, I., Kolb, A., & Koch, R., 2010. Time-of-flight sensor calibration for accurate range sensing. Computer Vision and Image Understanding, 114(12), 1318-1328.

Luhmann, T., Stuart, R., Kyle, S., Boehm, J., 2013. Close-range Photogrammetry and 3D Imaging. De Gruyter, Berlin, pp. 702.

May, S., Surmann, H., Perv, K., & Augustin, D.-S., 2006. 3D time-of-flight cameras for mobile robotics. Proceedings of the IEEE/RSJ International Conference on Intelligent Robots and Systems, pp. 790–795. doi:10.1109/IROS.2006.281670

Mittet, M.-A., Grussenmeyer, P., Landes, T., Yang, Y. and Bernard, N., 2013. Mobile outdoor relative localization using calibrated RGB-D cameras. In: *8th International Symposium on Mobile Mapping Technology*, pp. 2–7.

Pagliari, D., Menna, F., Roncella, R., Remondino, F., & Pinto, L., 2014. Kinect Fusion improvement using depth camera calibration. In: *International Archives of the Photogrammetry, Remote Rensing and Spatial Information Sciences*, 40(5), 479-485.

Remondino, F., & Stoppa, D., 2013. TOF Range-Imaging Cameras. Springer, 240 pages.

Reshetyuk, Y., 2009. Terrestrial laser scanning: error sources, self-calibration and direct georeferencing. VDM publishing house ltd, 161 pages.

This contribution has been peer-reviewed.
doi:10.5194/isprsarchives-XL-5-W4-93-2015

INVESTIGATING 3D RECONSTRUCTION METHODS FOR SMALL ARTIFACTS

V. Evgenikou, A. Georgopoulos

National Technical University of Athens, School of Rural and Surveying Engineering, Lab. of Photogrammetry
Zografou Campus, Heroon Polytechniou 9, 15780, Zografou, Athens, Greece
vevge@hotmail.com, drag@central.ntua.gr

Commission V, WG V/4

KEY WORDS: Laser Scanning, SfM, 3Dmodels, Artifacts

ABSTRACT:

Small artifacts have always been a real challenge when it comes to 3D modelling. They usually present severe difficulties for their 3D reconstruction. Lately, the demand for the production of 3D models of small artifacts, especially in the cultural heritage domain, has dramatically increased. As with many cases, there are no specifications and standards for this task. This paper investigates the efficiency of several mainly low cost methods for 3D model production of such small artifacts. Moreover, the material, the color and the surface complexity of these objects id also investigated. Both image based and laser scanning methods have been considered as alternative data acquisition methods. The evaluation has been confined to the 3D meshes, as texture depends on the imaging properties, which are not investigated in this project. The resulting meshes have been compared to each other for their completeness, and accuracy. It is hoped that the outcomes of this investigation will be useful to researchers who are planning to embark into mass production of 3D models of small artifacts.

1. INTRODUCTION

Modern technological advancements have enabled the production of three dimensional models of small objects and artifacts of cultural heritage value. At the same time the demand for such three dimensional reconstructions is increasing as they are used in a variety of applications in archaeological studies, education, tourism and many more. Today many different techniques are available for completing this task effectively. These techniques range from image based methods to scanning hardware and software using either laser or common light. Their implementation cost also spreads over a wide price range and so do the results as far as accuracy, completeness and reliability are concerned. The success rate of all these methods depends on the equipment used, on the methodology itself and, of course on the object properties. By this its shape complexity, surface color and material are implied. This issue has not been extensively investigated in the past albeit for larger objects (Toschi et al. 2014); hence this paper is trying to add another piece of relevant information to the puzzle of the multitude of methods and of hardware and software capable of producing three dimensional models of artifacts of different shape and nature.

2. MOTIVATION

As digital technologies advance rapidly, digital libraries for cultural heritage assets have evolved, e.g. Europeana (www.europeana.eu). However, up to now only 2D material, i.e. images are stored in these repositories. The first ever 3D object uploaded in Europeana is the 3D model of a Byzantine Church in Cyprus (Sofocleous et al. 2006). In the last years all the more museums and cultural heritage repositories of any kind demand the mass 3D digitization of monuments and artifacts. This is by no means a trivial task, as all these items present a variety of shapes, a multitude of surface colors, a diversity of materials.

In addition, for many years terrestrial laser scanning has shown a rapid progress and its use spread rapidly over many applications including documentation of cultural heritage assets (Gonzo et al., 2007; Koska & Kremen, 2013; Mandferdini & Galassi, 2013). However, in the last decade photogrammetric algorithms have been automated with the contribution of the computer vision community and hence image based techniques have gained the lost ground in the competition for point cloud generation. It is believed that perhaps they already have certain advantages over scanning methods as they are more cost effective, sometimes faster, equally efficient and accurate and they are capable of providing texture to the 3D models (Skarlatos & Kiparissi 2012). Hence, when it comes to mass production of 3d models of artifacts, users are facing a dilemma as to which method is most suitable to use.

Figure 1: The collection of different artifacts tested (from left to right: Brass, Marble, Plaster, Mahogany, Wax, Wood, Porcelain, Bronze, Plastic, Metal, Clay)

For that purpose ten different artifacts were carefully selected presenting a multitude of geometry and shape complexity, of texture and a variety of materials (Figure 1). They included small ornamental objects with surfaces made of metal, plastic, wax, wood, marble, plaster and other materials or combination thereof. Some of them had shiny surfaces, a fact which presented additional problems. The main purpose of the project was to investigate the efficiency of the various methods tested to produce the three dimensional surface. Hence image texture was not included in the investigation as it is mainly related to the properties of the digital camera. Except for the marble base in case an object was standing on a base of different material, this was excluded from the investigation.

3. METHODOLOGY

3.1 3D data acquisition methods

Nowadays 3D reconstruction usually implies point cloud generation and processing. Technological advances have enabled the development of a lot of methods for 3d data acquisition. The methods available today for producing three dimensional models are many, diverse and mainly could be categorized into two main groups. Passive methods record the electromagnetic energy, i.e. visible light, emitted by the objects to be documented. Active methods, on the other hand, emit electromagnetic energy, e.g. laser beams, infrared light etc., and record its reflection form the object (Remondino & El-Hakim, 2006; Bianco et al, 2013).

Several of these methods, representative of the wide range have been selected for the test. Laser and structured light scanners, web based services for image based scanning, a commercial software applying Structure-from-Motion (SfM) algorithm and a contemporary total station with scanning capabilities were employed.

Seven different methods were used to generate 3D point clouds of the artifacts chosen. Those methods belong to two main categories. The first category includes the image based methods (IBM) i.e. passive methods, which are either automated or semi-automated, while the second category includes a variety of active scanners. Specifically:

- *Automated image based methods (IBM)*: 123DCatch software by Autodesk® (now replaced by ReCap), Arc3D web service, MS Photosynth software
- *Semi-automated image based methods (IBM)*: Photoscan software by Agisoft®
- *Structured Light Scanner*: SL2 by XYZRGB® Inc.
- *Triangulation based 3D laser scanner*: Next Engine®
- *Time-of-flight 3D laser scanner*: Topcon® IS imaging station

All objects were scanned in such a way in order to produce the three dimensional surface in the most efficient way for the comparison to be objective

3.2 Image Based Methods

Theoretically, the generation of a point cloud can be achieved faster and cheaper through the image based methods (IBM), while the quality and accuracy of the result depends on image scale and pixel size. Practically, image based methods are proven to be more efficient while facing light-absorbing and dark colored materials, but less efficient while facing items with low textured surfaces (Remondino & El-Hakim, 2006). Specular and/or very smooth materials have negative effect to the quality of the point cloud, regardless of the method used. In an attempt to overcome those problems, most objects were sprayed with chalk and/or talc powder in order to create texture and to reduce shine and disturbing reflections.

Image acquisition was very carefully planned, in order to ensure best results and objectivity. The object was placed on an elevated stool and two Bowens studio flashes were employed in order to achieve the best lighting conditions (Figure 2). The images were taken with a Canon EOS 1D MIII full frame DSLR with a resolution of 21Mpixel and a fixed lens of 50mm nominal principal distance. Also, the image overlap was 80-90%, the photographed object appeared focused in the biggest part of each capture, the angle between the shots was less than 10 degrees and the surrounding area remained stable. For ensuring even illumination of the objects a set of two Bowens® studio flashes was also employed.

Figure 2: Image acquisition

The number of the captured images for each object varies and depends on the characteristics of each artifact. Blurry images were excluded and the rest were loaded to the software for processing. The results were obtained a few hours later. An example of the procedure described above appears in Figure 3.

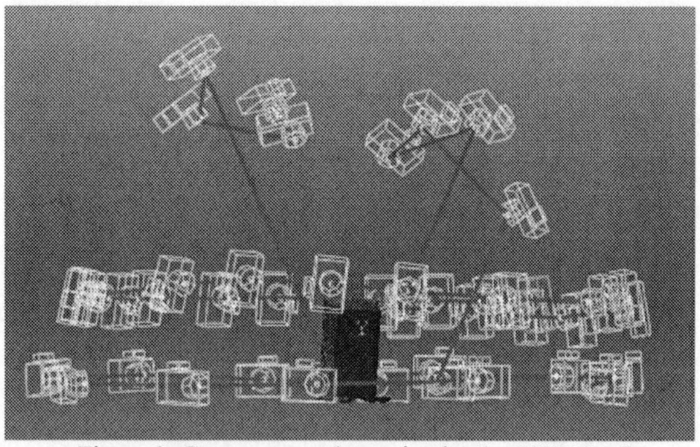

Figure 3: Camera pose determination by 123DCatch

The images acquired for each artifact were uploaded to the 3 web based services, i.e. 123DCatch, Arc3D and Photosynth. The point cloud registration was performed by the service itself, as the user has practically no control. Similarly, in the case of

This contribution has been peer-reviewed.
doi:10.5194/isprsarchives-XL-5-W4-101-2015

Agisoft Photoscan, the images were aligned by the software and the best possible point cloud and dense mesh were produced.

3.3 Scanning

As already mentioned three different scanners were used for data acquisition. In each case the system parameters were set in such a way as to achieve the highest possible accuracy and density of the generated clouds.

3.3.1 SL2 XYZRGB® Structured Light Scanner

The SL2 structured laser scanner by XYZRGB® (Valanis et al. 2009, Valanis et al. 2010) is composed by a projector and two cameras (DSLR or Machine Vision) and is connected to a computer running the proprietary software (Figure 4). The cameras were placed at the desired positions, i.e. so that a base-to-distance ratio of approximately 1:3 was ensured. The whole system needs to be calibrated for each different setup. Due to the different artifact sizes, two different setups were required and hence two calibrations were performed.

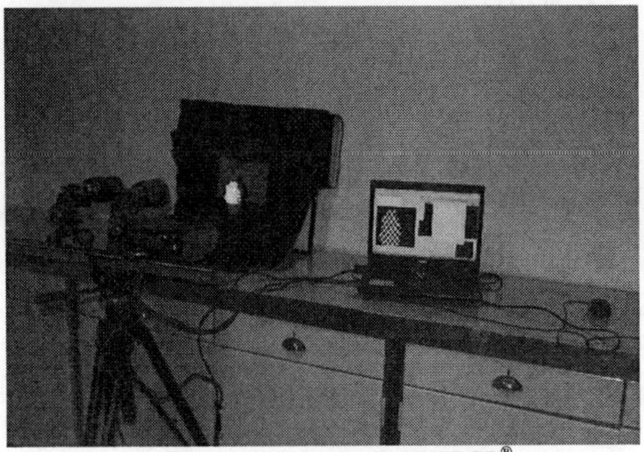

Figure 4: SL2 scanner by XYZRGB®

Each artifact was placed in front of a black background, as structured laser scanners in general cannot determine points on black surfaces. The camera parameters, i.e. exposure values for the DSLR's and brightness, contrast for the machine vision cameras etc. were specified. In order to scan the whole surface of each object, it was manually rotated in between the scans. The scanning process for every object lasted 3 hours. Meshes were created from the generated point clouds using the system's software. These meshes were subsequently exported for further processing in Meshlab.

3.3.2 NEXT ENGINE 3D Laser Scanner

The Next Engine 3D laser scanner is composed of the main scanner hardware and a connected to the scanner rotating base, on which the objects are placed (Figure 5). For each scan several necessary parameters should be specified, like distance, object brightness, cloud density, positions and number of scans. Each object was scanned in approximately 20 different rotated positions. All the generated meshes were cleaned from unwanted faces and afterwards they were aligned both semi automatically. The merged model was exported to Meshlab for further processing. The scanning, repairing and merging of each model was completed within 5 hours approximately.

Figure 5: The next Engine scanner

3.3.3 TOPCON IS Imaging Station

The Topcon IS total station (http://www.topcon positioning.com/products/total-stations) is one of the first such instruments to incorporate imaging and scanning capabilities. The manufacturer claims distance accuracy 1mm±(10mm± ±10ppm), 1.3MP camera and 20pts/sec scanning speed. Considering the imaging station's limited measurement and accuracy capability to obtain a high density 3D point cloud, this method was used for demonstration purposes and it was decided to scan only a single side of each artifact. The Topcon IS was placed 2.20m away from the objects, an arbitrary local coordinate system was created and the scanning parameters were set in order to generate an accurate and dense point cloud to the best of the instrument's capabilities. The data acquisition for each object lasted approximately 40min.

3.4 Data processing

3.4.1 Data preparation

Depending on the used method, the result was a point cloud (Topcon IS), a group of point clouds (Photosynth), a group of meshes (SL2 XYZRGB, Next Engine) or a merged mesh (123DCatch, Arc3D, Agisoft Photoscan). Topcon IS and Photosynth proved unable to produce a mesh. Hence only point clouds were processed from these two 3D data sources.

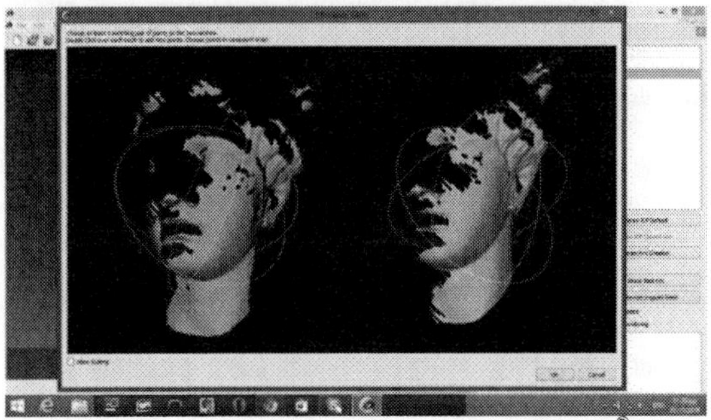

Figure 6: Processing of point clouds in Meshlab®

For all other methods meshes were produced and were subsequently processed in Meshlab (Figure 6). This software has been chosen because it is freeware and it offers a large variety of processing options (http://meshlab.sourceforge.net).

This contribution has been peer-reviewed.
doi:10.5194/isprsarchives-XL-5-W4-101-2015

Before that step, all meshes that contained more than 3 million points were subsampled in the Geomagic Studio v.12 software, as Meshlab cannot accommodate point clouds larger than that. This subsampling has been very carefully performed in order not to discard valuable information of the surface.

3.4.2 Mesh processing

The following processing stages were performed in Meshlab software:

- *Cleaning*. The various methods generated points not only from the surface of the artifacts but also from items in the surrounding area. Those points were selected and cropped.
- *Alignment*. Points were picked and matched manually in order to align the meshes or in order to improve the images' alignment whenever possible. From the Image Based Methods only 123DCatch provided this option. The alignment was optimized automatically.
- *Merging*. All the partial meshes of each method were merged into a single mesh in order to create a unique model.
- *Surface Reconstruction*. The points and normals of a cloud were used to build a surface using the Poisson Surface reconstruction approach. The parameters were fixed at the following values as, after experimentation, it was proven that they lead to the most accurate results that Meshlab can produce within a reasonable amount of time.
    - Octree Depth: 10,
    - Solver Divide: 8,
    - Samples per Node: 1,
    - Surface offsetting: 1.

  This approach creates a surface with no holes, so the extra parts ought to be cropped.
- *Repairing*. The models were repaired by the removal of duplicate and zero area faces duplicated and unreferenced vertices and faces from non-manifold edges.

4. EVALUATION OF RESULTS

The resulting three dimensional models were then compared and evaluated for their qualitative and quantitative properties in order to attempt to draw some conclusions relating to the efficiency and accuracy capabilities of each different 3D reconstruction method in conjunction with the material and the geometric complexity of each object. In addition some proposals for the amelioration of the resulting three dimensional models are also attempted.

For the quantitative comparison, highly accurate measurements on the objects were performed with specialized instrumentation. These measurements were also performed on the three dimensional models and the results are presented, discussed and evaluated (Table 1). For comparison the three dimensional models specialized freely available software, i.e. CloudCompare® was used (www.cloudcompare.org). It has been established that MS Photosynth and the Topcon total station are rather poor in performance for producing any of the three dimensional models. On the other hand the structured light scanner, contrary to the initial expectations failed in certain cases to produce complete point clouds. However it was rather successful in many other cases tested. The point clouds produced with the NextEngine® scanner are acceptable. Specific difficulties were presented by highly complex objects. The web services, i.e. Arc3D and 123DCatch, performed satisfactorily with many of the objects, while presenting certain weaknesses in shiny objects and artifacts with no texture. This was also the problem of the commercial software and in certain cases it was overcome by spraying the objects with chalk powder in order to create some kind of texture.

4.1 Qualitative evaluation

The qualitative evaluation was performed through the careful observation of the created 3D models. The parameters taken into account were:

- Total or partial point cloud generation
- Surface distortions
- The amount and the size of the existing holes
- Surface roughness

The following conclusions were reached for each method of 3D point cloud acquisition.

4.1.1 123DCatch

The service failed to create a complete model only in the case of the wooden artifact. All the produced meshes have deformations that appear to be significant on the 3D models of the plastic and the wax item due to low texture quality. On the other hand, the majority of the details were modeled and no holes were located.

4.1.2 Arc3D

The images of the plastic and the marble object did not lead to the creation of a model. Also, the created models were full of large holes and had great distortions.

4.1.3 Photosynth

The low density of the point clouds that were produced with this method characterizes it as incapable for visualizing the details of the object (Kersten, 2012). The surface construction was possible with post processing, but it was very time-consuming and required the implementation of many different software and algorithms.

4.1.4 Agisoft Photoscan

The Photoscan software created successfully a 3D model for every artifact. The surface roughness appears to be lower than it should. Also, tiny holes and slight deformations are detected all over the models. The alignment of the images of the metal and the wax artifact led to meshes of even lower quality with great deformations. Nevertheless, it was the only method that generated an acceptable model of the candle.

4.1.5 Next Engine

The Next Engine scanner creates high-quality meshes with minimum noise. Great deformations are detected only on the 3D model of the wooden object. Also, the point cloud of the brass artifact contained a lot of noise that led to high surface roughness. Additionally, black texture inserts noise to the cloud. The scanning of the candle led to the acquisition of a very low-dense point cloud that basically contained just noise.

This contribution has been peer-reviewed.
doi:10.5194/isprsarchives-XL-5-W4-101-2015

4.1.6 SLS

Generally, the method leads to great results. The scanning of objects with too dark color (not entirely black) and/or reflective surfaces, results to the creation of 3D models with plenty of imperfections, such as deformations, large holes and misaligned meshes. The black objects cannot be scanned, so the method failed to acquire a point cloud from the mahogany item. It also failed to generate a model from the candle.

4.1.7 Topcon IS

The generated point clouds deviate significantly from the form of the objects regardless of the material and has such low density that it cannot be converted to a mesh to produce any of the 3D models.

Representative examples of what is described above are the meshes that were acquired from the clay artifact (Figure 7).

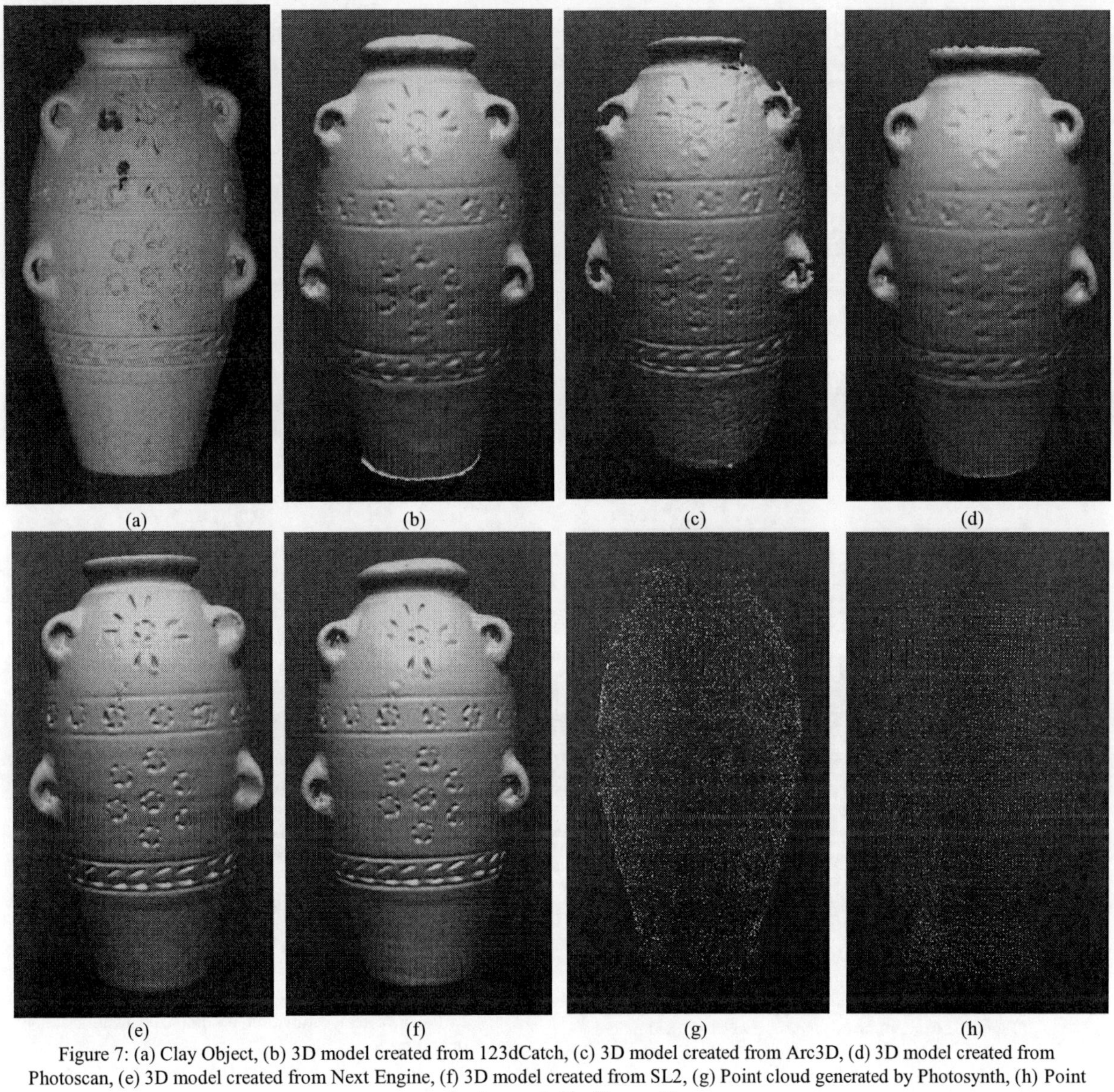

Figure 7: (a) Clay Object, (b) 3D model created from 123dCatch, (c) 3D model created from Arc3D, (d) 3D model created from Photoscan, (e) 3D model created from Next Engine, (f) 3D model created from SL2, (g) Point cloud generated by Photosynth, (h) Point cloud generated by Topcon IS

4.2 Quantitative evaluation

To evaluate the accuracy and the model scale of the created models highly accurate measurements were performed on the artifacts. For that purpose, a digital caliper with accuracy of 0.01mm was used to measure approximately fourteen control distances on each object. The respective distances were also digitally measured on three dimensional models. Those distances varied in length and were dispersed on the surfaces of the artifacts. Due to the low density of the point clouds, the distances could not be measured on the clouds generated by Photosynth and Topcon IS.

This contribution has been peer-reviewed.
doi:10.5194/isprsarchives-XL-5-W4-101-2015

The first step was to calculate the scale of each model from equation [1]. Regardless of the fact that scanners generate scaled models, every model's scale (K) was calculated and amended in order to reach more objective conclusions.

$$K = (S1/S1'+S2/S2'+......+S(n-m)/S(n-m)')/(n-m) \quad [1]$$

S: distance measured using the digital caliper
S': distance measured within CloudCompare
n: the amount of the control measurements
m: the amount of the distances that cannot be accurately located due to model deformations.

Additionally, the mean error of each model was estimated from equations [2] and [3], and the standard deviation was also calculated. The results are presented in Table 1 in mm. Some values are missing because the method failed to produce a point cloud.

Error: $E_i=S_i-S_i'*K$ [2]
Mean model error: $M.E=(E1+E2+......+En)/n$ [3]

	123DCatch		ARC3D		PhotoScan		Next Engine		SL2	
	Mean	St.D	Mean	St.D	Mean	St.D	Mean	St.D	Mean	St.D
PLASTER	0,17	1,21	**0,09**	**1,07**	0,02	1,60	0,14	1,34	0,39	1,19
WOOD	**-0,30**	**1,70**	-1,10	2,81	-0,36	2,47	0,35	4,66	-1,16	2,38
PORCELAIN	-0,43	1,85	-2,47	4,73	-0,36	2,43	0,17	2,08	**0,09**	**2,00**
CLAY	0,52	0,95	0,50	1,24	-0,58	1,38	**-0,13**	**0,64**	-0,34	0,82
MAHOGANY	**0,04**	**0,89**	-0,38	0,95	0,10	1,03	-0,17	0,68	-	-
METAL	0,40	0,66	-0,20	1,24	1,26	1,83	**-0,30**	**0,58**	0,63	1,36
MARBLE	0,02	1,08	-	-	**0,00**	**0,73**	-0,09	0,70	-0,16	1,17
BRASS	0,12	0,93	-0,15	0,92	0,13	0,64	**0,10**	**0,92**	0,33	1,11
BRONZE	-0,32	1,42	0,11	1,87	0,03	1,33	-0,19	1,11	**0,08**	**0,46**
PLASTIC	-0,76	1,18	-	-	0,78	1,21	**0,32**	**0,54**	0,41	0,65

Table 1: 3D Models' Mean Error and Standard Deviation (in mm)

Figure 8: Computing distances between aligned models

After the identification of the models with the greater accuracy in combination with the most favorable standard deviation (highlighted in Table 1), they were compared with all other models the same object that were created using different methods. Within CloudCompare® the 3D models were scaled, aligned and the distances between the aligned models were computed in mm (Figure 8).

Control distances were could not be measured either on the candle or on the clouds acquired from it, because of the almost complete lack of distinguishable features. However, from the qualitative evaluation of the models, it was clear that the model generated by Photoscan was the only acceptable, so it was used as base for the quantitative comparisons. The scales of the created models were adjusted automatically.

Scale adjustments were also performed for the Photosynth's and Topcon's point clouds, although, in the case of clouds acquired with the Topcon IS, it was rarely successful. Hence the relevant values are not shown in Table 2. These adjustments justify the low values shown on table 2.

The values computed by the Cloud Compare (Table 2) were slightly different from the expected. Those divergences are justified due to the following reasons:

- The characteristics of each artifact set limitations on the choice of the control measurements which could have led to the incorrect estimation of the scale factors and the mean errors.
- The surface distortions and holes definitely affected the results
- The boundaries of the objects were not identical.

This contribution has been peer-reviewed.
doi:10.5194/isprsarchives-XL-5-W4-101-2015

	123DCatch		Arc3D		PhotoScan		Photosynth		Topcon IS		Next Engine		SLS	
	MD	St.D.	MD	St.D.	MD	St.D.	MD.	St.D.	MD	St.D.	MD	St.D	MD	St.D
PLASTER	-0,10	0,54	B	B	0,12	0,79	0,10	0,68	-0,17	3,25	0,09	0,61	-1,31	0,64
WOOD	B	B	0,10	0,92	-0,07	1,54	-0,01	1,15	-0,04	3,86	0,24	1,89	0,09	0,89
PORCELAIN	0,70	0,35	3,00	1,41	0,27	0,27	0,00	0,68	-	-	0,40	0,24	B	B
CLAY	1,96	0,74	1,19	0,32	-0,63	0,48	0,04	0,95	-	-	B	B	0,36	0,32
MAHOGANY	B	B	0,26	0,47	-0,16	0,45	-0,07	1,17	-	-	0,18	0,43	-	-
METAL	0,43	0,26	-0,32	0,57	0,99	1,00	0,03	0,46	-	-	B	B	0,63	0,60
MARBLE	0,07	0,12	-	-	B	B	-0,07	0,56	-	-	-0,11	0,53	0,15	0,54
BRASS	0,04	0,24	-0,12	0,39	0,25	0,29	0,06	0,57	-	-	B	B	1,12	0,59
BRONZE	-0,16	0,47	0,94	2,34	-0,01	0,40	0,10	0,82	-	-	-0,18	0,42	B	B
PLASTIC	-0,02	0,38	-	-	0,02	0,30	0,00	0,74	-	-	B	B	0,00	0,15
WAX	0,01	1,08	-0,03	3,04	B	B	-0,38	1,92	-0,69	5,08	-	-	-	-

Table 2: Mean distance and standard deviation between each model and the base model (B), as computed in CloudCompare

5. CONCLUDING REMARKS

As expected, there is no clear advantage for any method investigated for being suitable for small artifact 3D modelling given the variety of materials included in this investigation. Moreover, the list of parameters taken into account in this project was by no means complete. It has been endeavoured to include in this research the materials which appear more often in such cases. In addition the methods used are not all low cost, but they constitute a representative sample of methods that are available today for this very important task.

Geometric problems due to the complex surface of the objects causing occlusions are apparent in almost all artifacts with such properties. However, this problem may easily be resolved using Image based methods, as one can easily take suitable pictures to image points in the occluded area.

In Table 3 the characteristics of the methods used are presented giving some measure of general evaluation. Acquisition and processing time are approximate estimates, as they varied a little from object to object. User friendliness is based on subjective criteria. It could differ if other users were involved.

	123DCatch	Arc3D	Photosynth	Photoscan	Topcon IS (one side)	Next Engine	SLS
Cost of equipment ($)	-	-	-	3500	30000	3000	80000
Acquisition time (h)	1.5	1.5	1.5	1.5	1	2	3,5
Processing time (h)	4	6	1	6	0.5	6	8
User-friendliness	4/5	5/5	5/5	4/5	3/5	5/5	2/5
Average cloud density (before decimation)	100000	180000	13500	160000	6500	6500000	4000000

Table 3: Characteristics of the methods used

From the analysis presented above, which was based on the qualitative and quantitative evaluation, an overall ranking was attempted. In Table 4 the relevant marks are presented. These marks are based on both the objective and the subjective assessment.

	123DCatch	ARC3D	Photosynth	Photoscan	TOPCON IS	Next Engine	SLS	MEAN
PLASTER	8,5	7,5	3	9,5	1	**10**	8,5	6,9
WOOD	5	5	2	**8,5**	1	5,5	6	4,7
PORCELAIN	6,5	0	2	6	0	**8,5**	7,5	4,4
CLAY	6	3,5	1,5	6,5	0,5	8	**8,5**	4,9
MAHOGANY	6,5	5,5	2	7	0,5	**9,5**	-	5,2
METAL	7,5	5	2	5,5	1	**8**	5,5	4,9
MARBLE	8,5	-	2,5	9	0	**9,5**	7	6,1
BRASS	8,5	8	2	8	0,5	**10**	5	6
BRONZE	6	2	1	8	0	**8,5**	7,5	4,7
PLASTIC	3,5	-	0	7,5	0	**8**	7,5	4,4
WAX	5,5	3	1	7	0,5	-	-	3,4
MEAN	6,5	4,4	1,7	7,5	0,5	8,6	7	

Table 4: Overall assessment of the methods used

This table depicts the best method for each different material and summarizes in the last row the overall evaluation of each method for the different artifacts. In the last column of Table 4, the behavior of each different material is shown as a mean of the assessment of all methods used, or all methods that produced an acceptable result. This, of course, is just an indication of the mean reaction of the specific material to all scanning methods.

The conclusions of this project could assist a potential user for the selection of the most suitable methodology for the three dimensional reconstruction of small artifacts taking into account the special properties of the objects, such as their complex geometry and shape and their material and color properties. In addition the cost of equipment and related instrumentation should also be taken into account. Moreover the cost and time necessary for data acquisition and processing are also a decisive factor.

REFERENCES

5.1.1 References from Books:

5.1.2 References from Other Literature:

Bianco, G., Gallo, A., Bruno, F., Muzzupappa, M., 2013, A Comparative Analysis between Active and Passive Techniques for Underwater 3D Reconstruction of Close-Range Objects, *Sensors* 2013, *13*, 11007-11031; doi:10.3390/s130811007

Gonzo, L., Voltolini, F., Girardi, S., Rizzi, A., Remondino, F., El-Hakim, S.F., 2007, Multiple Techniques Approach to the 3D Virtual Reconstruction of Cultural Heritage, Eurographics Italian Chapter Conference 2007.

Kersten, Th., Lindstaedt, M., 2012: Automatic 3D Object Reconstruction from Multiple Images for Architectural, Cultural Heritage and Archaeological Applications Using Open-Source Software and Web Services. Photogrammetrie, Fernerkundung, Geoinformation (PFG) 2012 (6).

Koska, B., Kremen, T., 2013. The combination of laser scanning and structure from motion technology for creation of accurate exterior and interior Orthophotos of St. Nicholas baroque church. International Archives of the Photogrammetry, Remote Sensing and Spatial Information Sciences, Volume XL-5/W1, 2013. 3D-ARCH 2013 - 3D Virtual Reconstruction and Visualization of Complex Architectures, 25 – 26 February 2013, Trento, Italy

Manferdini, A., Galassi, M., 2013, Assessments for 3D Reconstructions of Cultural Heritage using Digital Technologies, International Archives of the Photogrammetry, Remote Sensing and Spatial Information Sciences, Volume XL-5/W1, 3D-ARCH 2013 – 3D Virtual Reconstruction and Visualization of Complex Architectures, Trento, Italy.

Remondino, F., EL-Hakim, S., 2006, Image-Based 3D Modelling: A Review, Swiss Federal Institute of Technology (ETH), Zurich, National Research Council, Ottawa, Canada, The Photogrammetric Record 21(115): 269-291, September 2006.

Skarlatos, D., Kiparissi, S., 2012, Comparison of Laser Scanning, Photogrammetry and SFM-MVS Pipeline Applied in Structures and Artificial Surfaces, Dept. of Civil Engineering and Geomatics, Cyprus University of Technology, ISPRS Annals of the Photogrammetry, Remote Sensing and Spatial Information Sciences, Volume I-3, 2012, XXII ISPRS Congress, Melbourne, Australia

Sofocleous, E., Georgopoulos, A., Ioannides, M., Ioannidis Ch., 2006. "The Geometric Documentation of the Asinou Church in Cyprus", VAST 2006, Lefkosia Cyprus, Oct-Nov. 2006, in: The e-volution of Information Communication Technology in Cultural Heritage (eds. M. Ioannides et al.), p. 138-144 ISBN-10: 963 8046 75 9.

Toschi, I., Capra, A., De Luca, L., Beraldin, J.-A., Cournoyer, L., 2014. On the evaluation of photogrammetric methods for dense 3D surface reconstruction in a metrological context. ISPRS Annals of the Photogrammetry, Remote Sensing and Spatial Information Sciences, Volume II-5, 2014 ISPRS Technical Commission V Symposium, 23 – 25 June 2014, Riva del Garda, Italy.

Valanis, A., Georgopoulos, A., Sofronidou M., Hadzilacos Th., 2009. "Scanning for Microns" Proceedings XXII CIPA Symposium, October 11-15, 2009, Kyoto, Japan.

Valanis, A., Georgopoulos, A., Ioannidis, Ch., 2010. "Assessing The Performance of a Structured Light Scanner", International Symposium of ISPRS Comm. V, Newcastle 21-25 June 2010.

5.1.3 References from websites:

https://www.nextengine.com (last visited on January 2015)
http://www.topconpositioning.com (last visited on January 2015)
http://meshlab.sourceforge.net (last visited on January 2015)
www.cloudcompare.org (last visited on January 2015)
http://www.123dapp.com/catch (last visited on January 2015)
http://www.arc3d.be (last visited on January 2015)
https://photosynth.net (last visited on January 2015)
http://www.xyzrgb.com (last visited on January 2015)
http://www.geomagic.com (last visited on January 2015)
http://www.agisoft.com (last visited on January 2015)

This contribution has been peer-reviewed.
doi:10.5194/isprsarchives-XL-5-W4-101-2015

MIRROR IDENTIFICATION AND CORRECTION OF 3D POINT CLOUDS

P.-F. Käshammer and A. Nüchter

Informatics VII – Robotics and Telematics
Julius-Maximilians University Würzburg, Germany
andreas@nuechti.de

Commission V, WG V/4

KEY WORDS: mirror identification, jump edge detection, panorama creation, plane detection, 3D point cloud correction

ABSTRACT:

In terrestrial laser scanning (TLS), the surface geometry of objects is scanned by laser beams and recorded digitally. This produces a discrete set of scan points, commonly referred to as a point cloud. The coordinates of the scan points are determined by measuring the angles and the time-of-flight relative to the origin (scanner position). However, if it comes to mirror surfaces laser beams are fully reflected, due to the high reflectivity. Mirrors do not appear in the point cloud at all. Instead, for every reflected beam, a incorrect scan point is created behind the actual mirror plane. Consequently, problems arise in multiple derived application fields such as 3D virtual reconstruction of complex architectures. The paper presents a new approach to automatically detect framed rectangular mirrors with known dimensions and to correct the 3D point cloud, using the calculated mirror plane.

1 INTRODUCTION

Scanning reflective surfaces with a 3D laser scanner yields incorrect 3D point clouds, since the emitted laser light is reflected by the surface and a wrong distance is determined. Fig. 1 presents an example, which has won the LiDAR-as-art-contest in 2013. The problem has impact on many post processing steps, e.g., automatic scene modeling and object recognition. In the area of mobile robotics, laser scanners are commonly used for system navigation and robotic mapping or simultaneous localization and mapping (SLAM), i.e., mobile robots have to be able to locate themselves and to identify obstacles. Faulty sensor values of the environment lead to unwanted behavior of the robot and to inconsistent maps.

The scientific contribution of this paper is the implementation of a new approach of solving this problem for the special case of framed rectangular mirrors whose dimensions are known. Thereby, an identification criterion is created, which allows to break down the problem of mirror detection into several subproblems, which are solved by methods of image processing and mathematics. Up to our knowledge, for the first time, a tool has been created to identify mirrors in 3D laser scans and to correct the corresponding 3D point clouds. Throughout the paper, we will demonstrate the algorithms using 3D data acquired with a pulsed Riegl VZ-400 3D laser scanner in multiple mirror situations. The evaluation shows, that the resulting algorithms are applicable without modification to 3D scans acquired using the phase-shift principle. First, we describe the detection of potential mirror contours in a 2D panoramic representation of the point cloud based on a jump edge detection and 3D contour extraction algorithm. Second, we show how actual mirror contours are identified, how the mirror plain is calculated using the Principal Component Analysis (PCA) and also how the point cloud is corrected.

2 RELATED WORK

Not much research has been done dealing with the problem of faulty laser scanner measurements on mirrors and windows. There is only one approach by Shao-Wen Yang and Chieh-Chih Wang Yang and Wang (2008, 2011). In their work, problems of both mirrors and windows are addressed and integrated into the robot's localization, mapping, and navigation framework. They introduced a sensor fusion technique to detect potential obstacles us-

Figure 1: Left: The image depicts how our robot Irma3D sees itself in a mirror. The laser looking into itself creates distortions as well as changes in intensity that give the robot a single eye, complete with iris and pupil. Thus, the image is called "Self Portrait with Duckling". Image courtesy Jan Elseberg. Right: Scanned scene with two mirrors.

This contribution has been peer-reviewed.
doi:10.5194/isprsarchives-XL-5-W4-109-2015

ing sonar sensors and a laser scanner. However, their approach works in 2D only. Mirrors are finally described as obstacles in a 2D occupancy grid map. In contrast to that, this paper describes more a tool to find the exact 3D mirror position and to correct 3D point clouds after the recording of the laser scan.

Only a small number of researchers have looked into the issue of detecting specular surfaces and correcting the errors automatically. A related problem is the detection of transparent or translucent objects. The related works in the area range from 3D point cloud processing of kinect data to environment perception and object reconstruction methods Foster et al. (2011); Ihrke et al. (2010); Klank et al. (2011); Albrecht and Marsland (2013).

3 PANORAMA CREATION

To find potential mirror contours we detect jump edges. First, a 2D panoramic range image of the point cloud is created using the *equirectilinear* projection method Hamidreza Houshiar and Nüchter (2013). When creating panorama images from laser scans, each point of the 3D point cloud is projected onto a 2D array of a certain resolution. Therefore the scanned environment of the laser scanner is seen as the projection of a sphere around the scanner. Given the 3D point in spherical coordinates (θ, ϕ, r) the projection method determines on which 2D pixel with the coordinates (u, v) a 3D point is projected. If multiple 3D points fall into a pixel, e.g., by full-wave-analysis or resolution mismatches, they are stored in a vector. The distance r of one 3D point defines the value of a pixel in the range image version of the *equirectilinear* projection. While many projection methods are available (Hamidreza Houshiar and Nüchter (2013)) for this work, the projection method *equirectangular* is used which is defined by the simple projection rule:

$$u = \theta$$
$$v = \phi$$

with longitude θ and latitude ϕ.

Depending on the resolution of the panorama image, multiple points of the point cloud are usually projected on the same pixel (u, v). If this occurs, there are two possibilities: The map method *FARTHEST* takes the point with biggest range r as represented pixel; the map method *NEAREST* the one with smallest range r. Depending on the point cloud and the mirror situation both methods have different effects, so it has to be chosen individually for each scan, cf. Fig 2.

4 JUMP EDGE DETECTION AND CONTOUR EXTRACTION

Jump edges are detected in the panoramic range image and a corresponding jump edge image of same resolution is created. Between two adjacent pixels of an image, there is a jump edge if the difference in their value is greater than a certain threshold. The sign of the difference describes which of the two pixels is closer to origin, i.e., scanner pose. A jump edge is only represented in the jump edge image, if the considered pixel is closer to origin than the one compared to. This is important so that the detected contour describes only the mirror frame and contains no pixels within the actual mirror plane. Otherwise 3D contours cannot be extracted properly.

Extracting the detected 2D contours of the jump edge image is a known problem of image processing and is done by using the corresponding OpenCV-function `findContours` in `CvRetrExternal`

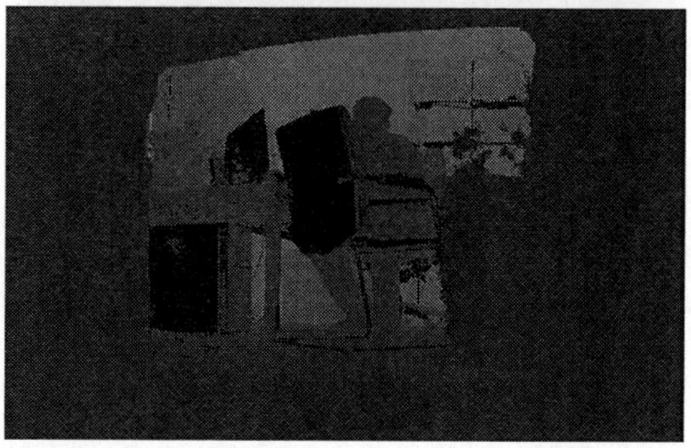

Figure 2: Part of a range image of a mirror. Undefined pixel appear black. Without fillZero-function. Top: Method farthest. Bottom: Method nearest. Not ideal case with 3D points on the mirrors plane.

mode. Subsequently the corresponding 3D contours are extracted by testing which 3D point is projected on the considered 2D contour pixel, which is the reverse process of creating the range image. Note: The chosen map method changes the outcome of the extracted 3D contour. To get the frame point as 3D contour point, the map method has to be *nearest.*

In the ideal case, all 3D mirror points are laser beam reflection and are therefore further away from the scanner and need to be corrected. However, in some cases 3D points occur that describe the actual mirror surface due to not perfect reflections or detection using the full-wave-analysis, cf. Fig. 2 bottom picture. The 2D contour will then be shifted to the inside of the mirror and will not describe the mirror frame anymore. Consequently, 3D contours might not seem to be connected what leads to the need of filtering these contours or erroneous calculated plane. In those cases the map method *farthest* is chosen to avoid this problem (Fig. 2 top picture).

5 IDENTIFICATION OF MIRROR CONTOURS

So far we describes how potential 2D and corresponding 3D contours are extracted. Now the actual mirror contours have to be identified among all detected ones.

In a first step a connected filter is applied on all detected contours. It calculates the distances of all 3D contour points towards their next neighbor. If the distance between two neighbored contour points is larger than a certain threshold, the considered contour is seen as unconnected and therefore filtered. If a considered 3D mirror contour contains points that do not describe the mirrors frame due to previous problems in detecting the jump edges, this filter decreases in efficiency because the used threshold has to be

This contribution has been peer-reviewed.
doi:10.5194/isprsarchives-XL-5-W4-109-2015

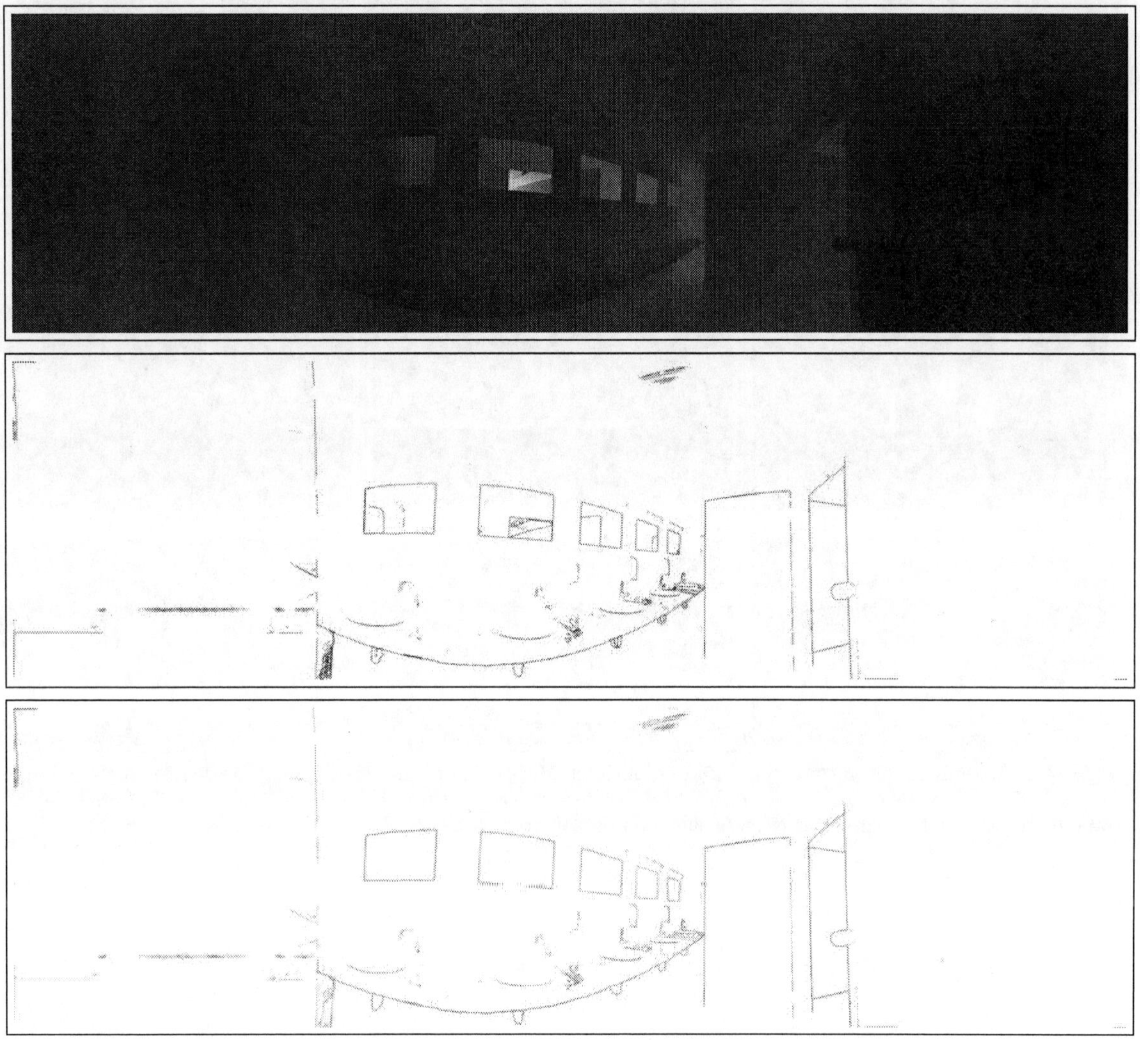

Figure 3: Top: Range image of scanned bathroom environment with five mirrors of same size and resolution of 3600×1000. Middle: Corresponding jump edge image with a threshold of 0.1 m. Bottom: Extracted 2D contours of picture above.

This contribution has been peer-reviewed.
doi:10.5194/isprsarchives-XL-5-W4-109-2015

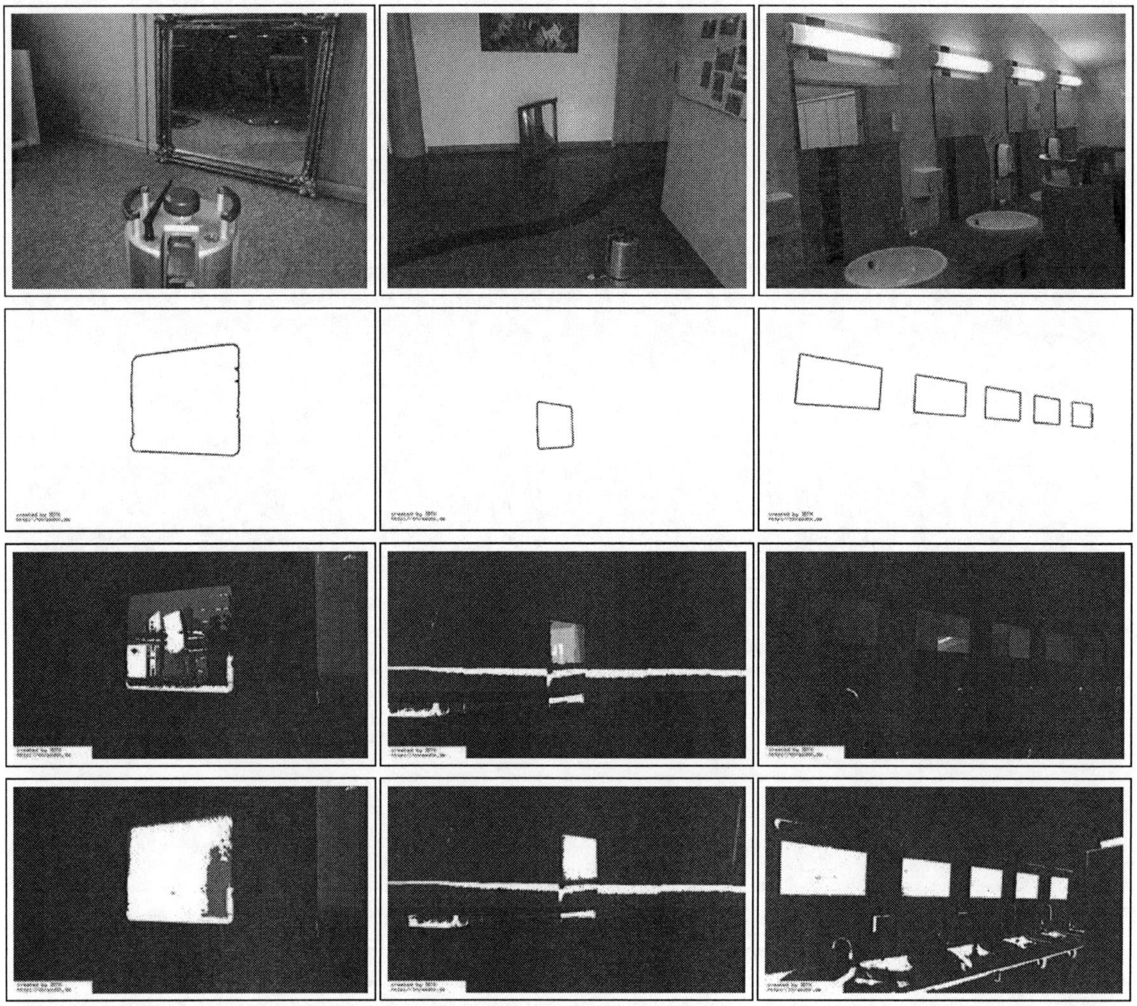

Figure 4: Left: Mirror of size 56×64 cm in an office environment. Middle: Same mirror placed in an office corridor. Right: 5 mirrors of size 40×60 cm in a bath room. Top row: Photo of the scenes. Second row: Extracted 3D contours. Third row: Screenshots in a 3D point cloud viewer without mirror correction. Bottom: Corrected 3D point clouds.

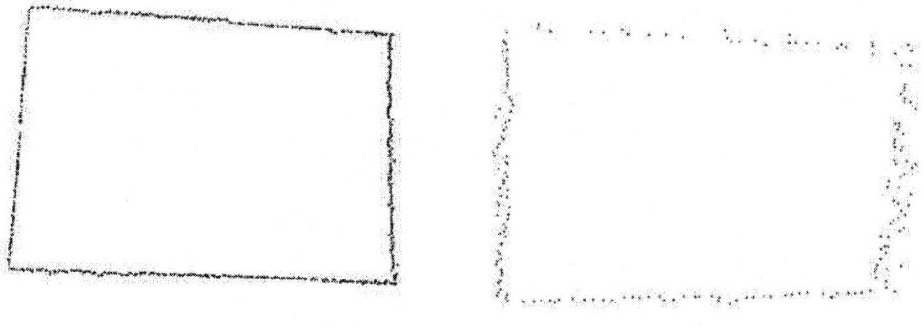

Figure 5: Two extracted mirror contour as 3D point clouds of mirrors. The right side features always more noise, since the incidence angle is much smaller theren.

This contribution has been peer-reviewed.
doi:10.5194/isprsarchives-XL-5-W4-109-2015

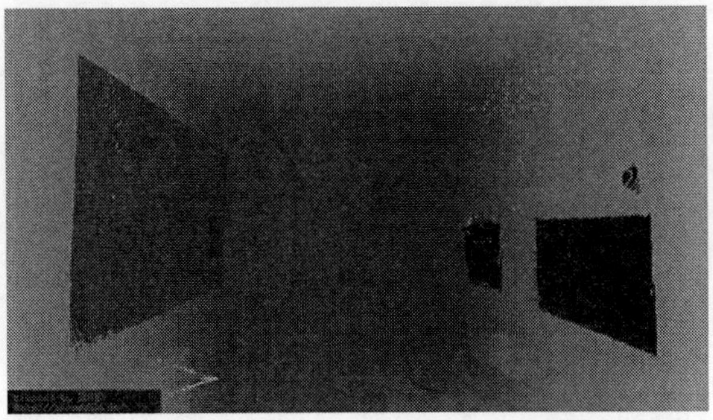

Figure 6: Screenshot of corrected point cloud acquired by Z+F 5006 Scanner in bathroom environment.

increased to not filter the actual mirror contours. Consequently more non-mirror contours pass this filter. In a second step, for all remaining 3D contours, the dimensions as well as the best fit plane through the contour points, which is the mirror plain for a mirror contour, are calculated using the Principal Component Analysis (PCA) Pearson (1901).

In a second filtering process, contours having the specified mirror dimensions within a certain margin of error are identified. For an identified mirror contour, the original point cloud is corrected at hand the corresponding 2D contour as well as the calculated mirror plane. To this end, the OpenCV function `pointPolygontest` is used to find all 3D points which are projected within the 2D mirror contour. Those are then reprojected on the mirrors plane, stored in a separate 3D point cloud and deleted from the original one.

6 DEALING WITH UNDEFINED PIXEL

There are many challenges in dealing with real-world data. E.g., dealing with undefined pixel in the created panoramic range image which originate mainly due to highly reflecting surfaces in the scanned environment, e.g., when scanning water. This means there are pixel in the panoramic image on which no 3D point is projected and therefore do not posses a range value.

Since these undefined pixel cause false jump edges and so creating incorrect contours, a method for dealing with those has been developed. Therefore a `fillZeros` function assigns values to undefined pixels by copying the closest properly measured pixel from top-left in the 2D image coordinates. This algorithm often leads to shifting of the 2D mirror contour either towards the inside of the mirror or towards the outside. The second case is not a problem because normally the contour will still describe the frame or at least the mirrors plane. In case of a shifting towards the inside, a second algorithm called `findFramePoint` fixes this by searching for 3D points which lie in the opposite direction of the mirror contours geometrical mean.

7 EXPERIMENTS AND RESULTS

The mirror identification software is implemented in C/C++ based on the Open Source software *3DTK — The 3D Toolkit*. Andreas Nüchter et al. (2015). The software has been tested with a Riegl VZ-400 terrestrial laser scanner and the Zoller+Fröhlich Imager 5006 in the same environment.

For both scanners all mirrors have been identified and thus the corresponding point clouds were corrected. The only case where

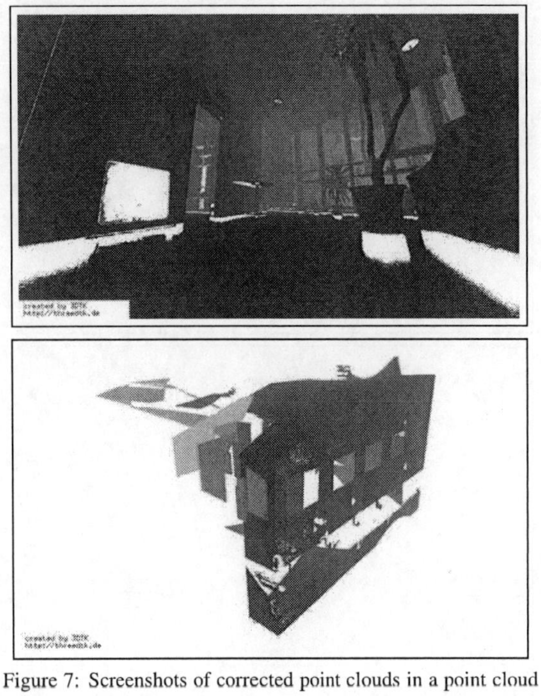

Figure 7: Screenshots of corrected point clouds in a point cloud viewer.

mirrors cannot be identified at all is if mirrors in the panoramic representation are cut of by the panorama frame. This is because the mirror contour cannot be detected as connected in the current software version. In future work, we will mind the scanners starting position with respect to the mirror. For all tested point clouds, 3D mirror contour errors are up to 9 cm in x and y dimension and up to 3.5 cm in z dimension, which is due to the discretization induced by the (relatively low resolution of the) panorama image. There a two main effects that increase the 3D mirror contours and consequently also effect the error of the mirror plane and corrected points. With decreasing incidence angle between laser beams and mirror plane, noise in the extracted frame increases, cf. Figur 5. If mirror plane and laser beams were parallel, the mirror would not appear at all in the point cloud.

Figure 8 shows the result of the automatic correction on data acquired with a Riegl VZ-400, a plused laser scanner. The top row and bottom row show that the 3D points are projected to the correct location. Figure 7 (bottom) gives a 3D view of the bottom row of Figure 8. The remaining misalignment can easily be resolved scan matching methods, such as ICP (iterative closest points). The middle row of Figure 8 demonstrates, how our software can be used to look around the corner using a mirror.

All presented algorithms are also applicable to scans of a Zoller+ Fröhlich Imager 5006, which measures according to the phase-shift principle. Nevertheless there is one difference in the outcome. The corrected point clouds of the Zoller+Fröhlich Imager contain several points "flying around" which look like 3D mirror contours but shifted relative to the scanner, cf. Figure 6. Those points arise from the different principle of operation. Due to the finite spot size, if the laser beams hits the mirrors frame, the measurments are averaged between the frame point and the mirrored point. In contrast to the Riegl, this scanner is not able to distinguish several targets by Full Wave Analysis. Since those faulty points are reprojected by the correction algorithm. Figure 6 whos a 3D point cloud with corrected mirrors and erronous contours.

This contribution has been peer-reviewed.
doi:10.5194/isprsarchives-XL-5-W4-109-2015

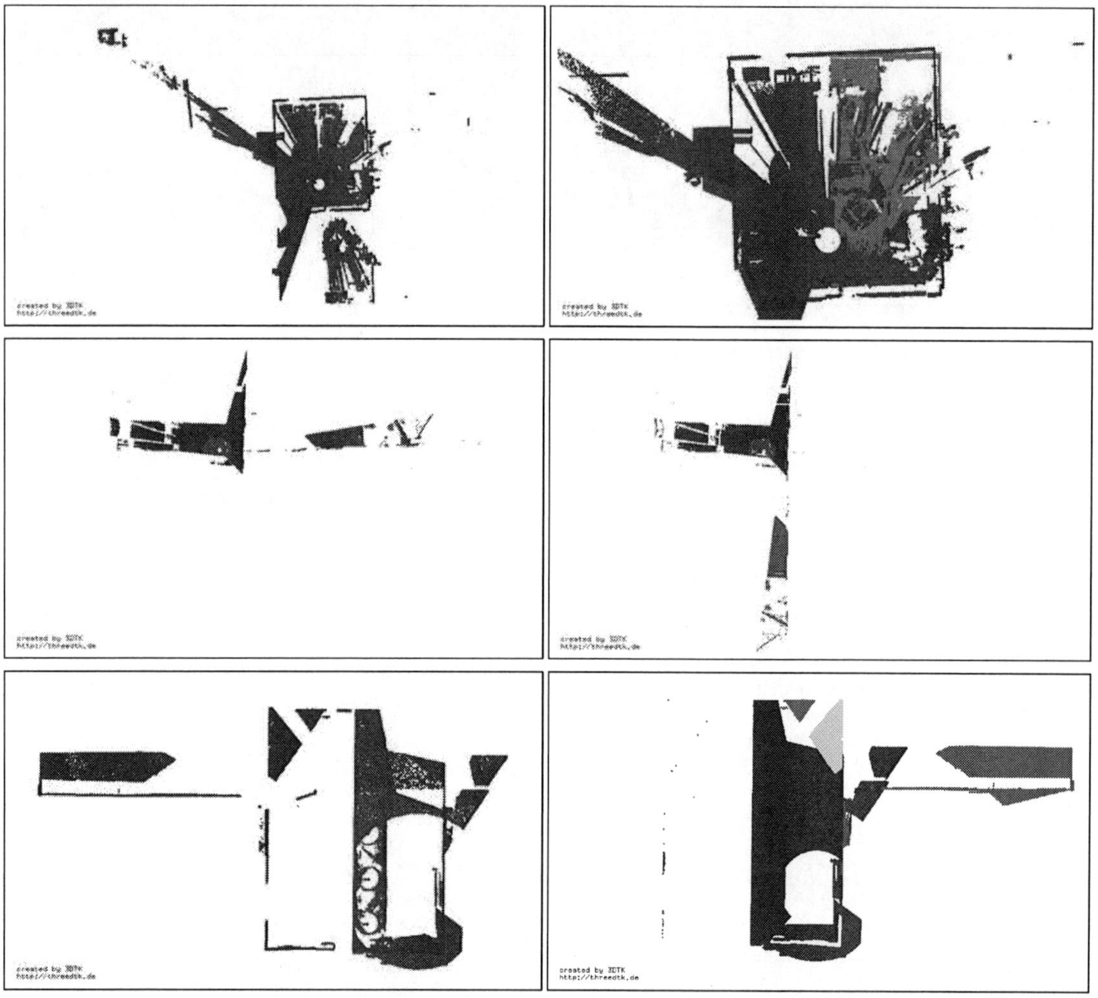

Figure 8: Screenshots of 3D point clouds in bird eye view. Left: Uncorrected point clouds. Right: Corrected point clouds. Top: Office environment. (Note that right picture is zoomed) Middle: Office corridor. Bottom: Bathroom with 5 mirrors.

8 CONCLUSIONS AND OUTLOOK

This paper describes a complete solution to correct point clouds that contain a mirror of know size. It has been demonstrated to work well in various environments and with pulsed and phse-shift scanners.

Needless to say, a lot of work remains to be done. In future work, will concentrate on arbitrary mirrors, i.e., without knowing the mirror size in advance. Furthermore, we will integrate this technology into our registration and mapping methods and aim to develop a method to distingish mirrors from windows. Overall, we aim at reducing the time needed for manually correcting and post-processing of 3D point clouds.

References

Albrecht, S. and Marsland, S., 2013. Seeing the unseen: Simple reconstruction of transparent objects from point cloud data. In: Robots in Clutter (RIC) Workshop at Robotics: Science and Systems (RSS '13), Berlin, Germany.

Andreas Nüchter et al., 2015. 3DTK — The 3D Toolkit. http://slam6d.sourceforge.net/.

Foster, P., Sun, Z., Park, J. J. and Kuipers, B., 2011. Visagge: Visible angle grid for glass environments. In: Proceedings of the IEEE International Conference Robotics and Automation (ICRA '11), Karlsruhe, Germany, pp. 2213–2220.

Hamidreza Houshiar, Dorit Borrmann, J. E. and Nüchter, A., 2013. Panorama based point cloud reduction. In: Proceedings of the 16th IEEE International Conference on Advanced Robotics (ICAR '13), Montevideo, Urugauy.

Ihrke, I., Kutulakos, K., Lensch, H., Magnor, M. and Heidrich, W., 2010. Transparent and specular object reconstruction. Computer Graphics Forum 29(8), pp. 2400–2426.

Klank, U., Carton, D. and Beetz, M., 2011. Transparent object detection and reconstruction on a mobile platform. In: Proceedings of the IEEE International Conference Robotics and Automation (ICRA '11), Shanghai, China, pp. 5971–5978.

Pearson, K., 1901. On lines and planes of closest fit to a system of points in space. The London, Edinburgh, and Dublin Philosophical Magazine and Journal of Science 6(2), pp. 559–572.

Yang, S.-W. and Wang, C.-C., 2008. Dealing with laser scanner failure: Mirrors and windows. In: IEEE International Conference on Robotics and Automation (ICRA), Pasadena, California.

Yang, S.-W. and Wang, C.-C., 2011. On solving mirror reflection in lidar sensing. IEEE/ASME Transactions on Mechatronics 16(2), pp. 255–265.

This contribution has been peer-reviewed.
doi:10.5194/isprsarchives-XL-5-W4-109-2015

A NEW APPROACH FOR SEGMENTATION-BASED TEXTURING OF LASER SCANNING DATA

Z. Lari [a, *], A. Habib [b]

[a] Dept. of Geomatics Engineering, University of Calgary, 2500 University Drive NW, Calgary, AB, Canada, T2N 1N4 – zlari@ucalgary.ca

[b] Lyles School of Civil Engineering, Purdue University, 550 Stadium Mall Drive, HAMP 4108 West Lafayette, IN 47907 – ahabib@purdue.edu

Commission V, WG V/4

KEY WORDS: Laser scanning, Texturing, Segmentation, Boundary detection

ABSTRACT:

In recent years, laser scanning systems have been recognized as a fast and accurate technology for the acquisition of high density spatial data. The advent of these systems has reduced the cost and increased the availability of accurate 3D data for mapping, modelling, and monitoring applications. The original laser scanning data does not explicitly provide meaningful information about the characteristics of the scanned surfaces. Therefore, reliable processing procedures are applied for information extraction from these datasets. However, the derived surfaces through laser scanning data processing cannot be effectively interpreted due to the lack of spectral information. To resolve this problem, a new texturing procedure is introduced in this paper to improve the interpretability of laser scanning-derived surfaces using spectral information from overlapping imagery. In this texturing approach, individual planar regions, derived through a laser scanning data segmentation procedure, are textured using the available imagery. This texturing approach, which aims to overcome the computational inefficiency of the previously-developed point-based texturing techniques, is implemented in three steps. In the first step, the visibility of the segmented planar regions in the available imagery is checked and a list of appropriate images for texturing each planar region is established. An occlusion detection procedure is then performed to identify the parts of the segmented regions which are occluding/being occluded by other regions in the field of view of the utilized images. In the second step, visible parts of planar regions are decomposed into parts which should be textured using individual images. Finally, a rendering procedure is performed to texture these parts using available images. Experimental results from real laser scanning dataset and overlapping imagery demonstrate the feasibility of the proposed approach for texturing laser scanning-derived surfaces using images.

1. INTRODUCTION

Over the past few years, laser scanning systems have been established as the leading technology for the rapid collection of high density 3D point cloud. The original laser scanning point cloud usually does not provide semantic information about the types and characteristics of the scanned surfaces. The collected point cloud consequently needs to be processed to extract the required information for different applications (e.g., terrain mapping, environmental monitoring, 3D city modelling, transportation planning, cultural heritage documentation, structural health monitoring, industrial site modelling, and forest mapping). The derived surfaces through laser scanning data processing surpass the quality of the reconstructed surfaces using other methods such as photogrammetric reconstruction. However, those surfaces cannot be easily and effectively interpreted due to the lack of descriptive (spectral) information. On the other hand, the imagery data collected by high quality passive imaging systems provides rich descriptive information regarding the scanned surfaces. Therefore, the complementary characteristics of the images can be utilized to improve the interpretability of the laser scanning-derived surfaces.

The synergistic properties of laser scanning systems and images can be fully exploited only when they are successfully registered relative to a common reference frame (Habib and Schenk, 1999; Habib et al., 2004). Once the images and laser scanning data are registered to a common reference frame, the spectral and positional surface information from both datasets can be linked together through a texturing procedure. Traditionally, the link between laser scanning and image data is established through point-by-point projection of the dense point clouds onto the images(Lamond and Watson, 2004; El-Omari and Moselhi, 2008; Koch and Kaehler, 2009; Kim and Habib, 2009). Due to the large volume of scanned points, the texturing of a complete laser scanning dataset using this approach will be very time and memory consuming. Furthermore, the laser-scanned surfaces may not be homogenously textured due to local point density variations. In order to overcome the second drawback of the point-based texturing techniques, different surface-based texturing approaches have been introduced in recent years. In these approaches, the laser scanning point cloud first is converted into continuous surfaces using a surface modelling technique. The surface modelling is performed by fitting smooth surfaces (Vosselman et al., 2004), fitting basic geometric shapes (Guelch, 2009), or triangulation/meshing (Kazhdan et al., 2006) of laser scanning point cloud. Once the continuous surfaces are constructed, the vertices of the individual surfaces, within the generated mesh, are projected onto the image space using collinearity equations (Kraus, 1993). The textures within the projected surfaces onto the image are then mapped to the mesh components to produce a photo-realistic appearance for the generated mesh (Wang et al., 2001;

* Corresponding author

This contribution has been peer-reviewed.
doi:10.5194/isprsarchives-XL-5-W4-115-2015

El-Hakim et al., 2003; Alshawabkeh and Haala, 2005; Bannai et al., 2007; Xu et al., 2010; Chen et al., 2012). These approaches also suffer from the computational inefficiency of point-based texturing approaches due to point-by-point projection of the scanning point cloud onto the image/images. The other shortcoming of point-by-point projection – in both of the introduced point-based and surface-based texturing approaches – is the occlusion problem. This problem occurs when two laser scanning points, at the vicinity of abrupt surface changes, are competing for the same image location. In order to identify and resolve instances of this problem during the texturing procedure, different point-based visibility analysis techniques have been proposed (Ahmar et al., 1998; Kuzmin et al., 2004; Habib et al., 2007). The majority of these approaches suffer from the computational inefficiency and cannot accurately identify instances of occlusion problems in complex scenes.

In order to overcome the drawbacks of the aforementioned techniques, a new segmentation-based texturing approach is introduced in this paper. This approach improves the computational efficiency of the texturing procedure and provides photo-realistic textures along the laser scanning-derived surfaces (segments). Moreover, a region-based occlusion/visibility analysis procedure is proposed to resolve instances of the occlusion problem. Figure 1 shows the outline of the proposed method for texturing laser scanning-derived surfaces using overlapping images.

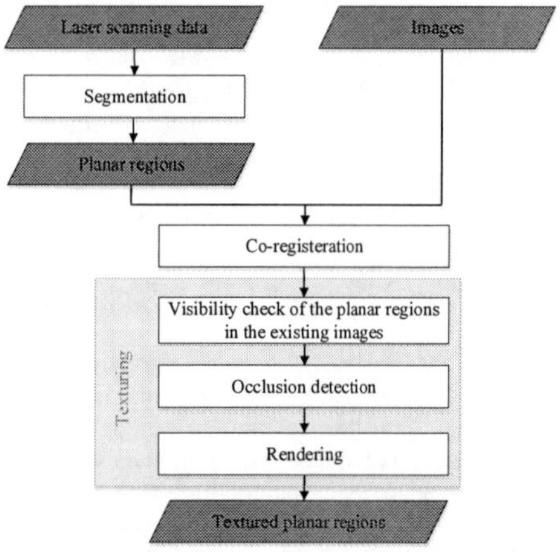

Figure 1. The outline of the proposed texturing approach

This paper starts by the introduction of the utilized segmentation approach for the extraction of planar regions from laser scanning point cloud. The next section introduces the proposed approach for texturing of segmented planar surfaces using imagery. The newly-developed visibility analysis and occlusion detection approaches are also described in this section. Afterwards, experimental results with real data are presented to verify the feasibility of the proposed approach for texturing laser scanning-derived surfaces using overlapping images. Finally, concluding remarks and recommendations for future research work are summarized.

2. PLANAR FEATURES SEGMENTATION

In this paper, the parameter-domain segmentation approach proposed by (Lari and Habib, 2014) is utilized for the extraction of planar regions from laser scanning data. In the first step of this segmentation approach, a Principal Component Analysis (PCA) of local neighborhoods of individual laser scanning points is performed to detect and model potential planar neighborhoods within the laser scanning point cloud (Pauly et al., 2002). Afterwards, geometric attributes which describe the detected planar neighnorhoods are estimated using an adaptive cylindrical buffer defined based on the PCA outcome (Figure 2). This cylindrical buffer is a modified version of adaptive cylinder neighborhood proposed by Filin and Pfeifer (2005). The diameter of this cylindrical buffer is specified by the distance between the query point and its nth-nearest neighbor, where n is the number of points required for reliable plane definition while considering the possibility of having some outliers. The orientation of the cylinder axis is adaptively changing to be aligned along the normal to the plane through the majority of the neighboring points and the height of this cylindrical buffer – maximum normal distance between a neighboring point and best fitted plane – is usually determined based on the noise level in the point cloud coordinates.

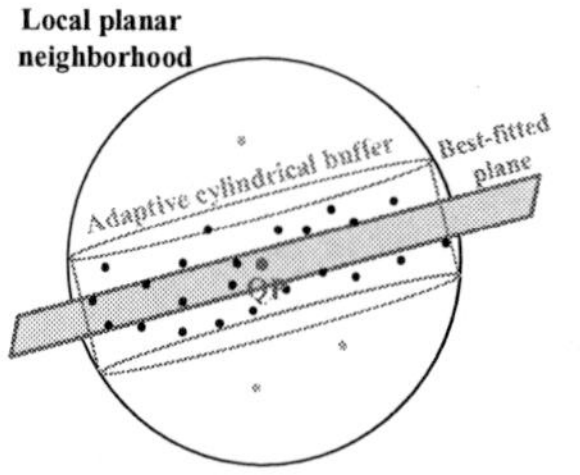

Figure 2. Adaptive cylindrical buffer of a query point (QP) belonging to a planar neighborhood

The implemented cylindrical buffers are then employed for the estimation of local point density variations along the involved planar surfaces. The estimated local point density indices are utilized to refine the characteristic attributes for planar features segmentation. In order to speed up the segmentation process, the classification and representation of local planar neighborhoods is followed by a region growing procedure to aggregate all the points which belong to planar features into groups based on the estimated local point density indices. Further processing steps are carried out for the individual groups of planar surfaces rather than the entire points classified as being part of these surfaces.

The segmentation attributes associated with the individual laser points, which belong to local planar neighborhoods, are then defined as the coordinates of the origin point's projection onto the best-fitted plane (X_0, Y_0, Z_0) to that planar neighborhood. Afterwards, a clustering approach is utilized for the detection and extraction of the accumulated peaks in the attribute space. In this segmentation approach, the estimated attributes are firstly organized in a kd-tree structure, which is manipulated for peak detection. A novel peak detection approach is then implemented, for which an appropriate extent of the cluster is estimated while considering acceptable deviation (angular and spatial) among the attributes of coplanar points. All the points in the parameter domain are checked to count the number of their neighboring points within the established cluster extent. The attribute point with the highest neighbors' count is chosen as the peak. The points in the spatial domain whose attributes are the constituents of the detected peak are then considered as one segment and removed from the parameter domain. The search for the highest count peaks is continued until the number of detected attribute points in the identified peak is less than the size of the smallest region to be segmented. The disadvantage of such a segmentation procedure is that the points belonging to coplanar but spatially disconnected planes will be segmented into the same cluster. In order to resolve this ambiguity, a neighborhood analysis is conducted through boundary detection

doi:10.5194/isprsarchives-XL-5-W4-115-2015

of the clustered points using the modified convex hull algorithm (Jarvis, 1977).

3. TEXTURING

In order to take advantage of the complementary characteristics of laser scanning data and images for the interpretation of segmented planar surfaces, a new texturing procedure is introduced in this section. The objective of this procedure is to texture the outcome from the laser scanning data segmentation (i.e., the planar features) using single/overlapping images. Rather than assigning the texture on a point-by-point basis, which is how it was implemented in the previously-developed texturing approaches, the proposed procedure will be based on polygonal image-to-object rendering.

To ensure that both the laser scanning data and the images can be linked together with their fully synergistic properties, prerequisite procedures such as camera calibration and co-registration of both datasets should be performed to provide the Interior Orientation Parameters (IOPs) of the utilized imaging sensor and the Exterior Orientation Parameters (EOPs) of the available images. The proposed texturing procedure is then initiated by investigating the partial/full visibility of the segmented planar regions in the available imagery. Once the visibility of the segmented regions in the available imagery is checked, an occlusion detection procedure is performed to check if a planar region is occluding/being occluded by other regions in the field of view of existing images. In the next step, the segmented planar regions are decomposed based on their visibility within the available images. Finally, a rendering procedure is implemented to visualize the textured planar regions on the screen. The detailed explanation of these steps will be presented in the following subsections.

3.1 Visibility check

In this subsection, a visibility analysis procedure is introduced and implemented to check if a segmented planar region is fully or partially visible in the available imagery. One should note that this procedure does not consider whether the region of interest is occluded by other regions in the field of view of a given image. In the first step of this procedure, the suitability of available images for the texturing of a planar region is investigated while considering the angle between the region surface normal and the optical axis of the image. For a given region, the image in question is considered to be appropriate for the texturing procedure only if the region surface normal makes an acute angle with the optical axis of the image.

For the images which satisfy the abovementioned condition, the image footprint (the four corners of each image) is then projected onto the infinite plane encompassing the planar region of interest using collinearity equations, while enforcing the plane equation of the region of interest. The overlap area between the projected image footprint and the planar region of interest is then identified to determine if the planar region is fully or partially visible within the image in question. The overlap area between two planar regions is identified using the Weiler-Atheron algorithm (Weiler and Atherton, 1977).

The planar region of interest is fully visible in an image if it is completely covered by the projected image footprint (Figure 3.a). On the other hand, the region of interest is partially visible in an image if more than the pre-specified percentage of the region area is covered by the projected image footprint (Figure 3.b). Once this visibility check is carried out for all planar regions in all images, a list of the appropriate images that could be used for the texturing of a given region is established.

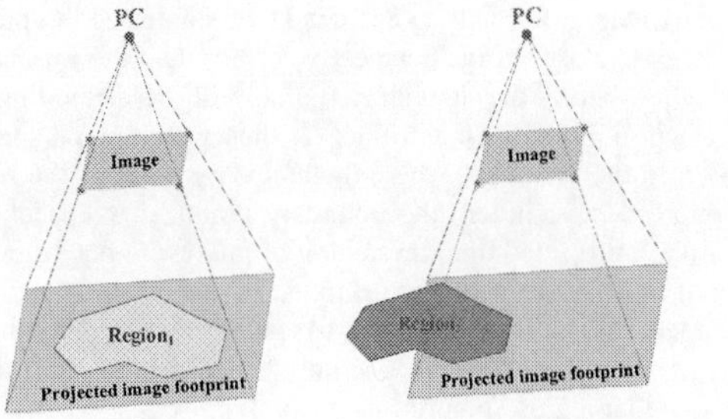

(a) (b)

Figure 3. Representation of (a) a fully visible region and (b) a partially visible region in an image

3.2 Occlusion detection

The objective of this step is to investigate whether the visible part of a planar region in an image is occluding/being occluded by the other regions in the field of view of that image. An option for conducting this test is to check whether or not every point in the overlap area between the region of interest and the footprint of the image in question is visible in that image. A point will be visible in a given image if the line of sight connecting the perspective center of the image to that point does not intersect any other region. A problem with this option is that we can derive an infinite number of points in the overlap area between a given region and the projected image onto that region. Therefore, the visibility analysis of these points in the given image will be time and memory consuming.

In order to overcome these drawbacks, an alternative occlusion detection approach will be adopted in this section. The conceptual basis of this alternative process is based on the boundaries (inner and outer boundaries) of the segmented region as well as their intersection with the footprint of the image in question. This procedure is implemented by checking whether or not a boundary point is visible in a given image. This visibility analysis is performed by intersecting the line segment connecting the perspective center of a given image and the boundary point in question with all the regions that are visible in that image. If the intersection point is outside the investigated region, the boundary point is not occluding/ being occluded by that region within the image in question (Figure 4.a). Alternatively, if the intersection point happens to be inside the investigated region, the boundary point is either occluding or being occluded by the intersection point (Figure 4.b).

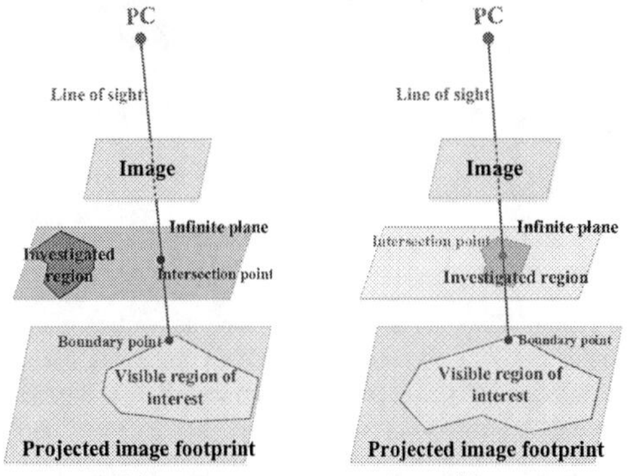

(a) (b)

Figure 4. Visibility analysis of a planar region's boundary point in a given image, (a) intersection point outside of region boundary (b) intersection point inside region boundary

This contribution has been peer-reviewed.
doi:10.5194/isprsarchives-XL-5-W4-115-2015

The boundary point will be occluded by the intersection point if the latter is closer to the perspective center than the former. On the other hand, the boundary point will be occluding the intersection point if the former is closer to the perspective center. In the first case, where the boundary point of the region of interest is occluded, the boundary points of the occluding region are projected onto the region of interest to determine the part of that region which is occluded in the given image. This occluded part is identified by intersecting the planar region of interest with the projected occluding region onto the infinite plane of that region (Figure 5).

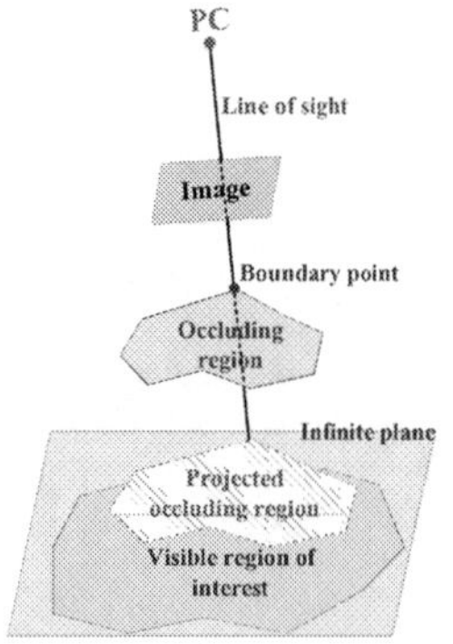

Figure 5. Identification of the occluded part of the region of interest (projection of boundary points of the occluding region onto the region of interest's infinite plane)

In the second case, where the region of interest is occluding another region, two different situations may happen:

1. All the boundary points of the region of interest are projected inside the occluded region (Figure 6.a).
2. A portion of the boundary points of the region of interest are projected inside the occluded region (Figure 6.b).

The first situation can be solved by projecting the boundary points of the region of interest onto the occluded region and the second situation will be handled when the occluded region becomes the region of interest.

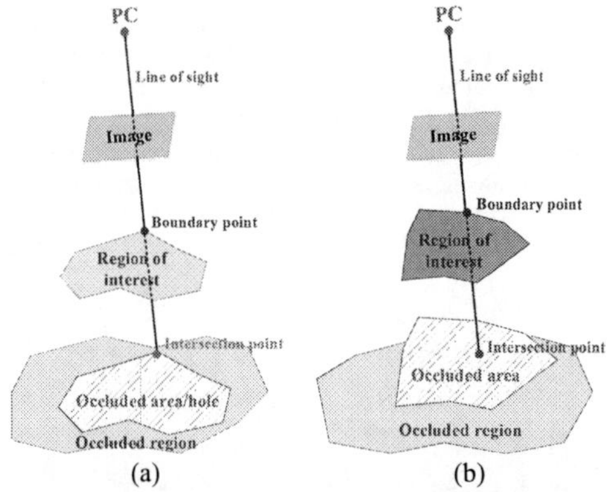

(a) (b)

Figure 6. The occlusion of other regions by the region of interest (a) all the boundary points of the region of interest are projected inside the occluded region and (b) a section of the boundary points are projected inside the occluded region

For each of the visible regions in an image, one can sequentially trace the visible boundary points, the points along the image footprint intersection with the region, if they happen to be inside the region, and the internal points that are occluded by the boundary points of the other regions. The outcome of this procedure, for a given region, will be the sub-regions that are visible in different images.

3.3 Decomposition of planar regions according to their visibility within available images

Once the visible part/parts of the planar regions within available images are identified through the proposed procedures in the previous two subsections, the segments of these parts which should be textured using the individual images should be identified. In this section, a decomposition approach is introduced to determine these segments within planar regions. In other words, the objective of this procedure is to identify which parts of each planar region will be textured as best as possible using individual images. In this case, we are dealing with one of the following three scenarios:

1. In the first scenario, the entirety of a given region is visible in a single image. In order to render this region, all of its boundary points are projected onto that image and the part of the image within the projected boundary will be rendered onto the given region.
2. In the second scenario, the entirety of a given region is simultaneously visible in several images. In this case, the texturing of the region of interest is carried out using the image that has the best homogenous sampling distance along that region. The image which has the best homogenous sampling distance is defined as the image which is closest to the region of interest and has the smallest angle between the surface normal and the optical axis of the image.
3. In the third scenario, the region of interest is composed of sub-regions which are visible in several images (Figure 7). The rendering of such a region is carried out by decomposing the visible sub-regions into the parts which are visible in single or multiple images. In order to identify these parts, the visible sub-regions in the individual images first are intersected together using the Weiler-Atheron algorithm. One should note that the occluded parts of the visible sub-regions in each image were identified in the previous step. Figure 8 shows the visible sub-regions in each image and Figure 9 shows the outcome of intersecting these sub-regions. The intersected parts of the visible sub-regions represent the segments which are visible in multiple images ($segment_{12}$ and $segment_{23}$). The overlap areas between the sub-regions (i.e., the segments that are visible in multiple images) are then subtracted from the individual sub-regions to identify the segments which are visible in a single image (i.e., $segment_{11}$, $segment_{22}$, and $segment_{23}$).

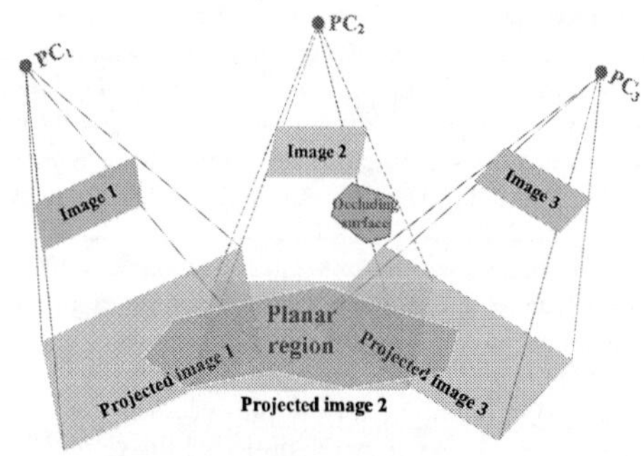

Figure 7. A planar region which is partially visible in the overlapping images

This contribution has been peer-reviewed.
doi:10.5194/isprsarchives-XL-5-W4-115-2015

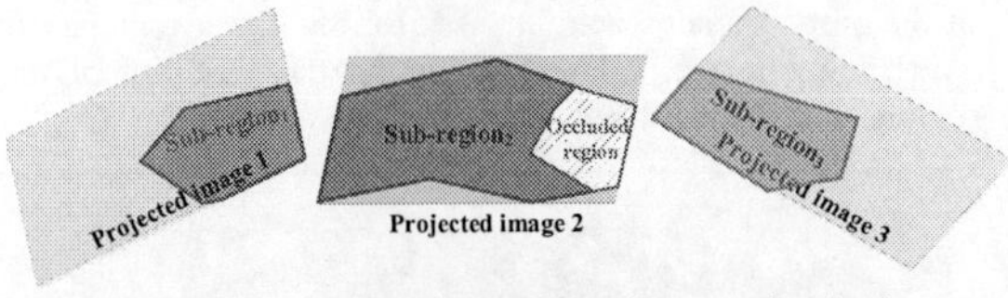

Figure 8. Visible sub-regions of the planar region in the overlapping images

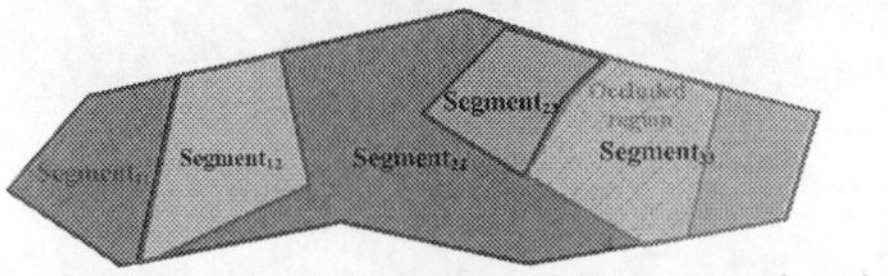

Figure 9. The intersection between visible sub-regions in each image

For the segments which are visible in a single image, the rendering procedure is carried out by projecting their boundary points onto that image and selecting the part of the image within the projected boundary for texturing that segment (Figure 10). However, for the segments which are visible in multiple images, the rendering procedure is carried out using all the images encompassing those segments. Accordingly, the boundary points of such a segment are projected onto all of the images enclosing that segment. In this case, the projected boundary onto the image which has the best sampling distance along that segment is selected as the texture master. In order to incorporate the spectral information from other enclosing images into the master texture, the correspondence between the conjugate pixels within the projected boundary onto the images in question is established using a 2D projective transformation. The assigned color to a pixel within the master texture is determined by averaging the colors of its conjugate pixels in all images enclosing the projected boundary in question. The intersected segment then will be textured using the established master texture (Figure 11).

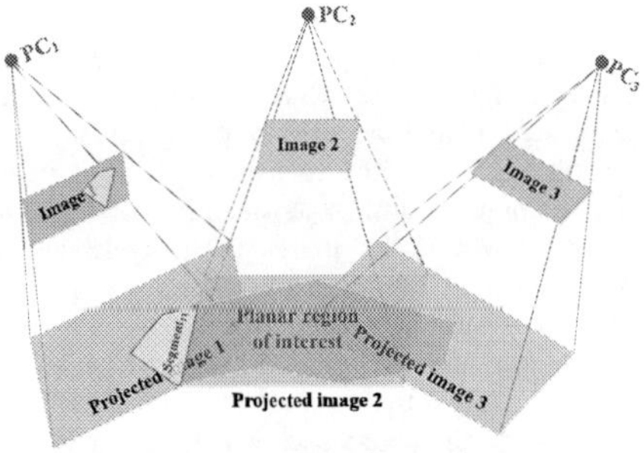

Figure 10. Rendering a segment which is visible in a single image (segment$_{11}$)

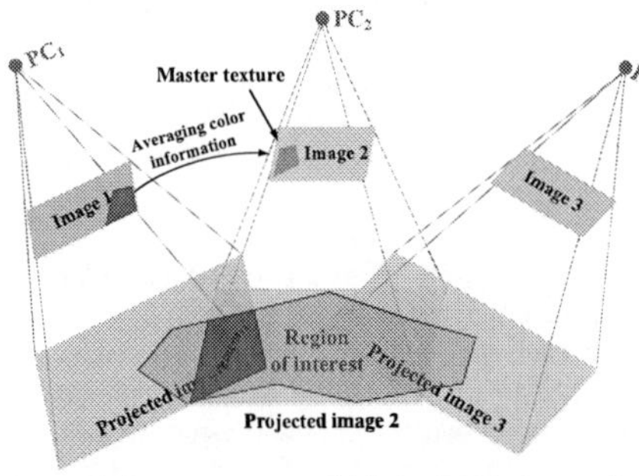

Figure 11. Rendering a segment which is visible in two images

3.4 Rendering and visualization

Once the parts of planar regions are decomposed according to their visibility within individual images, a rendering procedure should be implemented to render and visualize them on the screen. The objective of this procedure is to apply the identified textures, within the images, to the segmented planar surface in 3D space and visualize the textured surfaces on the 2D screen. More specifically, in rendering procedure, the clipped textures are rendered onto the segmented planar surface in 3D object space, which is then mapped to the destination image (on the screen) using projective transformation.

In this research, the transformation between object space and texture space was already established using the collinearity equations in the previous section while the transformation between the object space and screen is carried out using the Open Graphics Library (OpenGL) software interface (Shreiner, 2009). In order to optimize the performance of the rendering procedure, OpenGL is only able to render convex and solid planar surfaces onto 2D screen space. However, in reality, we are also dealing with concave planar surfaces, planar surfaces with holes, and complex planar surfaces. In order to handle the projection of theses surfaces onto a 2D screen, they should be tessellated into simple convex polygons. Therefore, the Delaunay triangulation algorithm (Bowyer, 1981; Watson, 1981) is utilized to subdivide concave, hollow, and complex polygons into easier-to-render convex polygons (triangles). Since the correspondence between texture space and object space has already been established through the texturing procedure, the vertices of the derived triangles, in the object space, are identified on the assigned texture to the planar region of interest to determine the parts of the texture that belongs to those triangles. The clipped textures are then rendered onto the triangles and visualized in 2D screen space.

One should note that the identified textures and the objects to be textured, on the screen, are rarely the same size in pixels. Therefore, filtering techniques should be applied to determine how each pixel, in the original texture image, should be expanded or shrunk to match a screen pixel size (Figure 12). In this case, the color information for each pixel on the screen should be simulated or interpolated based on the utilized texture image. If the quality of the texture image is less than the screen resolution, the color information for each pixel in the filtered texture image is sampled using the color information from the nearest neighboring pixel in the original texture image. However, if the quality of the texture image is higher than the screen resolution, the assigned color to each pixel in the filtered texture is derived by linear interpolation of its neighboring pixels in the original texture image.

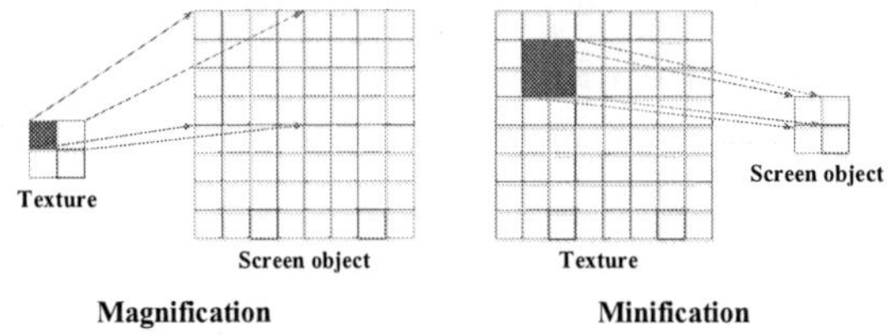

Figure 12. Filtering methods for adapting a texture image to the screen resolution

4. EXPERIMENTAL RESULTS

In this section, experiments using the segmented planar regions from an airborne laser scanning data and overlapping images are conducted to verify the feasibility of the proposed texturing approach. This texturing procedure is performed to facilitate the interpretation of the laser scanning-derived surfaces. The utilized dataset for this experiment includes six overlapping airborne laser strips and eight aerial images captured over an urban area in Calgary, Canada. The airborne laser scanning data was acquired by an Optech ALTM 3100 laser scanner and aerial images were collected using an RC30 analogue camera. This dataset was acquired from a flying height of 770m, with a ground sampling distance of 6cm for the images and an average point density of 2pts/m^2 for the laser scanning data. Figures 13.a and 13.b show one of the aerial images and the laser scanning data (displayed in different colors according to the height) over the area of interest, respectively.

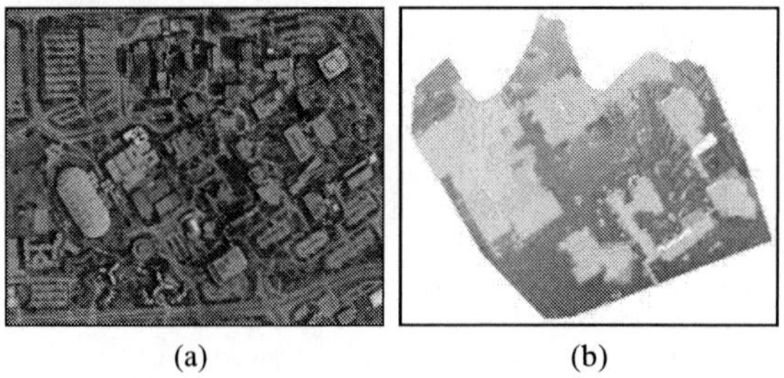

(a) (b)

Figure 13. (a) An aerial photo and (b) and airborne laser scanning data colored according to height over the area of interest (University of Calgary campus)

The utilized laser scanning point cloud is initially segmented to establish the individual planar regions which will be textured using the images. This dataset includes 421,974 points, where 387,643 of them have been segmented into 72 planar regions. Figures 14.a shows the planar features segmentation outcome for this point cloud. The boundary points of the segmented planar regions (inner and outer boundary points) are then detected and utilized for texturing these regions using the available images. Figures 14.b shows the boundaries of the segmented planar regions within the utilized laser scanning point cloud. The segmented planar regions are finally textured using the texturing approach introduced in the previous section. The laser scanning points which had not been segmented into planar regions were then individually projected onto the image space and colored using the spectral/color information of the nearest neighboring pixel to the projected image coordinates, after visibility analysis using the Z-buffer approach.

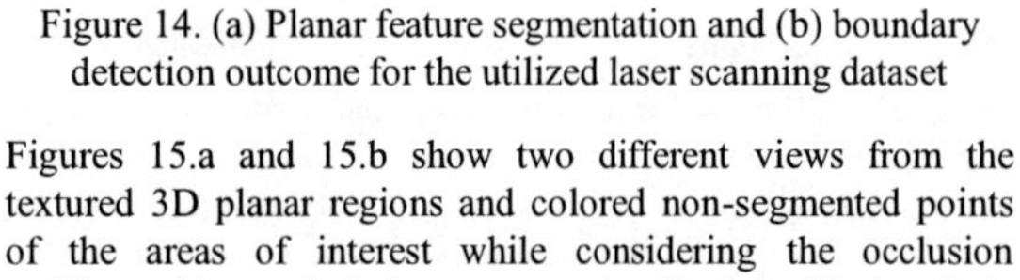

(a) (b)

Figure 14. (a) Planar feature segmentation and (b) boundary detection outcome for the utilized laser scanning dataset

Figures 15.a and 15.b show two different views from the textured 3D planar regions and colored non-segmented points of the areas of interest while considering the occlusion problems (the occluded areas are visualized in black color). Qualitative evaluation of the derived texturing results through visual inspection of Figures 15.a and 15.b verifies the feasibility of the proposed approach for the texturing planar regions using overlapping images while considering the visibility of those regions within the images.

(a)

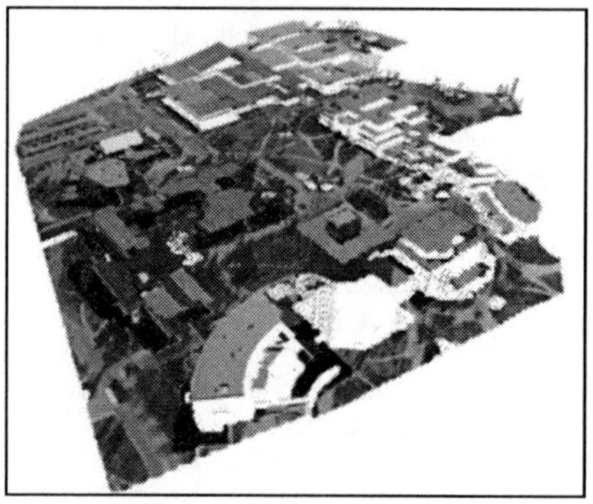

(b)

Figure 15. (a, b) Two different views from the textured planar regions and individually colored non-segmented points

5. CONCLUSIONS AND RECOMMENDATIONS FOR FUTURE RESEARCH WORK

This paper introduced a new region-based texturing procedure for the establishment of the link between the laser scanning positional and the images' spectral information. This texturing procedure is implemented while investigating the visibility of the segmented planar regions in the provided images. This investigation is conducted in two steps: 1) a new visibility analysis technique is utilized to identify the visible parts of a given region within the individual images, and 2) a new occlusion detection approach is implemented to check whether the visible part/parts of a region in an image is occluding/being occluded by the other regions. The proposed visibility analysis technique overcomes the computational inefficiency of existing approaches (i.e., the approaches which are implemented based on point-based texturing procedures) by checking the visibility of the boundary points of a given region rather than its entire points. Once the visibility of the segmented regions within the existing images is checked, a decomposition technique is applied to identify the parts of the segmented planar regions which should be textured using individual images. A rendering procedure is finally performed in an OpenGL environment to apply the 2D texture images onto the segmented planar regions. Qualitative evaluation of the derived segmentation results verified the feasibility of the proposed approach for texturing of laser-scanning derived surfaces.

In conclusion, the introduced texturing procedure resolves the computational inefficiency of the point-based texturing technique and provides continuous spectral information along the segmented planar regions. The textured 3D surfaces can be

This contribution has been peer-reviewed.
doi:10.5194/isprsarchives-XL-5-W4-115-2015

ultimately utilized for different surface interpretation, object extraction, modelling, and monitoring applications. Future research work will be focused on the extension of the proposed approach for texturing non-planar surfaces (e.g., linear/cylindrical features, cones, spheres) to provide completely textured scenes. In addition, the assigned textures to the segmented planar regions can be processed using image processing techniques (e.g., image segmentation) to identify and quantify characteristics of individual planar features (e.g., possible deteriorations or cracks that have not been properly represented in the laser scanning point cloud).

ACKNOWLEDGEMENTS

The authors would like to thank Tecterra and the Natural Sciences and Engineering Research Council of Canada Discovery and Strategic Project Grants (NSERC) for financial support of this research work.

REFERENCES

Ahmar, F., Jansa, J., Ries, C., 1998. The generation of true orthophotos using a 3D building model in conjunction with a conventional DTM, in: *International Archives of Photogrammetry, Remote Sensing and Spatial Information Science*. 32 (Part 4), pp. 16–22.

Alshawabkeh, Y., Haala, N., 2005. Automatic multi-image photo-texturing of complex 3D scenes, in: *Proceedings of CIPA 2005 XX International Symposium*. Torino, Italy.

Bannai, N., Fisher, R.B., Agathos, A., 2007. Multiple color texture map fusion for 3D models. *Pattern Recognition Letters*. 28, 748–758.

Bowyer, A., 1981. Computing Dirichlet tessellations. *Computer Journal*. 24, 162–166.

Chen, Z., Zhou, J., Chen, Y., Wang, G., 2012. 3D texture mapping in multi-view reconstruction, in: *Advances in Visual Computing*, Lecture Notes in Computer Science. Springer, Berlin, Germany, pp. 359–371.

El-Hakim, S.F., Remondino, F., Voltolini, F., 2003. Integrating techniques for detail and photo-realistic 3D modelling of castles. *GIM Int* 22 21–25.

El-Omari, S., Moselhi, O., 2008. Integrating 3D laser scanning and photogrammetry for progress measurement of construction work. *Automation in Construction*. 18, 1–9.

Filin, S., Pfeifer, N., 2005. Neighborhood systems for airborne laser data. Photogramm. Eng. Remote Sens. 71, 743–755.

Guelch, E., 2009. Advanced matching techniques for high precision surface and terrain models, in: *Proceedings of Photogrammetric Week 2009*. Heidelberg, Germany, pp. 303–315.

Habib, A., Ghanma, M., Mitishita, E.A., 2004. Co-registration of photogrammetric and LiDAR data: methodology and case study. *Brazilian Journal of Cartography*. 56, 1–13.

Habib, A., Kim, E.M., Kim, C., 2007. New methodologies for true orthophoto generation. *Photogrammetric Engineering and Remote Sensing*. 73, 25–36.

Habib, A., Schenk, T., 1999. New approach for matching surfaces from laser scanners and optical sensors, in: *International Archives of Photogrammetry, Remote Sensing and Spatial Information Sciences*. 32(3-W14), pp. 55–61.

Jarvis, R. A., 1977. Computing the shape hull of points in the plane. in: *Proceedings of the IEEE Computing Society Conference on Pattern Recognition and Image Processing*, pp. 231-241.

Kazhdan, M., Bolitho, M., Hoppe, H., 2006. Poisson surface reconstruction, in: *Proceedings of Symposium on Geometry Processing 2006*. pp. 61–70.

Kim, C., Habib, A., 2009. Object-based integration of photogrammetric and LiDAR data for automated generation of complex polyhedral building models. *Sensors* 9, 5679–5701.

Koch, M., Kaehler, M., 2009. Combining 3D laser-scanning and close-range Photogrammetry: An approach to exploit the strength of both methods, in: *Proceedings of the 37th International Conference of Computer Applications in Archaeology*. Williamsburg, Virginia, USA, pp. 1–7.

Kraus, K., 1993. Photogrammetry, Vol.1, Fundamentals and Standard Processes. Dummler Verlag, Bonn.

Kuzmin, Y.P., Korytnik, S.A., Long, O., 2004. Polygon-based true prthophoto generation, in: *International Archives of Photogrammetry, Remote Sensing and Spatial Information Science*. 35. pp. 529–531.

Lamond, B., Watson, G., 2004. Hybrid rendering - a new integration of photogrammetry and laser scanning for image based rendering, in: *Proceedings of Theory and Practice of Computer Graphics*. pp. 179–186.

Lari, Z., Habib, A., 2014. An adaptive approach for the segmentation and extraction of planar and linear/cylindrical features from laser scanning data. *ISPRS Journal of Photogrammetry and Remote Sensing*. 93, 192–212.

Pauly, M., Gross, M., Kobbelt, L.P., 2002. Efficient simplification of point-sampled surfaces, in: *Proceedings of the Conference on Visualization (VIS) 2002*. IEEE Computer Society, Washington, DC, USA, pp. 163–170.

Shreiner, D., 2009. OpenGL Programming Guide: The Official Guide to Learning OpenGL, Versions 3.0 and 3.1 (7th Edition). Addison-Wesley Professional, Indianapolis, IN, USA.

Vosselman, G., Gorte, B.G.H., Sithole, G., Rabbani, T., 2004. Recognising structure in laser scanner point clouds, in: *The International Archives of the Photogrammetry, Remote Sensing and Spatial Information Sciences*, 8-W2. pp. 33–38.

Wang, L., Kang, S.B., Szeliski, R., Sham, H.Y., 2001. Optimal texture map generation using multiple views, in: *Proccedings of Computer Vision and Pattern Recognition 2001*. pp. 347–354.

Watson, D.F., 1981. Computing the n-dimensional Delaunay tessellation with application to Voronoi polytopes. *Computer Journal*. 24, 167–172.

Weiler, K., Atherton, P., 1977. Hidden surface removal using polygon area sorting, in: *Proceedings of the 4th Annual Conference on Computer Graphics and Interactive Techniques*, SIGGRAPH '77. ACM, New York, NY, USA, pp. 214–222.

Xu, L., Li, E., Li, J., Chen, Y., Zhang, Y., 2010. A general texture mapping framework for image-based 3D modeling, in: *Proceedings of 17th IEEE International Conference on Image Processing (ICIP)*. pp. 2713–2716.

This contribution has been peer-reviewed.
doi:10.5194/isprsarchives-XL-5-W4-115-2015

3D POINT CLOUD MODEL COLORIZATION BY DENSE REGISTRATION OF DIGITAL IMAGES

N. Crombez, G. Caron, E. Mouaddib

University of Picardie Jules Verne
MIS Laboratory
33 rue Saint Leu
80039 Amiens CEDEX 1
{nathan.crombez, guillaume.caron, mouaddib}@u-picardie.fr

Commission V, WG V/4

KEY WORDS: Point clouds, Virtual and Visual Servoing, Images Registration, Colorization

ABSTRACT:

Architectural heritage is a historic and artistic property which has to be protected, preserved, restored and must be shown to the public. Modern tools like 3D laser scanners are more and more used in heritage documentation. Most of the time, the 3D laser scanner is completed by a digital camera which is used to enrich the accurate geometric informations with the scanned objects colors. However, the photometric quality of the acquired point clouds is generally rather low because of several problems presented below.
We propose an accurate method for registering digital images acquired from any viewpoints on point clouds which is a crucial step for a good colorization by colors projection. We express this image-to-geometry registration as a pose estimation problem. The camera pose is computed using the entire images intensities under a photometric visual and virtual servoing (VVS) framework. The camera extrinsic and intrinsic parameters are automatically estimated. Because we estimates the intrinsic parameters we do not need any informations about the camera which took the used digital image. Finally, when the point cloud model and the digital image are correctly registered, we project the 3D model in the digital image frame and assign new colors to the visible points.
The performance of the approach is proven in simulation and real experiments on indoor and outdoor datasets of the cathedral of Amiens, which highlight the success of our method, leading to point clouds with better photometric quality and resolution.

1. INTRODUCTION

Cultural heritage contains important remnants of our past and has to be carefully safeguarded. That is why patrimony documentation represents an important research issue. Cultural heritage includes monuments, works of art and archaeological landscapes. These sites are frequently faced with environmental conditions, human deterioration and can be victim of their age. The E-Cathedrale[1] research program was born in this context of cultural heritage preservation. It is dedicated to the digitization of the cathedral of Amiens, in France, to get a complete and precise model never obtained. This cathedral reaches an internal height of 42.30 meters for 200,000 cubic meters interior volume. Among French cathedrals of which the construction is over, the one of Amiens is the tallest and has the greatest interior volume. The cathedral was built between 1220 and 1270 and has been listed as a UNESCO World Heritage Site since 1981.
Recent technological advances have enabled the development of tools and methods, like 3D laser scanners which allow to quickly and accurately scan the spatial structure of monuments. The scan result is a dense and panoramic 3D point cloud of the scene. The first three scanning campaigns of the cathedral have allowed to obtain billions of points (Fig. 1) with approximately 2 millimeters sampling resolution on the details and less than 5 millimeters on the other parts. A digital camera is included into the laser scanner to assign color informations to each acquired 3D point. However, the photometric quality of the acquired point clouds model is generally rather low because of several problems. First of all, the scanner camera has a low resolution compared to the 3D scanner acquisition itself, because of that a 3D point and its nearest neighbors are colorized with the same digital image pixel. This phenomenon creates a blur effect on the point cloud model which are not representative of the real visual object aspect. Secondly, we do not recover the real-world luminance nuances because of the low photometric dynamic of the scanner camera. Finally, to entirely cover a large scene, the laser scanner has to be placed at different geographical positions. The obtained point clouds are acquired at different time of the day, so at different sun exposures. The merging of these point clouds leads always to visual inconsistencies (Fig. 2). Even if the geometric precision is really accurate the visual aspect of the virtual model is clearly unrealistic. This paper proposes a method to colorize a 3D point cloud model using digital images acquired from any viewpoints.

Figure 1: View of the Amiens cathedral point clouds model

[1]https://mis.u-picardie.fr/E-Cathedrale/

This contribution has been peer-reviewed.
doi:10.5194/isprsarchives-XL-5-W4-123-2015

Figure 2: Color problems resulting of the point clouds merging and the scanner camera low quality

2. RELATED WORK

In recent years, the use of 3D technologies has become the major method to accurately measure architectural or historical objects because of their high performance. These measured "objects" can be small as the Sidamara grave (Altuntas et al., 2007) and huge as a desert palace in Jordan (Al-kheder et al., 2009) or as the Todai-ji in Japan, the world biggest wooden edifice (Ikeuchi et al., 2007). In all cases, the aim is to obtain an accurate documentation with a high level of realism for protection, conservation and for possible future restoration or archaeologists studies. Several measurement acquisition methods exist: with terrestrial laser scanners (Haddad, 2011), terrestrial or aerial photogrammetric methods or by close range photogrammetry (Moussa et al., 2012).
It seems interesting to use the laser scanner and the digital camera separately. Indeed, terrestrial laser scanner measurements can be combined with Close Range Photogrammetry method or with digital images to improve the texture quality and resolution with a photometric realism (Al-kheder et al., 2009) (Henry et al., 2012). These methods require the registration of the data sets which are collected by the laser scanner and the digital images acquired by the camera. This registration is a crucial step for a good colorization, this is why it has long been studied and many approaches has been proposed. These approaches have been categorized by M. Corsini and al. (Corsini et al., 2012).
The first category contains the simplest methods, when the two device poses are known relatively to each other. This method is used in (Abmayr et al., 2004) where 3D point clouds and 360 degrees high resolution images are acquired from a same geographical position. Therefore, transforming both coordinate systems into each-other is easily done. However, to cover a large environment the scanner has to be placed far from the scene. The camera is far from the scene too, therefore the acquired digital images have a lack of visual details.
Registration can be also done using corresponding features from both data sets. These corresponding features can be points, lines or else. These can be manually chosen or automatically detected. In (Adan et al., 2012), points are manually chosen. Point clouds are used to generate ortho-images and corresponding points are manually selected in the virtual ortho-images and the digital one. A transformation matrix is computed using these sparse sets of corresponding points to register images and the 3D point cloud. Points can be automatically detected using SIFT, ASIFT or other point features extraction methods. In (Moussa et al., 2012), features are extracted using SIFT in the digital image and in a point-based environment model. In some cases (Alshawabkeh and Haala, 2004), manually or automatically selected, lines may be more appropriate given the straight geometry of the scanned environment. Laser scanners assign to the acquired points a reflectance value. This reflectance is sometimes used to make the data of the two systems more similar (Mastin et al., 2009) and facilitate the features detection.
Registration can be done comparing 3D model shapes and objects contours detected in the digital image (Belkhouche et al., 2012). Some works are using several images to obtain a sparse 3D point cloud, for example by Structure from Motion (SfM). Then, this point cloud is registered with the 3D target model. This 3D/3D registration performed, the images used for the SfM can be aligned to the model and used to colorize it. However, these methods require a large number of digital images.
Finally, other methods try to minimize mutual informations between digital images and 3D model geometric features. Different features like normals, reflection, ambient occlusions are compared in (Corsini et al., 2009). The results of their study show better image-to-geometry registrations using normals maps and a combination of normals and ambient occlusion.
The huge size of the cathedral leads us to use the automatic features extraction methods. The color problems presented above make difficult the features detection and registration. Thanks to a point clouds colors homogenization, we are able to automatically detect and match point features between a virtual image of the 3D scene and a digital one. A camera pose is computed using these sparse sets of 2D/3D corresponding points to register the digital image and the 3D point cloud model. However, these kind of registrations are not enough accurate, the digital and the virtual images are not correctly aligned. We propose to correct the obtained camera pose using the entire images intensities as features. This dense photometric feature has the advantage of avoiding geometric features detection and matching and has the potential to reach a higher registration precision. The photometric feature has been used for visual tracking of planes (Benhimane and Malis, 2004). In the latter work, image intensity variations are related to the plane motion in the image, expressed as the so called homography, a 2D motion model. In our case, we propose a method to directly use image pixel intensities to estimate the extrinsic and intrinsics camera parameters in a virtual 3D space. For this reason, taking inspiration of (Benhimane and Malis, 2004), we need to relate an image intensity variation to the camera pose variation. This theoretical framework was first tackled for the photometric vision based robotic system control, that is the photometric visual servoing (Collewet and Marchand, 2011). For over twenty years, visual servoing allows to control a robot using various visual features (points, lines, moments etc...). It has been previously extended to Virtual and Visual Servoing (VVS) (Marchand and Chaumette, 2002), aiming to compute a camera pose using image measurements, by non-linear optimization, and exploiting visual servoing knowledge. Our approach extends the photometric visual servoing (Collewet and Marchand, 2011) to the VVS involving the photometric feature.

This paper is organized as follows. First, Section 3.1 compares different similarity measures to determine the most suitable cost function for our data. Section 3.2 describes the photometric virtual and visual servoing. Then Section 4. presents some results of simulations to validate our method. Section 5. shows results of point clouds colorizations using our method. Finally, conclusions and future works are presented in Section 6..

3. PROPOSED APPROACH

The formation of a two-dimensional representation of a three-dimensional world has been extensively studied. A perspective camera transforms a 3D point $\mathbf{X} = (X\,Y\,Z\,1)^T$ in the image plane at homogeneous coordinates $\mathbf{x} = (x\,y\,1)^T$ by the perspective pro-

This contribution has been peer-reviewed.
doi:10.5194/isprsarchives-XL-5-W4-123-2015

jection :

$$\mathbf{x} = (x\,y\,1)^T = pr(\mathbf{X}) \quad \text{with} \quad x = f\tfrac{X}{Z} \quad \text{and} \quad y = f\tfrac{Y}{Z} \qquad (1)$$

where f is the camera focal length. To obtain a visual representation of the image, homogeneous coordinates image points are converted in pixel coordinates. A pixel $\mathbf{u} = (u\,v\,1)^T$ is obtained by $\mathbf{u} = \mathbf{Kx}$ where the matrix $\mathbf{K}$ contains the camera intrinsic parameters :

$$\mathbf{K} = \begin{pmatrix} \alpha_u & 0 & u_0 \\ 0 & \alpha_v & v_0 \\ 0 & 0 & 1 \end{pmatrix} \qquad (2)$$

α_u and α_v are scale factors which relate pixels to distance and u_0 and v_0 represent the principal point. The focal length f is integrated in α_u and α_v, thus the complete 3D point $\mathbf{X}$ transformation to pixel point $\mathbf{u}$ is expressed as :

$$\mathbf{u} = \mathbf{K}pr(\mathbf{X}) \qquad (3)$$

A digital image taken in the real world depends on the extrinsic camera parameters (position and orientation in the world) and the intrinsic camera parameters $\mathbf{K}$.

We note $P_{i,j}(\mathbf{X}, I)$ the jth 3D point from the point cloud acquired by the station i where $\mathbf{X} = (X\,Y\,Z\,1)^T$ is its coordinates and I its intensity. The point cloud acquired by the station i is noted $\mathbf{PC}_i = \{P_{i,j}(\mathbf{X}, I)/\forall j \in [\![0, N_i - 1]\!]\}$ with N_i the number of points acquired by the ith station. The complete model $\mathbf{PCm}$ contains the N registered point clouds: $\mathbf{PCm} = \{\mathbf{PC}_i/\forall i \in [\![0, N-1]\!]\}$.
A virtual image is generated by the projection of 3D points $P_{i,j}(\mathbf{X}, I)$ to 2D image points by the complete perspective projection (3). We expressed the formation of a virtual image of a 3D point cloud by $\mathbf{I_v}(\mathbf{PCm}, \mathbf{K}, {}^c\mathbf{M}_o)$ where $\mathbf{K}$ contains the intrinsic virtual camera parameters and ${}^c\mathbf{M}_o$ is the virtual camera pose in the 3D point cloud. The matrix ${}^c\mathbf{M}_o$ represents the transformation from the $\mathbf{PCm}$ (object (o)) frame to a camera (c) frame :

$${}^c\mathbf{M}_o = \begin{pmatrix} {}^c\mathbf{R}_{o(3\times3)} & {}^c\mathbf{t}_{o(1\times3)} \\ \mathbf{0}_{(3\times1)} & 1 \end{pmatrix}_{(4\times4)} \qquad (4)$$

This matrix ${}^c\mathbf{M}_{o(4\times4)}$ is formed by a rotation matrix ${}^c\mathbf{R}_{o(3\times3)}$ and a translation vector ${}^c\mathbf{t}_{o(1\times3)}$.

To register a digital image $\mathbf{I}$ on a point cloud $\mathbf{PC}_m$, we need the extrinsic parameters of the camera which took $\mathbf{I}$ expressed in the $\mathbf{PC}_m$ frame (the transformation matrix ${}^c\mathbf{M}_o$) and the intrinsic camera parameters $\mathbf{K}$.

3.1 Photometric cost function comparisons

This section aims to determine the most suitable cost function in our study case for a dense visual servoing. We are looking for a robust measure of similarity to handle the color problems previously mentioned (Figure 2). The different similarity measures compared are the Zero-mean Normalized Sum of Squared Differences (ZNSSD) and the Zero-mean Normalized Cross Correlation (ZNCC) between digital images and virtual one in grayscale. We also compare these results with the Mutual Information (MI) between digital images and normal maps and reflexion maps. Normal and reflexion maps have been employed in (Corsini et al., 2009), we use the same methods to create our renderings and compute Mutual Information.
We compute these different measures around a virtual camera pose close to a digital image pose varying two degrees of freedom of the virtual camera. Camera motions are made along its x and y axes. We use the intrinsic camera parameters of the camera which took the digital image.
This study compares cost function shapes obtained with these correlation criteria (Figure 3), to facilitate the comparison we plot the opposite ZNCC and MI. We can observe that the ZNSSD and the ZNCC cost functions have a more convex shape than the others and a more pronounced minimum. The MI cost functions are more noisy than the others. These observations indicate that it is preferable to use the ZNSSD or ZNCC or the mutual information based on the intensities in order to converge the pose of the virtual camera. Finally, the low complexity of the ZNSSD calculation whether for the criteria minimization/maximization or for the interaction matrix computation, confirms the relevance of the choice of the ZNSSD.
Centering and normalizing the intensities of both virtual and digital images permits to handle the color problems related to the merged point clouds. The error to minimize is computed on these intensities. That is why the cost function shape is so clean, convex and has a clearly distinguishable minimum. Moreover, the ZNSSD is faster to compute than de MI.

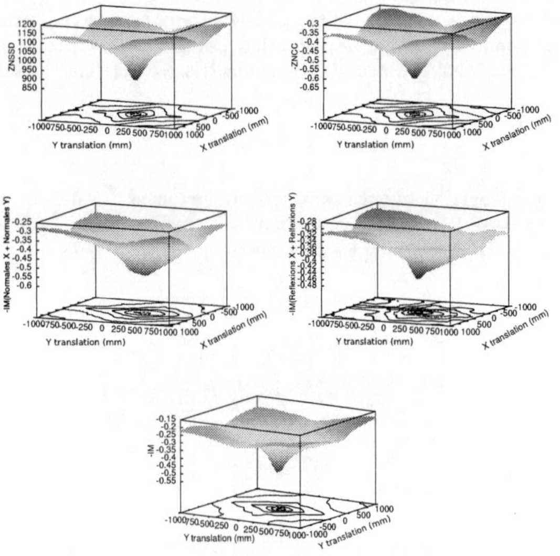

Figure 3: Example of cost functions for five correlation criteria

3.2 Photometric virtual and visual servoing

We express the extrinsic and intrinsic camera parameters estimation problem as an optimization problem under the visual servoing framework. However, unlike the photometric visual servoing (Collewet and Marchand, 2011), our method does not try to minimize the difference between two real images from the same vision sensor. Our method minimizes the difference between a desired digital image $\mathbf{I}^*$ and a virtual image $\mathbf{I_v}(\mathbf{PCm}, \mathbf{K}, {}^c\mathbf{M}_o)$. For this, the criterion to minimize is:

$$\mathbf{e}(\mathbf{r}) = ||\mathbf{I}^* - \mathbf{I_v}(\mathbf{PCm}, \mathbf{K}, {}^c\mathbf{M}_o)|| \qquad (5)$$

where $\mathbf{r} = (t_X, t_Y, t_Z, \theta u_X, \theta u_Y, \theta u_Z, \alpha_u, \alpha_v, u_0, v_0)$. The first six elements of $\mathbf{r}$ represent the current virtual camera pose ${}^c\mathbf{M}_o$ by 3 translation and 3 rotation elements (considering the rotation is represented as an angle and a unit vector) and the last four are the intrinsic camera parameters.
The virtual camera motion control law is computed thanks to a

This contribution has been peer-reviewed.
doi:10.5194/isprsarchives-XL-5-W4-123-2015

Levenberg-Marquardt based one:

$$\dot{\mathbf{r}} = \lambda(\mathbf{H} + \mu diag(\mathbf{H}))^{-1}\mathbf{L}_I^{\mathrm{T}}(\mathbf{I}^* - \mathbf{I}_\mathbf{v}(\mathbf{PCm}, \mathbf{K}_\mathbf{v}, {}^c\mathbf{M}_o)) \quad (6)$$

with $\mathbf{H} = \mathbf{L}_I^{\mathrm{T}}\mathbf{L}_I$, considering $\mathbf{L}_I$ is the interaction matrix related to luminance of image $\mathbf{I}$. This interaction matrix is also named the image Jacobian, i.e. the matrix linking the image feature motion to the camera pose variation. The interaction matrix is derived under the constant illumination consistency (see (Collewet and Marchand, 2011) for details). In the iterative process, the pose ${}^c\mathbf{M}_o$ is updated with the pose increment $\mathbf{r}_6$ composed by the first six elements of $\dot{\mathbf{r}}$ using the exponential map of the special euclidean group SE(3).

$$\begin{aligned} {}^{c^{[t+1]}}\mathbf{M}_o = (e^{(\mathbf{r}_6)})^{-1} \times {}^{c^{[t]}}\mathbf{M}_o = \\ ({}^{c^{[t]}}\mathbf{M}_{c^{[t+1]}})^{-1} \times {}^{c^{[t]}}\mathbf{M}_o = \\ {}^{c^{[t+1]}}\mathbf{M}_{c^{[t]}} \times {}^{c^{[t]}}\mathbf{M}_o \quad (7) \end{aligned}$$

The intrinsic camera parameters are incremented with the last four elements of $\mathbf{r}$.
The optimization stops when the residual error becomes constant. The optimization is made more robust using an M-Estimator (Comport et al., 2003) which is based on robust statistics (Huber, 1981).

4. SIMULATION

In this Section, simulation results are presented to validate our approach. We use a 3D scene composed by four simple geometric meshed objects as test environment (figure 4).

Figure 4: Virtual 3D test environment

A virtual camera is placed inside the 3D environment and a virtual image is generated from this pose. The intrinsic virtual camera parameters are known. We consider this virtual image as the desired image $\mathbf{I}^*$ (figure 5a).
The virtual camera is randomly moved inside a sphere of 0.2 meters around the desired pose. The camera orientations are randomly incremented by an angle between -5 and 5 degrees around its three axes. The intrinsic camera parameters are randomly incremented by a number between -20 and 20 pixels. A virtual image is generated with these new camera extrinsic and intrinsic parameters. This image is considered as the initial image. We use the photometric virtual and visual servoing from this initial pose. Figure 5b shows the image differences between the desired and the initial image. Figure 5c shows the image differences between the desired and the image generated at the end of the servoing. We can see that the virtual camera has converged to the desired camera pose (figure 5d) by minimizing the residual error between the desired image and the images generated during the servoing. The intrinsic camera parameters have also been correctly estimated as shown by the figure 5e.
This experiment has been repeated one hundred times with randomly extrinsic an intrinsic parameters. The virtual camera parameters have successfully converged to the desired parameters

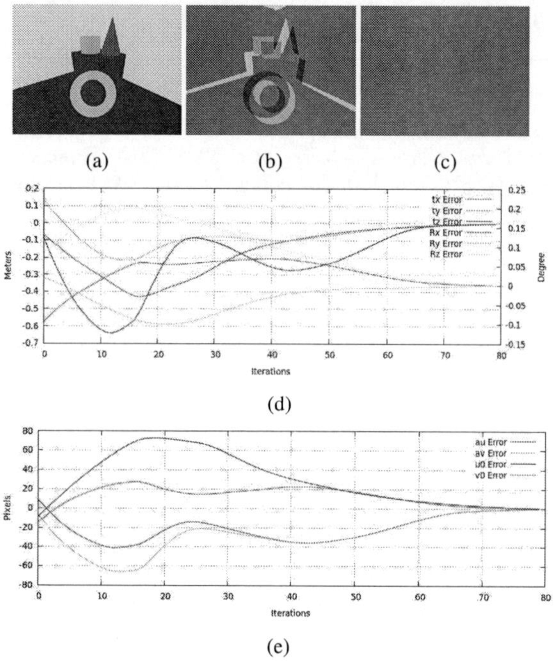

Figure 5: Simulated experiment : (a) desired image, (b) differences between desired and inital image, (c) differences between desired and final image, (d) translations and rotations error, (e) intrinsic camera parameters error

for 95 experiments. The 5 others experiments have failed because the overlap of the geometric objects projection in the desired image and in the initial image was not enough.

5. COLORIZATION RESULTS

5.1 Methodology

The estimated pose allows us to transform each point of the cloud into the digital image frame. The 3D point cloud data are initially expressed in object reference frame. We use the ${}^c\mathbf{M}_o$ from our pose estimation to express the point cloud in the camera reference. Then, using the perspective projection (3) with the estimated intrinsic camera parameters $\mathbf{K}$, we express the points into the 2D digital image plane $\mathbf{I}^*$. Our points have now real coordinates included in the digital image dimensions. We use a bilinear interpolation to assign a new color value to each 3D point. This RGB color value is computed with the weighted average of the RGB values of the four neighboring pixels in the digital image.
In most cases, the colorization or the texturing is done when the point cloud is meshed. In our case, we want to colorize directly the original point cloud. Because of that, we can not expect good colorization results without handling the occlusion issues. Indeed some points coming from parts of the scene which are invisible from the camera position may be considered as visible and colored. This can occur in two situations: when two (or several) points from different depths are on the same projected ray and so are projected into a same pixel or when every points, projected in a pixel, come from invisible parts of the scene. In these two cases, points, which are supposed to be invisible, receive a false new color. To fix these issues, before the colorization we apply the Hidden Point Removal (HPR) (Katz et al., 2007) on the point cloud to remove the 3D points which should be invisible as viewed from the estimated pose ${}^c\mathbf{M}_o$.

This contribution has been peer-reviewed.
doi:10.5194/isprsarchives-XL-5-W4-123-2015

5.2 Experimental results

First Experiment : we use our approach to colorize the point cloud of the south portal shown in figure 2 with the digital image shown in figure 6a. The camera is calibrated, so its intrinsic parameters are known.

(a)

(b)

(c)

Figure 6: Colorization of the south portal : (a) desired digital image, (b-c) colorized point cloud, (e) evolution of the residual error during the servoing

We use our method to determine from where the digital image has been taken (in the 3D scene which contains the point cloud). The virtual and visual servoing moved the virtual camera by computing iteratively its six degrees of freedom and reached a stable state after 17 iterations. Figure 7 shows the evolution of the residual error during the servoing. Figure 8 represents the difference between the digital image that we want to reach and the virtual one at the first (a) and the last (b) iterations. The distance between the initial virtual camera pose and the optimal one is 674.50mm.

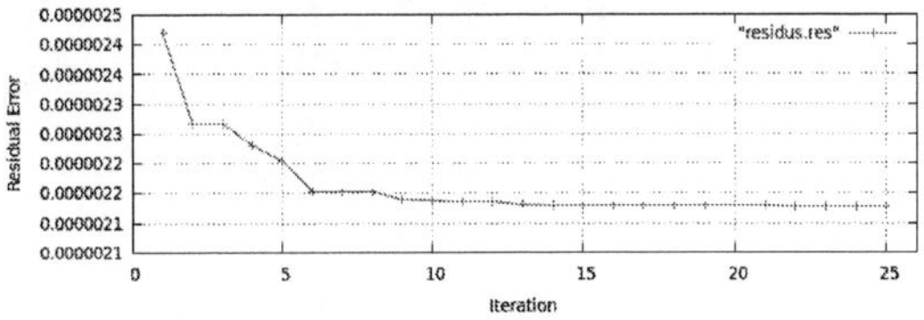

Figure 7: Evolution of the residual error during the servoing

With the estimated camera pose ${}^{c}\mathbf{M}_o$, the visible points of $\mathbf{PC_m}$ are projected in the digital image $\mathbf{I}^*$ (figure 6a) and are colorized. Figure 6b and figure 6c show the point cloud $\mathbf{PC_m}$ after its colorization. We obtain a point cloud which seems perfectly colored, with much more uniform colors but if we closely look at the cloud (figure 9), we can see that we need to have several digital images for covering the portal entirely. We can also see the lack of accuracy on the statue face because of a bad digital image point of view choice. To improve the detail visibilities we need a better image resolution for the colorization. Therefore, we have to take images from nearest positions from the statue.

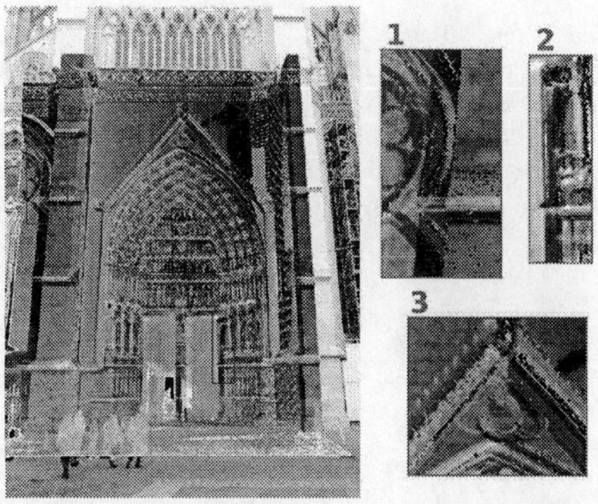

(a)

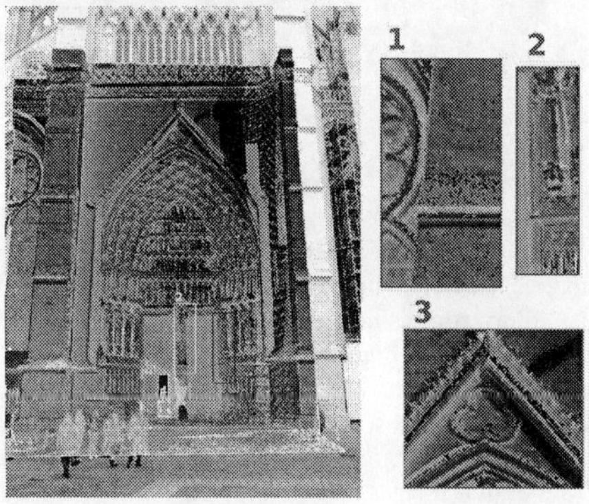

(b)

Figure 8: Differences images between the desired digital image and (a) the intial image (b) the final image

Figure 9: The south portal point cloud colorized. The points recolorized in red are invisible from the digital image pose

Second experiment : we use our pose optimization method to entirely colorize a smaller part of the south portal : the Virgin Mary statue. Our goal is to obtain an accurately colorized point cloud with photo-realistic quality. To obtain this, ideally, the result would be to have each 3D point visible from the camera pose

This contribution has been peer-reviewed.
doi:10.5194/isprsarchives-XL-5-W4-123-2015

projected in one different pixel. Knowing the intrinsic camera parameters and the point cloud resolution, we computed the distance needed between the object and the camera to ensure that a pixel receives a unique projected 3D point. Empirically, we calculated a four meters distance needed between the statue and our camera to project each visible 3D point in one different pixel in the 2D image. Three digital images have been taken $\mathbf{I}_1^*$, $\mathbf{I}_2^*$,$\mathbf{I}_3^*$ around the statue for a maximum covering (figure 10). The three

Figure 10: Images of the Virgin Mary used to the colorization (These digital images have a 4032x3024 pixels resolution)

camera poses are estimated with the pose estimation presented before and we use our complete colorization process. Right now, the points colorization is done one image after the other. Points are projected in the first digital image, points covered by this one are colored and the others are labelled as not yet colored. Then, points are projected in a second digital image and only the covered and labelled points are colorized. This operation is repeated with all the digital images or when all the points are colorized. A comparison between the Virgin Mary point cloud before and after our colorization is shown in Figure 11. We took another

(a)

(b)

Figure 11: The Virgin Mary point cloud (a) before and (b) after our colorization.

digital image $\mathbf{I}_4$ of the Virgin Mary from a different view angle for a visual validation. This image has not been used for the point cloud colorization. Figure 12 shows the Virgin Mary point cloud from the same camera pose than the new image after and before our colorization process. As we can see, the good colorization results are clearly undeniable. The two images focused on the Virgin's face (Figure 12) show us the really accurate color resolution of the point cloud after its colorization. The digital image $\mathbf{I}_4$ has been taken 5 months (different season) after we took the three images $\mathbf{I}_1^*$, $\mathbf{I}_2^*$,$\mathbf{I}_3^*$ used for the colorization. This explains the colorimetry slight differences between $\mathbf{I}_4$ and the colored point cloud.

(a) (b) (c)

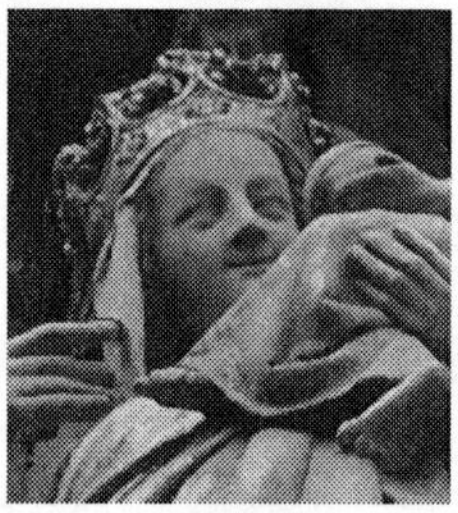

(d)

(e)

(f)

Figure 12: (a) A Virgin Mary statue image from a low-angle shot. (b) The original point cloud of the Virgin Mary from the same camera pose. (c) The colored point cloud of the Virgin Mary from the same camera pose. (d-f)A focus on the color resolution of the Virgin's face.

Third Experiment : we also used the colorization process on a point cloud of the painting in the chapel of Saint-Sebastian.

This contribution has been peer-reviewed.
doi:10.5194/isprsarchives-XL-5-W4-123-2015

Figure 13 shows a comparison between the point cloud colors acquired by the scanner (bottom area) and after its colorization (top area).

Figure 13: Painting in the chapel of the Saint-Sebastian before (bottom) and after (top) its colorization.

Fourth Experiment : thanks to the intrinsic camera parameters estimation, we can also use digital images without any information about the camera which took them. We colorize a part of the point cloud of the western faade portal using a digital image from the internet (figure 14a).

(a)

(b)

Figure 14: (a) Digital image of a part the western façade portal from the internet. (b) Point cloud of the western faade portal with points colored using the digital image

6. CONCLUSIONS AND FUTURE WORKS

We proposed an alternative method for automatically registering digital images over 3D point cloud. The existing manual methods used to detect the correspondence features of the two data sets can provide good results. However, these methods can be really tedious to handle an important size of data. Automatic detections of geometric features can be delicate because they require to have similarities between the digital images and the virtual images generated with the 3D point clouds.

In our approach, rather than using geometric image features, we use the entire image intensities which make the registration more robust and precise than by manually selecting points. In the future, it seems interesting to use the RGB information rather than only the intensities.

The optimization of the intrinsic camera parameters allows to use digital images without any informations about the camera which took them. The presented results show that our process provides point clouds with a realistic texture quality. Future works will be focused on the research of the optimal positions needed to obtain digital images to colorize a point cloud.

REFERENCES

Abmayr, T., Hartl, F., Mettenleiter, M., Heinz, A., Neumann, B. and Fr02hlich, C., 2004. Realistic 3d reconstruction - combining laserscan data with rgb color information.

Adan, A., Merchan, P. and Salamanca, S., 2012. Creating Realistic 3D Models From Scanners by Decoupling Geometry and Texture. International Conference on Pattern Recognition (ICPR) pp. 457–460.

Al-kheder, S., Al-shawabkeh, Y. and Haala, N., 2009. Developing a documentation system for desert palaces in Jordan using 3D laser scanning and digital photogrammetry. Journal of Archaeological Science 36(2), pp. 537–546.

Alshawabkeh, Y. and Haala, N., 2004. Integration of digital photogrammetry and laser scanning for heritage documentation. IAPRS pp. 12–23.

Altuntas, C., Yildiz, F., Karabork, H., Yakar, M. and Karasaka, L., 2007. Surveying and documentation of detailed historical heritage by laser scanning. XXI International CIPA Symposium pp. 01–06.

Belkhouche, Y., Buckles, B., Duraisamy, P. and Namuduri, K., 2012. Registration of 3d-lidar data with visual imagery using shape matching. Int. Conf. on Image Processing, Computer Vision, and Pattern Recognition pp. 749–754.

Benhimane, S. and Malis, E., 2004. Real-time image-based tracking of planes using efficient second-order minimization. IEEE/RSJ International Conference on Intelligent Robots and Systems pp. 943–948.

Collewet, C. and Marchand, E., 2011. Photometric Visual Servoing. IEEE Transactions on Robotics 27(4), pp. 828–834.

Comport, A., Marchand, E. and Chaumette, F., 2003. A real-time tracker for markerless augmented reality. In: ACM/IEEE Int. Symp. on Mixed and Augmented Reality, ISMAR, Tokyo, Japon, pp. 36–45.

Corsini, M., Dellepiane, M., Ganovelli, F., Gherardi, R., Fusiello, a. and Scopigno, R., 2012. Fully Automatic Registration of Image

Sets on Approximate Geometry. International Journal of Computer Vision 102(1-3), pp. 91–111.

Corsini, M., Dellepiane, M., Ponchio, F. and Scopigno, R., 2009. Image-to-geometry registration: a mutual information method exploiting illumination-related geometric properties. Computer Graphics Forum 28(7), pp. 1755–1764.

Haddad, N. A., 2011. From ground surveying to 3D laser scanner: A review of techniques used for spatial documentation of historic sites. Journal of King Saud University - Engineering Sciences 23(2), pp. 109–118.

Henry, P., Krainin, M., Herbst, E. and Ren, X., 2012. Rgb-d mapping : Using kinect-style depth cameras for dense 3d modeling of indoor environments. The International Journal of Robotics Research 31(5), pp. 647–663.

Huber, P. J., 1981. Robust Statistics. John Wiley and Sons.

Ikeuchi, K., Oishi, T. and Takamatsu, 2007. The great buddha project: Digitally archiving, restoring, and analyzing cultural heritage objects. Int. J. Comput. Vision 75(1), pp. 189–208.

Katz, S., Tal, A. and Basri, R., 2007. Direct visibility of point sets. ACM Trans. Graph. 26(3), pp. 24.

Marchand, E. and Chaumette, F., 2002. Virtual visual servoing: A framework for real-time augmented reality. In: Eurographics Conference Proceeding, Vol. 21(3), pp. 289–298.

Mastin, A., Kepner, J. and Fisher, J., 2009. Automatic registration of lidar and optical images of urban scenes. IEEE Conference on Computer Vision and Pattern Recognition pp. 2639–2646.

Moussa, W., Abdel-Wahab, M. and Fritsch, D., 2012. An automatic procedure for combining digital images and laser scanner data. ISPRS - International Archives of the Photogrammetry, Remote Sensing and Spatial Information Sciences pp. 229–234.

This contribution has been peer-reviewed.
doi:10.5194/isprsarchives-XL-5-W4-123-2015

AUTOMATIC TEXTURE AND ORTHOPHOTO GENERATION FROM REGISTERED PANORAMIC VIEWS

U. Krispel[a,d], H. L. Evers[b], M. Tamke[b], R. Viehauser[c], D. W. Fellner[d,e]

[a] Fraunhofer Austria, Graz, Austria - ulrich.krispel@fraunhofer.at
[b] Centre for Information Technology and Architecture (CITA), Copenhagen, Denmark - (heve, martin.tamke)@kadk.dk
[c] Institute for Computer Graphics and Vision (ICG), TU Graz, Austria
[d] Institute of ComputerGraphics and KnowledgeVisualization (CGV), TU Graz, Austria
[e] GRIS, TU Darmstadt & Fraunhofer IGD, Darmstadt, Germany

Commission V, WG V/4

KEY WORDS: spherical panorama, image registration, texturing, point clouds, machine learning, automated generation of semantic information

ABSTRACT:

Recent trends in 3D scanning are aimed at the fusion of range data and color information from images. The combination of these two outputs allows to extract novel semantic information. The workflow presented in this paper allows to detect objects, such as light switches, that are hard to identify from range data only. In order to detect these elements, we developed a method that utilizes range data and color information from high-resolution panoramic images of indoor scenes, taken at the scanners position. A proxy geometry is derived from the point clouds; orthographic views of the scene are automatically identified from the geometry and an image per view is created via projection. We combine methods of computer vision to train a classifier to detect the objects of interest from these orthographic views. Furthermore, these views can be used for automatic texturing of the proxy geometry.

(a) (b)

Figure 1: The data acquisition equipment: For range data acquisition, a Faro Focus 3D Scanner was used (a); additional color information was obtained using a customized adapter connecting the base tripod of the 3D scanner with a HDRI camera (b).

1. INTRODUCTION

Many 3D scanners provide more information than range data, as for instance color information. The scanners are equipped with a camera that takes photographs during the scanning session. These photographs are subsequently stitched in order to create a panoramic photo sphere at the scanners position. Such panoramic views can then be used to apply color information to pure range data.

Sole point clouds are unstructured and have no semantic information. This challenges their integration into existing work flows of stakeholders, such as engineers and architects (Tamke et al., 2014). The interplay of geometric and color information carries meaningful aspects that provide so far unused semantic information to 3d scan data. Objects that are hard to detect on a geometrical level, might be revealed in a visual inspection of the dataset and provide the information needed for semantic processing. In case of scanned building data this information can be stored alongside the data e.g. to make future renovations and reconstructions more efficient. Within the DURAARK project [1], the presented tool is already used to enrich the initially unstructured point cloud data with this semantic information on electrical installations.

The problem with wall mounted power sockets is that they usually stick only 3-4 mm out of the surrounding surface. This makes them hard to detect in pure range data, as they disappear in the geometric noise of a 3D scan. A recognition using a purely image based (2D) approach, with the photographs obtained during the scanning, will not succeed in all cases, as the perceived objects might be severely distorted. Therefore, it is necessary to combine the geometric information of the point cloud scans with the images of the session in order to create orthographic views of planar elements in the scene (e.g. walls) that contain the desired perceived objects (e.g. sockets). Computer vision methods can be applied using these views to train and perform object detection.

In this paper, we present such a work flow that identifies and creates orthographic views. These views are created from registered panoramic images and a proxy geometry that was derived from the point clouds.

2. RELATED WORK

Spherical panoramic imaging is a widely used technique to acquire photographs of a scene using a complete field of view, i.e. capture all surroundings of a specific viewpoint. These spherical panoramic images are typically created by the fusion of a number of photographs, taken from different directions at the same position. Alternatively, special hardware exists that acquires a panoramic view directly.

[1] `http://duraark.eu`

This contribution has been peer-reviewed.
doi:10.5194/isprsarchives-XL-5-W4-131-2015

The work of D'Annibale et al. (D'Annibale and Fangi, 2009) has been concerned with spherical photogrammetry, i.e. using spherical panoramic images for surveying and reconstruction. Phototexturing of 3D models from surveyed data has been in the focus of the work of Grammatikopoulos et al. (Grammatikopoulos et al., 2007) by combining 3D models from laser scanning and multi-image texture projection.

The automatic creation of texture maps from geometry has also been a relevant research topic, as this will aid in in many applications that involve the creation of textures for 3D models, for example in the entertainment industry. The work of Levy et al. (Lévy et al., 2002) for example creates an automatic texture atlas (or an atlas of charts) for an given object. A similar approach that obtains an atlas based on surface features was presented by Zhang et al. (Zhang et al., 2005).

Furthermore, the same set of problems has to be addressed when doing purely image-based reconstruction, as was the case in the work of Furukawa et al. (Furukawa et al., 2009) for indoor scenes, or the work of Pitzer et al. (Pitzer et al., 2010) which used a robot for a fully automated approach for indoor scene reconstruction. The work of Waechter et al. (Waechter et al., 2014) was concerned with the problem of large scale texturing for achieving photoconsistency over multiple images in image-based reconstruction.

3. METHOD OVERVIEW

Our method is designed for the acquisition of indoor scenes, therefore, we assume that most of the surfaces will be planar (floor, walls etc). The method identifies planar, rectangular regions in the scene and produces an orthographic view (image) of desired resolution. An overview of the work flow is shown in Figure 2.

In the data acquisition phase, the scene is scanned using a laser scanning device, which yields a point cloud and a low resolution panoramic image. Furthermore, a high resolution panoramic image is acquired using a camera.

The preprocessing phase consists of creating a 3D surface model of the scene using a semiautomatic approach, and the alignment of the high resolution panoramic image to the point cloud scan.

In the last step, the rectangular planar regions (patches) are identified from the surface model, and an orthographic view is created per patch, by projecting the color information from the high resolution panoramic image onto the patch at a desired resolution.

4. DATA ACQUISITION AND PREPROCESSING

After the physical data acquisition, the data needs to be preprocessed for the ortho view generation. The developed workflow consists of publicly available software components, which are partially interfaced with custom scripts. Used tools are: Faro Scene [2] point cloud software, CloudCompare [3] point cloud software, ptGUI Pro [4] panoramic image software and Rhinoceros 5[5] / Grasshopper [6] 3D modeling software. The point cloud is converted into E57 Format (Huber, 2011).

4.1 Measuring Equipment

The data acquisition has been done using a terrestrial laser scanner (Faro Focus 3D), as shown in Figure 1a. The acquired data

[2]http://www.faro.com/
[3]http://cloudcompare.org/
[4]http://www.ptgui.com/
[5]http://www.rhino3d.com/
[6]http://www.grasshopper3d.com/

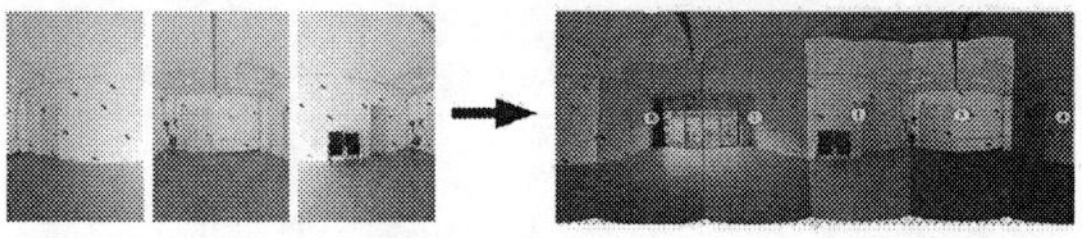

Figure 3: The stitching of Canon DSLR Images to panoramic HDRI was carried out in ptGUI Pro (a).

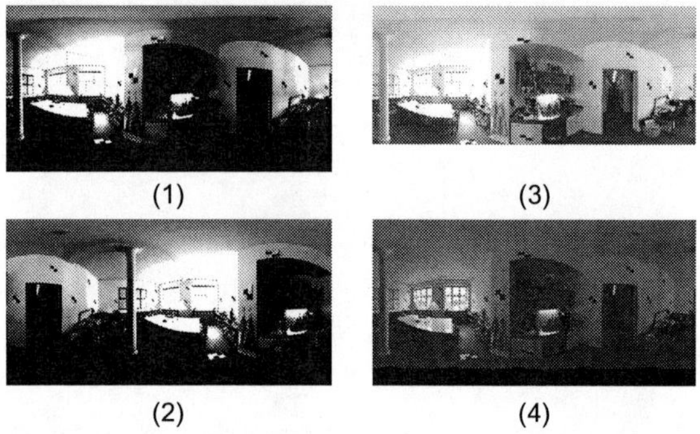

Figure 4: Image-based registration: the normalized panoramic image taken with the DSLR (2) has to be aligned to the normalized panoramic image which was exported from the laser scanning unit (1). The automatic alignment described in Section 4.2 produces an aligned high-resolution panoramic image (4) which is aligned to the panoramic image from the laser scanning unit (3).

is a 3D point cloud (Figure 5a) and a set of images, which can be stitched into a single spherical panoramic image. First experiments showed that the images acquired from the inbuilt camera in the Faro Focus 3D often exhibit overexposed or underexposed areas. Hence, a series of 6 sets of 9 bracketing photos was aditionally acquired. These constitute a spherical high dynamic range (HDR) panoramic image. In order to obtain the panoramic image at the same position as the 3D scanner, an adapter for the base tripod of the 3D scanner has been developed, see Figure 1b. This allows a professional Canon 500D digital single-lens reflex (DSLR) camera to be mounted and take panoramic images with the help of a Nodal Ninja 3.

4.2 Panoramic Image Registration

A special focus in the development of the work flow was to deliver panoramic images that are precisely registered to the point cloud. At first, the images from the Canon DSLR are stitched together to a spherical HDRI panoramic image within the ptGUI Pro panoramic image software, see Figure 3.

Although the panoramic image was taken at the scanner location, the image returned from the stitching software will not be aligned azimuth-wise (a rotation about the up direction) in general. We resolve this degree of freedom by image based method; the high-resolution panoramic image is aligned to the panoramic image of the Faro Focus 3D, which registered to the point cloud, as shown in Figure 4.

Typically, such an image registration task is based on feature detection and matching (Szeliski, 2004). However, due to the fact that there is only one degree of freedom, we can use a much simpler method: both images are converted to grayscale and normalized (subtract grayscale mean and divide by standard deviation). Then, both images are rescaled to the same size. The alignment is resolved by an exhaustive test using a sliding window along

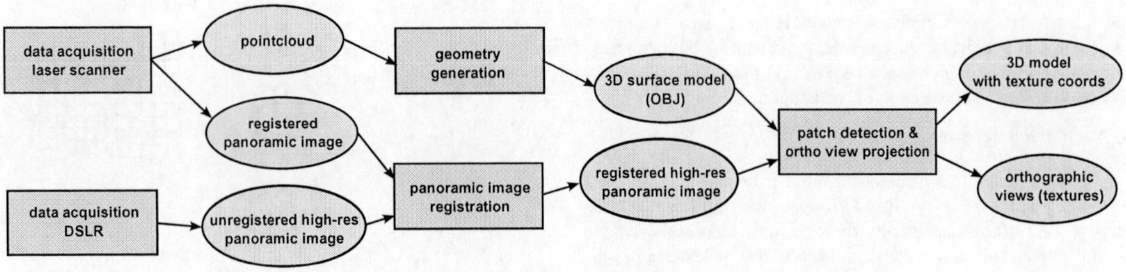

Figure 2: Overview of the whole work flow: boxes depict methods and ellipses reflect (intermediate) results. First, the acquired input data is registered and preprocessed (geometry generation and panoramic image registration). Afterwards, the patch detection and ortho view projection step identifies rectangular patches in the input geometry and generates an orthographic image for each patch, as well as texture coordinates for the 3D surface model.

the azimuth axis for each pixel position, and finding the minimal SAD (sum of absolute differences). In the rare cases where the lighting conditions between the laser scan and the DSLR photo acquisition were substantially different, it was necessary to examine several local minimums of this error function, as the global minimum might not correlate to the optimal position.

4.3 Geometry generation

Several approaches to model the base 3D model for the creation of the orthographic images exist and have been discussed and used within the project:

The base 3d model can be**manually modeled** with an underlying 3d point cloud as reference. The need for consistency in the test data prohibited this approach.

A desirable **automated process of geometry generation** can be achieved through a point cloud reconstruction software similar to the Poisson Surface Reconstruction algorithm by (Kazhdan et al., 2006). This, for example, is integrated in the open source CloudCompare. These approaches results often in high amounts of faces and topological complexity, as preliminary tests showed. This is generally not wanted and conflicted with the projects initial limitation for quad-based geometries.

An **alternative automatic generation** of a simplified 3D base model can be achieved by extracting the architectural spaces within a point cloud like described in (Tamke et al., 2014). While this method would work very well, it is based on the extraction of several spaces in point clouds with multiple scans. The approach described in this paper uses at the moment only single scans.

A **semi-automated approach** ensures the consistency and overcome both human in-accuracies and topological complexity. The employed approach uses a random sample consensus (RANSAC) algorithm (Schnabel et al., 2007), which detects planes within the point cloud Figure 5b. The resulting clusters are imported into Rhinoceros 5 through a custom IronPython[7] script in the visual programming plug-in Grasshopper. This reads the clusters and separates them into walls, ceiling and floor by orientation and area thresholds, and organizes them in a polar coordinate space. From this, a simple quad based mesh model of the room is created (Figure 5c). This is delivered as OBJ geometry to the next step in the work flow.

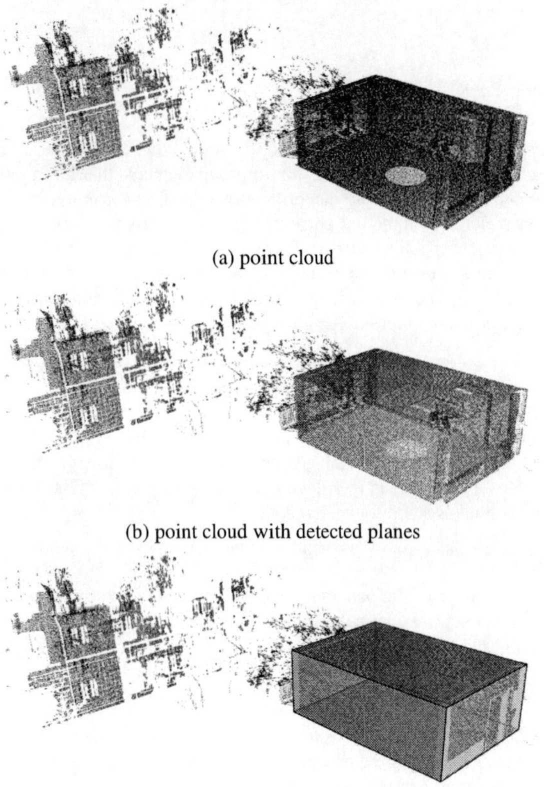

(a) point cloud

(b) point cloud with detected planes

(c) point cloud with final geometry

Figure 5: A coarse 3D model is created from the point cloud (a) using algorithmic clustering of the point cloud into walls, ceiling and floor (b), from which a simplified model with quadrilateral faces is created (c).

5. ORTHO VIEW GENERATION

From the acquired and processed data (geometry and registered panoramic images) we obtain orthographic images using the following approach: First, the method processes the input geometry

[7] http://ironpython.net/

This contribution has been peer-reviewed.
doi:10.5194/isprsarchives-XL-5-W4-131-2015

to identify elements (triangles, quads) lying in the same plane. For a group of such elements an oriented bounding box with minimal area is fitted, which we call a *patch*. For each patch, an ortho view is created via projection of the panoramic image.

5.1 Patch detection

In order to generate a meaningful number of patches, the input geometry has to be analyzed and grouped. We begin by first clustering the surface elements by their normal vector, this is done using a mean shift (Comaniciu and Meer, 2002) clustering in the space of normalized normal directions. We use a flat kernel with size 0.3.

This yields a group of k clusters that correspond to similar directions. We denote these clusters $d_0 \cdots d_k$ and their normal directions $\vec{n}_0 \cdots \vec{n}_k$. However, this clustering step will not group elements that lie in a similar plane, but have opposite normal directions.

In order to group the elements with opposite normal directions, we perform a second clustering step on the rows of the $k \times k$-matrix A^n, whose row and column indices i correspond to a cluster d_i. The elements a_{ij} of A^n correspond to a 1 if

$$|\langle \vec{n}_i, \vec{n}_j \rangle| \geq 0.95$$

and 0 otherwise. The equivalent rows of A^n correspond to clusters of principal normal directions.

After grouping the elements into the main directions, we need to perform another clustering step to group elements that lie in the same plane. We use normal projection to project the midpoint of each element on the line corresponding to the principal direction, and perform a third clustering step on the length of the projected vector. Again, we use meanshift, this time with a flat kernel of size 0.1m. The resulting clusters correspond to the elements with similar main direction, lying in the same plane.

When these final clusters have been found, an arbitrary orthonormal basis is identified using the approach of Frisvad (Frisvad, 2012), where the Z direction corresponds to the cluster plane normal. Finally, a bounding rectangle of the elements in the plane with minimal area is obtained, by evaluating an exhaustive search over 90 orientations in 1-degree steps, which yields a final local coordinate system for each patch.

5.2 Image data projection

As the pose of the panoramic sphere corresponds to the pose of the scanner, we obtain this information directly from the exported point cloud in E57 format. The pose consists of an Euclidean transformation, which is expressed by a position vector and a rotation quaternion.

The system now proceeds to create an orthographic view of each patch, given a desired resolution in mm/pixel, using a simple projection approach which is shown in Figure 6: For each pixel p of the patch, the system creates a ray from the corresponding 3D position to the center of the panoramic sphere. The intersection p' of the ray and the sphere is then transformed into the local spherical coordinate system of the panoramic sphere, which yields the azimuth angle ϕ and the elevation angle θ. These angles are used to acquire the color value of this pixel from the panoramic photograph.

Finally, the system also creates texture coordinates for the input geometry. The texture coordinates are obtained by projecting the vertices of each surface element into the local patch coordinate system, and normalizing the result to texture coordinate range. The textured model is written as .OBJ file to disk.

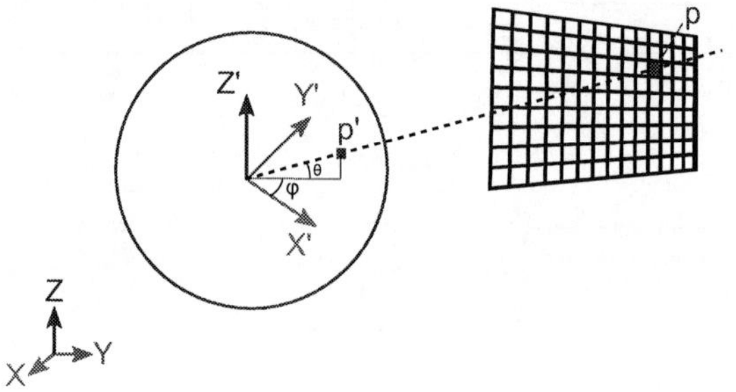

Figure 6: A rectangular patch in 3D space is sampled at a specific resolution, e.g. 1 pixel/mm. Each pixel p is transformed into the local coordinate frame (X', Y', Z') of the panoramic sphere in spherical coordinate angles azimuth ϕ and elevation θ to determine the color value in the panoramic image.

6. APPLICATION AND RESULTS

We evaluated the method on scans of several indoor scenes. The resulting orthographic views were used in an computer vision pipeline to detect electrical appliances, i.e. sockets and switches.

6.1 Reconstructions

We applied the described pipeline to 6 scans of indoor scenes. Two examples can be seen in Figure 7. A third example, also showing the generated ortho views, is shown in Figure 8. A room with slightly more complex geometry can be seen in Figure 9.

The clustering was able to group suitable elements in all datasets. For example, the input geometry of the scene depicted in Figure 7 (left column) consists of 218 triangles and 76 quadrilateral faces, which have been grouped into 3 main directions and 15 ortho patches.

6.2 Electrical Appliance Detection

The extracted ortho views are well suited for image processing methods in order to detect different wall-mounted objects. Exploiting the known relation from pixel to the actual wall geometry, the scale of the searched objects in the image is fixed to 1mm per pixel. Thus, object detection can be performed by a simple sliding window approach: For each pixel $\mathcal{P}(i)$ of an ortho view, the detection probability of a object class is determined by analyzing a small image patch centered around $\mathcal{P}(i)$.

Detecting power sockets and light switches in arbitrary indoor scenes form a reasonable application example. However, these objects can be very challenging to classify: they are usually untextured and designed to be unobtrusive, but on the other hand they exist in many variations regarding shape and colors. In general, sockets and switches are mostly defined by their silhouettes and mostly homogenous coloration of varying colors (e.g. different brands).

It is therefore practical to make use of both gradient and color information in the search window. Thus, we form the following feature descriptor pool:

- The '"Histograms of oriented gradients" (HoG) descriptor by Dalal and Triggs (Dalal and Triggs, 2005). In recent years, it has been a very successful approach for identifying objects by their global gradient appearance.

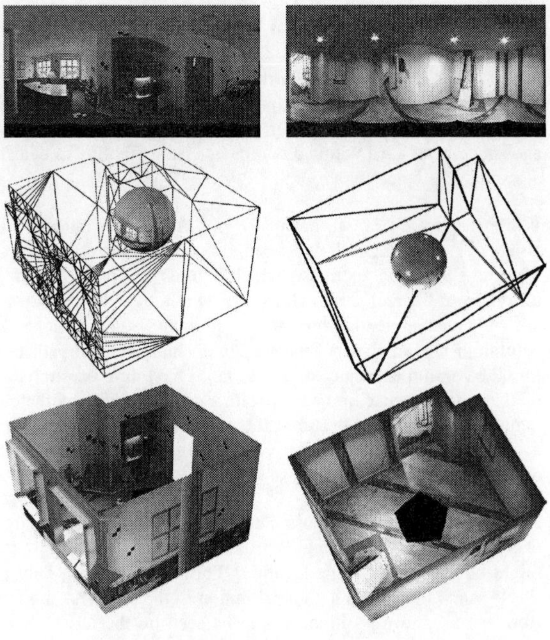

Figure 7: Two indoor scenes in different stages of reconstruction. The acquired high-resolution input panoramas (top row) were aligned to the geometry that was created from the laser range scan. The middle row shows the input geometry, and the aligned panoramic spheres. In the bottom row we see the same geometry using the generated orthographic views as textures, the ceiling has been omitted for a better visualization.

- A self developed descriptor that models the distribution of image gradients differently than the HoG descriptor. For each window pixel where the local gradient magnitude is larger than a threshold, we calculate the unit vector $\vec{v}(\mathcal{P}(i))$ pointing in the gradient direction φ_i: $\vec{v}(\mathcal{P}(i)) = (1, \angle\varphi_i)$. Projecting $\vec{v}(\mathcal{P}(i))$ onto a set of 4 fixed unit vectors, each 45° apart of its neighbors. yield a 4 dimensional value that uniquely and continuously describe the vectors direction. By considering only the absolute values of the projections, the descriptor becomes invariant to contrary object- and wall-intensity values. The final descriptor entries are build by comparing the mean orientation values of randomly sized and located sub-regions, similar to Haar-like features from (Viola and Jones, 2001).

- In order to model the color distribution, the color channels (RGB) of the image patch are separated. Again, differences in mean intensity values of randomly sized and located sub-region pairs over all 3 channels form the descriptor entries.

All these descriptors are combined by concatenating their entries to one feature vector.
According to its values, a pre-trained random forest classifier (Breiman, 2001) retrieves the probability of each class. In order to train the classifier, we created a large training set of labeled image patches representing 3 object classes: power socket, light switch and background. After classification, a subsequently applied non- maxima suppression on the class probability maps yield the final detection results.

(a) panoramic input image

(b) generated orthographic views (patches)

(c) textured geometry

Figure 8: The spherical panoramic input image (a) is transformed into orthographic views (b) for each quadrilateral patch that was automatically identified from the input geometry. (c) shows a rendering of the textured geometry.

This contribution has been peer-reviewed.
doi:10.5194/isprsarchives-XL-5-W4-131-2015

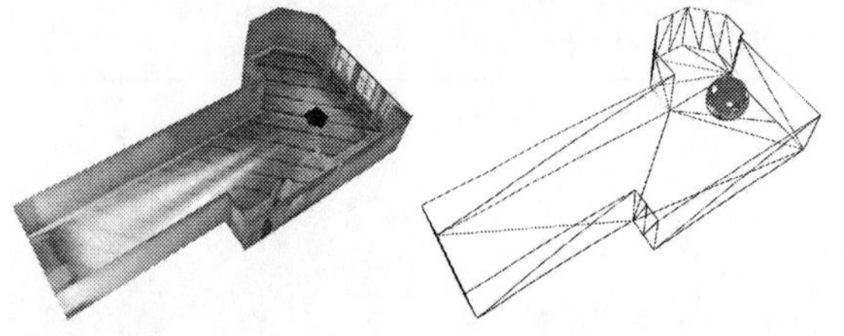

(a) (b)

Figure 9: Our workflow is able to handle rooms with non-orthogonal geometry, as long as all important aspects can be scanned from one position.

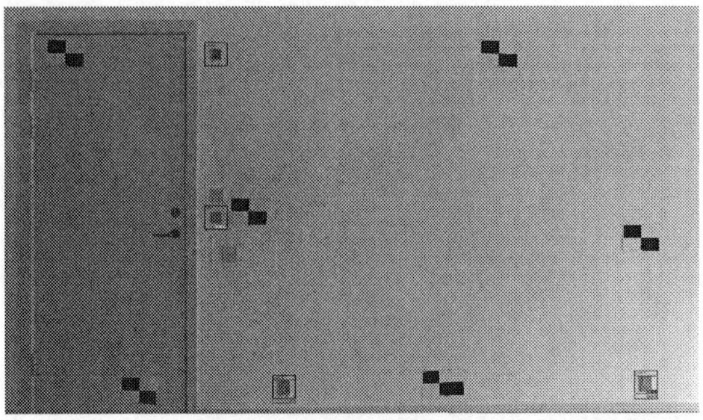

(a)

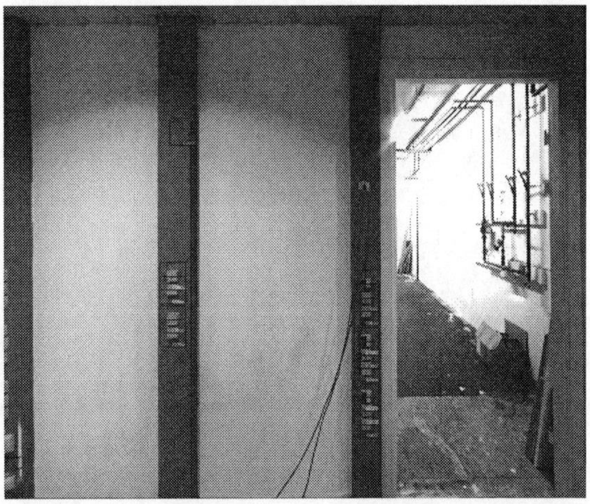

(b)

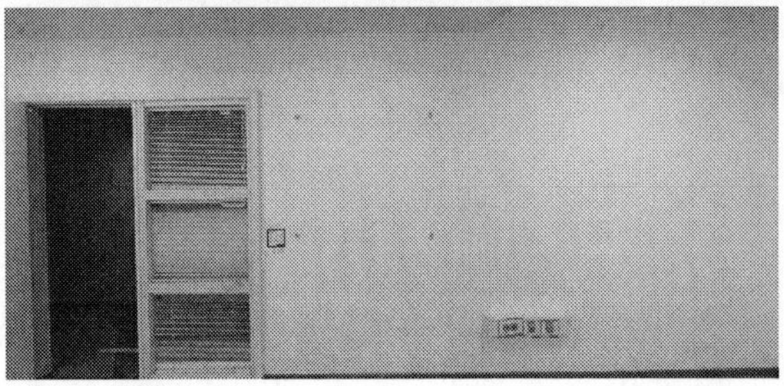

(c)

Figure 10: We use computer vision methods to train a classification system and detect electrical appliances, in this case switches. The two trained classes correspond to switches (violet rectangles) and sockets (orange rectangles).

7. CONCLUSION AND FUTURE WORK

In this paper, we presented a work flow for the semiautomatic extraction of orthographic views for indoor scenes from laser range scans and high resolution panoramic images. The resulting images have been used within a computer vision pipeline to detect electrical appliances in a room.

While the images are sufficient for our application purpose of object detection, the ortho views might contain projection errors where the supplied geometry, which is often an approximation, differs from the real scene which is shown in the panoramic image. As an example, the room shown in Figure 7 on the left has a column inside which is not reflected in the input geometry, therefore the column is projected on the wall. One future research direction is therefore to find ways to either create less approximated input geometry, or make use of the point cloud information e.g. to filter out pixels in an ortho view whose projected depth lies outside the ortho view plane.

However, it might not be possible to scan all important contents of a room using only a single scan, as not all parts of the room might be visible from one location. Furthermore, using only a single scan might yield a poor resolution of the parts that are far from the scanning position, as can be seen on the left of Figure 9. Another important aspect of future work is therefore to research methods to integrate the information of multiple scans and panoramic images into a single model.

8. ACKNOWLEDGEMENTS

This work was partially funded by the European Community's Seventh Framework Programme (FP7/2007-2013) under grant agreement no. 600908 (DURAARK - Durable Architectural Knowledge), 2013-2016.

REFERENCES

Breiman, L., 2001. Random forests. Machine learning 45(1), pp. 5–32.

Comaniciu, D. and Meer, P., 2002. Mean shift: A robust approach toward feature space analysis. IEEE Trans. Pattern Anal. Mach. Intell. 24(5), pp. 603–619.

Dalal, N. and Triggs, B., 2005. Histograms of oriented gradients for human detection. In: Computer Vision and Pattern Recognition, 2005. CVPR 2005. IEEE Computer Society Conference on, Vol. 1, IEEE, pp. 886–893.

D'Annibale, E. and Fangi, G., 2009. Interactive modeling by projection of oriented spherical panoramas. the case of ad deir-petra. In: Proceedings of 3D-ARCH 2009: 3D Virtual Reconstruction and Visualization of Complex Architectures, Vol. XXXVIII-5/W1.

Frisvad, J. R., 2012. Building an orthonormal basis from a 3d unit vector without normalization. Journal of Graphics Tools 16(3), pp. 151–159.

Furukawa, Y., Curless, B., Seitz, S. M. and Szeliski, R., 2009. Reconstructing building interiors from images. In: International Conference on Computer Vision, pp. 80–87.

Grammatikopoulos, L., Kalisperakis, I., Karras, G. and Petsa, E., 2007. Automatic multi-view texture mapping of 3d surface projections. In: Proceedings of 3D-ARCH 2007: Virtual Reconstruction and Visualization of Complex Architectures, Vol. XXXVI-5/W47.

Huber, D., 2011. The astm e57 file format for 3d imaging data exchange. In: Proceedings of the SPIE Vol. 7864A, Electronics Imaging Science and Technology Conference (IS&T), 3D Imaging Metrology, Vol. 7864A.

Kazhdan, M., Bolitho, M. and Hopper, H., 2006. Poisson surface reconstruction. In: K. Polthier and A. Sheffer (eds), Eurographics Symposium on Geometry Processing (2006).

Lévy, B., Petitjean, S., Ray, N. and Maillot, J., 2002. Least squares conformal maps for automatic texture atlas generation. ACM Trans. Graph. 21(3), pp. 362–371.

Pitzer, B., Kammel, S., DuHadway, C. and Becker, J., 2010. Automatic reconstruction of textured 3d models. In: IEEE International Conference on Robotics and Automation, ICRA 2010, Anchorage, Alaska, USA, 3-7 May 2010, pp. 3486–3493.

Schnabel, R., Wahl, R. and Klein, R., 2007. Efficient ransac for point-cloud shape detection. Computer Graphics Forum 26(2), pp. 214–226.

Szeliski, R., 2004. Image alignment and stitching: A tutorial. Technical Report MSR-TR-2004-92, Microsoft Research.

Tamke, M., Blümel, I., Ochmann, S., Vock, R. and Wessel, R., 2014. From point clouds to definitions of architectural space - potentials of automated extraction of semantic information from point clouds for the building profession. In: E. M. Thompson (ed.), Proceedings of the 32nd eCAADe Conference, Vol. 2, pp. 557–566.

Viola, P. and Jones, M., 2001. Rapid object detection using a boosted cascade of simple features. In: Computer Vision and Pattern Recognition, 2001. CVPR 2001. Proceedings of the 2001 IEEE Computer Society Conference on, Vol. 1, IEEE, pp. I–511.

Waechter, M., Moehrle, N. and Goesele, M., 2014. Let there be color! — Large-scale texturing of 3D reconstructions. In: Proceedings of the European Conference on Computer Vision, Springer.

Zhang, E., Mischaikow, K. and Turk, G., 2005. Feature-based surface parameterization and texture mapping. ACM Trans. Graph. 24(1), pp. 1–27.

This contribution has been peer-reviewed.
doi:10.5194/isprsarchives-XL-5-W4-131-2015

ACQUISITION AND REPRODUCTION OF SURFACE APPEARANCE IN ARCHITECTURAL ORTHOIMAGES

A. Martos[a]*, R. Cachero[a]

[a]Dogram, Parque Tecnológico de Asturias, 33428 Llanera, Spain - (martos, rocio)@dogram.es

Commission V, WG V/4

KEY WORDS: Orthoimage, Multi-View Stereo, Photometric Stereo, Surface Appearance, BRDF Capture, Physically Based Render

ABSTRACT:

Software tools for photogrametric and multi-view stereo reconstruction are nowadays of generalized use in the digitization of architectural cultural heritage. Together with laser scanners, these are well established methods to digitize the three-dimensional geometric properties of real objects. However, the acquired photographic colour mapping of the resulting point clouds or the textured mesh cannot differentiate the proper surface appearance from the influence of the particular illumination present at the moment of the digitization. Acquisition of the actual *surface appearance*, separated from the existing illumination, is still a challenge for any kind of cultural heritage item, but very specially for architectural elements. Methods based on systematic sampling with commuting light patterns in a laboratory set-up are not suitable. Immovable and outdoor items are normally limited to the existing and uncontrolled natural illumination.

This paper demonstrates a practical methodology for appearance acquisition, previously introduced in (Martos and Ruiz, 2013), applied here specifically for the production of re-illuminable architectural orthoimages. It is suitable for outdoor environments, where the illumination is variable and uncontrolled. In fact, naturally occurring changes in light among different images along the day are actually desired and exploited, producing an enhanced multi-layer dynamic texture that is not limited to a frozen RGB colour map. These layers contain valuable complementary information about the depth of the geometry, surface normal fine details and other illumination-dependent parameters, such as direct and indirect light and projected self-shadows, allowing an enhanced and re-illuminable ortoimage representation.

1. INTRODUCTION

1.1 State of the art

Nowadays the use of diverse software tools for photogrammetric and image-based dense 3D reconstruction is widely generalized in cultural heritage digitization. There are quite effective methods to automatically acquire and reproduce the shape and colour texture of real cultural heritage objects. The so-called Multi-View Stereo (MVS) or Image Based Modelling (IBM) methods depart only from multiple images, normally taken under existing (passive) illumination conditions. When properly applied they can typically reconstruct the original geometry, reaching a considerable level of detail and accuracy, competing with other (active) methods such as laser scanning. Other active scanning methods such as structured light scanning are of limited use when outdoors or applied for large architectural elements, mainly due to practical limitations on the intensity of the light.

In any of these cases the colour of the surfaces is intended to be reproduced by means either of coloured point clouds or static texture maps. Both are generated by projection of the images. In some cases the colour is quite directly extracted from the photographs by projecting the pixels of each image onto the 3D model surface polygonal mesh, creating high resolution blended colour textures. When using a laser scanner, complementary photographs are typically taken on the spot as well, using a separate or an integrated camera. This provides either a coloured point cloud by back-projecting the points onto the images. Both coloured models can be used to straightforwardly produce conventional orthoimages by simply rendering a projection of the 3D model onto an orthogonal plane of choice.

*Corresponding author

One obvious limitation of these methods resides in the fact that even when the objects are accurately photo-textured, and regardless of the method used to map the colours, the reproduced surface texture of the orthoimage will be a static snapshot frozen onto the surface of the object, resembling only the existing conditions at the moment of the acquisition. When the images are taken in a cloudy day the variation of illumination among different images is normally small, favouring thus the matching and reconstruction. However the resulting representation of colour texture is perceived as flat, lacking much of the potential contrast and detail due to the limited shading (Fig. 1).

Figure 1: Portada de la Oliva, Asturias, Spain. Orhtoimages of the same object generated from photographs taken either in cloudy conditions (left) or under direct sunlight (right).

Whereas under direct sunlight the surface shading and self-projected shadows will be much vivid and informative. Sharper shading may significantly help in the intuitive interpretation of the image, by giving better appreciation of volumes and fine relief details. Ordinarily all the images in each set should be taken under the same fixed illumination conditions to reduce mismatches caused by the moveable shadows due to changes in sun position, and to avoid the appearance of artificial seams in the blending of multiple images. This requires the selection of the proper time of the day and also completing the field work in a relatively short period of time, which might not be a practical approach for large and complex elements.

Even when all the images were taken with similar shading, the final resulting appearance will not be able separate the proper surface properties from the fixed effects of shading caused by the existing illumination. Hard shadows will be frozen into the model texture, sometimes even hiding relevant details. In some cases faces oriented north (in the northern hemisphere) may never receive enough sunlight at the appropriate angle.

It is not always feasible to avoid changes in natural light when doing outdoor projects. Nevertheless the intended goal in appearance acquisition is actually to characterize how the surface appears to reacts to changes in illumination, turning these variations in the source images into something indeed necessary to complete the documentation. In a similar way, previous approaches to solve appearance acquisition in the laboratory do actually need to produce variable illumination patterns to characterize the Bidirectional Reflectance Distribution Function (BRDF).

1.2 State of the art in appearance acquisition

The acquisition and realistic reproduction of the surface appearance properties of cultural heritage elements is still an open topic addressed only recently. Several previous efforts have been made in order to model and capture the properties of the surfaces, producing digital models that can be observed and illuminated from any angle, and to try to achieve a fully dynamic photo-realistic quality. Some of these approaches are specific for the capture of bi-directional reflectance information of cultural heritage homogeneous materials. The MERL BRDF database includes samples of several materials recorded by the Rochester Institute of Technology in the USA while only as homogeneous and isotropic samples (Chen et al., 2011).

To acquire a non-homogeneous texture mapping of the surface properties, most of the approaches proposed so far require active and variable illumination set-ups, such as domes (Klein and Schwartz, 2010), (Schwartz and Klein, 2012) or goniometer style arms (Haindl and Filip, 2013) and/or rotating tables. These typically require a very large amount of photographs, taken from different known camera positions and commuting multiple point light sources. While these set-ups have been already successfully into acquiring appearance for small objects that can fit inside the working volume, they are still highly impractical or impossible to use for medium or large size objects such as sculptures or architectural items. These devices cannot be used out of laboratory conditions, nor with unmovable objects or for outdoors acquisition. Due to the size of the objects controlling an active illumination completely is no longer a viable option, severely limiting thus the possibilities of these acquisition processes.

Additionally these approaches aim to very exhaustively record all the information required to completely reproduce the very multidimensional sampling space of the (BRDF) texture. This a very systematic but not completely practical approach, requiring very exhaustive field data and heavy data processing, together with accurate prior knowledge of geometric and radiometric calibration, as well as a very controlled and closed light environment.

Other approaches successfully combine dense geometry with smart photographic mapping from multiple images, overcoming thus some of these limitations (Callieri et al., 2008). The resulting visualization is often considered photo-realistic and has enough freedom of movement for the observation point, as long there are no relative movements between the item and the light. However the actual limitation to achieve a completely realistic dynamic visualization lies in the fact that, at best, all the texture/color information is still frozen onto the object surface, not reacting to any potential changes in the color or position of lights or the environment.

Another interesting approach for the reproduction of reflectance properties of cultural heritage objects is Polynomial Texture Mapping (PTM), a computational photographic method that captures a subject's surface shape and color and enables the interactive re-lighting of the subject from any direction (Malzbender et al., 2001). Each RTI resembles a single, two-dimensional (2D) photographic image, but unlike a typical photograph, it can be re-lighted using a "virtual" light. This changing interplay of light and shadow in the image discloses fine details of the subject's 3D surface form. Digitization procedures are flexible and allow the digitalization of larger artworks, but does not capture the 3D model of the object. Instead it represents a texture image which, unlike a typical photograph, can be re-lighted using a virtual light source. All images must be taken from the same position and since the method does not produce true reconstruction it does not allow for an arbitrary choice of the point of view for orthoimage generation.

The approximate capture of the combined illumination and reflectance model was recently addressed with fixed, but unknown light environmental maps (Palma et al., 2012), using a Phong model with clustering of materials. This work shows that effective field acquisition of some properties of the illumination map and surface appearance properties is feasible even with partially specular objects, and that very simple reflectance models can clearly enhance the appearance perception of the rendered model.

The approach proposed in (Tunwattanapong et al., 2013) approximates the illumination conditions using continuous spherical harmonics, using a complete 3D geometrical model and a set of combined images of the object . This method allows the visualization of a variety of objects difficult to acquire with other techniques.

We focused in developing methods able to simultaneously determine the three-dimensional shape of the object and to map the surface appearance and environment illumination simultaneously as a lightmap, determined indirectly, instead of a set of well known point sources. This would allow to work with hand-held digital photographs taken under variable but flexible illumination (including natural light) and would define an approach to appearance capture as practical and flexible as standard MVS, which can be used successfully in a variety of situations and environments, very specifically for architectural and outdoor objects.

1.3 Field and outdoor acquisition

Working outdoors with large architectural elements or sculptures means that illumination can rarely be controlled much, if at all. Even modelling the light-map is not a trivial task. While the size, position and movement of the sun is absolutely predictable given accurate geo-location and time, this is not the case for the indirect light coming from partly overcast sky, ground and nearby buildings. This indirect illumination will still play a very relevant role in the capture process since it is a significant component of the actual light map (see Fig. 3. To illustrate the adaptation of the appearance capture technique described in (Martos and Ruiz,

This contribution has been peer-reviewed.
doi:10.5194/isprsarchives-XL-5-W4-139-2015

2013) to outdoors illumination, we will show following examples, taking advantage of naturally occurring variations along the day.

In practice it is often possible to find more than sufficient variation in the illumination pattern along in a single day. With some proper planning and depending on the weather, it is even possible to perform the acquisition taken the pictures under several significantly different light-maps, in just a few minutes.

Figure 2: Ortoimage generated from the virtual model with appearance capture simulating completely overcast sky. Shadows are very diffuse and there is limited volumetric appreciation.

Figure 3: Ortoimage generated from the virtual model with appearance capture simulating direct sunlight with partly overcast sky. Shadows are slightly less sharp than in the original images.

The door of Santa Maria de la Oliva in Asturias, Spain, was photographed at a few different times in the day: at morning, at midday and then again at dusk. Partly cloudy conditions on a windy day in an urban environment provided a large variety of different illumination changes. The most useful set however was obtained in the scope one hour at dusk, when there were rapid changes in sunlight were casting onto nearby structures. Overcast photographs where useful to initialize the texture maps while finer details in texture where obtained from sunny and partly overcast sets of photographs, for which Photometric Stereo methods worked the better.

In any case the selected sets of images should represent a reasonable variation in light. When images are taken under variable illumination conditions, if they are properly grouped and sorted this will not affect matching significantly since differences will not be very significant. However the resulting composition of a single orthoimage from multiple images would exhibit discontinuities in different patches, or result at best in a blended image.

Photographs taken when completely overcast cannot provide much fine texture detail information, but still worked robustly for usual IBM tasks such as feature matching, bundle adjustment, camera calibration and surface reconstruction (Fig. 2). While the resulting mesh was not highly detailed, the additional details are provided by photometric stereo methods with the more direct sunlight images.

Direct sunlight (Fig. 3) also caused saturation in some areas due to the limited dynamic range of the photographs (12 bits). To minimize these problems the photographs where taken with bracketing methods to enhance dynamic range. The light-map estimation has to be calculated in this case with increased resolution (256x256 pixels) and dynamic range (32 bit).

2. METHODOLOGY

In this paper we present new methods for the acquisition and reproduction of surface appearance properties of immovable and outdoor objects, based only on naturally changing illumination conditions. For this acquisition method the sky and surrounding environment will be approximated as a variable and initially unknown light-map, for which changing daily illumination is expected to cause enough variability in the illumination pattern to reproduce the apparent response of the surfaces to the changing illumination.

To illustrate this method we will show how it is was applied to the generation of reilluminable orthoimages of the *Portada de la Oliva* and the door of *Peñalba de Santiago* both architectural outdoor elements with a significant cultural heritage value and interest in detailed representations.

The whole process is achieved departing only from photographs. Several snapshots of the doors where taken using a calibrated digital camera with some particular settings (fixed focus and focal length, manual exposure, low ISO setting, etc...). A basic three-dimensional dense geometry reconstruction is performed by means of dense stereo MVS and SfM methods. This provides dense depth information (Fig. 4) to an scale in the millimetre range which is mapped as a UV texture map (Fig. 5).

A few more photographs where taken within a few hours intervals with varying environmental lighting conditions, including variable sunlight and overcast levels. Using this information a rough illumination model is built and some properties of the BSDF (Bidirectional Scattering Distribution Function) of the surfaces are deduced from the different lighting settings. This allows to produce an specially built bi-directional "pixel shader" function

This contribution has been peer-reviewed.
doi:10.5194/isprsarchives-XL-5-W4-139-2015

Figure 4: Depth map of the densely reconstructed model. Each pixel represents a different depth w.r.t the chosen orthogonal view plane.

Figure 5: Colour coded UV map, to arbitrarily reassign 3D model coordinates to the 2D texture coordinates of the multiple maps.

to reproduce any new lighting condition including arbitrary new angles for sun incidence.

Additionally fine details in the surface relief are revealed by a normal map (Fig.6) produced by a simplified Photometric Stereo reconstruction method. Direct and indirect shadows, as well as Indirect light scattering are also modelled to properly estimate all illumination contributions in acquisition and to provide a higher degree of realism in the final render. This will provide unified and convenient information about illumination conditions to construct complex ortho-image documents, to enhance depth perception and to provide the volumetric information in the representation that is often missing from conventional planar orthoimages.

We used Patch-based Multi-view Stereo [PMVS] (Furukawa and Ponce, 2010) to reconstruct a dense point cloud with normal mapping, from the set of images and previously calibrated camera internal parameters. The output, a large set of oriented point, is filtered and interpolated with a subsequent matching and hole covering pass, averting the creation of a polygonal (or a mesh) model. This produces both the 3D coordinates and the surface normals estimated at each oriented point which is sufficient for displacement map rendering. However the normal map obtained by this method was shown insufficient for the purposes of detailed surface appearance acquisition. Instead

A natural improvement was to take an initial step where a Lambertian, matte or pure diffusive reflective model is assumed for the surfaces. In this case the apparent color does not depend on the observation angle but only on the incidence of light. It is possible then to determine the light-map by an iterative optimization process based on Photometric Stereo (PS) methods (Basri et al., 2007), used to recover the reflectance properties simultaneously to the refinement of the light-map. Lighting map calculated includes arbitrary combination of diffuse, point and extended sources. General lighting conditions are represented using low order spherical harmonics discretized in the light-map, using simple optimization in a low-dimensional space. This approach is limited when the surfaces are significantly glossy of specular. While this is not the case in this case study, however

Figure 6: Normal map of the reconstructed model obtained by Photogrammetric triangulation. It was later refined at render time by small corrections acquired by Photometric Stereo methods. Red/blue colours code horizontal/vertical normal components.

this is considered in further refinement stages and the methodology has been tested as well with significantly specular reflections.

These case-studies will show the feasibility of working with field calibration and initial rough approximations for the camera model and light-maps, addressing thus the flexibility required for practical field documentation in museum environments or outdoors. The potential for diffusion will be shown with the use of open source software tools for enhanced visualization. The presented capture methods are integrated with the specific adaptation of open-source GPU-based render engines to produce two flavours of 3D inspection/visualization tools with proper relighting capa-

This contribution has been peer-reviewed.
doi:10.5194/isprsarchives-XL-5-W4-139-2015

bilities, able to reveal very subtle details: An quasi-real time realistic engine (Blender Cycles), which is also the basis for the capture process and is focused on realistic reproduction, and a real-time version based on customized *pixel shaders*, for the real-time visualization of lightweight models on web browsers and other interactive applications.

This is achieved not only by reconstructing the shape and projecting colour texture maps from photographs, but also modelling and mapping the apparent optical response of the surfaces to light changes, while also determining the variable distribution of environmental illumination of the original scene. This novel approach integrates Physically Based Render (PBR) concepts in a processing loop that combines capture and visualization. Using the information contained in different photographs, where the appearance of the object surface changes with environmental light variations, we show that it is possible to enhance the information contained in the usual colour texture maps with additional layers. This enables the reproduction of finer details of surface normals and relief, as well as effective approximations of the Bi-directional Reflectance Distribution Function (BRDF). The appearance of the surfaces can then be reproduced with a dedicated render engine providing unusual levels of detail and realism due to enriched multi-layer texture maps and custom shading functions.

The development of these digitalization methods was inspired by several concepts borrowed from relatively recent, but already well developed, 3D rendering methods: The separate capture and modelling of object geometry, reflectance textures and environmental light. The optimizations used by the realistic rendering algorithms, are not just convenient ideas to increase realism under highly constrained render times. Common render pipelines and capture algorithms share fundamental analogies, closing the cycle in order to realistically reproduce digitalized models.

In contrast, we developed a flexible methodology to produce effective surface appearance approximations with a single moving camera and uncalibrated extent lightmaps instead of discrete point sources. This is strongly based on the same principles of free moving camera/station determined by Bundle Adjustment, which is well known for Photogrammetric or Structure from Motion (SfM) methods. The idea is extended here to capture not only the object geometry by triangulation of rays, but similarly, to determine the position of individual light sources which can be approximated with similar principles by uncalibrated stereo-photometric methods. The second key concept exploited is the additivity of light in the image formation process: Light coming from multiple light sources, including indirect or secondary sur face reflections in this consideration, is added in the final image, being thus linearly separable. It is thus possible to produce multiple layers of texture and environmental map. This process can be iteratively refined and solved until convergence of the difference between the original and rendered images.

2.1 Observations and limitations

This methodology is characterized by the practical combination of some differential key concepts and methods:

1. The common use of Physically Based Render *shaders* both for surface appearance acquisition and render.
2. Separation of the relevant surface properties in multiple 'render-oriented' layers such as depth, normal, direct and indirect shading, calculated with different methods.
3. The use of an extent light-map, determined indirectly from the images, instead of known point sources.

Figure 7: Map with the ambient occlusion shading produced by the diffuse component of the light-map.

Figure 8: Projected shadows layer, showing areas that will not receive direct light from the sun if set in an arbitrarily given direction. Notice diffuse borders due to sun angular extension.

4. Normal maps calculated by Stereo Photometric methods, more detailed than those calculated by triangulation.

The involved techniques are individually known already and applied separately in cultural heritage digitization and visualization, although not often together. There are actually existing open-source or well documented algorithms for each stage that can be applied to extend the usual MVS pipeline work, by the integration also of PBR and PS methods.

Physically Based Render methods are not only important to provide a more realistic look to the appearance characterization. In-

This contribution has been peer-reviewed.
doi:10.5194/isprsarchives-XL-5-W4-139-2015

Figure 9: Composition of all the estimated illumination components of one of the images, diffuse, direct and shadow.

Figure 10: Final orthoimage, made by composition of layers with alternative position for the sun w.r.t Fig.3.

direct illumination is indeed a relevant factor in acquisition which cannot be neglected to achieve good results, as shown in figure (Fig.13). Here some surfaces in the object cast light onto nearby surfaces, especially in salient features. This is specially strong when original images have been taken under direct sunlight, producing also hard shadows.

It has been noted as well that concentrated light sources such as the LEDs used in other approaches for appearance acquisition, do actually produce very sharp self-casting shadows. The sharp discontinuity in shadows produce situations like Fig.14 which is actually difficult to remove completely when creating the texture maps, specially with only a few images, leaving some undesirable halos in the texture. However this is not a practical issue

Figure 11: Final render of the fully composed orthophoto with a synthetic slanted illumination that highlights the fine relief details of the texture. Enhanced detail is achieved thanks to the photometric stereo method.

Figure 12: Synthetic drawing created from the processing and composition of multiple layers, removing color information and adding edge and shading, with the purpose showing only the details in shape and fine relief.

since most sources, including the sun on a clear day, are actually extensive to some degree. There is also normally some significant indirect illumination from the environment or other parts of the object itself. Best results are actually achieved in a partially clouded day.

This contribution has been peer-reviewed.
doi:10.5194/isprsarchives-XL-5-W4-139-2015

Figure 13: Indirect self-illumination showing the influence of the second bounce of the light in he model, where close surfaces of the own model cast light onto nearby surfaces.

Figure 14: A synthetic render where there is a single point light source, to illustrate the hardness of the shadows when there is no significant environmental indirect light, in contrast with Fig.3. This pattern may occur at night with an artificial spot beam.

3. CONCLUSIONS

We have demonstrated a practical methodology for appearance acquisition which was previously introduced in (Martos and Ruiz, 2013), and is specifically applied here for the production of reilluminable architectural orthoimages. It is suitable for outdoor situations, where the illumination can be variable and uncontrolled. Actually the naturally occurring changes in light among different images are desired and exploited to produce and enhanced multilayer color texture, not limited just to a fixed RGB map. These layers contain separated information about the depth of the geometry, surface normal fine details and other illumination-dependent parameters such as direct and indirect illumination, or projected self-shadows.

Figure 15: One of the original photographs (up) of the door of Santiago de Peñalba, Leon, Spain and an synthetic view (down) of the 3D model with acquired appearance, rendered using physically based methods for a different viewpoint and illumination.

The enhanced texturing enables a dynamic appearance response to the light when using realistic render methods, as well as the possibility to add, adjust or fully remove shading and projected shadows at will. This is useful not just for achieving a realistic visualization but also for the production of enhanced drawings, with highlighted edges and volumetric shading.

Some of the intermediate layers, specially those containing depth and shading information can be used also for alternative representations providing the look and feel of artistic rendition of volumetric drawings. By processing, including edge detection at different levels, filtering and adjusting relative weights of the various bitmaps, it is possible to generate automatic compositions that try to emulate the process that an artist would do interpreting volumes, shading, and projections of shadows, also standing out further those edges with stronger changes in relief and depth. The result, generated automatically from the re-illuminable orthoimage is show in Figs.12 and 17, and can be used as alternative representations with improved volumetric interpretation.

Finally the whole documentation methodology for outdoor items was demonstrated to be quick, flexible and practical, basically requiring just a digital camera, while still taking advantage of natural or induced variations in environmental light. In both case-studies, *La Oliva* and *Santiago de Peñalba*, photographs were

This contribution has been peer-reviewed.
doi:10.5194/isprsarchives-XL-5-W4-139-2015

Figure 16: Door of Santiago de Peñalba, Leon, Spain. Detail of the orthoimage(down) with slightly different slanted illumination to highlight relief details.

Figure 17: Automatic volumetric drawing generated by processing and composing some of the multiple layers with different edge detection methods (base RGB color information is removed).

taken under variable yet uncontrolled illumination conditions (ie, different times of the day) in a few hours or event minutes, without much preparation and without significantly increasing field work when properly planned. As a result the surface appearance could be still acquired and rendered successfully with complete flexibility to re-illuminate the object at convenience.

REFERENCES

Basri, R., Jacobs, D. and Kemelmacher, I., 2007. Photometric stereo with general, unknown lighting. International Journal of Computer Vision 72(3), pp. 239–257.

Callieri, M., Cignoni, P., Corsini, M. and Scopigno, R., 2008. Masked photo blending: Mapping dense photographic data set on high-resolution sampled 3d models. Computers & Graphics 32(4), pp. 464–473.

Chen, T., Taplin, L. A. and Berns, R. S., 2011. Munsell color science laboratory artist material database brdf fitting. Munsell Color Science Laboratory Technical Report, www.art-si.org.

Furukawa, Y. and Ponce, J., 2010. Accurate, dense, and robust multi-view stereopsis. IEEE Trans. on Pattern Analysis and Machine Intelligence 32(8), pp. 1362–1376.

Haindl, M. and Filip, J., 2013. Visual Texture - Accurate Material Appearance Measurement, Representation and Modeling. Advances in Computer Vision and Pattern Recognition, Springer.

Klein, R. and Schwartz, C., 2010. 3D-Coform. 2nd year report. [Online; accessed 06-May-2013].

Malzbender, T., Gelb, D. and Wolters, H., 2001. Polynomial texture maps. In: Proceedings of the 28th annual conference on Computer graphics and interactive techniques, ACM, pp. 519–528.

Martos, A. and Ruiz, B., 2013. Realistic virtual reproductions. image-based modelling of geometry and appearance. In: Digital Heritage International Congress (DigitalHeritage), 2013, Vol. 1, IEEE, pp. 127–134.

Palma, G., Callieri, M., Dellepiane, M. and Scopigno, R., 2012. A statistical method for SVBRDF approximation from video sequences in general lighting conditions. Eurographics Symposium on Rendering 31(4), pp. 1491–1500.

Schwartz, C. and Klein, R., 2012. Acquisition and presentation of virtual surrogates for cultural heritage artefacts. In: EVA 2012 Berlin, Gesellschaft zur Förderung angewandter Informatik e.V., Volmerstraße 3, 12489 Berlin, pp. 50–57.

Tunwattanapong, B., Fyffe, G., Graham, P., Busch, J., Yu, X., Ghosh, A. and Debevec, P., 2013. Acquiring reflectance and shape from continuous spherical harmonic illumination. ACM Transactions on Graphics (TOG) 32(4), pp. 109.

This contribution has been peer-reviewed.
doi:10.5194/isprsarchives-XL-5-W4-139-2015

REMOTE VISUALIZATION AND NAVIGATION OF 3D MODELS OF ARCHEOLOGICAL SITES

M. Callieri, M. Dellepiane, R. Scopigno

Visual Computing Lab, ISTI-CNR, Via G. Moruzzi 1, 56124 Pisa (PI), Italy
Web: http://vcg.isti.cnr.it/ Contact: *surname*@isti.cnr.it

ABSTRACT:

The remote visualization and navigation of 3D data directly inside the web browser is becoming a viable option, due to the recent efforts in standardizing the components for 3D rendering on the web platform. Nevertheless, handling complex models may be a challenge, especially when a more generic solution is needed to handle different cases. In particular, archeological and architectural models are usually hard to handle, since their navigation can be managed in several ways, and a completely free navigation may be misleading and not realistic. In this paper we present a solution for the remote navigation of these dataset in a WebGL component. The navigation has two possible modes: the "bird's eye" mode, where the user is able to see the model from above, and the "first person" mode, where the user can move inside the structure. The two modalities are linked by a point of interest, that helps the user to control the navigation in an intuitive fashion. Since the terrain may not be flat, and the architecture may be complex, it's necessary to handle these issues, possibly without implementing complex mesh-based collision mechanisms. Hence, a complete navigation is obtained by storing the height and collision information in an image, which provides a very simple source of data. Moreover, the same image-based approach can be used to store additional information that could enhance the navigation experience. The method has been tested in two complex test cases, showing that a simple yet powerful interaction can be obtained with limited pre-processing of data.

1. INTRODUCTION

The remote visualization (via Web) of complex geometry has become feasible only in the last couple of years. The WebGL framework allowed to overcome the proprietary plugins issue, and three-dimensional content is slowly becoming a usual type of content for web pages.
The issues about the visualization of 3D models are related not only to the necessity to handle complex data, but also to the interaction with them. Navigating 3D environments is not trivial, due to both the nature of data and the fact that most of the users are not used to that. This is especially true when dealing with 3D models of complex environment (terrains with buildings and ruins, often found in the archeological field); even more when 3D models come from sampling technologies (3D scanning or 3D-from-photo), which produce high-resolution, unstructured triangulated models. What is the best way to interact with them? A completely free navigation may be hard to handle, while extremely constrained approaches may limit their potentials. While it is possible to borrow ideas from the entertainment world (i.e. videogames industry), the best solution would be to find a flexible, lightweight system needing only a small amount of preprocessing.
This paper presents a method for the navigation of complex archeological 3D environments, especially tailored for web visualization. The main contributions include:

- An intuitive navigation paradigm which includes two possible modes: the *bird's eye* view, where the user can explore the model from top; and the *first person* mode, where the user can walk inside the environment in a walk-through fashion. The two modes are linked by the presence of a *point of interest*, which helps the user in the navigation from top, and permits an intuitive passage between the two modes.

- An image-based encoding of the constraints for navigation. Each image channel contains the data needed for realistic interaction (i.e. height from ground, collision), without the need of the presence of proxy 3D models or similar. Moreover, the other channels of the image can encode also other information which can be used for the navigation (i.e. location or invisible hotspots). The image can be generated with a semi-automatic approach starting from the geometry, or by hand with the usual image processing tools.

The approach is especially suited to archeological environments made by one or more "levels", but it could be adapted to other environments. It was tested on two cases: an entire insula in Pompei acquired with terrestrial laser scanners (where location information was used to integrate the 3D model with existing documentation, linked through a website) and a medieval village whose model was acquired with UAV. In both cases, a very intuitive navigation was obtained without the need of complex pre-processing operations.

2. RELATED WORK

Remote visualization of complex architectural environments deals with several general issues. The main ones are the integration of three-dimensional contents on the web, and the navigation of complex environments. While a comprehensive overview of all the related approaches goes well beyond the scope of the paper, we provide an overview of the state of the art in the following subsections.

2.1 3D and the Web

Three-dimensional data have always been considered part of multimedia content, but their role in the context of web pages hasn't been a major one until a few years ago. This was mainly due to the difficulty to handle complex data, but also to a lack of standardization. Hence, their visualization and use was supported through the use of embedded software components, such as Java applets or ActiveX controls (ACT, n.d.).

This contribution has been peer-reviewed.
doi:10.5194/isprsarchives-XL-5-W4-147-2015

Some research effots were devoted to define a common data format (Raggett, 1995, Don Brutzmann, 2007), but 3D scene visualization was still delegated to external software components.
Only an initiative by the Khronos Group (Khronos Group, 2009a), the creation of WebGL standard (Khronos Group, 2009c), was able to generate a remarkable change. WebGL is a mapping of OpenGL|ES 2.0 specifications (Khronos Group, 2009b) in JavaScript. Direct access to the graphics hardware by the web browsers allows to fully exploit the potentials of rendering (Evans et al., 2014).
Following this standard, several actions were made to provide a sort of interface between the low-level OpenGL commands and the data visualization and navigation. We can see two different directions of research: the first one is more dedicated to a declarative approach, based on the concept of *scenegraph*. Two examples of declarative programming solutions are X3DOM (Behr et al., 2009) and XML3D (Sons et al., 2010). In alternative to the declarative approach, several other actions went in the direction of an imperative approach, where a more direct interaction is possible. Several libraries have been developed (most of them based on Javascript as a basic language), ranging from scene-graph-based interfaces, such as Scene.js (Kay, 2009) and GLGE (Brunt, 2010), to more programmer-friendly paradigms, such as SpiderGL (Di Benedetto et al., 2010), WebGLU (DeLillo, 2009), and Three.js (Dirksen, 2013). Several examples of integration of 3D data on the web are appearing now. Among the practical issues that have to be solved to provide usable tools, there are the necessity to integrate the models with other types of data (Jankowski and Decker, 2012, Callieri et al., 2013), but also the need for methods to handle the visualization of single complex objects (Behr et al., 2012, Callieri et al., 2013). The Smithsonian X3D explorer (http://3d.si.edu) is an alternative example where 3D models are associated to additional content, but the structure and flexibility of the proposed system are not known.

2.2 Camera control in Computer Graphics

The issue of 3D camera control is well known in the context of Computer Graphics. Please refer to the work of Christie (Christie et al., 2008) for a comprehensive overview.
The main solution developed to interactively observe and inspect a 3D model is the virtual trackball (Chen et al., 1988, Bell, 1988, Shoemake, 1992, Henriksen et al., 2004), that allows 3D rotations with just 2 Degrees of Freedom input devices, such as the mouse. This interaction works very well when the aim is to rotate around an object, but it tends to fail when the environment is more complex. For example, when the user needs to explore the inside of an object, or when its shape cannot be approximated to a sphere, this approach usually fails.
On the other side, a totally free navigation may be difficult to implement, and it could lead to a *lost-in-space* effect, where the user is not able to find the orientation in the context of a scene. Moreover, there are important object-related issues that should be taken into account (i.e. in an architectural environment, the user does not want go below the ground).
Solutions have been proposed to constrain and guide the trackball paradigm (Fitzmaurice et al., 2008, Hachet et al., 2008), but in the case of more complex environments (i.e. the architectural ones) other methods have been implemented. They could be based on a preliminary analysis of the scene to calculate pre-defined paths (Andujar et al., 2004), or constraining the camera with force-fields to avoid collisions (Wan et al., 2001, McCrae et al., 2009). Another possible solution is to select a set of possible points of view, and constrain the navigation to these points, which can be connected by the means of pre-computed paths (Hanson and Wernert, 1997). The choice of the points of view describing a complex environment is a very complex task (Scott et al., 2003), and authoring is usually necessary (Burtnyk et al., 2002, Burtnyk et al., 2006).
Following this approach, image-based techniques have been used to remove limitations on scene complexity and rendering quality for interactive applications. Please refer to the recent work by Di Benedetto (Di Benedetto et al., 2014) and its bibliography for a comprehensive overview. Nevertheless, the aim of our work was to provide a solution supporting the free navigation of the model, trying to define implicit constrains and taking advantage of the strong points of different navigation paradigms.

3. EFFICIENT RENDERING

The first problem related to the visualization scheme was to render effectively the complex 3D model representing the area of interest. As we stated in the introduction, we are focusing towards the visualization of complex 3D models of archeological/historical areas, created by sampling the real environment (by 3D scanning or 3D-from-photos). These models are not a "simple" terrain, that can be easily managed using 2.5D methods, or well known terrain-specific level-of-detail approaches, but require a full 3D management. Additionally, we have to work with extremely complex 3D models, made of millions of triangles.

While not the main contribution of this work, it is nevertheless necessary to present the solution which we have used in order to stream these dataset over the net and render them inside a web browser.

3.1 WebGL and SpiderGL

Our visualization framework is based on WebGL (Khronos Group, 2009c), a component of HTML5. Thanks to this, the 3D content works natively inside the browser, without the need of plugins on most modern browsers (Firefox, Chrome and Internet Explorer) on all platforms. WebGL provides direct access to the GPU of the PC, enabling a programmer to write extremely optimized low-level code, as one would do in standard software programming. However, since WebGL is very low-level, its direct use is not really common, and many developers use support libraries, able to provide higher level functionalities, data structures and helper functions. To this aim, we decided to use SpiderGL (Di Benedetto et al., 2010), a JavaScript support library oriented to Computer Graphics programming.

Using these components we built a simple, yet flexible framework for the creation of 3D presentation of high-res models on the web. Using a declarative-like approach, it is possible to define a simple scene. The framework takes care of asynchronous data loading, event management, rendering and animation. All the components of the scene (3D models, trackball, camera) are configurable and extensible, making it possible to create simple visualization webpages with ease, but at the same time, when a more complex visualization is needed, it is possible to fully exploit the modular/configurable capabilities of the framework. This is what we have done in this work: by tweaking the behavior of the renderer and of the navigation/interaction components, we were able to create a specialized visualization webpage.

One of the core modules of this framework is the management of multiresolution 3D models, a feature that we have used to stream and display the huge 3D models shown in the examples (Section 5.).

3.2 The Multiresolution engine

Displaying high resolution models on a web browser is not just a matter of optimizing the rendering speed, but it also involves considering the loading time and network traffic caused by transferring a considerable amount of data over the network. Loading a high-resolution model as a whole through the web requires to transfer a single chunk of data in the order of tens, if not hundreds, of megabytes: this causes a lot of band usage, and the user has to wait for the transmission to end before seeing any visual result.

Multiresolution techniques provide the solution for both rendering and data transfer. These schemes generally split the geometry in smaller chunks; for each chunk, multiple levels of detail are available. Transmission is *on demand*, requiring only to load and render the portions of the model strictly needed for the generation of the current view. While this approach is key to be able to render very large models at an interactive frame rate, it is also helpful to optimize the data transfer over the network, since the geometry will be divided in small chunks and only transferred when needed. The model is immediately available for the user to browse it, even though at a low resolution, and it is constantly improving its appearance as new data are progressively loaded. On the other hand, since refinement is driven by view-dependent criteria (observer position, orientation and distance from the 3D model sections), only the data really needed for the required navigation are transferred to the remote user.

Figure 1: The multiresoluton Engine. On left: the patches used , smaller and more detailed near the camera, larger and at a lower-resolution farther away. Middle: when rendering, the geometry appear as a continous surface. On right: the multiresolution easily manage the whole 27 million triangles model of the Pompeii Insula.

We implemented one of those multiresolution scheme, Nexus (Cignoni et al., 2005) (http://vcg.isti.cnr.it/nexus/), on top of the SpiderGL library (Di Benedetto et al., 2010), obtaining very good performances. In the two examples shown in this paper, we are dealing with high-resolution 3D models: the Pompeii insula is more than 27 millions triangles (Figure 1), while the San Silvestro Hill is around 22 millions (Figure 7). Both datasets are displayed at interactive rate, with almost immediate startup at the page loading, and the progressive refinement provides good visual results even on slower networks. This makes possible a quick, seamless exploration of the entire dataset, and supports instant jumping from one part of the area to another.

4. NAVIGATION

As we stated, navigation is the most important component of the visualization. Our aim was to implement a navigation simple to use, but able to fully explore mixed ground-buildings 3D dataset like the Pompeii insula or the San Silvestro Hill.

Instead of trying to implement a single all-purpose navigation method, which would result in an extremely complex interface, we decided to implement two different but complementary navigation methods: a "bird's eye" view, to explore the dataset from above, and a "first-person" view to explore the area in a more immersive way. In this way, each navigation mode is easier to use and more specialized but, by combining both navigation methods, it is possible to fully explore the 3D dataset with greater flexibility.

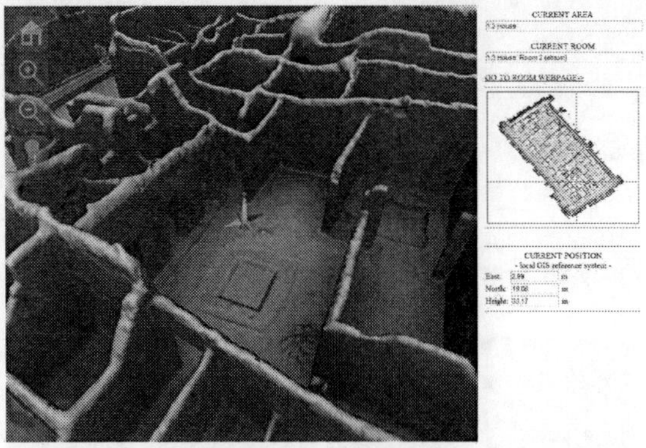

Figure 2: The visualization webpage interface. On left: the 3D view, in bird's eye view mode, with the current point of interest shown by the 3D locator. On top right: the current area and room name, tracked using the location channel of the map. On middle right: the minimap, graphically showing the current position. On bottom right: the current position of the user, expressed in this specific case as relative coordinates with respect to a local origin defined in a GIS tool

4.1 Bird's eye mode

Bird's eye view mode mimics flying above the environment or, with a more precise metaphore, looking from above at a maquette of the area. The user viewpoint hovers the area, looking down at specific details (see Figure 2).

This is, however, not a completely unbound navigation (like in the free-camera paradigm), but is similar to what is available in applications such as Google Earth. Basically, the camera always refers to a *point of interest*, located on the ground of the environment. The user may move this point, panning across the map, and change the camera orientation (vertical tilting, orbiting, and zooming). We did not consider the use of a completely free camera because it is really difficult to be controlled by non-experts (especially in a web browser environment), without providing a significative increase of flexibility. Even considering its constrains, this navigation mode enables the user to easily reach every portion of the environment. This is especially true in the Pompeii dataset, but also in most of the archeological sites: since most of the roofs are missing, it is possible to see almost all of the walls and details of the structure.

The *point of interest* is shown on the 3D view with a locator (a simple 3D model, visible in Figure 2); this helps the user in having a clear feedback on its position inside the area. The locator is nearly human-size, to better understand the scale of the area, and show at its base the cardinal points (the red arrow points towards north).

The implementation of bird's eye view is quite easy. Since we are dealing with a terrain-like environment, it is simply a matter of defining an absolute XY positioning on the ground (X axis aligned with East-West axis and Y axis on the North-South one). Beside the extents of the dataset area, it is necessary to have a way to follow the ground geometry when moving across the area. Most datasets presents irregular, sloped, uneven terrain; and having a way to accurately position the point of interest is crucial

This contribution has been peer-reviewed.
doi:10.5194/isprsarchives-XL-5-W4-147-2015

to cope with these irregularities. We will explain this point in Section 4.4.1. No collision is necessary in this navigation mode, since it would impose unnecessary constrains in user movement, making the navigation of the whole area more difficult.

Bird's eye view is the starting navigation mode of the visualization webpage. The user may, at any time, switch to the first-person view (see next Section). While switching to first person view, location and orientation are preserved, i.e. the point of interest of the bird's eye will become the location of the first-person camera, and the view direction of the bird's eye will be the same of the one used by the first-person (see Figure 3). In this way, it is much easier to go back and forth between views, allowing for a seamless exploration of the dataset. The same correspondence will be preserved when switching back from first-person to bird's eye.

Figure 3: Switching from bird's eye to first-person preserves the user position and orientation, as it is visible in these images.

4.2 First person mode

Even if, in an environment like Pompeii, with most of the ceilings missing, the bird's view is more or less sufficient to explore the entire map, it is sometimes useful/necessary to "jump into" the 3D model, to explore the environment like it would be done in person.

The first-person view mode lets the user moves around the environment at ground level (the camera is at a human-compatible height). Our implementation follows the same mechanism used in many similar tools: the first-person lets the user look in all direction (using the mouse) and move around on the ground (using the mouse or the WASD keys, generally used in videogames).

Similarly to the bird's eye mode, also in the first-person mode it is necessary to follow the ground geometry when moving around. There is, however, an additional constrain often found in similar navigation modes: the collision with walls and other geometries. Implementing these features for a first-person view is always complex, given the unstructured nature of the kind of 3D models we are using, generated from 3D scanning. In most videogames or interactive visualization, the 3D model used comes from manual modeling; it has a lower resolution and, more important, it is highly structured (walls and floors are explicitly marked, and often a collision mesh is also manually created). This helps a lot the navigation, but it is not easily obtainable when dealing with this kind of 3D models.

For the terrain-height following, we have used the method described in Section 4.4.1 , while in Section 4.4.2 we explain the method used for managing walls collision.

4.3 Minimap

In order to simplify the exploration, we also added a mini-map to the visualization page (see Figure 2). The minimap has two purposes: to give an immediate feedback on the current position, thus avoiding users to "get lost" while zooming or when in first-person view, but also to have a way to instantly jump to another position.

The position of the point of interest is shown using a crosshair drawn over the image, dynamically updated at every user movement. Conversely, when the user clicks on a point on the minimap, the point of interest is instantly moved to that location. Both these actions exploit the fact that there is a perfect match of the area covered by the 3D dataset, the minimap and the one covered by the data map (see next section). Locations are passed between these three elements using relative coordinates (0.0-1.0), to gather independence from image resolution.

4.4 The "Navigation Image" encoding

As we stated in the previous section, in order to provide a easy-to-use navigation, it is necessary to implement some mechanism able to follow the terrain height and to manage collisions. Additionally, tracking the user position is another need often present in this kind of visualization. In all cases, it all boils down to knowing, for a given position in the dataset area, a specific information: which is the terrain height in this point? Is there a wall in this point, or is it free-ground? In which room/zone is this point?.

The idea is to have all these location-dependent information stored in a single, easy-to-access structure; knowing the location of the user (the point-of-interest for the bird's eye, and the viewer position in first-person), it is easy to retrieve the location-specific data, and have the visualization page react accordingly. We chose to store all the data in an image file, where each channel encodes a specific information. PNG format is used, given its lossless compression. An example of such image is shown in Figure 4. This "Navigation Image" is read at the page loading; then, it is drawn on a hidden canvas and read back in memory in an array. At this point, the pixels are directly addressable by Javascript code. This structure is accessed using relative coordinates, thus ensuring independency from the image resolution, and an easier mapping between coordinates.

4.4.1 Red channel - Terrain Height Most of the 3D datasets built using sampled data have a high-res non-flat terrain; moving across this terrain requires then to alter the camera height to follow the geometry of the ground. This is generally obtained by creating a low-res collision ground, which is then tested as the user moves. Given that this lower-res collider is generally smooth, encoding its height in a map is not a problem, and makes much easier its use. The red channel of our map is used to store the base height of the terrain, i.e. a smoothed height-field representing the ground without the walls and other occluders. By using this map, the camera always follows the terrain when moving in the bird's eye and first-person view. The height map is smoothed to avoid useless camera shaking, and the walls and small colliders are not included, to have the camera follow the "real" ground level. Its use is straightforward: every time the XY position of the user changes, its Z coordinate is modified according to the value encoded in this map.

4.4.2 Blue channel - Collision Another feature which is often sought in this kind of visualization is the collision with walls and other large geometries. This is a non-trivial task, since it requires, in most game engines, the manual creation of collision meshes (often, with specific geometrical-topological constrains, to obtain better performances), and a complex management of the collision check. The blue channel of our map is used to store the collision/validity map: each pixel encodes if the corresponding area is free, occupied by an occluder (mostly, walls), or impassable (the area outside the 3D model, or not suitable for first-person navigation). The collision map is used mostly when the

This contribution has been peer-reviewed.
doi:10.5194/isprsarchives-XL-5-W4-147-2015

user is in first-person view; every time the user XY position has to be updated, the map is checked: if the new XY position would cause a collision, the motion is culled at the wall, preventing the collision. Bird's eye view does not use the collision map, since it may be useful for the user to place its point of interest anywhere in the map and, beside this, ignoring the walls and unsurmountable terrain does allow a quicker exploration of the map. The collision map is, however, used in bird's eye view when clicking on the minimap: if the user has asked do go in an "impossible" position, its position is snapped to the closest "safe" position. The same mechanism is used when the view mode is switched from bird's eye to first-person: if the point of interest is currently inside a wall, the user is moved to the closest open ground, to avoid having the user being trapped.

4.4.3 Green channel - Location A situation often occurring in this kind of datasets is the need to determine the exact location of the user, not just in terms of absolute XY coordinates on the map (which we already have), but in terms of areas, sub-areas, rooms, or any other semantical subdivision of the map. Again, this information is easily encoded in an image map. The green channel of our map is used to determine the position of the user point of interest. Each pixel simply stores the ID of the area; the 255 available values are enough to encode the areas for most datasets.
Every time the user position changes, the ID of the pixel under the user XY position is fetched and used to access a JSON (JavaScript Object Notation, an open standard for representing simple data) structure with all the information on the areas: a globally accessible variable is updated and a call-back function is called when the ID changes, in order for the webpage to react accordingly.

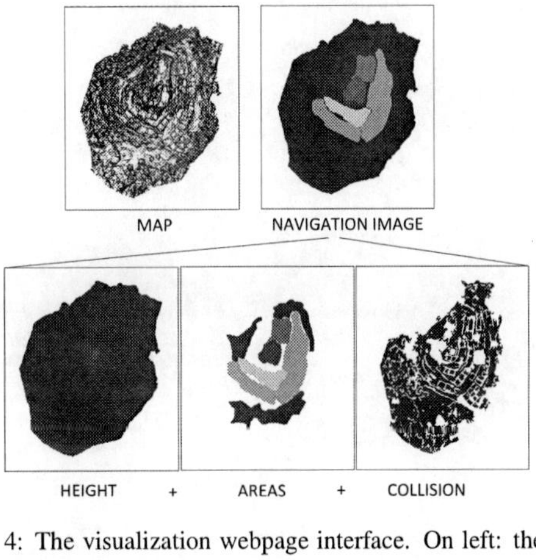

Figure 4: The visualization webpage interface. On left: the 3D view, currently in bird's eye view mode, with the point of interest shown by the 3D locator. On top right: the current area and room name, tracked using the location channel of the map. On middle right: the minimap, graphically showing the current position. On bottom right: the current position of the user, expressed in this specific case as relative coordinates with respect to a local origin defined in a GIS tool

4.4.4 Other channels and encoding Since we are using PNG, it is possible to also have an Alpha channel; at the moment, it is not used. It may contain additional IDs of areas (if more than 255 areas have to be encoded); or could be used in conjunction with the red height channel, to have more resolution on the vertical axis, when dealing with a steeper terrain; or to encode any other location-based data. The assignment of the channels is completely arbitrary, and may be easily altered to cope with specific characteristics of the dataset. It is also possible, if running out of space, to add another image map, adding three/four more channels.

While there are other ways to encode the same type of information, this access strategy exploits already existing and highly optimized mechanisms of the browser. Additionally, given the data contained in the channels, it behave very well when compressed. The most valid alternative would be to keep all this data in a vectorial format; this would surely work, but with a limited impact on the size of the data (depending, for example, on how complex the geometry of the walls is), and, more importantly, it would require a much more complex authoring and editing.

Considering the Pompeii dataset, the PNG image used is only 500kb in size, for a 1142 x 980 resolution, with a physical size of each pixel around 10cm, more than enough to provide an accurate management for terrain height, walls collision and location tracking.

4.4.5 Authoring navigation images From the start, we wanted a method for managing the aforementioned features (ground height, wall collision, area detection) which was simple both when using it in realtime *and* when creating the data needed for its use.

The height channel may be easily created from the 3D model: a simple color-coded (Z value to red value) orthographic rendering from above (or below) is enough to generate this channel. Then, to have a more continuous height, it is possible to smooth the rendered image, it is easier and have better results than starting from a smoothed geometry. We stated that it is better to not consider, in the height map, the walls and other non-terrain geometries. These areas may be manually found and eliminated in the 3D model or in the rendered image (and then smoothly filled). It is also possible to use, on the 3D model, some heuristics to automatically select walls and similar geometries, and then cut them. This automatic approach is viable and time saving; in the cases where the detection is not really accurate, a small manual intervention would help obtaining a good result in a shorter time, with respect to the completely manual approach.

The collision channel is slightly more complex. While it is an option to manually mark all the walls and colliders using an image-editing software, this may be time consuming. The same heuristics mentioned for the height channel may also be applied to find the walls for the collision channel. Also in this case, the result of this automatic detection may be further refined with a limited manual intervention, resulting in a much cleaner result. This is the procedure used to generate the collision maps shown in Figure 4 and Figure 5.

Both of these maps (height and collision), can also be created from GIS raster or vector data: this method is easy and time-effective, making it possible to exploit existing data.

Unfortunately, for most applications, the "location" channel will necessarily should be created manually, since there is not a straightforward way to automatically subdivide a 3D model in an archeological-sensible way. However, creating this image from an existing GIS vector/raster layer is a viable (and recommended) option.

4.4.6 Page interconnection Especially when dealing with complex archeological environment, it is often possible to have existing databases or documentation-rich websites. It would be interesting to connect the visualization page with these existing repositories. This would mean having a way, on one hand, to open from the existing website the 3D visualization page with the point of interest and view parameters already focusing on a specific area

This contribution has been peer-reviewed.
doi:10.5194/isprsarchives-XL-5-W4-147-2015

and, on the other hand, from the current point of interest, go back to the existing website, directly accessing the section related to the current 3D location.

This is indeed possible, thanks to the parametric nature of the visualization page. In order to open the visualization page, the *Query String* mechanism can be used. It is possible to specify completely the position and view parameters; for example, the following url query string opens the visualization with the point of interest located at 36.0 meters East, 12.6 meters North, with camera looking at 45 degree with respect to north direction, 60 degree azimuthal angle, zoom level 2.0:

```
http://aaa.com/pompeii3D.html?StartNS
   =36.0&StartEW=12.6&StartPhi=45.0&
       StartTheta=60.0&StartZoom=2.0
```

It is also possible to ask the webpage to go to one of the "areas" or "rooms" defined in the JSON structure mentioned in section 4.4.3. This structure contains a list of areas/subareas, each one with an associated ID, name, description, and URL (see next paragraph) and parameters for setting point of interest and view direction to frame it. By passing the name of the desired area to the URL of the 3D visualization webpage, the page will open by framing the desired area, fetching the appropriate parameters in this JSON structure. The following url query string opens the visualization with the point of interest location and camera orientation stored in the JSON entry for the "house of Torello di Bronzo":

```
http://aaa.com/pompeii3D.html?
     StartRoom=1.7_house_torello
```

Going in the other direction is also easy. Since at all times, the webpage knows where the user point of interest is located, in absolute coordinates or at area/room level (thanks to the location map), it is possible to open back the existing website at the desired section. This can be implemented by using the same JSON structure just mentioned: the URL stored in each area/room is used for this specific purpose.

5. RESULTS

The proposed rendering and navigation method was applied on two real datasets, in order to test the usability and the possible uses of the image map encoding. The first one represented a nearly planar but slightly sloped, complex architectural structure, while the second one represented a less planar, irregularly urbanized medieval borough on top of an hill.

5.1 The Pompei Insula V.1

The first test case was the model of the Insula V.1 of Pompei. This wide area (1330 square meters), was acquired in the context of a project coordinated by the Swedish Archeological School (Dell Unto et al., 2013) using phase shift laser scanners. The acquisition campaigns and the subsequent processing generated several complete models of the insula at different resolutions. The model (which was composed by 27M triangles, with a resolution of around 3cm) was also referenced w.r.t. the reference system of the archeologists.
The original map of the insula was used as the minimap, while the height and collision were created in a semi-automatic fashion. The walls were removed from the 3D model, using an automatic selection function of MeshLab, and the collision map was obtained by producing a snapshot of a blue colored model from the same point of view as the minimap (see Figure 5,right).
The height map was obtained through filling the missing portions of the geometry via Poisson reconstruction, and encoding the height from the ground in the red channel using a per-vertex color function of MeshLab tool (see Figure 5,left). Some examples of the navigation of the Pompei dataset are shown in Figures 2 and 3. In this case, the green channel of the image map was used to link the 3D model to an already existing website that contained the documentation (text and images) for each single room of the Insula. By encoding in the green channel the Id of each room, it was possible to access directly from the navigation context the corresponding page on the Pompei Project website.
Figure 6 shows an example of the link between the position of the focus point and the corresponding data sheet on the website. A possible future extension of this system would be a perfect integration with the website, where the user could jump in and out of the two websites in an integrated fashion.

Figure 5: The height and collision maps for the Pompei example

Figure 6: Linking the navigation with a data sheets website.

5.2 San Silvestro village

San Silvestro was a miners medieval village which is currently part of a minerary park in Tuscany, Italy. The village was the target of an acquisition campaign using a UAV and Multi-view stereo matching techniques. Several different flights around the village allowed to extract more than 400 frames that were used for obtaining a complete multi-view stereo reconstruction.

This contribution has been peer-reviewed.
doi:10.5194/isprsarchives-XL-5-W4-147-2015

The model of the village can be extremely valuable for several aspects, and the possibility to freely navigate a three-dimensional model gives the chance to overcome its peculiar placement on top of an hill. The final 3D model (with color) was made of 22M triangles, with a resolution of nearly one point every 5 cm.
As in the previous example, the model was scaled to real measures, and geo-referenced using the documentation provided by the park authorities. Then, it was processed to be used with multiresolution visualization. The mini map (see Figure 4,top-left) for navigation was obtained through a non photo-realistic rendering of the model from top. The complete image map can be seen in Figure 4 with all the channels encoded: the height on the red channel, the location on the green one, and the collision in the blue one.

The height channel was obtained in an automatic fashion by

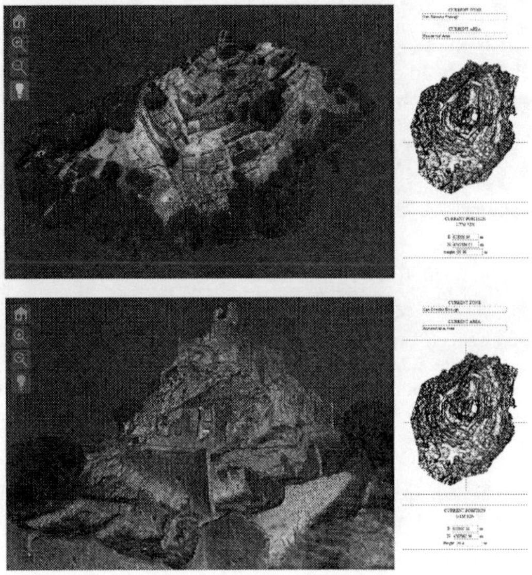

Figure 7: Two snapshots of bird's eye navigation of San Silvestro model

encoding the color w.r.t. the height from the ground using a per-vertex color function of MeshLab tool, and generating a snapshot from the top view. The collision channel was obtained in a semi automatic fashion, by first removing the vertical walls from the geometry, and then modifying by hand the blue colored remaining geometry. In this case, the collision surface was less easy to define, due to the presence in the model of architectural remains, rock and vegetation. The location channel was easily produced by hand starting from the minimap. It defines the different (i.e. residential, industrial, church) areas of the village.
Figure 7 shows two points of view that can be reached using the bird's eye modality: it's easy to get a view from the top, but also to explore portions of the village. On the top right of each snapshot, the peculiar area associated to the focus point is shown. Moreover, the UTC coordinates are available (bottom right).

Figure 8 shows two points of view that can be reached with first person navigation: one that can re-create a real visit (the one from the ground), the other providing a point of view not available to visitors (it is currently not possible to climb on top of the tower). The test on this model shows that it is possible to prepare a dataset where the ground is not planar, with only a minimal intervention in the creation of the image maps, and providing an extremely flexible yet intuitive navigation.

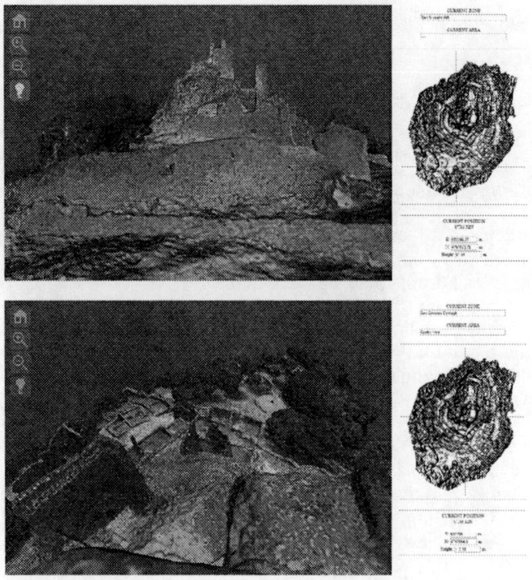

Figure 8: Two snapshot of first person navigation of San Silvestro model. Top: on top of the tower. Bottom: from the industrial area outside the village.

6. CONCLUSIONS AND FUTURE WORK

This paper presented a solution for the navigation of 3D environments on which the usual trackball paradigm for navigation cannot be applied. The system integrates two different visualization modalities (linked by the concept of "point of interest"), that could account for different needs in the exploration of a complex architecture.
Moreover, the important information needed to ensure a realistic experience are stored in a image map, that could be used also to store additional information (i.e. the location of the current point of interest). The image map can be produced with a limited effort by the user, and it's easily integrated in the navigation environment.
The main limitations of the current approach are: the need for data that are not organized on different levels (although the navigation could "switch" from one layer to another when needed), the use of bird's eye view when the model has roofs (although this could be overcome by using ad-hoc visualization, like x-ray-like shaders), and the necessity to produce the image map "by hand" (although the procedure is quite straightforward and could be automatized, for example using a MeshLab script).
Nevertheless, the system could be directly applied to a number of possible test cases in the field of Cultural Heritage, from archeological excavation to city navigation, or even for the exploration of artifacts with small detail (like engraving or small decorations).
Regarding future improvements of the system, three possible directions of work can be outlined: the implementation of simple interaction functionalities (measurement, sections, simple annotation), the creation of authoring tools to help in preparing new scenes, and a better integration with existing web visualization systems.

ACKNOWLEDGEMENTS

The research leading to these results was funded by EU FP7 project ICT Harvest4D (http://www.harvest4d.org/ , GA n. 323567) and EU INFRA Project Ariadne (GA n. 313193, http://www.ariadne-infrastructure.eu/). We'd like to thank the University of Lund,

Scuola Archeologica Svedese, and Andrea Berton, for their support.

REFERENCES

ACT, n.d. Microsoft ActiveX Controls. http://msdn.microsoft.com/en-us/library/aa751968(VS.85).aspx.

Andujar, C., Vazquez, P. and Fairen, M., 2004. Way-finder: guided tours through complex walkthrough models. Computer Graphics Forum 23(3), pp. 499–508.

Behr, J., Eschler, P., Jung, Y. and Zöllner, M., 2009. X3dom: a dom-based html5/x3d integration model. In: Proceedings of the 14th International Conference on 3D Web Technology, Web3D '09, ACM, New York, NY, USA, pp. 127–135.

Behr, J., Jung, Y., Franke, T. and Sturm, T., 2012. Using images and explicit binary container for efficient and incremental delivery of declarative 3d scenes on the web. In: Proceedings of the 17th International Conference on 3D Web Technology, Web3D '12, ACM, New York, NY, USA, pp. 17–25.

Bell, G., 1988. Bell's trackball. Written as part of the "flip" demo to demonstrate the Silicon Graphics (now SGI) hardware.

Brunt, P., 2010. GLGE: WebGL for the lazy. http://www.glge.org/.

Burtnyk, N., Khan, A., Fitzmaurice, G. and Kurtenbach, G., 2006. Showmotion: camera motion based 3d design review. Proceedings of the 2006 symposium on Interactive 3D graphics and games pp. 167–174.

Burtnyk, N., Khan, A., Fitzmaurice, G., Balakrishnan, R. and Kurtenbach, G., 2002. StyleCam: Interactive Stylized 3D Navigation using Integrated Spatial & Temporal Controls. Vol. 4Number 2, ACM, pp. 101–110.

Callieri, M., Leoni, C., Dellepiane, M. and Scopigno, R., 2013. Artworks narrating a story: a modular framework for the integrated presentation of three-dimensional and textual contents. In: Web3D, 18th International Conference on 3D Web Technology.

Chen, M., Mountford, S. J. and Sellen, A., 1988. A study in interactive 3-d rotation using 2-d control devices. In: Proceedings of the 15th annual conference on Computer graphics and interactive techniques, SIGGRAPH '88, ACM, New York, NY, USA, pp. 121–129.

Christie, M., Olivier, P. and Normand, J.-M., 2008. Camera control in computer graphics. Computer Graphics Forum 27(8), pp. 2197–2218.

Cignoni, P., Ganovelli, F., Gobbetti, E., Marton, F., Ponchio, F. and Scopigno, R., 2005. Batched multi triangulation. In: Proceedings IEEE Visualization, IEEE Computer Society Press, Conference held in Minneapolis, MI, USA, pp. 207–214.

DeLillo, B., 2009. WebGLU: A utility library for working with WebGL . http://webglu.sourceforge.org/.

Dell Unto, N., Ferdani, D., Leander, A.-M., Dellepiane, M., Callieri, M. and Lindgren, S., 2013. Digital reconstruction and visualization in archaeology case-study drawn from the work of the swedish pompeii project. In: Digital Heritage 2013 International Conference, IEEE, pp. 621–628.

Di Benedetto, M., Ganovelli, F., balsa Rodriguez, m., Jaspe Villanueva, A., Gobbetti, E. and Scopigno, R., 2014. exploremaps: Efficient construction and ubiquitous exploration of panoramic view graphcs of complex 3d environments. In: Computer Graphics Forum, 33(2), 2014. Proc. Eurographics 2014, To appear.

Di Benedetto, M., Ponchio, F., Ganovelli, F. and Scopigno, R., 2010. Spidergl: a javascript 3d graphics library for next-generation www. In: Proceedings of the 15th International Conference on Web 3D Technology, Web3D '10, ACM, New York, NY, USA, pp. 165–174.

Dirksen, J. (ed.), 2013. Learning Three.js: The JavaScript 3D Library for WebGL. Packt Publishing.

Don Brutzmann, L. D., 2007. X3D: Extensible 3D Graphics for Web Authors. Morgan Kaufmann.

Evans, A., Romeo, M., Bahrehmand, A., Agenjo, J. and Blat, J., 2014. 3d graphics on the web: A survey. Computers & Graphics 41(0), pp. 43 – 61.

Fitzmaurice, G., Matejka, J., Mordatch, I., Khan, A. and Kurtenbach, G., 2008. Safe 3d navigation. In: Proceedings of the 2008 symposium on Interactive 3D graphics and games, I3D '08, ACM, New York, NY, USA, pp. 7–15.

Hachet, M., Decle, F., Knodel, S. and Guitton, P., 2008. Navidget for easy 3d camera positioning from 2d inputs. In: 3D User Interfaces, 2008. 3DUI 2008. IEEE Symposium on, pp. 83 –89.

Hanson, A. J. and Wernert, E. A., 1997. Constrained 3d navigation with 2d controllers. In: Proceedings of the 8th conference on Visualization '97, VIS '97, IEEE Computer Society Press, Los Alamitos, CA, USA, pp. 175–ff.

Henriksen, K., Sporring, J. and Hornbæk, K., 2004. Virtual trackballs revisited. IEEE Transactions on Visualization and Computer Graphics 10(2), pp. 206–216.

Jankowski, J. and Decker, S., 2012. A dual-mode user interface for accessing 3d content on the world wide web. In: Proceedings of the 21st international conference on World Wide Web, WWW '12, ACM, New York, NY, USA, pp. 1047–1056.

Kay, L., 2009. SceneJS. http://www.scenejs.com.

Khronos Group, 2009a. Khronos: Open Standards for Media Authoring and Acceleration.

Khronos Group, 2009b. OpenGL ES - The Standard for Embedded Accelerated 3D Graphics.

Khronos Group, 2009c. WebGL - OpenGL ES 2.0 for the Web.

McCrae, J., Mordatch, I., Glueck, M. and Khan, A., 2009. Multiscale 3d navigation. In: Proceedings of the 2009 symposium on Interactive 3D graphics and games, I3D '09, ACM, New York, NY, USA, pp. 7–14.

Raggett, D., 1995. Extending WWW to support platform independent virtual reality. Technical Report.

Scott, W. R., Roth, G. and Rivest, J.-F., 2003. View planning for automated three-dimensional object reconstruction and inspection. ACM Comput. Surv. 35(1), pp. 64–96.

Shoemake, K., 1992. Arcball: a user interface for specifying three-dimensional orientation using a mouse. In: Proceedings of the conference on Graphics interface '92, Morgan Kaufmann Publishers Inc., San Francisco, CA, USA, pp. 151–156.

Sons, K., Klein, F., Rubinstein, D., Byelozyorov, S. and Slusallek, P., 2010. Xml3d: Interactive 3d graphics for the web. In: Proceedings of the 15th International Conference on Web 3D Technology, Web3D '10, ACM, New York, NY, USA, pp. 175–184.

Wan, M., Dachille, F. and Kaufman, A., 2001. Distance-field based skeletons for virtual navigation. In: Proceedings of the conference on Visualization '01, VIS '01, IEEE Computer Society, Washington, DC, USA, pp. 239–246.

This contribution has been peer-reviewed.
doi:10.5194/isprsarchives-XL-5-W4-147-2015

ADAPTIVE TEXTURE SYNTHESIS FOR LARGE SCALE CITY MODELING

G. Despine[a], T. Colleu[a]

[a] SIRADEL, Department of Digital Map Data Production, Research and Development Service, 35760 Saint-Grégoire, France - (gdespine, tcolleu)@siradel.com

Commission V, WG V/4

KEY WORDS: Generic Texture, Window Detection, Texture synthesis, City Modelling, Façade Parsing

ABSTRACT:

Large scale city models textured with aerial images are well suited for bird-eye navigation but generally the image resolution does not allow pedestrian navigation. One solution to face this problem is to use high resolution terrestrial photos but it requires huge amount of manual work to remove occlusions. Another solution is to synthesize generic textures with a set of procedural rules and elementary patterns like bricks, roof tiles, doors and windows. This solution may give realistic textures but with no correlation to the ground truth. Instead of using pure procedural modelling we present a method to extract information from aerial images and adapt the texture synthesis to each building. We describe a workflow allowing the user to drive the information extraction and to select the appropriate texture patterns. We also emphasize the importance to organize the knowledge about elementary pattern in a texture catalogue allowing attaching physical information, semantic attributes and to execute selection requests. Roofs are processed according to the detected building material. Façades are first described in terms of principal colours, then opening positions are detected and some window features are computed. These features allow selecting the most appropriate patterns from the texture catalogue. We experimented this workflow on two samples with 20 cm and 5 cm resolution images. The roof texture synthesis and opening detection were successfully conducted on hundreds of buildings. The window characterization is still sensitive to the distortions inherent to the projection of aerial images onto the facades.

1. INTRODUCTION

1.1 Context and objectives

One activity of the Siradel Company is to produce virtual 3D cities textured with aerial images for architectural or urban planning projects. These models are well suited for bird-eye navigation or to evaluate any urban project but generally the image resolution does not allow pedestrian navigation or detailed view. On the other hand Siradel is leading the Probesim3D research project aiming to produce massive 3D data for military training in simulators. These kinds of simulators usually offer virtual environments close to video game. For this use case the look and feel is as important as the geographic truth: the level of details and the resolution of the textures are essential to improve the immersive feeling of the learner and finally the acceptance of the learning tool.

One solution to face this problem is to increase the resolution by completing aerial images with terrestrial photos. This can be done with a Mobile Mapping System or even with a simple camera. According to our experience, this solution involves huge amount of manual work to remove occlusions (trees, poles, wires…) and to perfectly fit textures on buildings, even with automatic process workflow. This is an expensive and time consuming solution that can't be applied to large scale data bases.

Another solution, and this is the object of this paper, is to artificially improve the grain and the level of detail with an adaptive texture synthesis. The idea is to use a procedural modelling strongly constrained with information extracted from aerial low resolution images. The goal here is not to model the exact ground-truth but to offer a believable low-cost environment that is as close as possible to the real scene.

1.2 Related works

(Alizadehashrafi and Abdul-Rahman 2011) use a Dynamic Pulse Function for texture enhancement. It is based on facet situation and observation for tessellating the textures. Missing faces of polyhedron and missing objects can be solved.

(Loya, 2008) presents a technique for generating abstract representations of buildings in modern urban areas by determining the dominant colours and primary periodic features of a building from textures of facades. A parametric model of building facades as waveforms, based on Fourier series, is used to approximate the facade structure. (Dai, 2013) also details another interesting method for synthesizing realistic façade textures from a simple example. The sample image is parsed and described as semantic components that can be tilled to generate some new facade textures with different dimensions or with occluded parts inpainted.

(Ripperda and Brenner, 2006) propose a formal grammar to derive a structural façade description based on aerial imagery. This description process is guided by statistic distributions of façade structures (e.g. distribution of widows sizes, building heights, number of stairs …), distributions that are explored using a reversible jump Markov Chain Monte Carlo. This approach is very interesting and close to our problem but the creation of the statistic distribution set is a heavy task and is only valid for a given region of interest.

This contribution has been peer-reviewed.
doi:10.5194/isprsarchives-XL-5-W4-155-2015

Procedural modelling based on grammar rules allows fast producing of cost effective generic cities (Parish and Müller, 2001). This process has been improved by an automatic determination of the architectural style (Mathias 2011). A classifier is first trained on a set of façades with identified style then, for each new façade, some descriptors are calculated and a distance to each candidate class is computed. The knowledge of the style allows to choose the appropriate modelling rule.

(Müller, 2007) describes algorithms to automatically derive 3D models of high visual quality from single façade images of arbitrary resolutions. They combine a procedural modelling pipeline of shape grammars with image analysis to derive a meaningful hierarchical façade subdivision.

Texture synthesis usually involves elementary texture patterns representing a small sample of wall bricks or roof tiles. These patterns are repeatable in X and Y covering the whole roof or façade but creating unaesthetic periodic effect. (Cohen 2003) explains a simple stochastic system for non-periodically tiling based on a small set of Wang Tiles (Wang 1965).

This topic is very active and well documented with high quality results. Nevertheless, in most cases, examples show few building configurations and images are of good quality with a low distortion due to perspective. Our industrial needs faces the challenges of large-scale modelling. It led us to develop this intermediated solution between façade comprehensive description and fully generic textures.

2. METHOD PRINCIPLES AND PRODUCTION FRAMEWORK

As a digital map data producer Siradel has at its disposal many 3D virtual cities. In these databases, buildings ground outlines are captured with an accuracy of 0.5m and the roofs' main shapes are modelled in 3D. Walls and roofs are textured with aerial imagery with a resolution varying from 5cm to 20cm (Figure 1)

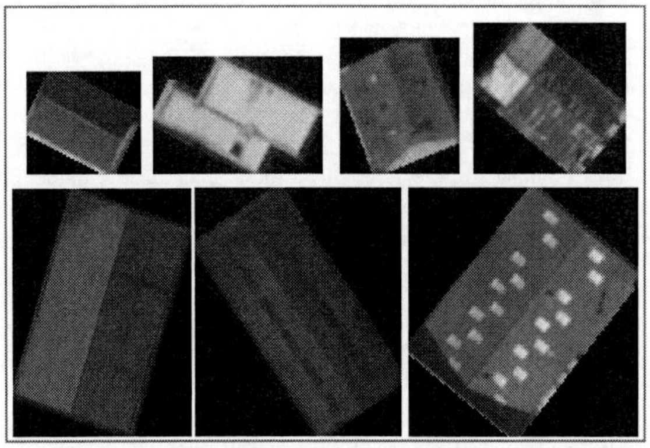

Figure 1. Examples of 20 cm resolution aerial images in real size. Metal, cement and slate roofs: three different building materials very similar even for human eye.

2.1 General principles

Our method is based on three steps:

1- Detection of information: this step enables to extract information from the image and also geometrical features from the available 3D model. Various image analysis tools are involved to extract information such as building material (tile, bricks, concrete…), principal colours of the facades or the position and size of punctual objects like the doors and windows.

2- The Interpretation step is performed by a set of user defined rules allowing refining the raw information from the first step. Here the user can express his knowledge of construction rules and local architectural styles to complete information: by example on low resolution images, it is difficult to distinguish natural slate, bitumen or metal (Figure 1). However the building shape can give indication on the most appropriate material: flat roofs of high buildings are generally made of concrete or bitumen, industrial buildings are covered with metal. The position of windows can also be corrected by this step: roof windows generally do not cross ridge and façade windows are often aligned to each other.

3- The synthesis step is the composition of the final texture. An empty image covering the entire façade or roof area is created with the desired resolution. Then a texture pattern corresponding to the building material is chosen and repeated in X and Y in order to cover the whole façade. For roof we take into account the slope of each roof section to orient correctly the pattern. After that, the process places linear pattern representing objects like cornices or roof-ridges and finishes with punctual pattern like doors or windows. The choice of the patterns is guided by the information found in the two previous steps: a request is performed on a texture catalogue, if multiple patterns are returned the final choice is done randomly. All the problem in this process is to preserve visual coherence between the chosen patterns to make the resulting image credible: if old scraped ceramic roman tiles are chosen for the roof surface, the same type of tile should be chosen for ridges. The obtained results are often flat, uniform with an artificial feeling. A radiometric adjustment is then performed. It changes the dynamics so that the average and variance of each channel from the synthesized picture becomes identical to the aerial image. This operation has two advantages: it makes the final texture closer to the reference image and it brings diversity between buildings sharing the same texture pattern.

2.2 Production framework and user interactions

In our context of industrial production we should model cities with very different architectural styles. It is essential to offer a strong interaction with the user all along the production workflow. This has been achieved by developing a framework offering the following capabilities: a tool for texture patterns management and the ability to the user to write its own process rules.

2.2.1 Texture Catalogue Manager

Even for manual modelling a graphic designer should pay a great attention to the choice of textures. If he needs to add a window on a façade he must spend time to find the right image corresponding to the correct usage and architectural style. For this task he uses his knowledge and feeling to find the best looking pattern.

Texture synthesis is much more than just a detection problem because all this subjective knowledge and relationship between patterns must be formalized and accessible to automatic process.

This contribution has been peer-reviewed.
doi:10.5194/isprsarchives-XL-5-W4-155-2015

For this reason we developed the Texture Catalogue Manager, a tool handling the following aspect:

- Elementary pattern images storage
- Physical information like texture real size or texture tilling mode:
    - Surface textures are repeatable in 2 directions (X and Y). They represent building materials like tiles or bricks.
    - Linear textures are repeatable in 1 direction (X or Y): they represent linear objects like cornices, gutters or ridges.
    - Punctual objects are not repeatable: they represent local details like windows or doors.
- Semantic information (attributes) like the texture usage (roof, wall…), the building material or its cleanliness, the type of building it is suited for.
- Logical links between textures allowing grouping them by family or style (e.g. a Renaissance style window should be linked to a cut stoned wall texture pattern). These links are very important to ensure a consistent architectural style.
- Capability to make requests on this database. The selection criterion can mix information such as texture usage, object real size, attributes or logical links.

2.2.2 Interactive modelling rules

Procedural tools usually are based on a grammar allowing writing of modelling rules that can lead to very complex reconstruction process. Each rule can use a stochastic behaviour and the chaining of the rules is not deterministic. It is this randomness which creates diversity and makes that no building is similar to another.

Grammar formalization and development of a grammar interpreter are heavy tasks. In our case, even if random takes an important place to create lacking information, the diversity is not the main goal and the modelling process is driven by the aerial image analysis. Thus we decided to simplify the rule management and to restrain the user interactions to three specific tasks:

- In the first task the framework offers to the user various tools to extract information on building shape and texture features. The user can activate or deactivate tools depending on the architectural style of its region of interest. These tools are mainly image classification, pattern recognition and geometry information extraction about buildings.

- The user can then write its own rules to refine the building material: e.g. slates and metal roof are frequently confused by automatic classification but industrial buildings with soft slope are generally covered with metal.

- The last task concerns pattern selection: according to extracted information, the user can write requests to find the most appropriate pattern in the texture catalogue. The requests are based on the following selection criterion :
    - Texture usage (surface, linear or punctual)
    - Building material
    - Function (windows, door…)
    - Any user defined attributes or link (building usage, construction period, architectural style)

Interactions between the user and the framework are implemented with the Angel Script engine[1]. The script allows the user to write its own rules and to interact with hard coded modules.

Using this framework we now present two applications. The first experiment concerns the production of roof texture synthesis from aerial images. We then extend our experiment to tackle the problem of façade textures synthesis.

3. ROOF TEXTURES SYNTHESIS

3.1 Detection of material and objects

First, the material of roofs is analysed. Four materials are considered: ceramic tiles, metal or cement, slates, and rusty metal. To find the image best fitting material, we use a SVM classification applied on colour descriptors. The first descriptor is the minimum difference between the red channel and blue or green channel of the roof's main colour. This descriptor is good for ceramic tiles or bricks where the red channel is dominant. Since rusty metal roofs are also locally red, we use a second descriptor which is the standard deviation of the hue channel around the red value. To finish, slates, metal or cement are very similar and generally difficult to discriminate even for human eye (Figure 1). However, since their saturation is usually very low we used the S channel of HSV space colour as the third descriptor. Figure 2 shows a visual example of the SVM training step with two descriptors.

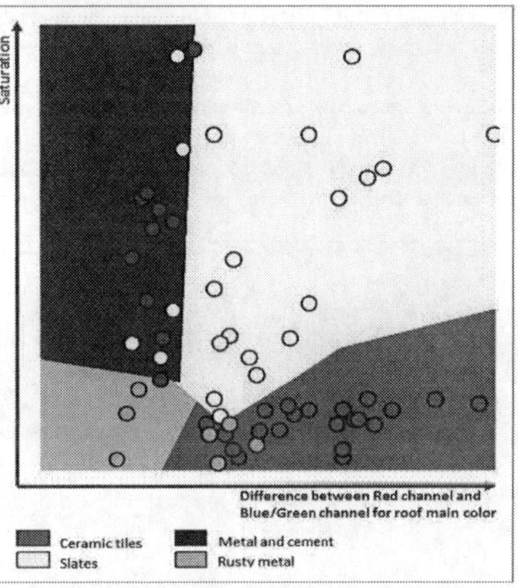

Figure 2. Example of SVM roof classification with two descriptors: errors mainly concern confusion between slates and metal classes. 4 rusty roofs are also labelled as Ceramic tiles because of their dominant red colour

Then, the roof-windows are detected. They are generally very dark or very bright compared to the surrounding. Therefore we posterize the image using K-means algorithm with 4 clusters and extract cluster contours. We finally select the candidate contours with geometric criterion like minimal or maximal area as well as compactness and convexity.

---

[1] http://www.angelcode.com/angelscript

Finally, to get a more natural rendering we also introduce material defaults like scratches or fluid flows and light variations by merging the original image with 30% of transparency to the obtained texture (figure 3).

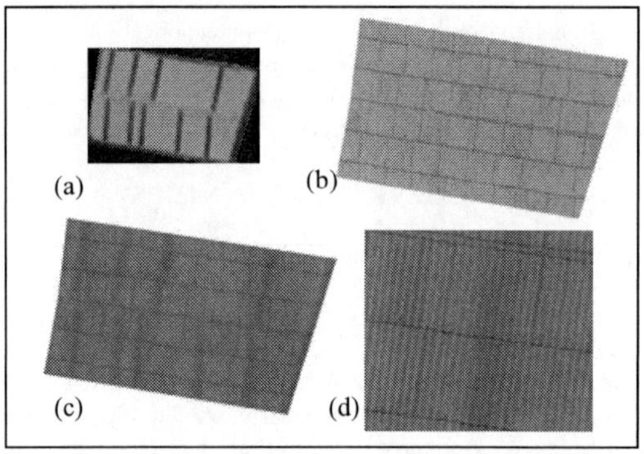

Figure 3. Original texture (a), raw synthesized texture (b) and mixed images with transparency (large and detailed views c, d).

3.2 Results

We tested our algorithms on the Saint-Jacques database, a small city in the suburb of Rennes in north-west of France. The aerial images have a resolution of 20 cm. The testing area presents the following building types: industrial, agricultural sheds, residential areas and apartment or office flats. The roofs are mainly covered with metal for industrial, with bitumen for flats and with natural slates for private houses. For the latter, the main ridge is often made of red ceramic tiles.

The following snapshots present some of our results (Figure 4). The 503 buildings of our testing area were processed in less than 2 minutes on a very standard computer to produce 2 cm resolution textures.

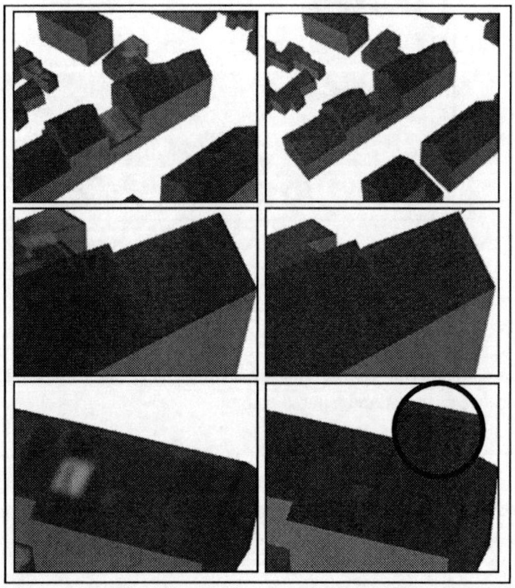

Figure 4. Texture improvement on a residential area. General view and details.

One observed problem in the tested data (Figure 5) is related to the lack of 3D superstructures in the geometric models. Indeed, some roof superstructures like dormers are detected but they can hardly be represented by a simple 2D texture pattern. Instead they should be replaced by a specific 3D model.

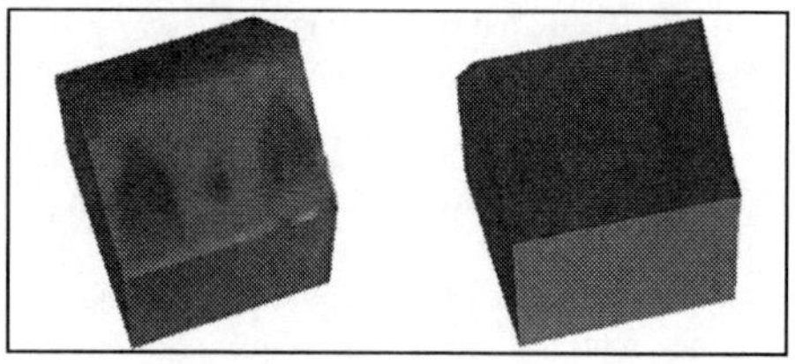

Figure 5. Error in object interpretation: the two glass walls are actually dormers, a 3D structure that cannot be represented by a simple 2D texture

4. FAÇADE TEXTURE SYNTHESIS

4.1 Façade principal colours

The first step while performing façade image reconstruction is to find the background colour of this façade. Some buildings have more than one colour, sometimes the ground floor is darker or for modern buildings some stripes of paint can separate the floors. This colour distribution is generally eye catching and is essential for building recognition by humans. We now explain how we detect these colour variations (Figure **6**).

For this study we only consider the horizontal colour transitions which are the most common cases but similar approach can be used to find vertical stripes. The underlying problem of façade colour description is to answer to the question of "what is wall surface and what is not?". In an image, wall surface can be defined as some clusters of connected pixels with the same colour. The cumulated surface of these clusters should be significant compared to the surface of the whole façade.

We used two simplifications to face our problem: the first is to cut our image into elementary slices and the second is to reduce the colour number for each slice.

For the *slice cutting,* horizontal edges are first computed with a Sobel filter. This gradient image is stacked horizontally to emphasize the colour transitions. The stack is then thresholded and each local maximum is kept and give the position for the slice cutting. After that, each slice is processed independently. We used the k-means algorithm to posterize the slice and reduce the number of colours to 3. This operation transforms the image slice in a set of adjacent patches of uniform colour. These patches are first filtered with an area threshold then they are referenced in a structure storing the following information for each colour:

- Number of patches for this colour
- Cumulated area of patches with this colour
- Most representative (greatest) patch of this colour

The local colour for the façade is then the colour which covers the greatest area (*cumulated area*) with the fewer patches and with a good compactness for these patches. Note that a simple selection based on the most present colour gives no satisfaction because it can represent other objects than walls like balconies or windows (Figure **6** d2).

This contribution has been peer-reviewed.
doi:10.5194/isprsarchives-XL-5-W4-155-2015

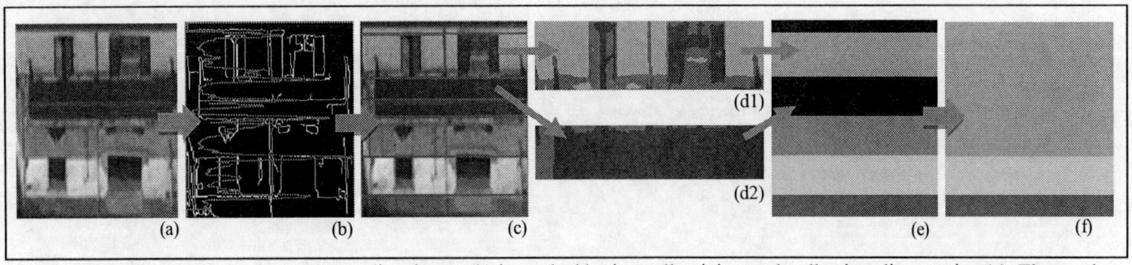

Figure 6. Façade principal colours: the gradient image (b) is stacked horizontally giving peaks allowing slices cutting (c). The number of colour is reduced (d) and the distribution of colour patches analysed to find the façade colour (d1) or set it to *undef* if too scrappy (d2). Missing colours are deduced from neighbouring slices and close colours merged to give the final image (f)

If the slice's patches are too scrappy the façade colour is set to *undefined*. Therefore, when all the slices are processed a second step is performed to complete these *undefined* colours. The average colour of the *undef* slice if compared to the surrounding slice (one beside and one bellow) and the closest facade colour is set to the current slice.

Finally a corrective step is performed to fuse similar slices and only keep the most representative colours.

4.2 Openings detection

This section now details the detection of facade openings from aerial images. The word "opening" is used for windows, doors, loggias and balconies (the latter is not an opening but indicates the presence of such a one). The proposed method follows the idea of (Martinovic, A. et Al., 2012) in the sense that no shape grammar is imposed to the façade. Instead, openings are detected and refined using weak architectural principles such as size and position. Moreover a hypothesis of row/column structure helps to improve the detection.

The proposed detection consists in two steps. The first one is a pixel-based detection of the facades using machine learning. The second is an object-based detection using architectural constraints. As a post-process, images are first normalized to increase contrast and filtered using mean-shift clustering to reduce noise and enhance edges.

4.2.1 Pixel-based detection using machine learning

The goal of this process is to analyze the images using only pixel information and a set of training images. To do so, the multi-class segmentation algorithm implemented in the Darwin library is used (Gould, S, 2012) version 1.8.0. First, a set of training images is manually selected and labeled. These images are used to train a conditional random field model (CRF). Several features are computed to train the model: color, texton filters, histogram of oriented gradients (HOG), local binary patterns (LBP), and average over rows and columns. The labeled objects are: *background, building, ground, roof, vegetation and opening*. At this stage, only windows and doors are labeled as openings in the training images because they have rather regular size and appearance. On the other hand, loggias and balconies are too noisy because they often contain objects like tables, plants, dryers or antennas. Therefore they will be taken into account only in the next step. The output of the pixel-based detection is shown in the second column of Figure 7. The facade in the first example contains few windows and a door. They are all detected although the precision of the contours is not perfect. The second façade example contains dozens of loggias. They are not detected by the algorithm.

4.2.2 Object-based detection using architectural constraints

Architectural constraints are now used to both refine detected openings and detect new openings.

Refinement of detected openings. In the previous step, no information about the façade structure is used. As a result, false detections occur, contours are noisy and openings are not aligned. In order to refine the detection, the following constraints are used: Openings and buildings have often a rectangular shape. Thus they are adjusted to their bounding boxes. Openings have usually width and height in certain intervals of values. Thus, too small or too large openings are removed. Openings are often repeated in rows and columns. Thus, clusters of openings that are almost aligned in rows or columns and have almost the same height/width are refined to be well aligned in positions and size. Doors are usually close to the ground. Thus openings close enough to the ground with a minimum height are classified as door openings.

Detection using repetitions. Openings include a wide variety of shapes and textures. This is particularly true with large buildings made of loggias and balconies. As explained in the pixel-based detection step, only windows and doors were detected. This part now assumes that repeated templates in a façade are likely to be openings, thus the goal is to find repetitions and validate or not the opening hypothesis. Images are processed as follows:

- Compute a mask of opening hypotheses (Figure 8): Horizontal gradients are accumulated vertically and vertical gradients are accumulated horizontally. Local accumulation peaks are detected. The peaks are used to create masks of rows and columns. They are multiplied together to obtain the final mask of openings hypotheses. The third column of Figure 7 shows the resulting mask of hypotheses.
- Opening hypotheses are matched pair-wise to compute a correlation score. A high matching score indicates that the opening is repeated. If the same opening appears at least three times, then the hypothesis is validated.
- Fuse these newly detected openings with previous opening of section 4.2.1.
- Validate by rows and columns: In a row/column of opening hypothesis, if enough openings have been detected, then other hypotheses of the same row/column and same size are validated. The last column of Figure 7 shows the output of this detection using repetitions. In the second row, all the loggias are detected. Note that the detection result is different than the grid of hypotheses because of the matching step and the size constraints that have removed some hypotheses.

This contribution has been peer-reviewed.
doi:10.5194/isprsarchives-XL-5-W4-155-2015

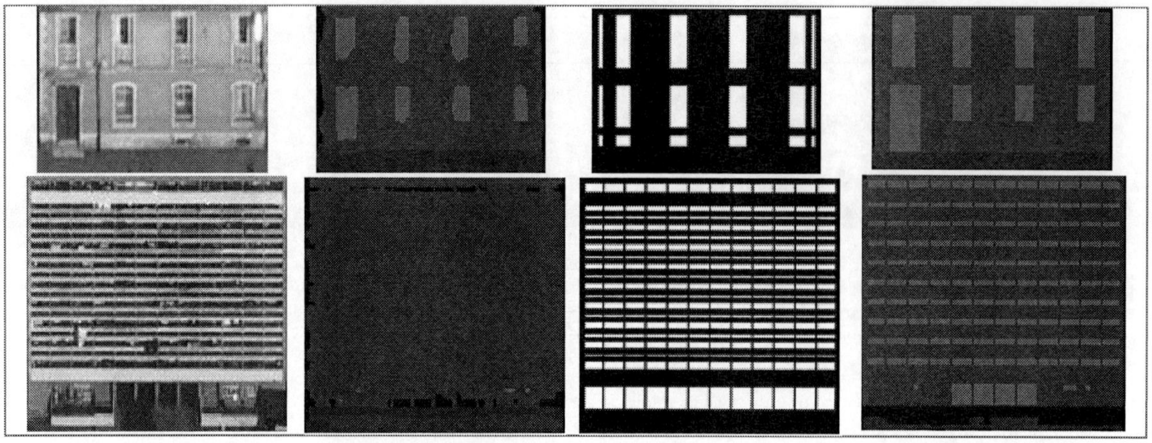

Figure 7: Original image; Pixel-based detection; Opening hypothesis using accumulation of gradient; Object-based detection

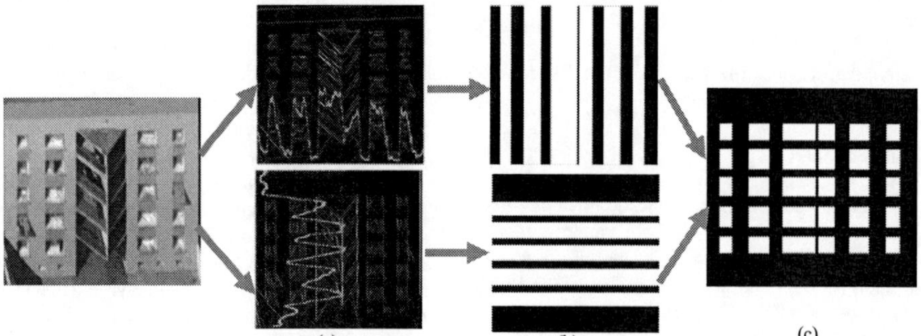

Figure 8: Mask of opening hypotheses. (a) Horizontal gradients are accumulated vertically and vertical gradients are accumulated horizontally. (b) Local accumulation peaks are used to create masks of rows and columns. (c) The masks are multiplied to obtain the final mask of opening hypotheses.

In this process the mask of hypotheses computed from accumulation of gradients gives a nice structure of hypotheses which may not be error-free but provides a convincing visual appearance. Therefore, if the previous detection results in a low ratio of openings compared to the grid of hypotheses, then this grid is used rather than sparse distribution of openings.

4.3 Windows characterization

In the previous section the labelled mask gives the positions and sizes of openings. We now need to find more information about the window type and architectural style.

The resolution of the image sample is not good enough to allow a comprehensive description of the window features. Moreover the perspective introduces important distortion in the windows shape (Figure 9). In these conditions we decided to limit the investigation to very simple tests and reduce the number of windows types to 3 (Figure 10):

- Old style windows
- Recent windows
- Loggia (i.e. balcony which does not extrude the façade)

Each type is represented in our catalogue by several family of pattern.

We used a set of simple image procedures. Each procedure will validate (or invalidate) an *indicator* and the combination of these indicators will help to determine the type of the window. As a procedure can conclude to no result each indicator can take three values: *true, false* or *undefined.*

The first indicator is given by the shape of the mask which is more or less equivalent to the windows shape.

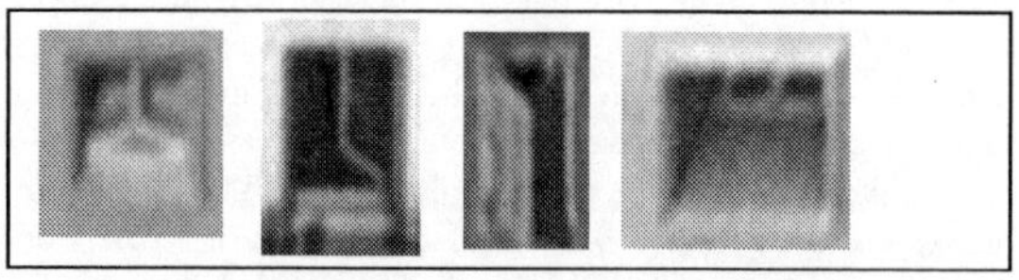

Figure 9. Low resolution, open windows, closed shutters and important distortion due to perspective are obstacle to window characterization

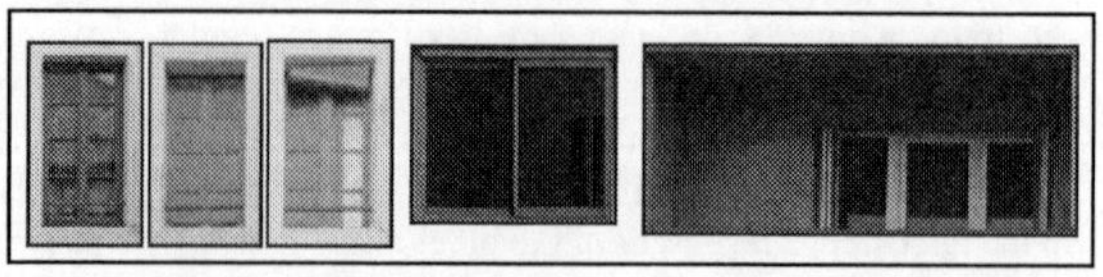

Figure 10. Extract of the texture catalogue with the 3 types of windows (old, new and loggia).

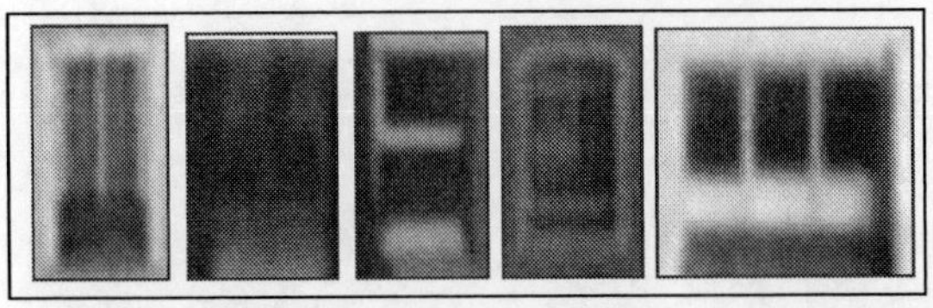

Figure 11. Windows with 1, 2 or 0 "clear" vertical bar

The second indicator is related to the presence of a clear vertical bar at the middle of the windows. It gives indication on the number of glasses. The notion of "clear" is really relative as it can be seen in Figure 11: in the best case the bar is as tall as the window's frame, white and surrounded with dark glasses but sometimes it can be brown, visible only in the upper part of the image or can be occulted by closed shutters. In some cases there is more than one vertical bar or it can be horizontal. As we are looking for a vertical structure the image can be reduced to a single row averaging the corresponding columns. We then research a local maximum near the image centre.

The third indicator is related to the presence of open shutters. Only image wider than 1.5m are considered by this test. The colour space is reduced to 3 levels using the k-means algorithm, then image is divided in 3 parts: the left and right parts are ¼ wide of the total image width. If left and right parts contain shutters they should have a uniform colour. Moreover the colour should be identical for both parts. This predominance is usually obvious in the histogram as it can be seen in Figure 12. In comparison the central part of the window image shows a great variation in colour distribution, some areas are bright and other very dark.

All these indicators are combined with simple heuristics to find the window style (*old/new/loggia*):

- Shapes wider than 4m are considered as loggia.
- Generally windows higher than wide are ancient, and windows wider than high are recent.
- Wide windows with only one glass are usually recent.
- Windows with panel shutters are old style.

Style, dimension, shutters and number of glasses are combined to request the texture catalogue and find the most appropriate window patterns. If several pattern candidates are returned, one is chosen randomly. For the moment a façade can accept only one type of window which is a limitation to realism.

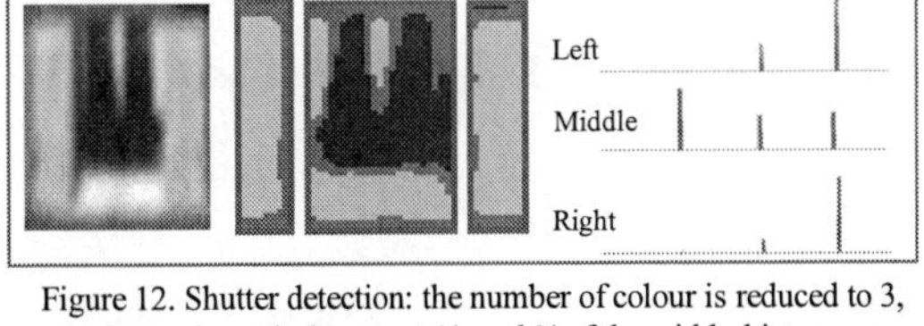

Figure 12. Shutter detection: the number of colour is reduced to 3, the image is cut in 3 parts at ¼, and ¾ of the width; histograms show a clear dominance at the same value for both shutters

4.4 Façade texture synthesis results

We conducted our experiment on our Nimes database, a city in the south of France, with an image resolution of 5 cm. Façade characterization is more demanding than roof characterization and the previous database with a resolution of 20 cm was not sufficient.

For the opening detection method, the training set is made of 75 façade images with labels (training labels were manually edited). The test set is composed of 28 façade images. The facades have different architectural styles and size in order to represent the diversity of the city (e.g. individual houses, classical buildings, tall modern buildings). Over a total of 457 openings in the test set, 431 openings have been detected including 47 false detections and 71 openings have not been detected. In terms of classification accuracy, this results in a precision of 89% and a recall of 84%. In addition, a larger set of 1000 images as been processed in about 35 minutes.

The detection obtained with this algorithm gives satisfying results considering the low quality of the images and the great variety of facades. Concerning the assumption that repeated patterns are likely to be openings, only few errors were observed due to repeated patterns that were actually not openings. As shown in Figure 13, one problem comes from precision errors on the position and size of the openings (although alignment constraint is used). To finish, some improvement could be considered: the detection of more objects like balconies and shop windows; automatic choice of thresholds using statistics on the training set; Mask of building occlusions in the images. Moreover more tests will be performed on other cities and with other resolutions.

Figure 13: Precision errors on the position and size of the openings

The results of texture synthesis (Figure 14) are promising but we are still facing problems to generalize the usage of façade texture synthesis to a whole data base. For our point of view the principal limitations are related to these two points:

- The conception of the texture catalogue needs a lot of attention. On one hand a deep reflection must be carry out to find relevant attribute and relations between patterns. On the other hand, some strong graphic designer skills are required to ensure visual consistency in the resulting image.

- In this study, the architectural style (old/recent) research is mainly driven by window characterization. It would be interesting to take into account other façade elements like cornices, balconies, ornaments or even to consider the façade as a whole like in (Mathias 2011). Geometrical information such as building outlines or roof shape can also give clues to its architectural style.

5. CONCLUSION

We present in this study a method to artificially increase the textures quality for 3D virtual cities textured with aerial images. The idea developed here is to use adaptive procedural modelling to add details to the texture. Instead of using pure random modelling we constrain this process by the information that could be extracted from the original aerial images

.

This contribution has been peer-reviewed.
doi:10.5194/isprsarchives-XL-5-W4-155-2015

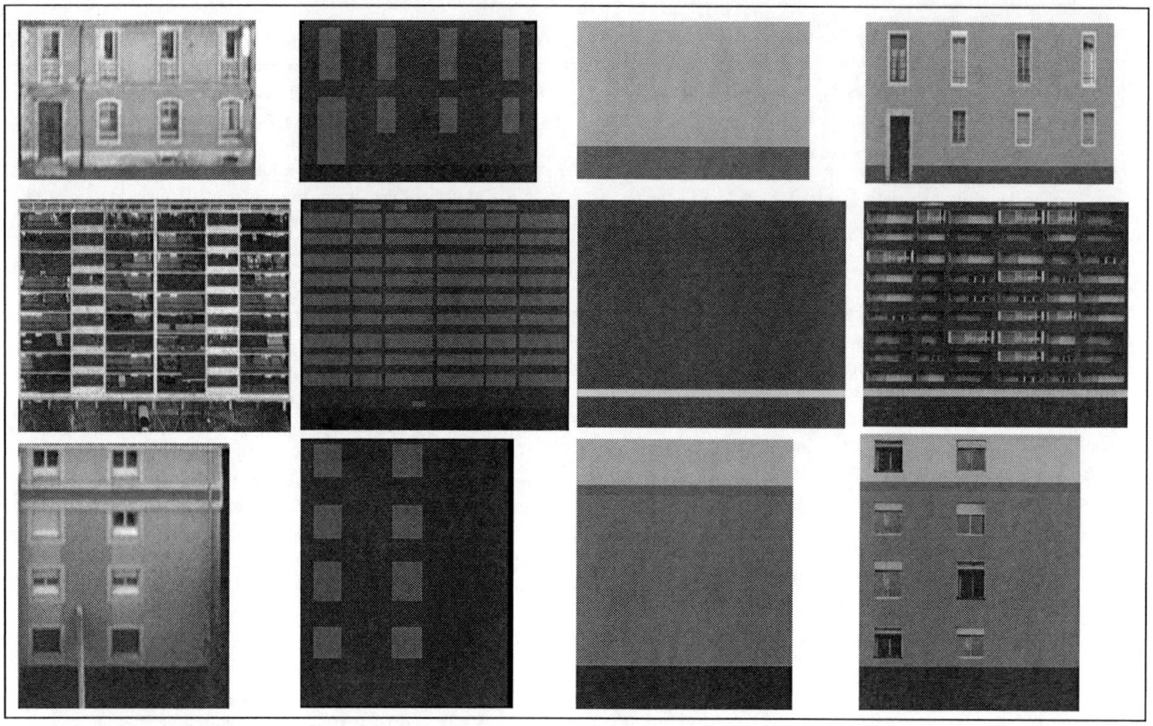

Figure 14. Texture synthesis results for facades

Initially thought as a pure detection and recognition problem it also appears to be an information management problem: correct texture reconstruction needs good knowledge of the available high resolution patterns and how they should be used together. It is then essential to add semantic information to these patterns, to link them by family or style and to automatically access this information by requests.

The conducted tests are relevant for roof texture modelling and industrial exploitation can now be considered. The façade texture reconstruction is promising but due to façade complexity some complementary works must be achieved, especially in the resolution of architectural style.

REFERENCES

Alizadehashrafi , A. and Abdul-Rahman, A., 2011. Enhancing textures of 3D building façade. In: *10th International Symposium & Exhibition on Geoinformation (ISG 2011) & ISPRS Commission II/5 & II/7 Conference* , Shah Alam, Malesia.

Cohen, M. et Al., 2003. Wang Tiles for image and texture generation. In *ACM Transactions on Graphics (TOG) - Proceedings of ACM SIGGRAPH 2003,* San Diego, USA, Volume 22 Issue 3, pp 287-294.

Dai, D. et Al., 2013. Example-based facade texture synthesis. In: *IEEE International Conference on Computer Vision (ICCV)*, Sydney,Australia, pp 1065-1072.

Gould, S, 2012. DARWIN: A Framework for Machine Learning and Computer Vision Research and Development. In: *Journal of Machine Learning Research (JMLR)*, pp: 3533−3537.

Loya, A. et Al., 2008. A Practical Approach to Image-Guided Building Façade Abstraction. In: *Computer Graphics International*, Istanbul, Turkey.

Martinovic, A. et Al., 2012. A Three-layered Approach to Facade Parsing. In: *Proceedings of the 12th European Conference on Computer Vision*, Florence, Italy, Vol. VII, pp 416-429.

Mathias, M. et Al., 2011. Automatic Architectural Style Recognition. In: *ISPRS - International Archives of the Photogrammetry, Remote Sensing and Spatial Information Sciences*, Volume XXXVIII-5/W16, 2011, pp.171-176.

Müller, P. et Al., 2007. Image-based Procedural Modeling of Facades. *ACM. Transactions on Graphics*, Vol. 26, pp. 85.

Parish, Y. and Müller, P., 2001. Procedural modeling of cities. In: *SIGGRAPH '01 Proceedings of the 28th annual conference. on Computer graphics and interactive techniques*, Los Angeles, USA, pp: 301-308

Ripperda, N. and Brenner, C., 2006. Reconstruction of Façade Structures Using a Formal Grammar and RjMCMC. In: *Pattern Recognition, 28th DAGM Symposium*, Berlin, Germany, Lecture Notes in Computer Science Volume 4174, pp 750-759.

Wang, H. 1965. Games, logic, and computers. In: *Scientific American (November)*, pp: 98-106.

This contribution has been peer-reviewed.
doi:10.5194/isprsarchives-XL-5-W4-155-2015

ADVANCES IN MULTI-SENSOR SCANNING AND VISUALIZATION OF COMPLEX PLANTS: THE UTMOST CASE OF A REACTOR BUILDING

J.-F. Hullo [a, *], G. Thibault [a], C. Boucheny [a]

[a] EDF R&D, EDF Lab
1 avenue du Général de Gaulle, 92141 Clamart, France
(jf@hullo.fr, guillaume.thibault@edf.fr, christian.boucheny@edf.fr)

Commission V, WG V/4

KEY WORDS: 3D as-built reconstruction, multi-source data, terrestrial laser scanner, panoramic images, 2D and 3D CAD models, virtual mock-up, complex scenes visualization

ABSTRACT:

In a context of increased maintenance operations and workers generational renewal, a nuclear owner and operator like Electricité de France (EDF) is interested in the scaling up of tools and methods of "as-built virtual reality" for larger buildings and wider audiences. However, acquisition and sharing of as-built data on a large scale (large and complex multi-floored buildings) challenge current scientific and technical capacities.

In this paper, we first present a state of the art of scanning tools and methods for industrial plants with very complex architecture. Then, we introduce the inner characteristics of the multi-sensor scanning and visualization of the interior of the most complex building of a power plant: a nuclear reactor building. We introduce several developments that made possible a first complete survey of such a large building, from acquisition, processing and fusion of multiple data sources (3D laser scans, total-station survey, RGB panoramic, 2D floor plans, 3D CAD as-built models). In addition, we present the concepts of a smart application developed for the painless exploration of the whole dataset. The goal of this application is to help professionals, unfamiliar with the manipulation of such datasets, to take into account spatial constraints induced by the building complexity while preparing maintenance operations. Finally, we discuss the main feedbacks of this large experiment, the remaining issues for the generalization of such large scale surveys and the future technical and scientific challenges in the field of industrial "virtual reality".

1. INTRODUCTION

Dense scanning methods are used for more than 20 years in the industry for different specific needs related to maintenance to describe the actual shape of a facility, with its obstacles and free spaces, to help handling, storage, replacement or change of important components of the plant.

Figure 1 illustrates the history of challenges in acquisition of as-built data at EDF in the last twenty years (1993-2014).

The contributions of this paper are the following:

- A state of the art of tools and methods for acquisition and processing of as-built data in complex indoor facilities, based on a twenty years experience of as-built data production within our company,
- Results and analysis of a large scale experiment: a first dense multi-sensor scanning survey of a whole nuclear reactor building (1000+ stations),
- Recommendations and examples for developing dedicated applications of virtual visits of complex architectures using multiple data types,
- Listing of the remaining bottlenecks and challenges for the perspective of a generalization of large and dense multi-sensor scanning surveys.

2. STATE-OF-THE-ART OF RECONSTRUCTION AND VISUALIZATION OF COMPLEX INDUSTRIAL SCENES

2.1 Specific aspects of the scanning of industrial indoors

The characteristics of industrial environments, in their shapes (congested, with a huge number of objects, whose manufactured geometries correspond to a set of 3D primitives like cylinders, boxes, etc.), their surfaces (objects sometimes very reflective or with very dark albedos) and accessibility (on-site restricted access time, difficult conditions of temperature and vibrations) have led to the development of dedicated tools for acquisition and processing, such as Terrestrial Laser Scanning in the 1990's, cf. [Pot et al., 1997] and Figure 1.a.

Unfortunately, and because of the specificities of industrial environments, many developments intended for the scan of non-industrial scenes (natural, outdoor, less congested or unstructured environments) have not provided answers for all bottlenecks for dense scanning of large-scale industrial scenes: accuracy of the raw points for a large spectrum of albedos, accuracy of registration and referencing as well as productivity of 3D reconstruction.

* Corresponding author.

This contribution has been peer-reviewed.
doi:10.5194/isprsarchives-XL-5-W4-163-2015

2.2 State of the art of large scale dense scanning surveys

Recent breakthroughs allowed the perspective of a first very large-scale and multi-sensor scanning, from acquisition and procession to data integration in a dedicated application.

Regarding the acquisition, photogrammetric and lasergrammetric tools and methods have significantly increased in quality and speed of acquisition in the past few years. Motorized heads with full frame DLSR can generate panoramas of half-billion of pixels in 5 minutes [Kopf et al., 2007] with a high automation rate in stitching and HDR rendering. Phase based laser scanners can capture roughly the same time a 50 million point cloud of mid-range measurements (< 1 m to 50 m). On most objects surfaces, depth measurements in this range are under 1 cm in that range, [Barras et al., 2013]. However, surfaces of some industrial objects do not correspond to the calibration standard ranges (albedo, reflectance and incidence angle) and leads to significant measurement errors [Voegtle and Wakaluk, 2009]. A part of these errors can be filtered automatically (comet-tailed effect, outliers) though another, sometimes non negligible part, can only be removed through manual segmentation (multi-reflections on specular surfaces in particular).

Regarding registration of data, indoor passive geolocation systems do not offer centimetric accuracy for portable devices. Today, best lightweight sensors can lead to a 1 % tolerance of the trajectory, which is still not accurate enough for laser scans and need to be frequently reset on known positions to reduce shift error (though this step can be done automatically using existing maps and checkpoints : doors, ladders, stairs). However, these sensors can be used either for a first approximation of position and orientation of laser scans, either for positioning roughly the panoramic images.

Fine registration of laser dataset is an open and active research topic. For complex buildings, cloud to cloud approaches such as ICP and variants, that have a great automation rate, cannot be used for several reasons. First, due to the proximity of objects, a small relative change in scanner position induces strong differences on point clouds. Thus cloud to cloud constraints cannot, by definition, lead to accurate and confident results. Secondly, these approaches cannot take into account datum constraints for data referencing without a great loss of automation. Finally, these approaches are nor efficient neither robust for very large datasets (dozens of hundreds of stations). Currently, only the target-based registration and referencing, paired with a total station survey for datum constraints, allow the fine registration of hundreds of laser scans (3σ = +/- 2 cm) at the scale of a 10-floors building.

Finally, recent improvements in semi-automatic tools for 3D reconstruction from large point clouds (dozens of billions) have made possible the 'as-built' reconstruction of a full mock-up. Without recent out-of-core algorithms, processing billions of 3D points would be impossible [Elseberg and al., 2013]. But since high density can be used, semi-automatic algorithm (region-growing, RANSAC) for primitive fitting performs better in a reasonable time. These algorithms are also valuable in quality monitoring thanks to their statistics (standard deviation of fitting, number of points used, etc.) and are generraly preferred to more manual reconstruction tools (free adjustment or point snapping).

Despite these improvements, the review of the state of the art for complex indoor dense scanning surveys and reconstruction leads to the statement that both current tools and methods available today have not been designed for very large dataset; that's why scaling up still remains a great challenge.

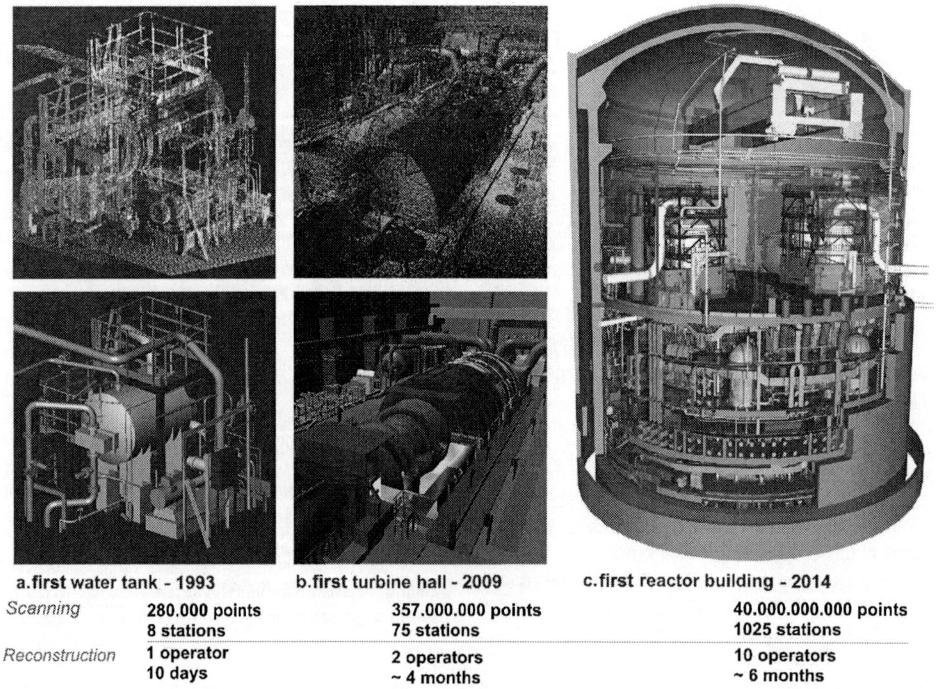

Figure 1: Review of 3 breakthroughs projects (1993-2014) of as-built reconstruction from laser scanning data of industrial facilities at EDF.

This contribution has been peer-reviewed.
doi:10.5194/isprsarchives-XL-5-W4-163-2015

2.3 Challenges in the acquisition and visualization of multi-sensors, multi-users, multi-sources and multi-epochs dataset

But scaling-up the size of scanning survey is not the only challenge for the industry: the increasing number and profiles of end-users of as-built data is another one. Indeed, the set of data (floor plans, 3D models, panoramic photographs, laser scans) required to truthfully represent the complexity of an industrial facility is too large and complex to be shared without care to non-expert professionals. Consolidation of all these data in a sole "database" requires a lot of semi-automatic processes, smart enough to ensure consistency as well as to detect differences between epochs. Once the data is consolidated, the development of user interfaces might be carried out with respect to the experience of different end-users, profiles and needs (users have heterogeneous skills for visualization and exploration of 2D/3D topographic representations of a complex environment).

This statement led to the design of dedicated software tools (see § 5) to help professional users explore and exploit these data, with annotations, distance measurements, specific rendering, path finding, etc.

3. THE MOST COMPLEX BUILDING OF A 1300 MW POWER PLANT AS AN UTMOST CASE STUDY: THE REACTOR BUILDING

3.1 Context of the experiment

In the wake of several research projects that contributed to the development of tools and methods of dense scanning [Girardeau-Montaut et al., 2005], [Bey et al., 2011] or [Hullo et al., 2012] as well as tools for the visualization of complex dataset [Boucheny, 2009], EDF carried out a first very large-scale experiment in the most complex building of a 1300 MW nuclear power plant: a reactor building.

The goal was to assess the effective performance (quality, costs and speed) of current tools and best-practices through an important research project launched in 2013, in order to state on the generalization of multi-sensor scanning of power plants and underline remaining bottlenecks. Another goal was to maximize the benefit of this dataset for the company, especially by sharing it to as many users as possible to help them in their daily work.

Thus, to meet the needs of many jobs of maintenance and operations, a dense multi-sensor survey (total stations, laser scanning and panoramic RGB) was carried out during summer 2013.

3.2 Description of the building architecture

That reactor building, whose shape is cylindrical, consists in 11 floors with complementary ground heights in each of them. Moreover, the main part of the 100+ rooms of the building is particularly congested and many are also only accessible through ladders.

Regarding the environmental conditions, exploring and scanning such environments is constrained by poor lightning, restrictions of access and congestion due to the equipments of the plant itself. Key numbers of the multi-sensor scanning survey are shown in Figure 2.

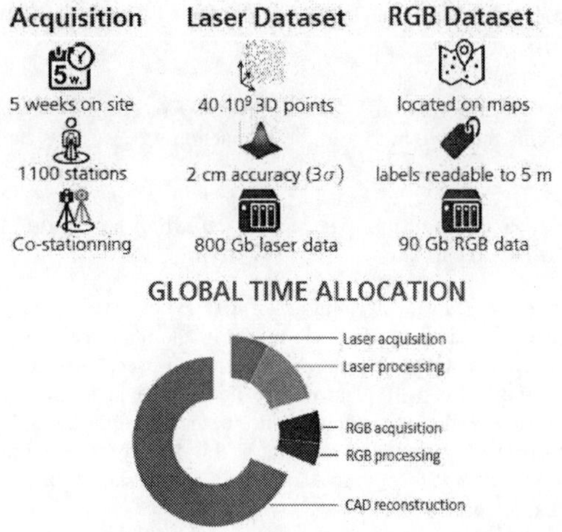

Figure 2 : Key numbers of the multi-sensor scanning survey of a nuclear reactor building; the main order of magnitude that may be highlighted is number of stations: more than 1000 stations (for more details see [Hullo and Thibault, 2014]).

As detailed in [Hullo and Thibault, 2014], thanks to some specific evolutions and developments of the tools and methods to take into account the scaling up, and with an important number of operators (5 laser scanning surveyors and 3 photographers) this acquisition phase has shown the feasibility of such surveys. This experiment also underlined some brakes on the generalization of current post-processing tools and methods in terms of qualities, times or costs (fine registration of large laser scanner dataset, quality monitoring and validation tools, issues in formats, storage and data sharing).

To illustrate the complexity of fine registration of laser scans in complex architectures, we show in Figure 3 a view of the 3D network used in the adjustment. This experiment has also underlined the lack of robustness of weighted least squares in such large blocks and the need to work with 3D topographic networks instead of 2.5D. Unfortunately, though at the state of the art, the registration tool we have used implements a basic version of least squares algorithm and lacks of tools for blunder detection, error propagation and quality monitoring.

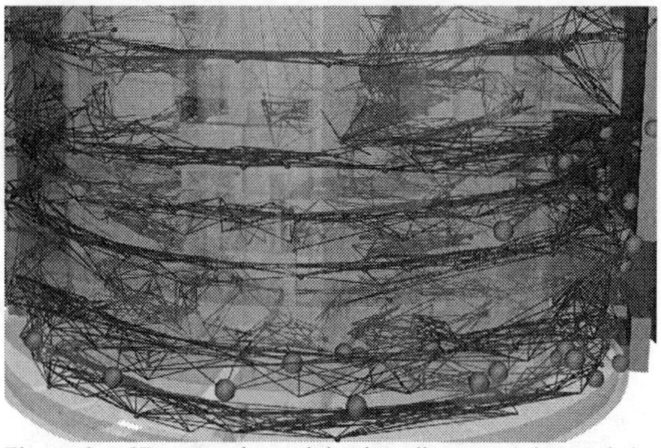

Figure 3 : 3D network used in the adjustment phase of the registration of laser scans, composed of scan stations (red), measurement of registration spheres from these stations (blue lines) and datum spheres for referencing (green).

4. IMPROVING PRODUCTIVITY OF THE AS-BUILT CAD RECONSTRUCTION: QUALITIES, TIMES, COSTS

As shown in bottom of Figure 2, the CAD reconstruction still represents the main budget in the production process of as-built data and

4.1 Impact of building size in task definition and planning for 3D reconstruction

A large dataset of 40 billions of 3D points was used to reconstruct an as-built 3D CAD model of the plant (see Figure 4). In order to fulfil the many needs of end-users, the quality levels of produced 3D data had to be defined in accordance with their intended uses: maintenance tasks planning (including the associated logistics: easements access, scaffolds, handling areas), workers safety, virtual tour for low accessibility rooms by inexperienced professionals.

The consequences of this multiplicity of needs led to a detail specification for reconstruction task depending on object type: fitting tolerances, naming, methods (cloud least-square fitting, region growing cloud-shape fitting, cloud snapping, copies). Tolerances for reconstruction vary from +/- 2 cm on door edges to +4 cm / -50 cm on shapes representing obstacles on pathways (+ outside / - inside maximum error in the reconstructed shape).

The use of panoramic RGB images was very valuable for the reconstruction process to help understanding complex areas.

In order to face the huge amount of work and to reduce time of data delivery, the 3D CAD reconstruction had to be split and parallelized by sectors, and then allocated to a semi-dozen of CAD operators for several months. To reach and certify the quality of reconstructed as-built model, a level-2 check procedure was carried out by two independent operators and led to iterations and re-work. This validation was performed thanks to mesh-cloud distance computations, for example using a software originally implemented by EDF R&D and Telecom ParisTech (CloudCompare, [Girardeau-Montaut et al., 2005]) and by using 'out-of-core' technologies to display a maximum amount of points for the visual inspection, guided by the analysis of a SQL database storing the standard deviations of each reconstructed geometry (~60.000 objects).

4.2 A level-2 procedure for quality control of large datasets

We mentioned the use of a procedure for quality control of large as-built dataset (laser scans, RGB panoramic or 3D CAD model). These level-2 procedures are a standard for quality monitoring and to detect non-conforming materials. Our experiment has shown their usefulness. In several breakpoints, we have thus implemented it in the following manner:

- Level 1: data creator checks 100 % of the dataset and document it
- Level 2: project management performs spot checks both on areas of interest and randomly on the whole dataset.

Due to the important rate of human actions in the processes and the size of the dataset, at least 2 iterations are generally needed to reach quality levels.

4.3 2D maps update and analysis as a "graph database" of the plant

A global model of the plant is a key data for designing the structure of the dataset, unify the names of the objects and perform analysis on it. By constructing a graph of the various objects and some relationships (adjacency, verticality, inclusion, etc.), it makes possible to answer questions like these:

- What are the panoramic images taken in the rooms adjacent to a specific one?
- Can I access to a specific location in the building without taking ladders?

This entire graph database was built by using 2D floor plans, updated using the 3D dataset: walls, grounds, ladders, stairs and many others, see Figure 5.a. These as-built floor plans were then automatically processed to extract all the required information and build a "topological graph" describing several properties of the installation (shape, location, names, types, navigation and access, etc.) as well as render dedicated basemaps. A previous example of this kind of approach can be seen in [Whiting, 2006]. This graph is then used within the applications to answer queries on the semantic, geometric or topological properties of the buildings and its components, cf. Figure 5.b.

Figure 4: Dense data sources used for as-built reconstruction of the nuclear reactor building from point clouds. Each type of 3D object is reconstructed with specific rules (fitting tolerances and naming codification).

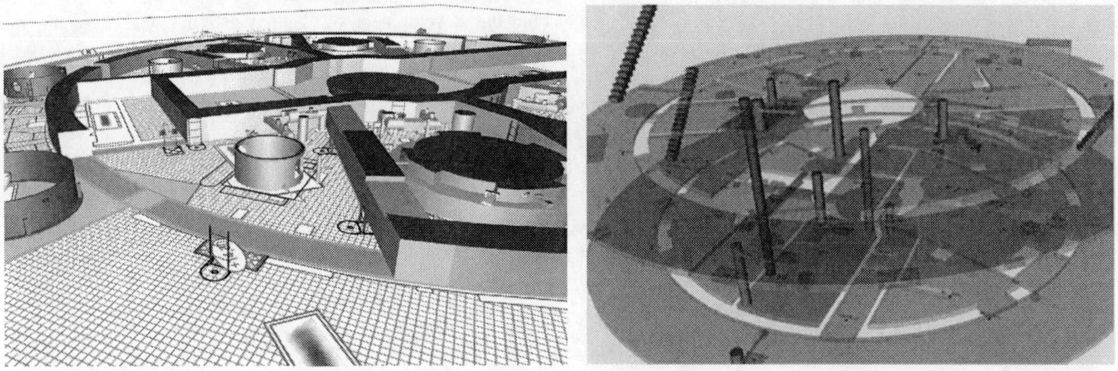

a. Floor maps are updated using as-built dataset (panoramic images, laser scans and 3D models)

b. 3D visualization of the "topological graph" automatically created using floor maps. The red cylinder indicates a ladder and all the red floors are connec--ted to this ladder.

Figure 5: 2D as-built floor maps (a.) updated using the 3D as-built dataset and (b.) processed to create a semantic, geometric and topological representation of the building.

4.4 Recommendations and remaining bottlenecks

Even for a company with over 20 years experience in the field of as-built documentation, a multi sensor and dense scanning survey at the size of a whole 10-floors building remains a challenges for which tools and methods have not been designed for.

In addition to the several recommendations and choices that were exposed in the previous paragraphs, our experiment highlighted some specific aspects of large and multi-sensor scanning surveys of complex architectures. For these projects that answers the growing need of as-built data for professionals, we recommend to:

- Specify in detail the needs, requirements and constraints. At the scale of a building, every misunderstanding or fuzzy specification may have a strong impact on costs, qualities or durations.
- Document every step of the process and perform quality monitoring from the beginning to both help fixing non-conformal data and enrich the dataset for future use.
- Parallelize as much as possible the tasks (acquisition and post-processing). When the data production lengthens, the number of non-qualities increases greatly.

Regarding remaining bottlenecks, we have mentioned several technical brakes to generalization of current tools and methods for large and complex buildings. We have also discussed the 3 axis of the 3D reconstruction: costs, durations and qualities. Actually, only the third one is a goal: the first two only are an answer to the third. By performing rigorous quality monitoring and control, each dataset can be produced with a great level of confidence.

Indeed, the main drawback for reaching expected quality levels is the difficult inter-validation of data sources. To perform such a control procedure, the dataset needs to be merged and visualised all together. But due to current tools and formats limitations, this fusion and exhaustive check of multi-sources dataset remain very complex with poor automation and loss rates in conversions.

5. HELPING THE GREATEST NUMBER OF WORKERS BY SMART MULTI-DATA VISUALIZATION

In the previous sections, we have detailed the challenges for the production of data representing a complex building. Due to that complexity and the intended uses, multiple types of data are required: high resolution panoramic RGB, laser scans, 3D as-built CAD model and 2D floor plans. Once the dataset is completed, issues remain in the visualization and exploration, taking into account that users may not be familiar with this various data types.

5.1 State-of-the art of multi-data visualization for complex architectures

By nature, each data source represents only one aspect of the reality of the plant and fulfills some needs among many. The co-visualization of the multiple data sources is therefore required to assist and help workers finding answers to their daily jobs. Over the past years, several solutions have been proposed for the problem of multi-source as-built data visualization:

- virtual tours with floor plans and panoramic RGB [Koehl et al., 2013],
- navigation through several laser spherical views,
- navigation and path calculation in 3D environment, for instance in 3D videogames or 2.5D cartography services such as Google Maps indoor.

However, none of these solutions integrates all the data sources required to represent some complex buildings such as a nuclear reactor building and easily navigate across the large dataset. Indeed, apart from the constraints of technical integration, the complexity of the plant itself is an issue for virtual navigation (multiple levels with multiple floors) and requires a specific interaction design to handle it.

5.2 Rules for the development of virtual visits of a complex plant

Our goal is to develop an application which can be valuable to many people working in nuclear plants, taking into account that

This contribution has been peer-reviewed.
doi:10.5194/isprsarchives-XL-5-W4-163-2015

the targeted building is rarely accessible. Potential values include the following scenarios:

- Improvement of the efficiency of maintenance operations via a consolidated preparation stage that takes into account the spatial constraints of the environment; this requires the collaboration of different teams around a shared representation of the environment (e.g. mechanical workers defining their needs in scaffoldings).
- Improvement of the accuracy and reduction time in engineering studies to prepare modifications of the plant, with limited needs for the teams to go into the facility.
- Ability for many recently hired workers to be familiarized more rapidly with their working environment through dedicated training sessions, including tutoring stages.

These few scenarios illustrate the variety of user profiles that should be considered for the common visualization we wish to design. Many of them are not familiar with the handling of complex datasets, and their use of the tool may be very occasional, which enforces the need for user friendly and simple interfaces.

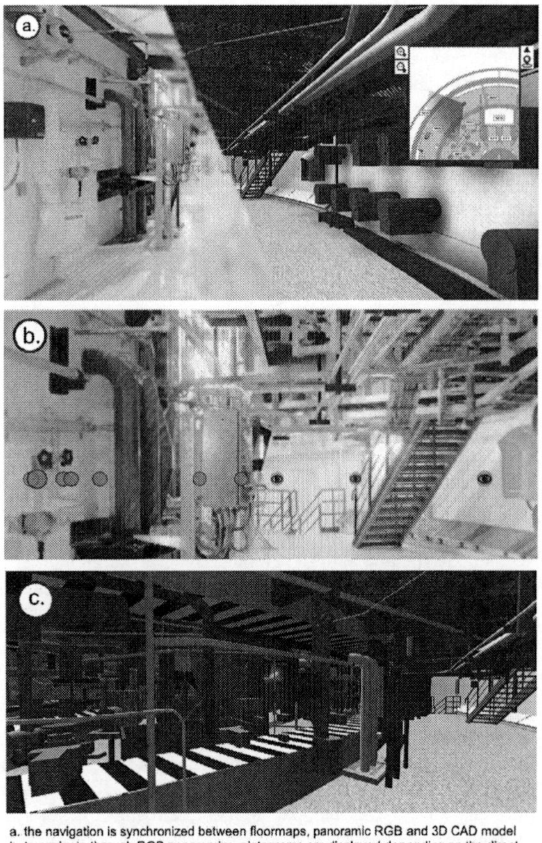

a. the navigation is synchronized between floormaps, panoramic RGB and 3D CAD model
b. to navigate through RGB panoramics, pictograms are displayed depending on the direct visibility or the horizontal/vertical neighbourhood
c. other tools such as real-time wall cutting help unfamiliar users to understand the complexity of the building

Figure 6: Multi-source data exploration and navigation in a complex building: example of key principles for the development of an application

To address this problem, several principles have been selected for the development of the navigation interface within our application (as illustrated in the Figure 6):

- Multi-view synchronization across multi-source dataset, with the possibility at any time to move from one view to another without losing user position/orientation (shared, unique position of the user in the environment).
- Ability to explore the building in its vertical dimension, which is required by tasks such as the handling of heavy equipments.
- Fine processing and registration of 2D-3D data sources for high legibility and fluent user experience, with ensuring the consistency of the various sources.

For instance, we estimate that in the synchronized muti-view experience the spherical photographs need to be positioned in the virtual environment with a tolerance of 50 cm and 5°. This range of accuracy seems sufficient to help users focusing on their task, by not being disturbed by some inaccuracies of the transitions when switching between views. Furthermore, the verticality of the panoramic images must be known with an error under 5° otherwise the navigation becomes very uncomfortable.

In addition to the navigation features, we provide specific interactions with the dataset to the users. Each feature has been designed accordingly to the user's needs. To implement them, we carried out several iterations with the users to overcome the technical feasibility and genuinely create powerful and smart tools. Among these, we can mention distance measurements on 3D CAD models (perpendicularly to the normals, vertical constrained or free), annotations and snapshots in every dataview, interactive section of walls or project management. In all these developments, a large number of feedbacks have been collected in a dynamic development process to reach an optimal user benefit.

6. CONCLUSION & FUTURE WORK

Professional uses of as-built documentation have recently increased in a significant manner, inducing two major challenges: scaling up dense and multi-sensors scanning surveys to a whole building and share this dataset to a very large audience.

In this paper, we have presented a large scale experiment in a 10+ floors 1300 MW nuclear reactor building. From acquisition of 1000+ co-stations (RGB panoramic and laser scans) to 3D as-built reconstruction, the project has involved many contributors for almost a year and a half. In every step, quality requirements have stressed out tools and methods. A first application has been developed to help many professionals in their daily job by helping them exploring virtually the plant through many data types.

We can summarize the contributions of our work in 3 points:

- Assessing the feasibility, as well as underlining the current complexity, of scaling-up state-of-the art tools for multi-sensor scanning of industrial environments (1000+ stations). Best practices still need to be improved to carry out the survey of a whole nuclear reactor building with a great level of confidence in the quality, costs and duration of such surveys.

This contribution has been peer-reviewed.
doi:10.5194/isprsarchives-XL-5-W4-163-2015

- Highlighting weak steps in the production workflow: registration and processing of large laser scanner datasets, quality monitoring, quality control and inter-validation of multiple data sources.
- Sharing our feedback on the development of multi-source data visualization and navigation applications in complex buildings with recommendations for general principles applied to virtual tours.

On the base of these findings, future work is planned in order to reduce the remaining issues:

- Improving data registration/referencing procedures and algorithms to ensure a confident level of quality, for both laser scans and HD panoramic images.
- Developing new tools for data validation and consistency across a large multi-source dataset.
- Collect user feedbacks on our first prototypes of applications and still better use consistency of dataset to enrich user experience.

However, some serious bottlenecks remain and seem farther away from the current state of the art:

- How could we increase significantly (benefit >50% of current cost) productivity of 3D as-built CAD models from very large point clouds (hundreds of billions ?)
- Which procedures and resources should be employed to update a large multi-sensor dataset and detect errors or inconsistencies between two epochs?

References

Barras V., Delley N. and Chapotte G. (2013) Analyses aux limites des scanners laser terrestres. Revue de l'Assiocation Française de Topographie, XYZ number 137, pages 19-26.

Bey A., Chaine R., Marc R., Thibault G., Akkouche S. (2011) Reconstruction of consistent 3D CAD models from point cloud data using a priori CAD models. Proceedings of the ISPRS Laser Scanning 2011, Calgary, Canada.

Boucheny, C. (2009). Interactive scientific visualization of large datasets: towards a perceptive-based approach. PhD Thesis, Université Joseph Fourier (in French only).

Girardeau-Montaut D., Roux M., Marc R., and Thibault G. (2005). Change detection on points cloud data acquired with a ground laser scanner. International Archives of the Photogrammetry, Remote Sensing and Spatial Information Sciences, 36, 30–35.

Elseberg J., Borrmann D. and Nüchter A. (2013) One billion points in the cloud – an octree for efficient processing of 3D laser scans. ISPRS Journal of Photogrammetry and Remote Sensing. Volume 76, February 2013, Pages 76–88.

Hullo J.-F. And Thibault, G. (2014). Scaling up close-range surveys: a challenge for the generalization of as-built data in industrial applications - The International Archives of the Photogrammetry, Remote Sensing and Spatial Information Sciences, ISPRS Technical Commission V Symposium , Riva del Garda, Italy.

Hullo J.-F., Grussenmeyer P., Landes T. and Thibault G. (2012). Probabilistic Feature Matching Applied to Primitive based Registration of TLS Data. XXII ISPRS Congress - ISPRS Annals of the Photogrammetry, Remote Sensing and Spatial Information Sciences, Volume I-3, Melbourne, Australy, August 25 - September 1st 2012.

Kopf J., Uyttendaele M., Deussen O. and Cohen M. (2007). Capturing and viewing gigapixel images. ACM Transactions on Graphics (Proceedings of SIGGRAPH 2007), volume 26 issue 3, article 93.

Koehl M., Scheider A., Fritsch E., Fritsch F., Rachebi A. and Guillemin S. (2013) Documentation of historical building via virtual tour: the complex building of baths in Strasbourg. ISPRS Archives, Volume XL-5/W2, XXIV International CIPA Symposium, pages 385-390.

Pot J., Thibault G., & Levesque P. (1997). Techniques for CAD reconstruction of 'as-built' environments and application to preparing for dismantling of plants. Nuclear Engineering and Design, Issue I, 135-143.

Thibault, G., Pasqualotto, A., Vidal, M., Droulez, J., & Berthoz, A. (2013). How does horizontal and vertical navigation influence spatial memory of multifloored environments? Attention, Perception & Psychophsics, 75, 10-15.

Voegtle T. and Wakaluk S. (2009) Effects on the measurements of the terrestrial laser scanner HDS 6000 (Leica) caused by different object materials. Laser scanning 2009, IAPRS, Vol. XXXVIII, Part 3/W8 – Paris, France, September 1-2, 2009.

Whiting E. (2006). Geometric, topological & semantic analysis of multi-building floor plan data, Thesis (S.M.)--Massachusetts Institute of Technology, Dept. of Architecture, 2006.

This contribution has been peer-reviewed.
doi:10.5194/isprsarchives-XL-5-W4-163-2015

AUTOMATIC ANALYSIS AND CLASSIFICATION OF THE ROOF SURFACES FOR THE INSTALLATION OF SOLAR PANELS USING A MULTI-DATA SOURCE AND MULTI-SENSOR AERIAL PLATFORM

L. López[a*], S. Lagüela[a, b], Inmaculada Picon[a], D. González-Aguilera[a]

[a] Department of Cartographic and Land Engineering, University of Salamanca, Hornos Caleros, 50. 05003 Ávila. Spain

[b] Applied Geotechnologies Research Group, University of Vigo. Rúa Maxwell s/n, Campus Lagoas-Marcosende, 36310 Vigo, Spain.

Commission V, WG V/4

KEY WORDS: 3D reconstruction, aerial trike, photogrammetry, infrared thermography, point cloud, buildings, solar influence, solar panel.

ABSTRACT:

A low-cost multi-sensor aerial platform, aerial trike, equipped with visible and thermographic sensors is used for the acquisition of all the data needed for the automatic analysis and classification of roof surfaces regarding their suitability to harbour solar panels. The geometry of a georeferenced 3D point cloud generated from visible images using photogrammetric and computer vision algorithms, and the temperatures measured on thermographic images are decisive to evaluate the surfaces, slopes, orientations and the existence of obstacles. This way, large areas may be efficiently analysed obtaining as final result the optimal locations for the placement of solar panels as well as the required geometry of the supports for the installation of the panels in those roofs where geometry is not optimal.

1. INTRODUCTION

Several techniques have been applied so far for the calculation of the solar incidence on roofs (Agugiaro et al., 2012; Agugiaro et al., 2011; Hofierka and Suri, 2002; Nguyen and Pearce, 2010) trying to find the most suitable for the installation of solar panels with optimum performance. These techniques generally use simplified models and approximations of the positions and orientations of the roofs, which could imply errors in the calculation of the solar incidence on the surfaces of the roofs causing large variations in their productivity. In order to avoid this, the use of geo-referenced 3D models of roofs will allow the performance of accurate analysis of areas, orientations and slopes of the roofs. Furthermore, if these data is complemented with thermal information, which provides direct knowledge about the temperatures of the roofs and consequently about the influence of the solar radiation on them, the removal of false positive surfaces that could have the ideal geometric conditions to install solar panels will be possible. For instance, those surfaces containing elements that could prevent their installation like skylights or chimneys will be possible.

This work proposes and tests a methodology for capturing data with different devices, useful for the performance of the automatic classification of the roofs with respect to their suitability to harbour solar panels. The methodology consists on the processing of visible and thermographic images acquired from a low-cost aerial trike equipped with a multi-sensor platform (MUSAS-MUltiSpectral Airborne Sensors) towards the generation of 3D point clouds of roofs, followed by their classification into different categories according to their suitability for the installation of solar panels.

The paper has been structured as follows: after this introduction, section 2 includes a detailed explanation of the materials and methods used for the data acquisition and processing for the automatic segmentation and classification of the roof surfaces for the installation of solar panels. Section 3 is devoted to be an explanation of the methodology presented through its application to an urban area selected as case study; finally, section 4 establishes the most relevant conclusions of the proposed approach.

2. MATERIALS AND METHODS

2.1 Equipment

2.1.1 Aerial trike

The aerial trike is a driven low-cost aerial system equipped with a MUSAS platform able to accommodate different types of sensors, from imaging sensors (RGB and thermographic cameras) to navigation systems (GNSS/IMU). The aerial trike is an experimental “Tandem Trike AIRGES” (Table 1), with a weight capacity up to 220 kg, flying altitudes up to 300 m according to the current Spanish legislation. The main advantages of this platform over the usual driven aerial vehicles are its low cost, ease of use and the possibility of flying below 300 meters, consequently obtaining better GSD (Ground Sample Distance) with the imaging sensors. This increase in the resolution of the RGB images allows the generation of dense point clouds, as well as improves the use of the thermographic images, given that the thermographic sensor presents very limited resolution implying a high GSD that, for altitudes higher than 300 m could prevent the differentiation of characteristic elements of roofs. Regarding the UAV’ (Unmanaged Aerial Vehicle) platforms, the aerial trike is able to transport greater weights, which implies the possibility of using more and higher quality equipment with longer autonomy than copter-type platforms.

* Corresponding author. E-mail address: luisloez89@usal.es (L. López).

This contribution has been peer-reviewed.
doi:10.5194/isprsarchives-XL-5-W4-171-2015

Parameter	**Value**
Empty weight	110 Kg
Maximum load	220 Kg
Autonomy	3.5 h
Maximum speed	60 Km/h
Motor	Rotax 503
Tandem paraglide	MACPARA Pasha 4
Emergency system	Ballistic parachutes GRS 350
Gimbal	Stabilized with 2 degrees of freedom (MUSAS)
Minimum sink rate	1.1
Maximum glide	8.6
Plant surface	42.23 m^2
Projected area	37.8 m^2
Wingspan	15.03 m
Plant elongation	5.35
Central string	3.51 m
Boxes	54
Zoom factor	100%

Table 1.Technical specifications of the manned aerial platform, aerial trike.

Sensors are supported by a specific gyrostabilized platform (MUSAS) (Figure 1) allowing a full coverage of the study area with an appropriate ground sample distance (GSD). This device includes two servomotors arranged on the x and y axes to maintain the vertical position of the camera along the flight path with precision. The servomotors are controlled by an Arduino board, which incorporates an IMU with 6 degrees of freedom: 3 accelerometers with a range of ± 3.6 G m s^{-2}, a double-shaft gyroscope (for pitch and roll control) and an additional gyroscope for yaw control (both gyroscopes have a measurement range of ±300°/s). The software developed for the control of the device was based on Quad1_mini V 20 software, with DCM (Direction Cosine Matrix) as the management algorithm for the IMU (Premerlani, 2013) This aerial platform is used to obtain good quality georeferenced images from a zenithal point of view, allowing better spatial perception given the absence of obstacles between the camera and the object. The navigation devices allow the geolocation of each datum for the posterior generation of geo-referenced point clouds with real and thermographic texture and the corresponding orthophotos and orthotermograms.

Figure 1: Aerial trike and MUSAS platform used in this study; top-right and bottom-left corner: detail of the gyrostabilized platform for the installation of sensors.

2.1.2 RGB camera

RGB cameras are used for the acquisition of images towards the reconstruction of 3D point clouds as well as to provide visual information of the state of the roofs. The visible camera selected for this work is a full frame reflex camera Canon 5D mkII. This camera has a CMOS sensor which size is 36x24 mm with a resolution of 21.1 megapixel and is equipped with a 50mm focal length lens. The size of the image captured with this sensor is 5616x3744 pixel with a pixel size of 6.4 µm.

The camera is calibrated prior acquisition in order to allow the correction of the distortion and perspective effects from the data collected and the 3D reconstruction in the photogrammetric process. The calibration of cameras working in the visible band of the electromagnetic spectrum (Table 2) is performed through the acquisition of multiple convergent images of a geometric pattern (known as calibration grid), with different orientations of the camera. The adjustment of the rays ruling the position of the camera and the image in each acquisition allows the determination of the inner orientation parameters of the camera (focal length, format size, principal point, radial lens distortion and decentering lens distortion). The calibration is processed in the commercial photogrammetric station Photomodeler Pro5©, which performs the automatic detection of the targets in each image and computes and adjusts the orientation of each image, resulting in the computation of the calibration parameters of the camera.

Focal length (mm)	Value	50.1
Pixel size (µm)	*X* Value	6.2
	Y Value	6.2
Principal point (mm)	*X* value	17.2
	Y value	11.5
Radial lens distortion	K_1 value (mm^{-1})	6.035546 x 10^{-5}
	K_2 value (mm^{-3})	-1.266639 x 10^{-8}
Decentering lens distortion	P_1 value (mm^{-1})	1.585075 x 10^{-5}
	P_2 value (mm^{-1})	6.541129 x 10^{-5}

Table 2. Interior orientation parameters of visible camera Canon 5D mkII, result of its geometric calibration.

2.1.3 Thermographic camera

The thermographic camera selected for this study is the FLIR SC655. This device has been specially developed for scientific applications. It allows the capture and recording of thermal variations in real time, allowing the measurement of heat dissipation or leakage. Its sensor is an Uncooled Focal Plane Array (UFPA). 640x480 size, capturing radiation with 7.5 to 13.0 µm wavelengths and measuring temperatures in a range from -20 °C to 150°C. The Instant Field of View (IFOV) of the camera is 0.69 mrad, and its Field of View (FOV) is 25° (Horizontal) and 18.8° (Vertical) with the current lens of 25 mm focal length.

Thermographic cameras capture radiation in the infrared range of the spectrum, in contrast to photographic cameras that work in the visible range. For this reason, the geometric calibration (Table 3) of the camera is performed using a specially designed calibration field, presented in (Lagüela et al., 2012) (Figure 2), which is based on the capability of the thermographic cameras

This contribution has been peer-reviewed.
doi:10.5194/isprsarchives-XL-5-W4-171-2015

for the detection of objects that are at the same temperature as being at different temperatures if they present different emissivity values. This calibration field consists on a wooden plate with black background (high emissivity) on which foil targets are placed (low emissivity).

Focal length (mm)	Value	25.1
Format size (mm)	Value	10.9 x 8.2
Principal point (mm)	*X* value	5.3
	Y value	4.1
Radial lens distortion	K_1 value (mm^{-1})	5.281 x 10^{-5}
	K_2 value (mm^{-3})	7.798 x 10^{-7}
Decentering lens distortion	P_1 value (mm^{-1})	1.023 x 10^{-4}
	P_2 value (mm^{-1})	-3.401 x 10^{-5}

Table 3. Interior orientation parameters of thermographic camera FLIR SC655, result of its geometric calibration.

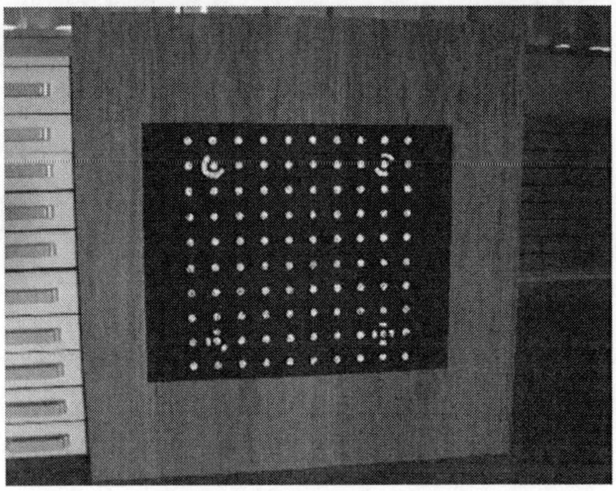

Figure2: Calibration grid for the geometric calibration of thermographic cameras.

2.2 Methodology

2.2.1 Flight planning and data acquisition

Proper flight planning is important to ensure that the data captured allows the performance of the study, optimizing available resources, ensuring a high quality of the images and minimizing capture time. The geomatic information required for the flight planning can be obtained free of charge from the National Center of Geographic information in Spain (CNIG), from its National Aerial Orthoimage Plan (PNOA, 2009) with a GSD of 0.25 m and a Digital Terrain Model (DTM) with a 5m grid resolution. The flight planning was carried out based on the classical aerial photogrammetric principles (Tait, 1994) but adapted to the new algorithms and structure from motion (SfM) strategies (Agarwal et al., 2011), ensuring image acquisition with forward and side overlaps of 70% and 30%, respectively. Given the format difference between the thermographic and RGB sensors, time between shots is different in order to ensure these overlaps (500 ms for the RGB sensor and 160 ms for the thermographic sensor for a flight speed of 50 Km/h).

The gyrostabilized platform ensures the theoretical geometric setup of a photogrammetric aerial image capture in each shot, which stablishes the optical axis of the camera should be nadiral. The theoretical definition of scale in digital aerial photogrammetry is related to the geometric resolution of the pixel size projected over the ground (GSD). This parameter can be calculated by considering the relationship between flight altitude over the ground, the GSD, the focal length of the sensor and the pixel size (eq. 1)

$$\frac{f}{H} = \frac{pixelsize}{GSD} \quad (1)$$

Where f = focal length of the sensor
H = flight altitude over the ground
GSD = Ground Sample Distance
$pixelsize$ = pixel size of the sensor

Due to the lower resolution of thermographic images, the flight planning is performed according to the characteristics of the sensor (Figure 3), considering the planning completely valid also for the RGB sensor. The flight altitude over the ground selected was 160m allowing a GSD of 11cm and 2cm for the thermographic and RGB sensors respectively.

Figure 3: Example of flight planning: (Top) Areas captured by each image shot for the thermographic sensor. (Middle) Flight planning for the navigation of the aerial trike. (Bottom) GPS track after the aerial trike flight with the position of each image shot.

This contribution has been peer-reviewed.
doi:10.5194/isprsarchives-XL-5-W4-171-2015

2.2.2 3D point cloud reconstruction

The image-based modelling technique based on the combination of photogrammetry and computer vision algorithms allows the reconstruction of dense 3D point clouds. The absolute orientation (position and attitude) of each image is known because the position of the imaging sensors is registered with respect to the GNSS/IMU navigation devices of the aerial trike, and data acquisition is synchronized with the navigation. For this reason, only the dense matching process is required, taking as basis the semi-global matching technique (SGM) (Deseilligny and Clery, 2011; Gehrke et al., 2010; Hirschmuller, 2005). This procedure allows the generation of a dense and scaled 3D model (Figure 4) resulting from the determination of the 3D coordinates of each pixel using homologous points between images. This process is performed using the RGB images because their higher resolution provides a 3D point cloud with higher point density than the point cloud generated from the thermographic images.

On the other hand, the transmission of the thermographic texture to the point cloud is developed through the computation of the spatial resection (Kraus, 1993) of the thermographic images using as reference the 3D point cloud obtained from the RGB images, identifying homologous entities (key points) between each image and the point cloud. The identification of homologous entities entails great complexity since the radiometry of the thermographic images represents a temperature value not directly related to the intensity level represented in the point cloud. For this reason, the identification of the maximum number of significant features possible in each thermographic image is essential to minimize the error in the computation of the external orientation parameters. The result obtained is a 3D dense point cloud with the thermographic information mapped of the study case from a bird's eye view (Figure 5).

Figure 4: Example of 3D point cloud (2.179.610 points) generated from images captured with the RGB camera. Area: 23040m^2. Resolution: 100 points/m^2.

Figure 5: Previous 3D point cloud (2.179.610 points) with the thermographic texture mapped on the roofs.

In order to generate additional metric products to support the study from the dataset captured orthoimages which integrated thermographic information were generated. The process to generate these orthothermograms consists on the projection of the texture of the images to the mesh generated from the point cloud. Then the texture of the mesh is projected in a horizontal reference plane using an orthogonal projection, obtaining an image with geometric scale and no distortion, in such way that the measurement of a distance on the image, multiplied by the scale, results in the measurement in the reality.

2.2.3 Automatic planes segmentation

Once the 3D point cloud is generated from the RGB images and the thermographic texture is mapped, the following procedure is the segmentation of the roofs. This will be performed in different steps using the Point Cloud Library (PCL) (Rusu and Cousins, 2011), open source and licensed under BSD (Berkeley Software Distribution) terms, which includes a collection of state-of-the-art algorithms and tools useful for 3-D processing, computer vision and robotic perception. First, ground and vegetation are removed using a pass through filter with a Z coordinate restriction. A pass through filter performs a simple filtering along a specified dimension removing the elements that are either inside or outside a given range. In this case the filter gets the minimum Z value of the point cloud and removes all points with a Z value close to the minimum Z. The distance threshold is established as a parameter set for the user, being 5 m a recommendable value, established experimentally by the authors after the segmentation of several point clouds. In addition, although the presence of points belonging to the facades is minimal due to the vertical configuration of the capture, a conditional filter based on the angle between the normal vector of each point and the vertical axis is applied to remove these points. Then, the point cloud representing all roofs is segmented using a Euclidean cluster segmentation (Figure 6) allowing better and faster results for the subsequent extraction of the different planes of each roof (Gallo et al., 2011; Hulik et al., 2014) by dividing the point cloud in individual roofs. This way, RANSAC (Random Sample Consensus) algorithm is applied to each roof individually for the extraction of the composing planes, with an important reduction in the number of iterations given that the number of planes per roof is limited. RANSAC is an iterative method used to estimate parameters of a mathematical model from a dataset (Fischler and Bolles, 1981). The RANSAC algorithm assumes that all the data is comprised of both inliers and outliers. Inliers can be explained by a model with a particular set of parameter values, while outliers do not fit that model under any circumstance. The input to the RANSAC algorithm is the dataset to evaluate, a parameterized model which can explain or be fitted to the dataset and some confidence parameters to adjust the results of the algorithm to the desired output. RANSAC achieves its goal by iteratively selecting a random subset of the original data that will be hypothetical inliers and this hypothesis is then tested as follows: A model is fitted to the hypothetical inliers, i.e. all free parameters of the model are reconstructed from the inliers. All other data are then tested against the fitted model and, if a point fits well to the estimated model, it is also considered as a hypothetical inlier. The estimated model is reasonably good if there are a considerable number of points classified as hypothetical inliers. Then, the model is recalculated using all the hypothetical inliers, improving the results of the initial computation performed from the initial set of hypothetical inliers. Finally, the model is evaluated by estimating the error of the inliers relative to the model. This procedure is repeated a

This contribution has been peer-reviewed.
doi:10.5194/isprsarchives-XL-5-W4-171-2015

fixed number of times, each time producing either a model which is rejected because the number of points classified as inliers is too limited or a refined model together with a corresponding error value. The process finishes when the mathematical model fits the dataset according to the parameters specified.

Once each roof is clustered in different planes (Figure 7), and the coefficients that describe each surface in a Cartesian coordinate system by the general equation of the plane are known (eq. 2), the geometric evaluation is performed, resulting in the orientation, slope and energy loss values for each surface.

Figure 6: Result of the extraction of each roof using the Euclidean cluster extraction algorithm.

Figure 7: Results of the extraction of each planar surface using the RANSAC algorithm applied in different roofs.

$$Ax + By + Cz = D \quad (2)$$

From the components A, B and C of the general equation of the plane that match with the values of the vector normal to the plane (Eq. 3) we can proceed to compute the orientation and slope values. To obtain the orientation of the surface is necessary calculate the angle between the projection of the vector $\vec{v}$ on the horizontal plane and the Y Cartesian axis (Eq. 4) evaluating the quadrant to which it belongs to obtain de orientation angle (Eq. 5 to 8) (Figure 8).

$$\vec{v} = (A, B, C) \quad (3)$$

$$\theta = arcTan(A/B) \quad (4)$$

$$\text{1st Quadrant: } Or = \theta - 180^{\circ} \quad (5)$$

$$\text{2nd Quadrant: } Or = -\theta \quad (6)$$

$$\text{3rd Quadrant: } Or = \theta \quad (7)$$

$$\text{4th Quadrant: } Or = 200 - \theta \quad (8)$$

Where: $\vec{v}$ = vector normal to the plane.

A, B and C = parameters from the general equation of the plane.

θ = Angle between the horizontal projection of the normal vector and the Y Cartesian axis.

Or = Orientation of the plane.

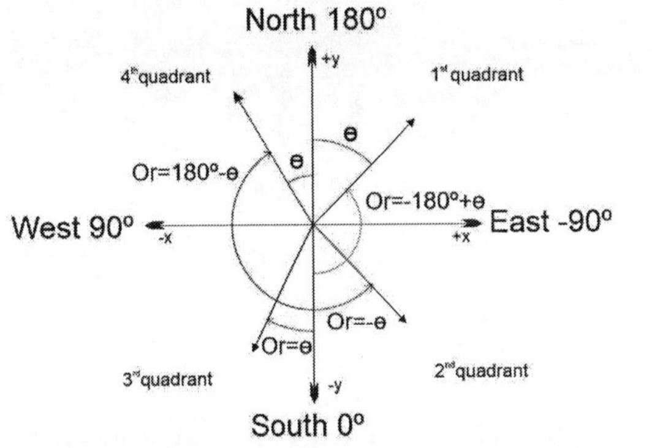

Figure 8: Absolute orientation of the slope from the angle between the projection of the vector $\vec{v}$ on the horizontal plane and the Y Cartesian axis.

The slope of the surface is calculated by a simple trigonometric process in a right-angled triangle where the normal vector is considered the hypotenuse and its projection on the Z = 0 plane and C value the adjacent and opposite legs respectively (Figure 9) (Eq. 9 to 11).

This contribution has been peer-reviewed.
doi:10.5194/isprsarchives-XL-5-W4-171-2015

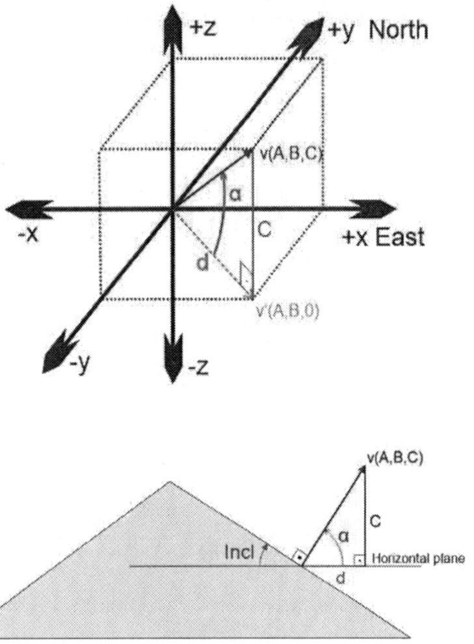

Figure 9: Inclination of the slope from the values of the normal vector of the plane.

$$d = \sqrt{A^2 + B^2} \quad (9)$$

$$\alpha = arcTan\left(\frac{C}{d}\right) \quad (10)$$

$$Incl = 90^{\circ} - \alpha \quad (11)$$

Where: $\vec{v}$ = vector normal to the plane.

A, B and C = parameters from the general equation of the plane.

d = Module of the vector $\vec{v}$ projected on the horizontal plane (Z = 0).

α = Angle between the horizontal plane (Z = 0) and the normal vector of the surface.

$Incl$ = Inclination of the plane.

2.2.4 Geometric analysis and classification

Once the different planes of the roofs are detected and segmented, its area, slope and orientation is analysed in order to perform their geometric classification. The roofs with an area smaller than the necessary for the installation of solar panels and those with North orientation are discarded. The remaining roofs are classified in different groups according to their theoretical productivity due to their inclination and orientation (Figure 10 and 11), taking into account the possibilities of integrating the solar panels in the roof and installing them in configurable supports. The CTE sets the South orientation as the optimum position for these elements, with a slope equal to the latitude where they are installed. However, the limits on the inclination can be computed according to the minimum efficiency allowed for the orientation of the surface using the method explained by the IDAE (Institute for Diversification and Saving of Energy) (Eq. 12 and 13) (IDAE, 12 Jan. 2015). These limits allow the evaluation of the suitability of the surface under study and the computation of the ideal geometry of the support platforms if needed, taking into account that losses cannot exceed the 20% if solar panels are installed using the overlapping which implies the installation of solar panels directly over the roofs not changing the angle of solar incidence regarding to the slope of the surface where they are installed technique and the 10% using the general method which implies the installation of solar panels in supports to modify the angle of solar incidence regarding to the slope of the surface where they are installed.

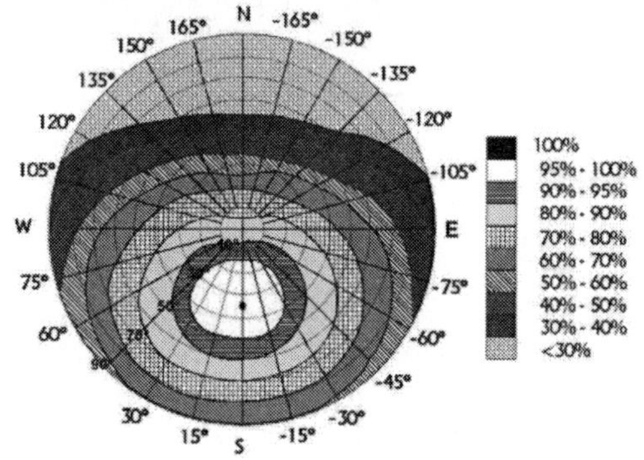

Figure 10: Diagram for the estimation of solar losses by orientation and inclination (Latitude 41°) (IDAE). The radial lines set the orientation of the surface. The concentric circles sets the slope of the surface. The texture found in the intersection of these parameters gives the losses by inclination and orientation of the roof (Source: IDAE).

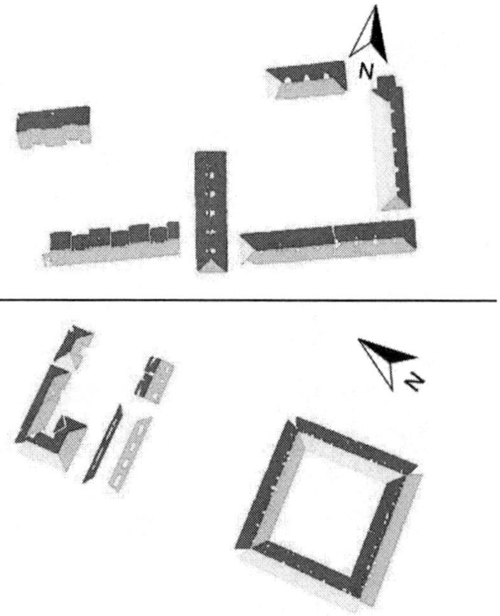

Figure 11. Details of the previous point cloud segmented and classified by its suitability to install solar panels. (Green) Surfaces suitable to install solar panels without supports. (Yellow) Surfaces suitable to install solar panels using a support to modify the solar incidence angle. (Red) Surfaces not suitable to install solar panels.

This contribution has been peer-reviewed.
doi:10.5194/isprsarchives-XL-5-W4-171-2015

If $\beta > 15$

$$Losses(\%) = 100 \cdot \left[1.2 \cdot 10^{-4} \cdot (\beta - \varphi + 10)^2 + 3.5 \cdot 10^{-5} \cdot \alpha^2 \right] \quad (12)$$

If $\beta < 15$

$$Losses(\%) = 100 \cdot \left[1.2 \cdot 10^{-4} \cdot (\beta - \varphi + 10)^2 \right] \quad (13)$$

Where β = slope
α = azimuth
φ = latitude.

2.2.5 Thermographic refinement of surfaces

Once the geometric classification is performed, the thermographic data allows the location of elements that could be an impediment to the installation of solar panels in order to avoid shadows or obstacles that could reduce their productivity allowing the analysis to find the optimum location within each roof. The existence of obstacles and anomalies in the roofs will involve the existence of different materials with different emissivity values, so the temperature detected by the thermographic images will be different even if the solar radiation received was the same. On the other hand, the existence of obstacles that prevent direct sunlight also will be manifested by changes in the roof surfaces temperatures. This facts will allow the location of the surfaces of the roofs affected by these anomalies performing a statistical study of the mean and the standard deviation of the temperatures. A point will be considered as an obstacle when the difference between its temperature value and the mean temperature of the roof is higher than the standard deviation of the temperatures of the surface that is being analysed. Considering that a shadow produced by an obstacle will move around this following the path of the sun, a perimeter around each obstacle will be marked to facilitate further analysis (Figure 12).

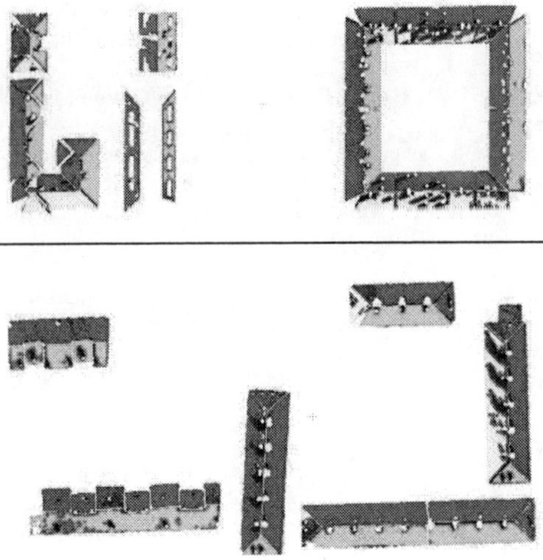

Figure 12. Details of the previous point cloud segmented and classified after the thermographic analysis. (Black) Points removed by the statistical thermographic analysis. (Grey) Perimeter defined by a user parameter (distance) around the obstacles detected to help the optimum location to place the solar panels.

3. EXPERIMENTAL RESULTS.

The proposed methodology has been validated in an wide urban area in the city of Avila (Lat 40° 38', Lon 4° 41'), with an extension of 237.250m^2 in a rectangular shape of 365m x 650m (Figure 13), chosen by the existence of roofs with different geometric characteristics allowing the test of the methodology. The study area includes 34 roofs which have been automatically segmented into 146 planar surfaces and evaluated according to their orientation, slope, surface and the radiometry of the thermography. The final results are geo-referenced point clouds of the surfaces of the roofs classified by their theoretical productivity regarding solar energy catchment with the elements that could reduce their productivity located and surrounded by an area that could be affected by its influence along the day and a plane text file for each surface with its geometric information (area, slope, orientation ...) and the optimum inclination of the supports if they were necessary (Table 4). This information allows the location of the optimum areas for the installation of solar panels in a wide urban area in a fast an accurate way, without the need to consult the technical documentation of each building.

Thanks to that during all processing the geometric information of the 3D point clouds has been preserved, perform measurements directly on the point clouds of the segmented planes will be possible allowing decision-making processes about issues related with the dimension or type of solar panels that could be installed and calculating the actual productivity of the selected components.

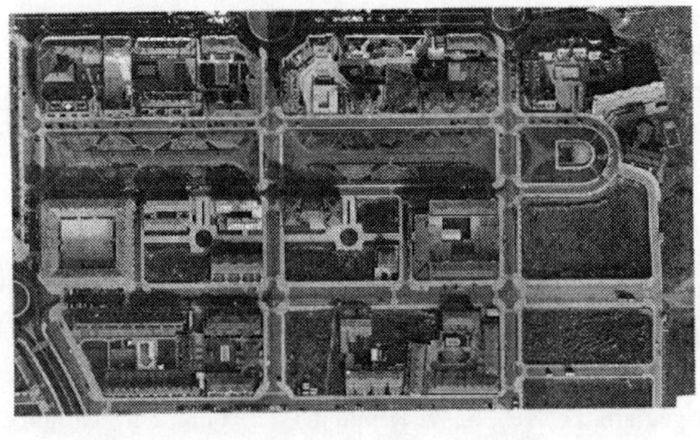

Figure 13. RGB Orthoimage of the study area.

Surface	1	2	3	4
Orientation	166.20°	-14.12°	-103.86°	75.97°
Slope	19.43°	19.26°	19.07°	19.54°
Energy losses	-	2.08%	-	21.5%
Suitable	NO	YES	NO	YES
Angle of supports	-	-	-	20.46°
Surface	5	6	7	8
Orientation	165.90°	-13.89°	103.99°	75.91°
Slope	19.35°	19.43°	19.06°	13.36°
Energy losses	-	2.02%	-	21.53%
Suitable	NO	YES	NO	YES
Angle of supports	-	-	-	20.64°

Table 4. Example of the results after the geometric analysis for a roof of the dataset.

This contribution has been peer-reviewed.
doi:10.5194/isprsarchives-XL-5-W4-171-2015

4. CONCLUSIONS

This article presents a methodology for the generation of thermographic point clouds and metric images from a multi-data source and multi-sensor aerial platform, as well as the automatic plane segmentation of the roof surfaces, its classification according to its theoretical productivity derived from its geometric characteristics (orientation, slope and surface) and a final refinement according to the thermographic radiometry that allows the location of anomalies or obstacles that could prevent the installation of solar panels or reduce their productivity. In particular, the hybrid product obtained provides complete thermographic and metric information of the different roof surfaces, which enables a better detection, spatial location and interpretation of the suitability of each surface to harbour solar panels than the traditional methods that use simplified models or in-situ visual inspections of the buildings analysed.

What is more, the methodology proposed is also valid for the processing of datasets captured with different airborne sensors like LIDAR (Laser Imaging Detection and Ranging) or any other RGB and Thermographic sensors transported by any manned or unmanned aerial vehicle.

This project opens new trends for future work both from a sensorial and methodological points of view. Concerning the first, the calibration of the thermographic equipment regarding the positioning device will allow to obtain the external orientations of each thermographic image allowing the automation of the thermographic texture mapping and the removal of the tedious process of identifying homologous entities between thermographies and point clouds. From the software point of view, the integration of the sun path effects to perform a better location of the shadow areas that could reduce the productivity of the solar panels will be the next milestone which will improve considerably the reliability of the decision-making processes from the processing results from this methodology.

REFERENCES

Agarwal, S., Furukawa, Y., Snavely, N., Simon, I., Curless, B., Seitz, S.M., Szeliski, R., 2011. Building rome in a day. Communications of the ACM 54, 105-112.

Agugiaro, G., Nex, F., De Remondino, F., Filippi, R., Droghetti, S., Furlanello, C., 2012. Solar radiation estimation on building roofs and web-based solar cadastre. ISPRS Annals of Photogrammetry, Remote Sensing and Spatial Information Sciences 1, 177-182.

Agugiaro, G., Remondino, F., Stevanato, G., De Filippi, R., Furlanello, C., 2011. Estimation of solar radiation on building roofs in mountainous areas. IAPRS&SIS 38, 155-160.

Deseilligny, M.P., Clery, I., 2011. Apero, an open source bundle adjusment software for automatic calibration and orientation of set of images, Proceedings of the ISPRS Symposium, 3DARCH11, pp. 269-277.

Fischler, M.A., Bolles, R.C., 1981. Random sample consensus: a paradigm for model fitting with applications to image analysis and automated cartography. Communications of the ACM 24, 381-395.

Gallo, O., Manduchi, R., Rafii, A., 2011. CC-RANSAC: Fitting planes in the presence of multiple surfaces in range data. Pattern Recognition Letters 32, 403-410.

Gehrke, S., Morin, K., Downey, M., Boehrer, N., Fuchs, T., 2010. Semi-global matching: An alternative to LIDAR for DSM generation. International Archives of the Photogrammetry, Remote Sensing and Spatial Information Sciences, Calgary, AB, 38 (B1).

Hirschmuller, H., 2005. Accurate and efficient stereo processing by semi-global matching and mutual information, Computer Vision and Pattern Recognition, 2005. CVPR 2005. IEEE Computer Society Conference on. IEEE, pp. 807-814.

Hofierka, J., Suri, M., 2002. The solar radiation model for Open source GIS: implementation and applications, Proceedings of the Open source GIS-GRASS users conference, pp. 1-19.

Hulik, R., Spanel, M., Smrz, P., Materna, Z., 2014. Continuous plane detection in point-cloud data based on 3D Hough Transform. Journal of Visual Communication and Image Representation 25, 86-97.

IDAE, <http://www.idae.es/uploads/documentos/documentos_5654_FV_pliego_condiciones_tecnicas_instalaciones_conectadas_a_red_C20_Julio_2011_3498eaaf.pdf>. (12 Jan. 2015)

Kraus, K., 1993. Advanced Methods and Applications. Vol. 2. Fundamentals and Estándar Processes. Vol. 1. Institute for Photogrammetry Vienna University of Technology. Ferd. Dummler Verlag. Bonn.

Lagüela, S., González-Jorge, H., Armesto, J., Herráez, J., 2012. High performance grid for the metric calibration of thermographic cameras. Measurement Science and Technology 23, 015402.

Nguyen, H.T., Pearce, J.M., 2010. Estimating potential photovoltaic yield with r.sun and the open source Geographical Resources Analysis Support System. Solar Energy 84, 831-843.

Premerlani, W., Bizard, P, 2013 Direction Cosine Matrix IMU: Theory.
<http://gentlenav.googlecode.com/files/DCMDraft2.pdf>

Rusu, R.B., Cousins, S., 2011. 3d is here: Point cloud library (pcl), Robotics and Automation (ICRA), 2011 IEEE International Conference on. IEEE, pp. 1-4.

Tait, D.A., 1994. Photogrammetry, vol. 1. Fundamentals and standard processes: Karl Kraus. Ferdinand Dümmels Verlag, Bonn, 1993, 402 pp., DM 58.00, ISBN 3-427-78681-1. ISPRS Journal of Photogrammetry and Remote Sensing 49, 42-43.

This contribution has been peer-reviewed.
doi:10.5194/isprsarchives-XL-5-W4-171-2015

VIRTUAL MODELING FOR CITIES OF THE FUTURE. STATE-OF-THE ART AND FUTURE CHALLENGES.

J. Valencia[a, *], A. Muñoz-Nieto[b], P. Rodriguez-Gonzalvez[b]

[a] PhD candidate Department of Cartographic and Land Engineering; High School of Ávila, University of Salamanca, Ávila, Spain – javi.valencia.m@gmail.com
[b] Department of Cartographic and Land Engineering; High School of Ávila, University of Salamanca, Ávila, Spain – (almuni, pablorgsf)@usal.es

Commission V, WG V/4

KEY WORDS: Cartography, Rendering, City, Three-dimensional, Spatial Infrastructures, Visualization, GIS, Virtual Reality

ABSTRACT:

3D virtual modeling, visualization, dissemination and management of urban areas is one of the most exciting challenges that must face geomatics in the coming years. This paper aims to review, compare and analyze the new technologies, policies and software tools that are in progress to manage urban 3D information. It is assumed that the third dimension increases the quality of the model provided, allowing new approaches to urban planning, conservation and management of architectural and archaeological areas.
Despite the fact that displaying 3D urban environments is an issue nowadays solved, there are some challenges to be faced by geomatics in the coming future. Displaying georeferenced linked information would be considered the first challenge. Another challenge to face is to improve the technical requirements if this georeferenced information must be shown in real time. Are there available software tools ready for this challenge? Are they useful to provide services required in smart cities?
Throughout this paper, many practical examples that require 3D georeferenced information and linked data will be shown. Computer advances related to 3D spatial databases and software that are being developed to convert rendering virtual environment to a new enriched environment with linked information will be also analyzed. Finally, different standards that Open Geospatial Consortium has assumed and developed regarding the three-dimensional geographic information will be reviewed. Particular emphasis will be devoted on KML, LandXML, CityGML and the new IndoorGML.

1. INTRODUCTION

Three-dimensional rendering of complex architectural environments is nowadays, one of the most attractive challenges in Geomatics.
Advances in geographic information acquisition and development of hardware and software, are opening new perspectives for the management of the third dimension, making possible a new outlook as a result of the ability to include Virtual and Augmented Reality.

Thanks to advances in technologies such as LiDAR, Mapping Mobile Systems (MMS), unmanned aerial vehicles (UAV), etc. (Niemeyer et al., 2014) (Haala et al., 2008) (Puente et al., 2013) (Nex and Remondino, 2014) and linked to the progress of new theories about Big Data, Smart Cities, Building information modeling (BIM), (Volk et al., 2014) etc., a need has been detected to develop new data models able to work with the third dimension and related new 3D features. Also, the temporal component involves an additional challenge, because allows a better understanding of the temporary changes of urban environment. In this sense, they are also necessary databases equipped to cope with these models and software tools for their management and visualization.

Although there are some initiatives emerging to overcome weaknesses in management and publishing three-dimensional geographic information, we can consider that the adaptation of the third dimension is in an initial stage, with many challenges, due to the growth of applications in several fields of engineering and architecture (Yin and Shiode, 2014).

On the one hand we have many tools able to acquire 3D information and, on the other hand, the solutions for rendering and displaying this information are in a point of great maturity. But, what happens when, in addition to display three-dimensional environments, we display information associated with them? And what about technical requirements if this georeferenced information should be shown in real time? Are software tools ready for this challenge? Do they have functionality for this kind of services? Through this paper, a review, comparison and analysis of several solutions that have emerged to solve these requirements is carried out.

The paper is structured as follows: In section two we will overview different elements and types of modeling. Section three is devoted to discuss problems associated when modeling cities and other large areas. In section four we will deal with the 3D information publishing and dissemination. Finally the conclusions are given in section five.

2. 3D MODELING OF LARGE AREAS

2.1 Overview

Third spatial dimension has become a real need for many technical and scientific tasks (Scianna and Ammoscato, 2010). Frequently it has been used as an attribute or as additional

* Corresponding author.

information attached to geographic features to enable further spatial analysis.

Getting three-dimensional information no longer reflects the technological and financial effort that it used to take few years ago, but today is already possible to obtain significant results with less effort and resources (van Oosterom et al., 2008). From a theoretical point of view, the lack of the third spatial dimension decreases objectively the quality of the cartographic models (Varela García et al., 2007). We should also note the importance of the third spatial dimension from the practical point of view since an increasing number of scientific fields, requires 3D dataset and analysis in order to improve their research i.e. medicine, industrial engineering, urban planning, geology, architecture, etc. (Calle et al., 2010).

Rendering three-dimensional environments is a task well solved by different software tools able to design, create, and model spatial structures, both simple and complex, effortlessly and with a high degree of realism. Basically the starting point is a 3D model, from CAD files or modeling software, and a set of measurements that are added to its features and proprieties. In order to achieve photorealistic effect, textures and lighting models are also applied (Bénitez and Baillard, 2009)

In most cases, the aim of 3D rendering is to show some infrastructure that already exists and, due to its importance in the study area, worth to be modeled. In other cases, for the infrastructures to be built in the future, the aim is to see how would be their appearance. That is the reason because urban planning, architecture or heritage conservation increasingly demand 3D rendering (Dellepiane et al., 2011). Since modeling large areas, such as cities, countries or even regions is not feasible from the process described above, information coming from Geographical Information Systems (GIS), together with other datasets acquired by LiDAR, radar, photogrammetry, remote sensing, Mobile Mapping, etc. is been used nowadays.

The current problems lie in store and manage all the enormous amount of data. In this sense, a powerful database manager (van Oosterom et al., 2008) is essential. Examples of such situations could be the attempts that several countries are developing in order to achieve 3D cadasters (van Oosterom et al., 2011) is. Another factor of complexity that can be added to the above 3D scenarios is the need to link information with the various elements of our three-dimensional model. Problems related with exchanging data format also must be solved.

2.2 Modeling cities in 3D

Three-dimensional city models are already considered as a very important resource for urban planning and for decision-making in urban policies (van Oosterom et al., 2008). Many of them are based, as already mentioned, on information stored in GIS. The simplest but effective modeling strategy is the extrusion of 2D building shapes getting heights from other sources of information such as LIDAR. These models can be created quickly, providing an easy way to 3D city displaying. This first modeling approach has been used successfully over the Internet, either with browser or plugins developed for this purpose (e.g. Google Earth) (Figure 1), or using webs developed based on OpenGL.

Even though different algorithms are being developed to automate the rendering process is important to highlight that good results with LiDAR point clouds and Poincaré-TEN algorithm have been achieved (Rebecca et al., 2008). However, this modeling strategy has poor capabilities for spatial analysis since it works with two really simple independent datasets. On the one hand the terrain models DTM and on the other hand the building model (DBM).

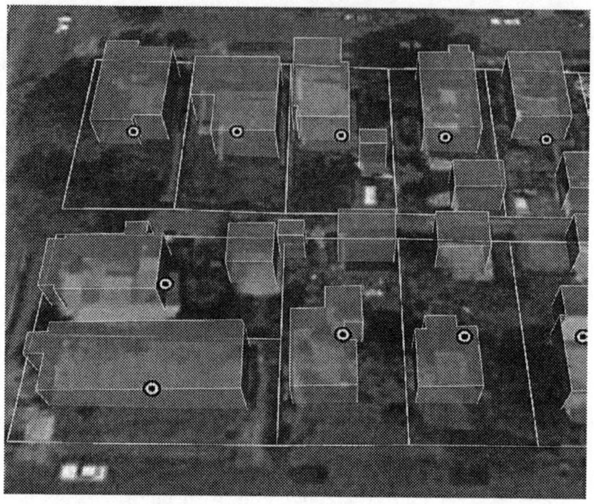

Figure 1. Buildings extruded and uploaded in Google Earth.

This kind of models can be improved by adding information from aerial or terrestrial pictures (Figure 2). In this way, the quality of the model would be higher, since we can get the photographic textures of the buildings. It would be also possible to model and place in their right position the street furniture. Thus we are getting valuable information for spatial analysis of traffic zones or pedestrian areas (Slingsby and Raper, 2008).

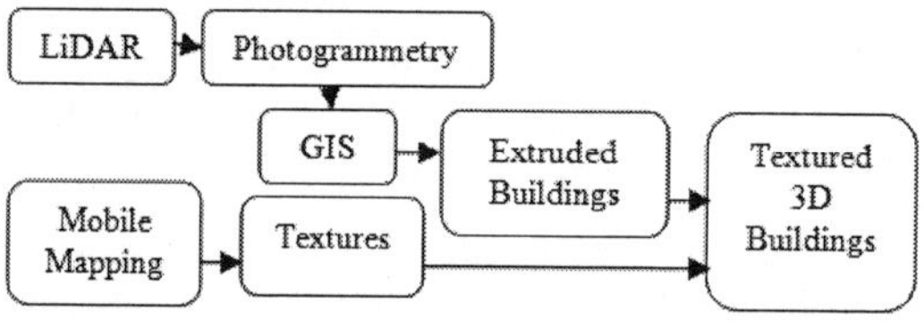

Figure 2. 3D model production workflow.

From these improved models it can be derived a lot of interesting analysis such a 3D topological study of human mobility. Extracting relevant information we can map obstacles for disabled pedestrians. Other possible uses deal with the ability to calculate real distances and routes, both for pedestrian and drivers. Determine urban visual exposure (often used in marketing) will be also possible (Suleiman et al., 2011).

Currently there are different software tools to create and manage these models such as CityEngine, ESRI, Infrastructure Design Suite, Autodesk, Bentley Map, Bentley and GeoMedia 3D Hexagon Geospatial. They have in common a set of options that make them very versatile for generating such models (Figure 3). Uploading information to the city area (DTM, BIM, CAD, 3D models, etc.) is the first step that will be completed with the other elements that make up the model (street furniture and vegetation). Finally aesthetic elements related to the lighting and atmosphere are incorporated (Engel et al., 2012). Although a comprehensive modeling is a high time consuming task, once the city model is created, it becomes very important due to their capability to interact with other3D geographic information.

Figure 3. 3D model made with Esri CityEngine. Figure courtesy of Esri.

As seen in the following table (Table 1) not all the software options are equal efficient for related functionalities.

Functionality	Esri	Autodesk	Bentley	Geomedia
Import CAD and GIS	●●●	●●●	●●●	●●
Support Geo DBMS	●●●	●●●	●●●	●●●
Editing tools	●●	●●●	●●●	●●
3D Modeling & Texturing Tools	●●●	●●●	●●	●●●
Terrain and Maps	●●●	●●●	●●●	●●●
Basic 2D and 3D Format Support	●●●	●●	●●	●●●
OpenStreetMap Import	●●●	○	○	○
3D Web Scene	●●●	○	○	○
KML Support	●●●	●●●	●●●	●●●
CityGML Suport	●●●	●●●	●●	●●
3D Libraries	●●●	●●●	●●●	●
Advanced 3D Format Support	●●	●●●	●●	○
3D Analysis	○	●●	●	●●●
3D Navigation	●●●	●●	●●●	●●●
BIM interaction	●	●●●	●●●	●
Virtual flights films	●●	●●	●	●●
Scripting Interface	●●●	●●	●	●
Simulation	●●●	●●●	●●	●●

Table 1. Qualitative comparison of software functionality (○: none, ●: bad, ●●: regular, ●●●: good)

As seen in previous comparative, CAD and GIS tend to converge in many functionalities. However, everything related to the interaction with BIM formats or other advanced 3D formats are best handled in the software that comes from CAD world. Something similar occurs with advanced editing of models. Meanwhile, GIS-derived software has better rendering performance, resulting in much lower resources consumption than the derived CAD software. As a consequence, we can conclude that geomatics and BIM disciplines are converging at the moment (Curtis, 2008), since BIM provides answers to the problem of building modeling and geomatics science, with a multidisciplinary geospatial scope, supply these issues.

Similar tools, currently developed, as RCP (VirtuelCity, 2010) or DbMAP Flyer 3D (GeoMind, 2013) show a worse interaction among various sources of geographical information offering a smaller amount of possibilities for information import. But, it should be taken into account, that since the software complexity and functionalities are reduced, the final city model time also decreases with consequently poor results.

Finally, we have to consider that mention software tools are desktop based, but there are also a cloud based tools option for users. An example of this tools is 5D Smart City (Cityzenith, 2013), which is focused on the creation and management of geospatial components for smart cities (Figure 4).

Figure 4. 3D model of Barcelona by 5D Smart City. Figure courtesy of 5D Smart City.

2.3 Modeling large areas in 3D

Three-dimensional modeling of large areas (countries, regions, states) implies additional problems. The main one is the amount of information to process, not only for their analysis and management, but also for their possible dissemination or publication on the Web (Elberink et al., 2012). This scenario is undoubtedly one of the main challenges to be addressed by Spatial Data Infrastructures (SDI). SDI have been one of the main drivers of the technological development in geomatics in recent years, but until now they have not a comprehensive answer to the problem of three-dimensionality (Basanow et al, 2008).

The large amount of information to process involves dealing with algorithms, software, analysis and management methods from the Big Data (Amirian et al., 2014). There are two key elements to consider: the data model and the database manager. For to the data model, before addressing this type of problem, a proper definition of the required level of detail is fundamental. As well as, the different guidelines and/or standards provided by the organization and committees (OGC, FGDC, ISO,...). Regarding to the database manager, its capability to deal with the 3D information is required. In this sense, is worth to mention Oracle and PostGIS developments (Herreruela, 2012) to allow not only storing spatial information, but also associated alphanumeric information, and also performing several topological and geospatial analysis functions (Walenciak et al., 2009).

All advances in the context of SDI and the management of Big Databases, involve linking spatial DBMS with 3D models using Web protocols (Zlatanova et al., 2004).

This contribution has been peer-reviewed.
doi:10.5194/isprsarchives-XL-5-W4-179-2015

In this way many kind of projects can take advantage of this union, such as 3D cadaster (Figure 5), control and monitoring of people in the area of Geospatial Intelligence or the control and management of supply networks.
However, in several of these projects an added problem has to be considered; the online publication of 3D models with associated information.

Figure 5. 3D cadastral sketch. Figure courtesy of Luis Ignacio Virgós Soriano and José Miguel Olivares García, General Direction of Cadastre of Spain.

3. 3D MODELING WITH ASSOCIATED INFORMATION

One of the most complex issues addressed in this paper is three-dimensional models with linked alphanumeric information. When a complex 3D model must incorporate alphanumeric data, a new data model is needed. The problem is not to combine the two data models solving storage or management problems, but the increasingly demand to serve or publish these two types of information.

The great diversity of file formats for the 3D information storage, such as, STL, OBJ, DAE, 3DS, DWG, DGN, JSON... is increasing since, virtually, every 3D modeling software has its own format created to serve a specific goal, in spite the import/export capabilities. However, none of the mention file formats are able in a comprehensive way to incorporate alphanumeric information associated with geometric elements.

The incorporation of alphanumeric information linked with geometric features is essential for the city modeling (Pinto et al., 2013) this way; it is possible move for 2D GIS to 3D GIS, allowing us to address the 3D SDI (Figure 6). The 3DGIS provides, besides the proper methods for geometric representation, the possibility to link semantic information. In this way, the current research lines are focused on filling the gap between the urban simulation and urban space management (Ying et al., 2012).
At that point is worth to mention the integration of three-dimensional data in X3D format with other data labelled in XML, such as temporal data (Hetherington et al., 2006).

The concurrence of geographic and alphanumeric data involves, for software tools, the ability to store, manage and process them simultaneously (Karki et al., 2010).This practically forces to manage complex scenarios from a database manager that is capable to work with both types of information together. It usually implies to join different computational technologies.
Due to the numerous possibilities that may arise regarding the information needs of these projects, data models are often developed based on the Levels of Details (LoD). The concept of LoD could be summarized as how much detail is included in the model element (BIMForum, 2013), and should not be confused with spatial resolution (interrelated term). For example, a building includes constructive elements, rooms and other objects (walls, doors, windows, floor surfaces, ceilings), depending on the degree of detail. Moreover, the accuracy of the shapes of these objects can be described and classified using different codes; e.g. the German geo-information system ATKIS (Authoritative Topographic-Cartographic Information System) based on standardized data models for an object-based description of the Earth's surface (AdV, 2013).

The precision code and LoD allows us to overcome the classical management of geographic information on scales. This way when the client requests a piece of land to a given LoD, the database manager can filter data through code accuracy to provide the information requested.

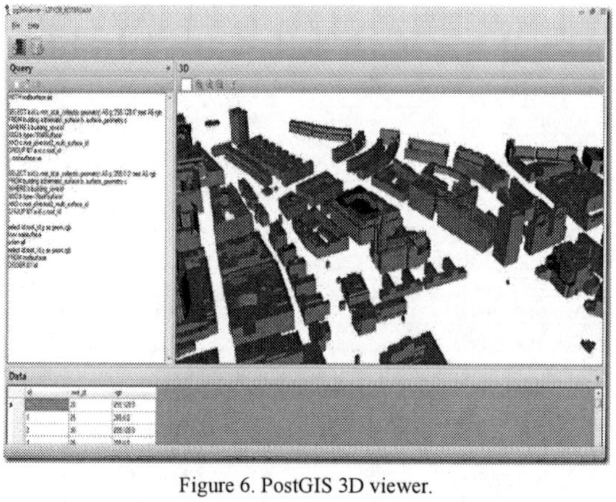

Figure 6. PostGIS 3D viewer.

However, a further step when publishing or sharing such models is necessary. In most cases these direct connection to three-dimensional databases is not considered because of the high transactional cost involved and because most of the 3D viewers have no direct connection to these databases.
In client / server structures, it is impossible to have direct connection among all possible exchange information formats. The communication both thin clients (web browsers) as fat clients (GIS software), works in an encrypted file using a known standard (Figure 7). This situation has already happened in the 2D world, when trying to publish geographic information, since when a user connects to a WMS service, the client receives from the server an image in a standardized format (Jpeg, Tiff, etc.). For Web Feature Service (WFS) both geometries and associated data is received in standardized file format as GML (Himsolt, 2010), regardless of the format in which the geographical information is stored on the server.

Currently, one of the strategies most commonly used is the combination of databases and files in formats that can incorporate data and 3D geometries. It is very common to go from database to any standardized formats that are widely used by most 3D viewers, such as KML/KMZ or CityGML (OGC, 2012).

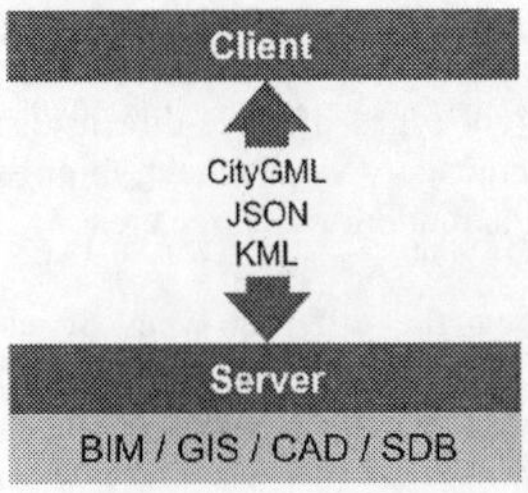

Figure 7. Client/Server architecture for 3D information interchange.

In this context major advances are occurring for the exchange of 3D spatial information. And it will be the core of the next point.

4. PUBLISHING 3D INFORMATION. OGC STANDARDS

It is a well known fact that publish and disseminate information through the Internet is on the basis of a successful policy for geographic information. In this sense, some innovative and lively ways to publish 3D models are being developed currently (Figure 8).
Today we found a situation where most of 3D modeling software has its own viewers which can be desktop plugins or web based. Although import and export functions are able to perform communication between systems, this way is not the most suitable. Therefore a new strategy is required according to the criteria of interoperability

To achieve this goal is very important to define a data model which provides a common framework between different modeling and 3D visualization tools. There are different data models that meet this assumption, such as KML, CityGML, LandXML or IndoorGML. All of them are based on XLM and have already become standards for the OGC.

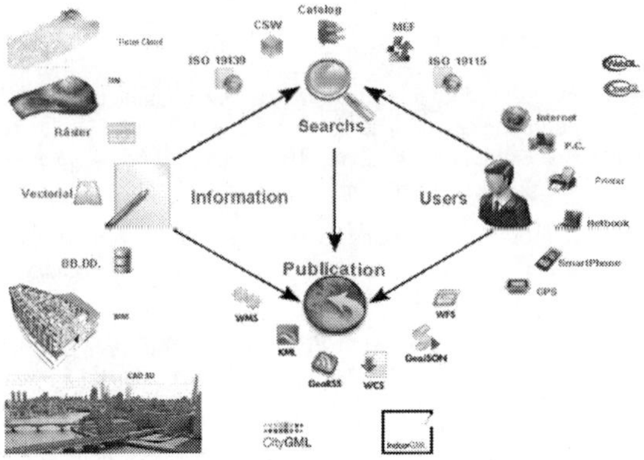

Figure 8. Publishing of 3D models with interoperability.

The KML model has practically become a standard due to its popularity and use. It is currently in its 2.2.0 development version. It allows publishing in the same file many different data such as vector, image, text, links, etc. This format also includes navigation capabilities along different points controlling the speed. As a consequence KML could be considered as a simple and well-structured data model.

CityGML was designed to store and exchange virtual 3D city models. It is another standard for the exchange of spatial data developed by the OGC and ISO TC211. Its main objective is to become the reference format for the exchange of basic entities, attributes and relationships of city models.
In spite of files become more memory demanding than in KML data model it has a greater capacity to store alphanumeric information.

It is important to highlight that CityGML is able to work with the LoD concept. For this purpose five different levels of detail are covered (Figure 9) according to the Open Geospatial Consortium (OGC, 2012):

- LoD0: terrain-only
- LoD1: extruded polygons upon a terrain
- LoD2: LoD1 with roof structures and roof textures
- LoD3: LoD2 with external architectural details
- LoD4: LoD3 with internal rooms (LoD4)

Figure 9. Different LoD in CityGML. Figure courtesy of OGC.

LandXML is another standard based on XML. It is the oldest of all and was developed for the exchange of civil engineering projects, with its associated data and measurements.

Finally IndoorGML (Nagel et al., 2010) is a topological model that has recently become a standard adopted by the OGC. Its primary mission is to serve as a method of transferring information on indoor location. The initiative to create IndoorGML arises from East Asian countries, showing great interest in developing this type of location technologies and indoor navigation due to the high growth of technologies such as RFID (Radio Frequency IDentification) and an enormous development of their telecommunications networks.
IndoorGML provides a detailed description and topologically structured navigable spaces of a building (Figure 10), but also allows easy integration of external navigation systems. These properties make this data model a good alternative for risk prevention projects, emergencies, security, etc.

This contribution has been peer-reviewed.
doi:10.5194/isprsarchives-XL-5-W4-179-2015

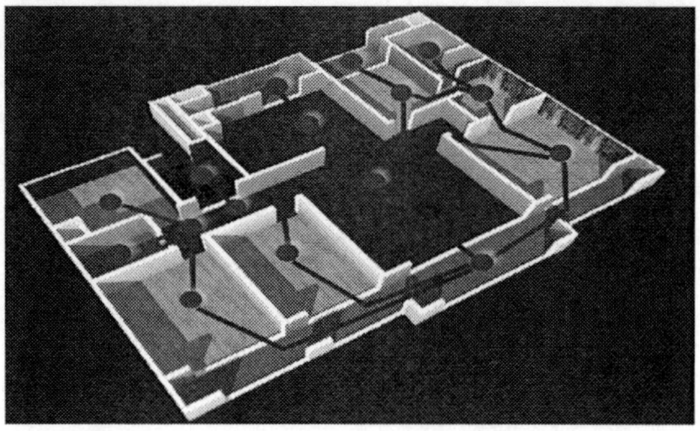

Figure 10. Model of IndoorGML schema. Figure courtesy of OGC.

As shown, each model has different properties even though all are based on interoperability, universality, accessibility and exchange criteria. The idea is that you should be able to access 3D models for the purpose needed at the time, with any computer, in any part of the world, at any moment of the day and independently of the software you are working.

There are many desktop software capable of manage some kind of referred model and visualize it in 3D. However, the challenge is to publish the models directly on the Web, i.e. to work from a web browser with 3D models without having to install anything. Current technologies such as standard formats, libraries and plugins for 3D visualization are part of the so-called 3D Web. The most significant are (Prieto et al., 2012):

- WebGL: a cross-platform open standard API for low 3D graphics. It can be accessed as a DOM interface.
- X3DOM: experimental open source framework for integrating HTML5 specification and 3D content. It is based on Javascript and WebGL.
- HTML5: the canvas in which to embed 3D graphics. X3D thus include HTML objects.

These three technologies allow to publish on the corresponding data model files on X3D format from W3DS service. The aim is to watch the 3D information present in models, in different Web applications.

Three-dimensional models on the Web requires consideration of the spatial data and information associated with it (Rodrigues et al., 2013). So it is necessary to develop a protocol for transactions both spatial data and alphanumeric data. The way to do this is shown in the following figure (Figure 11)

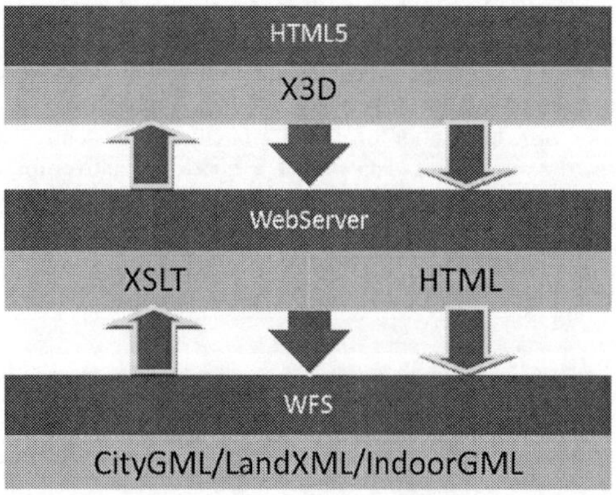

Figure 11. Transactions of geographical and alphanumeric data schema.

5. CONCLUSIONS

Throughout this paper complex scenarios that we can find when modelling different types of three-dimensional elements were studied, from elements of small extent, through the modeling of cities and even the rendering of large areas.

It has been shown the different types of alternatives for 3D modeling cities, which has been compared in order to assess their functionalities. This paper has attempted to analyze the different situations that we can face when modeling 3D geographic information, weighing up alternatives and showing solutions that are being discussed right now. The importance to link associated information to graphic 3d models has been highlighted too.

Finally we address the issue of web based 3D models due to the importance of publishing and disseminating geographical information policies, according to the interoperability criteria.
The virtual modeling faces many challenges ahead, not only in the city modeling, but also in the complex architectures and cultural heritage, with their inherent difficulties.
As discussed, the BIM and Geomatics, as a researching areas, have obvious synergies that feed themselves, demanding common strategies and policies, and predicting interesting advances in the coming years.

REFERENCES

References from Books:

Dellepiane, M., Callieri, M., Corsini, M., Scopigno, R., 2011. Using digital 3D models for study and restoration of cultural heritage artifacts. *Digital imaging for cultural heritage*. CRC Press, New York, USA, pp. 39-70.

Pinto, N. N., Tenedório, J. A., Antunes, A. P., Cladera, J. R., 2013. *Technologies for urban and spatial planning: Virtual cities and territories*. IGI Global, Hershey, USA.

References from Journals:

Elberink, S. O., Stoter, J., Ledoux, H., Commandeur, T., 2013. Generation and dissemination of a national virtual 3D city and landscape model for the Netherlands. *Photogrammetric engineering & remote sensing*, 79(2), pp. 147-158.

Hetherington, R., Farrimond, B., Presland, S., 2006.Information rich temporal virtual models using X3D. *Computers & graphics*, 30(2), pp. 287-298.

Nex, F. Remondino, F., 2014. UAV for 3D mapping applications: a review. *Applied geomatics*, 6(1), pp. 1-15.

Niemeyer, J., Rottensteiner, F., Soergel, U., 2014. Contextual classification of Lidar data and building object detection in urban areas. *ISPRS journal of photogrammetry and remote sensing*, 87, pp. 152-165.

Prieto, I., Egusquiza, A., Delgado, F. J., Martínez, R., 2012. CityGML como modelo de datos para la representación, intercambio y visualización de información sobre el patrimonio arquitectónico. *Virtual archaeology review*, 3(5), pp. 48-52.

Puente, I., González-Jorge, H., Martínez-Sánchez, J., Arias, P., 2013. Review of mobile mapping and surveying technologies. *Measurement*, 46(7), pp. 2127-2145.

This contribution has been peer-reviewed.
doi:10.5194/isprsarchives-XL-5-W4-179-2015

Volk, R., Stengel, J., Schultmann, F., 2014. Building information modeling (BIM) for existing buildings - Literature review and future needs. *Automation in construction*, 38, pp. 109-127.

Yin, L., Shiode, N., 2014. 3D spatial-temporal GIS modeling of urban environments to support design and planning processes. *Journal of urbanism: international research on placemaking and urban sustainability*, 7(2), pp. 152-169.

Zlatanova, S., Rahman, A. A., Shi, W., 2004. Topological models and frameworks for 3D spatial objects. *Computers & geosciences*, 30(4), pp. 419-428.

References from Other Literature:

Amirian, P., Basiri, A., Winstanley, A., 2014. Evaluation of data management systems for geospatial big data. *Lecture notes in computer science*, 8583, pp. 678-690.

Basanow, J., Neis, P., Neubauer, S., Schilling, A., Zipf, A., 2008. Towards 3D spatial data infrastructures (3D-SDI) based on open standards—experiences, results and future issues. *Advances in 3D geoinformation systems*, pp. 65-86.

Benitez, S., Baillard, C., 2009. Automated selection of terrestrial images from sequences for the texture mapping of 3D city models. In: *The international archives of the photogrammetry, remote sensing and spatial information sciences*, Rostock, Germany, Vol. XXXVIII, Part 3/W4, pp. 97-102.

Calle, M. D. L., Gómez-Deck, D., Olaya Ferrero, V., 2010. Desarrollo personalizado de aplicaciones SIG 3D. In: *IV jornadas SIG libre*, Girona, Spain.

Curtis, E., 2008. Serving CityGML via web feature services in the OGC web services-phase 4 testbed. *Advances in 3D geoinformation systems*, pp. 331-340.

Engel, J., Pasewaldt, S., Trapp, M., Dollner, J., 2012. An immersive visualization system for virtual 3D city models. In: *20th international conference on geoinformatics*, pp. 1-7.

Haala, N., Peter, M., Kremer, J., Hunter, G., 2008. Mobile Lidar mapping for 3D point cloud collection in urban areas – A performance test. In: *The international archives of the photogrammetry, remote sensing and spatial information sciences*, Rostock, Germany, Vol. XXXVII, Part B5, pp. 1119-1124.

Herreruela, J., 2012. Deploying 3D city models for urban and metropolitan planning. In: *Oracle spatial user conference*, Washington, DC, USA.

Himsolt, M., 2010. GML: A portable graph file format. Technical report, Universität Passau.

Karki, S., Thompson, R., McDougall, K., 2010. Data validation in 3D cadastre. *Developments in 3D geo-information sciences*, pp. 92-122.

Nagel, C., Becker B., Kaden, R., Li, K.J., Lee, J., Kolbe, T. H., 2010. Requirements and space-event modeling for indoor navigation, OGC 10-191r.

Rebecca, O. C., Gold, C., Kidner, D., 2008. 3D City modelling from LiDAR Data. *Advances in 3D geoinformation systems*, pp. 161-175.

Rodrigues, J. I., Figueiredo, M. J., Costa, C. P., 2013. Web3DGIS for city models with CityGML and X3D. In: *17th international conference information visualisation*, pp. 384-388.

Scianna, A., Ammoscato, A., 2010. 3D GIS data model using open source software. In: *The international archives of the photogrammetry, remote sensing and spatial information sciences*, Melbourne, Australia, Vol. XXXVIII, part 4-8-2-W9, pp. 120–125.

Slingsby, A., Raper, J., 2008. Navigable space in 3D city models for pedestrians. *Advances in 3D geoinformation systems*, pp. 49-64.

Suleiman, W., Joliveau, T., Favier, E., 2011. 3D urban visibility analysis with vector GIS data. In: *19th annual GIS research UK*, Portsmouth, UK.

Van Oosterom, P., Stoter, J., & Lemmen, C., 2005. Modelling of 3D cadastral systems. In: *28th cadastral seminar*, Busan, Korea, pp. 594-606.

Van Oosterom, P., Stoter, J., Ploeger, H., Thompson, R., Karki, S., 2011. World-wide inventory of the status of 3D cadastres in 2010 and expectations for 2014. In: *FIG working week*, Marrakechh, Morocco, pp. 18-22.

Varela García, F. A., Hernández Ibáñez, L. A., Taibo Pena, J., Seoane, A., 2007. Ampliación de las capacidades de visualización de un SIG libre mediante la comunicación con un navegador 3D. In: *I jornadas de SIG libre*, Girona, Spain.

Walenciak, G., Stollberg, B., Neubauer, S., Zipf, A., 2009. Extending spatial data infrastructures 3D by geoprocessing functionality-3D simulations in disaster management and environmental research. In: *International conference on advanced geographic information systems & web services*, pp. 40-44.

Ying, S., Guo, R., Li, L., He, B., 2012. Application of 3D GIS to 3D cadastre in urban environment. *3rd international workshop on 3D cadastres: Developments and practices*, Shenzhen, China.

References from Websites (Last accessed on 2015, Jan. 24):

AdV. 2013. Digital basic-landscape model (Basic-DLM). http://www.adv-online.de/icc/exteng/broker.jsp?uMen=84d3021b-2c20-c221-d5ad-8f138a438ad1.

BIMForum, 2013. Level of development specification. http://bimforum.org/wp-content/uploads/2013/08/2013-LOD-Specification.pdf.

Cityzenith, 2013. 5D smart city. http://www.cityzenith.com/city.html.

GeoMind, 2013. http://www.geomind.it/dbmap-3d/.

OGC, 2012. OpenGIS® city geography markup language (CityGML) encoding standard, version 2.0. http://www.opengeospatial.org/standards/citygml.

VirtuelCity, 2010. http://www.virtuelcity.com/en/3d-engineering.

This contribution has been peer-reviewed.
doi:10.5194/isprsarchives-XL-5-W4-179-2015

Low-cost surveys of the *Domus* of Stallius Eros in Pompeii

A. Bosco [a, b, *], M. Barbarino [b], R. Valentini [b], A. D'Andrea [b]

[a] PhD student, University of Salerno, Department of Cultural Heritage, Salerno - Italy
[b] Centro Interdipartimentale di Servizi di Archeologia, Università degli studi di Napoli L'Orientale, Napoli – Italy.
*Corrisponding autor e-mail: a.bosco14@studenti.unisa.it

Commission V, WG V/4

KEY WORDS: 3D Model, Photogrammetry, 3D-ICONS, Total Station, Pompeii

ABSTRACT:

In the framework of the Project 3D-ICONS (www.3dicons-project.eu), different large archaeological models were acquired in Pompeii. In particular, the 3D Model of the *House* of *Stallius Eros* (*Regio* I, *Insula* VI, 13 - 14) was realized. This house conserves a rich vertical stratigraphy covering a large chronological period and offers interesting information about the development of this side of the ancient city. To implement a detailed survey of the house and to reconstruct the original structure of the *domus*, different types of survey were carried out, in order to provide a complete 3D model of the house. Three data-acquisition methodologies were used and their final results were compared in order to measure their accuracy, efficacy and velocity. At first, the *domus* was surveyed by an electronical total station, then two different interventions with un-calibrated photogrammetry were planned. The paper presents some preliminary results obtained by the three methodologies and the integration of the three models. The computational environments reproducing the ancient Domus of Stallius Eros allows to propose new archaeological interpretations and hypothesis about the reconstruction of the ancient House.

1. INTRODUCTION

In the framework of the Project 3D-ICONS, addressed to digitize a series of architectural and archaeological masterpieces with particular reference to European culture, different large archaeological models were acquired in Pompeii by the CISA, (Centro Interdipartimentale Servizi di Archeologia), of the University of Naples "L'Orientale". In particular the work focused on less known archaeological areas, but likewise relevant for the analysis of the different masonries and stratigraphical superimpositions. Among these areas particular attention has been dedicated to the *House* of *Stallius Eros* (*Regio* I, *Insula* VI, 13 - 14), which conserves a rich vertical stratigraphy covering a large chronological period and offers interesting information about the development of this side of the ancient city (see Figure 1).

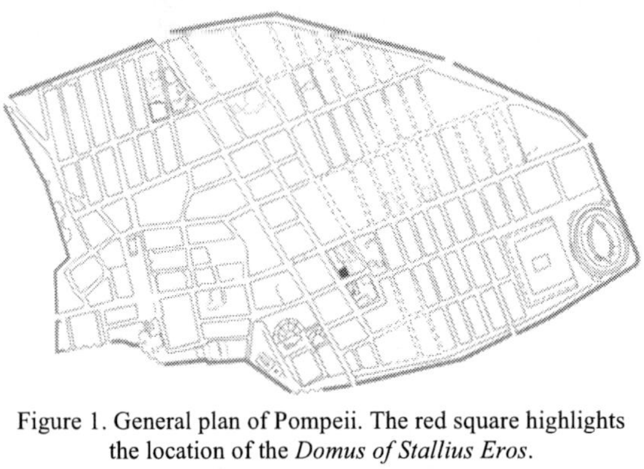

Figure 1. General plan of Pompeii. The red square highlights the location of the *Domus of Stallius Eros*.

It is evident that a complex site such as that of Pompeii, consisting of several structures of different types and with different levels of details, requires more techniques and methods for 3D modelling.

Even if the total station is an instrument quite frequent in the contemporary archaeological excavations, some limitations reduce the use. In particular the main limit is represented by the amount of time need for the acquisition of the data due to the quite manual process of data-capture. Both massive use of reference targets and, as alternative, manual application of a reflecting prism in each recorded location are time consuming operations. On the other hand, the precision with which this survey operations are carried out are strongly operator dependant and affect the high level of precision of the instrument. Finally, the survey obtained with the total station has no information about colours and materials of the surveyed structures.

For these reasons, during last decades several new techniques were introduced in the archaeological field in order to compensate these deficiencies (Eisenbeiss et al.,2005).

A promising technique is the un-calibrated photogrammetric survey. This is increasingly present among the techniques used to facilitate archaeological mapping and research, but still clashes with the prejudices of some researchers regarding the accuracy of the survey. As regards the aerial photogrammetry, however, the difficulties are increased because of costs and management of drones equipped "ad hoc", but the growing demand for UAVs in various sectors (from the agrarian to surveillance in football stadiums), produced in the recent years cost reduction and improvement in the quality of the vehicles.

The following sections show a short historical and archaeological description to highlight the issues of the site and the reasons for the choice of the techniques applied to the case study. Then, the surveying techniques are shortly summarized and some of acquired and processed data are shown. Finally, the results obtained by means of each method are compared, to give some preliminary conclusions and to indicate future developments.

This contribution has been peer-reviewed.
doi:10.5194/isprsarchives-XL-5-W4-187-2015

2. HISTORICAL AND ARCHAEOLOGICAL CONTEXT

The *domus* was excavated between 1926 and 1927, by the archaeologist Amedeo Maiuri (Maiuri, 1929), but no precise stratigraphic data have been registered and surveyed. It had two levels currently witnessed by two stairs. It was relatively small for roman standards, with about 300 m^2 for the first floor. The rooms open onto the west side of the *tuscanic* atrium, while the access to a small garden was on the north side. Such zone (*tablinum*) had a floor in *cocciopesto* decorated with white tiles forming a pattern in meanders and rhombus. A complex hydraulic system was set up before the house was destroyed. Restructuring operations were certainly performed few years before the eruption of 79 AD. Some coins found during the excavation suggest a date about 65 AD.

According to Maiuri hypothesis, the house was a complete ruin before the final eruption of Vesuvius. He concluded that the large pile of sand in the rooms around the front hall rendered this house uninhabitable and indicated that it had been adapted as a private storage area for building material. Figure 2 clearly shows below the pumice stone due to the eruption a dark deposit representing the state of abandonment of the house already before the eruption.

In the framework of the project "Pompeian Households: An On-line Companion", Allison included this house in a group of 30 houses so-called atrium houses. All, except one, have at least one garden as well as the defining front hall, or atrium (Allison P.M., 2004).

After the excavation of the twenties, no other in-depth analysis has been done in this area. In the last months, the study of the life the house has been taken in the framework of a MSc Thesis at the University of Naples L'Orientale.

On the other hand, it is worth to note that a strong attention is paid to correctly use survey techniques for the site of Pompeii. This attention was magnified by recent several damages. Indeed, the interventions including the use of 3D models are contained in the preservation program for the "Grande Progetto Pompei" activities.

The first aspect to be investigated for preserving these kind of historical monuments is the right interpretation of the building phases. For example in this case, simple observation of the walls of the house show that several heterogeneous structural interventions over time. Many architectural techniques are recognizable, some of the oldest known to the city (by frame and block) alternating with newer techniques.

All these information could be included in a comprehensive 3D model, which can replace the traditional record by unifying the geometrical information (from survey) with the structural information (from photos and investigations).

Figure 2. Historical photography of the excavation of 1926

3. THE SURVEYS

Since June 2014, the team of University of Naples "L'Orientale", under the supervision of Prof. F. Pesando, started the cleaning of the domus with the aim to carry out a detailed survey of the house and reconstruct the original structure of the Domus. During this intervention, several surveys were carried out in order to provide a complete 3D model of the house. Different methodologies and technologies were used and the final results compared in order to measure their accuracy, efficacy and velocity. At first, the domus was surveyed by an electronical total station, which is commonly used for archaeological investigation. Then, further two different interventions, both with un-calibrated photogrammetry, were planned to render the 3D model of the house.

The case study is particularly interesting because just after this survey the site was closed due to technical reasons. Therefore, the study of the site can continue based on the data recorded also through the 3D models.

3.1 Survey with total station

The total station is commonly used tool for "indirect" survey in archaeology; this methodology has become cheap and the software of the instruments particularly user-friendly also for no skilled operators. For the Stallius Eros survey, we planned the acquisition of points necessary to build a general plan of the house with the addition of few selected points for quoting the height and the uneven floors. Figure 3 shows the results of the survey after the processing in CAD software of the data obtained by means of the total station. This kind of operation has required a working day of two archaeologists with the need to make several movements of the total station and subsequent alignment and recalibration of the instrument. The result is the acquisition of enough points to outline the dimensions of the walls, potholes and drains water of the *domus*, and to lay down the averages heights and depths.

Even if it is possible to use the total station also to capture in detail the characteristics of the vertical stratigraphy of the walls of the house, this operation would require a long time and inevitable approximation of the smallest details.

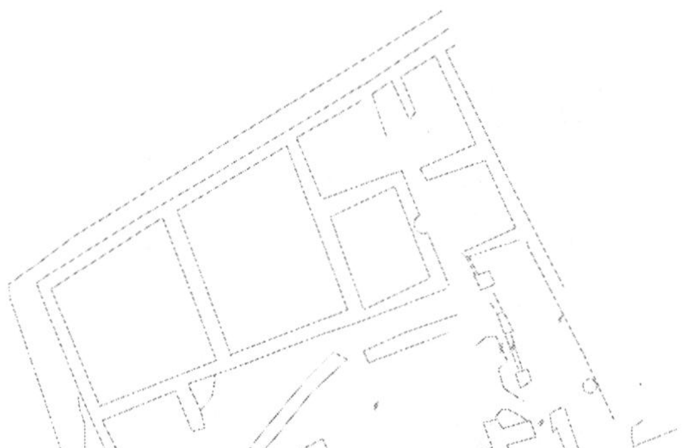

Figure 3. Planimetry of *Domus* of Stallius Eros - Elaboration by total station and CAD

3.2 Terrestrial and aerial photogrammetry

In order to solve some problems given by the total station, two different un-calibrated photogrammetical surveys were planned. The first work was carried out with a terrestrial "traditional" technique. About 1500 photos were acquired by Nikon D90 and some targets were measured by Total Station to define the geometry of the archaeological structure and to align the final model to the traditional survey. In terms of time-acquisition the work on the field required just few hours and two archaeologists. Figure 4 shows a snapshot of the model obtained by terrestrial photogrammetry with the location and the direction of each photo.
A second intervention was realized with a low cost drone (Apollo IdeaFly) equipped with a small digital camera (Canon PowerShot SX 260). In order to gain time to have a longer flight, and consequently a major number of shots, the drone was lightened and equipments reduced. Furthermore in order to reduce the vibrations of the quadric-copter, the camera was fastened directly to the drone in a perpendicular position under the drone by means a light attachment. For this reasons we didn't use a tool (Gimbal) for the fulfilment of the vertical axe of the camera in relationships to the vibrations of the quadric-copter. Also for this second aerial intervention some targets were acquired by the total station to assure the superimposition among the different surveys.

Figure 4. Terrestrial photogrammetry - location and direction of photos for 3D model

Some issues, typical for the surveys of archaeological buildings only standing partially, were dealt with the photogrammetical approach. The use of terrestrial and aerial shots was needed to integrate the two different perspectives from the bottom and top views. Stairs and poles were used for the acquisition of the higher part of the walls, while the roofs and the top part of the walls were acquired by the Drone. Furthermore, as the area is opened to tourists we were obliged to scheduled the flight before the opening of the archaeological area to the public, generally after 8.30 a.m. when the light conditions are not optimal. To overcome the scarce light of the dawn, we modified sensibly the ISO, setting it to 400, to gain visibility. The shutter speed is set to maximum, ie 1/2000, due to the movement of the quadric-copter, and continuous shooting has guaranteed a number of shots greater than necessary. This redundancy of photos allowed the most high quality photos to be selected without losing overlap in pre-processing. The results obtained were very satisfying, bearing in mind that it was not possible to set a path before the flight either know the height reached by the quadricopter. During the 8 minutes flight, about two hundreds photos were taken from a height of about 15 meters.

4. DATA PROCESSING

Data processing has been carried out by the software for photogrammetry "Agisoft Photoscan" and some work on the mesh was made with software such as Geomagic Studio® and MeshLab. Despite of the various, not quite conventional, adjustments to the quadricopter, the result of the flight was quite satisfactory. The photographs, after a needed careful selection (at the end about 200 photos have been used), showed a very good superimposition and a good resolution which allowed to create a model of about 21.600.000 faces. Figure 5 shows a view of a textured mesh.
The quality of the photos from the ground, was extremely higher than the quality of the photos taken by the drone.
All photos were processed in a single chunk, and have produced a dense cloud of about 98.000.000 points.
For this reason, finally, a model of more than 225.000.000 faces was created. However, such a complete model is not manageable by normal and also high performance computers. Therefore, during the mesh of the model some simplifications were applied, and the final model contains about one-tenth of the initial information. Figure 6 shows a top view of a textured mesh obtained by this technique. This final model was chosen after an accurate comparison of different versions with more or less accuracy, based on a sound balance between the manageability of the model and the level of accuracy necessary for the scopes of the survey.
In order to compare the models was chosen a wall very interesting for the analysis of the development of the house, because it shows one of the oldest construction techniques used in Pompeii, the "opera a telaio". This wall (namely the east wall of the atrium) is 6.25 m long, according to the reading obtained from the total station, while the same surface is 6.28 m in the model derived from terrestrial photogrammetry. Errors with the use of the total station are generally due to the instrument calibration and manual imprecision in the position of the targets necessary to record the position on the points to be acquired. This error is generally lower than 0.5% of the total length (less than 3 cm on 6.25 m). For this reason the technique provides very reliable measures for linear walls and therefore is considered reliable for archaeological surveys at medium distance. Moreover the difference between the measures obtained by the total station and that obtained by the terrestrial un-calibrated photogrammetry is approximately 3 cm (6.25 vs 6.28 respectively). Therefore the two techniques seems to have the same reliability.

This contribution has been peer-reviewed.
doi:10.5194/isprsarchives-XL-5-W4-187-2015

Figure 5. 3D model of *Domus* of Stallius Eros obtained by drone photogrammetry (top view)

Figure 6. 3D model of *Domus* of Stallius Eros obtained by terrestrial photogrammetry (top view)

This contribution has been peer-reviewed.
doi:10.5194/isprsarchives-XL-5-W4-187-2015

5. METHOD COMPARISON

The primary objective of the surveys was to provide detailed and as accurate as possible support to archaeological research, with an intervention as possible low cost and bearing in mind the management issues that often involves a three-dimensional survey. The use and the integration of the results obtained by three allowed to highlight advantages and limitations of each techniques.

The use of the total station to obtain digital maps and phases plans is currently a routine on the archaeological excavation. Normally this task involves a simplification compared to the existing, because what archaeologists want to know are the dimensions of the built and its location in space. The details are commonly assigned to the cards, photos, sections, etc. By creating a three-dimensional model all this information may be extracted easily from the reconstruction. By setting maps and sections it is possible to obtain a detailed plan of the spaces allowing an easy analysis of the deterioration of a wall (figure 7).

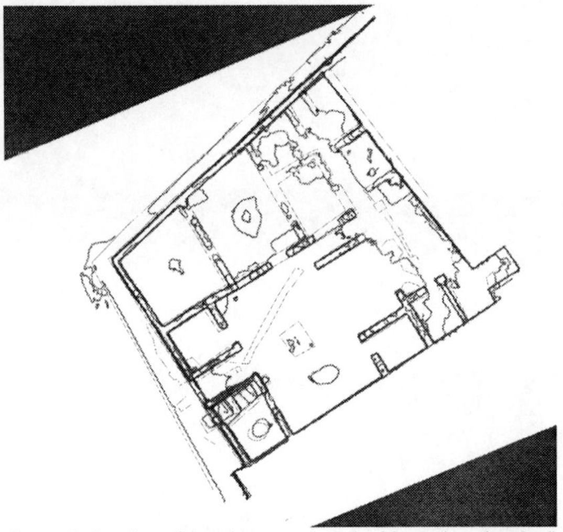

Figure 7. Overlap of the plants produced with total station (red) and with the 3D model (black)

As it is possible easily extracting by the 3D Model ortho-photographs (Bornaz et al., 2006) and other rectified maps characterized by a detailed and coloured description of the surface, the archaeologist have available at any time a digital resource precise and measurable (scale 1:1) to identify stratigraphic masonry techniques and recognize the standing remains of the building.

This technique allows also to remove, during the processing phase of the reconstruction of the 3D model, any modern detail which impedes a complete reading and interpretation of the archaeological feature. Furthermore it is possible to extract features according to different phases or period in order to propose possible alternative or later reconstructions.

In the case of the examined *domus* it was impossible to have an overall view of the outer west wall, as it overlooks a narrow street (figure 8). A photomosaic straightened would certainly have created distortions resulting unreliable.

The model obtained by photogrammetry from the ground allows the extraction of this detailed information.

Figure 8. Ortho-photography of the west wall outside, extracted from the 3D model from terrestrial photogrammetry

The aerial photogrammetry low cost has not obtained a detail such as to allow the extraction of information of detail, but it has been found useful for a vision planimetric and to obtain an image of context (the roof, the top of the standing walls and the narrow streets surrounding the house).

6. CONCLUSIONS

The absence of stratigraphic information makes primary the correct reading of the vertical stratigraphy, as the only information source about the changes of the house (possible extension, reduction, different alignments). The classical approach of the archaeological study of masonry structures provides a catalogue with several photos. These photos have often a metric reference. In other cases these photo are rectified. Very rare are the ortophotos.

Orthophotos, plans and sections extracted from the 3D model become an essential tool for the study and understanding of the archaeological data. Furthermore the model allows the analysis of the state of the conservation of the monument and the study of the masonries can be used for a monitoring of the *domus*.

No standards are available to evaluate and measure the accuracy of 3D acquisition. To date, the final models obtained through different techniques with decimation and simplifications of the surfaces can be clearly compared only by extracting some specific and partial basic geometrical information (such as the height, length and depth). Further comparisons of the models could demonstrate that integrating different techniques is particularly useful for the surveying of parts of the standing structures. This integration not only solves visibility problems (a niche can not be seen from above, while a not accessible roof is not acquirable in terrestrial photogrammetry), but could improve the definition and measurement of the vertical and horizontal planes of the walls to reduce any distortion. The estimation of resolution in pixel or the number of polygons is not sufficient to demonstrate the reliability of a model, nor its chromatic properties. The increasing number of applications of terrestrial and aerial photogrammetry should provide data for a more detailed analysis of the limits of these techniques and a large scale use of these methodologies for rapid 3D acquisition.

ACKNOWLEDGEMENTS

The authors would gratefully acknowledge Professor Fabrizio Pesando, for encouraging this work and providing input to the research and Dr. Marco Giglio for providing survey data through the total station.

REFERENCES

3D Sistem – Geomagic Solution, 2012. Geomagic Studio 2012.1.1, User guide

Agisoft 2014. Agisoft PhotoScan User Manual Professional Edition, Version 1.0.3

Allison P. M., 2004. "Pompeian Households: An On-line Companion is a publication of The Stoa: A Consortium for Electronic Publication in the Humanities", Ross Scaife, ed. http://www.stoa.org/projects/ph/house?id=5

Bitelli, G., Tini, M.A., Vittuari, L., 2003. "Low-height aerial photogrammetry for archaeological orthoimaging production". ISPRS Archives Vol. XXXIV, Part 5/W12, pp. 55-59.

Bornaz L., Dequal S., Lingua A., 2006. "L'ortofoto solida di precisione: un prodotto innovativo per la rappresentazione e la gestione dei dati tridimensionali". Atti Conferenza nazionale SIFET, Taranto.

Chiabrando F., Lingua A., Maschio P., Rinaudo F., Spanò A. (2012). Mezzi aerei non convenzionali a volo autonomo per il rilievo fotogrammetrico in ambito archeologico. In: Una giornata informale per i 70 anni del Prof. Carlo Monti - 3 Maggio 2012, Milano, 3 maggio 2012. pp. 1-12

Eisenbeiss, H., Lambers, K., Sauerbier, M., Zhang, L., 2005. "Photogrammetric documentation of an archaeological site (Palpa, Peru) using an autonomous model helicopter". ISPRS Archives Vol. XXXIV-5/C34, CIPA, Torino, Italy ,pp. 238-243.

Maiuri, A. 1929. Pompei - Relazione sui lavori di scavo dall'aprile 1926 al dicembre 1927. In *Notizie degli Scavi di Antichità*, Serie 6, n.. 5, pp. 354-476.

This contribution has been peer-reviewed.
doi:10.5194/isprsarchives-XL-5-W4-187-2015

MULTI-SENSOR RADIOMETRIC STUDY TO DETECT PATHOLOGIES IN HISTORICAL BUILDINGS

S. Del Pozo [a, *], J. Herrero-Pascual [a], B. Felipe-García [b], D. Hernández-López [b], P. Rodríguez-Gonzálvez [a], D. González-Aguilera [a]

[a] Dept. of Cartographic and Land Engineering, High School of Ávila, University of Salamanca, Ávila, 05003, Spain - (s.p.aguilera, sabap, pablorgsf, daguilera)@usal.es
[b] Regional Development Institute-IDR, University of Castilla-La Mancha, Albacete, 02071, Spain - bfelipe@jccm.es, david.hernandez@uclm.es

Commission V, WG V/4

KEY WORDS: Architecture, Cultural Heritage, Camera, Multispectral, Laser Scanning, Remote Sensing, Close Range, Classification

ABSTRACT:

This paper presents a comparative study with different remote sensing technologies to recognize pathologies in façades of historical buildings. Building materials deteriorate over the years due to different extrinsic and intrinsic agents, so assessing these diseases in a non-invasive way is crucial to help preserve them. Most of these buildings are extremely valuable and some of them have been declared monuments of cultural interest. In this way through close range remote sensing techniques, it is possible to study material pathologies in a rigorous way and in a short duration field campaign.

For the investigation two different acquisition systems were applied, active and passive methods. The terrestrial laser scanner FARO Focus 3D was used as active sensor, working at the wavelength of 905 nm. For the case of passive sensors, a Nikon D-5000 and a 6-bands Mini-MCA multispectral camera (530-801 nm) were applied covering visible and near infrared spectral range. This analysis allows assessing the sensor, or sensors combination, suitability for pathologies detection, addressing the limitations according to the spatial and spectral resolution. Moreover, the pathology detection by unsupervised classification methods is addressed in order to evaluate the automation capability of this process.

1. INTRODUCTION

Historical buildings and monuments are extremely valuable constructions for the area where they are placed. The degradation of their construction materials is caused mainly by environmental factors such as pollution and meteorological conditions. Water increases the ability of air contaminants to degrade stone by combining with them to produce a hard and blackened layer on the surface of the rock (Marszałek, 2004). In addition, results of previous uncontrolled restoration techniques applied in some cases influence the degradation (Price, 2011). For that reason the use of non-contact and non-destructive technologies to study stone damages is important for the preservation of buildings and for the choice of the best technique for restoration (Fort et al., 2002; Weritz et al., 2009).

Terrestrial laser scanner and digital cameras are two different technologies that are suitable for these studies. They are non-destructive and non-invasive sensors that allow researchers to acquire huge geometric and radiometric information across the building with high accuracy and in a short duration acquisition. The geometrical information provided by laser scanner technology has been successfully applied in a large number of fields such as archaeology (Lamberts et al., 2007), civil engineering (González-Aguilera et al., 2008), geology (Buckley et al., 2008) and geomorphological analysis (Armesto et al., 2009). On the other hand, radiometric information, provided by the intensity measurements in the case of laser technology and by the digital levels in the case of digital cameras, is used less frequently. Even so, it has demonstrated its high potential for classification tasks and recognition of different materials. Nowadays, in the literature, one can find works related to this issue: from methodologies of radiometric calibration (Kaasalainen et al., 2008), to corrections of intensity values (Franceschi et al., 2009; Höfle and Pfeifer, 2007) and to applications of the intensity data (Lichti, 2005). Spectral classification methods are based on the fact that each specific material has wavelength dependent reflection characteristics. There are many classification methods, which vary in complexity. These methods include hard and soft classifiers, parametric and non-parametric methods and supervised and unsupervised techniques (Mather and Tso, 2003). There are several works related to the application of these techniques to the identification of damage on building surfaces (Lerma, 2001; Lerma, 2005; Ruiz et al., 2002).

The main goal of this study was to compare three different sensors to detect damages in facades of historical buildings. Additional goals were to detect different building materials and also increase the degree of automation in the pathology detection of façades. To succeed these objectives, the paper is divided into the following sections: Section 2 gives the details and specifications of the equipment employed and thoroughly describes the methods employed in the workflow methodology. Section 3 shows the classification results for both unsupervised and supervised algorithms, closing with Section 4 which summarizes the conclusions drawn from the study.

2. MATERIAL AND METHODS

Sensors and their role in the workflow developed are outlined in Figure 1.

This contribution has been peer-reviewed.
doi:10.5194/isprsarchives-XL-5-W4-193-2015

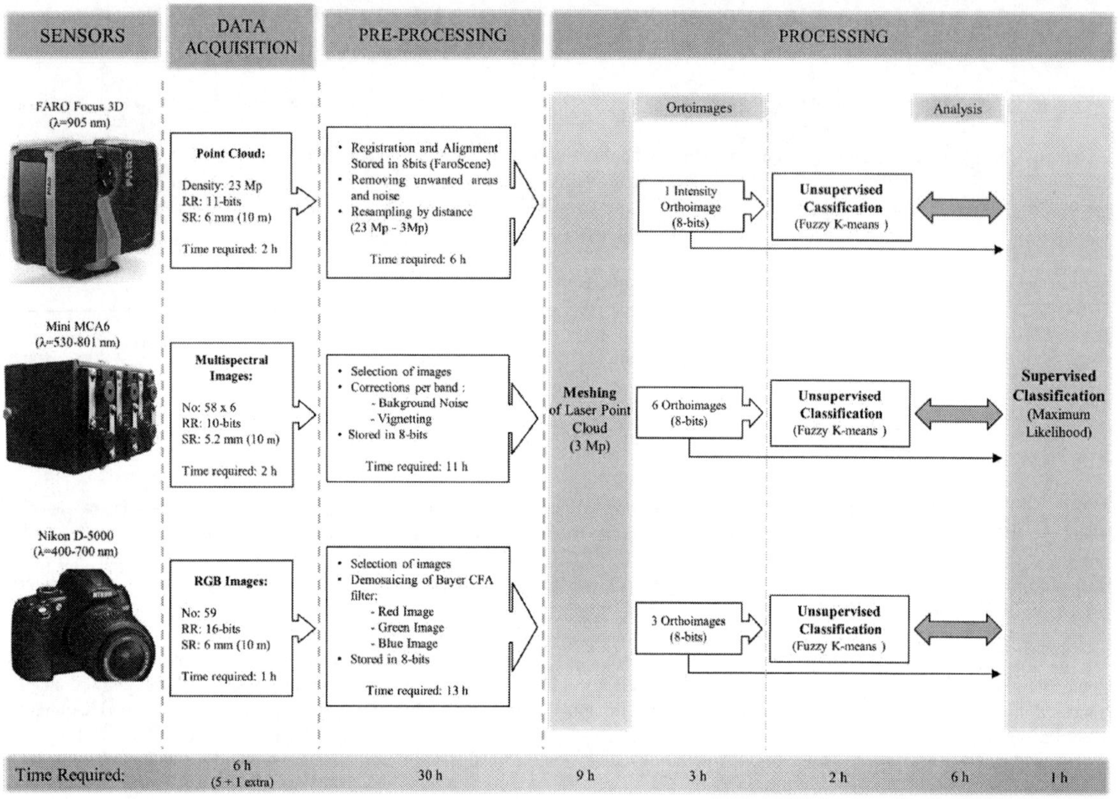

Figure 1. The workflow of the methodology presented. RR= Radiometric resolution, SR= Spatial resolution, Mp= Million points.

2.1 Equipment

For the documentation of the façade three different sensors were used with different characteristics and data acquisition principles, one active terrestrial laser scanner and two passive digital cameras (a multispectral camera and a single lens reflex-SLR digital camera). Figure 1 shows the different wavelength sensors used in this research.

The FARO Focus 3D laser scanner is based on the principle of continuous wave, with a wavelength of 905 nm. This device measures distances in a range of 0.60–120 m with a point measurement rate of 976,000 points per second. It has an accuracy of 0.015° in normal lighting and reflectivity conditions and a beam divergence of 0.19 mrad, equivalent to 19 mm per 100 m range. The field of view covers 320° vertically and 360° horizontally with a 0.009° of angular resolution and the returning intensity is recorded in 11 bits. This laser scanner includes, in addition, a double compensator in the horizontal and vertical axis that can be used as constraint for the scan alignment.

For the multispectral data acquisition, a lightweight Multiple Camera Array (MCA-6, Tetracam) was employed. This low-cost sensor allows versatility in data acquisition, however requires the radiometric and geometric corrections to ensure the quality of the results (Del Pozo et al., 2014). It includes a total of 6 individual sensors with filters for the visible and near infrared spectrum data acquisition. More concretely, the individual bands of 530, 672, 700, 742, 778 and 801 nm were used. The longest wavelength was chosen taking into account that the multispectral sensor is not external cooled. In spite of its 1280 x 1024 pixels of spatial resolution, the camera has a radiometric resolution of 10 bits. The focal length of 9.6 mm and the pixel size of 5.2 µm yield a façade sample distance (FSD) of 5.4 mm for a distance of 10 m, which should be taken into account for the pathology detection performance in small elements. The main limitation of this camera is the field of view (38° x 31°), so several captures were needed to keep the FSD.

The Nikon D5000 is a high-resolution digital SLR camera chosen to acquire information relating to the visible range of the spectrum with a resolution of 4288 x 2848 pixels. This camera uses RBG CMOS (Complementary Metal-Oxide-Semiconductor) sensor with a pixel size of 5.5 µm. The focal length chosen for the study is 18 mm in order to capture high field of view.

2.2 Data acquisition

2.2.1 Range data: For the laser scanner, intensity data at 11-bit resolution was collected at an average distance of 10 m. Scanning positions were selected according to the different technical specifications of the scanner for an spatial resolution of 6 mm at the working distance. The laser network was adapted and optimized due to the presence of obstacles that hinders a single station data acquisition.

This contribution has been peer-reviewed.
doi:10.5194/isprsarchives-XL-5-W4-193-2015

2.2.2 Image data: Illumination was a crucial parameter for both sensors in the image data acquisition since they were passive sensors and also several shot positions per sensor were required to cover the object of study. For that reason take fewer number of photos and farther was prioritized ensuring the greatest resolution. Images were collected at 10-m distance with a spatial resolution of 5.4 mm for the worst case (multispectral images).

2.3 Pre-Processing

2.3.1 Optimization and alignment of the point clouds: The raw TLS data were filtered and segmented in order to remove those points that were not part of the object of study (adjacent building, artificial elements, trees, etc.). The individual point cloud alignment was done by a solid rigid transformation by the use of external artificial targets (spheres). The spheres were stationed in tripods in the vertical of ground control points (GCP) surveyed by GNSS. The laser local coordinate frame could be transformed to a global coordinate system (UTM30N in ETRS89), allowing the geo-reference of the subsequent classification for a global analysis e interpretation. This proposed workflow allowed a final relative precision of the GCP coordinate of 0.01 m and an absolute error of 0.03 m after post-processing. As a result, a unique point cloud in a local coordinate system with 11-mm precision (due to the error propagation of inherent error sources of TLS (Reshetyuk, 2009) and the error associated to the definition of the coordinate frame) was generated.

2.3.2 Multispectral images corrections: Low-cost sensors are more likely to be affected by different noise sources so that the actual value of radiation collected by them is altered (Equation (1)) (Al-amri et al., 2010). The Mini MCA6 was affected by two different sources errors: a background noise and a vignetting effect (Del Pozo et al., 2014). Both errors were studied under laboratory controlled conditions for each wavelength band.

The background noise is a systematic error caused by the sensor electronics of the camera. It was analysed in a completely dark room in the absence of light determining the noise per band and exposure time. For this study, the maximum background error was for the 801-nm band and involved a 1.07% increment of the actual digital level value. Regarding the vignetting effect, the radially attenuation of the brightness was studied taking images of a white pattern with uniform lighting conditions. Digital levels of each multispectral image were corrected for these two effects through a script developed in Matlab to improve the data quality before the orthoimages generation.

$$DL_{raw} = DL_{radiance} + (DL_{bn} + DL_{v}) \quad (1)$$

where DL_{raw} = digital levels of the raw images
$DL_{radiance}$ = digital levels from the radiance component
DL_{bn} = digital levels from background noise
DL_{bn} = digital levels from vignetting component

2.3.3 Demosaicing of RGB images: The Nikon D5000 uses a Bayer filter over the CCD (Figure 2). It is a colour filter array (CFA) for arranging the red, green and blue colour filters on a rectangular grid of the sensor (Lu and Tan, 2003). Two filtered green, one blue and one red were obtained from each square of four pixels. There is double number of pixels for the green because the human eye is more sensitive to this colour. Each image is a mosaicked image. The process consists of estimate the three colours with the only one spectral measurement stored per pixel according to the neighbouring colours (Li et al., 2008). We make use of the open-source dcraw (Coffin, 2011) computer application to convert these images. As result, the demosaiced images enclose the half spatial resolution of the original ones.

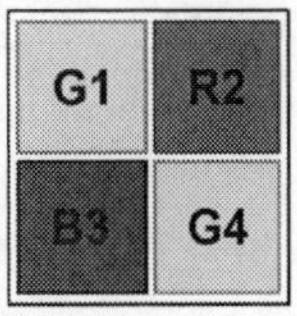

Figure 2. Bayer-pattern block CFA.

2.4 Processing

In this section data from the different sensors were joined to achieve the orthoimages of each wavelength and the subsequent classifications maps. The pre-selected images from both cameras and the optimized laser scanner point cloud are the input data of the following procedures.

2.4.1 Meshing: Once the final point cloud was generated a triangulation was applied to create the digital surface model (DSM). This step was required in order to generate continue 2D products (in the form of orthoimages) and carry out the pathology detection by the classification process. For the DSM generation the incremental triangulation Delaunay algorithm was applied (Bourke, 1989). The output was refined to avoid artifact, meshing gaps, and other errors (Attene, 2010).

2.4.2 Orthoimages generation: Orthoimages are highly demanded products that offer many benefits: metric accuracy and radiometric information useful to analyse quantitatively and qualitatively different kind of information.

For the orthoimage generation, there is necessary to known the external orientation of the images with respect to the coordinate system of the laser point model. For that purpose an average of 20 corresponding points between the point cloud and images were establish. The image projection is characterised by a rigid transformation (rotation and translation) together with the internal camera parameters.

Orthoimages were generated based on the anchor point method (Kraus, 1993). This method consists of applying an affine transformation to each one of the planes formed by the optimised triangular mesh, which was obtained from the point cloud determined by the laser. Through the collinearity condition (Albertz and Kreiling, 1989), the pixel coordinates of the vertices of the mesh are calculated, and the mathematical model of the affine transformation directly relates the pixel coordinates of the registered image and of the orthoimage.

This contribution has been peer-reviewed.
doi:10.5194/isprsarchives-XL-5-W4-193-2015

2.4.3 Orthoimages classifications: In order to categorize the orthoimages in different informational classes a firstly unsupervised and a posterior supervised classification were performed. The unsupervised classification was based on the Fuzzy K-means clustering algorithm where each observation can concurrently belong to multiple clusters (Bezdek, 1981). For a set of n multidimensional pixels, the automatic management in k classes iteratively minimizes the Equation (2) (Kannan et al., 2009):

$$J_m = \sum_{i=1}^{n}\sum_{l=1}^{k} u^{m}_{i,l}\left\|x_i - c_j\right\|^2 ; 1 \le m < \infty \qquad (2)$$

where m = any real number greater than 1
x_i = the i-th of d-dimensional measured data
$u_{i,l}$ = degree of membership of x_i *in* the cluster l
c_l = d-dimensional centre of the cluster
$\|**\|$ = any norm expressing the similarity between any measured data and the centre

Fuzzy partitioning is carried out through an iterative optimization of the objective function shown above, with the update of membership and the cluster centres by Equation (3).

$$u_{i,l} = \frac{1}{\sum_{k=1}^{C}\left[\frac{\left\|x_i - c_l\right\|}{\left\|x_i - ck\right\|}\right]^{\frac{2}{m-1}}} ; \; c_l = \frac{\sum_{i=1}^{N} u^{m}_{il} x_i}{\sum_{i=1}^{N} u^{m}_{il}} \qquad (3)$$

This iteration will stop when $\max_{il}\left\{\left|u_{il}^{(k+1)} - u_{il}^{(k)}\right|\right\} < \varepsilon$

where ε = termination criterion between 0 and 1
k = the iteration steps

After this classification, a fist approach of the spectral classes and different construction materials was obtained. With a subsequently supervised classification and applying the expert knowledge of some classes, the final results improve.

Furthermore, this classification will serve as reference to discuss which sensor is the ideal for detecting materials and pathologies in façades.

In this case, a maximum likelihood (ML) classification algorithm (Richards and Richards, 1999) was applied. The ML classifier quantitatively evaluates both the variance and covariance of the category spectral response patterns when classifying an unknown pixel. The resulting bell-shaped surfaces are called probability functions (Equation (4)), and there is one such function for each spectral category (Lillesand et al., 2004).

$$\Pr(k/g) = -\ln\left|\sum k\right| - (g-\mu_k)^T \sum k^{-1}(g-\mu_k) \qquad (4)$$

where $\sum k$ = covariance matrix
w_k = mean vector

3. EXPERIMENTAL RESULTS

The study area was the Shrine of San Segundo declared World Cultural Heritage in 1923 (García, 2006) (Figure 3). This Romanesque shrine is located in the west of the city of Ávila (Spain) and was built in the 12th century with unaltered grey granite plinths and walls with the alternation of granite blocks with different alteration degrees. The unaltered granite is mainly present in the blocks of low areas because of its high compressive strength and resistance to water absorption.

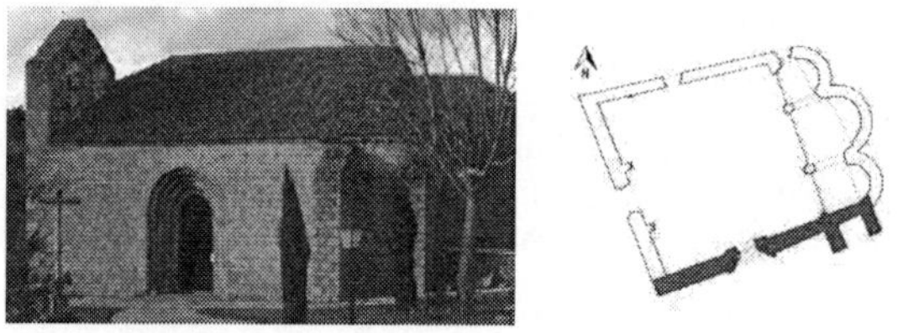

Figure 3. South façade of the Church of San Segundo in Ávila (Spain).

Figure 4. Orthoimages of the southern façade from the 3 analysed sensors.

This contribution has been peer-reviewed.
doi:10.5194/isprsarchives-XL-5-W4-193-2015

The field work was carried out on 27 July 2012 around the southern façade of the church (Figure 3), the most interesting façade from a historical point of view because it preserves the Romanesque main front. The five archivolts and capitals are decorated with plant and animal motifs. A total of 3 stations for the case of laser scanner were performed to cover the façade at a distance of 10 m. The resolution of the data capture of the FARO Focus 3D was 6 mm. Moreover, the façade was photographed at the same distance with both passive sensors, Nikon D5000 and Mini MCA6-Tetracam. A total of 3 RGB images and 9 multispectral images were used for the orthoimages generation. The total volume of information generated amounted to 10.7 GB, where the great part was due to the meshes and orthoimages generation projects. Figure 4 shows the set of the 10 final orthoimages.

In order to compare the discrimination capability of the three sensors to distinguish rock and pathologies a first unsupervised classification for the orthoimages of each sensor was performed (Figures 5, 6 and 7). A final supervised classification with the complete set of 10 orthoimages was carried out. This last classification serves as a reference against which compare each individual unsupervised classification. The steps followed by the workflow are shown in Figure 1.

3.1 Unsupervised classifications

Ten predefined clusters were used in each case for the unsupervised classification algorithm. In all of them, the resulting map showed the existence of affected areas. Post-analysis reduced the number of clusters. The number of clusters decreased from 10 initial clusters to 6 thematic classes with real meaning: 1) unaltered granite, 2) altered granite, 3) wood (door of the church), 4) areas with moisture evidences (caused by capillarity or filtration water), 5) mortar between blocks and 6) biological colonization areas.

Figure 5. FARO Focus 3D map for the 6–classes unsupervised classification.

Figure 6. Mini MCA6 map for the 6-classes unsupervised classification.

Figure 7. Nikon D5000 map for the 6-classes unsupervised classification.

It is noteworthy that data are not as satisfactory as could be due to large variability in lighting conditions during the data acquisition. As mentioned at the beginning of Section 3, the fieldwork took place on July 27, with a 6-hours total acquisition time.

Comparing the results with a visual inspection, results corresponds quite well to reality for both types of existing granites (unaltered and altered) by two well differentiated clusters in all classification maps (Figure 5, 6 and 7). Note that only 6 clusters of the final 6 pre-set informational classes were distinguished for the case of laser scanner (Figure 5). Regarding pathologies detection, it was not possible to draw conclusions with these first unsupervised classifications. However, this process served to perform a better defined supervised classification.

Table 1 summarizes the values of the overall accuracy for the Fuzzy K-means unsupervised classification for each data set evaluated.

Overall accuracy (%)	
Sensor	Fuzzy K-means algorithm
FARO Focus 3D (905 nm)	78.96
Mini MCA6 (530-801 nm)	86.20
Nikon D5000 (Visible)	75.92

Table 1. Unsupervised classification accuracies achieved for each data set.

The best classification accuracy was 86.20%, achieved for the Mini MCA6 unsupervised classification (Table 1). This indicates that the best correlation between the number of pixels correctly classified and the total number of pixels occurred for this sensor.

3.2 Supervised classification

With the full set of the 10 orthoimages a supervised classification of the façade was performed (Figure 8) taking into account the two existing variants of granite and their pathologies derived primarily from moisture and biological colonizations.

In order to compare the reference classification (based on training areas) with each sensor data unsupervised classification Table 2 was performed. The table shows the sum of pixels belonging to each class for each of the classifications performed. The count is expressed as a percentage of the total number of classified pixels.

Count of pixels (%)	Reference map	Laser map	Multispectral map	Visible map
▪ Unaltered granite	29.29	7.77	22.17	31.98
▪ Altered granite	14.69	34.02	25.04	18.97
▪ Wood	30.65	5.72	0.00	0.00
▪ Moisture	21.81	12.67	34.23	35.04
▪ Mortar	1.71	39.80	10.85	5.98
▪ Biological colonization	1.84	0.03	7.71	8.04

Table 2. Pixels computation belonging for each thematic class.

In a quantitative analysis, intensity data from laser scanner is the furthest from the reference. Results show a high number of pixels classified as mortar in the case of laser map (38.09% higher with respect to the reference map) and few pixels classified as unaltered granite offset by the altered granite count. For the multispectral and visible passive sensors results are quite similar. For both cases it was not possible to discretize wood from biological colonization so these two classes were exempted from being part of the multi-sensor analysis. Note that in the case of the multispectral camera, moisture classification was closer to reality. Since best results for moisture detection were achieved by the multispectral camera and it is the most comprehensive sensor with results closer to the reference it is concluded that the best of the three sensors to detect pathologies and construction materials differences is the Mini MCA6, which also had the best classification accuracy.

To evaluate the separability between classes the transformed divergence was used as a quantitative estimator (Davis et al., 1978) being the most widely used estimator for this purpose (Tolpekin and Stein, 2009). Table 3 shows the separability between the final 7 classes (taking into account the background or null cluster).

A high separability was achieved for all 7 classes, highlighting the good separability between the two granite types. The worst case was for moisture and unaltered granite. This fact is explained because most of the moisture was part of the unaltered granite. This granite type was constructively situated in lower areas to support the loads of the building (also in buttress), coinciding with the most likely areas to be affected by filtration and capillarity of water. This radiometric misunderstanding did not occur in the case of altered granite since being part of the centre of the façade hardly had moisture.

	Unaltered granite	Altered granite	Wood	Moisture	Mortar	Biological colonization
Altered granite	1.99	-	-	-	-	-
Wood	2.00	2.00	-	-	-	-
Moisture	1.51	1.99	2.00	-	-	-
Mortar	1.99	1.92	2.00	1.99	-	-
Biological colonization	1.99	2.00	2.00	1.97	1.99	-
Background	2.00	2.00	2.00	2.00	2.00	2.00

Table 3. Transformed divergence for the supervised classification.

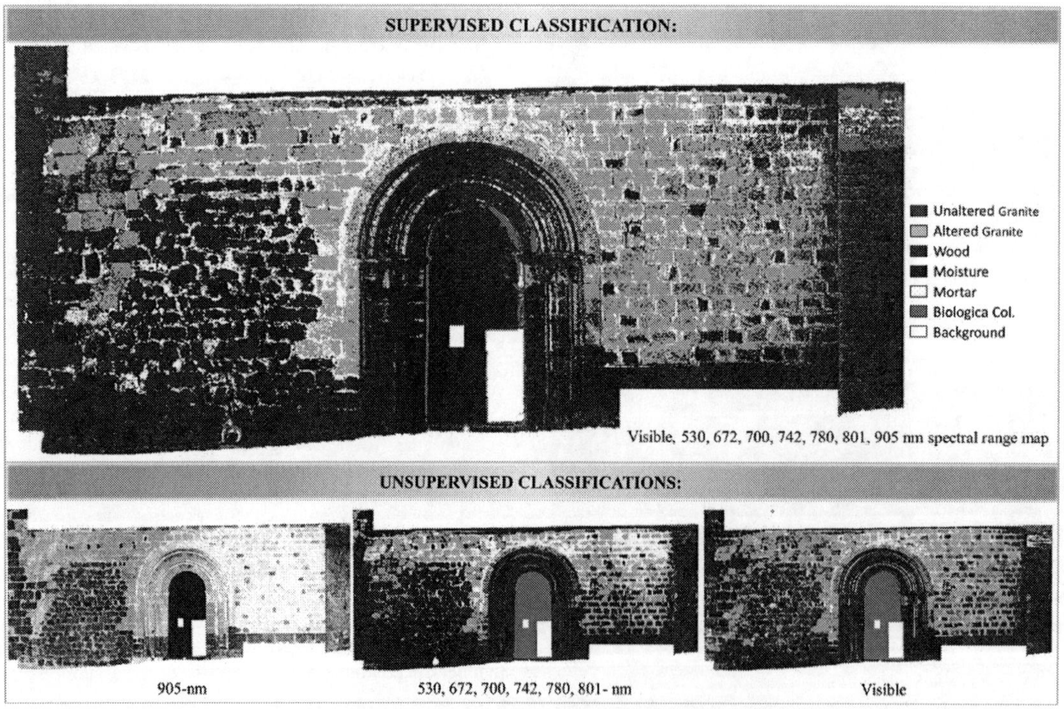

Figure 8. Multisensory map for the 6-classes unsupervised classification.

This contribution has been peer-reviewed.
doi:10.5194/isprsarchives-XL-5-W4-193-2015

4. CONCLUSIONS

The work presented in this paper shows a comparison of the classification results of different sensors in the detection of pathologies in materials of historical buildings façades. By combining the use of two different data acquisition techniques (active and passive), three sensors were examined: a 3D laser scanner, a multispectral camera and a digital SLR camera. The results show the different radiometric responses of the ashlars with different damages (mainly moisture and biological colonization) levels. The classification algorithms used for the classification processes were the Fuzzy K-means and the maximum likelihood (ML) classification algorithm.

A complete description of the workflow followed: data acquisition, pre-processing, orthoimages generation, and the application of two classification algorithms to assess the final results was outlined. Our results show that the most comprehensive sensor for which the best results were obtained is the MCA6. This is possibly due to the advantage of having the highest spectral resolution of the three analysed sensors. However, for the limited registration to a single wavelength, the results from the FARO laser scanner were quite good. Furthermore, geometric models of the study object can be derived thanks its data capture. With these models, physical pathologies (such as fissures, desquamations, etc.) could be analysed and both these and chemical pathologies could be quantified. Therefore, by consider all that issues, it is concluded that the best solution for pathology detection and quantification is a sensor combination where laser scanning as a primary choice. By adding the intensity information to visible or multispectral information results of classification improves in a quantitative and a qualitative way.

As future perspective, additionally to the proposed workflow, the possibility of extend the pathological analysis by the geo-referenced data, will allow analysing global pattern in areas with high density of cultural heritage elements. Therefore, according to the inherent characteristics of each sensor a specific radiometric calibration of each one will allow working in reflectance values instead of digital levels. The final results will be improved due to variable light conditions will not affect them.

REFERENCES

References from Books:

Albertz, J., Kreiling, W., 1989 Photogrammetrisches Taschenbuch. Wichmann.

Bezdek, J. C., 1981. Pattern recognition with fuzzy objective function algorithms. Kluwer Academic Publishers.

Davis, S. M., Landgrebe, D. A., Phillips, T. L., Swain, P. H., Hoffer, R. M., Lindenlaub, J. C., Silva, L. F., 1978. Remote sensing: the quantitative approach. McGraw-Hill, New York.

Fort, R., López de Azcona, M.C., Mingarro, F., 2002. Assessment of protective treatment based on their chromatic evolution: limestone and granite in the Royal Palace of Madrid, Spain. In: Protection and conservation of the cultural heritage of Mediterranean cities. Balkema Publishers, pp. 437-441.

García, F. A. F., 2006. La invención de la iglesia de San Segundo. In: Cofrades y frailes abulenses en los siglos XVI y XVII. Institución Gran Duque de Alba.

Kannan, S. R., Sathya, A., Ramathilagam, S., Pandiyarajan, R., 2009. New robust fuzzy C-Means based gaussian function in classifying brain tissue regions. In: Contemporary computing, Springer Berlin Heidelberg, pp. 158-169.

Kraus, K., 1993. Photogrammetry .Fundamentals and standard processes Vol 1. 4th edition. Dümmler. Bonn, Germany.

Lillesand, T. M., Kiefer, R. W., Chipman, J. W., 2004. Remote sensing and image interpretation. John Wiley & Sons Ltd.

Marszałek, M., 2004. Deterioration of stone in some monuments exposed to air pollution: a Cracow case study. In: Air polution and cultural heritage. Balkema, London, pp. 151-154.

Mather, P., Tso, B., 2003. Classification methods for remotely sensed data. CRC press.

Price, C. A., Doehne, E., 2011. Stone conservation: an overview of current research. Getty Publications.

Richards, J. A., Richards, J. A., 1999. Remote sensing digital image analysis (Vol. 3). Springer, Berlin.

References from Journals:

Al-amri, S.S., Kalyankar, N.V., Khamitkar, S.D., 2010. A comparative study of removal noise from remote sensing image. International journal of computer science issues, 7(1), pp. 1002-1148.

Armesto, J., Ordóñez, C., Alejano, L., Arias, P., 2009. Terrestrial laser scanning used to determine the geometry of a granite boulder for stability analysis purposes. Geomorphology, 106(3-4), pp. 271-277.

Attene, M., 2010. A lightweight approach to repairing digitized polygon meshes. The Visual Computer, 26(11), pp. 1393-1406.

Buckley, S. J., Howell, J. A., Enge, H. D., Kurz, T. H., 2008. Terrestrial laser scanning in geology: data acquisition, processing and accuracy considerations. Journal of the Geological Society, 165(3), pp. 625-638.

Del Pozo, S., Rodríguez-Gonzálvez, P., Hernández-López, D., Felipe-García, B., 2014. Vicarious radiometric calibration of a multispectral camera on board an unmanned aerial system. Remote Sensing, 6(3), pp. 1918-1937.

Franceschi, M., Teza, G., Preto, N., Pesci, A., Galgaro, A., Girardi, S., 2009. Discrimination between marls and limestones using intensity data from terrestrial laser scanner. ISPRS journal of photogrammetry and remote sensing, 64(6), pp. 522-528.

González-Aguilera, D., Gómez-Lahoz, J., Sánchez, J., 2008. A new approach for structural monitoring of large dams with a three-dimensional laser scanner. Sensors, 8(9), pp. 5866-5883.

Höfle, B., Pfeifer, N., 2007. Correction of laser scanning intensity data: Data and model-driven approaches. ISPRS journal of photogrammetry and remote sensing, 62(6), pp. 415-433.

This contribution has been peer-reviewed.
doi:10.5194/isprsarchives-XL-5-W4-193-2015

Kaasalainen, S., Kukko, A., Lindroos, T., Litkey, P., Kaartinen, H., Hyyppa, J., Ahokas, E., 2008. Brightness measurements and calibration with airborne and terrestrial laser scanners. IEEE transactions on geoscience and remote sensing, 46(2), pp. 528-534.

Lambers, K., Eisenbeiss, H., Sauerbier, M., Kupferschmidt, D., Gaisecker, T., Sotoodeh, S., Hanusch, T., 2007. Combining photogrammetry and laser scanning for the recording and modelling of the late intermediate period site of Pinchango Alto, Palpa, Peru. Journal of archaeological science, 34(10), pp. 1702-1710.

Lerma, J. L., 2001. Multiband versus multispectral supervised classification of architectural images. The photogrammetric record, 17(97), pp. 89-101.

Lerma, J. L., 2005. Automatic plotting of architectural façades with multispectral images. Journal of surveying engineering, 131(3), pp. 73-77.

Lichti, D. D., 2005. Spectral filtering and classification of terrestrial laser scanner point clouds. The photogrammetric record, 20 (111), pp. 218-240.

Lu, W., Tan, Y. P., 2003. Color filter array demosaicking: new method and performance measures. IEEE transactions on image processing, 12(10), pp. 1194-1210.

Tolpekin, V. A., Stein, A., 2009. Quantification of the effects of land-cover-class spectral separability on the accuracy of Markov-random-field-based superresolution mapping. IEEE transactions on geoscience and remote sensing, 47(9), pp. 3283-3297.

Weritz, F., Kruschwitz, S., Maierhofer, C., Wendrich, A., 2009. Assessment of moisture and salt contents in brick masonry with microwave transmission, spectral-induced polarization, and laser-induced breakdown spectroscopy. International journal of architectural Heritage, 3(2), pp. 126-144.

References from Other Literature:

Bourke, P., 1989. An algorithm for interpolating irregularly-spaced data with applications in terrain modelling. In: Pan Pacific computer conference, Beijing, China. 6 pages.

Li, X., Gunturk, B., Zhang, L., 2008. Image demosaicing: a systematic survey. In: Proceedings SPIE 6822, Visual communications and image processing, pp. 68221J-68221J.

Reshetyuk, Y., 2009. Self-calibration and direct georeferencing in terrestrial laser scanning. PhD thesis, KTH - Royal Institute of Technology, Stockholm, Sweden.

Ruiz, L. A., Lerma, J. L., Gimeno, J. 2002. Application of computer vision techniques to support in the restoration of historical buildings. In: The international archives of the photogrammetry, remote sensing and spatial information sciences, Graz, Austria, Vol. XXXIV, Part 3A+B, pp. 227-230.

References from Websites (Last accessed on 2015, Jan. 24)**:**

Coffin, D. 2011. Raw digital photo decoding. https://www.cybercom.net/~dcoffin/dcraw/.

This contribution has been peer-reviewed.
doi:10.5194/isprsarchives-XL-5-W4-193-2015

3D VIRTUALIZATION BY CLOSE RANGE PHOTOGRAMMETRY INDOOR GOTHIC CHURCH APSES. THE CASE STUDY OF CHURCH OF SAN FRANCISCO IN BETANZOS (LA CORUÑA, SPAIN)

A. Pérez Ramos [a] *, G. Robleda Prieto [a]

[a] Dept. of Technology and Graphical Representation, University School of Technical Architecture – University of La Coruña , Rúa da Fraga 27, Campus da Zapateira 15008, La Coruña, Spain – araceli.perez.ramos@udc.es

Commission V, WG V/4

KEY WORDS: Cultural Heritage, Close Range, Photogrammetry, Low cost photogrammetric techniques, Point cloud, Virtualization.

ABSTRACT:

Virtualization using low cost photogrammetric techniques, is often replaced by Terrestial Laser Scanning inside churches. Especially in the case of Gothic churches where light penetrates the interior of the building difficulting shooting in proper condition to perform their restitution. The need to use Terrestial Laser Scaning for indoor virtualization is a significant increase in the final surveying cost. In these cases, the Terrestial Laser Scanning is used to generate dense point clouds that can produce high resolution models. However, many Terrestial Laser Scanners are not able to provide color images or can not reach the quality of images which can be obtained through a semiprofessional camera. So, digital photogrammetry is often used to make these models high resolution textures that Terrestial Laser Scanner based methodology is not capable of providing. This article aims to solved the problem posed by virtualizating Gothic churches indoors. Making that task more affordable exclusively by low cost photogrammetric techniques. The proposed methodology allows obtaining photographs in such a good conditions for virtualizing the target by point cloud. In order to verify the usefulness of the method, It has been decided to apply it to Gothic apse of the church of San Francisco in Betanzos (La Coruña). The equipment used is inexpensive and easy to carry: DSLR camera with 18-135 mm lens, tripod, lights and total station.

1. INTRODUCTION

Nowadays, heritage virtualizing is used as basis for multiple applications: documentation, monitoring, tours and virtual museums, augmented reality (AR) …All them have in common the need of pre-generate a 3D model. Demanded requirements of these models will vary depending on the application, particularly model resolution (Visitini, 2009).

Data acquisition techniques commonly used are laser scanning and photogrammetry. These techniques are often mixed within a project in order to obtain results required or appropriate to the subsequent application of 3D model.

Inside buildings, similar to the case presented, the process commonly used involves using Terrestial Laser Scanners to generate volumes and geometries. However, laser scanner can not provide color images or image quality is not enough if we pursue a high resolution model. So, semi-professional cameras are used to obtain high quality images and build model texture. Methodology that combines laser scanning and photogrammetry provides good results (Visintini, 2009). Nevertheless, laser scanning dependence make it a high cost tecnique not available to many professionals. Furthermore, using both techniques entails doing work twice (Tucci, 2003).

Using photogrammetry as unique technique to survey allows for volumen and texture within a single workflow. (Pérez, 2011) Not to mention costs, as any profesional could afford equipment used.

The main problem in trying to virtualize a Gothic building indoors, or similar features, exclusively by photogrammetric techniques lies in difficulty to find proper lighting conditions, which should be as uniform as possible to get adequate restitution images, especially around windows and gaps opened in building walls (Ginovart, 2014).

This article aims to solved the problem posed by virtualizating Gothic churches indoors. Making that task more affordable exclusively by low cost photogrammetric techniques. The proposed methodology allows obtaining photographs in such a good conditions for virtualizing the target by point cloud. In order to verify the usefulness of the method, It has been decided to apply it to the apse. Since the apse is one of the most complex interior part of these kind of buildings.

2. CHURCH OF SAN FRANCISCO

The church of San Francisco, located in Betanzos, is in the english way to Santiago de Compostela. The convent patron was Fernán Pérez de Andrade “O Boo”, who chose it as nobility pantheon. The only original part that remains of the church is the apse, the rest has undergone many changes througout history. It was part of a former convent that was attached to north wall and had been built in the late fourteen century. Gothic building with many Romanesque details like buttresses or capitals. Materials used are common in the area: granite quarries close to Betanzos, chestnut wood and tile.

This contribution has been peer-reviewed.
doi:10.5194/isprsarchives-XL-5-W4-201-2015

Figure 1. South view of San Francisco church.

Figure 2. Interior view of San Francisco church.

3. METHODOLOGY

Restitution of any item requires good fieldwork planning in order to get a good result optimizing resources. Reducing time spent by caring fieldwork photography strategy and thus, time dedicated to fieldwork by reducing number of images and therefore time image processing by software.

To create a photorealistic indoor 3D model apse, geometry and texture generated must be accurate to reflect ornamental richness in detail.

Materials used are total station Leica TCR370, laser distancemeter Leica DISTO D410, Cannon EOS 550D SLR with 18-135 mm lens, tripod and lighting equipment.

3.1 Pre-planning.

Fieldwork was carried out according to pre-planning. Careful planning ensures success of shooting strategy, reducing processing time and avoid having to repeat visits for data collection. Photographed strategy is planned taken into account geometries and obstacles.

Calibrating camera in advance worths. In this case it has begun with camera calibration using template and it has been improved with later calibrations on 3D models. Camera parameters calculation have been adjusted gradually (on-the-field camera calibration). Best accuracies are achieved by calibrating the camera using a 3D object. Each focal length employed in the field must be calibrated.

3.2 Fieldwork or data collection.

3.2.1 Work enviroment recognition: First, place or select those points that serve as reference to establish our coordinate system and precision control points. Defects or damage from stones as checkpoints and coordinate system. It requires Topographical measurements.

Then, several measures are taken with laser distancemeter to ensure shooting strategy previously established conforms to reality.

3.2.2 Lighting conditions: Second, proper lighting settings are choosen availing lighting equipment. Trying to find uniformity and reduce shadows where posible.

Figure 3. Lighting conditions settings (general).

This contribution has been peer-reviewed.
doi:10.5194/isprsarchives-XL-5-W4-201-2015

Figure 4. Lighting conditions settings (detail).

3.2.3 Photo Capture: When adequate lighting conditions are set, we can begin to take photograhs according to previously established strategy. Canon 550D SRL (18 megapíxels) with 18-135 mm lens is used. Converging shots are much more effective than divergent, so in transition zone wall-dome we ensure covering entire surface with converging shots. Shooting in RAW format with 5184 x 3456 pixels resolution. RAW format allows us to correct the photograph parameters photograph without losing quality

Photos were taken in 18 mm focal lengh, ISO-100, f/11 aperture value and 1/5 s shutter speed. Under these conditions using tripod is necessary to prevent camera shake and noise and to ensure pictures sharpness.

The images acquired for 3D virtualization should overlap, at least, around 60% in order to ensure automatic matching between homologous points.

Photo captured is a crucial point in photogrammetric technique process since it depends largely on the accuracy and finish of the model. Establish the most appropriate shooting conditions for each case requires practice.

3.2.4 Topographical measurements: Uniformly distributed points are measured with total station Leica TCR370 to set coordinate system and control model accuracy.

Unlike methodology that combines laser sacnning and photogrametry, photogrammetry do not need a lot of points to fit images and 3D geometry (Tucci, 2003). Few points (10 in this study case) are enough to set the coordinate system (3 points out of 10) and control model accuracy.

3.3 Office Methodology or information processing.

3.3.1 Selecting and Developing of images: Before starting restitution tasks, those images that will be used are selected avoiding blurs and trepidations. Once selected, photos in RAW format are developed. In such cases, where lighting conditions vary between different areas of item to virtualize, reducing differences in development process is tried (color and exposure), paying attention to details. A suitable development contributes to improve image matching whether automatic or manual.

Figure 5. RAW development example.
Before(left) and after(right).

3.3.2 Restitution software: Data processing includes dense point clouds generating (3.641.273 points), scaling, rotating and translationing of model, building surfaces (635.932 faces) and texturing.

Figure 6. Mesh view with no texture (635.932 faces).

This contribution has been peer-reviewed.
doi:10.5194/isprsarchives-XL-5-W4-201-2015

Figure 7. High resolution model. Level of detail view.

Agisoft PhotoScan Professional 1.0.4.1847 has been used to processing point cloud, mesh generation and textures application. The workflow within PhotoScan is easy, even for professionals who are not used to working with photogrammetry software.

Figure 8. Front apse view.

4. RESULTS

Colored dense point cloud and Digital Surface Model are obtained following described methodology. These two products allow viewing on multiple 3D media. In addition, It provides alternative products: elevations, plans and sections as orthoimagery or vector drawing.

40 photograhs were enough to virtualize the apse. However, increasing the number of photograhs allowed us to eliminate those elements that do not belong to building as imagery, flowers, wires, lamps, speakers, furniture…

Obtained model was successful since it represents in detail ornamental elements of the apse and its texture. In this way, it constitutes a valuable document that registrates current apse conditions useful for future interventions or studies.

The best resolution that can be obtained in virtualized model is equivalent to pictures resolution. In this case the model has a minimun resolution around 8 pixels/cm. It increases in areas where photos were taken from a shorter distance.

5. DISCUSSION

The interior of Gothic buildings has features that makes virtualizing a complex and laborious task. Light entering through windows and other wall holes involves not optimal conditions for photo to restitution. In apses cases, space shape is an added difficulty. Thus, using lighting equipment and careful development become key points in this sort of work.

We have tried to find indoor light conditions as homogeneous as possible. However, it has been impossible to avoid some shadows as It can be seen in textured model.

One of the main drawbacks of this study case was impossibility to remove furniture and other items. This forced us to increase the number of photographs and therefore working time in data preparation and processing.

This contribution has been peer-reviewed.
doi:10.5194/isprsarchives-XL-5-W4-201-2015

Figure 9. Apse vault general view. (Above)
Figure 10. High resolution nerve detail. (Right)

During meshing process, small gaps are automatically closed by software so it does not fit to reality in some areas. Thus, capitals top are not shown in photos leaving a point cloud hole closes automatically by software. This drawback is shared with laser scanning technique so, areas not reached by laser beam are also hollow in point cloud.

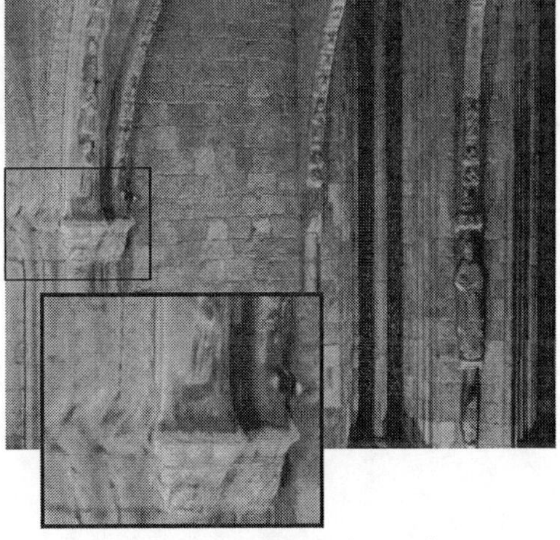

Figure 11. Areas do not fit to reality.

Something similar happens in texture generating, each triangle gets it from photo with the most perpendicular visual. Those areas are not visible in images taken so they have lower quality texture than others.

Using photogrammetry techniques can generate 3D geometry model and its texture within same workflow and within one software. Images adjust automatically to 3D model without making more tasks. Laser scanning would need extra tasks if high resolution model is pursued.

Data capturing by laser scanner is faster indoors, as few stations are required to generate a point cloud inside buildings with these characteristics (Sofocleous, 2006). Although point cloud is generated in real time, it requires cleaning. Furthermore, laser scanning usually requires images taken with an external camera and stablishing common points between images and 3D geometry for generating high resolution models.

Figure 12. Shadows that could not be avoided.

Photogrammetry requires more stations to cover the whole area to virtualize and more time for taking images and generating point clouds by software although, time required for cleaning is lower. Additional images for generating high resolution models are no needed.

In laser scanner working, point cloud density is defined at data captured time and it can be reduced later while, in photogrammetric techniques point cloud density is defined during office work according to needs. Files generated by laser scanning are huge and heavy so, they usually need to be reduce in order to handle it easily.

6. CONCLUSIONS

The documentation inside Gothic buildings is a complex task. Using photogrammetric techniques for virtualizing historic buildings indoors requires planning more careful than laser scanning, enabling surface and texture generation in a single workflow. In addition, photogrammetry with non-metric cameras, as presented , enables any professional could afford to do these works without look to expensive equipment that would never be amortized.

Photogrammetry make possible point cloud density similar to providing by laser scanning. Futhermore, the ones generates by laser scanning not always can reproduce objects color.

Additional lighting is crucial for indoors Gothic virtualization, especially around windows and wall holes. A careful development is also important if we want to optimize work results.

The result of the methodology has been satisfactory so It could be extended to any indoor area of the church and applied to future virtualization projects. Restitution by point cloud through photogrammetric techniques combining accuracy on models and resolution quality textures.

ACKNOWLEDGEMENTS

The authors gratly thanks Liubov Rozhkova (Agisoft Sales manager) and Diana Ovod (Agisoft Tehcnical Support) for providing PhotoScan license. Thanks for fully support are given to Jorge Tomás Cueli López. In addition, we wish to thank Department of Representation and Architectonic Theory (University of La Coruña) for the usage of laser total station and laser distancemeter and Department of Technology and Graphical Representation for putting at our disposal all available resources.

REFERENCES

Ginovart, J.L., Toldrá, J.M., Costa, A., Coll, S., 2014. Close Range Photogrammetry and Constructive Characterizacion of Masonry Gothic Vaults. *Journal of Construction* 13(1),pp. 47-55. http://www.scielo.cl/pdf/rconst/v13n1/art06.pdf (07 Nov. 2014)

Pérez, A., Cachero, R., Navarro, S., Jordá, F., López, D., Lerma, J. L., Martos, A., 2011. Close Range Photogrammetry and Terrestrial Laser Scanning: High Resolution Textured 3D Model of the Capel of The Kings in the Palencia Cathedral as a Case Study. In: *XXIIIrd International CIPA Symposium. Prague, Czech Republic. Journal Geoinformatics FCE CTU.* http://cipa.icomos.org/fileadmin/template/doc/PRAGUE/079.pdf (15 May 2014)

Sofocleous, E., Georgopoulos, A., Ioannides, M., Ioannidis, Ch., 2006. The geometric documentation of the Asinou Church in Cyprus. In: *7th Symposium on Virtual Reality, Archeology and Cultural Heritage, Lefkosia, Cyprus.* http://users.ntua.gr/drag/files/Download/The%20Geometric%20documentation%20of%20the%20Asinou%20Church.pdf (15 May 2014)

Tucci, G., Algostino, F., Borona, V., Chiabrando, F., 2003. 3D Modeling and restoration: From metric to thematic survey. The case study of San Francesco Al Prato in Peruga. In: *XIX International CIPA Symposium. Antalya, Turkey.* http://cipa.icomos.org/fileadmin/template/doc/antalya/142.pdf (15 May 2014)

Visintini, D., Siotto, E., Menean, E., 2009. 3D Modelling of the St. Anthony Abbot Church in S. Daniele del Friuli (I): From laser scanning and photogrammetry to vrml/x3d model. Report on the Joint ISPRS Commission WG/V4 Workshop "3D Virtual Reconstruction and Visualization of Complex Architectures", Trento, Italy http://www.isprs.org/proceedings/XXXVIII/5-W1/pdf/visintini_etal.pdf (22 May 2014).

This contribution has been peer-reviewed.
doi:10.5194/isprsarchives-XL-5-W4-201-2015

TOOTEKO: A CASE STUDY OF AUGMENTED REALITY FOR AN ACCESSIBLE CULTURAL HERITAGE. DIGITIZATION, 3D PRINTING AND SENSORS FOR AN AUDIO-TACTILE EXPERIENCE

F. D'Agnano *, C. Balletti, F. Guerra, P. Vernier

Università Iuav di Venezia, S. Croce 191, 30135 Venezia, Italia - (fdagnano,balletti, guerra, pvernier)@iuav.it

Commission V, WG V/4

KEY WORDS: cultural heritage, documentation, prototiping, low-cost sensor, 3D modelling, augmented reality

ABSTRACT:

Tooteko is a smart ring that allows to navigate any 3D surface with your finger tips and get in return an audio content that is relevant in relation to the part of the surface you are touching in that moment.
Tooteko can be applied to any tactile surface, object or sheet. However, in a more specific domain, it wants to make traditional art venues accessible to the blind, while providing support to the reading of the work for all through the recovery of the tactile dimension in order to facilitate the experience of contact with art that is not only "under glass."
The system is made of three elements: a high-tech ring, a tactile surface tagged with NFC sensors, and an app for tablet or smartphone. The ring detects and reads the NFC tags and, thanks to the Tooteko app, communicates in wireless mode with the smart device. During the tactile navigation of the surface, when the finger reaches a hotspot, the ring identifies the NFC tag and activates, through the app, the audio track that is related to that specific hotspot. Thus a relevant audio content relates to each hotspot.
The production process of the tactile surfaces involves scanning, digitization of data and 3D printing. The first experiment was modelled on the facade of the church of San Michele in Isola, made by Mauro Codussi in the late fifteenth century, and which marks the beginning of the Renaissance in Venice.
Due to the absence of recent documentation on the church, the Correr Museum asked the Laboratorio di Fotogrammetria to provide it with the aim of setting up an exhibition about the order of the Camaldolesi, owners of the San Michele island and church. The Laboratorio has made the survey of the facade through laser scanning and UAV photogrammetry. The point clouds were the starting point for prototypation and 3D printing on different supports.
The idea of the integration between a 3D printed tactile surface and sensors was born as a final thesis project at the Postgraduate Mastercourse in Digital Architecture of the University of Venice (IUAV) in 2012. Now Tooteko is now a start up company based in Venice, Italy.

1. TACTILE EXPLORATION FOR THE VISUALLY IMPAIRED

The loss of sight that characterizes people with visual deficit involves a redistribution of the perception of reality to other senses. In particular, blind people perceive the world mostly from touch and hearing. The tactile exploration of a blind person, regularly called haptic exploration, has some similarities with sight, and Pierre Villey once stated: "La vue est un toucher à longue portée, avec la sensation de couleur en plus ; le toucher est une vue de près avec la couleur en moins, et avec la sensation de rugosité en plus" (Villey, 1954). Tactile perception especially when strengthened from a young age, has the possibility of distinguishing a large number of details even in very small dimensions. The transmission of knowledge for blind people occurs prevalently through touch, through the exploration of tangible objects and the use of alphabetical forms, such as the Braille alphabet.
Simultaneously the perception of the world also occurs through hearing which can give content information (text reading, sonic exploration, etc.) but also spatial exploration. For example the presence of a tray on a table is described in the following way by Augusto Romagnoli: "At the height of my forehead my voice was reflected by a large and non-halogen obstacle, but prominent in the middle. [...] I immediately understood that was the hearing of a circular form, which curve was heard by my hearing thanks to the absence of secondary resonances, which come from well pronounced cuts and from the informal stretching of sound without changes in direction as it happens with the diversity of faces in an angular body." (Romagnoli, 1924). In fact, tactile and hearing perception often help the formulation of mental images of the blind, compensating for sight.
Despite this evidence, the examples of simultaneous union of audio and tactile information are quite rare in communication devices for the visually impaired. In particular, the experiments in synesthetic translation of color through sound operated by the Audiograph (Alty, Rigas, 1998), and the studies on human interfaces based on sounds (Gaver, 1986).
The theme is of high interest because technological help, able to combine touch and hearing to allow a better perception of reality for blind people, can lead the visually impaired to a more complete life and an autonomous exploration of the world.
This need is most relevant in the presence of objects that cannot be touched because of the scale or the value of the piece (such as a church's facade or a painting).

* Corresponding author. This is useful to know for communication with the appropriate person in cases with more than one author.

As it is obvious that fruition of cultural heritage should not be object of discrimination tied to temporary or permanent disabilities, the elimination of barriers for personal disabilities involving some physical operations (removal of stairs, obstacles) could be achieved, while it is much more difficult to reduce the barriers that are mainly perceptive such as those encountered by the people with visual deficiencies.
An ulterior restriction is due to the choice of operating only on works with a tridimensional component (sculptures, objects and architectural works) and not considering two dimensional works of art (drawings, paintings, engravings, etc) which would need a translation from two to three dimension and which goes outside this article's focus.

2. DESCRIPTION OF THE SYSTEM

The system allows, through replicas of works of art made through the use of 3d printing, along with a special ring which is able to read sensors on the replica, to have contextual audio information, relevant to the part of the object that is being touched in a specific moment. In this way an itinerary in museums is offered for visually impaired visitors, but since the possibility of "touching the work of art" is created, the systems becomes useful also for other categories of visitors that favor the tactile dimension, such as children, with the adaptation of content for their needs. The system's goal is to intensify and simplify the relationship with the visitor, always more inclined to using interactive technologies even in the cultural field. Through making available different engineered reproductions of the works for the users, the visitor will benefit from a versatile, multilingual guide instrument, with different levels of depth studying, able to provide for all kinds of visitors, starting from the visually impaired, as an addition to the reading of the work, favoring the experience of a direct contact, even if solely through a 3D reproduction, with the art work.
The system consists therefore in the realization of a technology that transforms mute replicas of a work of art in speaking models, integrating the direct tactile experience with the fruition of localized audio content. This technology, called Tooteko, allows touching a smaller scale model of a museum's work of art, feeling the materials, distinguishing the details with fingertips and listening at the same time, during the tactile recognition, the connected audio illustrations through headphones.
Tooteko puts together touch and hearing in such a way that both hands can explore the object at the same time, while the related audio content is being received by the person that is exploring the surface. A unique object such as a smartphone may detect NFC sensors and give an audio content in return, but the two things cannot happen at the same time while allowing the use of both hands on the surface. Whereas the ring keeps the hand movement free and communicates in Bluetooth mode with a smartphone, a tablet, a pc, an IWB, etcetera.

The system needs some fundamental components such as: the work of art replica to be explored with touch, special NFC sensors that can be inserted inside the object or directly on the surface, an NFC reader able to decode the touched data, and finally a personal audio reproduction system (headphones) that can stream the audio associated with the touched sensor.

Since the audio content is hosted on a cloud, Tooteko offers the opportunity to access content in more languages, to access content tailored for different users, to access music, sounds, or anything that runs on the Internet.

Tooteko technology offers integration between audio and touch, not available elsewhere, thus providing accessibility to audio information by touching any surface that has previously been tagged. Tooteko is a low cost technology that can apply to any three-dimensional surface and make it smart.

The system is made of three elements: a high-tech ring, a tactile surface tagged with NFC sensors, and an app for tablet or smartphone.

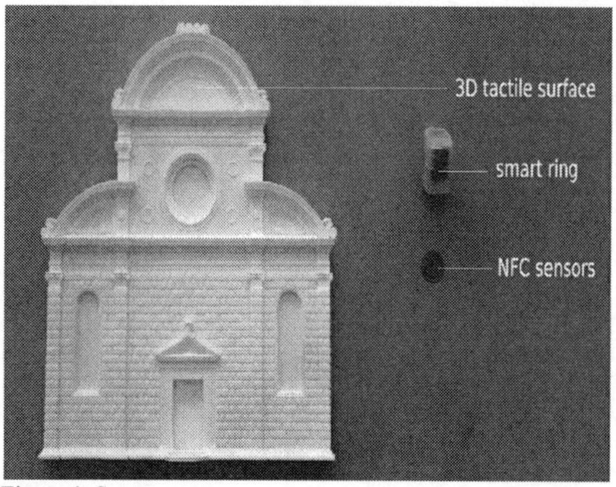

Figure 1. System parts

The ring detects and reads the NFC tags and, thanks to the Tooteko app, communicates in wireless mode with the smart device. During the tactile navigation of the surface, when the finger reaches a hotspot, the ring identifies the NFC tag and activates, through the app, the audio track that is related to that specific hotspot. Thus a relevant audio content relates to each hotspot. The production process of the tactile surfaces involves scanning, digitization of data and 3D printing.

Figure 2. Ring prototype

This contribution has been peer-reviewed.
doi:10.5194/isprsarchives-XL-5-W4-207-2015

3. DATA ACQUISITION

3.1 The case study

Tooteko first applied to the model of the facade of the church of S. Michele in Isola by Codussi, which represents the beginning of the Venetian Renaissance. The model is the result of an integrated detection achieved by the Laboratory of Photogrammetry at the Iuav University of Venice (Balletti et al, 2012).

A facade's survey is a traditional operation in architecture, a field of application for new technologies and instruments as well as an opportunity to verify consolidated methodologies. The creation of 3D models requires a powerful methodology able to capture and digitally model the fine geometric details of such architecture (Remondino, 2011). The point cloud, a very useful instrument in the realization of vectorial drawings and of tridimensional models, in this case derives from both the laser scanning of the application of multi-image photogrammetry and the images obtained through an Unmanned Aerial Vehicle (UAV).

3.2 The reality based survey

The first part of the survey was achieved with a laser scanner at phase difference. The achievement of the facade was reached with the Faro Focus 3D instrument, well known on the market (Balletti C. et al, 2013, Giannattasio et al., 2013) and characterized by a high speed of acquisition (up to 900.000points/sec) and high precision (± 2 mm between 10 and 25 m).

The scan acquisition scheme recalls the monoscopic convergent photogrammetry scheme that requires a central station, one from the right and one from the left to minimize shadow zones (at least those with horizontal trend).

It was chosen to make the scans at highest resolution to obtain a very dense point cloud and therefore permit the extraction of a high quality pseudo-orthophoto. Four scans were taken from the external space in order to contextualize the prospect of the church. Furthermore, four scans were taken from inside the church to investigate on the relationship between the facade and the inside of the church.

In this context, the instrument is particularly effective as it allows to achieve the whole spherical cap, with a horizontal coverage of 360 degrees and a vertical coverage of 270 degrees.

Figure 3 The laser scanning point cloud of the facade of San Michele in Isola

Figure 4. DSM of the facade of the church with "shadow" zone due to entablatures

The sum of the four scans has allowed to have, for the sole facade, a cloud of around 45 million points with a step in the lower part lower than around 1 mm and in the upper part around 5 mm.

The registration of the point clouds was achieved topographically. The targets, recognized semi automatically by the dedicated software Faro Scene, were detected with a total station to guarantee the verticality of the local reference system and for the management of all the scans – both of the inside and the outside – in one unique reference system. The georeferencing was also tuned up by adding some geometrical links. For example, some planes with horizontal and vertical arrangement were identified and used as tie-plane between the different scans.

First result of the survey was therefore the point cloud (fig. 3-4) obtained through laser scanner, in which are obvious the deficiencies determined by the facade's overhangs. In the zones underlying the entablatures you can in fact see a section in which no data was obtained (around 0.4 was the gap of data due to the first entablature and up to 1.2 on the second).

In the case of San Michele in Isola this effect is particularly evident since the church's parvis has a contained depth (round 8 meters) and it is therefore not possible to carry out the achievement of data from larger distances. The characteristics of the facade make these deficiencies particularly serious. There is no way to obtain a complete representation of the shells on top of the lateral aisles.

As a remedy for the deficiencies shown by the point clouds, generally, there is an attempt to utilize mechanical vehicles to raise the point of view (forklifts or mobile platforms) and carry out the photogrammetric photos or other laser scans. In particular environments like in the Venetian one, using such systems becomes complex both for the movement of the vehicles, and because the bearing capacity of the walking surface is quite limited. This last aspect in fact requires the use of lighter systems that do not allow reaching great heights and that are relatively unstable.

For all these reasons, for this detection, the final choice was to use UAV systems (Unmanned Aerial Vehicles) that in the last few years have had a remarkable development in terms of maneuverability and the possibility of stationary flight in accordance with electronic development (Nex et al, 2014).

This contribution has been peer-reviewed.
doi:10.5194/isprsarchives-XL-5-W4-207-2015

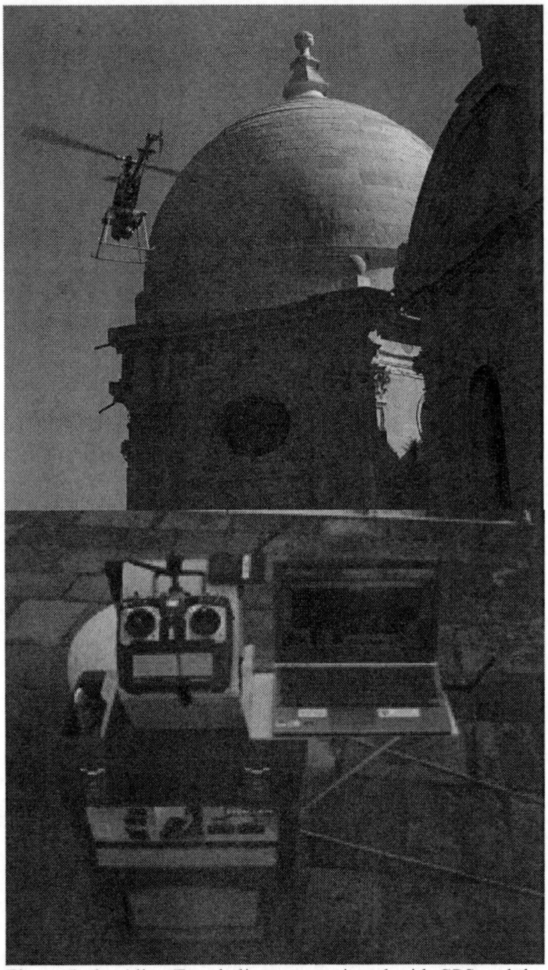

Figure 5. the Align Trex helicopter, equipped with GPS and the Sony Nex camera and its ground control

These systems are characterized by their transportable weight limit, but digital cameras for high angle pictures can be mounted easily.

In a second phase the acquisition was carried out through high angle pictures achieved with low cost and amateur equipment: the Align Trex helicopter (fig. 5), equipped with GPS and the Sony Nex camera with a 24mm lens. The aircraft was flown from the ground following the trajectories that could allow the achievement of the aerial stips suitable for restitution with a multi-image system. The path and the camera's photos were followed and recorded through a pc.

In this case, having to fly so close to the building, manual control of the helicopter was preferred over the autopilot in order to avoid trajectory errors that could be caused by potential signal losses, which could damage the facade. For the photos, working without a pre-set shot time was preferred, observing the angle from the ground and shooting with the remote to have better control of the coverage and the framed subject. The operations, conducted throughout three flying sessions of around 7 minutes each, were conducted within the period of the morning, clearly much less time than what any other system would take.

Figure 6. The photogrammetric model of the upper decoration of the facade in Photoscan

Figure 7. The STL model divided into 4 parts

At the end of the process, the over 500 images taken were used to add to the data already obtained. Among the different possibilities, it was chosen to use a multi-image photogrammetry to obtain a point cloud that would go to complement the already existent one (lo Brutto et al., 2012).

The frames were elaborated with the Photoscan software by Agisoft to obtain a tridimensional model of the facade's upper side, in particular of the zone that was presenting a gap in the model obtained by the laser scanners. The software uses tough algorithms that allow the orientation of the images even though obtained with a configuration that did not correspond with the photogrammetric principles.

For the modelling of the facade, various models were created to complete the various deficiencies. For example, in the model of the part just above the architraves that also includes the shells, 21 images were used. The point model obtained (fig. 6) was scaled not by using a clear measure, but going back to the points achieved with the laser scanner. In this way, the point cloud obtained through photogrammetry was not only scaled but also georeferenced in respect to the laser data. The alignment of the two datasets was occurred with 3mm error.

Other models were made for the facade's details such as the rosettes on top of the curved gables and the statue. At the end of the surveying campaign a unique point cloud was produced, in which the laser scanner data was integrated with the data obtained through the dense stereo matching.

3.3 The 3d printing

In order to proceed with the solid printing it was necessary to build a tridimensional model. The abundance of points and the integration of the extracted photogrammetrical data allowed the realization of a mesh model for the whole facade. Therefore inevitable small deficiencies and zones with higher noise were still compatible with the scale chosen for printing.

Figure 8. The final resin model

The process for the realization of the model is the one already well known in literature.
After the external processing of the data to get all the clouds (laser scanner and photogrammetrical ones) in the same reference system, every scan was decimated and cleaned void of outliers and noise until achieving a one unique point cloud.
The triangulation, the closing of the holes and the essential cleaning operations of the mesh allowed reaching the printing-ready model.
The solid print is made with the Ultra 2 printer by Envisiontech, based on the technology of image projection on curing resins. The model is divided into very thin slices (50 micron width) that are then printed individually and end up composing the whole model. Since the Ultra 2 has a reduced print dimension and the chosen scale was 1:50, the model had to be printed into 4 parts (fig.7), later glued together thanks to resin (fig.8).

3.4 Nfc tags

After having created an exact replica of the architecture as previously described, it is necessary to insert inside the object a series of hotspots that, when touched, stream the unique identifying code that characterizes them. NFC (Near Field Communication) technology is characterized by a passive component (tag) and an active component (sensor). Through the proximity of these two elements (normally less than 3cm) an electromagnetic field is created by the two loop antennas. This electromagnetic field is able to provide enough energy for the tag (passive component) to wake up and read and write the data inside the internal chip. NFC operates at 13.56 MHz on ISO/IEC 18000-3 air interface and at rates ranging from 106 kbit/s to 424 kbit/s.
NFC tags (passive components) are then inserted inside the model, even though an irreversible operation, since they do not require maintenance or batteries.
The NFC tag decoding happens through an extremely compact module based on a Philips PN532 processor for contactless communication at 13.56MHZ. The PN532 supports different operating modes and the system uses a Reader/writer mode supporting ISO 14443 / MIFARE.

Figure 9. NFC tag and sensor

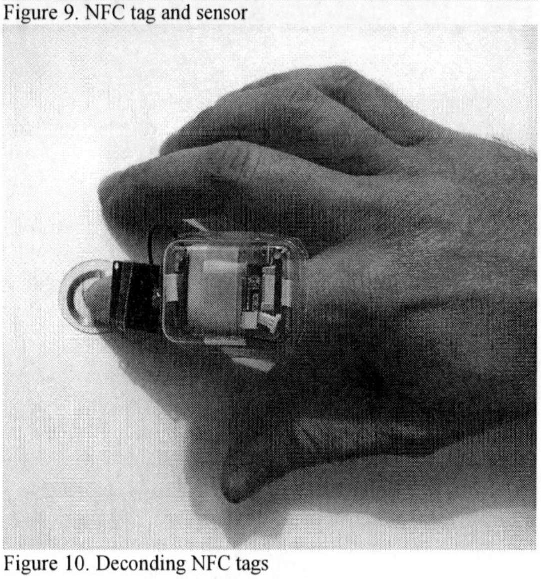

Figure 10. Deconding NFC tags

This contribution has been peer-reviewed.
doi:10.5194/isprsarchives-XL-5-W4-207-2015

The PN532's internal transmitter part is able to drive a reader/writer antenna designed to communicate with ISO14443A /MIFARE, guaranteeing a reliable demodulation and decoding circuitry for signals.

3.5 Processing and sending collected data

When the PN532 reader collects an external tag compatible with ISO 14443A it begins the decoding through the NFC antenna of the tag's content. Inside the tag is inserted a unique ID string that contains the following data: museum ID/work of art ID/part of the artwork touched ID, with an XXX;YYY;ZZZ format.
This information is elaborated by an Atmega32u4 processor by Atmel Corporation. The Atmega32u4 is a processor of small dimension that works at 8 bit with 16 MHz, featuring 32KB self-programming flash program memory.

Figure 11. Hotspots

The string that results from the reading of the tag is then sent via Bluetooth to the device that expects the reception (smartphone, tablet or audio guide). To avoid interference with other rings present inside in the museum, another string containing the ring's unique ID is sent simultaneously. This way the system expecting the reception can filter the arriving data and allow a virtual pairing, without access control, which makes the relation between the ring and the touched tag unique.

3.6 The App

The audio reproduction system is made up by a common smartphone or tablet, equipped with a specific app that has the task of receiving the Bluetooth LE signal, filtering the information, entering the database (internal or remote depending on the situation) and streaming the audio relative to the received string. In substance, the portable device bypasses the compatibility problem of NFC decoding (which is not a standard since most portable devices on the market are still not equipped with an NFC reader or still hasn't added it as a standard, as is the case for Apple products) giving the electronic ring the task of working as a bridge between NFC and portable devices through Bluetooth transmission, which is on the contrary a consolidated standard.
The string transmitted by the ring therefore contains the data relative to the user's identification, to the museum that owns the work of art, to the artwork and to the specific hotspot.
This string activates a remote audio file (which can be provided in audio streaming in the case of personal devices) or locally (in case of an audio guide) that takes into consideration the user's language and his target demographic (adolescent, expert visitor, common visitor or other) to allow a selection of information based on the user's needs.

Figure 12. Data Transmission

4.1 Conclusion

3d printing is experiencing a quick evolution, and allows, through digital data collection and the translation to a point cloud and successfully a mesh, the reproduction of works of art and architectural works in tangible objects at a low price. The following engineering of these models can allow the creation of objects that can be touched and will then stream multimedial information (audio and video) relative to specific points and guarantee an experience of augmented reality.

In a more specific domain, Tooteko wants to make traditional art venues accessible to the blind, while providing support to the reading of the work for all through the recovery of the tactile dimension in order to facilitate the experience of contact with art that is not only "under glass."
Tooteko offers the enjoyment of the work of art in its context, and not in a museum of replicas for the blind, and at the same time it offers the possibility for other categories that favor the tactile dimension, such as children, to approach the work of art, adapting the content to their needs.

This augmented reality can be of particular effectiveness to also permit the fruition from objects of cultural heritage to people who suffer from sensorial deficiencies such as the visually or hearing impaired, allowing a sensible reduction of the barriers.

This contribution has been peer-reviewed.
doi:10.5194/isprsarchives-XL-5-W4-207-2015

Figure 13. Accessible cultural heritage).

ACKNOWLEDGEMENTS

Istituto Ciechi di Milano, Serena Ruffato and Gilda Lombardi, partners of Tooteko.

REFERENCES

References from Journals

Nex, F.; Remondino, F., 2013. UAV for 3D mapping applications: A review. *Appl. Geomat.* **2013**, *6*, 1–15.doi: 10.1007/s12518-013-0120-x

Remondino, F, 2011. Heritage Recording and 3D Modeling with Photogrammetry and 3D Scanning. Remote Sensing, 3(6), pp.1104–1138.

References from Other Literature

Alty, jL, Rigas, D. Comunicating graphical information to blinds using music: the role of contex. Loughborough University Institutional Repository, in Proceedings of the SICHI conference on human factors in computing systems, Los Angeles California, April 18-23/ 1998.

Balletti, C., Adami,A., Guerra, F., Vernier, P., 2012. Dal rilievo alla maquette. Il caso di san Michele in Isola. Archeomatica, 2/2012, pp 24-30.

Giannattasio, C., Grillo, S. M., and Vacca, G., 2013. Interdisciplinary studi for the knoledge and dating of the San Francesco convent in Stampace, Cagliari – Italy (XIII-XXI century). ISPRS Ann. Photogramm. Remote Sens. Spatial Inf. Sci., II-5/W1, pp.139- 144

Lo Brutto, M. L., Meli, P., 2012. Computer vision tools for 3D modelling in archaeology. International Journal of Heritage in the Digital Era, 1, pp. 1-6.

References from Books

Villey, P., Le Monde des aveugles, Corti, 1954

Romagnoli A, 1924. Ragazzi Ciechi, Zanichelli, 1924

AA.VV., Toccare l'arte. L'educazione estetica di non vedenti e ipovedenti. a cura di Andrea Bellini, ed. Armando, Roma, 2000.

Gaver, W., Using Sound in Computer Interfaces, Human Computer Interaction, Vol. 2, No. 2, (1986), 167-

References from Website

Tooteko. http://www.tooteko.com/

Arduino. http://www.arduino.cc/

This contribution has been peer-reviewed.
doi:10.5194/isprsarchives-XL-5-W4-207-2015

3D INTEGRATED METHODOLOGIES FOR THE DOCUMENTATION AND THE VIRTUAL RECONSTRUCTION OF AN ARCHAEOLOGICAL SITE

C. Balletti *, F. Guerra, V. Scocca, C. Gottardi

Università Iuav di Venezia, S. Croce 191, 30135 Venezia, Italia - (balletti, guerra, vscocca, cgottardi)@iuav.it

Commission V, WG V/4

KEY WORDS: Archaeological Site, Documentation, Laser Scanning, Photogrammetry, 3D Modelling, 3D Reconstruction

ABSTRACT:

Highly accurate documentation and 3D reconstructions are fundamental for analyses and further interpretations in archaeology.
In the last years the integrated digital survey (ground-based survey methods and UAV photogrammetry) has confirmed its main role in the documentation and comprehension of excavation contexts, thanks to instrumental and methodological development concerning the on site data acquisition.
The specific aim of the project, reported in this paper and realized by the Laboratory of Photogrammetry of the IUAV University of Venice, is to check different acquisition systems and their effectiveness test, considering each methodology individually or integrated.
This research focuses on the awareness that the integration of different survey's methodologies can as a matter of fact increase the representative efficacy of the final representations; these are based on a wider and verified set of georeferenced metric data. Particularly the methods' integration allows reducing or neutralizing issues related to composite and complex objects' survey, since the most appropriate tools and techniques can be chosen considering the characteristics of each part of an archaeological site (i.e. urban structures, architectural monuments, small findings).
This paper describes the experience in several sites of the municipality of Sepino (Molise, Italy), where the 3d digital acquisition of cities and structure of monuments, sometimes hard to reach, was realized using active and passive techniques (rage-based and image based methods).
This acquisition was planned in order to obtain not only the basic support for interpretation analysis, but also to achieve models of the actual state of conservation of the site on which some reconstructive hypotheses can be based on.
Laser scanning data were merged with Structure from Motion techniques' clouds into the same reference system, given by a topographical and GPS survey.
These 3d models are not only the final results of the metric survey, but also the starting point for the whole reconstruction of the city and its urban context, from the research point of view. This reconstruction process will concern even some areas that have not yet been excavated, where the application of procedural modelling can offer an important support to the reconstructive hypothesis.

1. INTRODUCTION

Today's technologies and methodologies in archaeology documentation allow to record the position, size and shape, as well as generate very accurate and realistic 3D models both in terms of geometry and textures, that are required in any project related to the conservation of CH, forming an important element of the documentation and analysis process (Patias *et al.*, 2006; English Heritage, 2011).
The methods and tools used in the survey of archaeological sites cover a wide range according to the characteristics of the site and of the required precisions considering the different scale of representation. The survey becomes a very complex operation both from the methodological and operational point of view. It requires critical interpretation abilities for a correct comprehension of the surveyed object from which the correct strategies can be derived to obtain the best result from the survey operations (Guidi *et al.* 2004; Cignoni *et al.*, 2008; Remondino *et al.*, 2010; Brunetaud *et al.*, 2012; Chiabrando *et al.*, 2014; Balletti *et al.*, 2014).
Several scenarios could appear, each one requiring an adequate planning to obtain the best results from the survey operations. Particularly, the actual spectrum of methodologies and instruments is extremely wide: it goes from the direct survey, a simpler but fine and useful method, to the more sophisticated known as indirect measurement. Often, the multi-resolution approach and the application of different techniques and methodologies (such as photogrammetry, laser scanning, topographic survey, etc.) provide better results in terms of accuracy and optimisation of the final product (2D or 3D).

2. THE STUDY CASE

The archaeological investigations that have been conducted on several occasions in the area of Sepino since 1820 and that are still in progress, have outlined a strong topographical fragmentation of the cantonal settlements and, especially in recent decades, have focused on the shaping of their respective and articulated forms of integration. The structural evidence of Altilia (and later the municipal urban plant of *Saepinum*), the fortified settlement of Terravecchia and the sanctuary of San Pietro di Cantoni have always been considered the keystones of a sophisticated occupation, extended and stable in the only apparent contradiction between small settlement with the function of magnetizer and temporary centre of gravity, and sprawling and scattered settlement.
The easy and transparent identification of Terravecchia (Fig. 1a), at a height of 953m above sea level, with the ancient city of

* Corresponding author. This is useful to know for communication with the appropriate person in cases with more than one author.

This contribution has been peer-reviewed.
doi:10.5194/isprsarchives-XL-5-W4-215-2015

Saepinum, at the centre of important events during the third Samnite war, has always raised curiosity, suggested and urged cognitive interventions through surveying, recognition and excavation actions (Colonna, 1962; Scerrato, 1981). However, the difficulties of carrying out the interventions due to the steep slopes, the absence of useful connections, the spontaneous reforestation and in short, due to the marginality of the place, have not allowed an extensive, detailed and systematic exploration until recent years.

The sanctuary of San Pietro di Cantoni (Fig. 1b) is located along the north-eastern side of the Matese, at a low hill (altitude 666m above sea level), and is configured as a point of balance and also as a topographical point, between high areas used for defence and valley areas destined to the trade and the production within the community and the canton of *Saepinates*.

Figure 1. Terravecchia (a) and San Pietro di Cantoni (b)

The complex is located on a large artificial terrace, with a polygonal wall inside of which there are the remains of an Italic monumental temple (prostyle tetrastyle *in antis*), built around the middle of the second century B.C. (Matteini Chiari, 2004) and accessible by means of a paved structure culminating in a staircase located along the North-eastern angle of the podium and orthogonal to it.

Saepinum (Fig. 2), the city of the plain (altitude 553m above sea level), is born, as suggested by the same name, from the root of the verb *saepio* ("fence"), as a place to store and park the transhumant flocks along the *tratturo* Pescasseroli Candela. Enclosed in a quadrangular area of about 12 acres from the city walls, made in *opus quasi reticulatum* and with a total length of 1,270 m., is, according to what is documented by the investigations of the excavation performed several times within the city limits, largely vital and functionally organized since the second century. B.C. as it was a small proto-urban community centre. However, the city lives its period of maximum building fervour and consistent urban definition only during the Augustan age, a few decades after the bestowal of the municipal statute within the *Regio IV augustea* (Matteini Chiari *et al.*, 1993).

3. THE SURVEY CAMPAIGN

A first survey campaign, which involved the use of topographic instruments and terrestrial laser scanners and of a multi-rotors hexacopter for the acquisition of aerial photogrammetric photos took place in July 2013 in the archaeological site of Terravecchia di Sepino. The intervention, characterized by absolutely positive outcomes despite the difficulties dictated by the limited accessibility of the place, covered three different areas of the excavations: the church of San Martino, located behind the Northern slopes of the settlement, the so-called rectangular tower, located in the vicinity of the acropolis, and a long stretch of walls between the Porta Medievale, one of road access to the medieval village, and Postierla del Matese, a covered pedestrian gate inside the mighty samnitic walls in cyclopean masonry with double curtain. Unlike the laser scanning instrumentation, used with positive results in all areas of the programmed intervention, it was not possible to use the drone in the area occupied by the church of San Martino because of the extremely low hedges of trees that reduced the room for manoeuvre of the drone itself and therefore made the flight operations unsafe. The creation of an adequate aerial photo coverage, to ensure the integration of ground-acquired data, made of nadiral and oblique photogrammetric photos taken in succession, was assured by the use of the drone in the area of the rectangular tower and in between the Porta Medievale and Postierla del Matese by taking five subsequent flight sessions.

For the topographic support of the scans and photogrammetric photos, both aerial and ground-based, we did a special survey, simultaneously and in parallel with the scanning sessions, using for this purpose 14 topographical network stations, used for the entirety of the excavation operations, constructed from a pair of cemented cornerstones, georeferenced with GPS instrumentation in a planimetric and altimetric (global type) reference system, and adopting a flat -coordinate system Gauss-Boaga, functional to georeference all excavation data on cartographic support CTR at 1: 5000.

The use of the same acquisition techniques has also marked the survey campaign, carried out in the summer of 2013 and 2014, in the area of San Pietro di Cantoni: here we carried out simultaneously ten scans with laser scanner and five different flying sessions with the drone with the aim of achieving a detailed photographic documentation of the main structures in the sanctuary.

In this case the almost complete absence of trees allowed to achieve an almost complete photographic documentation of the sacred area through the acquisition of aerial, nadiral and oblique images, through sequences of contiguous strips. The simultaneous survey with GPS instrumentation also allowed to place laser scanning, aerial photogrammetric and terrestrial surveys in a unique reference system.

Figure 2. The archaeological site of *Saepinum*.

A similar survey campaign, carried out once again together with the Superintendence for Archaeological Heritage of Molise (*Soprintendenza per i Beni Archeologici del Molise*) and with the University of Perugia- Urban Planning and Surveying, in the first edition of the Summer School of Surveying and 3D Modeling, held in Sepino between the 20te and 30th July 2014, has affected a large portion of urban and extra-urban areas of the Roman town of *Saepinum*. The aim is to provide a 3D model of the entire urban plan that represents the actual condition and that suggests a documented and possible reconstruction in its various stages of life. The result of this work, started for cognitive and informative reasons, aims at the

This contribution has been peer-reviewed.
doi:10.5194/isprsarchives-XL-5-W4-215-2015

preparation of future digital papers fully accessible to the visitors of the archaeological site and aimed at facilitating the reading and understanding of the structures autonomously with the aid of mass mobile devices. For this reason we have undertaken the survey of the major urban infrastructures, which is to say the forensic area, the theatre, the commercial and residential buildings alongside the *decumanus* as well as the layout of the two main urban roads (*Cardo* and *Decumanus Maximi*).
The surveying activities were carried out also in two of the four main gates of the town, Porta Boiano and Porta Benevento, and also just outside the city walls on two tombs which are the *ara* mausoleum of *Publius Numisius Ligus*, the tholobate mausoleum of *Caius Ennius Marsus* and also the blocks of grade plans of the base plinth of a third *ara* funerary monument which was discovered during the excavations of the last century just a few metres away from the mausoleum of *Caius Ennius Marsus*.

4. THE REALITY BASED 3D DIGITIZATION

4.1 Terrestrial photogrammetry

In the survey of the archaeological sites, in order to allow future analysis and interpretations, it is fundamental to have a thorough and precise documentation of monuments. The photogrammetric technique is the right answer to this as it allows to acquire precise metric and radiometric data. With a digital camera and software based on *ad hoc* algorithms it is possible to carry out the survey of the analysed objects and to reconstruct the 3D model in a digital environment (Gonizzi Barsanti *et al.*, 2013).
The applications of the photogrammetric survey can range from a big to a small scale: in this case we have decided to use photogrammetry both for the survey of archaeological monuments and the digital acquisition of some smaller finds that are in the Museo Provinciale Sannitico of Campobasso and in the Museo Città di Cannara (PG). In order to obtain complete metric data, some monuments were surveyed through laser scanning and the digital terrestrial photogrammetry. Merging the photogrammetric and the laser scanning data nowadays is an established methodology and it is used not only because of the thoroughness of the geometric survey, but also to improve final radiometric data. Using a digital camera and software based on algorithms of Structure from Motion (SFM) and of Dense MULTI VIEW 3D Reconstruction (DMVR) it is possible to estimate the parameters of the internal and external orientation of the photograms and thus re-create the 3D model of the surveyed object (Balletti *et al.*, 2014). The SFM software used during the Summer School of Surveying and 3D Modeling and for the subsequent data elaborations, were mainly Agosoft Photoscan and 3DF Zephir Pro.
For the acquisition of the photograms of monuments we have used the digital camera Canon EOS 5D Mark with a focal length of 16mm. The camera has a CMOS Full Frame sensor of 36x24 mm with a maximum resolution of 5616x3744 pixel. The images, in this case, have been acquired with a resolution resampled to 2000x3000 pixel that has permitted to obtain not too heavy photograms (file-dimension wise), nonetheless maintaining a good quality for the texture of the model.
Due to the geometrical complexity of the analysed objects, in order to cover the entire surface, it was necessary to acquire a high number of images: the Mausoleum of *Caius Ennius Marsus* with 27 photo (Fig. 3) and of *Publius Numisius Ligus* (48 photo), the Porta Benevento (71 photo) and Boiano (125 photo) have been acquired with the photogrammetry. Images have been elaborated using one of the main commercial software of Structure from Motion: Agisoft Photoscan.

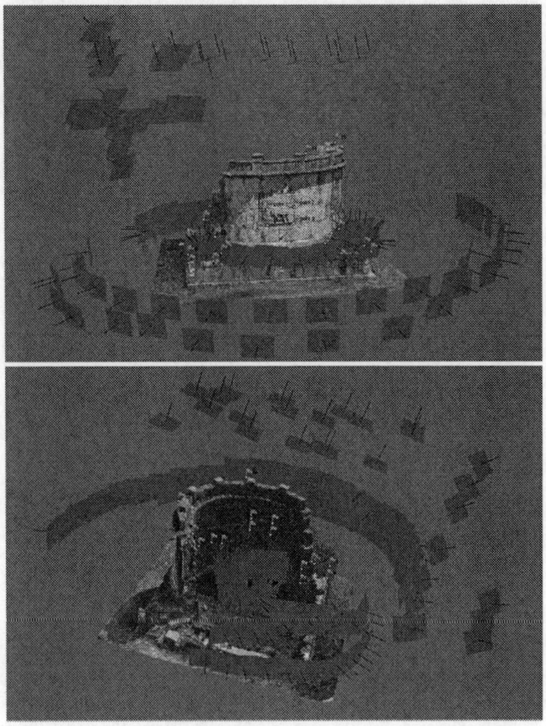

Figure 3. Photogrammetric model of *C. E. Marsus's mausoleum.*

Through an algorithm that matches the same points in the images, it is possible to trace the position of the camera and the geometry of the scene; in this way it is possible to visualize a sparse cloud, composed of all the points used in the matching phase. Right afterwards, the dense cloud and the triangle mesh have been created thus obtaining the 3D models of the two gates and of the mausoleum. For the phase of absolute orientation, i.e. to georeference the models in the reference system of the survey of the entire archaeological site, we have identified some points on each of the four monuments to which we have associated the coordinates obtained through topography.
After having processed the photograms, the precision of the oriented final models was less than one centimetre, therefore they were adequate for a detailed architectural representation.
The first results obtained through the applied procedures have been on one side the creation of orthophotos – extremely important for the bi-dimensional redesign – and on the other the creation of a 3D polygonal mesh, useful for rebuilding the whole archaeological site in its three dimensions in a virtual environment.
The photogrammetry technique has been used also to analyse in depth some smaller findings. In this case, we have used a digital camera Nikon D5100 with a sensor CMOS 23.6x15.7mm and a maximum resolution of 4928x3264 pixel. For the acquisitions we have used a focal length of 18 mm. Given the geometry of the objects, it has been used a scheme of radial photo with convergent optical axes. Moreover, in order not to have areas of shadow in the final model, the images have been acquired at three different heights, covering the whole surface object of the analysis.
In this case, the objects of the survey were some findings from the site of the sanctuary of San Pietro di Cantoni of Sepino, mainly of terracotta and bronze. In particular, the survey was

This contribution has been peer-reviewed.
doi:10.5194/isprsarchives-XL-5-W4-215-2015

carried out on two small votive bronzes of goddesses, a *thymiaterion,* a male head of terracotta, a bronze *flabellum* handle and a bronze jug with a high handle (Fig. 4), some tanagrines and a two-faced herms made of Luni marble. In this case the images have been processed with Agisoft Photoscan, with the only difference that, since it was not possible to acquire the coordinates of certain points topographically, the objects have been scaled on a measured distance.

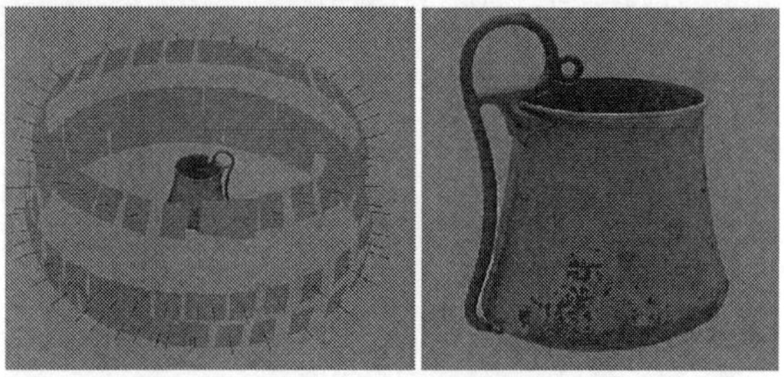

Figure 4. Photogrammetric model of a bronze jug.

The 3D models that resulted from this could be particularly useful for a virtual museum as they would allow a single visitor to interact independently with the exhibited objects. Thanks to the digital fruition of data, it will be possible to reposition in the original place the most important findings, read the historical information and analyse their shape and plasticity in the three dimensions.
The introduction of multimedia contents in a museum allows the user to interact autonomously with the information given, according to his interests and needs. The three-dimensional graphic and the virtual reality have increased the perceptive universe of fruition of information and has created new ways of approach and interaction with the contents. The need of thinking about new solutions for the communication in the museum goes together with the development of recent methodological and technological tools, whereas digital technologies can, in many cases, give answers appropriate to the needs of the museums to adapt to the changes of the society and of the communication methods. The virtual fruition of the archaeological data, widely debated in the academic world, has an increasingly important role in the establishment and encoding of communication processes that facilitate the scientific divulgation and in the planning of tools aiming at a more effective, intuitive and captivating communication of information. The creation of reproductions and maquette could allow the trial of alternative ways of divulgation, prerogative of a non-specialized public or of specific groups of visitors with certain handicaps. An alternative to the traditional plaster casts is the prototyping and the solid printing, that allow to produce very accurate copies and reproductions of the selected models, without the need of creating a direct contact with the source, thus avoiding the chances of damaging very delicate and fragile surfaces. The possibility of integration between the traditional museums and the digital tools available at very competitive prices thanks to the technological development can become a way to improve the cognitive processes and the scientific divulgation – which is actually the final aim of each research project.

4.2 UAV photogrammetry

In the site of *Saepinum* various tests have been carried out on the monuments, following different work-flows. The data obtained have been confronted with those acquired through the more traditional system for the aerial close range photogrammetry (cameras calibrated on professional drones) and/or laser scanners.

In the past few years, the development of photogrammetry with drone has developed the applications in the close range, managing to combine the aerial photogrammetric survey with the terrestrial one, introducing new low-cost possibilities compared to the traditional aerial technique. The development of software solutions for the extraction of point-clouds from groups of non-oriented images resulted in a great level of attention from the geomatic scientific community and also from external users, such as archaeologists. In the surveying campaign we have adopted two devices for the aerial-photogrammetric acquisition (Fig. 5): two professional UAV (traditional type), created by the company Neutech s.r.l, with calibrated cameras and two experimental Parrot Ar. Drone drones, normally used for recreational purposes, that are less expensive and easy to use through applications on tablets and smartphones. On the Parrots we have installed a Go Pro Hero3 Black Edition camera (Ballarin, 2014).

Figure 5. The two different UAV's systems used.

The first professional drone used is a hexacopter with a flight autonomy of 10-12 minutes, which can obviously vary according to the wind and the weight of the load lifted (max 250 – 300 gr). The UAV uses a GPS sensor with a planar array and other integrated sensors (a triaxial accelerometer MEMS, three MEMS gyroscopes and a triaxial magnetic sensor and a pressure sensor at 12 bits) to determine the position and the balance. The system allows to register the telemetric data on a micro SD on the drone and also to transmit to the ground-station the speed of flight, the altitude, the distance from the pilot, the GPS position and above all the battery level. Moreover, it is possible to transmit in real time to the ground-station what the camera focuses, in order to facilitate the operator both in the navigation and during the manual acquisition of the photograms. The second drone is a quadricopter with eight rotors, counter-posed two by two and characterised, like in the previous model, by GPS and other integrated sensors. The recording of the flight data is carried out in the same way as for the hexacopter.
On both drones, a Canon Powershot S100 was installed, a compact and light camera (198 gr) with a good resolution (12Mpx). The camera has CMOS 1/17'' sensor, with dimensions of 7.44 x 5.58 mm, pixel of 1.86 m. The 5x zoom allows to obtain an equivalent focal length that varies in a range between 24 and 120 mm.
As far as the low-cost system is concerned, Parrot + Go Pro, the action cameras are more widespread and have been used more and more during extreme activities. Their photogrammetric use though has not been easy since they could not supply high resolution images and videos and additionally their geometry is far away from the theoretical model of central projection due to their wide angle or fish eye lenses. Several approaches have recently tested the use of the GoPro Hero 3 camera, particularly for the recording and/or photogrammetric use (Balletti *et al.*, 2014).
The methodology of acquisition with UAV photogrammetry has been applied to all the archaeological sites analyzed, with the

This contribution has been peer-reviewed.
doi:10.5194/isprsarchives-XL-5-W4-215-2015

exception of the church of San Martino of Terravecchia, due to adverse environmental conditions.
The schemes of photos adopted in all the study cases envisage the acquisition of sequences of photogrammetric strips characterized by two different structures of view, nadiral and inclined. The choice of this last method is due to the necessity of obtaining the correct correspondence with the data acquired on the ground both with the laser scanner and the photogrammetry. In this way, integrating the different surveying methods, it is possible to obtain complete and accurate metric data, reducing the possible presence of shadows.
For each monument, we have done different flight sessions in order to obtain a redundant number of images: while for the terrestrial photogrammetry we carried out a real planning of the photos using the classical photogrammetric schemes. In the instance of the UAV photogrammetry, we have used the photo's schemes that were more appropriate for the shape of the buildings. As a consequence, in the phase of data processing, we selected the photograms useful for the creation of the different elaborations.

Figure 6. Photogrammetric model of *Saepinum*'s *Forum*.

The choice of the method to use to set up the single structures was influenced each time by their configuration and the conservation status. Whilst for plain structures, such as the forensic area (Fig. 6) and the main road axes, the aerial photogrammetry was preferred to the laser scanning acquisitions on the ground, for more articulated buildings with high surfaces, as for example *Porta Boiano* (Fig. 7), it was fundamental to use different techniques together to obtain a complete survey. The final photogrammetric model, which resulted from the integration of aerial and terrestrial data, has a precision on Ground Control Points between 4mm and 6mm, according to the elaborations.

Figure 7. Photogrammetric model of *Saepinum*'s Porta Boiano.

4.3 Laser scanning survey

In the field of archaeology, or generally in CH, laser scanners (triangulation or time-of-flight) and structured-light systems were very successful, making easier the acquisition process of data related to the geometry and shape of both simple and more complex structures (Balletti *et al.* 2014; Tucci *et al.* 2013; Bonfanti *et al.* 2013).
Generally, active sensors can provide directly and in very short time a huge amount of 3D data that are normally returned as a dense point cloud to which can also be added the radiometric value for each single point.
In the sites of Molise, two laser scanners Faro Focus 3D phase shift (www.faro.com) were used as they are particularly suitable for fast acquisitions (up to 976,000 points per second) and they have a very high precision (±2 mm in a range between 0.6 and 120 m).
The scans were acquired with an angular pitch of 0.035°, in order to guarantee the acquisition of a point every 6mm at 10m of distance, that is to say with a final density appropriate for the ratio 1:50. Thanks to the sensor integrated to the system, it has been possible to acquire the geometry of the structures and also the respective radiometric values.

Figure 8. Point clouds of the *Saepinum*'s theatre

The acquisition of data with laser scanner has been carried out for all the monuments described in the previous sessions. The complexity of the structures required a high number of scans, for example with the Theatre of *Saepinum* (Fig. 8), we carried out 45 scans in total, each one characterized by approximately 24 million points. The high number of scans is due to the high geometrical complexity of the building of which all the elements have been surveyed, including the external ambulatory.
To orientate the scans, we have used the chequered targets, whose coordinates have been acquired topographically. The only exception concerned the targets on the ambulatory walls. In this case the scans have been orientated using the method known as "cloud on cloud": only the two targets located the two opposite sides of the ambulatory have then been linked to the general net through topographic collimation. In this specific case the average precision on the alignment of the scans was of ± 4mm.

4.4 Topographic survey

Due to the "scattering" typical of the location of the analysed sites in the area of Sepino and of the single monuments and architectures of the city of *Saepinum*, it was necessary to create a topographic network that allowed to georeference the different surveys, obtained through total station and GPS. The aim was to define a homogenous reference system where it is possible to

This contribution has been peer-reviewed.
doi:10.5194/isprsarchives-XL-5-W4-215-2015

orientate the acquired data in relation to each surveyed structure.

To georeference the single monuments of the archaeological site of Sepino, we created a local reference system with GPS in RTK mode, using a GPS system Topcon Hiper Pro (Fig. 9). The master station was positioned on a cornerstone on top of one of the main gates of the city (Porta Boiano), while the rover was used to determine the natural target points for the photogrammetric support.

The laser scans were orientated with a topographic support made with the total station Leica TCR 1103 (Fig. 9) and afterwards they were roto-traslated in the local system of coordinates identified with the RTK survey.

Figure 9. The topographic instruments used.

For the orientation of the photogrammetric models, such as the one of the *forum*, realized with the experimental low-cost system, we have identified natural points on the angles of the floor plates of the forensic area, surveyed with GPS. For the forum, contrarily to what was done for other monuments, we wanted to test the accuracies also in the borderline case where the geometry of the object is extremely difficult to reconstruct and the Ground Control Points are not easy to collimate.

5. RESTITUTION AND RECONSTRUCTION HYPOTHESIS

The reconstruction of the historical landscape and of urban contexts is an open question both in the strictly archaeological field and also in Virtual Archaeology; thanks to the computer graphics' potentialities the virtual systems offer an easier comprehension of contexts usually difficult to understand (Guidi *et al.*, 2013; Kontogianni *et al.*, 2013).

The contribution offered by GIS systems and by procedural modelling software was really useful especially due to the past researches focused on Bologna, *Pompeii* and Rome (De Luca *et al.*, 2013; Müller *et al.*, 2009; Dylla *et al.*, 2009). In fact, the GIS systems allow an integrated and functional management of information in the same environment, whereas the procedural modelling software allows in part to automate the modelling process thanks to the application of fitting algorithms.

For this reason we decided to start a similar project on the ancient city of *Saepinum*, starting from the data acquired from the laser scanning and photogrammetric survey and also from the available archaeological documentation. This new research is based on a scientifically accurate and philologically virtual reconstruction of the entire urban complex. For this first part of the project, we decided not to describe the city's development from a diachronic point of view but to focus the research on a specific historical phase: the Augustan age. The archaeological investigations carried out in different areas have demonstrated the construction and the restoration of the most important buildings of the ancient city, such as the forensic area, the *basilica*, the thermal bath, the *macellum*, the sewer, the theatre with the complex of *campus piscina porticus* in the background and the city walls; in this particular case the documents have shown a direct intervention of the two princes of the imperial dynasty, *Tiberius* and *Drusus* (*CIL*, IX, 2443), between the 2 b.C. and 4 a.D. century.

We decided to use the procedural modelling software Esri's CityEngine® for the project's realization; this work will also be supported by GIS systems (Esri ArcGis®) for the integrated management of geographical and archaeological data and also by 3D solid modelling software (3dsMax® di Autodesk) for the modelling, rendering and animation phases.

The test on CityEngine®, specifically created for the reconstruction of urban complexes, allows us to take advantage of the procedural modelling prerogatives. In fact, it will be possible to generate different architectonic models in an almost automatic way; the operator can set up and edit the procedural rules depending on the characteristics of urban areas.

Due to the co-operation of multidisciplinary knowledge, the reconstruction of the entire ancient city will be possible only thanks to the data elaborated from the documentation acquired on-site, confirmed by some hypothesis and by some historical and archive documentation.

It was also necessary to plan different levels of detail for the reconstruction of the *forma urbis* and of the city's urban structure; in fact, the complexity of the city wasn't completely known from the archaeological point of view because of the extension of some urban sectors not yet examined. For example, for the surveyed structures the level of detail is extremely high.

The urban sectors with less historical documentation will mainly benefit from the contribution and potentiality of the procedural modeling. In fact, in this case, the use of automated processes and the generation of different types of buildings, thanks to the modeling procedures of the software, will be fundamental due to the lack of archaeological documentation and historical comparison.

The potentialities of CityEngine® grammar rules of CGA (computer generated architecture) is specifically based on the generation of different types of standardized buildings, that can be reproduced on a wide extension in an almost or automatic way.

6. CONCLUSIONS

This work aims at testing the use of digital technologies in the field of archaeological research, focusing on the definition, the testing and the development of the techniques of indirect survey for point clouds in the different contexts of application, deepening, to this end, different acquisition systems and verifying the effectiveness in isolated or integrated forms.

The systematic and experimental application of 3D survey techniques, undertaken during the campaigns held in Terravecchia, San Pietro di Cantoni and *Saepinum*, allowed in most circumstances a clear acceleration of the excavation and documentation operations, leaving the detailed surveys to a second stage after the operations in the field, as it generally involves a considerable amount of time. We hope to keep going in this direction and make the application and integration of photogrammetric techniques and laser scanning in archaeology a habit and not just an occasional fact, identifying the limits and enhancing its qualities and strengths.

The digital survey currently plays a central role in the documentation and understanding of the excavation contexts, especially in the light of the instrumental and methodological progress concerning the acquisition of on-site data. The deepening of different acquisition systems and the simultaneous

This contribution has been peer-reviewed.
doi:10.5194/isprsarchives-XL-5-W4-215-2015

testing of their effectiveness both in isolated or integrated form is at present a specific theme of research and testing at the Laboratory of Photogrammetry of the IUAV University of Venice, always being aware that the combination of different methodologies may in fact increase the representative effectiveness of the final work and may reduce or neutralize issues related to complex surveys.
The results obtained by the integrated application of laser scanning, photogrammetry and UAV's technologies, combined with more traditional and consolidated methods of documentation related to the archaeological research, can thus be expected to allow significant progress for research itself as well as result in the reduction of the time spent on acquisition of metrological data in the field and a considerable increase in precision, accuracy and adherence to the actual survey.
The broad and varied spectrum of digital tools applicable to the field of cultural heritage can therefore provide a valuable support during key moments of the archaeological investigation, which is why, from the definition of tools and chosen methods, it is essential to proceed to their direct, constant and systematic application on the excavation site, in order to test, verify and validate the actual effectiveness in each phase and of subsequent analysis, interpretation and processing of the acquired data.

REFERENCES

References from Journals:

Balletti, C., Brussa, N., Gottardi, C., Guerra, F., 2014. The documentation and reintegration of a lost past. In: *ISPRS Annals of the Photogrammetry, Remote Sensing and Spatial Information Sciences*, Riva del Garda, Italy, 23-25 June 2014, pp. 49-55.

Balletti C., Guerra F., Tsioukas V., Vernier P., 2014. Calibration of Action Cameras for Photogrammetric Purposes. In: *Sensors*, 14, pp. 17471-17490.

Bonfanti, C., Chiabrando, F., Rinaudo, F., 2013. TLS data for 2D representation and 3D modelling. Different approches tested in the case of San Giovanni in Salluzzo (CN) Italy. In: *ISPRS Annals of the Photogrammetry, Remote Sensing and Spatial Information Sciences*, Strasbourg, France, II-5/W1, pp. 37-42.

Brunetaud, X., De Luca, L., Janvier-Badosa, S., Beck, K., Al-Mukhtar, M., 2012. Application of digital techniques in monument preservation, *European Journal of Environmental and Civil Engineering*, 16 (5), pp. 543-556.

Chiabrando, F., Lingua, A., Noardo, F., Spanò, A., 2014. 3D Modelling of trompe l'oeil decorated vaults using dense matching techniques. In: *ISPRS Annals of the Photogrammetry, Remote Sensing and Spatial Information Sciences*, Riva del Garda, Italy, 23-25 June 2014, pp. 97-104.

Cignoni P., Scopigno R., 2008. Sampled 3D models for CH applications: a viable and enabling new medium or just a technological exercise. *ACM Journal on Computing and Cultural Heritage* 1(1), Article 2.

Colonna, G., 1962. Saepinum. Ricerche di topografia sannitica e medievale. In: *Archeologia Classica*, XV, pp. 80-107.

De Luca, D., Guidazzoli, A., Liguori, M. C., Spigarolo, M., 2013. Open Source e transmedialità. L'esperienza del Cineca nel campo del Virtual Heritage. In: *Storicamente*, n. 9, pp. 2-12.

Gonizzi Barsanti, S., Remondino, F., Visintini, D., 2013. 3D surveying and modeling of Archaeological sites - some critical issues. In: *ISPRS Annals of the Photogrammetry, Remote Sensing and Spatial Information Sciences*, Strasbourg, France, vol. II-5/W1, pp. 145-150.

Guidi G., Beraldin J.A., Atzeni C., 2004. High accuracy 3D modeling of Cultural Heritage: the digitizing of Donatello. *IEEE Trans Image Process* 13(3), pp. 370–380.

Guidi, G., Russo, M., and Angheleddu, D., 2013. Digital reconstruction of an archaeological site based on the integration of 3D data sources, In: *ISPRS Archives of the Photogrammetry, Remote Sensing and Spatial Information Sciences*, Trento, Italy, XL-5/W1, pp. 99-105.

Kontogianni, G., Georgopoulos, A., Saraga, N., Alexandraki, E., and Tsogka, K. 3D Virtual reconstruction of the middle *stoa* in the Athens ancient *agorà*, In: *ISPRS Archives of the Photogrammetry, Remote Sensing and Spatial Information Sciences*, Trento, Italy, XL-5/W1, pp. 125-131.

Patias, P., Sylaiou, S., Sechidis, L., Spartalis, I., Grussenmeyer, P., Meyer, E., Landes, T., Alby, E., 2006. A proposed low-cost system for 3D archaeological documentation. In: *Proceedings of the 7th International Symposium on Virtual Reality, Archaeology and Cultural Heritage*, Nicosia, Cyprus, pp. 145-149.

Remondino, F., Rizzi, A., 2010. Reality-based 3D documentation of Natural and Cultural Heritage sites – techniques, problems and examples, *Applied Geomatics*, Vol.2(3), pp. 85-100.

Tucci, G., Bonora, V., Guardini, N., 2013. Analyses of the factors affecting 3D models resolution – application to the recording of vaults in Sangallo's house. In: *ISPRS Annals of the Photogrammetry, Remote Sensing and Spatial Information Sciences*, Strasbourg, France, II-5/W1, pp. 307-312.

Scerrato, U., 1981. Ricerche di Archeologia medievale a Terravecchia di Sepino (notizia preliminare). In: *Almanacco del Molise*, pp. 109-122.

References from Books:

Dylla,K., Frischer, B., Mueller, P., Ulmer, A., Haegler, S., 2009. *Rome Reborn 2.0: A Case Study of Virtual City Reconstruction Using Procedural Modeling Techniques*. In: Frischer, B., Webb Crawford, J., Koller, D., (eds.), Making History Interactive. Computer Applications and Quantitative Methods in Archaeology (CAA), Oxford, pp. 62-66.

Matteini Chiari, M., Gaggiotti, M., De Benedittis, G., 1993. *Saepinum*, Campobasso.

Matteini Chiari, M., (ed.), 2004. *La Dea, il Santo, una Terra. Materiali dallo scavo di San Pietro di Cantoni di Sepino*. Spoleto, 2004.

Müller, P., Vereenooghe, T., Ulmer, A., Van Gool, L., 2005. *Automatic reconstruction of Roman Housing Architecture*. In: Recording, Modeling and Visualization of Cultural Heritage, Ascona, Switzerland.

This contribution has been peer-reviewed.
doi:10.5194/isprsarchives-XL-5-W4-215-2015

References from Other Literature:

Ballarin M., 2014. *Fotogrammetria aerea low-cost in Archeologia*. Tesi di dottorato, Corso di Dottorato in Ingegneria Ambientale e delle Infrastrutture, Dipartimento di Ingegneria Civile e Ambientale, Politecnico di Milano.

References from Websites:

English Heritage, 2011. 3D Laser Scanning for Heritage. Advice and guidance to users on laser scanning in archaeology and architecture. Available at www.english-heritage.org.uk

Faro Focus 3D, http://www.faro.com/it-it/home

This contribution has been peer-reviewed.
doi:10.5194/isprsarchives-XL-5-W4-215-2015

THE AVIGNON BRIDGE: A 3D RECONSTRUCTION PROJECT INTEGRATING ARCHAEOLOGICAL, HISTORICAL AND GEMORPHOLOGICAL ISSUES

M. Berthelot [a], N. Nony [a], L. Gugi [a], A. Bishop [a], L. De Luca [a]

[a] CNRS MAP (UMR 3495 CNRS/MCC), Marseille, France (nno, lgu, abi, ldl)@map.cnrs.fr

Commission V, WG V/4

KEY WORDS: 3D digitisation, data integration, virtual reconstruction, hypothetical representation

ABSTRACT:

The history and identity of the Avignon's bridge is inseparable from that of the Rhône river. Therefore, in order to share the history and memory of the Rhône, it is essential to get to know this bridge and especially to identify and make visible the traces of its past, its construction, its interaction with the river dynamics, which greatly influenced his life.
These are the objectives of the PAVAGE project that focuses on digitally surveying, modelling and re-visiting a heritage site of primary importance with the aim of virtually restoring the link between the two sides which, after the disappearance of the Roman bridge of Arles, constituted for a long time the only connection between Lyon or Vienna and the sea.
Therefore, this project has an important geo-historical dimension for which geo-morphological and paleoenvironmental studies were implemented in connection with the latest digital simulation methods exploiting geographic information systems. By integrating knowledge and reflections of archaeologists, historians, geomorphologists, environmentalists, architects, engineers and computer scientists, the result of this project (which involved 5 laboratories during 4 years) is a 3D digital model covering an extension of 50 km2 achieved by integrating satellite imagery, UAV-based acquisitions, terrestrial laser scanning and photogrammetry, etc.
Beyond the actions of scientific valorisation concerning the historical and geomorphological dimensions of the project, the results of this work of this interdisciplinary investigation and interpretation of this site are today integrated within a location-based augmented reality application allowing tourists to exploring the virtual reconstruction of the bridge and its environment through tablets inside the portion of territory covered by this project (between Avignon and Villeneuve-lez-Avignon). This paper presents the main aspects of the 3D virtual reconstruction approach.

1. INTRODUCTION

Despite its cultural and tourist notoriety, Avignon bridge still remains a relatively unknown architectural site. From a historical and archaeological point of view, very few research have focused on the story of its raising. The only knowledge about its history just underlines its legendary origins (Marié, 1953): "*...En 1177, a shephered called Benezet was asked by god to undertake the building of a bridge across the river...*" According to this story, the bridge was achieved 8 years later, in 1185.

The purpose of the PAVAGE project was to understand the real chronology of the bridge: when was it built? What was its original shape? Why was it finally abandoned, whereas some bridges built at the same period and in the same area are still active today?

The study of Saint-Benezet Bridge needed to expand the research fields beyond the traditional archaeological and architectural studies. Indeed, a bridge is a specific architectural element as it takes place into a landscape that determines its shape and evolution. Insofar as it takes a role of link between two territorial spaces, its analysis cannot be conducted without a study of its economic and politic context.

The issues raised by this study needed the collaboration of four research teams: geomorphologists for the landscape (CEREGE, Aix-en-Provence), historians for the social and economic context (CIHAM, Avignon), building archaeologists that focused on the bridge itself and the adjoining buildings (LAMMM, Aix-en-Provence), architects, engineers and 3D specialists for elaborating a rich and structured 3D representation of the bridge at several scales and temporal states (MAP, Marseille); plus the Avignon's Archives Service for gathering documentary resources on the topic. In this short paper we'll focus on the main aspects of our 3D virtual reconstruction approach, including the integration of multiple data coming from different domains, at different scales and in different formats.

2. REALITY-BASED 3D DIGITISATION AND RECONSTRUCTION

2.1 Collecting data in a web data base

The first task was to gather all the actual knowledge related to the bridge and its surroundings within a database shared between the whole research team.
The data gathering allowed to bring to light unpublished documents, both illustrated and textual, from medieval to contemporary times. In order to classify and share all the documentary sources according to relevant criteria, we built an on-line database based on PHP/MySQL (see figure 1) allowing researchers involved in the project to upload, document, comment and download documentary sources (texts, iconography, etc..) according to their specific study's object.

This contribution has been peer-reviewed.
doi:10.5194/isprsarchives-XL-5-W4-223-2015

The continuous access to these documents was very helpful for the 3D virtual reconstruction issues, as they described – even sometimes in a very precise way – the former appearance of the area at several scales and temporal states.

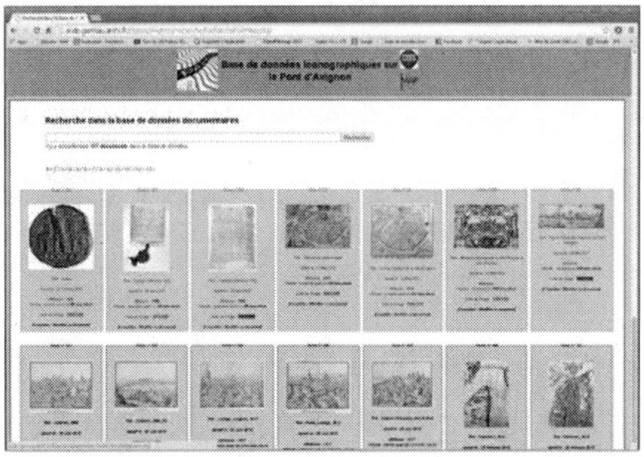

Figure 1. The developed on-line database for archiving and sharing all the documentary sources related to the Avignon's Bridge

Actual documents, such as IGN (Institut Géographique National) aerial orthoimages, topographical surveys from the CNR (Compagnie Nationale du Rhône), etc. were also added to this corpus.
These documents were digitised, geo-referenced and integrated into the used GIS system (ESRI ArcGIS). This overlapping of heterogeneous data (in terms of period, iconography and thematic nature), was a decisive step for programming and carrying out all the on-field investigations. As example the figure 2 shows the overlapping of an ancient map (Carte du Rhône à Villeneuve - J-F de MONTAIGU – 1685) with the CNR survey (in 1969) unveiling some inconsistences concerning the number of bridge's piers to be taken into account for the 3D virtual reconstruction.

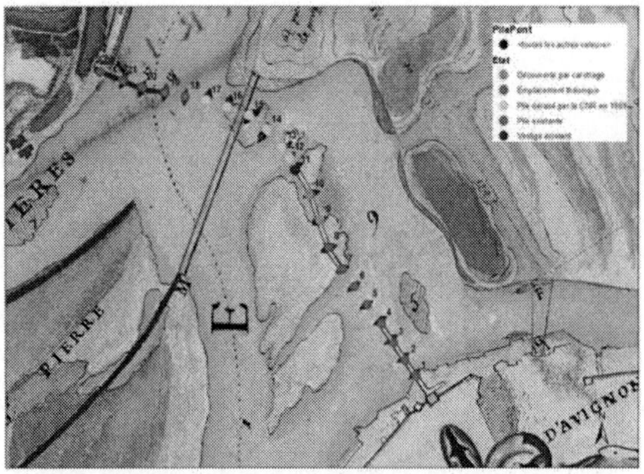

Figure 2. Overlapping of an ancient map (Carte du Rhône à Villeneuve - J-F de MONTAIGU – 1685) with the CNR survey of the piers remains in 1969

2.2 Multi-scales and multi-sensors surveying

With the aim of analysing and representing the environment surrounding the Bridge (50 square kilometres) various surveying techniques were integrated according to the morphological complexity levels to be observed and analysed: terrestrial 3D laser scanning, terrestrial and UAV-based photogrammetry, GPS, Coring, ERT surveying and acoustic river surveying.

2.2.1 3D laser scanning

Four 3D laser-scanning campaigns (based on time-of-flight and phase-shift technologies) were carried out on the main architectural elements: the bridge itself; the Saint-Benezet church on its second pier; the "Philippe le Bel" tower, on which the bridge ended up on Villeneuve side; the castle on the Avignon side. A GPS campaign allowed gathering these surveys into a unique geo-referenced system.

2.2.2 Photogrammetric survey

The photogrammetric campaign was carried out with the aim of providing archaeologists with precise and accurate 2D data. By starting from the application of accurate and rigorous acquisition protocols (Martin-Beaumont et al., 2013), the image processing was based on Apero/MicMac pipeline (Pierrot-Deseilligny et al., 2011) with the aim of generating dense and coloured point clouds (see Figure 3), orthoimages (see Figure 4) and depth maps of all the relevant buildings facades.

Figure 3. Coloured point clouds of Tour Philippe le Bel created by image matching with Apero/MicMac

Figure 4. Orthoimage of east façade of Saint-Bénézet Church, generated with Apero/MicMac starting from images acquired from a boat.

A UAV-based photogrammetric campaign was also leaded on the bridge surroundings, on Villeneuve side. The purpose was to generate a 3D model of the hill close to the former ending bridge (Figure 5).

This contribution has been peer-reviewed.
doi:10.5194/isprsarchives-XL-5-W4-223-2015

Figure 5. Sift point cloud and aerial image orientation of Tour Philippe le Bel surroundings, with Apero

2.2.3 Bathymetric surveying

The CEREGE team carried out a bathymetric surveying campaign (Vella et al., 2013) by basing on a CNR survey of the piers remains in 1969 (before their definitive destruction). Its purpose was to confirm the position (and the approximate shape) of these remains. This survey was made by acoustic camera at a resolution of 10 cm.

By starting from the elevations detected on the bottom of the river, a geo-referencing step allowed to define the relevant borders of the underwater remains of the bridge (see figure 6).

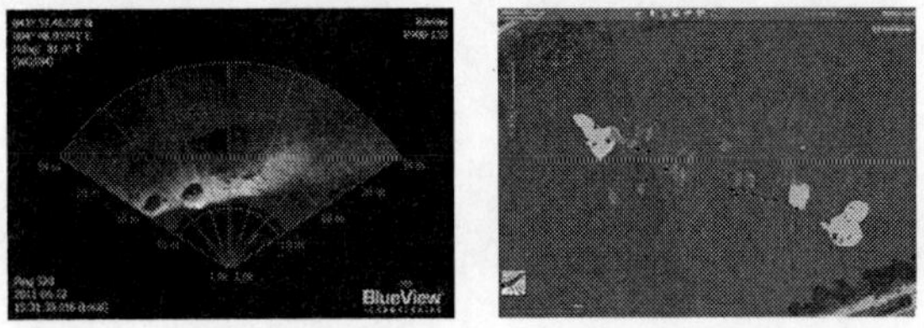

Figure 6. Acoustic camera results, showing the piers residues

2.2.4 Corings

The comparison between geo-referenced ancient maps - mainly Montaigu land registers – and actual ones permitted to identify areas in which corings would be carried out:

- A first campaign was carried on the estimated piers position. Four corings have been made, that confirmed the presence of two piers. A carbon dating made on wood residuals allowed to estimate the age of the layer around 1285.
- A drilling was also made on the Villeneuve side, in order to determinate the underground composition at the estimated former river bed

2.3 An integrated 3D reconstruction approach

The large amount of the gathered data, mainly 3D pointclouds, was post-processed in order to generate intelligible 3D models. The four remaining bridge arches, Saint-Bénézet chapel, as well as Philippe le Bel tower and Avignon castle was thus digitised, following some common interactive geometric modelling techniques (profile extraction, extrusion, primitive adjustment, etc.) for a point-cloud based 3D reconstruction.
This step was essential to provide archaeologists and architects with accurate and handleable geometric models. The models were detailed, according to the research advance and the objects relevance.

2.4 Data gathering in a GIS datawarehouse

All these data were geo-referenced and integrated into a H-GIS (Historical GIS). By integrating such heterogeneous data, in terms of space and time, this database permitted to enhance knowledge and analysis capacity of each document. Thus, the GIS became, besides a way to synthetize data, an important support for the scientific analysis.
The gathered documents were processed according to their nature. Ancient maps were vectorised (land use, urban weft, foot print of remarkable buildings, etc.). Ancient drawings were replaced in the estimated original position and orientation. This allowed to check, by comparing the elements represented on the drawing to the different maps and other drawings, the reliability of the artist.

3. HYPOTHETICAL RECONSTRUCTION

The hypothetical reconstruction focused on two periods: middle 14th and 17th century and covers a 50 square kilometres area around the bridge. This choice was made in order to show two different aspects of the bridge: 1350 corresponds to a (estimated) high activity period, whereas 1650 tally with its end of use (it is definitively abandoned in 1685).

3.1 Ancient Rhône and landscape restitution

The former landscape restitution was made by stacking former maps and actual. By adding the on-field acquired data during the coring campaigns, two DEM were built as depth maps. These models were then transformed into meshes, and post processed with digital sculpting (ZBrush) software (see Figure 7), in order to get a more realistic appearance and UV coordinates. Subsampled models extracted from original ones have finally been generated and added to the GIS.

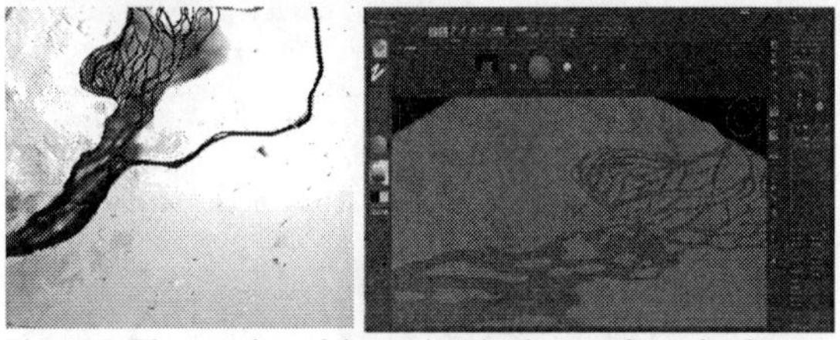

Figure 7. The creation of the ancient landscape, from depth map to 3D model

The missing parts of this study were extrapolated from available data and geologist analysis, the accuracy and density decreasing as moving away from the bridge.
This work has highlighted the modifications that affected the river during the last seven centuries: at the bridge origin, it was essentially made of meanderings, with low flow and small peeble islands – like Durance river today, close to Avignon. Its flow and level were then increased, and the meanderings changed into large channels separated by islands (Vella et al., 2013).

3.2 The Bridge

The research on the bridge itself has been splitted into two parts: its architecture and its path. In order to modify the geometric generation path, according to the archaeological research progress, without modifying the whole digital reconstruction, an animation technic was used: the bridge was skinned by means of 22 bones, each one controlling the position and orientation of one arch (Figure 8). Thus, moving one or more arches was very handy, as the whole bridge was connected to the moving parts. This allowed to better control the parametric relationships between parts and subparts of the overall geometry.

This contribution has been peer-reviewed.
doi:10.5194/isprsarchives-XL-5-W4-223-2015

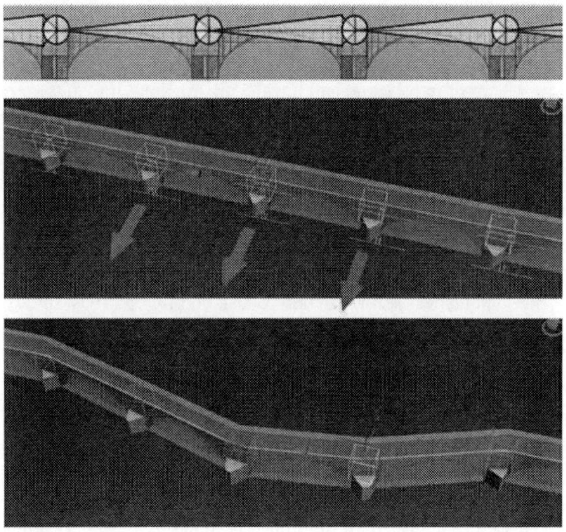

Figure 8. The bones system, created to control the bridge's path

3.3 Buildings

The surroundings buildings reconstruction was the subject of an intense collaboration between archaeologists and 3D modellers. This study was based on the 3D/2D surveying campaigns: 3D models of the actual state state and "pierre à pierre" surveys extracted from orthoimages. This data, combined to the large amount of ancient pictures, and many round trips between 3D models and archaeologists (Figure 8) permitted to build for each building 3 or 4 different temporal states, from the middle 14th century to the 18th.

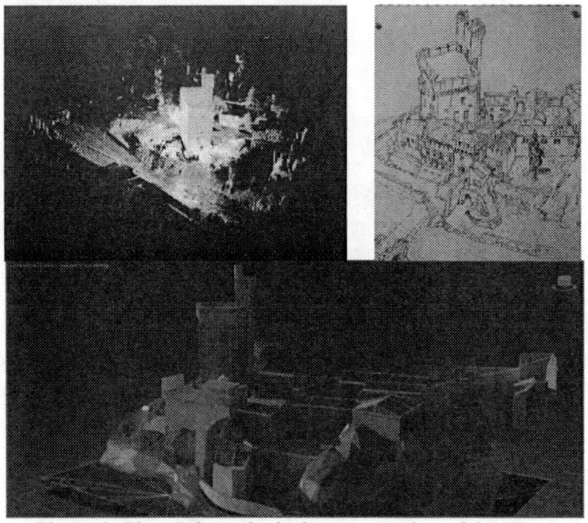

Figure 9. The 3D hypothetical reconstruction of the "Tour Philippe le Bel", based on laser scanner surveying and archaeological analysis

3.4 Urban spaces procedural modelling

Avignon and Villeneuve, as well as surrounding rural group settlements were represented using procedural modelling methods (City Engine; see Figure 9). This work was based on vectorised ancient maps. Urban weft was used as guides for the houses generation.

Geographers and historians established a series of criteria for defining morphological parameters by basing on the description of a set of basic typologies: roof inclination, window distribution, façade size, etc.

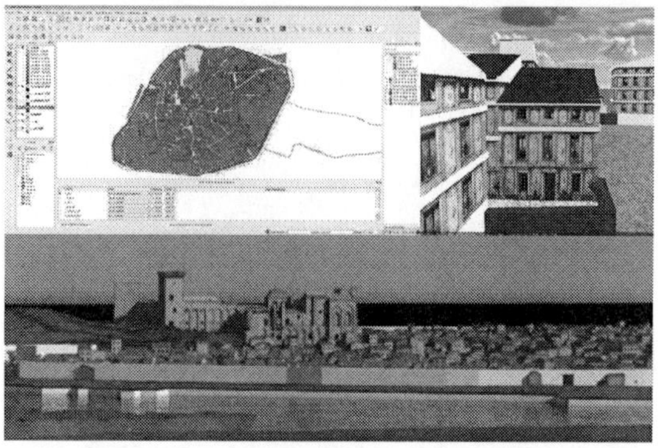

Figure 10. Procedural modelling of houses, based on urban wefts and architectural typology at the reconstruction time

3.5 Final 3D model

All this data, covering multiple representation scales and geometric levels of detail, were integrated within a unique referential using 3DsMax. This choice was essential for managing, visualising and processing a so complex 3D model. The main architectural elements were processed again in order to improve the final visual appearance of the surfaces: 3d details, textures, etc. Two overlapped 3D models were gathered, covering two periods (1350 and 1650).

4. FINAL OUTPUTS

Many 3D renderings were elaborated during the project, describing various aspects of the archaeological and geological reconstruction. Different rendering types were used, from realistic ones to wireframe, as well as some overlapping between ancient iconography and correspondent viewpoints on 3D models. All these images have been used as support for reasoning about the hypothetical reconstructions, as well as for improving the graphic environment of the final 3D scene by adding natural elements by post-production techniques.

The project achieved with a four minutes film, composed by two parts: one camera's path from Avignon to Villeneuve, and one in the opposite way. This video needed a full adaptation of the entire 3D scene in order to improve the photorealistic look of the final rendering. The buildings were detailed with high-resolution textures and various effects (flowing river, vegetation, etc.) were added (see figures 11 and 12). Due to the size and render settings (for the production of a HD video), the sequences of images were processed using renderfarms.

Finally, an augmented reality application (Avignon 3D), available for free on *AppleStore* and *Google Play* for smartphones and tablets, has been created by Art Graphique & Patrimoine starting from the above presented 3D scenes. It is based on 6 panoramic renders, located on both sides of the river, and on the Barthelasse island. The user can switch, for each point of view, between three different temporal states: 1350, 1650 and today. The application is improved by audio and video

This contribution has been peer-reviewed.
doi:10.5194/isprsarchives-XL-5-W4-223-2015

content, in which scientists involved in the project describe various technical aspects related to the project and the bridge itself.

Figure 11. The final render of the Bridge, from Avignon side, in its hypothetical state in 1650.

Figure 12. The final render of the Bridge, from Villeneuve side, in its hypothetical state in 1650.

5. CONCLUSIONS

This project, achieved in December 2014, required four research teams, during three years. It is a good example of how various research fields can cooperate and share information related to an exceptional heritage artefact. A lot of data was gathered and created, with different nature (more than 5 000 images, 150 laser scanner stations, two 1Gb 3D models, 300 archive pictures, etc.).

Various techniques, from on-field 2D/3D acquisition to geological searches have also been used, according to the different research teams involved.

The purpose, as explained in this paper, was both scientific and didactic: the results have been consistently shown to the public by means of various medias: exhibitions, movies, and so on.

On the other side, scientific results – a high definition 50 square kilometres 3D model and a webGIS – are still available for possible future researches.

ACKNOWLEDGEMENTS

On the memory of Michel Berthelot (1948-2014), the first author of this paper.

REFERENCES

Martin-Beaumont et al., 2013. Martin-Beaumont N., Nony N., Deshayes B., Pierrot-Deseilligny M., De Luca L. Photographer-friendly workflows for image-based modelling of heritage artefacts. Archives of the Photogrammetry, Remote Sensing and Spatial Information Sciences, XXIV International CIPA Symposium, Strasbourg, France, 2-6 September 2013.

Pierrot-Deseilligny et al., 2011. Pierrot-Deseilligny M., De Luca L., Remondino F. Automated Image-Based Procedures for Accurate Artifacts 3D Modeling and Orthoimage. Proceedings of XXIIIth International CIPA Symposium, Prague, Czech Republic, September 12 - 16, 2011

Vella et al., 2013. Vella M.A, Ghilardi M., Diouf O., Parisot J.C., Hermitte D., Provansal M., Fleury J., Dussouillez P., Sabatier-Delanghe D., Demory F., Hartmann-Virnich A., Delpey Y., Berthelot M., Bicket A. Géoarchéologie du Rhône dans le secteur du pont Saint-Bénézet (Avignon, Provence, France) au cours de la seconde moitié du deuxième millénaire apr. J.-C.: approche croisée de géographie historique et d'étude paléoenvironnementale. Géomorphologie, Relief, Processus et Environnement, 3, 287-310.

Marié, D.M., 1953. Le pont Saint-Bénézet, premier volume. Histoires et Réalité, 1953.

TESTING THE LOW-COST RPAS POTENTIAL IN 3D CULTURAL HERITAGE RECONSTRUCTION

M. Bolognesi [a, *], A. Furini [a], V. Russo [b], A. Pellegrinelli [a], P. Russo [a]

[a] Engineering Department, University of Ferrara, Saragat 1, 44122, Ferrara, Italy (marcello.bolognesi, alessio.furini, alberto.pellegrinelli, paolo.russo) @unife.it, [b] ing.valentinarusso@gmail.com

Commission V, WG V/4

KEY WORDS: photogrammetry, RPAS (UAV), 3D Digital Model, Accuracy, Low-cost systems, Cultural Heritage, GoPro

ABSTRACT:

In order to analyze the potential as well as the limitations of low-cost RPAS photogrammetric systems for architectural cultural heritage reconstruction, some tests were performed by a small RPAS equipped with an ultralight camera. The tests were carried out in a site of remarkable historical interest. A great amount of images were taken with camera's optical axis in vertical and oblique position. Images were processed by the commercial software PhotoScan of Agisoft and numerous models were realized, each of them was compared with an accurate TLS model used as a reference. The test, despite some problems found, has provided good results in terms of accuracy (average error <2cm) and reliability.

1. INTRODUCTION

In recent years there have been significant advances in Close Range Photogrammetry due to several well known factors, such as, for example, the application of Computer Vision techniques (Structure for Motion), the increase of the quality of low cost digital cameras and the significant development of photogrammetric software (Fonstad et al, 2013; Green et al, 2014). At the same time light RPAS (Remotely Piloted Aircraft Systems) have been equipped with navigation systems more and more advanced and reliable (Nex and Remondino, 2014; Turner et al, 2012; Mancini et al, 2013). The combination of these two technological developments led to a fast and widespread use of light RPAS in photogrammetric survey, with particular regard to Cultural Heritage (Hashim et al, 2012). By consequence a growing number of technicians and professionals from various fields not strictly belonging to the world of surveying, began to apply digital photogrammetry from RPAS. This trend points out the need for simple guidelines for the correct use of low cost photogrammetry performed from very light RPAS (weight less than 2 kg) in order to obtain sufficiently accurate 3D models of the surveyed objects (Koutsoudis et al, 2014). To achieve this goal it is important to test the low-cost technologies currently on the market. For such purpose in this paper we present the results of some experimental tests carried out by means of a Dji Phantom 2 equipped with a GoPro hero3 + Black Edition (Fig.1) and software as the PhotoScan of Agisoft.

Figure 1: RPAS used for the photogrammetric survey

The main advantage by operating with this kind of technology relates to the cost of the equipment, especially if compared with the standard of aerial photogrammetry. The choice of the equipment was made not only to save money, but also with a focus on convenience and ergonomics of the sensor and the RPAS.

The experiment consisted in performing some flights in a test site by acquiring several images with both vertical and oblique camera's optical axes. By means of the commercial software PhotoScan of Agisoft were created numerous models, each of which was compared with an accurate TLS reference model.

2. TEST SITE

The monument selected for the tests is the "Delizia Estense del Verginese" (Fig.2), a renaissance castle located in the province of Ferrara (Italy), already used in recently for assessing the accuracy of 3D models by RPAS and terrestrial photogrammetry in Cultural Heritage surveys (Bolognesi et al, 2014).

Figure 2: The test site, the "Delizia Estense del Verginese" castle (XVth century)

It is an ideal test-site for its historical and cultural importance, and for the possibility to survey it easily with different techniques.

* Corresponding author.

This contribution has been peer-reviewed.
doi:10.5194/isprsarchives-XL-5-W4-229-2015

In order to estimate the accuracy of the photogrammetric survey it is necessary to have a high accurate reference model of the castle. For this purpose, a TLS survey was performed by a Leica C10 laser scanner (figure 3).

Figure 3: The TLS point clouds of the Castle

The model of the castle was realized by several point clouds registered in the absolute reference system. Moreover about 400 control points belonging to the external walls of the building (GCP) were surveyed by a Leica total station from the vertices of a local high accurate geodetic network established around the castle. The reference model was validated by comparing it directly with the positions of GCPs.
As the average deviation between TLS clouds and Total station GCP resulted approximately 10-13 mm, the laser scanner model was considered as a reference in subsequent tests.

3. DATA ACQUISITION

The survey of the castle was performed entirely through DJI Phantom 2 and a GoPro camera. In total were carried out 5 different flights. A first flight at about 20 meters altitude with camera's vertical axis for the survey of the roof and towers. Other 4 flights were instead realized for the survey of the façades (one per side) with flight direction parallel to the façade and oblique camera's axes at a distance of about 5 - 7 meters from the walls. During the execution of the flights were acquired images also in correspondence of the 4 corners of the building in order to provide views which join together the façades (two by two) and part of the roof.
For an integrated processing of all the data such images have proved to be very useful, especially as regards the relative orientation of the frames automatically. The flights were performed in order to obtain an average overlap between frames of about 80%. Table 1 shows for each flight the number of images (IMG), the mean distance from the object and the average GSD.

ID flight/data set	Asset of the camera's axis	IMG	Object distance (m)	GSD (mm)
01	vertical	162	22,4	12,8
02	oblique - west	191	7,0	4,00
03	oblique - east	168	5,6	3,23
04	oblique - south	140	4,8	2,73
05	oblique - north	141	5,6	3,21

Table 1: flights/dataset performed, taken images, GSD and object distances

Furthermore, as already said, by means of the total station were detected about 400 natural points uniformly distributed on the four sides. Some of these points were also used in the early stages of orientation as GCPs.

4. DATA PROCESSING

Data processing was made with the commercial software PhotoScan of Agisoft ver.1.1.0. Given the large number of available images, it was decided to select a reduced data set to be used for processing, so as to optimize the computational time for the same accuracy of the geometrical model (Tab. 2).

ID flight/dataset	Asset of the camera's axis	IMG	IMG selected
01	vertical	162	114
02	oblique - west	191	175
03	oblique - east	168	152
04	oblique - south	140	105
05	oblique - north	141	124

Table 2: Images used for data processing

The datasets were used in different projects, from the basic up to more complex ones, varying the number of images, the camera calibration, the introduction of GCPs and their number, etc. This "modus operandi", step by step, allowed the control of the various phases of the processing. The default computation parameters of the software were assumed (Tab. 3). In a second step the parameters were changed to achieve better results.

WORKFLOW	
ALIGN PHOTO	
Accuracy	Medium
Pair preselection	Disabled
Key point Limit	40000
Tie point limit	1000
IMPORT GCP (GROUND CONTROL SETTINGS)	
Camera accuracy (m)	10
Marker accuracy (m)	0.005
Scale bar accuracy (m)	0.001
Projection accuracy (m)	0.1
Tie point accuracy (pix)	4
BUILD DENSE CLOUD	
Quality	Medium
Depth filtering	Aggressive

Table 3: Parameters and workflow used in Agisoft PhotoScan

STEP ONE: the first project was realized without specifying the type of camera, the calibration parameters and the location of the camera centres. In this phase it was maintained the division of flights, thus creating five separate models. One of these is shown in figure 4. In the figure it is observed a clear curvature of the model, the so-called "bowl-effect".

Figure 4: bowl-effect visible on first alignment made.

The result was actually to be expected since the photogrammetric software has no data to be able to eliminate the remarkable distortion of the GoPro camera. In this

This contribution has been peer-reviewed.
doi:10.5194/isprsarchives-XL-5-W4-229-2015

particular example were used only the photos taken with vertical axis. Differently in the single-side models (figure 5) the "bowl-effect" appears practically absent or greatly reduced.

Figure 5: results of alignment process in façade reconstruction

This result is due to the shorter distance from the object (about 7 m), the homogeneity of images scale and the greater degree of overlap. In addition, images taken with inclined axes at the two corners of the front have allowed a more accurate three-dimensional modelling

STEP TWO: in the second test it was decided to include some GCPs to try to eliminate or greatly reduce the "bowl-effect" also in the project with the vertical axes for the survey of the roof. In total were included 9 GCPs well spread on the ground around the castle. As seen in figures 6, the effect of the distortion continues to be significantly present, distortion also confirmed by residuals on the GCP on average resulting of 65 cm.

Figure 6: bowl-effect still visible in alignment with GCPs.

STEP 3: a further test was made under the assumption to know the calibration parameters of the camera. In this case, the camera was previously calibrated with the software Photomodeler, and the parameters obtained were incorporated into the project PhotoScan, not as rigid constraints but as approximated values, whose correction is estimated by PhotoScan during the processing. This choice is important and is due to the fact that the focusing system of the GoPro is automatic and may change slightly during the flight. In figures 7 and 8 are reported the models obtained respectively with photos with vertical axes (roof) and with oblique ones (single façades). In both models the bowl-effect seems to be eliminated or remarkably reduced. This result was confirmed in a second step in which a few GCPs were included in the computation obtaining the residuals shown in table 4.
Also in this project were obtained "dense clouds" models (figures 9 and 10) that have been compared with the reference TLS one.

Figure 7: alignment made using precalibrated camera (dataset 01)

Figure 8: alignment made using precalibrated camera (dataset 02)

ID flight/dataset	GCP	Error (m)	Error (pix)
01	11	0.057	0.778
02	7	0.022	0.465

Table 4: Residual error of GCPs estimated positions by PhotoScan

Figure 9: Point clouds obtained using precalibrated camera (dataset 01)

Figure 40: Point clouds obtained using precalibrated camera (dataset 02)

The "dense clouds" are very "noisy", especially the one obtained from vertical images. In addition there are missing data in the roof and the façades, caused by suboptimal alignments. However, comparing the clouds with the reference model (figures 11 and 12) are obtained deviations of the same magnitude of the residuals already reported in Table 4.

This contribution has been peer-reviewed.
doi:10.5194/isprsarchives-XL-5-W4-229-2015

Figure 11: Comparison between photogrammetric point clouds and TLS one. Comparison made by CloudCompare (dataset 01)

Figure 12: Comparison between photogrammetric point clouds and TLS one. Comparison made by CloudCompare (dataset 02)

The statistical analysis of such deviations shows that they agree with a Weibull type distribution. The correspondence between the distribution and frequency histograms obtained is in fact remarkably significant. Figure 13 shows the distributions of the two models.

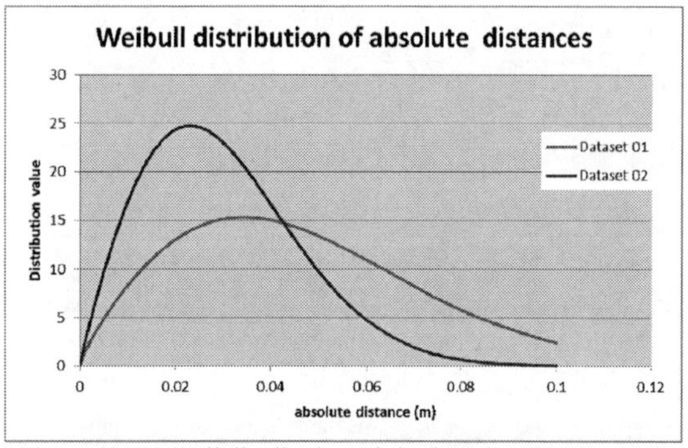

Figure 13: Weibull distribution of absolute distances between the photogrammetric point clouds and the TLS one. Project made from precalibrated camera.

The point clouds of the façades present deviations smaller than that the roof: in the latter model most of the errors is about 3-4 cm, while the models of the façades show average deviations slightly greater than 2 cm. Note how these values are consistent with the residuals obtained on GCP during the orientation phase. The differences between the two types of models are most likely related to the distance from the object: the images of the façades have been acquired at a lesser distance from the images of the roof (see table 1).

STEP 4: to improve results in this step the images have been previously corrected by the distortion using the software Photomodeler. For both models, the residuals of GCPs (table 5) resulted immediately acceptable being lower than the previous ones reported in table 4 (step 3).

ID flight/dataset	GCP	Error (m)	Error (pix)
01	8	0.021	0.253
02	6	0.007	0.191

Table 5: Residual error of GCPs estimated positions by PhotoScan

Even for these projects were created “dense clouds" models which were compared with the reference model. As shown in figure 14, the point clouds on the whole are very disturbed. This effect is probably due to the resampling operated on the images to reduce the lens distortion effects.

Figure 14: Point clouds obtained using undistorted photos (dataset 01)

Figure 15: Point clouds obtained using undistorted photos (dataset 02)

The model shown in figure 15 is better than that one in figure 14 even if also in this case the dense point clouds presents areas of high noise, but the façade of the building is correct, devoid of holes and missing data. Again this result is due to the shorter distance from the object and to the presence of photo taken with inclined optical axis. Also in this case the comparison with the reference model points out deviations that are in agreement with a Weibull distribution (fig. 16).

The values of the deviations are less than those in the previous steps, with significant differences in amplitude of the two graphs, confirming what already found on residuals of orientation of the GCPs (Values in the table 5 below those of table 4). This test with externally calibrated images from the project PhotoScan has therefore provided the best results, even if remains problems on dense clouds: noisy and with some sections incomplete.

STEP 5: During the course of the trial a new software release PhotoScan was made available dedicated to the use of cameras with a fisheye lens, such as the GoPro. Therefore it was made a new test with the original images and by setting the model fisheye. In Figures 17 and 18 are shown respectively the models obtained by the images with vertical and oblique axes.

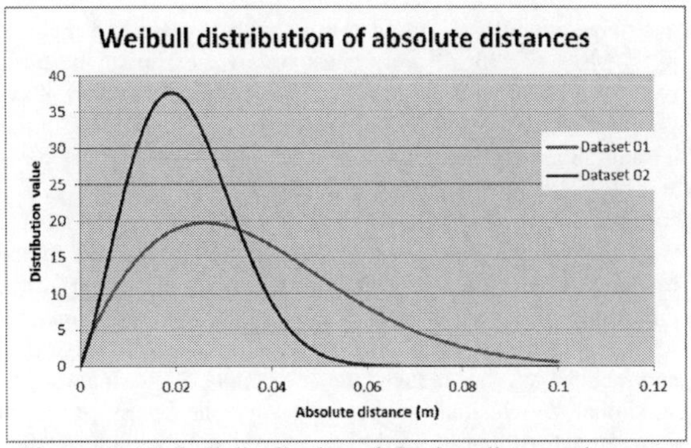

Figure 16: Weibull distribution of absolute distances between the photogrammetric point clouds and the TLS one. Project made from undistorted photos.

Figure 17: Point clouds and camera positions obtained using fisheye release (dataset 01)

Figure 18: Point clouds and camera positions obtained using fisheye release (dataset 02)

As can be seen qualitatively in the figures there is no trace of the bowl-effect. Table 6 shows the residuals of orientation calculated for some GCPs. More models made were made using the images with inclined axis: one for each side (from project 2 to project 5). The residuals obtained are the lowest among all the tests carried out so far. Also in this case were made the comparison with the reference model. The deviations found are shown in figures 20 and 21, while figure 22 illustrates the distribution of these deviations.

ID flight/dataset	GCP	Error (m)	Error (pix)
01	8	0.014	0.207
02	8	0.008	0.357
03	8	0.007	0.222
04	8	0.006	0.362
05	8	0.005	0.475

Table 6: Residual error of GCPs estimated positions by PhotoScan

You do not notice an improvement of the curves of deviation from the previous step, however, the models created are significantly more "clean", as shown in figure 23.

Figure 20: Comparison between photogrammetric point clouds and TLS one. Comparison made by CloudCompare (dataset 01)

Figure 21: Comparison between photogrammetric point clouds and TLS one. Comparison made by CloudCompare (dataset 02)

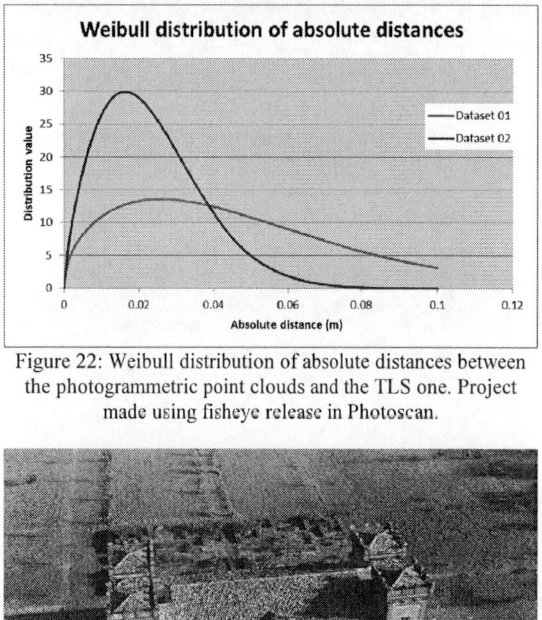

Figure 22: Weibull distribution of absolute distances between the photogrammetric point clouds and the TLS one. Project made using fisheye release in Photoscan.

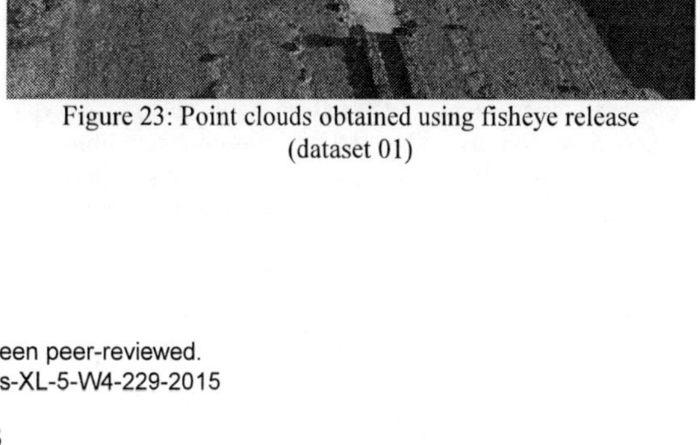

Figure 23: Point clouds obtained using fisheye release (dataset 01)

This contribution has been peer-reviewed.
doi:10.5194/isprsarchives-XL-5-W4-229-2015

STEP 6: using the latest version of the software PhotoScan able to work directly with "fisheye images" has made a final project in which they were introduced all the images, vertical and lateral, for a total of 670 images and 40 GCPs.
In figure 24 we observe the dense cloud obtained.

Figure 24: Point clouds obtained using fisheye release (project made using all dataset simultaneously)

Figures 25 and 26 show the deviations and their distribution, between global model obtained and the reference model. As can be seen, the results are absolutely acceptable.

Figure 25: Comparison between photogrammetric point clouds and TLS one. Comparison made by CloudCompare (project made using all dataset simultaneously)

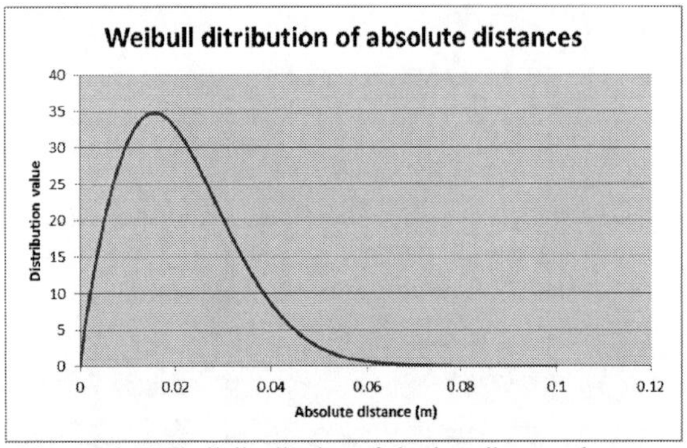

Figure 26: Weibull distribution of absolute distances between the photogrammetric point clouds and the TLS one. Project made using all dataset simultaneously.

5. CONCLUSIONS

At the end of the trial we can definitely say that micro RPAS equipped with low-cost sensors can be considered in all respects an effective tool for the survey of cultural heritage. To obtain accuracies comparable to those of photogrammetry systems "classical", is fundamental the knowledge of the parameters of optical distortion, as well as the use of a minimum number of GCPs (at least 8-10). Furthermore, it seems very important to maintain a close distance of the sensor from the object and to use simultaneously acquired images with vertical and inclined axes. More specifically, the latest tests have shown the good results obtained with the new version of PhotoScan that allows you to directly use images taken with fisheye sensor. In figures 27 and 28 are reported distributions relating to tests carried out so as to better understand the differences.
In figure 27 we observe that the deviations of the model created with the fisheye images (integrated model) compared to the reference model are on average better than the other three. Observing the residuals is known also another important feature, namely the simultaneous use of images with a vertical and inclined axis has reduced the uncertainty of the final model. A similar conclusion can be deduced from figure 28 which shows the distributions for the models of the individual façades.

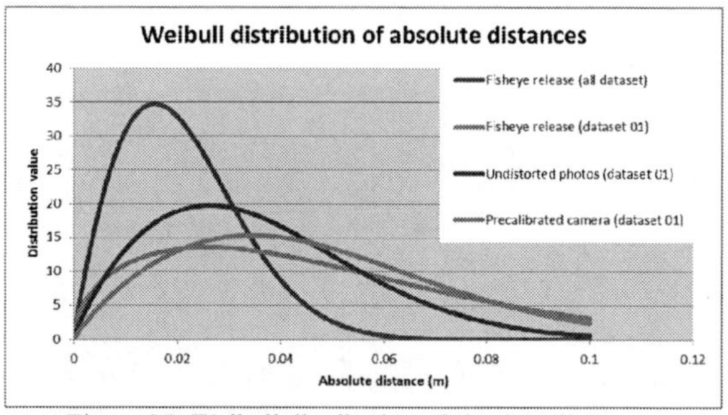

Figure 27: Weibull distribution of absolute distances. Comparison between integrated model and other ones made from dataset 01.

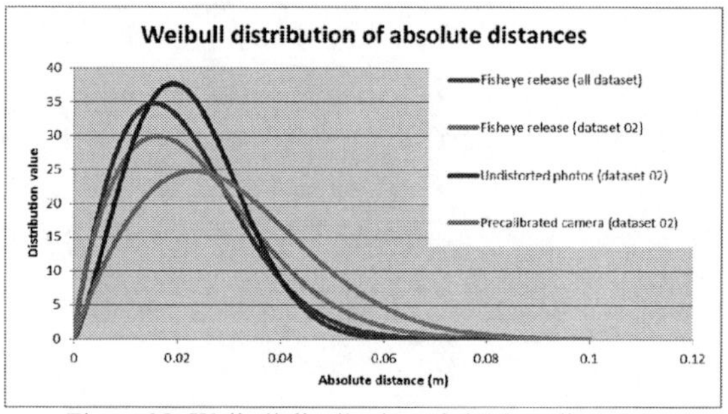

Figure 28: Weibull distribution of absolute distances. Comparison between integrated model and other ones made from dataset 02.

It is appropriate to conclude with a few comments on the modelling of architectural details. Note in figure 29 a detail of the cloud obtained from the integrated model.
Although architectural elements of the prospectus are clearly visible, a precise identification is unfortunately still difficult. Anyhow this type of problem is partly solved by varying the parameters of quality of the "dense cloud" that creates PhotoScan in the last stage of the processing. Figure 30 shows the same detail obtained with a higher density of points: it is easy to observe that the interpretation of the object, and thus the vectorization of the model, is definitely better.
Another important product is the orthographic projection of the model created. In figures 31 and 32 is reported the textured model necessary for creating orthophoto, while figure 33 shows a façade projected as orthophoto.

This contribution has been peer-reviewed.
doi:10.5194/isprsarchives-XL-5-W4-229-2015

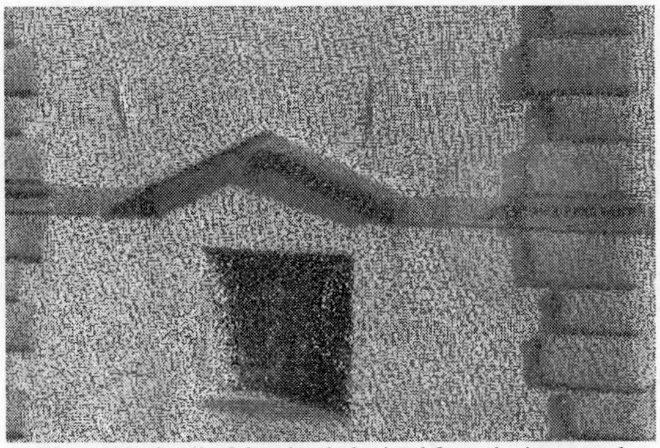

Figure 29: detail of the cloud obtained from the integrated model (medium quality)

Figure 30: same detail of the cloud obtained from the integrated model (ultra high quality)

Figure 31: Front view of the textured model of the castle.

Figure 32: Back view of the textured model of the castle.

Figure 33: Ortophoto of the front façade of the castle.

Figure 34 shows a zoom of the orthophoto: note the quality of the product and its easy vectorization.

Figure 34: a detail of the orthophoto in fig. 33

In conclusion, the potential offered by RPAS photogrammetric low-cost and user friendly systems, are very interesting and they will see a growing trend in the world of the cultural heritage survey. However in order to achieve an adequate level of precision is required compliance with some simple rules that were in part highlighted in this work.

REFERENCES

Bolognesi M., Furini A., Russo V., Pellegrinelli A., Russo P., 2014 Accuracy of cultural heritage 3D models by RPAS and terrestrial photogrammetry. In: *The International Archives of the Photogrammetry, Remote Sensing and Spatial Information Sciences*, XL-5, 2014, pp.113-119.

Green, S., Bevan A., Shapland, M., 2014. A comparative assessment of structure from motion methods for archaeological research. *Journal of Archaeological Science*, 46, pp. 173-181.

Fonstad, M. A., Dietrich, J., T., Courville, B. C., Jensen, J. L., Carbonneau P. E., 2013. Topographic structure from motion: a new development in photogrammetric measurement. *Eart. Surf. Process. Landforms,* 38, pp. 421-430.

Hashim, K. A., Ahmad, A., Samad, Abd. M., Nizam Tahar, K., Udin, W. S., 2012. Integration of low altitude aerial & terrestrial photogrammetry data in 3D heritage building modelling. In: *Proceedings – 2012 IEEE Control and System Graduate Research Colloquium, ICSGRC*, Shah Alam, Malaysia, Article number 6287166, pp. 225-230.

Koutsoudis, A., Vidma, B., Ioannakis, G., Arnaoutoglou, F., Pavlidis, G., Chamzas, C., 2014. Multi-Image 3D reconstruction data evaluation. *Journal of Cultural Heritage*, 15, pp. 73-79.

Mancini, F., Dubbini, M., Gattelli, M., Stecchi, F., Fabbri, S., Gabbianelli, G., 2013. Using unmanned aerial vehicles (UAV) for high-resolution reconstruction of topography: the structure from motion approach on coastal environments. *Remote Sensing*, 5 (12), pp. 6880-6898.

Nex, F., Remondino, F., 2014. UAV for 3D mapping applications: A review. *Applied Geomatics*, 6 (1), pp. 1-15.

Turner, D., Lucieer, A., Watson, C., 2012. An automated technique for generating georectified mosaics from ultra-high resolution Unmanned Aerial Vehicle (UAV) imagery, based on Structure from Motion (SFM) point clouds. *Remote Sensing*, 4 (5), pp. 1392-1410.

This contribution has been peer-reviewed.
doi:10.5194/isprsarchives-XL-5-W4-229-2015

ACCURATE 3D SCANNING OF DAMAGED ANCIENT GREEK INSCRIPTIONS FOR REVEALING WEATHERED LETTERS

A. I. Papadaki [a,*], P. Agrafiotis [a], A. Georgopoulos[a], S. Prignitz [b]

[a] National Technical University of Athens, School of Rural and Surveying Engineering, Lab. of Photogrammetry
Zografou Campus, Heroon Polytechniou 9, 15780, Zografou, Athens, Greece
al.i.papadaki@gmail.com, (pagraf, drag)@central.ntua.gr

[b] Berlin-Brandenburgische Akademie der Wissenschaften
Jägerstrasse 22/23, 10117, Berlin, Germany
sebastian.prignitz@gmx.de

Commission V, WG V/4

KEY WORDS: 3D Scanning, Structured Light Scanner, Structure from Motion, Weathered Letters Revealing, Ancient Inscriptions

ABSTRACT:

In this paper two non-invasive non-destructive alternative techniques to the traditional and invasive technique of squeezes are presented alongside with specialized developed processing methods, aiming to help the epigraphists to reveal and analyse weathered letters in ancient Greek inscriptions carved in masonry or marble. The resulting 3D model would serve as a detailed basis for the epigraphists to try to decipher the inscription. The data were collected by using a Structured Light scanner. The creation of the final accurate three dimensional model is a complicated procedure requiring large computation cost and human effort. It includes the collection of geometric data in limited space and time, the creation of the surface, the noise filtering and the merging of individual surfaces. The use of structured light scanners is time consuming and requires costly hardware and software. Therefore an alternative methodology for collecting 3D data of the inscriptions was also implemented for reasons of comparison. Hence, image sequences from varying distances were collected using a calibrated DSLR camera aiming to reconstruct the 3D scene through SfM techniques in order to evaluate the efficiency and the level of precision and detail of the obtained reconstructed inscriptions. Problems in the acquisition processes as well as difficulties in the alignment step and mesh optimization are also encountered. A meta-processing framework is proposed and analysed. Finally, the results of processing and analysis and the different 3D models are critically inspected and then evaluated by a specialist in terms of accuracy, quality and detail of the model and the capability of revealing damaged and "hidden" letters.

1. INTRODUCTION

During the recent years the significant progress in automating the photogrammetric processing and in popularizing close range scanning technology in conjunction with the effective incorporation of computer vision techniques has enabled scientists to capture accurate high resolution data of the three dimensional world and extract valuable information from them. Consequently this advancement has resulted in the evolution of many other research fields while broadening their horizons and modernizing them. For instance, the digital reconstruction of the surface of complex three dimensional objects with real texture and then the visual representation and dissemination of digital information allow the use of three dimensional models as an educational tool by museums (Gonizzi Barsanti and Guidi, 2013, Soile et al., 2013, Wachowiak and Karas, 2009). At the same time, recording and projecting cultural treasures enables the analysis by many different scientists providing them knowledge.

1.1 Related Work

In archaeology and cultural heritage documentation, a very important research field is the study of ancient stone inscriptions. Their documentation is a valuable tool for the epigraphists, in order to study and -sometimes- decipher them especially when the stone is difficult to read.

1.1.1 Study and Analysis of Ancient Inscriptions: So far, the study and the analysis of ancient inscriptions are mainly based on traditional manual techniques for their geometric documentation and the observations of the naked human eye in front of the stone. The most commonly used method for studying inscribed fragments includes the use of a special type of moisturized paper (squeeze) pressed into the stone, through which they attempt to create an impression of the surface of the inscription (Woodhead, 1992). Other materials used in the literature for the above process are latex or liquid rubber (Beck, 1963). Although the squeezes method produces reliable copies of inscribed surfaces, there are several drawbacks and problems that may rise during their use like the limited accessibility by many researchers, their preservation through time and their restrictive use in over-weathered surfaces. In some instances it is even not allowed to touch and tamper with the inscriptions, as they may be extremely fragile.

Hence non-contact and non-destructive techniques are sought in order to geometrically document them involving picture

* Corresponding author.

capturing under different lighting directions and camera viewpoints (Malzbender et al., 2001), video sequence or a set of 2D images for reconstructing a three dimensional scene (Lerma and Muir, 2014, Cornelis et al., 2001). However, the precision and detail of the obtained reconstructions is often not high enough for capturing details on inscribed surfaces such as letters and symbols. This is especially the case when the inscriptions are weathered and practically illegible. Other more sophisticated methods use laser scanners (Lerma and Muir, 2014, Landon and Seales, 2006), devices that can capture in a very detailed way the inscribed surface. A more cost-effective method that requires only the squeezes is presented by Barmpoutis et al. (2010). Through the aforementioned methods the epigraphists are called to answer questions and make decisions on specific issues including the ascription of fragments, the dating of the inscriptions, the identification of the place of origin, the identity of the letter cutter and the analysis of the lettering techniques. For the methods and the results presented in this paper, we have documented ancient Greek inscriptions carved in masonry or marble from Parthenon Inventory Lists written after 403 BC and we present extensive experimental results from their 3D reconstruction and their analysis.

1.1.2 Outline: The rest of the paper is outlined as follows. In Section 2, is presented the Data Acquisition Methodology including a detailed object's description. Section 3 is divided into several subsections describing the data processing steps that have been followed and the results (mesh generation and alignment in Sect. 3.1, meta-processing and analysis of 3D models along with the results and the evaluation presentation in Sect. 3.2). In Section 4 we conclude while in Section 5 we present our future work.

2. DATA ACQUISITION METHODOLOGY

2.1 Object's Description

Traditiones Parthenonis (handing over lists of the Parthenon) were inventory lists for the items stored in the Parthenon on the Athenian Acropolis. The magistrates of the Athenian democracy began in 434/3 BC, after the Parthenon was complete, to give accounts for the treasures of the Parthenon and Erechtheion. The lists with the items were inscribed on large stones, partially up to 2 m high and over 1 m wide, and stood on the Acropolis so that the accounts could be visible for the public. The last inventories were written around 300 BC. Later, when the stones came out of their primary context, they were reused as thresholds, doorposts and masonry blocks, sometimes they were recut for the secondary use or they fell down and broke. Therefore, only few stones are preserved completely, most of them came fragmentary to us, down to little fragments of only some centimeters diameter. The two main tasks for the editors of these kind of texts are to complete the puzzle by attributing fragments to each other and to decipher badly worn or weathered parts of the inscriptions as best as possible. These two steps are basis of any further research. The *Traditio Parthenonis* IG II^2 1425[1], dating to 368/7 BC, is written on a 2 m tall stone which was once built into the medieval western door of the Parthenon temple. The stone was taken out of the secondary use and is now stored in the Epigraphic Museum of Athens (inv. 7856/7). The account contains, as the other *Traditiones*, a list of items from the storerooms. One example (ll. 97-104):

θυμιατήριον ἐπίχρυσον ὑπόχαλκον, ἵνα [95] τὸ ἄλφα π<α>ρ[α]σεσήμανται, σταθμὸν σὺν τοῖς ἥλοις τοῖς χαλκοῖς XXXHHHH· θυμιατήριον ἐπίχρυσον ὑπόχαλκον, ἵνα τὸ βῆτα παρασεσήμανται, σταθμὸν XXXHꟿΔΔ χρυσίον ἐπίτηκτον ὑπάργυρον, [100] σταθμὸν ∴∵ χρυσίον ἐπίτηκτον ἀπὸ τῶν ἀκρωτηρίων τõ νεὼ τῆς Νίκης, σταθμὸν ∴∵ ΓΗΙΙΙ·χρυσίον ἐπίτηκτον ἀπὸ τῆς ἀσπίδος τῆς πρὸς τῶι νεώι, σταθμὸν ⊦⊦⊦IIIII·	*A censer from copper, overlaid with gold, [95] marked with an Alpha, weight with the bronze nails: 3400 drachmai. A censer from copper, overlaid with gold, marked with an Beta, weight: 3170 drachmai. Silver pieces overlaid with gold, [100] weight: 4 drachmai and 4 obols. A gilded piece from the acroteria of the Nike temple, weight: 6 drachmai and 3 obols. A gilded piece from the shield in front of the temple, weight: 3 drachmai and 5 obols.*

The stones with the *Traditiones* in some cases are "opisthograph", meaning that the letter cutter wrote on both sides to inscribe the complete text on one block of marble. Thus on the backside of the aforementioned stone, there is also text. There is even one more column of text than it was published in the last complete edition of 1927; but the letters of this fifth and last column are very faint and hardly readable. Therefore, as the first four columns of text are preserved relatively well and as this kind of text (lists) is – compared to other kinds of text – more "predictable" in its content, the unpublished column of text from the backside of this stone was our choice for 3D scanning.

2.2 Accurate 3D model Creation

The procedures applied and described below involve the 3D scanning of the inscriptions using image based techniques as the use of laser would deteriorate the accuracy. These involve the use of a Structured Light Scanner and Structure from Motion techniques. Since the use of structured light scanners is time consuming and requires costly hardware and software the alternative and cost/time effective SfM methodology for collecting 3D data of the inscriptions was also implemented for reasons of comparison. The result was the creation of an accurate 3D surface model, which would serve as a detailed basis for the epigraphists to try to decipher the inscription. This 3D model should preferably be non-coloured, i.e. without photographic texture in order not to mislead the archaeologists from misunderstanding and misinterpreting of the anaglyph.

2.2.1 Structured Light Scanner: As already stated, Structured Light Scanners are an alternative to laser scanners. In order to geometrically document the inscriptions, data were collected by using a Structured Light scanner (SL2 by RGBXYZ Inc.) consisting of two machine vision cameras (5MP) and a standard DLP projector.

The machine vision cameras are used for capturing 8 consecutive stereopairs of the projected light on the object for

[1] IG means *Inscriptiones Graecae*, the corpus for all ancient Greek inscriptions, published since 1815 by the Berlin Academy of Sciences. The Parthenon inventories written before 403 BC are published in IG I^3 292-362 (volume from 1981), the inventories after 403 BC in IG II^2 1370-1513 (volume from 1927).

doi:10.5194/isprsarchives-XL-5-W4-237-2015

the determination of its geometry, while the DLP projector was projecting the structured light patterns. These patterns are alternating black and white stripes with decreasing width. The two cameras were mounted on a rigid base with the distance between them being approximately equal to one third of the cameras to object distance. The maximum density capability equals to 150m and the maximum precision to 50m, according to the manufacturer (Georgopoulos et al., 2010). This specific model of structured light scanner, though, is slightly differentiated from the traditional structured light approach, mentioned above, as it consists of two cameras; the three dimensional coordinates are computed by triangulation between corresponding optical rays from the two machine vision cameras and not by triangulation between a ray and a plane. For achieving accurate and reliable results, a necessary setup calibration was applied. By this procedure are determined the interior orientation parameters of the cameras, their relative position and the scale of acquisition. For this purpose a custom calibration board, a simple checkered board, is imaged at various angles by both cameras (Figure 1).

Figure 1. System's Calibration Step

Intrinsic parameters of left camera:
Focal Length: fc_left = [3957.24341 3959.76799]
Principal point: cc_left = [606.10432 521.37367]
Skew: alpha_c_left = [0.0000]
Distortion: kc_left = [-0.13144 1.36164 0.00169 -0.00320 0.00000]
Intrinsic parameters of right camera:
Focal Length: fc_right = [3971.39481 3973.44483]
Principal point: cc_right = [638.80226 473.48252]
Skew: alpha_c_right = [0.00000]
Distortion: kc_right = [-0.12822 1.52410 0.00051 -0.00138 0.00000]

Extrinsic parameters (position of right camera wrt left camera):
Rotation vector: om = [-0.00958 0.43170 0.02699]
Translation vector: T = [-264.56873 -1.25983 64.33251]

Figure 2. The Structured Light system setup (Left) and an individual scan example (Right)

For the scanning of the inscriptions described in this paper, the system was placed at no more than 50cm from the inscription, having the capability of scanning 50-60cm^2 at a time (Figure 2 Left). Hence, 80 individual scans (Figure 2 Right) were performed considering a constant horizontal and vertical overlap of about 30%. In addition, the determination of the suitable imaging parameters of the two machine vision cameras was very crucial in order to acquire the data with best quality. The most well-known of these were pixel clock and exposure, which are related to the refresh rate of the pixels on the screen and the amount of light allowed falling on the photographic sensor.

2.2.2 Structure from Motion (SfM): SfM has a flexible camera model until calibration parameters are added to the workflow being used (Chandler and Fryer, 2013; Crandall et al., 2013). Through this procedure, a 3D point cloud could be generated assuming the matching overlap in a group of photographs. SfM can generate a 3D point cloud based on images taken from multiple viewpoints, otherwise known as Multi View Stereo (MVS), and corresponding geometry of shapes within a scene (Snavely et al., 2006; Crandall et al., 2013; Koutsoudis et al., 2013). For this approach, image sequences from varying distances were collected using a calibrated professional Canon EOS-1Ds Mark III full frame digital camera (5616x3744 pixels and 6.4μm pixel size) aiming to reconstruct the 3D scene through structure from motion (SfM) techniques in order to evaluate the efficiency and the level of precision and detail of the obtained reconstructed inscriptions.

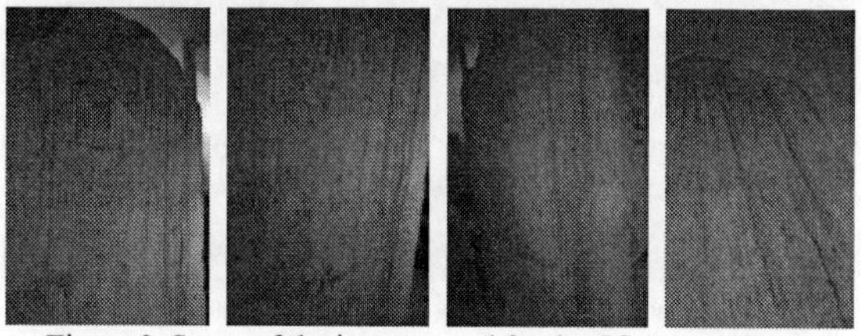

Figure 3. Some of the images used for the SfM procedure

A 24mm wide angle lens was used due to the small camera-to-object distance. The aforementioned characteristics lead to an average GSD of 120μm. 57 normal and slightly convergent RAW and JPEG images were acquired setting the sensitivity to a normal value, ISO 200. An approximate mean distance of 0.30m was used for the shots, fixing manual focus. As the data acquisition was an indoor process, a tripod was used to get non-blur images. The software used for this procedure was PhotoScan® by Agisoft.

2.3 Difficulties in the acquisition processes

Problems in the acquisition processes such as lighting conditions and camera-to-object relative positioning as well as difficulties in the alignment step and mesh optimization are also encountered without reducing the accuracy of the final results. The scanned inscription was not possible to move and was installed close to another inscription of the same size. In addition, due to their size, their weight and their damage, nothing could be done to acquire their top surface, producing therefore a complete model. Another problem occurred during the shooting, related to the logistic and to the blocked position of some objects, was the size of the system and the use of the right lens. In some cases, it was difficult to stand at the suitable position and height to acquire the required data from the two

This contribution has been peer-reviewed.
doi:10.5194/isprsarchives-XL-5-W4-237-2015

machine vision cameras of the Structured Light Scanner. In these cases the scanned surface was not a rectangle like the one presented in Figure 1b but a trapezium or another projection of the rectangle leading to occlusions. Finally, the illumination did not add any problems, due to the homogeneity in the colours of the object itself.

3. DATA PROCESSING AND RESULTS

The creation of the final accurate three dimensional model is a complicated procedure requiring large computation cost and human effort. It includes the collection of geometric data in limited space and time, the creation of the surface, the noise filtering and the merging of individual surfaces. It is important to note that in such cases, the detail of the surface may indicate the existence of a weathered letter, thus the noise filtering must be a carefully implemented procedure. The products of the scanning procedure were processed using various software packages, for the production of the final accurate 3D models of the inscriptions. In order to automate the various steps of the procedure and to achieve better results, various software, commercial and not, were combined. In the following the individual steps of this procedure and the used software are given.

3.1 Mesh Generation and Registration

3.1.1 Structured Light Scanner Data: In this section, the mesh generation procedure and their alignment is described for Structured Light Scanner data. The result of each scan is a point cloud. Each point cloud is triangulated into a dense (up to 150μm) mesh of triangles, with points on the objects surface, using Delaunay triangulation. The results from each individual scan are the mesh, which is exported in OBJ format for further processing, and an image in JPG format, which is responsible for the texture information. Initially, the aforementioned exported individual meshes were inserted in Geomagic Studio® for the registration and merging procedure, since each mesh refers to a different local system as well as for holes filling and noise filtering. For the meshes registration a combination of manual and automatic techniques was applied. After selecting 3 or more common points in unregistered meshes, the registration is made.

Figure 4. Individual scans after the registration process

Then, the ICP (Chen and Medioni, 1991) algorithm was used in order to refine the results of the manual registration. Through the manual step, values are calculated for the transformation, which are then optimised. Having completed the registration of all meshes, the final surface is processed as a whole and all the individual meshes are extracted separately in OBJ format so as to be georeferenced.

3.1.2 Structure from Motion (SfM) Data: In this section, the image alignment and mesh generation procedure is described for Structure from Motion (SfM) data. The necessary procedures were realized through the use of Agisoft's PhotoScan® software. Initially and after loading the photos, the calibration parameters of the system camera-lens were inserted. Subsequently, these photos need to be aligned. At this stage PhotoScan® finds the camera position and orientation for each photo and builds a sparse point cloud model. Alignment having been completed, computed camera positions and a sparse point cloud were displayed.

Figure 5. Reconstructed Mesh from Agisoft's Photoscan®

After an inspection of the alignment results, the generation of the dense point cloud model took place. At this stage, the program, based on the estimated camera positions calculates depth information for each camera to be combined into a single dense point cloud. At this stage, the desired reconstruction quality was specified as "ultra high" in order to obtain more detailed and accurate inscription surface (Figure 5). Due to some factors, like poor texture of some elements of the scene, noisy or badly focused images, there can be some outliers among the points. To sort out the outliers, several built-in filtering algorithms are applied. Since texture is not required in such applications in order not to mislead the archaeologists from misunderstanding and misinterpreting of the anaglyph, a generic texture mapping was applied for better visual quality of the final model and navigation through it. This Generic mapping mode allows the parameterization of texture atlas for arbitrary geometry. No assumptions regarding the type of the scene to be processed are made as well as the software tries to create as uniform texture as possible.

3.2 Meta-Processing of 3D Models and Results Analysis

As already mentioned, through the use of accurate 3D models and processing procedures the epigraphists can gain an additional possibility to analyze the inscriptions and make decisions on specific issues while trying to decipher the text. To this direction, meta-processing including mesh filtering and visual inspection of the 3D models on software providing the ability of changing light direction, is applied.

3.2.1 3D Model Selection: The resulting 3D models were visually inspected for their detail in letters description and mapping. The results indicated that the most suitable 3D model for the revealing of the weathered letters in the scanned inscriptions is the one created using the structured light scanner data. To this decision leads the fact that the 3D model created by the structured light scanner data depicts the letters in a more detailed way and this offers the capability of revealing better and more accurate "hidden" and destroyed letters (Figure 6).

Figure 6. Resulting surface from Structured Light data (Left) and SfM data (Right)

However, the implemented alternative methodology of SfM for collecting 3D data of the inscriptions has shown very promising results and further research in its implementation is planned at a later stage.

3.2.2 The Meta-processing Framework: In order to reveal the weathered letters, an efficient and easy to use from the epigraphists framework was developed and implemented requiring a minimal computational cost (Figure 7). The key idea for developing this framework is to smooth and normalize the surface model of this part of the already weathered inscription and then subtract it from the original scanned surface.

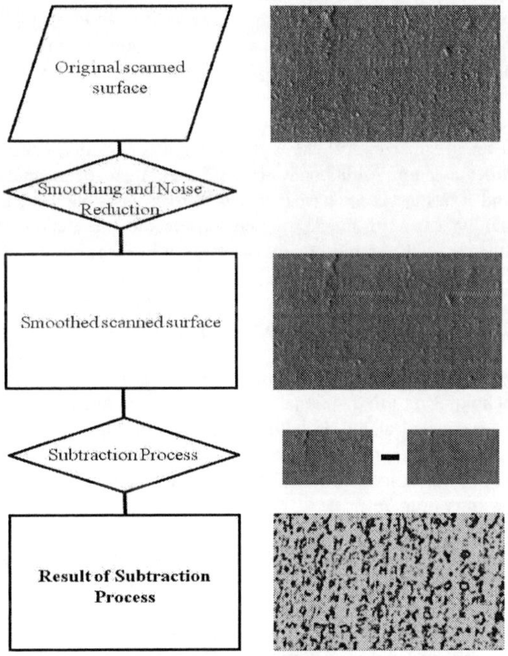

Figure 7. The proposed novel framework for revealing weathered letters

By this innovative procedure, any alterations of the original surface from the smoothed one will be a carving indicating a letter of a damaged part of the masonry. In that way, variations in the surface of inscription which are not recognized by the human eye are highlighted. Key point of this process is the level of the smoothing procedure as well as the threshold of the meshes subtraction. For this paper, the described framework was implemented through the use of commercial software leaving no possibilities to control all the parameters involved such as the smoothing algorithm and the subtraction process.

3.2.3 Results and Analysis of the Meta-processing: Figure 8 presents a non-weathered part of the scanned inscription while Figure 9 an extremely weathered part of the same inscription.

(a)

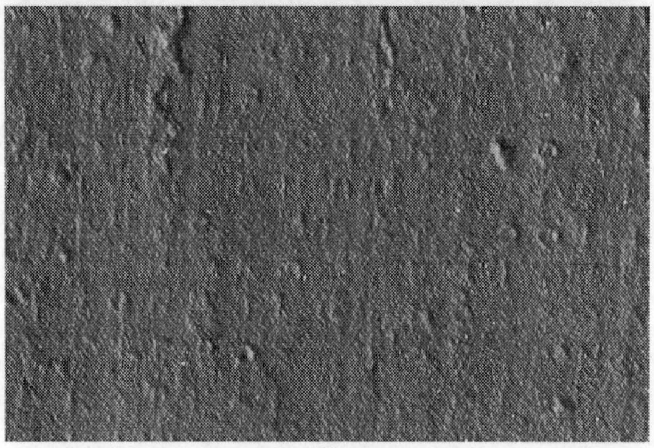

(b)

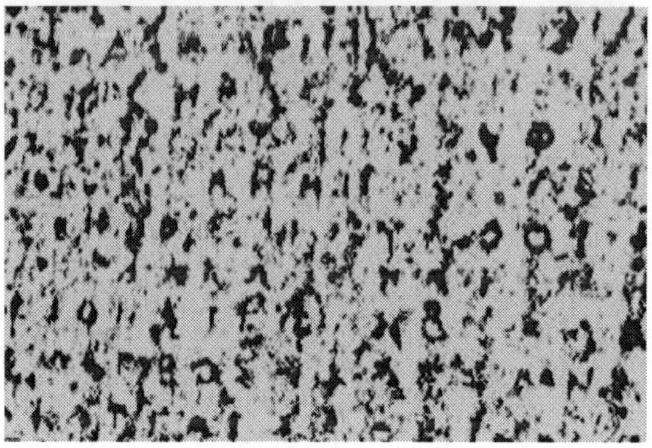

(c)

Figure 8. Textured (a) and non-textured (b) reconstructed surface of a non-weathered Parthenon Inventory List. (c) revealed letters as a result of processing

In the scanned part of Figure 8, letters are readable in the non-

This contribution has been peer-reviewed.
doi:10.5194/isprsarchives-XL-5-W4-237-2015

textured model (8b), the processed model (8c) and even in the textured model (8a). On the other hand, the textured and non-textured surface model on the pre-processing stage of an extremely weathered scanned part of the inscription, that are presented in Figure 9a and 9b respectively, make it clear that it is impossible for the archaeologists to observe horizontal and vertical rows of letters in order to analyse them. However, as it is observed in Figure 9, after meta-processing, can be recognized the stoichedon order (the letters are standing not only in horizontal lines, as usual in any inscription, but also in vertical rows) and some Omicrons are clearly obvious.

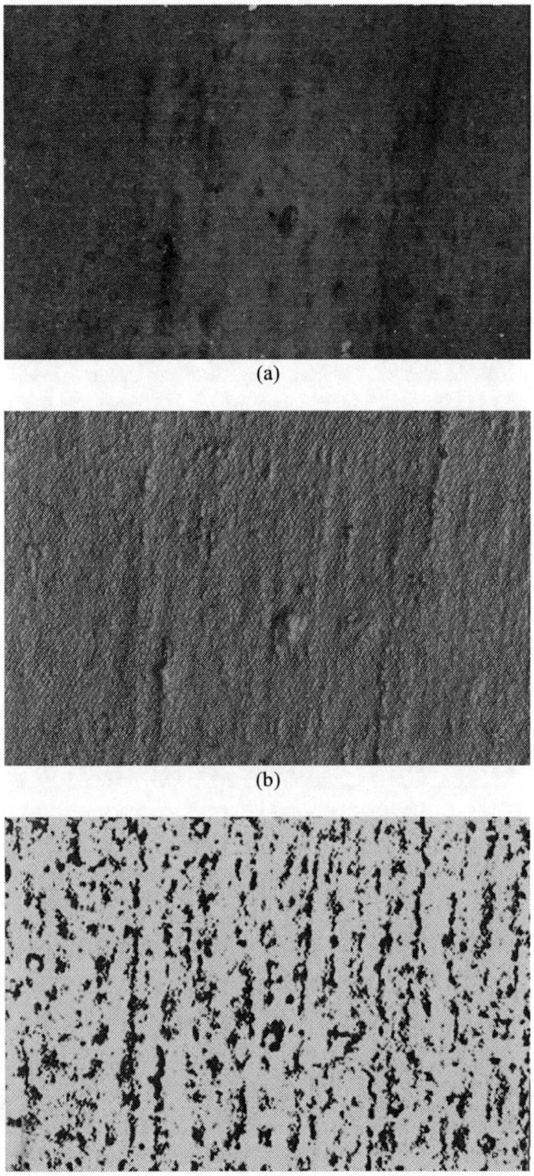

(a)

(b)

(c)

Figure 9. Textured (a) and non-textured (b) reconstructed surface of a very weathered Parthenon Inventory List. (c) revealed letters as a result of processing

3.2.4 Evaluation from an Expert: It is obvious that this novel proposed framework helps the epigraphists to follow the order of the letters both in horizontal and vertical direction. In addition, through this procedure, letters are appearing on weathered surfaces like the one presented in Figure 9. The results of this first attempt to deal with 3D models are encouraging. The chosen study area of the inscription is one of the most difficult tasks for deciphering: A column of text so weathered that it was not even recognized as a text column up to now. After the additional column was recognized, the first deciphering was done in front of the stone with a strong light. This version was now checked on screen with the 3D model. After meta-processing (Figure 9c), experts could already clearly see that (a) The whole area was once inscribed, and (b) that it was inscribed in the same *stoichedon* order as the other four columns of text.

With bare eyes and light, it was clear that there are some letters and that they comply with the size and the letter forms of the first four columns, but the electronic version shows traces of much more letters (even if not to decide which letter it has been in all instances), and even the whole *stoichedon* order that could be guessed but not assured with bare eyes. The proposed framework of meta-processing should be refined now in collaboration of 3D specialists and epigraphists.

For the revealing of up-to-now "hidden" letters by using the mesh, it was not only possible to recover the letters that were seen with the epigrapher's bare eyes, but also to make progress. This is mainly that we can better distinguish between letter and mere injury of the stone: one of the greatest problems in epigraphy is, in cases of badly preserved stones, whether an indentation is a casual cut in the marble or the trace of the letter cutter's chisel. An additional problem is colour change on the original stone that hinders the clear view. On a traditional squeeze, there are no different colours and in the imprint it can be better seen whether a trace comes from a chisel or not, but there are other issues, e.g. that a squeeze cannot be magnified. With a 3D model, the colour problem can also be dropped out, and by playing with the direction of light on screen, it can be better decided whether the cut is a regular cut done by a chisel or an irregular injury, like with turning a paper queeze around under a lamp. Additionally, the 3D model can be magnified. And it has to be emphasized again that it is possible to make a scan in cases of fragile surfaces where the production of a squeeze with its necessary mechanical pressure is perhaps sometimes inadvisable.

4. CONCLUSIONS

In this paper two non-invasive non-destructive alternative techniques to the traditional and invasive technique of squeezes are presented alongside with specialized developed processing methods, aiming to help the epigraphists to reveal and analyse weathered letters in ancient Greek inscriptions carved in masonry or marble.

The recommended techniques are based on creating a 3D Model of the carved surface of an inscription, through the process of which, it is possible to reveal "hidden" letters and decipher more parts of the inscription. In detail, the initially presented technique creates the 3D Model through Structured Light Scanner Data, though the second one through a Structure from Motion procedure. The results indicated that the most suitable 3D Model for the revealing of the weathered letters in the scanned inscription is the one created using the structured light

This contribution has been peer-reviewed.
doi:10.5194/isprsarchives-XL-5-W4-237-2015

scanner data, although time and effort consuming procedure. Consequently, an innovative meta-processing framework was developed and implemented on this model, consisting of subtracting a smoothed and normalized surface model from the originally scanned surface.

Through the presented techniques, is performed the acquirement of an accurate 3D Model of the inscription, which offers the capability of digitally storing of all the supplied information, even this which is not obvious with a bare eye. This fact as well as the non-invasive character of the presented techniques, offer the possibility not only of prevention from any damage, but also of accessibility, time and locality independent study, analysis and processing of any type of inscriptions, even the very weathered ones. Another contribution of the presented framework is the innovative meta-processing procedure that was developed through which the lettering schemes are efficiently recognized. The final product can be easily handled, is time saving and really reliable for accurate scientific conclusions and deciphering of inscriptions.

The proposed framework was applied to an ancient Greek inscription concerning the inventory of the Parthenon. The results of processing and analysis were inspected and evaluated by a specialist in terms of detail, accuracy and quality of the model, as well as the capability of revealing damaged or "hidden" letters and finally deciphering the inscription.

5. FUTURE WORK

Our future work, will include a further study and analysis on the proposed novel framework for revealing weathered letters. In more detail, state of the art smoothing algorithms will be implemented in order to evaluate the reliability and the accuracy of the results. Moreover, a future goal is the system to automatically calculate the parameters of the smoothing level according to the surface anaglyph and remaining letters' depth. Finally, the SfM technique will be further developed, in order to take advantage of its promising results.

ACKNOWLEDGEMENTS

The research leading to these results has been supported by European Union funds and National funds (GSRT) from Greece and EU under the project 3D ORO: Efficient and Effective 3D Computer Vision Tools for Improving the Performance of 3D Digitalization funded under the cooperation framework.

We also want to acknowledge for permission for scanning the Epigraphical Museum of Athens (Athanasios Themos and Elena Zavvou) and Prof. Constantin Papaodysseus and Dr. Athanasios Panagopoulos for their scientific cooperation.

REFERENCES

Barmpoutis, A., Bozia, E. and Wagman, R. S., 2010. A novel framework for 3d reconstruction and analysis of ancient inscriptions. Machine Vision and Applications, Vol. 21(6), pp. 989–998.

Beck, C. W., 1963. Synthetic elastomers in epigraphy. American Journal of Archaeology pp. 413–416.

Chandler, J., Fryer, J., 2013. Autodesk 123D catch: how accurate is it? Geomatics World, Vol. 2 (21), pp. 28-30

Chen, Y., & Medioni, G. (1992). Object modelling by registration of multiple range images. Image and vision computing, Vol. 10(3), 145-155.

Cornelis, K., Pollefeys, M., Vergauwen, M. and Van Gool, L., 2001. Augmented reality using uncalibrated video sequences. In: 3D Structure from Images SMILE 2000, Springer, pp. 144–160.

Crandall, D., Owens, A., Snavely, N., Huttenlocher, D., 2013. Discrete-continuous optimization for large-scale structure from motion. IEEE Trans. Pattern Anal. Mach. Intell. (PAMI). http://www.cs.indiana.edu/~djcran/papers/disco13pami.pdf (accessed 21.01.15).

Georgopoulos, A., Ioannidis, C. and Valanis, A., 2010. Assessing the performance of a structured light scanner. In: Commission V Symposium. International Archives of Photogrammetry, Remote Sensing and Spatial Information Sciences, Vol. 38(5), Citeseer.

Gonizzi Barsanti, S. and Guidi, G., 2013. 3d digitization of museum content within the 3dicons project. ISPRS Annals of Photogrammetry, Remote Sensing and Spatial Information Sciences II-5/W1, pp. 151–156.

Koutsoudis, A., Vidmar, B., Ioannakis, G., Arnaoutoglou, F., Pavlidis, G., Chamzas, C., 2013. Multi-image 3D reconstruction data evaluation. J. Cult. Herit., Vol. 15 (1), 73-79.

Landon, G. and Seales, W., 2006. Petroglyph digitization: enabling cultural heritage scholarship. Machine Vision and Applications, Vol. 17(6), pp. 361–371.

Lerma, J. L. and Muir, C., 2014. Evaluating the 3d documentation of an early christian upright stone with carvings from Scotland with multiples images. Journal of Archaeological Science, Vol. 46, pp. 311–318.

Malzbender, T., Gelb, D. and Wolters, H., 2001. Polynomial texture maps. In: Proceedings of the 28th annual conference on Computer graphics and interactive techniques, ACM, pp. 519–528.

Snavely, N., Seitz, S.M., Szeliski, R., 2006. Photo Tourism: Exploring Photo Collections in 3D. http://phototour.cs.washington.edu/Photo_Tourism.pdf (accessed 21.01.15).

Soile, S., Adam, K., Ioannidis, C. and Georgopoulos, A., 2013. Accurate 3d textured models of vessels for the improvement of the educational tools of a museum. ISPRS-International Archives of the Photogrammetry, Remote Sensing and Spatial Information Sciences, Vol. 1(1), pp. 211–217.

Wachowiak, M. J. and Karas, B. V., 2009. 3d scanning and replication for museum and cultural heritage applications. Journal of the American Institute for Conservation, Vol. 48(2), pp. 141–158.

Woodhead, A. G., 1992. The study of Greek inscriptions. Bristol Classical Press

This contribution has been peer-reviewed.
doi:10.5194/isprsarchives-XL-5-W4-237-2015

ACCURACY EVALUATION OF A MOBILE MAPPING SYSTEM WITH ADVANCED STATISTICAL METHODS

I. Toschi [a], P. Rodríguez-Gonzálvez [b], F. Remondino [a], S. Minto [a], S. Orlandini [c], A. Fuller [d]

[a] 3D Optical Metrology (3DOM) unit, Bruno Kessler Foundation (FBK), Trento, Italy - (toschi, remondino, minto)@fbk.eu
[b] Dept. of Cartographic and Land Engineering, High School of Ávila, University of Salamanca, Ávila, Spain - pablorgsf@usal.es
[c] Microgeo, Firenze, Italy - simone@microgeo.it
[d] RIEGL, Vienna, Austria

Commission V, WG V/4

KEY WORDS: Mobile, Accuracy, Precision, Statistic, Terrestrial, Photogrammetry, Laser scanning

ABSTRACT:

This paper discusses a methodology to evaluate the precision and the accuracy of a commercial Mobile Mapping System (MMS) with advanced statistical methods. So far, the metric potentialities of this emerging mapping technology have been studied in few papers, where generally the assumption that errors follow a normal distribution is made. In fact, this hypothesis should be carefully verified in advance, in order to test how well the Gaussian classic statistics can adapt to datasets that are usually affected by asymmetrical gross errors. The workflow adopted in this study relies on a Gaussian assessment, followed by an outlier filtering process. Finally, non-parametric statistical models are applied, in order to achieve a robust estimation of the error dispersion. Among the different MMSs available on the market, the latest solution provided by RIEGL is here tested, i.e. the VMX-450 Mobile Laser Scanning System. The test-area is the historic city centre of Trento (Italy), selected in order to assess the system performance in dealing with a challenging and historic urban scenario. Reference measures are derived from photogrammetric and Terrestrial Laser Scanning (TLS) surveys. All datasets show a large lack of symmetry that leads to the conclusion that the standard normal parameters are not adequate to assess this type of data. The use of non-normal statistics gives thus a more appropriate description of the data and yields results that meet the quoted *a-priori* errors.

1. INTRODUCTION

1.1 MMS technology

Mobile Mapping System (MMS) is nowadays an emerging technology, whose development began in the late 1980s and is constantly growing. From a technological point of view, MMS is a multi-sensor system, that consists mainly of three components: mapping sensors (active and/or passive 3D imaging systems), navigation/positioning sensors (IMU/GNSS) and a control unit, that synchronizes and integrates the acquisition of geometric/positioning information. All sensors are integrated on a rigid moving platform (e.g. vans, cars, trains, boats, snow mobile sledges, people, etc.), whose trajectory is computed and finally used to produce geo-referenced 2D/3D data. Land-based mobile laser scanners mounted on vans or cars represent the best and a cost-effective solution for capturing 3D point clouds of urban areas with an high recording rate, high point density/accuracy and remote acquisition mode. Thanks to these benefits and to continuous developments in both scanning and navigation technologies, MMSs are gaining more and more importance in many application fields, such as civil engineering and construction (Slattery et al., 2012), environment (Bitenc et al., 2011), pipeline design (Kawashima et al., 2012), road inventory (Pu et al., 2011) and cultural heritage (Ziparo et al., 2013). A general review of the different solutions available on the market and a comparison of their technical specifications can be found in Puente et al. (2013).

1.2 Literature review on approaches for accuracy evaluation of MMS

Terrestrial mobile mapping technology goes back to the 90's when the first experiments showed the potential of mobile acquisitions for mapping purposes (Bossler and Novak, 1993; El-Sheimy et al., 1995; El-Sheimy, 1996). MMSs have then developed from a typical academic research to commercially operating systems used for topographic surveying, 3D mapping of traffic arteries, city planning, visual street-level image and vector data acquisition, visualization, etc. The accuracy requirements for the acquired data is substantially different in each application, with cartographic mapping and road or rail infrastructure surveying being much more demanding in this particular respect. Nowadays there are different commercial MMS solutions and they show the best example of sensor integration and cost-effective acquisition of geo-referenced spatial data, with a combination of digital imaging devices, long-range laser scanners and GNSS/IMU positioning sensors (Tao and Li, 2007). Besides results published by system manufacturers and linked to internally-performed tests/procedures, there are few publications where MMS performances are examined with an external reference dataset and by an independent actor.

One of the most recent studies has been carried out within the European Spatial Data Research (EuroSDR) project "Mobile Mapping – Road Environment Mapping Using Mobile Laser Scanning" (Kaartinen et al., 2012). This action aimed at benchmarking the performance of various MMSs using a permanent urban test-field established around a shopping mall. A digital elevation model, some poles and building corners were adopted as reference objects. After comparing the collected datasets against the ground-truth and performing a 'gross error filtering', the accuracy was finally derived by computing the minimum, the maximum and the standard deviation values of the differences. Tests provided an elevation accuracy better than 3.5 cm for all professional systems properly calibrated (up to a range of 35 m), whereas the best system achieved a planimetric accuracy of 2.5 cm up to a range of 45 m.

Lim et al. (2013) performed a field test at Padre Island National Seashore (USA) in order to evaluate the accuracy of point clouds collected by a MMS under variable settings of vehicle speed and point density. Six vertical targets were GNSS-

This contribution has been peer-reviewed.
doi:10.5194/isprsarchives-XL-5-W4-245-2015

measured and adopted as references. Comparisons led to a mean target error of -0.222 m (East), -0.036 m (North) and -0.104 m (Height). Given the excellent conditions for GNSS provided by the test-area, discrepancies were ascribed to boresight errors and a parameter-domain approach was developed in order to adjust the lever arm parameters.
Prior to these studies, the StreetMapper mobile system received different attentions. Kremer and Hunter (2007) declared an accuracy range of 3-5 cm and reported some practical applications where the MMS could be used for mapping purposes. Barber et al. (2008) used GNSS measurements as references to evaluate precision and accuracy of laser scanning data collected in two test-sites, i.e. a residential and an industrial area. RMS errors in elevation were found to be in the order of 3 cm, whereas the planimetric accuracy was set around 10 cm. Haala et al. (2008) adopted an existing 3D city model of the city of Stuttgart to provide area covering reference surfaces. The investigation sought to assess the quality of data collected at building façades. After a semi-automatic selection of surfaces from the 3D city model, corresponding planar patches were estimated by least square adjustment from the MMS-collected data. Comparisons proved that an accuracy better than 3 cm (standard deviation of the differences between measured and reference data) can be achieved by the system in good GNSS conditions.

1.3 Paper objectives

This paper describes an approach designed to analyse the precision and the accuracy achieved by a commercial MMS system with advanced statistical methods. The evaluation studies so far performed were generally based on the assumption that errors follow a normal distribution. In fact, this hypothesis should be carefully verified in advance, in order to test how well the Gaussian classic statistics can adapt to data sets that are usually affected by asymmetrical gross errors. In those cases where even a proper process of outlier removal is not effective, robust and non-parametric methods for the derivation of accuracy measures should be preferred. This strategy can be further improved if the signed errors are computed, in order to analyse not only the dispersion values (which are related to the sensor noise), but also the bias effects on the object surfaces. Aim of the present paper is to test a rigorous statistical workflow for the evaluation of MMS data collected in a historic city centre. The site is selected in order to assess the system performance in a challenging urban scenario, that includes several building typologies with complex and diverse geometries/materials. As reference datasets we employ point clouds derived from a photogrammetric and a Terrestrial Later Scanner (TLS) survey. Where possible, an error budget computation is discussed, in order to develop an adequate understanding of the most important components of uncertainty affecting the measurements.

2. PROJECT DESCRIPTION

2.1 MMS data with RIEGL VMX-450

The MMS RIEGL platform (Figure 1 and Table 1) integrates two synchronously operated VQ-450 laser scanners, a portable control unit (VMX-450-CU) and IMU/GNSS navigation hardware. The system is able to measure up to 1.1 million points and 400 profiles per second, providing extremely dense and feature-rich data even at high driving speed. Furthermore, it exploits the RIEGL echo signal digitization technology and online waveform processing, resulting in a high penetration capability of obstructions (e.g. fences and vegetation). The platform is also equipped with the modular VMX-450-CS6 camera system, that complements the acquisition of geometric data with the recording of time-stamped images. Up to six industrial digital colour cameras with electronic shutters and 5 mm lens can be integrated. Camera calibration parameters are provided by the manufacturer, whereas exterior orientation parameters (i.e. time-stamped position and orientation of the cameras) are computed by RIEGL software in post-processing. The software includes several modules to manage each step of the processing pipeline, i.e. data acquisition (RiACQUIRE), data adjustment (RiPROCESS) and data geo-referencing (RiWORLD).

Figure 1. Configuration of the RIEGL VMX-450 system.

Sensor	VQ-450
Measuring principle	Time of Flight
Laser wavelength	Near infrared
Laser measurement rate	300 – 1100 kHz
Maximum range	140 – 800 m
Minimum range	1.5 m
Accuracy	8 mm, 1σ
Precision	5 mm, 1σ
Sensor	**IMU/GNSS**
Absolute position	0.020 – 0.050 m
Roll and pitch	0.005 °
True heading	0.015 °
Sensor	**VMX-450-CS6**
Resolution	5 Mpx
Sensor size	2452 x 2056 px
Pixel size	3.45 μm
Nominal focal length	5 mm

Table 1. Technical characteristics of the RIEGL VMX-450 system according to the manufacturer.

2.2 Area of study and data collection

The data acquisition took place on September 22nd, 2014. An area of about 700 m west to east by 500 m north to south was covered in the city centre of Trento (Italy). It represents a typical urban scenario, that can be divided into two test-sites:

- the Duomo square (ca. 80 m x 70 m, Figure 2) surrounded by some of the most outstanding historical buildings of the city. Among them, five houses were selected for further analyses and segmented from the collected data. They include the Cathedral of San Vigilio in typical Romanesque style, the Medieval Palazzo Pretorio and some Renaissance buildings characterized by precious frescoed façades.
- narrow streets (average width of 3-4 m) around the Duomo square, surrounded by painted private and public buildings. Among them, three houses were identified and further investigated, including the backside façade of Palazzo Pretorio, Santa Annunziata church and the old municipal building of the city.

The selected areas (Figure 3) represent a challenging test-field for MMS mapping, whose metric performance depends mostly on the accurate determination of the time-variant position and orientation parameters of the mobile platform. Therefore, multipath and signal loss effects caused by tall buildings on narrow spaces can deteriorate the GNSS conditions and, consequently the trajectory computation. The tight urban canyons of the surveyed city centre are a typical example of these unfavourable scenarios.

Since the methodology adopted by the study is aimed at validating the final "commercial" product, the collected data was internally-processed by RIEGL experts and finally delivered as point clouds (LAS file format) in WGS84 global coordinate system. No attempts to evaluate each step of the calibration and adjustment processes are thus here undertaken.

The point clouds are characterized by a mean spatial resolution of 2-3 cm and were exported separately for each of the two mounted laser scanners. Furthermore, three passes were collected in the Duomo square, resulting in six separate point clouds. This provided the possibility to assess the internal precision and consistency of the captured datasets.

Figure 2. Close-up view of the coloured 3D point cloud collected by MMS in the Duomo square of Trento (Italy).

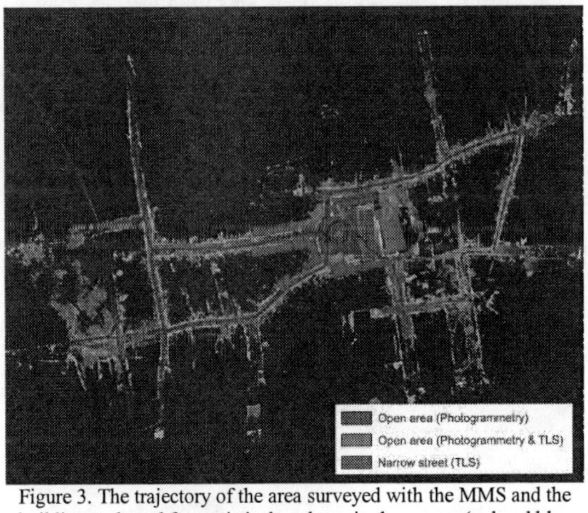

Figure 3. The trajectory of the area surveyed with the MMS and the buildings selected for statistical analyses in the square (red and blue symbols) and in the narrow streets (green symbols).

2.3 Reference data

2.3.1 Photogrammetric survey

The photogrammetric survey was carried out in the open square where the most notable historic buildings are located. This allowed an optimal design of the photogrammetric network, whose layout was selected in order to optimize the accuracy and completeness of the final 3D point cloud. Only the buildings in the square (red and blue symbols in Figure 3) were thus reconstructed.

The image acquisition phase was performed using a Nikon D3X digital camera (35.9 × 24.0 mm sensor size, 5.95 μm pixel size) equipped with two fixed focal length lenses (35 and 50 mm). A total of 359 images were collected at a mean camera-object distance of 25 m, achieving an average GSD of 3.5 mm. Using design equations and an *a-priori* image measurement accuracy of 1/2 pixel (Luhmann et al., 2006), the range uncertainty and lateral resolution on the final model were anticipated to be in the [4.0;4.2] mm and [2.9;4.2] mm ranges, respectively.

Both image orientation and dense image matching were carried out using Agisoft Photoscan. 18 natural GCPs, measured with a total station, were used as observed unknowns in the bundle adjustment for the image orientation phase. The re-projection error was 0.52 pixels. In order to achieve a reasonable trade-off between processing effort and resolution, the successive dense matching was performed using the second-level image pyramid. Thus the derived dense point cloud (Figure 4) features a mean spatial resolution of 7 mm and consists of more than 107 million points.

Figure 4. Photogrammetric dense reconstruction of the Duomo square in Trento (Italy).

2.3.2 Terrestrial Laser Scanner survey

A Leica HDS7000 TOF CW laser scanner was used to acquire ultra-dense point clouds of two symmetrical buildings in the square (Figure 5 - blue blocks in Figure 3) and of those in the narrow streets (green blocks in Figure 3). More than 15 different scans were necessary to achieve a complete 3D reconstruction of the scene, with a mean instrument-object distance of 10 m.

Figure 5. TLS ultra-dense reconstruction of two symmetrical buildings in the Duomo square of Trento (Italy).

This contribution has been peer-reviewed.
doi:10.5194/isprsarchives-XL-5-W4-245-2015

Adopting the error propagation formulated in (Reshetyuk, 2009) and computing the beam divergence according to (Lichti and Gordon, 2004), the *a-priori* error on the surface is expected to be 3 mm. The alignment of the TLS scans in a local coordinate system was carried out using both planar contrast targets and natural points. No absolute geo-referencing was performed. The final registration error was below 4 mm. This last value can be considered a more realistic representation of the noise level present in the final 3D point clouds. The final point cloud is characterized by a mean spatial resolution of 3 mm, resulting in more than 240 million points.

3. METHODOLOGY

3.1 Cloud-to-Cloud signed distance computation

In order to perform the statistical analyses with a reasonable computational effort and to highlight any effects dependent on the building typology, both MMS and reference data are segmented in separate blocks. The Duomo square is divided into five blocks (hereinafter termed “Cathedral”, “Palace”, “Café”, “Houses” and “Hotel” in accordance with their main present-day function), whereas in narrow streets three distinct buildings are segmented (hereinafter called “Garibaldi”, “Belenzani” and “Cavour” from the name of the corresponding street). After the removal of non-significant points (i.e. those belonging to time-variant elements such as pedestrians and vegetation), the MMS data is registered with the reference ones (both photogrammetric and TLS point clouds) using the ICP algorithm implemented in Trimble Realworks. For each segmented building, cloud-to-cloud signed distances are then computed adopting the method Multiscale Model to Model Cloud Comparison (Lague et al., 2013). This algorithm, implemented in CloudCompare v2.6, performs a direct comparison of point clouds in 3D, avoiding the preliminary phase of meshing or gridding. Two steps are sequentially performed. At first, surface normals are estimated and oriented in 3D at a user-selectable scale, that should be consistent with the local surface roughness. Then, for each selected core point, the local distance between the two clouds is extracted. A confidence interval depending on point cloud roughness and registration error is computed as well. In the presented tests, each point of the MMS data is selected as “core” and a 50 cm normal scale is adopted in accordance with the local spatial resolution and suggestions given in Lague et al. (2013).

3.2 Statistical analyses

Our accuracy tests are performed using ‘reference’ data derived from a photogrammetric and a TLS survey. When comparing point clouds acquired by different instruments, some issues should be considered:

- a proper error budget computation should be defined, since each system has its own sources of uncertainties and sensitivity to outlier presence.
- the compared 3D points are not exactly corresponding to each other.
- the object surfaces are not equally digitized as the acquisition positions are different.

Due to these reasons, the number of gross errors affecting the datasets increases, thus requiring a rigorous statistical approach to be applied. The workflow adopted in this study is summarized in Figure 6 and follows Rodríguez-Gonzálvez et al. (2014).

In order to prevent possible bias effects affecting the data processing, the normality assumption, i.e. the hypothesis that errors follow a Gaussian distribution, should be carefully checked. This can be carried out in various ways. Among them, the most common method is represented by the normality tests (D’Agostino, 1986). However, this approach does not work properly with very large samples (Rodríguez-Gonzálvez et al., 2014). In that case, a well-suited diagnostic should be preferred, such as the study of the graphical deviation in the Q-Q plot (Hasan et al., 2011; Hoehle, 2011). It reports the quantiles of the empirical distribution function with respect to the theoretical quantiles of the normal distribution function. If errors are distributed according to the Gaussian function, the Q-Q plot should be a diagonal straight line. Additional parameters, such as Skewness and Kurtosis can further support the normality assessment. The Skewness provides an indication of departure from symmetry in a distribution (asymmetry around the mean value) and is expressed as:

$$\text{Skewness} = \frac{\sum_{i=1}^{n}(x_i-\mu)^3}{(n-1)\sigma^3} \qquad (1)$$

The Kurtosis is a measure of whether the data are peaked or flat relative to a normal distribution and is expressed as:

$$\text{Kurtosis} = \frac{\sum_{i=1}^{n}(x_i-\mu)^4}{(n-1)\sigma^4} \qquad (2)$$

If the distribution is perfectly normal, Skewness and Kurtosis values of zero and three are obtained, respectively.

If the sample follows a Gaussian distribution, the error population can be characterized using the mean error (μ) and the standard deviation (σ). On the other hand, if the sample is not normally distributed, either due to the presence of outliers or because a different population’s hypothesis applies, a robust model based on non-parametric estimation should be employed. In this case, the median (m) and the median absolute deviation (MAD) are used as robust measures instead of the mean and standard deviation, respectively. The MAD is defined as the median (m) of the absolute deviations from the data’s median (m_x):

$$\text{MAD} = m(|x_i - m_x|) \qquad (3)$$

In this work, both Gaussian statistics and the robust ones are computed for each comparison. Furthermore, analyses are performed not only on the raw data, but also on those derived from a filtering process. The latter is performed by applying the classical 3σ rule for outliers removal (Nkurunziza and Vermeire, 2014).

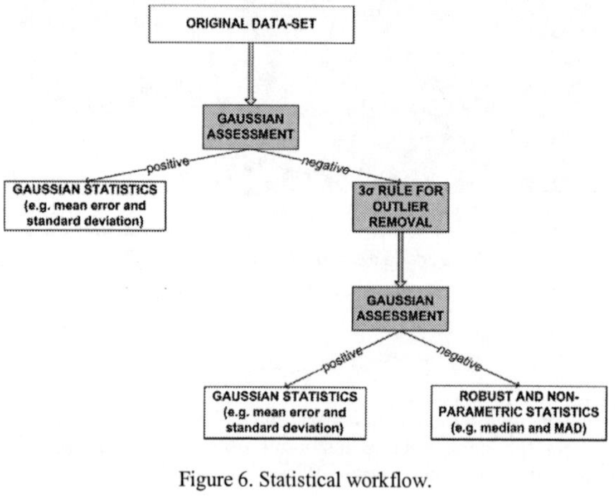

Figure 6. Statistical workflow.

This contribution has been peer-reviewed.
doi:10.5194/isprsarchives-XL-5-W4-245-2015

4. RESULTS

Following the methodology described in Section 3, a two-folds study is then carried out assessing precision and accuracy of the Trento's MMS data. The precision of the MMS point clouds is a measure of the scatter of observed values whereas the accuracy is meant as the agreement between measurement results and the corresponding true value.

The MMS data precision is determined by comparing the different scans captured by the two lasers scanners at the same time (i.e. same timestamps) and in two subsequent acquisitions (i.e. different timestamps) on the same common surface. The Cathedral façade is selected as test-area for this analysis. Of course, for perfect geo-referencing and calibration procedures, no discrepancies between the measures from different scanners and/or different epochs should be present. In fact, this problem may arise, as also previously revealed (Barber et al., 2008; Haala et al., 2008). Furthermore, several passes are usually required in order to achieve a complete reconstruction of complex buildings, especially when MMS mapping is applied to historic city centres. The quality of the post-processing adjustment should be thus assessed.

On the other hand, the MMS data accuracy is evaluated. Reference measures are derived from the photogrammetric and TLS surveys. The former is restricted to the buildings in the open square, where the photogrammetric results could be optimized due to a favourable camera network. On the other hand, TLS is exploited for the acquisition of ultra-dense point clouds in the narrow streets, where the GNSS quality is reduced and MMS acquisition angles are far from the optimal case. TLS measurements are acquired in the square too, in order to achieve a denser "ground truth". However, this is only feasible for two selected areas, due to the considerable effort for data collection.

4.1 Precision evaluation

Among the three subsequent passes collected in the square, two acquisition stripes are selected for the precision evaluation, since they both allow a good coverage of the Cathedral façade. At each of the two epochs (denoted as t_0 and t_1), two point clouds are recorded by the two laser scanners on the MMS (named LS_1 and LS_2). Four separate point clouds are thus recorded and compared. Data acquired by LS_1 at the epoch t_0 (named $LS_{1\text{-}t0}$) are selected as reference for all comparisons.

Following the workflow outlined in Figure 6 and given the large size of the samples, the Gaussian assessment is performed by evaluating the graphical deviation in the Q-Q plot. For each of the three comparisons, the statistical graphic is derived for the error population computed with the Multiscale Model to Model Cloud Comparison. The original distance dataset is then filtered with the 3σ criterion and the corresponding Q-Q plot is extracted again. Figure 7 shows the plots obtained from the comparison of $LS_{2\text{-}t0}$ point cloud against the reference one ($LS_{1\text{-}t0}$). The Q-Q plot shape extracted for the original (raw) dataset (Figure 7a) is significantly far away from the ideal normal hypothesis. On the other hand, the error sample derived from the outlier filtering is closer to the Gaussian distribution (Figure 7b), but not enough to allow a parametric error estimation. Both the parametric and non-parametric statistics are finally computed and summarized in Table 2.

As expected, Kurtosis and Skewness parameters agree with the outcomes of the graphical analyses (Q-Q plots): a significant percentage of outliers affects the original population and should be removed. Furthermore, by considering the robust error measures computed after the outlier removal, a mean bias between point clouds of 2.6 mm is observed. On the other hand, the robust measure of dispersion (MAD) reaches up to ± 5.8 mm, and increases up to ± 8.6 mm under a Gaussian hypothesis[1]. If comparisons performed with data acquired at the same epoch provide a means of assessment for errors related to LS noise and calibration uncertainty, the analyses on multi-epochs data account also for the post-processing error propagation. Values of median and MAD increase when point clouds with different timestamps are compared, highlighting some expected limitations of the geo-referencing procedure. The latter, in fact, was performed without including any ground control information within the trajectory adjustment procedure.

Finally, along with the numerical statistics, a spatial distribution of the discrepancies can be studied. The colour-coded maps of signed distances reveal the presence of typical error stripes, that are more or less coincident with scanning orientation. For instance, Figure 8 shows the errors computed from the comparison of $LS_{2\text{-}t0}$ point cloud against the reference one. The scale ranges from – 0.01 m (blue) to 0.01 m (red).

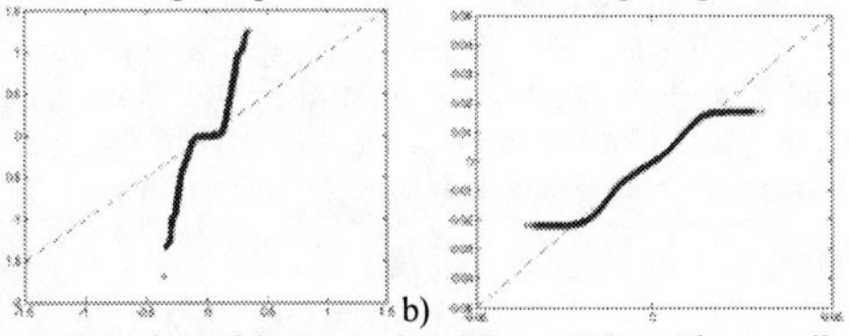

a) b)

Figure 7. Q-Q plots of the comparison $LS_{2\text{-}t0}$ vs. $LS_{1\text{-}t0}$. The quantiles of input sample (vertical axis) are plotted against the standard normal quantiles (horizontal axis). Pre-filtering (a) and post-filtering (b) data are shown.

Raw comparison	$LS_{2\text{-}t0}$	$LS_{1\text{-}t1}$	$LS_{2\text{-}t1}$
Number of points	3,651,128	829,741	661,362
Sample mean (cm)	-0.78	-0.06	-0.13
Standard deviation (cm)	± 6.83	± 4.96	± 4.92
Median (cm)	-0.27	-0.25	-0.26
MAD (cm)	± 0.43	± 0.62	± 0.65
Kurtosis (-)	47.24	69.59	75.46
Skewness (-)	-0.25	-0.01	0.57
Filtered comparison	$LS_{2\text{-}t0}$	$LS_{1\text{-}t1}$	$LS_{2\text{-}t1}$
Number of points	3,126,270	738,842	581,103
Sample mean (cm)	-0.25	-0.28	-0.36
Standard deviation (cm)	± 0.65	± 0.93	± 0.89
Median (cm)	-0.24	-0.27	-0.28
MAD (cm)	± 0.35	± 0.54	± 0.58
Kurtosis (-)	3.74	3.60	3.34
Skewness (-)	-0.08	0.00	-0.09

Table 2. Statistical results of the raw and post-processed MMS data analyses.

Figure 8. Colour-coded map of the comparison between $LS_{2\text{-}t0}$ vs. $LS_{1\text{-}t0}$. Differences range between – 0.01 m (blue) and 0.01 m (red). White areas represent no-data surfaces.

[1] The Gaussian estimation of the MAD is computed as (1.4826*MAD)

This contribution has been peer-reviewed.
doi:10.5194/isprsarchives-XL-5-W4-245-2015

4.2 Accuracy evaluation

To evaluate the accuracy of a single MMS point cloud against the reference data (photogrammetry and TLS), a pre-registration phase with ICP is carried out, in order to dismiss the uncertainties associated to the global coordinate geo-referencing. Only the errors affecting the 3D modelling and reconstruction of the architectural element are thus analysed.

4.2.1 Comparison with the photogrammetric reference data

All Q-Q plots prove that distances computed with the raw datasets are not suitable for a Gaussian estimation. As shown from the example given in Figure 9a, the input samples are far away from the ideal distribution. Gross errors should thus be filtered out, since they can corrupt the true statistical distribution of discrepancies. This evidence is verified again by the extremely high values of Kurtosis and Skewness parameters (see Table 3 - upper part).
After the filtering, the resulting datasets are more close to a normal distribution: this is confirmed by both graphical analyses (Q-Q plot in Figure 9b) and numerical ones (Kurtosis and Skewness parameters in Table 3 - bottom part).

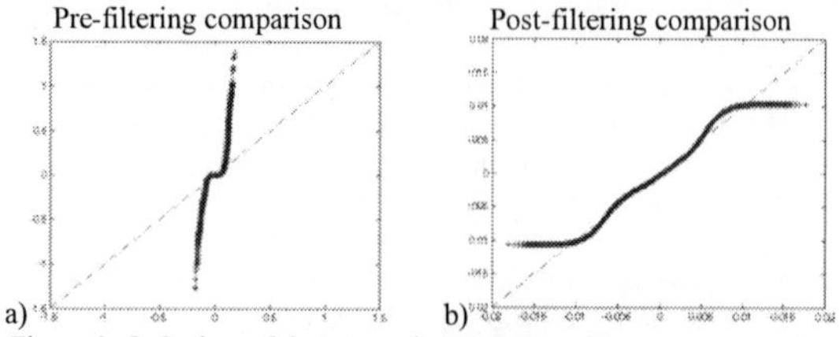

Figure 9. Q-Q plots of the comparison MMS vs. Photogrammetry for the Café dataset. The quantiles of the input sample (vertical axis) are plotted against the standard normal quantiles (horizontal axis). Pre-filtering (a) and post-filtering (b) data are shown.

However, a robust analysis of the MMS accuracy is here advised. Table 3 (bottom part) lists the derived statistical values, that allow a robust estimation of the error. This can be defined as median ± MAD and it is here estimated as -0.7 mm ± 4.0 mm. If a Gaussian measure of the MAD is computed, the error dispersion grows up to ± 5.9 mm. This second value can be compared against the *a-priori* value of uncertainty quoted for the reference dataset: results are consistent with both the photogrammetric lateral resolution and range uncertainty.

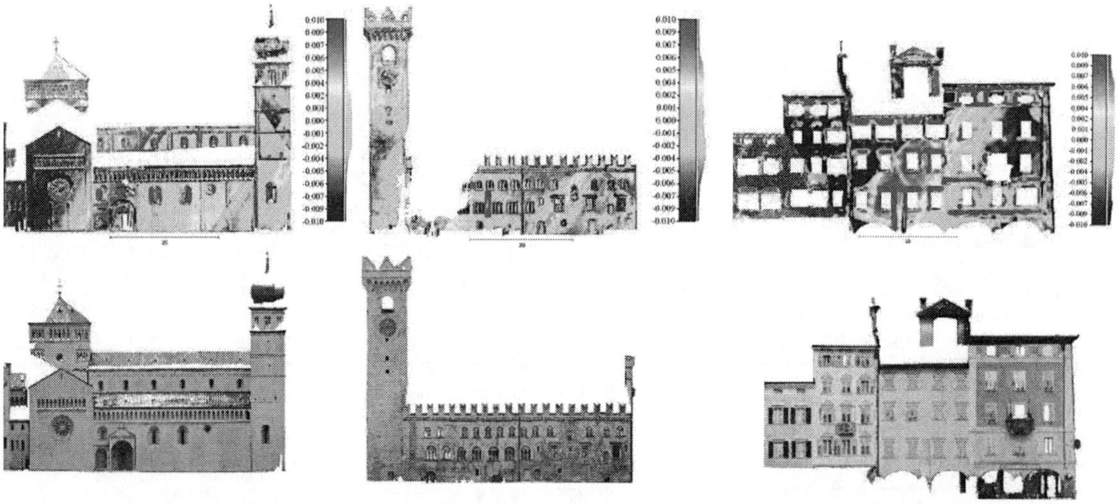

Figure 10. Colour-coded maps of the comparison analysis between the MMS and Photogrammetry data for the Cathedral (a), Palace (b) and Hotel (c) datasets. The scale ranges from - 0.01 m (blue) to 0.01 m (red). Corresponding photogrammetric point clouds are shown below.

Raw comparison	Café	Cathedral	Palace	Houses	Hotel
Number of points	249,368	3,257,278	687,885	1,372,294	695,607
Sample mean (cm)	-0.69	-0.12	0.25	-0.40	0.28
Standard deviation (cm)	± 6.42	± 3.51	± 3.95	± 8.68	± 3.86
Median (cm)	-0.03	-0.03	0.07	-0.36	0.09
MAD (cm)	± 0.26	± 0.25	± 0.34	± 0.92	± 0.60
Kurtosis (-)	48.78	118.65	65.49	30.01	84.76
Skewness (-)	-1.77	-0.05	1.62	1.06	3.21
Filtered comparison	**Café**	**Cathedral**	**Palace**	**Houses**	**Hotel**
Number of points	192,385	2,716,260	538,369	1,086,039	644,286
Sample mean (cm)	-0.04	-0.02	0.05	-0.32	0.06
Standard deviation (cm)	± 0.31	± 0.35	± 0.44	± 1.17	± 0.91
Median (cm)	0.00	-0.04	0.08	-0.31	0.07
MAD (cm)	± 0.19	± 0.20	± 0.26	± 0.71	± 0.55
Kurtosis (-)	3.39	3.79	3.42	3.67	3.31
Skewness (-)	-0.27	0.26	-0.23	-0.08	0.08

Table 3. Statistical results of the accuracy evaluation: MMS vs. Photogrammetry comparisons on the raw and filtered point clouds.

This contribution has been peer-reviewed.
doi:10.5194/isprsarchives-XL-5-W4-245-2015

Finally, along with the numerical error assessment, the spatial error distribution is highly relevant for assessing the MMS performance in modelling architectural façades. For this reason, three colour-coded maps of signed distances are showed in Figure 10, where the scale ranges from - 0.01 m (blue) to 0.01 m (red).

The Cathedral dataset covers a large area: 65 m in width and up to 50 m in height. The corresponding error distribution (Figure 10a) shows that the regions delivering the largest deviations are:

- areas acquired from high distances and with high acquisition angles, such as the left portion of the façade and the upper part of the tower (see also the trajectory in Figure 3);
- areas characterized by challenging materials (e.g. windows) and complex geometry (e.g. the main entrance).

Furthermore the same error patterns, previously highlighted by the MMS-MMS comparisons (see Figure 8), are here present too. The diagonal stripes may be linked to the direction of the scanning acquisition or probably due to internal calibration errors. These bands are characterized by a rapid change of the errors, that may cause a degradation of the final 3D reconstruction.

In the Palace dataset, beside a lack of data in the upper left corner of the building due to the presence of a large tree, systematic error stripes related to the LS orientation are found, together with an error increment in the tower (Figure 10b). The latter effect is again derived from the increased acquisition distance and scanning angle.

Finally, an area of buildings characterized by different materials and a less detailed surface is analysed (Café, Houses and Hotel datasets). The bottom part is removed from the investigation due to the presence of obstructions, such as cars, vegetation, people, etc. In addition, all time-variant elements are removed as well. For instance, a small area of 33 m x 15 m is showed in Figure 10c, corresponding to the Hotel dataset. Higher errors are found, as it is confirmed by the robust error dispersion (MAD), which reaches up to ± 7.1 mm for the Houses dataset (Table 3). The causes of this behaviour are under study and may be related to the material properties.

4.2.2 Comparison with the TLS reference data

Graphical analyses carried out with the point clouds acquired in the Duomo square (Cathedral - Figure 11 - and Café datasets) prove that the Gaussian model doesn't fit well with the acquired MMS raw data. Also in this case a filtering is necessary and a robust analysis of the MMS accuracy is then performed. The statistics reported in Table 4 agree with results achieved by adopting the photogrammetric model as reference.

Raw comparison	**Cathedral**	**Café**
Number of points	2,437,121	155,470
Sample mean (cm)	± 1.68	± 0.40
Standard deviation (cm)	8.76	8.26
Median (cm)	± 0.21	± 0.22
MAD (cm)	0.41	0.34
Kurtosis (-)	46.12	25.44
Skewness (-)	5.33	-0.84
Filtered comparison	**Cathedral**	**Café**
Number of points	1,876,985	122,113
Sample mean (cm)	0.12	0.18
Standard deviation (cm)	± 0.48	± 0.35
Median (cm)	0.17	0.18
MAD (cm)	± 0.26	± 0.23
Kurtosis (-)	3.73	2.80
Skewness (-)	-0.01	-0.20

Table 4. Statistical results of the accuracy evaluation: MMS vs. TLS for the square dataset on the raw and filtered point clouds.

Raw comparison	**Garibaldi**	**Belenzani**
Number of points	4,853,987	1,156,009
Sample mean (cm)	-0.24	-0.15
Standard deviation (cm)	± 5.03	± 5.56
Median (cm)	-0.12	-0.52
MAD (cm)	± 0.28	± 0.95
Kurtosis (-)	77.99	50.58
Skewness (-)	-0.66	2.82
Filtered comparison	**Garibaldi**	**Belenzani**
Number of points	3,995,458	989,354
Sample mean (cm)	-0.13	-0.42
Standard deviation (cm)	± 0.43	± 1.26
Median (cm)	-0.10	-0.53
MAD (cm)	± 0.21	± 0.79
Kurtosis (-)	4.08	3.38
Skewness (-)	-0.03	0.07

Table 5. Statistical results of the accuracy evaluation: MMS vs. TLS for the street dataset on the raw and filtered point clouds.

Figure 11. Colour-coded map of the comparison analysis between MMS and TLS data in the Cathedral dataset. The scale ranges from – 0.01 m (blue) to 0.01 m (red).

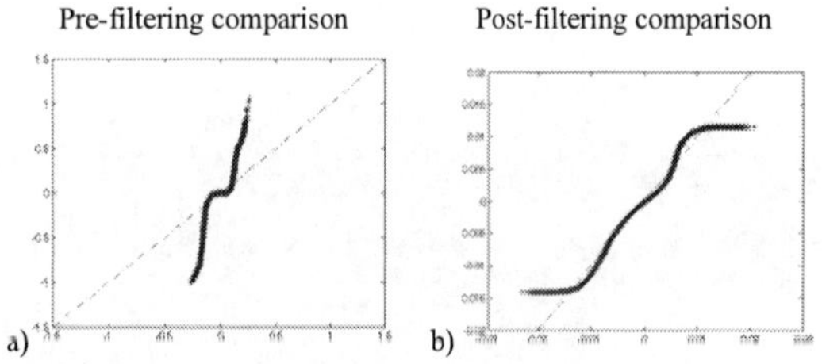

Figure 12. Q-Q plots of the comparison MMS vs. TLS for the Garibaldi dataset. The quantiles of input sample (vertical axis) are plotted against the standard normal quantiles (horizontal axis). Pre-filtering (a) and post-filtering (b) data are shown.

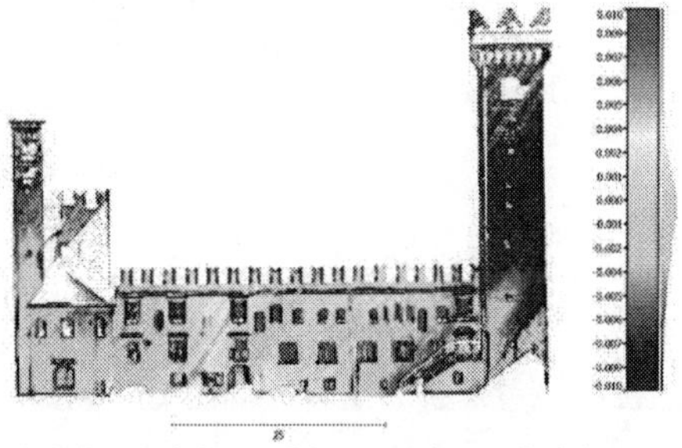

Figure 13. Colour-coded map of the comparison analysis between MMS and TLS data in the Garibaldi dataset. The scale ranges from – 0.01 m (blue) to 0.01 m (red).

This contribution has been peer-reviewed.
doi:10.5194/isprsarchives-XL-5-W4-245-2015

The study is then completed by analysing the signed distances computed for the three façades on the narrow streets (Garibaldi, Cavour and Belenzani datasets). Due to the limited street width and significant height of the buildings, the acquisition angles are expected to be more severe here, deteriorating the metric accuracy of the resulting point clouds. Likewise all previous comparisons, Q-Q plots show a significant departure from normality, that is reduced after the 3σ filtering (Figure 12).
Statistics are listed in Table 5; results achieved by the Cavour dataset are here omitted, since the significant amount of occlusions doesn't allow a proper analysis. An average robust dispersion (MAD) of ± 5.0 mm is outlined, that reaches ± 7.4 mm if a Gaussian estimation is derived. As expected, this value is higher than the one delivered by comparisons of the square datasets, but the gap is not statistically significant.
Worth to note is finally the spatial error distribution, shown as colour-coded map of distances. For the Garibaldi dataset (Figure 13), it is still affected by the systematic diagonal stripes. Furthermore, the error increases quickly on the tower, whose height requires an unfavourable acquisition direction.

5. CONCLUSIONS

This paper strives to provide advanced statistical methods specifically designed to assess precision and accuracy of a MMS solution. The inter-comparison methodology here adopted is based on reference 3D imaging techniques (both active and passive sensors), whose uncertainty is predicted in advance, using manufacturer specifications and design equations. The MMS point clouds (Figure 14 – Duomo square dataset) are then compared against these reference ones and a statistical analysis of the resulting error distribution is performed.
The suggested approach relies on a Gaussian assessment, followed by an outlier filtering process. Finally, non-parametric statistical models are applied, in order to achieve a robust estimation of the error dispersion. All datasets show a large lack of symmetry that leads to the conclusion that the standard normal parameters are not adequate to assess this type of data. The use of non-normal statistics gives thus a more appropriate description of the error population and yields results that meet what may be expected (*a-priori* errors) concerning the assessment of both precision and accuracy.
The comparisons between point clouds acquired by the two laser scanners at the same epoch deliver a Gaussian estimation of the MAD in the order of ± 5.2 mm, which is consistent with the precision (1σ value) quoted by the MMS datasheet. The error increases up to ± 8.6 mm if the acquisitions in two different epochs are compared. These differences can be traced back to geo-referencing errors and surface-related problems. The former may be mitigated by including control point information within the post-processing adjustment. Depending on the final metric requirements, this issue should be adequately dealt with, especially if the MMS is used for accurately mapping historic city centres. Indeed this requires complex trajectories and multiple passes, thus demanding an optimization of the geo-referencing process.
The accuracy assessment of one single point cloud highlights the good metric potentialities of the RIEGL mobile system. Errors estimated for the buildings in the Duomo square are characterized by an average dispersion of ± 5.9 mm (MMS vs. Photogrammetry) and ± 3.6 mm (MMS vs. TLS). These values are consistent with the uncertainty quoted by the MMS manufacturer (8 mm, 1σ). As expected, errors increase if MMS mapping is performed in narrow streets with high buildings and more challenging obstacles. However, the computed value of ± 7.4 mm in such areas still represents an acceptable trade-off.

Figure 14. Intensity (left) and elevation (right) map of the MMS data collected in the Duomo square of Trento (Italy).

REFERENCES

Barber, D., Mills, J., Smith-Voysey, S., 2008. Geometric validation of a ground-based mobile laser scanning system. *ISPRS Journal of Photogrammetry and Remote Sensing*, 63(1), pp. 128–141.

Bitenc, M., Lindenbergh, R., Khoshelham, K., Van Waarden, A.P., 2011. Evaluation of a LiDAR land-based mobile mapping system for monitoring sandy coasts. *Remote Sensing*, 3(7), pp. 1472–1491.

Bossler, J.D., Novak, K., 1993. Mobile Mapping System: new tools for the data collection of GIS information. *Canadian conference on GIS'93*, Ottawa, Canada, pp. 306-315.

This contribution has been peer-reviewed.
doi:10.5194/isprsarchives-XL-5-W4-245-2015

D'Agostino, R.B., 1986. *Tests for the normal distribution.* Marcel Dekker.

El-Sheimy, N., 1996. A mobile multi-sensor system for GIS applications in urban centers. In: *The International Archives of the Photogrammetry, Remote Sensing and Spatial Information Sciences*, Vienna, Austria, Vol. XXXI, Part B2, pp. 95-100.

El-Sheimy, N., Schwarz K.P., Gravel, M., 1995. Mobile 3-D positioning using GPS/INS/Video cameras. *The Mobile Mapping Symposium*, Ohio State, USA, pp. 236-249.

Haala, N., Peter, M., Kremer, J., Hunter, G., 2008. Mobile LiDAR mapping for 3D point cloud collection in urban areas – a performance test. *The International Archives of the Photogrammetry, Remote Sensing and Spatial Information Sciences*, Rostock, Germany, Vol. XXXVII, Part B5, pp. 1119-1124.

Hasan, A., Pilesjö, P., Persson, A., 2011. The use of LiDAR as a data source for digital elevation models – a study of the relationship between the accuracy of digital elevation models and topographical attributes in northern peatlands. *Hydrology and earth system sciences discussions,* 8, pp. 5497-5522.

Hoehle, J., 2011. The assessment of the absolute planimetric accuracy of airborne laserscanning. In: *The International Archives of the Photogrammetry, Remote Sensing and Spatial Information Sciences*, Calgary, Canada, Vol. XXXVIII-5W12, pp. 145-150.

Kaartinen, H., Hyyppä, J., Kukko, A., Jaakkola, A., Hyyppä, H., 2012. Benchmarking the performance of mobile laser scanning systems using a permanent test field. *Sensors*, 12(9), pp. 12814-12835.

Kawashima, K., Kanai, S., Date, H., 2012. Automatic recognition of piping system from laser-scale terrestrial laser scanner point cloud. *Journal of the Japan Society for Precision Engineering*, 78(8), pp. 722–729.

Kremer J., Hunter G., 2007. Performance of the StreetMapper mobile LiDAR mapping system in real world projects. *Photogrammetric week 2007*, Stuttgart, Germany, pp. 215-225.

Lague, D., Brodu, N., Leroux, J., 2013. Accurate 3D comparison of complex topography with terrestrial laser scanner: Application to the Rangitikei canyon (N-Z). *ISPRS Journal of Photogrammetry and Remote Sensing*, 82, pp. 10-26.

Lichti, D., Gordon, S., 2004. Error propagation in directly georeferenced terrestrial laser scanner point clouds for cultural heritage recording. *FIG working week*, Athens, Greece.

Lim, S., Thatcher, C. A., Brock, J. C., Kimbrow, D. R., Danielson, J. J., Reynolds, B. J., 2013. Accuracy assessment of a mobile terrestrial LiDAR survey at Padre Island National seashore. *ISPRS Journal of Photogrammetry and Remote Sensing*, 34(18), pp. 6355-6366.

Luhmann, T., Robson, S., Kyle, S., Harley, I., 2006. *Close range photogrammetry: Principles, methods and applications.* Whittles, Dunbeath, UK, 500 pages.

Nkurunziza, M., Vermeire, L., 2014. A comparison of outlier labeling criteria in univariate. *9th international conference on teaching statistics*, Flagstaff, Arizona (USA).

Pu, S., Rutzinger, M., Vosselman, G., Elberink, S., 2011. Recognizing basic structures from mobile laser scanning data for road inventory studies. *ISPRS Journal of Photogrammetry and Remote Sensing*, 66(6), pp. S28-S39.

Puente, I., González-Jorge, H., Martínez-Sánchez, J., Arias, P., 2013. Review of mobile mapping and surveying technologies. *Measurement*, 46(7), pp. 2127-2145.

Reshetyuk, Y., 2009. *Self-calibration and direct georeferencing in terrestrial laser scanning*. PhD thesis, KTH - Royal institute of technology, Stockholm, Sweden.

Rodríguez-Gonzálvez, P., Garcia-Gago, J., Gomez-Lahoz, J., González-Aguilera, D., 2014. Confronting passive and active sensors with non-gaussian statistics. *Sensors*, 14(8), pp. 13759-13777.

Slattery, K., Slattery, D., Peterson, J., 2012. Road construction earthwork volume calculation using three dimensional laser scanning. *Journal of Surveying Engineering*, 138(2), pp. 96–99.

Tao, V., Li, J., 2007. *Advances in mobile mapping technology.* ISPRS Book Series No. 4; Taylor & Francis: London, UK, 176 pages.

Ziparo, V.A., Zaratti, M., Grisetti, G., Bonanni, T.M., Serafin, J., Di Cicco, M., Proesmans, M., Van Gool, L., Vysotska, O., Bogoslavskyi, I., Stachniss, C., 2013. Exploration and mapping of catacombs with mobile robots. *IEEE international symposium on safety, security, and rescue robotics (SSRR)*, pp. 21-26.

This contribution has been peer-reviewed.
doi:10.5194/isprsarchives-XL-5-W4-245-2015

3D RECORDING, MODELLING AND VISUALISATION OF THE FORTIFICATION KRISTIANSTEN IN TRONDHEIM (NORWAY) BY PHOTOGRAMMETRIC METHODS AND TERRESTRIAL LASER SCANNING IN THE FRAMEWORK OF ERASMUS PROGRAMMES

T. Kersten [a, *], M. Lindstaedt [a], L. Maziull [a], K. Schreyer [a], F. Tschirschwitz [a] and K. Holm [b]

[a] HafenCity University Hamburg, Photogrammetry & Laserscanning Lab, Überseeallee 16, 20457 Hamburg -(thomas.kersten, maren.lindstaedt, kristin.schreyer, felix.tschirschwitz)@hcu-hamburg.de, Lena-Maziull@web.de
[b] NTNU - Norwegian University of Science and Technology, Division of Roads, Transport, and Geomatics, Høgskoleringen 7 A, NO-7491 Trondheim, Norway - knut.r.holm@ntnu.no

Commission V – WG V/4

KEY WORDS: 3D, CAD, meshing, model, reconstruction, virtual

ABSTRACT:

In this contribution the 3D recording, 3D modelling and 3D visualisation of the fortification Kristiansten in Trondheim (Norway) by digital photogrammetry and terrestrial laser scanning are presented. The fortification Kristiansten was built after the large city fire in the year 1681 above the city and has been a museum since 1997. The recording of the fortress took place in each case at the end of August/at the beginning of September 2010 and 2011 during two two-week summer schools with the topic "Digital Photogrammetry & Terrestrial Laser Scanning for Cultural Heritage Documentation" at NTNU Trondheim with international students in the context of ERASMUS teaching programs. For data acquisition, a terrestrial laser scanner and digital SLR cameras were used. The establishment of a geodetic 3D network, which was later transformed into the Norwegian UTM coordinate system using control points, ensured a consistent registration of the scans and an orientation of the photogrammetric images. The fortress buildings were constructed in detail from photogrammetric photographs and point clouds using AutoCAD, while the fortress area and walls were modelled by triangle meshing in Geomagic. The visualisation of the fortress was carried out 2013 with the software Cinema 4D in the context of a lecture in the Master study programme Geomatics. The 3D model was textured and afterwards presented in a video. This 3D model was finally transferred into the game engine Unity for an interactive 3D visualisation on 3D monitors.

1. INTRODUCTION

The Erasmus programme is a student exchange programme of the European Union. Although its name reminds of Erasmus of Rotterdam, a European formed philosopher of the Renaissance, ERASMUS is a backronym meaning "EuRopean Community Action Scheme for the Mobility of University Students" at the same time (DAAD 2012). Beside the 28 European Union member states also further states such as Norway participate in the world-wide largest funding programme of stays abroad at universities. Central components are the formal recognition of study achievements abroad based on the European Credit Transfer System (ECTS) and the financial support of exchange students, who regularly study at a participating university. Apart from study stays and foreign internships in the context of the study also training stays as well as advanced training of general university staff can be financially supported. Basis for the exchange of students and lecturer is an appropriate bilateral agreement between in-going and out-going universities.

The Photogrammetry & Laser Scanning Lab of HafenCity University Hamburg accomplished two-week summer schools in co-operation with NTNU (Norwegian University of Science and Technology, Department of Civil and Transport Engineering Division of Roads, Transport, and Geomatics) in Trondheim three times in the years 2009 to 2011 with the topic "Digital Photogrammetry & Terrestrial Laser Scanning for Cultural Heritage Documentation" in the context of ERASMUS Teaching programmes. In the practical part of the summer school historical buildings were recorded in each case by photogrammetry and terrestrial laser scanning, and their data were processed afterwards in the course. Bases of these summer schools are the courses Architectural Photogrammetry of the Bachelor study programme and the module Terrestrial Laser Scanning of the Master study programme Geomatics of the HCU Hamburg (see also Kersten 2005).

In the context of these two courses the Photogrammetry & Laser Scanning Lab of HCU Hamburg has already documented architecturally significant buildings in Northern Germany for 15 years by photogrammetry and terrestrial laser scanning (Kersten & Acevedo Pardo 2003, Kersten et al. 2003, Kersten et al. 2004, Kersten et al. 2006, Kersten & Lindstaedt 2012). Documentation of the following complex architectural objects are further examples of project works of HCU Hamburg in 3D modelling: The northern sluice of the ancient dam in Marib/Yemen (Kersten 2007), the historical peninsula in Istanbul (Kersten et al. 2009), the Almaqah temples in Sirwah/Yemen (Lindstaedt et al. 2008) and in Yeha/Ethiopia (Lindstaedt et al. 2010), as well as the Al Zubarah fortress in Qatar (Kersten & Mechelke 2013).

In this contribution the 3D object recording, 3D modelling and visualisation of the fortification Kristiansten in Trondheim (Norway) using terrestrial laser scanning and digital photogrammetry, is introduced. After a brief introduction of the fortress Kristiansten, the summer school is described. Afterwards the practical work is described with focus on the modelling and visualisation of the captured object.

* Corresponding author

This contribution has been peer-reviewed.
doi:10.5194/isprsarchives-XL-5-W4-255-2015

2. THE FORTIFICATION KRISTIANSTEN IN TRONDHEIM

Kristiansten Fortress (shown in Figure 1) is a fortification on the eastern side of the city of Trondheim. The fortress is built on a strategically important hill above the Old City Bridge, which was the main entrance to the city from north and east. From this hill it was possible to shoot into the city, and Swedish troops had controlled the city from this hill in 1564 and 1658. Plans for a fortification in this area were made in 1676, but nothing was done until after the big city fire in 1681.
When the two engineering officers Johan Caspar de Cicignon and Anthony Coucheron were sent to Trondheim to create a new city plan after the fire, they also built the fortress. The construction started 1682, and 1684 it was already finished (KAVLI 1987). The original plan was a symmetrical, star-shaped fortification around a central tower, the Donjon. However, the walls were adjusted to the terrain, and the shape became an irregular star with nine corners.
During the years several changes have been made. After being rebuilt in 1740, it has more or less kept its present shape. In the northern part, there are casemates, i.e. arched rooms connected to the inside of the outer walls. The casemates were used for accommodation, food storage, kitchen, prison, etc. They are well kept, and still in use for various purposes. There are three individual buildings on the fortress; the Commander's Residence, which is a modest one floor house, the guard house, standing close to the Commander's Residence, and the Donjon, the main tower.
The Donjon is a whitewashed stone building, appearing as a significant landmark of Trondheim. It covers 17m by 15m, the height is 20m and the wall thickness is four meters. The building has a cellar for storing gunpowder. The ground floor was originally used as residence for the commander. The next two floors were for cannons, which could be fired in all directions. The wooden floors are very robust, made to endure the weight and recoil from the guns. There is also a top floor, behind the roof, used as observation post.
The fortress has been in combat only once, during the great Nordic war. The Swedish king Karl XII sent General Carl Gustaf Armfeldt with ten thousand men to Trondheim in 1718. He was not able to capture the fortress, and thus he failed to capture Trondheim. After the death of Karl XII, Armfeldt and his men had a disastrous retreat over the mountains back to Sweden during the winter 1718/1719, where most of his men died.
In 1814 Norway was forced into a union with Sweden. Kristiansten was decommissioned by the Swedish/Norwegian king Karl III Johan 1816, and it was left to "deterioration by the time". However, the maintenance was continued because of its function as fire watch and fire alarm.
During the German occupation of Norway in World War II, a number of Norwegian patriots were executed at Kristiansten. After the war, the fortress was the execution place for convicted traitors and war criminals in Norway. A historical summary about the fortification Kristiansten is found e.g. in Bratberg (2008).
The fortress has been renovated during the last years. The Donjon is a museum. The Commander's Residence is a restaurant. A memorial chapel is established in one of the casemates. And the whole fortress is a popular area for people to visit for walking and enjoying the view (see Figure 1). Salutations by cannons are done from the fortress on the National Day (May 17), at royal visits, and at other significant occasions. The New Year fireworks of the municipality are sent up from the fortress.

Figure 1: Fortification Kristiansten in Trondheim – nadir aerial photo (top, Google Earth), view from the city to the defence tower Donjon (centre) and view to the city (bottom)

3. SUMMER SCHOOLS IN ERASMUS TEACHING PROGRAMMES

In the years 2009, 2010 and 2011 three two-week summer schools with the topic "Digital Photogrammetry & Terrestrial Laser Scanning for Cultural Heritage Documentation" in the context of the ERASMUS teaching programmes were performed at NTNU Trondheim. The content of each summer school was partitioned into a theoretical and practical part. In doing so, the lecture Architectural Photogrammetry (of the Bachelor study programme Geomatics) and the module Terrestrial Laser Scanning (of the Master study programme Geomatics) of HCU Hamburg formed the substantial content of the summer schools. In the first week, lectures on digital photogrammetry and terrestrial laser scanning were given in the theoretical part of each summer school. Focus areas in digital photogrammetry were

This contribution has been peer-reviewed.
doi:10.5194/isprsarchives-XL-5-W4-255-2015

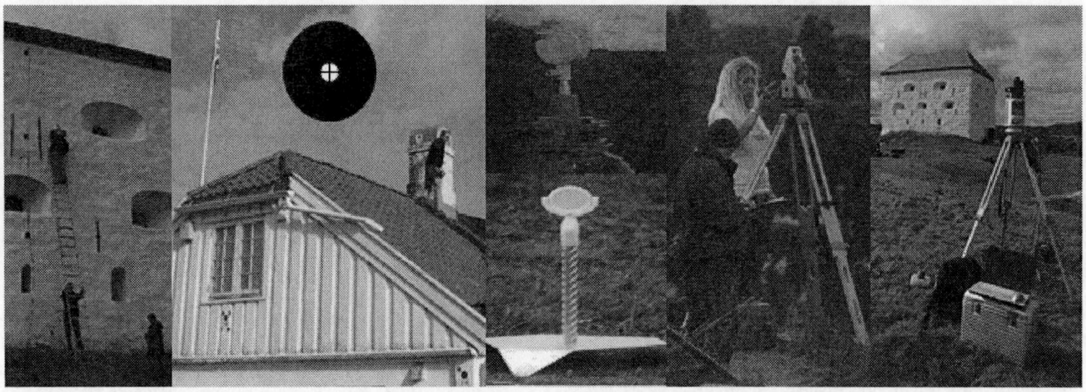

Figure 2: Control point signalisation at both fortress buildings, reflective targets for laser scanning, geodetic control point measurement and Riegl VZ-400 in use (f.l.t.r.)

digital camera systems, camera station configuration, project planning, control point determination, geodetic 3D network adjustment, image orientation and camera calibration by bundle block adjustment, 3D point determination in photos, CAD object reconstruction, modelling and visualisation. Terrestrial laser scanning was covered by the following topics: Measuring methods, systems, criteria for laser scanners, registration and geo-referencing, segmentation, filtering, geometrical accuracy tests, CAD object reconstruction from point clouds, modelling, 3D triangulation and meshing, and analysis of the results. For the practical part the participants worked mostly on a project in the second week. A total workload of 60 hours was defined for the summer schools in each case, which corresponds to two ECTS. The first summer school took place from 7 to 18 September 2009. As a practical project a section of the Nidaros cathedral inside and outside was photographed and scanned, since the NDR (Nidaros Domkirkes Restaureringsarbeider – The Restoration Workshop of the Nidaros Cathedral) was not only interested in data recording and evaluation procedures, but above all also in the data and results. Only in the second and third summer schools the participants were concerned with the fortification Kristiansten. In the second course, which took place from 30 August to 10 September 2010, primarily the fortress tower Donjon and parts of the internal area were recorded for object modelling, while in the third summer school from 29 August to 9 September 2011 the commandant's office, the internal area of the fort as well as the outside walls were recorded and processed. The number of participants of each summer school was between eight and ten students, which came from Norway, Germany, Poland and Finland.

4. DATA RECORDING AND 3D MODELLING OF KRISTIANSTEN FORTIFICATION

The data acquisition of the fortification was conducted on the third and fourth day of each summer school (01./02.09.2010 and 31.08./01.09.2011). In 1.5 days the following work was carried out (including preparation): measurement of a geodetic 3D network, signalisation and measurement of control points for photogrammetry and laser scanning, photogrammetric image recording and terrestrial laser scanning. The measurements of the geodetic 3D network were carried out with a total station Leica TCRP 1201 (see Figure 2), in which eight network stations were forced centring on stable tripods and were mutually measured, in order to ensure solid network geometry. The signalised control points (Figure 2) fixed at the two fortress buildings (Donjon and commandant's office) were measured from the network stations

Figure 3: Perspective view of the defence tower (Donjon) modelled from point clouds – manual texture mapping on meshes in Geomagic

This contribution has been peer-reviewed.
doi:10.5194/isprsarchives-XL-5-W4-255-2015

and from free-stationing. The 3D network and the control points were adjusted using the software PANDA of GeoTec (Laatzen). The network stations were determined with a precision of less than 1mm in both years, while the 28 signalised control points at the defence tower Donjon in 2010 and at the commandant's office in 2011 were adjusted with a precision of 2.5mm in XY and 4mm in Z. The control point and network station coordinates computed in PANDA were transformed into the UTM coordinate system using three identical points. The residuals in XY for two coordinates were about 12cm. The height was adapted by a systematic shift.

The defence tower Donjon was photogrammetrically recorded in 2010 with a Nikon D70 (3008×2000 pixel) and Nikkor lenses (focal length 18mm and 28mm) in 77 images as wraparound image block at eye level for multi-image triangulation. Using the same photogrametric multi image triangulation approach the commandant's office was recorded in 2011 with the Nikon D90 (4288×2848 pixel, focal length 20mm) in 92 images. The multi-image triangulation for image orientation and camera calibration was accomplished for the image block of the commandant's office with the software PICTRAN of technet GmbH, for which on the average 15.7 points were measured per photo and each object point in seven photos. The bundle block adjustment gave a standard deviation of 1mm for the control points and a measuring precision of 2.5μm in the photos. The image block of the defence tower was later used only for texture mapping of the 3D model.

The data recording of the fortress buildings and the structure inside and outside was realised with the terrestrial laser scanner Riegl VZ-400, in which the system was controlled with the software RiScanPro using a notebook (Figure 2 right). The defence tower Donjon was scanned from twelve scanner stations around the building in 2010. For the panorama scans of the fortress interior structure a scan resolution of 20mm at 25m distance was selected, while the building was scanned with a point spacing of 5mm at the building wall. For the subsequent registration of the scans a total of 26 reflectors with a diameter of 6cm (Figure 2 centre) were distributed in the scanned object area (14 in the field, four at the building and eight at the network stations). In 2011 the fortress walls and parts of the internal terrain were scanned from a total of 13 scan stations. 28 targets were set up around the object, and the coordinates were determined with the total station. The walls were scanned with a scan resolution of 20mm at the wall. After each scan panorama images were additionally recorded with a calibrated digital camera the Nikon D700 (Figure 2 right), which were firmly

Figure 4: Fortification Kristiansten as meshed 3D model including modelled canons (top) and 3D model textured in Geomagic (bottom)

This contribution has been peer-reviewed.
doi:10.5194/isprsarchives-XL-5-W4-255-2015

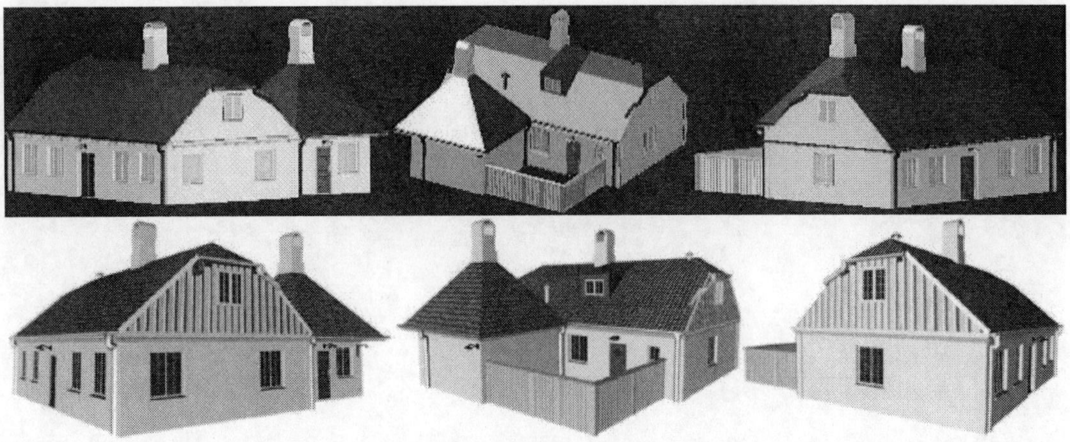

Figure 5: Commander's Residence of fortification Kristiansten – Constructed 3D CAD model from photogrammetric images (top) and 3D model textured in Cinema 4D (bottom)

installed on top of the scanner and which used a full format sensor with 36mm×24mm and a fixed Nikkor lens with 20mm focal length. The image data served for the RGB colouring of the scans. After registration, geo-referencing and colouring of the scans in RiScanPro, the following results for the 13 scans from 2011 were achieved. The scans could be registered with a standard deviation of 5mm for the control points, while a geo-referencing was obtained with a standard deviation of 8.9mm. The results of the 12 scans from 2010 were slightly better. Due to filtering of the scans in RiScanPro not only rough erroneous measurements but also the vegetation could be eliminated, so that as basis for 3D modelling in Geomagic the data of the terrain, the walls and the buildings were available.

As the first step the defence tower Donjon was modelled after further filtering (smoothing/data reduction) of the point cloud by triangle meshing (edge length 3cm) in Geomagic. However, some holes still needed to be closed after meshing, in particular on the roof sectors. A closed meshing with 2.9 million triangles was generated and textured in Geomagic with the photos from the photogrammetric image block (Figure 3). Subsequently, the internal terrain model and the walls of the fortification were modelled by meshing with 1.6 million triangles (edge length 15cm). Furthermore, two different cannon types were modelled as characteristic objects of the fortress (see Figure 4 bottom in red). However, some cannon parts had to be constructed due to incomplete laser scanning (Figure 4 top in the corner). The cannon in Figure 4 consists of 133.000 triangles. After the construction of the two cannon types, each type was copied seven times and placed on the appropriate positions in the model of the fortification (Figure 4 bottom).

As the only object of the fortress the commandant's office was not only photogrammetrically recorded, but also photogrammetrically processed. After image orientation and camera calibration the points necessary for a detailed construction of the building in AutoCAD were measured in different photos using PICTRAN. The CAD construction was carried out exactly the same way as in many other projects of the HCU Hamburg (Kersten et al. 2003, Kersten et al. 2004, Kersten 2005, Kersten et al. 2006), i.e. after determination of the 3D points these were imported into AutoCAD, there connected to 2D polylines, from which volume bodies were created. In CAD Boolean operations can be applied to two or more surfaces or volumes, whereby new object parts can be achieved using intersection, subtraction or union of the surfaces or volumes.

Figure 5 shows the commandant's office of the fortress Kristiansten constructed in detail in AutoCAD. The entire modelling of the fortification and the processing of the whole data from the two field campaigns in 2010 and 2011 were brought to completion as a semester thesis in the context of the ERASMUS student exchange programme in the winter semester 2011/2012 at NTNU Trondheim (Maziull 2011).

5. VISUALISATION

5.1 Video Sequence

In the following winter semester 2012/2013 the same student worked on the 3D model of the entire fortress Kristiansten generated at NTNU Trondheim in co-operation with a fellow student in the context of the module Project Visualisation in the Master study programme Geomatics. In order to generate a video sequence, the entire model was further edited and textured in detail using the software Cinema 4D from the company MAXON Computer GmbH. First the defence tower Donjon was revised in Geomagic by smoothing the meshing using filtering and by reducing the number of triangles to 2.4 million. The textures from Geomagic were not transferred when importing the data to Cinema 4D. Using the function texture mapping in Cinema 4D it turned out that the editing of complex, meshed models, such as the elevation model of the fortress, the walls and the defence tower Donjon, is clearly more labour intensive than the processing of 3D volume bodies (commandant's office). Since the meshed models consist of triangles and/or polygons, they must be attributed to defined surface areas in a Sisyphean task, before different textures can be manually assigned to them. However, automatic texture mapping of meshed models would also have been possible, if oriented image data of a calibrated camera was available in the same coordinate system as the 3D model (Kersten & Stallmann 2012). For the preparation of suitable and qualitatively better textures, photos were cropped and rectified in the free software GIMP. For optimal illumination of the model in Cinema 4D different sources of light were provided and placed, since special characteristics of the different objects thereby can be emphasized well just for the evening or night. In total six spotlights were positioned around the defence tower to light up the foundation walls (Figure 6 top left).

Figure 6: Cut-outs from the video of the fortress Kristiansten compiled in Cinema 4D – defence tower Donjon (top left) and commandant's office (top right) with lighting as well as perspective views on the entire fortress (bottom)

To generate a video sequence for this project, an object-oriented camera linked on a spline was selected, since it could be aligned to different objects on one hand and ran along the splines as camera path on the other hand. During the production of the image sequence single images were stored in the TIFF format after the rendering process with the standard renderer of Cinema 4D, in order to lose no data in case of a process abort, e.g. by a system crash of the computer. The rendering of 6000 Frames (240 seconds or 4 minutes of video length with a rate of 25 frames per second) on a standard computer was very time-consuming with 784 hours (which corresponds to 33 days). Two film sequences (walk-through and fly-through) were generated in the photo manager of Cinema 4D and stored in the video format AVI. For the unification of the two sequences the video cut software iMovie of the company Apple was used. After audio effects were added to the film, e.g. bird twittering and mood music, the video was exported as HD 720p, which corresponds to a resolution of 1280×720 pixel in the format MOV. This video shows a data volume of 261 MB, while the 3D model is 371 MB in the standard format of Cinema 4D (.c4d) and it uses 137 textures with a data volume of 192 Mbyte. The created video is available in the Internet in YouTube (https://www.youtube.com/watch?v=CtUh1El_exg). The expenditure of time for the video production was 175 hours without the time for rendering.

5.2 Interactive 3D Visualisation

Interactive visualisation is a representational form, where the viewing user selects the viewpoint, from which he would like to view the scene, and a game engine computes the selected perspective in real time and displays it. For this project the game engine Unity is used. It offers good performance and programmability using the language C#. A clear advantage in comparison to other common Engines is that it offers direct viewing capabilities on different platforms, such as PC (Windows, Mac), mobile devices (Android and iOS) and in the Web with a single click. This option is already available in the free version of Unity, which offers its use also within applications, for which no large investments are possible.

Since a complete 3D model of the entire fortification is available, due to data recording by laser scanning, the movement of the viewer in the model is translated as a walk-through. Thereby the user moves by foot over the terrain like a human being visiting the real imposing structure. As an interface between computer and user a control system is installed using computer mouse and keyboard as well as the controller of the Playstation3 (Dualshock3). Using buttons and pointers (joystick) both the changes of the position and the viewing direction can be communicated to the system. More details about different controllers in Unity are described in Tschirschwitz et al. (2014).

The output to the user can be carried out alternatively as simple perspective representation or in different forms of stereoscopic 3D. Adapted to different display devices the viewing can be effected using NVIDIA 3D vision for 3D monitors, or top-bottom or side by side stereo, e.g. for stereoscopic projectors.

Interactive visualisation can be performed using the textured model, which was basis of the video sequence rendered. Since no data reduction was carried out by producing a ÇAD model for the entire fortress, the representation (visualisation/display) can in this project only be offered on efficient hardware. Consequently, a presentation with smartphones, such as described in Tschirschwitz et al. (2014), must be relinquished here. The much to high point density of the triangle meshing exceeds the limitations of the processors and graphic chips used in modern smartphones. A strong data compression is not reasonable for this

This contribution has been peer-reviewed.
doi:10.5194/isprsarchives-XL-5-W4-255-2015

model, since high resolution is necessary for a realistic and attractive representation of this complex and challenging area.

6. CONCLUSION AND OUTLOOK

In this contribution it was shown that even larger projects in architectural photogrammetry could be successfully processed by the integration of practical projects into lectures. This is true for the integration into the Bachelor and Master Study programmes at HCU Hamburg as well as in lectures of international summer schools in the framework of ERASMUS lecturer and student exchange programs. The data acquisition and first working processes were realised in two summer schools for the example project fortress Kristiansten, in which students from different universities could intensively deal with the topic in lectures and practical project work within two weeks' time. The advanced and more in-depth processing of the project was then accomplished by a student in the context of an ERASMUS student exchange programme at NTNU Trondheim (Maziull 2011). In the following winter semester 2012/2013 the same student could continue working on the generated 3D model of the fortification Kristiansten in the course project visualisation of the Master study programme Geomatics. The entire model was arranged, further processed and textured very detailed in Cinema 4D, so that a video sequence could be generated, which is available at YouTube on the Internet. After the conversion of the 3D models from Cinema 4D into the game engine Unity further possibilities of immersive viewing for an interested user are given with interactive 3D visualisation on different output devices such as 3D monitors.

The integration of such practical projects in lectures of Bachelor and Master Study programmes promotes an increased motivation and willingness to learn, because the theoretical background and the practical projects can be linked well with each other. The preparation and execution of international summer schools demand an appropriate logistic and organisational expenditure, in addition to the necessary bureaucratic expenditure for ERASMUS programme. However, the benefit faces the expenditure by temporally limited, but intensive training, by the development of the personal network, the treatment of interesting projects and the extension of the own horizon by teaching and learning in an international environment.

REFERENCES

Bratberg, T.T.V., 2008. Trondheim Byleksikon. 2nd Edition, Kunnskapsforlaget, Oslo, ISBN 978-82-573-1762-1.

DAAD, 2012. ERASMUS (Special edition) - Happy birthday, ERASMUS! Die Erfolgsgeschichte der Europäischen Union feiert ihr 25-jähriges Bestehen. DAAD euroletter, Deutscher Akademischer Austauschdienst, August, 36 p. (Access at 19th January 2015), https://eu.daad.de/medien/eu/publikationen/sonderausgabe_euroletter.pdf.

Kavli, G., 1987. Norges festninger. Universitetsforlaget, ISBN 978-82-00-18430-0.

Kersten, T., Acevedo Pardo, C., 2003. Wie kommt das Ahrensburger Schloss in den Computer? - 3D-Gebäudeerfassung und Visualisierung durch digitale Photogrammetrie. *DenkMal! Schleswig Holstein, Zeitschrift für Denkmalpflege in Schleswig Holstein,* Landesamt für Denkmalpflege in Schleswig-Holstein (Hrsg.), Jahrgang 10, 2003, Westholsteinische Verlagsanstalt Boyens & Co., Heide, pp. 48-54.

Kersten, T., Eilmus, B., Lindstaedt, M., Acevedo Pardo, C., 2003. 3D-Erfassung und Visualisierung des Celler Schlosses durch digitale Architekturphotogrammetrie. *Photogrammetrie, Laserscanning, Optische 3D-Messtechnik - Beiträge der Oldenburger 3D-Tage 2003,* Th. Luhmann (Hrsg.), Wichmann Verlag, Heidelberg, pp. 213-222.

Kersten, T., Acevedo Pardo, C., Lindstaedt, M., 2004. 3D Acquisition, Modelling and Visualization of north German Castles by Digital Architectural Photogrammetry. *International Archives of Photogrammetry, Remote Sensing and Spatial Information Sciences,* Istanbul, Turkey, Vol. XXXV, Part B2, pp. 126-132.

Kersten, T., 2005. Digitale Architekturphotogrammetrie als ein Beispiel für die praxisorientierte Ausbildung an der HAW Hamburg. HAW Hamburg, Fachbereich Geomatik, Festschrift zur Verabschiedung von Prof. Dr. h.c. Jürgen Zastrau, 8. Juli 2005.

Kersten, T., Biebermann, M., Schneider, M., 2006. 3D-Erfassung und Modellierung des Duderstädter Westerturmensembles durch Kombination von digitaler Architekturphotogrammetrie und terrestrischem *Laserscanning. Photogrammetrie, Laserscanning, Optische 3D-Messtechnik - Beiträge der Oldenburger 3D-Tage 2006,* Th. Luhmann/C. Müller (Hrsg.), Wichmann Verlag, Heidelberg, pp. 254-263.

Kersten, T., 2006. Combination and Comparison of Digital Photogrammetry and Terrestrial Laser Scanning for the Generation of Virtual Models in Cultural Heritage Applications. *The 7th International Symposium on Virtual Reality, Archaeology and Cultural Heritage,* VAST (2006), M. Ioannides, D. Arnold, F. Niccolucci, K. Mania (Editors), Hilton Nicosia, Cyprus, Oct. 30 - Nov. 4, 2006, pp. 207-214.

Kersten, T., 2007. Virtual Reality Model of the Northern Sluice of the Ancient Dam in Marib/Yemen by Combination of Digital Photogrammetry and Terrestrial Laser Scanning for Archaeological Applications. *International Journal of Architectural Computing,* Special Focus on Cultural Heritage, 02 (05), Multi-Science, pp. 339-354.

Kersten, T., Büyüksalih, G., Baz, I., Jacobsen, K., 2009. Documentation of Istanbul Historic Peninsula by Kinematic Terrestrial Laser Scanning. *The Photogrammetric Record,* 24(126): pp. 122-138 (June 2009).

Kersten, T., Lindstaedt, M., 2012. Virtual Architectural 3D Model of the Imperial Cathedral (Kaiserdom) of Königslutter, Germany through Terrestrial Laser Scanning. In: *EuroMed 2012 - Int. Conference on Cultural Heritage,* Ioannides, M.; Fritsch, D.; Leissner, J.; Davies, R.; Remondino, F.; Caffo, R. (Eds.), Lecture Notes in Computer Science (LNCS), Volume 7616, Springer-Verlag Berlin Heidelberg, pp. 201-210.

Kersten, T., Stallmann, D., 2012. Automatic Texture Mapping for Architectural and Archaeological 3D Models. In: *International Archives of Photogrammetry, Remote Sensing and Spatial Information Sciences,* Melbourne, Australia, Vol. XXXIX, Part B5, pp. 273-278.

Kersten, T., Mechelke, K., 2013. Fort Al Zubarah in Katar – 3D-Modell aus Scanner- und Bilddaten im Vergleich. AVN - *Allgemeine Vermessungs-Nachrichten,* 2/2013, pp. 50-58.

Lindstaedt, M., Kersten, T., Mechelke, K., Götting, M., Heiden, R., 2008. Virtuelles 3D-Modell der antiken Tempelanlage in Sirwah/Jemen zur archäologischen Objektdokumentation durch terrestrisches Laserscanning und Photogrammetrie. *Publikationen der Deutschen Gesellschaft für Photogrammetrie, Fernerkundung und Geoinformation e.V.,* Band 17, Hrsg. E. Seyfert, 28. Wissenschaftlich-Technische Jahrestagung der DGPF, 23.-25. April 2008 in Oldenburg, pp. 59-68.

Lindstaedt, M., Mechelke, K., Schnelle, M., Kersten, T., 2011. Virtual Reconstruction of the Almaqah Temple of Yeha in Ethiopia by Terrestrial Laser Scanning. Proceedings of the 4th ISPRS International Workshop 3D-ARCH 2011: "3D Virtual Reconstruction and Visualization of Complex Architectures", Fabio Remondino, Sabry El-Hakim (Eds.), *International Archives of Photogrammetry, Remote Sensing and Spatial Information Sciences*, Trento, Italy, Volume XXXVIII-5/W16.

Maziull, L., 2011. 3D Modelling and Visualization of Kristiansten Fortress in Trondheim. Semester Thesis at NTNU Trondheim, December.

Maziull, L., 2013. Visualisierung der Festung Kristiansten in Trondheim, Norwegen in Cinema 4D. Semester Thesis at HCU Hamburg, March.

Tschirschwitz, F., Kersten, T., Zobel, K., 2014. Interactive 3D Visualisation of Architectural Models and Point Clouds Using Low-Cost-Systems. In: *Progress in Cultural Heritage. Documentation, Preservation, and Protection.* 5th International Conference, EuroMed 2014, Limassol, Cyprus, November 3-8, 2014, Ioannides, M., Magnenat-Thalmann, N., Fink, E., Zarnic, R., Yen, A.-Y., Quak, E. (Eds.), Lecture Notes in Computer Science (LNCS), Volume 8740, Springer International Publishing Switzerland 2014, pp. 268–278.

This contribution has been peer-reviewed.
doi:10.5194/isprsarchives-XL-5-W4-255-2015

MODELING AND ACCURACY ASSESSMENT FOR 3D-VIRTUAL RECONSTRUCTION IN CULTURAL HERITAGE USING LOW-COST PHOTOGRAMMETRY: SURVEYING OF THE "SANTA MARÍA AZOGUE" CHURCH'S FRONT

Robleda Prieto G. [a, *], Pérez Ramos A. [a]

[a] Dept. of Technology and Science of Graphical Representation. University School of Technical Arquitecture – University of La Coruña. Rúa de Rúa da Fraga 27, Campus da Zapateira, 15008 A Coruña, Spain – gustavo.robleda@udc.es

Commission V, WG V/4

KEY WORDS: Low-cost photogrammetry, Cultural Heritage, Architectonic Rehabilitation, Virtual Reconstruction

ABSTRACT:

Sometimes it could be difficult to represent "on paper" an architectural idea, a solution, a detail or a newly created element, depending on the complexity what it want be conveyed through its graphical representation but it may be even harder to represent the existing reality. (a building, a detail,...), at least with an acceptable degree of definition and accuracy. As a solution to this hypothetical problem, this paper try to show a methodology to collect measure data by combining different methods or techniques, to obtain the characteristic geometry of architectonic elements, especially in those highly decorated and/or complex geometry, as well as to assess the accuracy of the results obtained, but in an accuracy level enough and not very expensive costs. In addition, we can obtain a 3D recovery model that allows us a strong support, beyond point clouds obtained through another more expensive methods as using laser scanner, to obtain orthoimages.

This methodology was used in the study case of the 3D-virtual reconstruction of a main medieval church façade because of the geometrical complexity in many elements as the existing main doorway with archivolts and many details, as well as the rose window located above it so it's inaccessible due to the height.

1. INTRODUCTION

In the labour to collect data to represent architectonic elements, some authors (Scherer, 2002) reduce the acquisition of metric data for graphic representation of cultural and architectural heritage in general, to four methods (traditional, topographic, photogrammetric and scanning), while there are significant differences between them in terms of accuracy, cost and time employee in taking and/or data processing.

In this paper, we try merging the aforementioned methods to obtain a methodology that allows us to obtain results with high accuracy level as are obtained through sophisticated techniques as laser scanner, but less expensive than these: in our 3D-virtual reconstructions we mainly use the photogrammetric software "Agisoft – PhotoScan Professional Edition" (3,499.00 USD = 2,795.31€), a non-professional photographic camera "Canon 600D" including a 18-35mm lens and 32Gb SD-card (499.00 €). To this work equipment, needed tools should be taken into account to collect measure data on site to "scale, rotate and translationate" the virtual model: in our case we use a laser total station "Leica TCR-307" (nowadays phase-out model but similar, as "Leica TS-02", costs 7,018.00€). Thus, total amount goes from 3,294.00€ to 10,312.00€, depending on using the total station or more traditional techniques to set horizontal and vertical axis to obtain a local coordinate system and control measure differences between reference points and control points. Nevertheless, laser total stations can be rented from 40€/day, so topographical support can be easily assumable.

On the other hand, using laser scanner techniques is much more expensive than aforementioned ones: "Leica C-10" laser scanner ("time of fly" measure system) sale price is over 50,000€ and "Faro Focus 3D" ("phase difference" measure system) is around 34,000€. Point cloud treatment software should be added to these expenses. Renting option is also available in this case: "Leica C-10" rent is 600€/day but other 600€/day (at least one day) corresponding to a specialized operator belonging to the reseller have to be added (requirement imposed by the reseller consulted) including temporary license software, while "Faro Focus 3D" rent is 500€/day with point cloud treatment trial software available.

This cost analysis summarizes why are named "low-cost" the technique introduced by this paper. In addition, unlike laser scan technique, added value is obtained because of we can obtain photorealistic finishes into the same software where point clouds are generated. At the same time, this allows us a visual support in the process of vectorizing complex geometries, because we have the support of the images taken through our non-professional camera and not only coloured point clouds that could not be enough defined.

To show all of this, we use the study case of the 3D-virtual reconstruction of the Santa María de Azogue church's front in Betanzos, built between the late 14th and early 15th century in Gothic style. It has a main doorway with a semicircular arch richly decorated, above which there is an outstanding rose window. The church is located on English Santiago's Way, just in the cross of the two branches coming from Ferrol and from

* Corresponding author.

This contribution has been peer-reviewed.
doi:10.5194/isprsarchives-XL-5-W4-263-2015

La Coruña that led for many years pilgrims coming from England who landed in the two biggest northern seaports close to Santiago Cathedral. Because of its location, the church was affected by many architectonic influences which translated into a highly decorated and complex geometry that makes using traditional methods to collect data for its graphical representation unuseful so we thought advanced methods were better than traditional ones but without becoming unaffordable and at the same time with accuracy enough.

The 3D-virtual reconstruction accuracy has been assessed through GCP (ground control points) comparing the XYZ coordinates of the points measured on the "real model" with laser total station and the other ones supplied by the photogrammetric reconstruction on the "virtual model" for the homologous points. Previously we had to scale (putting in the same measure both real model and virtual model), rotate (stablishing the axis direction of the reference system) and traslationate (defining the origin of the reference system) the virtual model obtained through GRP (ground reference points) as well measured with laser total station. In this case, we got maximum error of 0.0047m (1.109 pixel), considered low enough to be admitted to obtain orthoimages to use in the surveying of the church.

2. METHODOLOGY AND RESULTS

Departing from the same hypothesis as another studies performed by us (Robleda Prieto G., 2014b) but changing the software used till now (Photomodeler Scanner), we try to validate a photogrammetric virtual recovery methodology by using mainly: a non-professional photographic camera (to take photographs), photogrammetric software running by PC's (to obtain a 3D virtual model), a total station (to obtain coordinates of some control and reference points) and a tablet (to define unmistakably points to be measured with total station).

In this case, we used an off-the-shelf camera "Canon 600D" with a 18-35mm lens, which was mounted on a tripod in order to obtain the maximum definition as possible and avoid the blurriness due to lack of stability with high shutter speed.

In order to get the most definition and sharpness, the project was divided in two subprojects:

2.1 Subproject "Doorway"

The main door is the most decorated and the major complex geometry element. So photos were taken from the closest distances as possible as to embrace the whole door.

The photos were taken from 9 tripod placements in a way that from each one of those, the camera was focusing towards the same direction which was approximately the doorway centre while the distances from each placement to the door were more or less the same (Cueli J.T., 2011). Thus, we covered all the elements (frontal and jambs) from different angles and, at the same time, the disposition of the tripod placements in arch form (on floor) allows to get more accuracy in the photogrammetric processes performed by the software, because of the convergence of the images (Esmaeili F et al., 2013).

The distance between each placement was approximately in a range of a ratio b/h = 0.10 - 0.15, being "b" the distance from a camera position to the next and "h" the distance to the object focused. This ensured us to get properly horizontal pairs of images to generate point clouds (Cueli J.T., 2011). Even more, from each settlement, two photos were shot from both positions: one in a height of the tripod used and the other one in a height of the tripod such as ratio "b/h" was kept in a 0.10-0.15 range, which means that distances between settlements were the same than distances between the two different photos taken from a only settlement. Thereby we could obtain not only horizontal pairs of images but also vertical pairs.

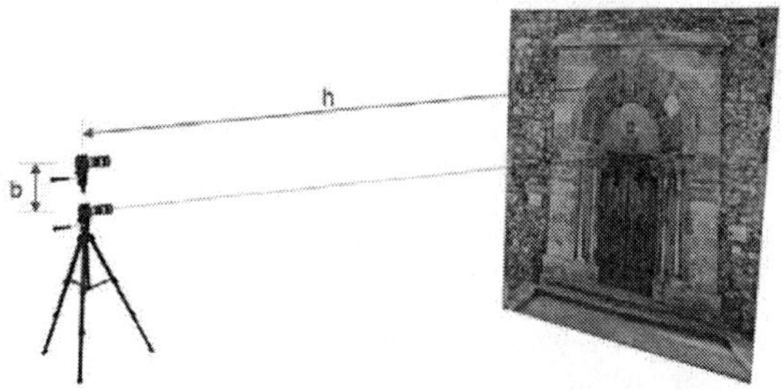

Figure 1. Ratio "b/h" in a tripod placement centered relative to the object, to obtain a vertical pair of images.

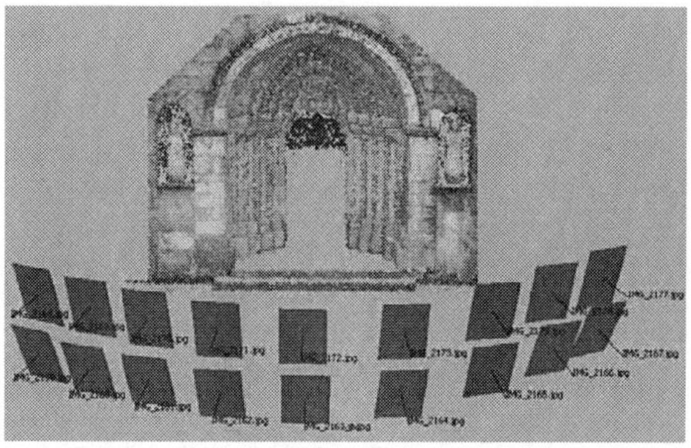

Figure 2. Sparse point cloud of the doorway with all the shootings needed to get both horizontal and vertical pairs.

Related on the shooting, next adjustments were set on manual mode camera (Robleda Prieto G., 2014a): focal length = 18 mm, ISO = 100 (to obtain the maximum sharpness our camera can supply), aperture value = f/11 (to get an optimum depth field without diffraction problems on images), shutter speed = 1/15seg (to balance the lighting level with the ISO choosen: the lower ISO, the more sharpness and the more light needed).

The photos were taken in maximum resolution as possible (5184x3456 pixel) and in RAW format to keep all their properties in the moment of the shoot instead of to use a compression format which could to remove some image information because of the compression processes. Once in the office, the photos were "printed" to JPEG format after to make some little corrections (colour, contrast, etc.) to try equalize image quality between them.
This images in JPEG format, were imported into the "Agisoft – PhotoScan Pro" software (Agisoft, 2015) to obtain point clouds from monoscopic digital multi-images (Hui, 2012) through the photogrammetric processing. But before, images must be

This contribution has been peer-reviewed.
doi:10.5194/isprsarchives-XL-5-W4-263-2015

aligned by the software, what means that software must solve how photos are located in the space and in relation with the object to be recovered. In this case we achieved to align all the images, so all the stereo-pairs could be used to get point clouds but software only uses the better stereo-pairs. After this process of aligning, a sparse point cloud with 242358 points was obtained.

2.2 Subproject "General Façade"

In those zones of the façade where geometry is less complex (rectangular ashlars) or in those other not showed into the scene of "Doorway Subproject" (like the Rosewindow)" with the same reasoning than before, 14 tripod placements were performed, all of them located at the same distance from de façade center (more or less) but enough to embrace the façade remaining. Perhaps, we could have subdivided the whole project in another subproject more for the Rosewindow only (in fact, we took more photos as to do it like this), but finally it was no necessary because this subproject including this singular element was considered sharpness enough.

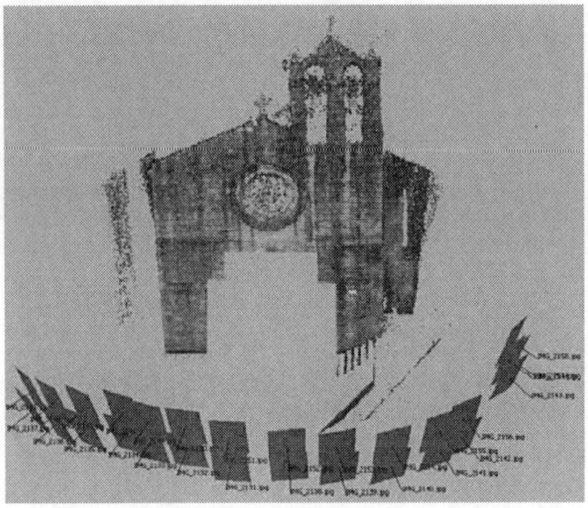

Figure 3. Sparse point cloud of the general façade (doorway zone excluded) with all the shootings needed: 14 tripod placements, two photo from each placement, so 28 images.

The camera settings were similar than the previous subproject: focal length = 18 mm, ISO = 100, aperture value = f/11, shutter speed = 1/13seg.

In a similar way than before, once images were converted from RAW to JPEG format, imported into the software and aligned, a second sparse point cloud was get in this subproject with 267382 points.

2.3 Merged Project

As a result of merging the sparse point clouds of the two subprojects, another sparse cloud was obtained with 507254 points: 242358 points concentrated in the doorway and 267382 belonging to the remaining front façade. Notice that 242,358 plus 267,382 is more (509,740) than 507,254 because of redundant points (very close between themselves) are removed in the merge process.

In the previous subprojects, before aligning images, masks were applied to photographs to avoid parts of the scenes (sky, obstacles, people, etc.) which could produce errors in the process of aligning photos and obtaining sparse point cloud, so the merge cloud hadn't to be barely cleaned.

Figure 4. Merged sparse point cloud of the whole project with the two sets of tripod placements and all the images (28+18).

In this step Dense Point Cloud could have been obtained but it was considered inopportune until knowing the reliability of the virtual model in terms of accuracy.

2.4 Accuracy assessment

On site, with a laser total station, measurement points were taken in order to assess the accuracy of the recovery model. The laser EDM allows to get distances directly to points located on a surface which can reflect the laser, so with the laser total station we can obtain coordinates X,Y,Z of any point on the façade.

In this case instead of using coded target, we used "natural targets" (any natural mark on the stone façade) to not damage the building. The chosen points must be unmistaken so those can be carefully chosen to be visible in the photographs and perfectly recognizable. To define this points a mobile device (tablet) was used: if a possible "natural target" can be seen in a photo taken with a mobile device it is sure than this point will be seen in a photo taken with a more resolution camera, so the target is marked on the mobile device.

Figure 5. Natural target nº 6: white dot on the granite façade.

This contribution has been peer-reviewed.
doi:10.5194/isprsarchives-XL-5-W4-263-2015

Once chosen all the targets, in the same order as defined in the mobile device, points measurements are taken with the laser total station.

Figure 6. Point measurements collected on site with laser total station aided with a mobile device.

This "special" points are usually named GCP (Ground Control Points) but other times those points can be named GRP (Ground Reference Points), depending on this points are used to scale and set the reference system or if they are used for assess the accuracy of the recovery model. In this study case, point n.1 and point n.2 were both considered GRP and these and another eight were GCP.

POINT ID.	RELATIVE COORDINATES (m)			LOCAL COORDINATES (m) [coord. relativas traslacionadas]			ROTATED LOCAL COORDINATES (m)		
	X	Y	Z	x	y	z	x'	y'	z'
1	0,000	17,578	2,197	0,000	0,000	0,000	0,000	0,000	0,000
2	4,790	18,013	3,726	4,790	0,435	1,529	4,810	0,000	1,529
3	0,692	17,470	6,013	0,692	-0,108	3,816	0,679	-0,170	3,816
4	2,800	17,699	6,964	2,800	0,121	4,767	2,799	-0,133	4,767
5	6,853	17,969	4,209	6,853	0,391	2,012	6,860	-0,230	2,012
6	-2,661	17,200	3,333	-2,661	-0,378	1,136	-2,684	-0,136	1,136
7	-1,437	17,317	7,251	-1,437	-0,261	5,054	-1,455	-0,130	5,054
8	1,715	17,599	7,360	1,715	0,021	5,163	1,710	-0,134	5,163
9	3,664	17,689	8,378	3,664	0,111	6,181	3,659	-0,221	6,181
10	6,816	18,070	12,288	6,816	0,492	10,091	6,833	-0,126	10,091

θ_{1-2}° ε°
84.81096 5.1890

Table 7. GRP (highlighted in orange) used to rotate the model and to set the origin of local coordinates (see next Fig.) and GCP used to scale the model and to assess the accuracy.

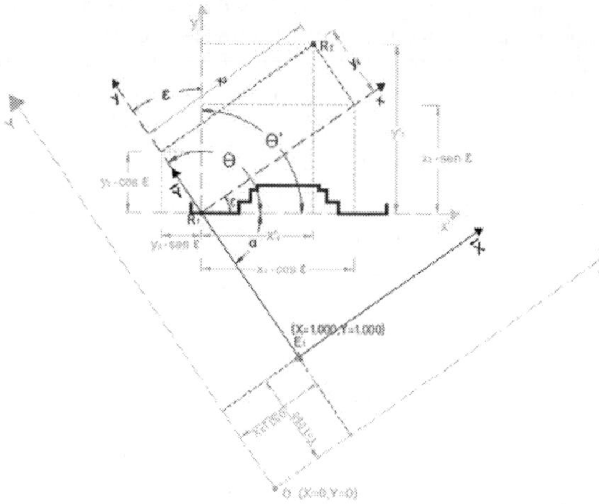

Figure 8. Scheme (floor plan) of arbitrary position and orientation of the total station reference system related with a building façade (cross section in blue).

It's not possible to place the total station on a point on the façade (the origin of the local reference system) or to place it perfectly orthogonal to the centre of the façade. So we placed the total station anywhere with arbitrary relative coordinates (for example x=1000, y=1000, z=1000) and unknown absolute coordinates (X=?, Y=?, Z=?), and a first point on the façade was chosen to set the origin of angles of total station, because this point will be the local coordinates origin (point n.1, x'=0, y'=0, z'=0) of the model. Then, a second point was also carefully chosen (point n.2) but this had to be a target in the same depth (y' point n.2 = 0) as the first because it was necessary to estimate the rotation angle: in this case ε = 5.1890° (see Tab. 7 and Fig. 8).

Thus, the GRP, in this case, were P.1 and P.2 (they aided us to reference the model) but other natural marks clearly visible on the façade were chosen as GCP (point 3 to point 10) to perform the assess of the accuracy. As GRP as GCP were imported into photogrammetric software to scale the model but before this, coordinates obtained from the total station had to be moved from placement station E1 to the local coordinates origin P.1 and after the local coordinate system had to be rotated towards direction P.1–P.2 (see Tab.7 and Fig.8) Then, the assessment of the accuracy were made through GCP, excepting those with a high error (probably due to a bad quality of the target chosen).

Markers	X (m)	Y (m)	Z (m)	Error (m)
1	0.000000	0.000000	0.000000	0.018064
2	4.810000	0.000000	1.529000	0.012264
3	0.679000	-0.170000	3.816000	0.008844
4	2.799000	-0.133000	4.767000	0.012628
5	6.860000	-0.230000	2.012000	0.000755
6	-2.684000	-0.136000	1.136000	0.008960
7	-1.455000	-0.130000	5.054000	0.008021
8	1.710000	-0.134000	5.163000	0.009369
9	3.659000	-0.221000	6.181000	0.007709
10	6.833000	-0.126000	10.091000	0.026668
Total Error				**0.008702**

Table 9. GRP and GCP imported into "Agisoft – PhotoScan Pro", used to scale the model and to assess the accuracy.

This contribution has been peer-reviewed.
doi:10.5194/isprsarchives-XL-5-W4-263-2015

2.5 Dense Point Cloud, Mesh and Texture

After to assess the accuracy the model was validated because the total error was only 8.7 mm (see Tab. 9), so the processes carried on with getting a Dense Point Cloud of the whole model (merged the two subprojects) 1with 2,423,632 points.

Figure 10. Orthoimage of the Dense Point Cloud of the whole project (2,423,632 points) and the GRP and GCP on it.

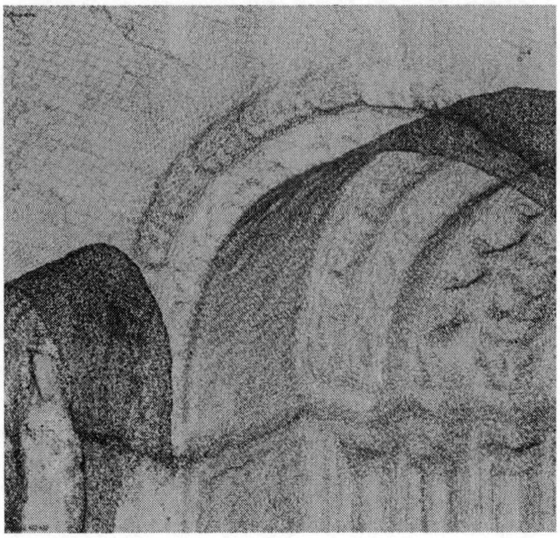

Figure 11. Dense Point Cloud detail in orthographic perspective

Next step was to triangulate the dense cloud points to get the Mesh. Which is needed to obtain faces, namely, the DSM (Dense Surface Model) on which photos are project to obtain the Textured Model.

Figure 12. Mesh (orthographic perspective) generated from the Dense Point Cloud: 218,200 vertex and 431,705 faces.

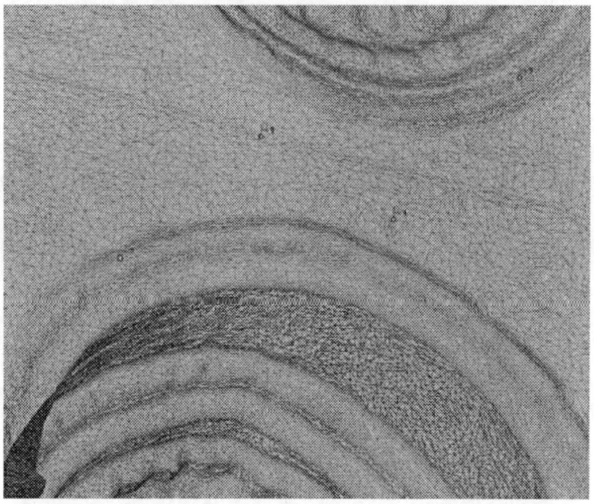

Figure 13. Mesh detail: vertices are closer each other in zones with more complex geometry so faces are smaller just there.

The triangles of the original mesh would have been very small because vertices coming from Dense Point Cloud were very close each other, so generated mesh was decimated to reduce a little bit the weight of the file and thus can manage the mesh in an easier way. For this, the number of the mesh vertex (218,200) are much less than points of the Dense Cloud (2,423,632). From this resultant final Mesh was get the DSM.

This contribution has been peer-reviewed.
doi:10.5194/isprsarchives-XL-5-W4-263-2015

Figure 14. Perspective of the Dense Surface Model (DSM).

At the end of the modelling process, the projection of the images on the Dense Surface Model was done. Having obtained a Mesh with a triangle size small enough on the face made by the DSM of each triangle, the software projects the image more orthogonal to each triangular face, so the smaller triangles and the more photos available, the better.

Figure 15. Orthoimage of the Textured Model obtained by projecting photos on the DSM.

To obtain the correct view (front, top, lateral), orthographic or perspective, is basic the process aforementioned of moving the coordinate system from relative to local as well as to rotate him parallel to an axis contained on the façade. If not, it couldn't export the correct view and even it couldn't assess correctly the accuracy of the model.

Figure 16. Doorway detail: orthographic perspective of the Textured Model getting from the DSM (from decimated Mesh).

Figure 17. Ortho-perspective: doorway top and rose window.

This contribution has been peer-reviewed.
doi:10.5194/isprsarchives-XL-5-W4-263-2015

2.6 Export model, export orthoimages

Having done correctly the process to convert the coordinates system from relative to local, the software used allows easily to export orthoimages of different projection planes.

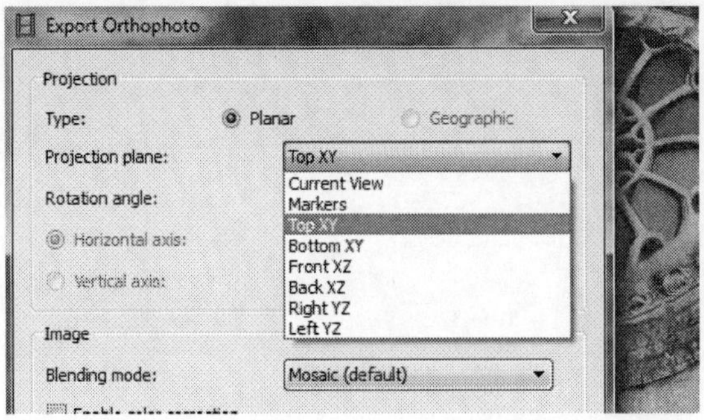

Figure 18. "PhotoScan Pro" Export Orthophoto window.

On the other hand, the model could be exported from "Agisoft PhotoScan Pro" to "Adobe PDF 3D" file which allows everybody to explore the model through zoom, rotate and move functions, as well as to get measurements directly from the model (that was exported in a properly scale).

Figure 19. PDF-3D model exported from "Agisoft PhotoScan Pro": measurement in first left column (annotation in green).

3. CONCLUSIONS

By combining different techniques, mainly topography and photogrammetry, it was possible to obtain dense point clouds of a complex architectonic element that by traditional methods would have been impossible, by inaccessible and complexity.

The reached accuracy was 0.008702m (8.702mm) with a medium error = 0.02972m and maximum error = 0.009369m. It was considered an accuracy enough: being greatest distance between GRP/GRC 7.19m (P.1 and P.9), the accuracy was 0.13% (0.009369/7.19 * 100) = 1.3‰.

The GCP n.10 had to be refused because of his high error (0.02668m), probably due to the long distance to him (at the base of the bell tower) it was necessary the point was visible on photos so this point was so big to obtain better accuracy.

Thus, this methodology can be considered a reliable tool for the graphical representation of architectonic elements (or others) with very complex geometry, and that another sophisticated methods would not be able to adequately solve or they could be unaffordable.

ACKNOWLEDGEMENTS

Liubov Rozhkova (Agisoft Sales manager) and Diana Ovod (Agisoft Tehcnical Support), the first for provide me a educational license of "PhotoScan Pro" and the second for reply all my questions.

Jorge Tomás Cuelli López, for his support, patience, numerous tips, resolution of doubts and lessons given to me along three (or more) last years about digital photogrammetry and photography for digital photogrammetry.

Department of Technology and Science of Graphical Representation (University of La Coruña), to which I belongs as a university teacher, for putting at my disposal available resources.

Department of Representation and Architectonic Theory (University of La Coruña) for lending me a laser total station and a laser distance meter.

REFERENCES

Agisoft LLC. Official website. Saint Petersbourg, Russia. http://www.agisoft.com (22 Jan. 2015).

Cueli López J.T. Fotogrametría práctica (Practical Photogrammetry). *Edidiones Tantín*. Santander 2011.

Esmaeili F., Varshosaz M., Mamid E., Displacement measurement of the soil nail walls by using close-range photogrammetry and introduction of CPD method. *Measurement*. 2013, n. 46, p. 3449-3459

Hui B et al. Accurate geometric camera calibration technique using multi-views of a non-metric planar grid. *Optic and Lasers in Engineering*. 2012, n. 51, p. 432-439.

Robleda Prieto G. Levantamientos Topográficos mediante técnicas fotogramétricas de bajo coste y corto alcance (Topographical Surveys through close-range and low-cost Photogrammetric Techniques). *Proceedings of Graphic Expression applied to Building, XII International Conference – APEGA2014*. Madrid, Spain.

Robleda Prieto G. Levantamiento Planimétrico de la portada de la Iglesia de Santiago en La Coruña mediante restitución fotogramétrica con métodos de bajo coste (Surveying of

This contribution has been peer-reviewed.
doi:10.5194/isprsarchives-XL-5-W4-263-2015

Santiago's Church Front in La Coruña, through Photogrammetric Restitution with low-cost methods). *Proceedings of Graphic Expression applied to Building, XII International Conference – APEGA2014*. Madrid, Spain.

Scherer M. About the synthesis of different methods in surveying. *Proceedings* of *CIPA 2002 XVIII International Symposium*. pp. 423-429 Postdam, Germany.

RECREATION AND SIMULATION OF PREINDUSTRIAL FLOUR PROCESSES IN THE MARGIN OF RIVERS

Juan J. Fernández[a], Jesús I. San José[a], José Martínez-Rubio[a], Javier Finat[b]

[a] Laboratory of Architectural Photogrammetry (LFA)
ETS Arquitectura, Universidad de Valladolid
Avda Salamanca s/n, 47014, Valladolid, Spain
(juanjo, lfa, jmrubio)@ega.uva.es
[b] Modelado, Biomecánica y Visualización Avanzada del Patrimonio (MoBiVAP)
Parque Científico de la Universidad de Valladolid
Paseo de Belén 11, 47010, Valladolid, Spain
jfinat@agt.uva.es

Commission V, WG V/4

KEY WORDS: Cultural Heritage, Augmented Reality, Preindustrial activities, Simulation of processes

ABSTRACT:

Manufacture and preindustrial activities have configured infrastructures and commercial development along several centuries. Both regional economic and social environment of the present days can be seen as a consequence of the secular interaction between available physical resources and social tissue along our history. Since ancient times, the economy of Castile and Leon has been based on livestock and agriculture, predominantly represented by cattle and large cereal extensions. Thus, flour industry plays an important role which has been reflected along the Medieval and Renaissance ages in a network of preindustrial installations involving mills and "aceñas" (which are also water-powered mills but set up as larger plants containing different kinds of grinders). In this work we have performed an architectural surveying of a bunch of those large-scale abandoned installations which precede the development of the flour factories that brought the industrial era. Our case of use focuses on the aceñas placed in a strech of the Duero river, between Tordesillas and Toro (Spain). Our work includes the virtual reconstruction of mechanisms and a simulation of the processes that milling involves by using the transformation of hydrodynamical forces into mechanical power. Furthermore the architectural container, our development allows the visualization of the milling machinery running, enriched with a simulation of some aspect of the involved hydrodynamic aspects.

1. INTRODUCTION

From the late nineties of 20th century there is an increasing interest about Industrial Cultural Heritage (ICH, in the successive) which has been extended to Pre-Industrial and Technological Cultural Heritage (PITCH) involving Renaissance and beyond. The PITCH involves large rural zones which were articulated in the past around transformation activities which have modified the physical and social landscape. It is necessary to engage citizens, business and communities to discover and be proud of its own identity. So, they will be able of maintaining, supporting and developing related activities which are specific of each zone, because each zone has its own history. The different social and economic agents must be protagonists of an integrated discourse, because they are the responsible of revitalizing their environment.

Intangible traditions involve to uses, costumes, routes, local celebrations ... whose origins are sometimes forgotten. Their preservation must be attached to real and/or virtual representations able of illuminating some keys for understanding them. Pre-industrial activities are linked to an organization of regional production and distributions in local markets. In particular, these activities were organized around small centers acting as intermediate between producers and consumers, as it occurs with the transformation of grain in flour. Hence, the architectural surveying of pre-industrial installations provides a physical and objective support to develop an integrated discourse able of incorporating different interpretations, and their cultural, economic and historical extensions which are strongly related between them. Thus, to begin with we provide a coarse description of some basic principles linked to a representation of architectural industrial heritage as a narrative tool.

In this work we introduce a methodology for surveying PITCH linked to flour mills along large rivers in a small zone of Castilla y Leon (Spanish region in the North plateau). Nevertheless the local character, our methodology can be extended to similar zones in other zones near to rivers or old commercial routes. Our goal is multiple and displays several complementary aspects such as contextualize the environment and the fabrics, architectural surveying of buildings, and simulation of processes involving grain milling. The main novelty concerns to the integration of all of them under a Mobile Augmented Reality (MAR) application. This integration is necessary for the mise-in-value of active usually neglected, and for their exploitation by local communities with the assistance of SMEs linked to the production of multimedia contents for Cultural Heritage.

Contextualize the surveyed object is very often neglected, and when it appears, it involves mainly to academic aspects linked to architectural styles or technological solutions. Instead, we propose another view which is based in the interactive discovery of physical landscape, historical, social and technological issues which provide the keys to understand the surveyed buildings as a whole in a common and integrated framework. With this approach we intend to explain how the pre-industrial and manufacture activities configure a local landscape, commercial routes and market activities which contribute to organize the socio-economic environment, besides architectural aspects of buildings and/or technological solutions. Obviously, this is long-term approach; thus, in this work, we present a preliminary contribution which sketches

This contribution has been peer-reviewed.
doi:10.5194/isprsarchives-XL-5-W4-271-2015

only some meaningful elements to be integrated and developed in following work to be accomplished.

Figure 1: Environmental approach from aerial views

The recreation of disappeared realities linked to PITCH, requires the introduction of virtual and augmented reality (VR/AR in the successive) models able of explaining the original context and how technologies were coupled between them. Along last years, there is an increasing development of VR/AR models and tools to understand Cultural Heritage contents, and its projection to enhance the feeling of an European citizenship. The European Commission is aware of recent advances of these technologies; currently, they are promoting the development of cooperation which is focused around the development of ICT innovative creative activities as a development motor for different stake holders. Main agents are integrated by multimedia industries (SMEs, e.g.) and academic partners.

Unfortunately, it is meaningful that along the FP6 of EU there is only a supported project CONSIST which is related to Industrial Cultural Heritage; in addition, this connection is a tangential one, because it is focused towards the Comparison of conservation materials and strategies for sustainable exploitation of immovable industrial cultural heritage made of iron and steel. This topic has no relation with the development of rural zones, neither the integration of less developed areas or the recovery of local and/or regional activities. In other words, it does not contribute to balance or to develop a specific cultural development.

More recently and along the FP7 of the EC, there have been several initiatives which are linked with particular phenomena linked to corrosion phenomena, damaged materials or another aspects linked to life of materials. Doubtless, these initiatives are very interesting for developing a collaboration between academic and industrial partners, including SMEs linked to Engineering Materials. However, they have no relation with the promotion of a citizenship based in the mutual knowledge and the mise-in-value of the own Cultural Heritage

Our aim is exactly the opposite one. We intend to promote a rediscovery of pre-industrial activities in less developed regions as a cohesion factor in the past, which explains the old and almost disappeared network of relationships between near zones. In this way and from a better knowledge and enjoy of these traditions, we intend to identify them as a way of articulating a richer social tissue, perform their reprojection towards the next future and to reinforce the consciousness of citizenship as participant in a joint project, and not a simple spectator of very large scale initiatives.

To illustrate this program, we have developed a case-of-use which is focused to explain the workflow of flour processes performed by large preindustrial installations, such as Aceñas, which are built in stone with a typical shape given as a ship with a counter-current prow; inside they contain are a large flour mill for grinding up the grain. They are very robust installations, which have been able of surviving large floods and they have worked till the nineteen century near to large rivers. Currently they are abandoned and it is not easy to understand workflows involving not only mechanical aspects, but also the intra-history around these objects.

To perform our modeling we have combined different aspects involving image- and range-based Architectural surveying, preindustrial devices for grinding as an antecedent of more advanced solutions in Mechanical Engineering, and a multilevel approach to their integration by using Computer Graphics simulations. In the next future, we shall incorporate more advanced rendering techniques for more realistic experiences by using advanced tools of Computational Dynamics of Fluids, in order to reproduce the behavior of involved mechanisms.

According to this scheme, we have organized our paper in four additional sections involving a description of landscape and architectural surveying which have been used for grinding grains fabrics, and their integration in a common model; this model provides a support to superimposed additional contents as successive layers. Next, we display a 3D reconstruction of the waterwheels and related mechanisms which were based in a reconversión of hydraulic power in mechanical power for flour mill. The performed reconstruction poses several problems involving simulation of mechanisms a some additional rendering issues which are discussed in an additional section. Last, we perform a discussion of results and present some conclusions and challenges to be solved in the next future.

2. ARCHITECTURAL SURVEYING OF PRE-INDUSTRIAL INSTALLATIONS

In this section we give a description of methods and techniques which have been performed for an architectural surveying of flour mills of abandoned installations along the Duero river from Tordesillas (Valladolid, Spain) to Toro (Zamora, Spain). Figure 1 shows the right choice for location of the flour mill of Pisones (Zamora) having into account the hidrographical characteristics of the river. Sometimes, installations are not enough secure and the visit is not always possible. Furthermore, most people are unable of understanding vestiges, identify components and articulate all of them in a coherent representation. It is necessary to understand the context, describe some assets (including basic principles of mechanisms) and the architecture as a common framework for the right interpretation of this pre-industrial landscape.

This contribution has been peer-reviewed.
doi:10.5194/isprsarchives-XL-5-W4-271-2015

Some recent related initiatives involve to surveying and fostering urban creative zones of interest from the ICH viewpoint. It is convenient to have an overall project strategy, which we have adapted from the methodology proposed in (Creative City Challenges: Industrial heritage in fostering urban creative zones, 2014). According to this approach, we select three stages labeled as (1) architectural mapping, (2) mental mapping, and (3) dialog and strategy / idea-development (in this work we are not concerned with description of potential and dialog among landowners). The goals to be achieved in the first stage relative to *Architectural Mapping* concern to (Creative City Challenges: Industrial heritage in fostering urban creative zones, 2014):

1. Identify, describe and evaluate areas with (pre-)industrial cultural heritage;
2. Register significant characteristics and assess the architectural qualities for further analysis;
3. Make material and results available for the upcoming phases

After identifying the object of study (mills of grain actuated by hydraulic power), next task concerns to registration aspects. Nevertheless, the lack of sources, the methodology proposed in (Creative City Challenges: Industrial heritage in fostering urban creative zones, 2014) can be extrapolated to our case of study. By following the flowchart proposed in page 9, this section is related to the subphase 2 (Description) which concerns to fieldwork and evaluation of available sources, involving to registering areas and markers in the field.

In our case, some additional troubles arise from scarce and disperse documentary sources for pre-industrial Cultural Heritage, and sometimes unaccessible places where pre-industrial installation are located; very often physical and/or technological resources are almost hidden by vegetation or partially destroyed by river floodings. Thus, the task to be performed has some archaeological component which is linked to the discovery and interpretation of PITCH vestiges. To achieve this goal, it is crucial to describe the context involving physical and cultural landscape, on one side, and human and technological resources on the other hand.

The understanding of relationships between landscape and resources in everyday life is performed on an architectural representation. Its interpretation makes possible to identify the role which is played by different social agents along large temporal periods. So, a better knowledge of interrelations between landscape and resources through its spatial representation, will allow the acknowledgement and mise-in-value of almost forgotten intangible resources. These topics concern mainly to the second stage which involve to *mental mapping* which is presented in the fourth section. All of them are linked to activities, routes, and different expressions of creativity involving economic and technological activities, and architectural surveying provides.

Very often the river course has been locally modified to obtain the necessary volume of flow for moving mills with hydraulic forces. Mill fabrics are usually located where the speed is highest, which includes non-convex "vertices" (opposite to meanders), which increase the dynamical forces to be used. To achieve this goal, keel-shaped stone constructions have been performed at such non-convex vertices, which reorient the flow volume along narrow stepped channels where the machinery is located. Aerial views are very useful, because they provide a global representation justifying the choice for the best locations of flour mills in our case.

Figure 2: Aerial view of Pisones Dam

Figure 2 provides a nearer approach which displays the relative location of some installations between them.

After contextualizing the environment from hydrographic viewpoint, it is necessary to identify the main characteristics of the environment. Very often, it is hidden by vegetation or partially destroyed by human actions and/or river floodings. Thus, aerial views do not provide the necessary information to give a local representation of the fabrics. Sometimes, fallen trees and other accidents make impossible to introduce laser scanners near to the objects to be surveyed. In these cases, it is only possible to take different views and try of performing a manual survey of the most outstanding characteristics. Figure 3 illustrates an approximate representation obtained of an almost inaccessible fabric [1] whose real look can be seen in the frontal view of the figure 4. In absence of a complete covering of the whole fabrics building, the integration of the captured information poses relevant challenges for a complete surveying of architectural objects.

Figure 3: Hand sketch of almost unaccessible mill fabrics

The construction of flour fabrics from milling of corn generates transformations in landscape which involves not only to the installations but to the network of paths and roads around the building, and a complex social structure involving offices and relationships between different agents. Currently, the abandon of old fabrics, a partial destruction by human and natural actions, and their coverage by vegetation poses additional challenges to recover these pre-industrial environments and to understand their role in the spatial articulation between 13th and 19th centuries.

3. WATERWHEELS AND RELATED MECHANISMS

Waterwheels and extensions of Archimedes screws have been used from the second century B.C. for the conversion of hydraulic power in mechanical energy. In presence of a large flow volume

[1] Sketch performed by Jesús I. San José

This contribution has been peer-reviewed.
doi:10.5194/isprsarchives-XL-5-W4-271-2015

Figure 4: Frontal view of Pisones flour fabric

or for low height, simple solutions based on some of both procedures provide amenable solutions which can be combined between them and adapted to different environments. However, in presence of high ground differences or unequal flow volumes, the problem becomes much harder.

Waterwheels are commonly used in Mediterranean (greek, roman, arabian) cultures and they are flourished in the medieval age following different formats with peculiar local adaptations, as it appears in medieval Christian Spanish kingdoms of Castilla and Aragón, also. A compendium and extension of all these contributions which incorporates a large amount of contributions arising from civil Italian and Spanish engineers of late Renaissance can be found at (Keller, 1998)

Perhaps the most sophisticated device for high ground differences along the sixteenth century is performed in Toledo by the Italian engineer Giovanni Torriani (Cremona 1500, Toledo 1585) called Juanelo Turriano in Spanish language. His goal was to lift the water of Tajo river by saving a high of ninety meters with a complicated orography and a large variety of slopes. The main technological problem to be solved was the conversion of a circular motion in rectilinear motion along the long path connecting the river course with the Alcazar and the city of Toledo. The original documentation of Torriani's solution is unfortunately lost and the last vestiges of civil engineering were destroyed at the end of the nineteenth century; in fact their solution was abandoned at the early 17th century because nobody was able of performing the right maintenance. Currently, there are two hypotheses about transmission principles, but we have not a complete understanding of his solution[2]

The Spanish word Aceña arises from an Arab expression which means hydraulic wheel. Hence, the terminology involves to the engineering principles which transform the hydraulic energy in mechanical energy for grinding any kind of corn by means of mills mechanisms. The coupling between wheels and axis in the perpendicular plane, plays the same role as the more modern turbines; both of them follow the same mechanical principles which have been used from more than eight centuries ago. Indeed, there are several types of wind and water mills, with their corresponding mechanisms to generate a multiplicative effect performed by grinding mechanisms.

Depth and flow volumes impose constraints for different kinds of mills. In a very simplified way, there are two main kinds of mechanical devices for conversion of hydraulic forces in mechanical efforts: For low depth and scarce flow volumes, it is necessary to use horizontal mechanisms based in wheels which are displaced by the water action. Hydraulic efforts for this second class of mechanisms are lesser than those which are required for vertical mechanisms. Horizontal motion is transmitted to a vertical axis which is coupled with another horizontal mechanisms. These simple mechanical considerations impose structural constraints about the architectural design which is necessary to achieve to obtain the right anchorage and a better performance of mechanical devices.

In our case, and nevertheless their morphological diversity, we have found both types of mechanisms. Vertical mechanisms are common in narrow channels between the keel-shaped stone constructions; in some cases, lifted water was used for another purposes linked to adjacent water deposits which were constructed as regulator elements or for fishing activities, e.g.. Horizontal mechanisms require less volume flow and they can be used in a complementary way in minor constructions or as an alternative solution when the water flow was not enough to move large vertical waterwheels. Figure 5 illustrates the basic horizontal mechanism used along the 16th century and before.

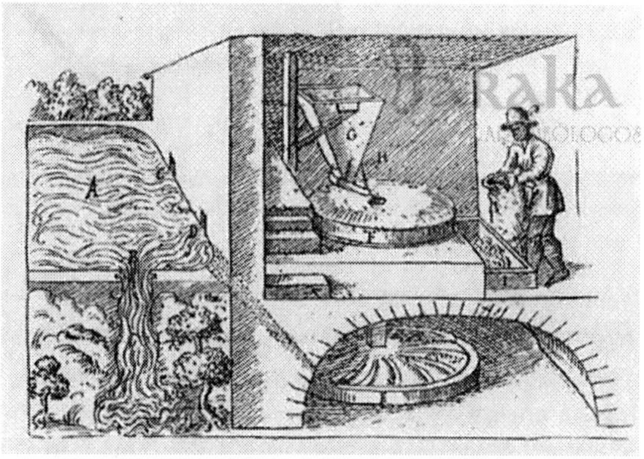

Figure 5: Horizontal mills or rodeznos

For low ground differences and nevertheless the unequal flow volume, the problem is easier because it suffices to perform a transformation between rotational efforts linked to wheels located in different planes. In this kind of mechanisms, the transmission depends on the relationships between diameters of wheels located at different planes and gear mechanisms connecting them, which is the case of mills activated by mechanical energy arising from hydraulic power. Nevertheless the architecture of installations must be enough robust to support dynamical forces arising from the water and embedded mechanisms in civil constructions.

From a larger socio-economic viewpoint, grinding processes have promoted and conditioned the development of villages which have become regional centers for transformations of agrarian exploitations. Some of these villages (Tordesillas, Toro) become the market centers for the distribution and resulting relationships have configured the physical landscape and a large amount of intangible CH along last seven centuries. For instance, Figure 6 is a gravure of the 19th century illustrating the architecture of mill fabrics adjacent to the Duero river passing throught Tordesillas.

Most installations for conversion of grains in flour by using water mills are composed by a dam, and a collection of buildings where the grinding processes are performed by using different mechanisms. Dams are constructed transversally to the river course, usually in non-convex corners where the flow is faster. Civil constructions are constructed to make smaller the the water courant

[2]A very nice simulation can be seen in `https://www.youtube.com/watch?v=MwU6m9tjM2A`

This contribution has been peer-reviewed.
doi:10.5194/isprsarchives-XL-5-W4-271-2015

Figure 6: Gravure of Tordesillas including flour fabrics

width and increase its speed; the resulting channeling in narrow corridors allows to transform the hydraulic power in a force acting along a horizontal ("rodezno" in Spanish language) or a vertical wheel ("noria" in Spanish language). The water action turns a wheel of palettes, and the circular motion is transmitted to another elements which are connected by means of a vertical or, alternately, a horizontal axis towards another wheels which move a conical grinder or crushing mechanism ("muela superior" in Spanish language) contained in a larger conical hopper which is connected with a smaller deposit ("tolva" in Spanish language) to refill flour sacks. The iteration of grinding processes would allow to develop a coarse-to fine result for resulting flour.

The basic mechanical principle is given by the transformation of hydraulic power in mechanical energy through a collection of coupled mechanisms which are disposed in perpendicular planes to optimize mechanical efforts (it is a very simple application of vector product in Mathematics). The right relative disposition of mechanisms and the design of dentate wheels to improve the multiplicative effects result in more efficient grinding processes. If there would be necessary, grinding process can be iterated to obtain finer flour. Figure 7 displays the current status of a wheel which has been used till the 19th century in regard with more modern uses as electricity production, also.

Risks linked to flooding (due to sudden changes in flow volume), and efforts supported by the corresponding transmission mechanisms require a very robust architecture. This requirement is translated in the construction of very robust docks (keel-shaped constructions) for re-orienting the river courant, able of supporting floods, and in civil engineering constructions to accommodate the machinery. Furthermore, there are some auxiliary constructions such as footbridges for connecting consecutive docks, and buildings to store grain, floor, or other materials linked with other tasks. In some cases, floor mills have been used as sawmills by coupling other mechanisms to cut out the wood. More recently and along the 19th century some of these installations have been used to produce electricity, where mechanical principles are similar ones to the described above.

4. AN INTEGRATED SIMULATION OF FLOUR GRINDING

In this section we illustrate perform a simplified representation of grinding process, contextualize the mechanical model inside a simplified representation of fabrics. In this way, we intend to provide a journey to a pre-industrial past going from the above realistic representation of the outdoor towards an increasingly virtualized of the indoor.

Resources based in VR/AR provide a mental mapping of physical space and technological resources involved in the flour production from grinding grain activities. According to the scheme

Figure 7: Original mechanism of waterwheels

proposed in (Creative City Challenges: Industrial heritage in fostering urban creative zones, 2014), the main purposes are the following ones:

1. Visibility, gathering, and identification of patterns (places and themes).
2. Generation of a report, and a map to make the material and results available for the upcoming phases.
3. Involvement and management, the focused dialog, further exploration and testing patterns.

In particular, we are mainly concerned with some aspects relative to the first item, which is the most meaningful from the academic viewpoint. In fact the report as a collection of materials relative to precedent sub-phases has been published already (I.SanJosé and J.J.Fernández-Martín, 2010), and the availability of maps with gathered patterns ordered after place; hence, its dissemination is a competence of Public Regional Administration to whom belong the materials. On the other hand, third point involves to the role to be played by different social agents in collaboration with professional in charge of promoting dissemination and the articulation of initiatives linked to the mise-in-value of PITCH. To ease the mise-in-value of these spaces it is necessary to develop some urgent interventions of conservation and restoration, because the current status is near to the ruin even for the best preserved buildings. Figure 8 illustrates the indoor of one the best preserved buildings.

To illustrate flour grinding processes we use very simplified VR models. Their incorporation to interactive AR applications on mobile devices follows the same methodology which has been

This contribution has been peer-reviewed.
doi:10.5194/isprsarchives-XL-5-W4-271-2015

Figure 8: Current state of indoor scenario

exposed in (Delgado et al., 2013). The main purpose is to improve the visibility, perform an open gathering, and identify some of the most common patterns involving the physical space and involved technologies. In this way, we intend to ease a greater acquittance with pre-industrial processes involving grinding grain and its conversion in flour. Far from being a consequence of conversion of hydraulic power in mechanical energy, the main goal for dissemination purposes would must be described in terms of a transformation of human work and its incidence on the nearest environment.

After a first approach to the local environment, it is convenient to have a three-dimensional model which can allow a more secure and interactive visit. To achieve these goals, we have used terrestrial and image- and range-based approaches, which we have completed with virtual models to include some details which are unfortunately lost. The architectural surveying includes different aspects concerning to location, outdoor and indoor visualization. Furthermore documentation aspects, we have developed 3D virtual models which can be displayed in videos to acquire a more faithful representation of the architectural representation. A static view of the waterwheel mechanism embedded into a simplified representation of building can be seen in 9.

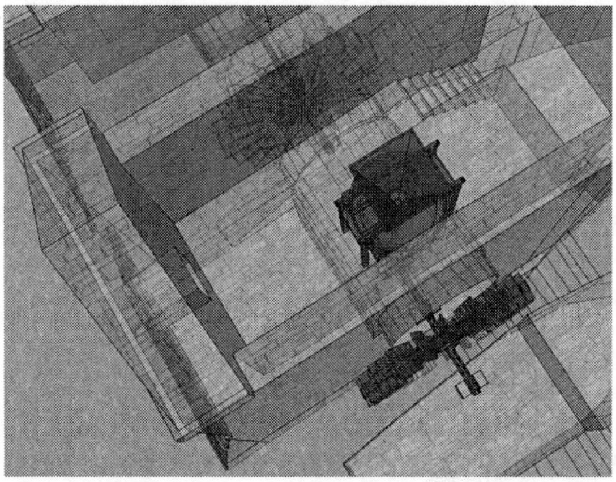

Figure 9: Simplified representation of the mill motion with inserted machinery

5. EVALUATION AND DISCUSSION

The most difficult problem is the integration of the architectural surveying of reconstructed environments with working mechanical devices in a realistic way. This task involves to a more realistic representation of Fluid Mechanics which is far beyond of the objectives of the work presented here. In the meantime, we restrict ourselves to display a model with different levels of details, involving the building and a decoupled representation of inserted mechanisms. The operation of coupled mechanisms involve to several wheels which are connected between them with orthogonal axes; its dynamical representation requires a careful design of gears, able of providing a realistic representation of motion which is illustrated in figure 10, by deleting the surrounding architectural environment.

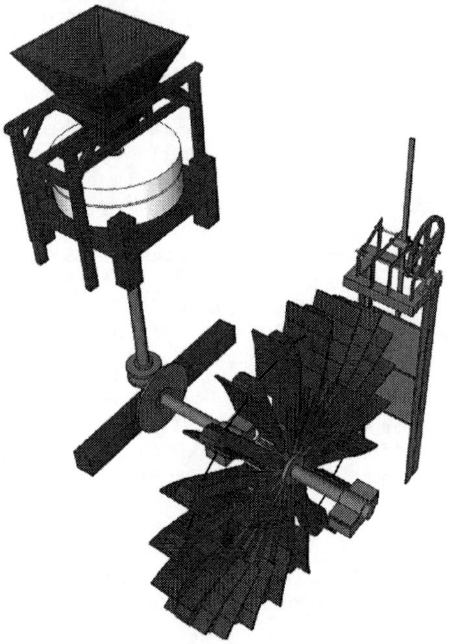

Figure 10: The grinding mechanism isolated from the mill for a better understanding of its operation

A more realistic representation can be useful for multimedia creation; it must include more advanced rendering aspects which involve to the Computational Fluid Dynamics. This broad reach goal is the most advanced challenge in ICT domain; its incorporation will allow to integrate currently available solutions for multimedia purposes in an interactive way; if resources would be available, it could be developed in the next future.

According to some recent tendencies, to illustrate the connection with socio-economic evolution of its environment, it is necessary to introduce human agents playing different roles able of interacting between them, as it occurs with avatars (by following similar patterns to those appearing in Second Life) and/or characters in videogames. These aspects concern to the exploitation of results to generate an economic return to these activities and we are not concerned with these topics.

6. CONCLUSIONS AND FUTURE WORK

In this work we have presented an architectural surveying of pre-industrial installations for grinding cereals and a multilayered approach to the simulation of grinding processes in flour mills. To illustrate these processes we have performed an approach in different steps including a simplified representation of the socio-economic and technological environment, an architectural surveying, a description of mechanisms and an illustration of their

This contribution has been peer-reviewed.
doi:10.5194/isprsarchives-XL-5-W4-271-2015

behavior in a simplified representation. This approach provides a support which can be useful not only for academic research linked to ICT contents, but to educational and entertainment purposes, also. In this way, we intend to contribute to the knowledge of past activities, the socio-economic articulation of a very rich historical and cultural environment, and the self-consciousness of an almost lost identity.

ACKNOWLEDGEMENTS

A precedent and much larger work was published as the book (I.SanJosé and J.J.Fernández-Martín, 2010) which corresponds to a research inside de PAHIS Program (Patrimonio Histórico), a project partially financed by the Junta de Castilla y León, Programa Operativo de Cooperación Transfronteriza España-Portugal (2007-2013), Fondo Europeo de Desarrollo Regional (FEDER), European Union. Aerial images arise from the National Plan of Aerial Ortophotography (PNOA) captured by the Instituto Tecnológico Agrario (ITACYL). Their contribution is grateful acknowledged.

REFERENCES

Creative City Challenges: Industrial heritage in fostering urban creative zones, 2014. Technical report, Hoje-Taastrup Kommune. Report co-funded by the Interreg IVB North-Sea Region Programme, the European Union and the Hoje-Taastrup Kommune.

Delgado, F. J., Martinez, R., Finat, J., Martinez, J., Puche, J. C. and Finat, F. J., 2013. Enhancing the reuse of digital resources for integrated systems to represent, understand and dynamize complex interactions in architectural cultural heritage environments. ISPRS Annals of Photogrammetry, Remote Sensing and Spatial Information Sciences 1(1), pp. 97–102.

I.SanJosé, J. and J.J.Fernández-Martín, 2010. Aceñas del Duero. Tordesillas, Toro y Zamora. Gráficas Lafalpoo. ISBN: 978-84-9718-620-9.

Keller, A., 1998. The twenty-one books of engineering and machines of Juanelo Turriano. Fundación Juanelo Turriano. A translation of the manuscript in the Biblioteca Nacional de España (Madrid) titled "Los veintiún libros de los ingenios y máquinas de Juanelo Turriano".

TOWARDS AN EASIER ORIENTATION FOR SPHERICAL PHOTOGRAMMETRY

Gabriele Fangi

Università Politecnica delle Marche – Italy
g.fangi@univpm.it

Commission V, WG V/4

KEY WORDS: Panoramic Spherical Photogrammetry, Block Adjustment, Trinocular Vision

ABSTRACT:

For architectural metric documentation, Spherical Photogrammetry (SP) has demonstrated its validity and efficiency in many projects already. The speed of surveying is high, the accuracy and completeness of the plotting are satisfactory. However, there are still many problems to be solved. The weakest point is the orientation procedure, which is rather difficult to perform, in the sense that only very experienced people can run it, and few people only make use of it. The old orientation steps are 1) model formation (limited to binocular panoramas couples); 2) link of all the models in a block adjustment with independent model triangulation; 3) block bundle adjustment with 4 parameters/pano (3 coord.+1 orientation bearing); 4) block bundle adjustment with 6 parameters/pano, say the previous 4 + 2 correction angles around the horizontal axes. The panoramas must be spherical and quasi-horizontal. In order to make easier the orientation, enabling more people to use SP, an improved approach has been set up. It consists in the combination of any possible model formed either by three and two panoramas. The trinocular vision, say the combination of three different panoramas to form a unique model, has the advantage to be much more robust in comparison to binocular vision in the sense that the trinocular model is likely to be more error-free than any of the three composing binocular models. It contains less model deformation, the model coordinates are validated by the mutual comparison of the three intersecting binocular models. In addition, the number of possible trinocular models is normally much larger than the one of binocular models. The steps for a semi-automatic orientation of a block of panoramas proceed as follows:

- Form any possible trinocular models by combination of the panoramas;
- in case that no trinocular model has been formed, form any possible binocular model;
- run a block adjustment with the algorithm of independent model, to link together the models in a unique reference system;
- run a block bundle adjustment with collinearity equations with 4 unknown parameters per panorama;
- run a block bundle adjustment with collinearity equations with 6 unknown parameters per panorama.

The control requirements are limited to one fixed control point and two perpendicular constraints.

Given a project with *n* panoramas, the trinocular possible models are the combination of n, 3 by 3, say (n*(n-1)*(n-2))/3!. As an example, in a project with 10 panoramas the trinocular possible models are 120, while the possible binocular models are (n*(n-1))/2= 45. Obviously not all the panoramas will be formed, because a minimum of four common tie points is required.

It is possible to select and filter out the best models by fixing a critical value on sigma naught for testing. The advantages are as follows:

- no need of approximate value for the unknown coordinates and parameters;
- easiness to find gross errors in the gradual block formation.

The epipolar geometry has been investigated, the expression of the curve described and assessed: it can be useful in the trinocular model formation. For any try of the model the operator, thus causing troubles to him, must approximately choose formation, up to now the starting value of the orientation bearing for any panorama. However, a new procedure, not requiring any approximate value of the orientation bearing, has been set-up and tested: the orientation bearing can vary stepwise by a given selected interval, until the model is successfully formed. The computation time increases significantly, but it is still negligible.

Up to now the control and tie points selection and observation is manual, making the orientation and the plotting to be slow and time consuming. Nevertheless, on the other end, the procedure is fully controlled and supervised by the operator, enabling the observation of the residuals and allowing the improvement of the orientation, and to perform some statistical quality analysis on the adjustment, which are not possible with automatic procedures based on Structure from Motion (SfM).

Another weak point of the SP is the speed of plotting. Commercial programs are now available, based on SFM algorithm, producing point clouds of good quality in an unknown scale and orientation. The correct scale and orientation can be therefore supplied by SP orientation with high accuracy. Examples are shown taken from two real projects.

1. INTRODUCTION

Panoramic Spherical Photogrammetry (PSP) makes use of spherical panoramas as information support. They are formed by overlapping images taken from the same point and stitched together on a sphere (Szelinsky and Shum, 1997) then mapped on a plane with the so-called latitude-longitude projection for the stitching for the type of cartographic projection (Snyder, 1993). The advantages of the spherical panorama consist in the Field Of View (FOV) that can reach 360°x180°, and in the high image resolution, image being omnidirectional. From the panorama image point, the two direction angles f and a towards the object point can be derived as they were taken with a theodolite (Figure 2). The coordinates of an arbitrary object point can be obtained by intersection of corresponding straight lines connecting the image point with the center of two or more spheres (orientation). The PSP can be suitable for fast and accurate recording of Architectural Cultural Heritage (Pisa et al., 2010). Attempts have been done for the automatic detection of corresponding points in the panoramas (Barazzetti et al.,2010). To overcome the lack of stereoscopy and enable complex surfaces to be plotted, the photo-modeling has been used; say the back-projection of the oriented panoramic spheres

This contribution has been peer-reviewed.
doi:10.5194/isprsarchives-XL-5-W4-279-2015

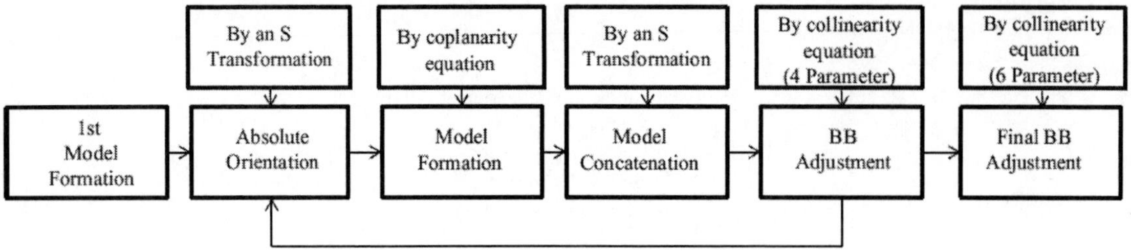

Figure 1: Flow-chart of the SP orientation procedure.

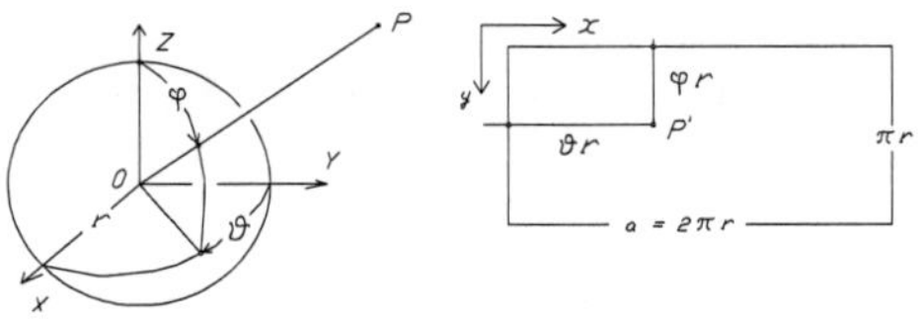

Figure 2: The sphere and its mapping.

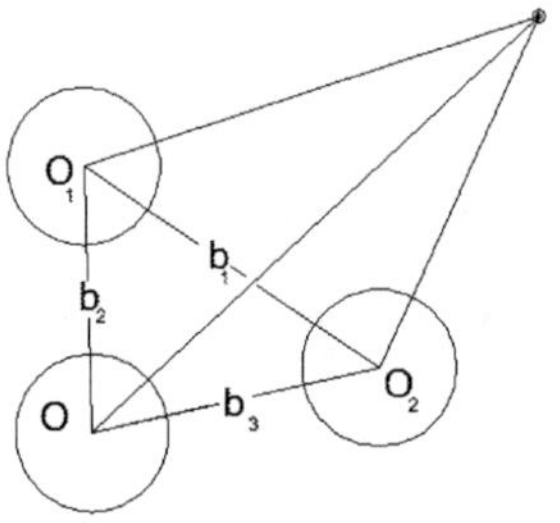

Figure 3: Triple intersection: the intersected planes.

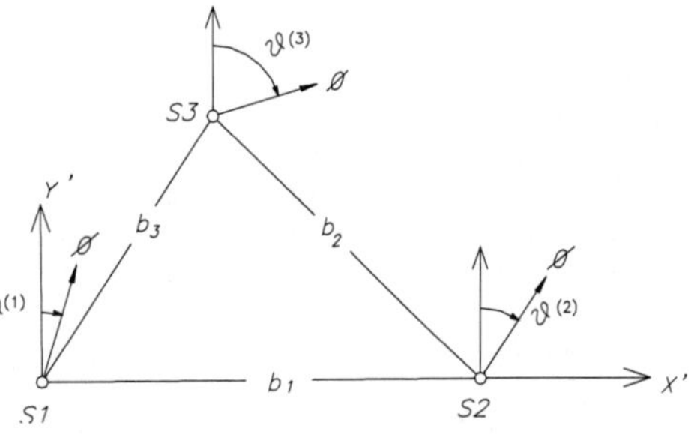

Figure 4: Selected reference system for trinocular model

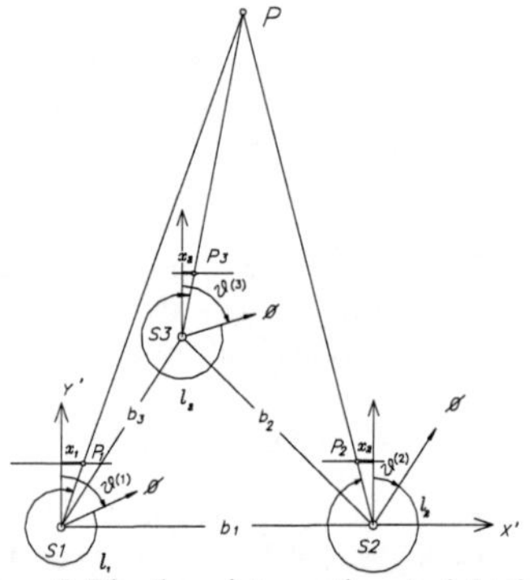

Figure 5: The three intersecting straight lines

over the rought model in a cad environment (d'Annibale, et al. 2009, Fangi, Piermattei, Wahbeh 2013).

The nature and algorithm of the epipolar curve has been investigated and shown (Fangi, Nardinocchi,2013). So far, the integration with point clouds produced by SFM algorithms has been fruitfully used in (Fangi, Pierdicca, 2012). Here the trinocular model for spherical panoramas is introduced.

2. THE ORIENTATION PROCEDURE OF THE SPHERICAL PANORAMAS

The orientation procedure (Fig. 1) for spherical panoramas has been explained (Fangi, Nardinocchi, 2013). This procedure takes place as follows:

- Relative Orientation by coplanarity ending with the models formation
- Block Independent Model Triangulation
- Block Bundle Adjustment

One model can be composed by two panoramas, having in common three or more tie points (binocular model). All the formed models by coplanarity, are linked together in a block adjustment with independent models (Independent Models Block Adjustment - IMBA). IMBA provides the approximated values for parameters and coordinates to be input in a Block Bundle Adjustment, this last using the collinearity equations. Block Bundle Adjustment can be performed in two steps: the first one with 4 parameters/per pano and the second final one with 6 parameters/per pano. The reason for that is the possibility to better find the gross errors. The coplanarity condition is:

$$g = x'^T M'^T b M'' x'' = 0 \qquad (1)$$

where $\boldsymbol{x'}$ and $\boldsymbol{x''}$ are the spherical coordinates of the image point P', M' and M'' the two rotation matrices, function of the angles $\varphi = y/R$ and $a = x/R + q_0$ being q_0 the zero bearing for the a panorama (figure 2), x and y the panorama image coordinates, $\boldsymbol{X}$ the spherical coordinates, $\boldsymbol{b}$ the base vector.

This contribution has been peer-reviewed.
doi:10.5194/isprsarchives-XL-5-W4-279-2015

$$[X' \quad Y' \quad Z']M'^{T}\begin{bmatrix} 0 & -bz & by \\ bz & 0 & -bx \\ -by & bx & 0 \end{bmatrix} M''\begin{bmatrix} X'' \\ Y'' \\ Z'' \end{bmatrix} = 0 \qquad (2)$$

For the absolute orientation of the block of the panoramas, a set of control points can be supplied, taken with traditional means (GPS, total station), as it is advisable for a professional project. As an alternative, the minimum requirements are one fixed point, one distance for the model size, together with three geometrical conditions in orthogonal directions. In (2) we can set by=0 and bx=1 and the unknowns reduce to three ϑ_0', ϑ_0'', and b_z. Since the coplanarity condition (2) is not a linear equation, it has to be linearized. Therefore it is necessary to input approximate values for the unknowns, for b_z it can be set to zero but for the two orientation angles ϑ_0', ϑ_0'', the operator must give approximated values, which are troublesome. But we made an improvement: the new computation procedure can start with arbitrary values for ϑ_0', ϑ_0''. In case of failure of the iterative computation, the next attempt will start with initial values differing from the previous ones by a preset value. The approximated orientation bearing can vary stepwise by a given selected interval, until the model is successfully formed. The computation time increases significantly, but it is still negligible.

The novelty of the here described orientation procedure consists in the trinocular model, which is the model formed by three panoramas (trinocular model – see Fig. 3,4,5). Trinocular model have been already investigated by many authors and by the A. (Fangi, 1998). Trinocular vision is used for machine vision in industrial application (Heinrichs et al.,2007) and medical application (Sechidis, Malian et al., 2000). The advantages of the trinocular models are in terms of amount of needed tie points, in terms of number of formed models, but mainly in their strength: while the binocular models of spherical panoramas can be easily deformed due to the orientation imperfections and bad estimation of the orientation parameters (Fangi, 2010), the trinocular models have less deformation, because of any tie point not one coplanarity condition only, but even three coplanarity hold.

2.1 The Trinocular Model

In case of a triple intersection, say intersection of the arbitrary point P from three panoramas S1, S_2 and S_3, the interested planes are three (figures 3,4,5), three independent stereo-pairs, i.e. left–right, left–center, and center–right. Per any observed point we can then write three condition equations. The unknowns are in this case:

- the three coordinates of a station (3)
- the two components of one of the three bases (2)
- the three components of the remaining two bases
  3x2= 6

We can select the type of orientation such as to set equal to zero the coordinates of one of the three stations, S1, then equal to zero the Y coordinate of S2. For quasi-horizontal panoramas, as they are in PSP, we can set the rotation angles α_x, α_y around the horizontal axes x and y, equal to zero, thus reducing the amount of unknowns, and enabling a better detection of gross errors. They are estimated later on, α_x, α_y , starting from initial zero value, in the last stage of the orientation, the block bundle adjustment.

We define a positive versus in the triangle sides. We define the components (with sign) of the three bases:

$$\begin{cases} b_x^{(i)} = X'(i+1) - X'(i) \\ b_y^{(i)} = Y'(i+1) - Y'(i) \\ b_z^{(i)} = Z'(i+1) - Z'(i) \end{cases} \qquad (3)$$

Because of the chosen system, $b_x^{(1)} = 1$ and $b_x^{(1)} = 0$.

To the coplanarity equations three conditions must be added, relative to the components in the three directions of the three bases.

$$\begin{cases} g_x = b_x^{(1)} + b_x^{(2)} + b_x^{(3)} = 0 \\ g_y = b_y^{(1)} + b_y^{(2)} + b_y^{(3)} = 0 \\ g_z = b_z^{(1)} + b_z^{(2)} + b_z^{(3)} = 0 \end{cases} \qquad (4)$$

Not all the 10 parameters of orientation are indeed independent: In fact, set

$b_x^{(1)} = 1$

in addition,

$b_y^{(1)} = 0$

from the (1) we have

$$b_x^{(3)} = 1 - b_x^{(2)} \qquad b_y^{(3)} = -b_y^{(2)}$$

and

$$b_z^{(3)} = -(b_z^{(1)} + b_z^{(2)}) \qquad (5)$$

Then the independent parameters are only the following seven:

$$\vartheta_1,\ \vartheta_2,\ \vartheta_3,\ b_z^{(1)},\ b_x^{(2)},\ b_y^{(2)},\ b_z^{(2)},$$

being ϑ_1 , ϑ_2, ϑ_3 the three zero bearing for any panorama, say the direction clockwise of the origin of the image with respect to the reference system. We can write three linearized coplanarity conditions. For the approximated values of all parameters, for the bearings we behave like for binocular models. We can set to zero bz components, $b_x^{(2)}=b_y^{(2)} = 0.5$. In case of failure of the computation we change the sign of $b_y^{(2)}$=-0.5.

To solve for the 7 unknowns, at least three points must be observed, with a redundancy equal to 2 (in the case of intersection only from two stations the redundancy would be zero).

With n observed points (m= observ. dir., c =equat. ; u= unknowns ; r =redun.:)

from 2 stations: $m=2n$ $u=2$, $c=n$, r = $n-2$ $m-2(n-1)$

from 3 stations: $m=3n$ $u=7$, $c=3n$, r = $3n-7$ $m=(3n-7)$

From the graph in Figure 6, one can see as follows: when $n \geq 5$ for increasing the redundancy, it is convenient to set up three pano-stations better than to enlarge the number of the observed points from two stations.

A third panorama helps to solve problems and provides additional constraints on panorama link, dramatically reducing the chance of gross errors. For the same project, the number of trinocular models is larger than the one of binocular models. Than it is necessary to filter out the worse ones. The filter can be performed by testing the final sigma naught with a preset value. Another test is the interval of admissible dimension for the model. Given d_{max} the distance to an arbitrary point from one of the three base point and b_{max} the value of the largest of the three bases, the ratio d_{max}/b_{max} should be included in a prefixed interval to avoid ugly configuration.

This contribution has been peer-reviewed.
doi:10.5194/isprsarchives-XL-5-W4-279-2015

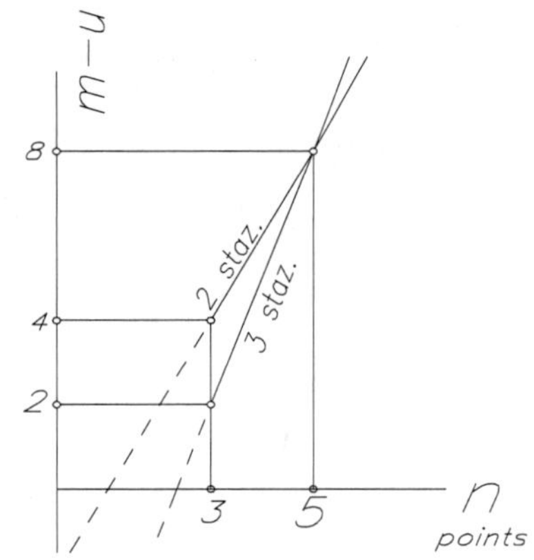

Figure 6: Relation between the number of the observed points and the number of observations.

The model formations can be a) fully automatic, b) select one panorama, c) select one couple, c) select a triplet.

a) The procedure tries to form trinocular model with any possible combination of existing panoramas;
b) The procedure tries to form trinocular models with any possible combination of two panoramas with the selected one;
c) The procedure tries to form trinocular model with any possible combination of the two selected ones with the remaining panorama;
d) The procedure tries to form the trinocular model formed by the selected triplet

From experience the most advisable choice of the four mentioned is the c). It permits to find gross errors as soon as they show up.

3. THE PSP IN COMBINATION WITH POINT CLOUD FROM SFM ALGORITHM

Another weak point of the PSP is the speed of plotting. Commercial programs are now available, based on SfM algorithms, producing point clouds of good quality in an unknown scale and orientation. The correct scale and orientation can be therefore supplied by PSP orientation with high accuracy. In the following, an example taken from a real project is shown. Santa Maria della Piazza façade, Ancona (Italy) has been surveyed with 9 spherical panoramas (figure 8). The possible triplets are (9x8x7/6!= 84). Choosing the limit for sigma naught value σ_0=0.001 rad, none model was formed, for σ_0=0.002 one model only, for s0=0.003 the formed triplets are 13, and finally for σ_0=0.005 the triplets successfully formed are 43. By increasing the value for sigma naught limit, increases the number of formed models, but in the same time increases also the possibility of undetected gross errors.

The orientation could provide nine control points well distributed on the surface. A point cloud project has been performed by means of a commercial software (Photoscan). The control points provided by PSP have been input in the Photoscan project, and the results are in Fig. 7. The average discrepancy is 1.08 cm while the maximum difference is 2.5 cm.

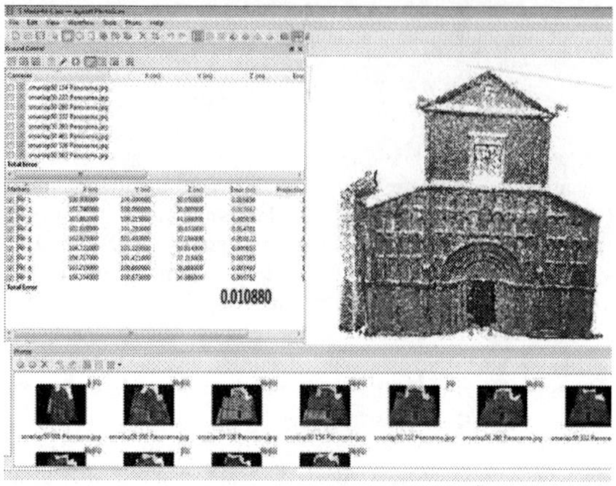

Figure 7: Errors of the control points produced by SP in a Photoscan Project – Santa Maria della Piazza façade, Ancona, Italy.

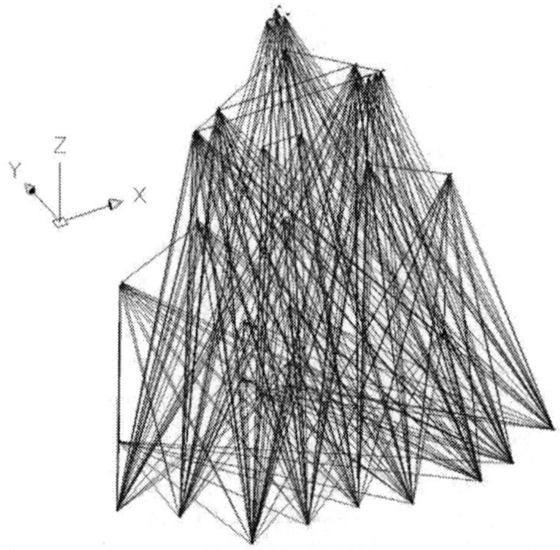

Figure 8: Santa Maria della Piazza façade.- The orientation network of the nine spherical panoramas.

4. CONCLUSIONS

The most difficult phase for PSP is the orientation phase. In order to make it easier some improvements have been set up in the procedure, the most significant one is the trinocular models. They are more robust in comparison with the binocular ones.
The combination with commercial software for the point cloud production like Photoscan has been tested, giving good results. PSP provides the needed control points for the point clouds.

REFERENCES

Barazzetti, L., Fangi, G., Remondino, F. and Scaioni, M., 2010. Automation in multi-image spherical photogrammetry for 3D architectural reconstructions. 11th International Symposium on Virtual Reality, Archaeology and Cultural Heritage (VAST), Paris, France. 75–81 (on CD-ROM).

D'Annibale, E. and Fangi, G., 2009. Interactive modelling by projection of oriented spherical panorama.International Archives of Photogrammetry, Remote Sensing and Spatial Information Sciences, 38(5/W1). 6 pages (on CD-ROM).

This contribution has been peer-reviewed.
doi:10.5194/isprsarchives-XL-5-W4-279-2015

Fangi G., 1998. The Coplanarity Condition For The Orientation In Surveying - ISPRS Arch. WG VI/3 Meeting "International Cooperation and Technology Transfer" Perugia Febbraio 1998 pg151, 166

Fangi G., 2010. Multiscale Multiresolution Spherical Photogrammetry With Long Focal Lenses For Architectural Surveys –ISPRS Mid-term Symposium NewCastle, June 2010 ISSN 1682-1777 ID 585

Fangi G., Nardinocchi C., 2013. Photogrammetric processing of spherical Panoramas - The Photogrammetric Record 28(143): 293–311, DOI: 10.1111/phor.12031 pg. 293- 311 © 2013 The Remote Sensing and Photogrammetry Society and John Wiley & Sons Ltd – ID 146702 2-s2.0-84883817879

Fangi, G. and Pierdicca, R., 2012. Notre Dame du Haut by spherical photogrammetry integrated by point cloud generated by multi view software. International Journal of Heritage in the Digital Era, 1(3): 461– 479.

Fangi G., Wahbeh W., Piermattei L., 2013. Spherical Photogrammetry as Rescue Documentation for The Reconstruction of some UNESCO Sites in Syria - International Journal of Heritage in the Digital Era – Publisher Multi Science Publishing – ISSN 2047-4970 – Volume 2, N.3 /September 2013 Pages 335- 342 – DOI 10.1260/2047-4970.2.3.335 – ID 156110

Heinrichs M., Rodehorst V.,Hellwich O., 2007. Efficient Semi-Global Matching For Trinocular Stereo - Stilla U et al (Eds) PIA07. International Archives of Photogrammetry, Remote Sensing and Spatial Information Sciences, 36 (3/W49A)

Luhmann, T. and Tecklenburg, W., 2004. 3D object reconstruction from multiple-station panorama imagery. International Archives of Photogrammetry, Remote Sensing and Spatial Information Sciences, 34(5/W16). 8 pages (on CD-ROM).

Malian A., van den Heuvel F.A., Azizi A., 2005. A Robust Photogrammetric System For Wound Measurement -The Photogrammetric Record Volume 20, Issue 111, pages 241–273, September 2005 DOI: 10.1111/j.1477-9730.2005.00319.x

Pedersini F., Sarti A., Tubaro S., 1996. A Multi-view Trinocular System for automatic 3-D Object Modeling and Rendering International Archives of Photogrammetry and Remote Sensing Vol. XXXI, part B5, Vienna 1996

Pisa, C., Zeppa, F. and Fangi, G., 2010. Spherical photogrammetry for cultural heritage. Proceeding of the Second Workshop on Heritage and Digital Art Preservation, Florence, Italy. 3–6.

Sechidis L., Tsioukas V., Patias P., 2000. An automatic Process for the Extraction of the 3d Model of the human Back Surface for Scoliosis Treatment - International Archives of Photogrammetry and Remote Sensing. Vol. XXXIII, Supplement B5. Amsterdam 2000.

Schneider, D. and Maas, H. G., 2005. Combined bundle adjustment of panoramic and central perspective images. International Archives of Photogrammetry, Remote Sensing and Spatial Information Sciences, 36 (5/W8). 4 pages.

Snyder, J. P., 1993. Flattening the Earth: Two Thousand Years of Map Projections. The University of Chicago Press, Chicago, Illinois, USA. 384 pages.

Szeliski, R. and Shum, H., 1997. Creating full view panoramic image mosaics and environment maps. Proceedings of SIGGRAPH. 251–258.

Torii, A., Imiya, A. and Ohnishi, N., 2000. Two- and three-view geometry for spherical cameras. 6th Workshop on Omnidirectional Vision, Camera Network and Non-classical Cameras, Beijing, China. 8 pages (on CD-ROM).

Torii, A. and Imiya, A., 2007. Computation of epipolar geometry and trifocal tensor from spherical images. Computer Vision Winter Workshop, St Lambrecht, Austria. 8 pages (on CD-ROM).

Volker S., Poyoc, 2000. Photogrammetric Measurements of Buildings and Architectural Models via Multi-View Stereovision eCAADe 1991 – PDF- Proceedings (conversion 2000)

Agisoft Photoscan - http://www.agisoft.com/

This contribution has been peer-reviewed.
doi:10.5194/isprsarchives-XL-5-W4-279-2015

RANGE AND IMAGE BASED MODELLING: A WAY FOR FRESCOED VAULT TEXTURING OPTIMIZATION

G. Caroti, I. Martínez-Espejo Zaragoza, A. Piemonte

Dipartimento di Ingegneria Civile e Industriale, Università di Pisa, Largo Lucio Lazzarino 1
56123 PISA, phone 050 221 7770, e-mail gabriella.caroti@ unipi.it

Commission V, WG V/4

KEY WORDS: Structure From Motion, Laser scanning, Photogrammetry, Cultural Heritage, Texture, Precision

ABSTRACT:

In the restoration of the frescoed vaults it is not only important to know the geometric shape of the painted surface, but it is essential to document its chromatic characterization and conservation status. The new techniques of range-based and image-based modelling, each with its limitations and advantages, offer a wide range of methods to obtain the geometric shape. In fact, several studies widely document that laser scanning enable obtaining three-dimensional models with high morphological precision. However, the quality level of the colour obtained with built-in laser scanner cameras is not comparable to that obtained for the shape. It is possible to improve the texture quality by means of a dedicated photographic campaign. This procedure, however, requires to calculate the external orientation of each image identifying the control points on it and on the model through a costly step of post processing. With image-based modelling techniques it is possible to obtain models that maintain the colour quality of the original images, but with variable geometric precision, locally lower than the laser scanning model. This paper presents a methodology that uses the camera external orientation parameters calculated by image based modelling techniques to project the same image on the model obtained from the laser scan. This methodology is tested on an Italian mirror (*a schifo*) frescoed vault. In the paper the different models, the analysis of precision and the efficiency evaluation of proposed methodology are presented.

1. INTRODUCTION

In geomatics, current surveying techniques and methodologies allow obtaining the shape of an object by means of dedicated hardware and software procedures. When applied in the field of historical-architectural heritage, such procedures can provide an exact geometric record of the de facto state of the object. Documentation for cultural heritage often requires 1:50, and sometimes 1:20, rendering. Such rendering scale allow for a detailed analysis of the morphological features of the object (Batini, 2014).
In time, many research groups have studied the accuracy and the precision of the obtainable 3-D models (Adembri et al., 2011; Benedetti et al., 2010; Biosca Taronger et al., 2007; Guidi et al., 2010; Guidi & Remondino, 2012; Juan Vidal & Martínez-Espejo Zaragoza, 2011; Merlo et al., 2013; Verdiani, 2011; Wenzel et al., 2013).
These new investigating technologies are extensively used both globally, for the survey of the historic architectural complex as a whole, and locally, for the survey of the decorative and architectural details. Among them, the painted decorations (frescoes, surface paintings in general, etc.) are perhaps the most significant, but possibly more fragile, elements, receiving the constant attention of all the figures that in various capacities are involved in the planning and implementation of restoration and conservation (architects, administrators, engineers, art historians, restorers, etc.).
While in the past restoration techniques for pictorial works were based on traditional visual inspection, nowadays, with the availability of new digital technologies it is possible to achieve objective, three-dimensional, geometrically accurate data, to which all results of different types of investigation can be referred. New technologies have therefore given a strong boost to digital cataloguing, documentation and monitoring of the preservation state of painted decorations, so that the availability of virtual models is now an essential tool in the development of heritage rendering.
It seems clear, therefore, that it is exactly in the transition phase from cognitive/investigative to operational that new technologies are required to interface and integrate with traditional techniques.
In 3-D renderings, chromatic data are often confined to the final popularization phase, i.e. colouring a simplified model. In fact, this appears as a demeaning of 3-D models as tool of scientific representation (Cipriani et al., 2014).
This research stems from the pressing need of restorers to have models high in precision and quality, in term of both geometry and texture colour. In this context, this work proposes a study of various techniques to apply high quality colour textures to laser scanner based models featuring high precision for topology and morphology.

2. STATE OF THE ART

2.1 Geometry survey

Laser scanning is currently an established surveying technique, whose accuracy and resolution are suitable for the rendering of historical, archaeological and architectural heritage aimed at classification and restoration (Bevilacqua et al., 2012; Caroti & Piemonte, 2008; Caroti & Piemonte, 2012; Dore and Murphy, 2012; Juan Vidal & Martínez-Espejo Zaragoza, 2011; Lerma et al., 2011; Marambio et al., 2009; Martínez-Espejo Zaragoza and Juan Vidal, 2012). An obvious advantage is the maximization of the ratio between the amount of data and survey duration. On the other hand, a major drawback is the high cost of both hardware and software, as well as the burdensome post-processing phase (nine days of in-house processing for each

This contribution has been peer-reviewed.
doi:10.5194/isprsarchives-XL-5-W4-285-2015

single day of field survey to achieve a high resolution textured model).
Over the last decade, a new methodology for 3-D surveying, by means of images, based on Structure from Motion (SfM) algorithms, has been gradually spreading. Using this computer vision technique the geometric structure of the 3D scene and the pose of the camera at each image can be quickly recovered from a series of overlapping, offset images. (Alsadik et al., 2013; Guidi et al., 2010; Guidi & Remondino 2012; Koska & Kremen, 2013; Lerma et al., 2010; Rodriquez 2012; Wenzel et al., 2013). This new approach to softcopy photogrammetry generates very dense 3D colour point clouds quite similar in size to the one produced from laser scanning surveys.

However, even if the software evolution in this field is very rapid and performance in terms of processing time, amount of manageable data and obtainable precisions is gradually improving, these proceduresmay not always be considered reliable. In fact, they suffer from variable precision, strongly dependent on the pattern present on surveyed objects, as well as the difficulty of having little control of the achievable accuracy at the geometric and morphological levels (Apollonio et al., 2014; Guidi & Remondino, 2012).

2.2 Texture mapping

In large scale 3-D models used as supporting documentation in restoration works, textures are not a mere aesthetic complement. In fact, in addition to supporting construction, material and chromatic studies, they also act as metric surveying tools, providing, once applied to the models, a guideline for measurements.
Therefore, if textures have to meet these requirements, their positioning accuracy must be consistent with the scale used, besides having the necessary chromatic precision.
Many laser scanners have built-in cameras, whose relative orientation is calibrated by the manufacturer, which allow direct true colouring of the point cloud. These textures are characterized by a high geometric accuracy, but the systems used for the photographic takes usually do not achieve good results in terms of resolution and colour fidelity (Apollonio & Remondino, 2010)
Simplified, realistic-looking models may not suffice for restorers, who require rigorous texture mapping for both morphology and colour information.
In these cases, it is essential to resort to a dedicated photographic campaign, performed with high quality cameras as regards optics, sensor size and post-processing graphics.
The traditional way to obtain textures from these photos is to derive camera orientation parameters from a single take, generate the UV map using so-called texture mapping of the projection of the single-frame model (Apollonio et al., 2010; Baldissini et al., 2010; Guidi et al., 2010) and finally merge the different partial UV maps to generate the texture of the model as a whole. Textures obtained in this way allow obtaining 3-high quality D models in terms of geometric, morphological and chromatic aspects (Fantini et al., 2012).
On the other hand, this procedure is very time-consuming, also requiring constant operator intervention in the collimation of the points on the model and on images. In addition, software and procedures commonly used quite often do not take into account the distortion present on images.
In the case of SfM software, the creation of models and textures is pretty much contextual, and the procedure usually involves self-calibration of the camera, which also takes account of characteristic distortion parameters. In these models, textures usually have good photographic quality, although, as stated in section 2.1, the overall morphological reliability of such models is not comparable to laser derived ones.

3. MATERIALS AND METHODS

Laser scanning models are quite accurate and detailed as regards geometry and morphology. On the other hand, the quality of pictures taken by the built-in camera is altogether substandard, hence textures need to be obtained by photographs taken with a reflex camera.
The traditional way to obtain a texture is to project the single-frame take on the model, and then merge the different partial UV maps to generate the texture of the overall model.
Models derived by the SfM methodology, although providing good rendering of the vault as a whole, cannot ensure homogeneous accuracy in rendering small morphological variations, due to their dependency on the vault surface pattern. Higher deviations coincide with cracks, the borders of collapsing plaster and gaps in frescoes, all important features for restorers. On the other hand, the colour quality of the texture is quite good, as it is in direct relation with the quality of the images used for texture application.

3.1 The case study

The case study is a frescoed vault on the ground floor of Palazzo Roncioni, in Pisa. The vault is a type rather common in Italy known as *a schifo* or mirror vault, having a not quite regular rectangular plan. The fresco, painted by Tuscan painter Giovan Battista Tempesti, is largely preserved, though it presents some obvious recent gaps due to environmental degradation.

3.2 Survey

This vault has been surveyed by Leica Geosystems' C10 ScanStation laser scanner, fitted with a built-in camera. The resolution of the survey was set at 4mm at 10m, resulting in a very dense cloud (on average 70pts/cm^2). In order to simplify the model where surfaces were smooth, a decimation of the points was performed, preserving in any case the geometrical information that, however small, is valuable for restorers (local deformation, cracks, ...).
In order to texture the derived model with more faithful colours, a dedicated photographic campaign was performed using a Nikon D700 camera fitted with a fixed 20mm objective. The shooting distance was on average 4.5m, which allowed single pixel coverage of about 2 mm. Furthermore, an additional photographic campaign, designed for subsequent three-dimensional modelling via SfM methodologies, was carried out. This is necessary because, while for texturing, as described above, the images are just required to cover the entire surface, ensuring a low mutual overlap, SfM software requires a large overlap between consecutive images to allow better matching results (in general better than 70%). The new photo campaign was carried out with the same camera and the same optical system as the previous one.

3.3 Laser scanner model texturing by means of individual images

In the photographic draft take the surface of the vault was divided in areas, characterized by almost constant curvature, in each of which photographic images were acquired, keeping the

This contribution has been peer-reviewed.
doi:10.5194/isprsarchives-XL-5-W4-285-2015

optical axis close to the direction of the radius of curvature of the centroid of each area. The images must ensure full coverage of the object taking into account that only the central portion of each image is used in order to avoid residual radial distortion at the edges.
The processing procedure is the same for each image to be projected. Considering the portions in common between the image and the 3-D model obtained from the laser scanner, a set of tie points was chosen, maintaining the distribution as uniform as possible. These points have been exported as coordinates in DXF format, using a point cloud management software of the laser model. These coordinates constitute the control points, which the photogrammetric software uses to orient the image. Photogrammetric software is used to derive the image coordinates of the chosen tie points; it is also possible to calculate the camera image features along with the precision with which the various parameters were determined.
Repeating this process for all the images yields the orientation parameters of the cameras, each of which is exported in a format compatible with the entertainment software used for texturing (e.g. Luxology Modo - format * .fbx).
In order to texturize the model, once it has been imported into the entertainment software, the so-called subdivision surfaces model must be created (Fantini, 2012) and the related UV map defined (Fig. 1).

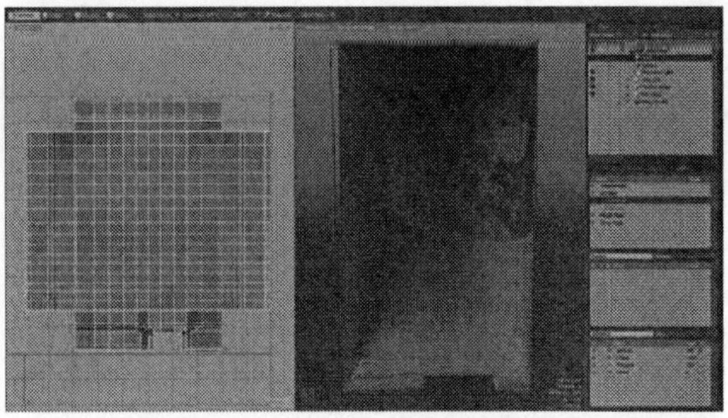

Figure 1. Projection of single image on the laser scanner model.

By projecting individual image partial UV maps are obtained for each one (Fig. 2 left), where colour is applied only to the framed part of the vault. At this point, all partial UV maps have to be reprocessed with an image editor (e.g. Photoshop) in order to merge them into a single image (Fig. 2 right). This operation requires processing the different images to produce uniform brightness, contrast and saturation and to select the portions of the partial UV map corresponding to the central portion of the respective image to reduce the influence of the residual radial distortion. Finally, the overall UV map is projected onto the laser model using entertainment software.

Figure 2: Partial UV map of a single image (left) and total overall UV map (right).

3.4 Model from SfM

A dedicated photographic campaign with high percentage of overlap along both axes has been performed in order to produce a set of images suitable for application of the SfM modelling methodology. The images have been processed via Agisoft's PhotoScan 1.0.0 software. This is quite common commercial software that enables performing the different steps needed for Image-Based Modelling: camera calibration, automatic detection of homologous points, image orientation, extraction of a dense point cloud, input of control points and absolute orientation of the model, surface generation and texture mapping (Manferdini & Remondino 2010).
The output is a 3-D model with good chromatic quality (Fig. 3 left). On the other hand, the precision of the underlying geometry (Fig. 3 centre) is not as good as that obtained with laser scanning-based surveys (Fig. 3 right).

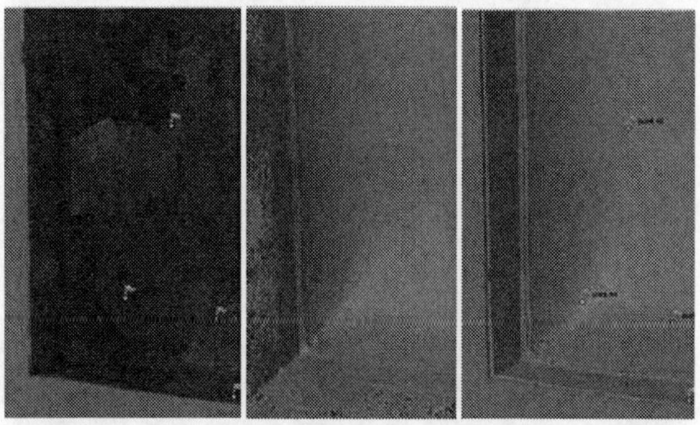

Figure 3. Texture from SfM (left); model from SfM (centre); model from laser scanning (right).

3.5 Laser scanner model texturing by means of SfM oriented images

The models obtained via laser scanning and SfM have been framed in the same reference system by means of 30 tie points evenly spread across the vault.
This has allowed using the external orientation of the cameras used for SfM-processed images to project the same images on the laser scanning model rather than on the SfM (Fig. 4).

Figure 4. Laser scanning model textured by images oriented with SfM technology.

This contribution has been peer-reviewed.
doi:10.5194/isprsarchives-XL-5-W4-285-2015

4. RESULTS AND DISCUSSION

The textured model was then checked for precision and quality. Effectiveness in terms of resources and time required for the different processing types has also been assessed.

4.1 Laser scanner model texturing by means of individual images

In order to apply the texture to the laser derived model, thirty tie points (on average) have been chosen on each image and on the model. The standard deviation following the image orientation procedure averaged 4 pixels. The texture applied to the model does not show any discontinuity in the transitions between images; moreover, its precision has also been checked by printing at a 1:1 scale on acetate paper and directly comparing them with the object (Fig. 5). This model can then be assumed as a reference also in assessing the positioning precision of the texture.

Figure 5. Precision check at 1:1 scale.

On the other hand, it must be noted that completion of the processing step, including image orientation and blending, took five person-days of work.

4.2 Geometrical comparison between laser- and SfM-derived models

The present research aims to use image orientation obtained by SfM to project the same images over a laser scanning derived model. Therefore, it was necessary in the first place to check the precision of the model obtained by SfM for an indirect assessment of image orientation.
Point-to-point deviations of the SfM model have been measured using the laser scanning model as a reference. This check performed on the whole model resulted in a standard deviation of about 3mm. These results show that image orientation is substantially accurate and confirmed the mean reprojection error involved with orientating each image via SfM to be 0.70 pixel with an average of 9000 tie points per image. On the other hand, maximum deviation values are in the range of 7-10mm and refer to cracks, plaster collapse borders and gaps in the fresco. Figures 6 to 11 show an overview of the fresco and some details on local deviations.

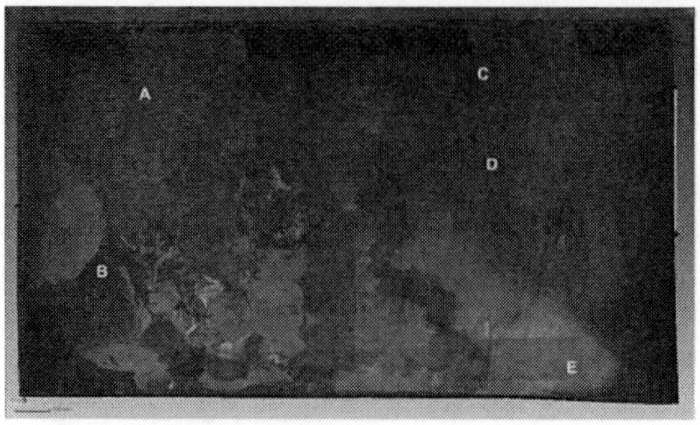

Figure 6. Regions checked for deviations between laser scanning- and SfM-derived models.

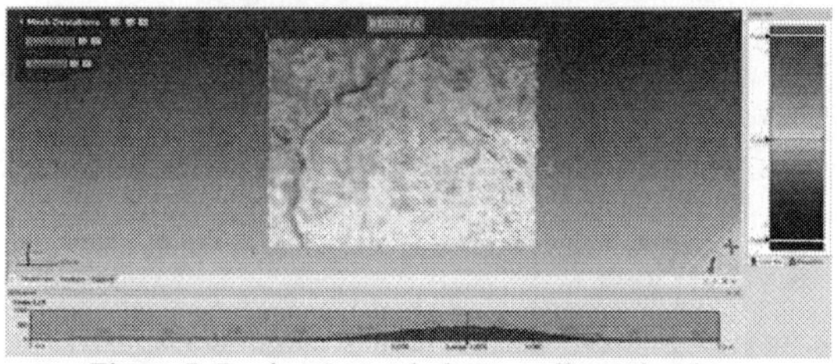

Figure 7. Region A: total plaster collapse borders.

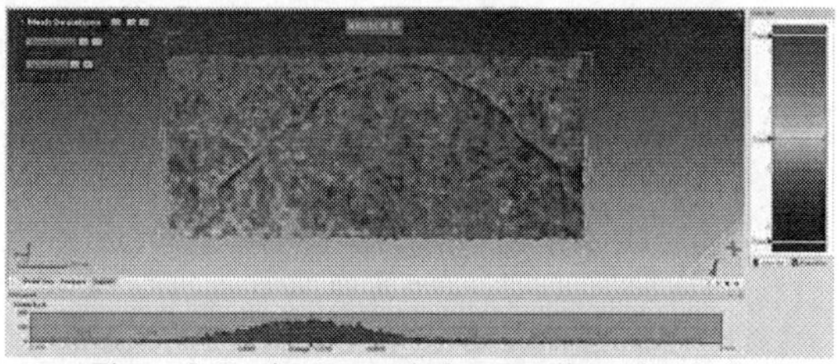

Figure 8. Region B: total plaster collapse borders.

Figure 9. Region C: gap in the fresco.

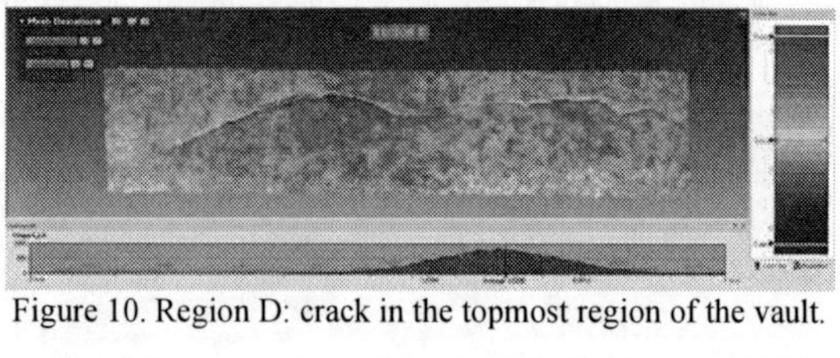

Figure 10. Region D: crack in the topmost region of the vault.

This contribution has been peer-reviewed.
doi:10.5194/isprsarchives-XL-5-W4-285-2015

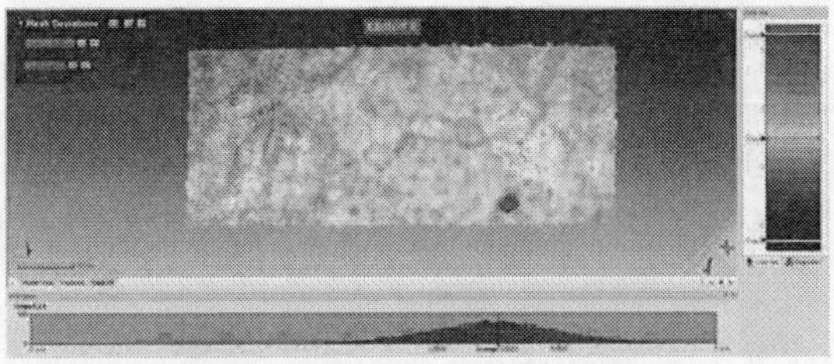

Figure 11. Region E: gap in the fresco.

Hence, the SfM-derived model is overall quite exact, but shows some flaws precisely in the regions of most interest to restorers. This processing methodology, on the other hand, has the advantage of significantly lower resource requirements: manual intervention is limited to inputting the support points to orient and scale the model, and the entire process required only person-day of work.

4.3 Texture precision of final model

The geometry of the final model is derived from the laser scanning survey. The texture of the model has been obtained via SfM software by importing the model and projecting the oriented images. The precision check for texture positioning was performed by measuring point-to-point deviations between easily identifiable features on the SfM and the laser scanner textures, which resulted in a standard deviation of about 3.5mm.

5. CONCLUSIONS

This paper presents a methodology that uses external orientation of images computed with SfM techniques to project the same images on the model derived from laser scanning.
For data integration purposes, the models generated by both methodologies are compared. The comparisons substantially confirm the good overall accuracy of camera orientation obtained by SfM. The comparison between the models is subsequently elaborated on through an precision check by means of tie points and analysis of the different processing methodologies.
The proposed methodology allows obtaining a three-dimensional model which retains the geometric precision of laser scanning surveys along with the texture quality obtainable with a photographic survey campaign.
The most interesting feature of the proposed methodology lies in the considerable reduction in the necessary time and computing resources as compared to the traditional methodology of orientation and projection of single images on the model. Finally, regarding texture positioning precision, both procedures have yielded results of the same order of magnitude.

REFERENCES

Adembri, B., Juan Vidal, F., Martínez-Espejo Zaragoza, I., 2011. "Hunting friezes of the Piazza d'Oro at the Hadrian's Villa: new hypothesis for a virtual reconstruction inside an integrated research strategy" in *Proceedings 16th International Conference on Cultural Heritage and New Technologies*. Museen der Stadt Wien – Stadtarchäologie. Vienna. ISBN: 978-3-200-02740-4, pp.140-153.

Alsadik, B., Remondino, F., Menna, F., Gerke, M., Vosselman, G., 2013. "Robust extraction of image correspondences exploiting the image scene geometry and approximate camera orientation" in *The International Archives of the Photogrammetry, Remote Sensing and Spatial Information Sciences*, Vol. XL-5/W1.

Apollonio, F.I., Baldissini, S., Manferdini, A.M., Remondino, F., 2010. "Rilievo 3D nell'area archeologica di Pompei: risultati e considerazioni." in *Modelli Digitali 3D in archeologia: Il caso di Pompei*. Edizioni della Normale. Scuola Normale Superiore Pisa, pp. 270-307.

Apollonio, F.I., Ballabeni, A., Gaiani, M.,Remondino, F., 2014. "Evaluation of feature-based methods for automated network orientation." in *The International Archives of the Photogrammetry, Remote Sensing and Spatial Information Sciences*, Vol. XL-5.

Apollonio, F.I. & Remondino, F., 2010. "Modellazione 3D da sensori attivi Pipeline con laser scanner" in *Modelli Digitali 3D in archeologia: Il caso di Pompei*. Edizioni della Normale. Scuola Normale Superiore Pisa, pp. 94-117.

Baldissini, S., Gaiani, M., Remondino, F., 2010. "Rilievo, gestione e mappatura del colore" in *Modelli Digitali 3D in archeologia: Il caso di Pompei*. Edizioni della Normale. Scuola Normale Superiore Pisa, pp. 200-235.

Battini, C., 2014. "Il colore come strumento di rilievo tridimensionale" in *Colore e Colorimetria. Contributi Multidisciplinari*, Vol. X A, Atti della Decima Conferenza del Colore, ISBN 978-88-916-0437-8.

Benedetti, B., Gaiani, M., Remondino, F. (ed.), 2010. Modelli Digitali 3D in archeologia: Il caso di Pompei. Ed. della Normale. Scuola Normale Superiore Pisa.

Bevilacqua, M.G., Caroti, G., Piemonte, A., 2012. "Rilievi integrati della Basilica romanica di San Gavino a Porto Torres", in *Proceedings 16° Conferenza Nazionale ASITA*. VICENZA, p. 391-396, ISBN 978-88-903132-7-1

Biosca Taronger, J.M., Navarro Tarín, S., Lerma García, J.L., 2007. "Modelado tridimensional de una bóveda barroca mediante la combinación de láser escáner y fotogrametría", in *7 Setmana Geomàtica*, Febrero 20–23, Barcelona, España.

Caroti, G. & Piemonte, A., 2007. "Laser scanner and photogrammetry for the survey of Monumental Cemetery in Piazza del Duomo, Pisa (Italy)." in *The International Archives of the Photogrammetry, Remote Sensing and Spatial Information Science*, vol. XXXVI part 5/C53; p. 201-206, ISSN: 1682-1750.

Caroti, G. & Piemonte, A., 2008. "An integrated survey for knowledge and preservation of a cultural heritage: the albanian fortified citadel of Elbasan.", in *The International Archives of the Photogrammetry, Remote Sensing and Spatial Information Sciences*, vol. XXXVII, part B5/V; p.373-378, ISSN: 1682-1750

Caroti, G. & Piemonte, A., 2012. "Studio relativo al rilevamento di dettaglio di una porzione del transetto nord della chiesa di San Paolo a Ripa d'Arno al fine di procedere al consolidamento strutturale", in *Proceedings Un incontro informale per i 70 anni del Prof. Carlo Monti*, pp. 1-6, Edizioni SIFET, ISBN 978-88-905017-6-1.

This contribution has been peer-reviewed.
doi:10.5194/isprsarchives-XL-5-W4-285-2015

Cipriani, L., Fantini, F., Bertacchi, S., 2014. "Affidabilità ed ottimizzazione del texturing del colore apparente dei modelli da sensori attivi e passivi: casi studio nel settore dei Beni Culturali" in *Colore e Colorimetria. Contributi Multidisciplinari*, Vol. X A, Atti della Decima Conferenza del Colore, ISBN 978-88-916-0437-8.

Dore, C. & Murphy, M., 2012. "Integration of Historic Building Information Modelling (HBIM) and 3D GIS for Recording and Managing Cultural Heritage Sites" in *18th International Conference on Virtual Systems and Multimedia (VSMM)*. VSMM 2012 Virtual Systems in the Information Society. 2 - 5 Septiembre 2012 Milan, Italia, pp. 369-376.

El-Hakim, S., Gonzo, L., Picard, M., Girardi, S., Simoni, A., 2003. "Visualization of frescoed surfaces: Buonconsiglio Castle – Aquila Tower, Cycle of the months" in *ISPRS Archives* – Vol. XXXIV-5/W10.

Fantini, F., 2011. "La modellazione in Displaced sub-D: Modelli a dettaglio variabile da scansione laser" in *Il ritorno all'immagine, nuove procedure image based per il Cultural Heritage*. ISBN 978-1-4452-9256-4.

Fantini, F., Rodríguez-Navarro, P., Di Tondo, S., 2012. "Il problema della mappatura del colore nei modelli digitali 3d a displaced subdivision surface da rilevamento laser scanner in ambito archeologico" in *Colore e colorimentria. Contributi multidisciplinari*. Vol. VIII A. Dip. Indaco. Politecnico di Milano,pp. 31-38.

Guidi, G. & Remondino, F., 2012. "3D Modelling from Real Data" in *Modelling and Simulation in Engineering*. Edited by Catalin Alexandru, InTech.

Guidi, G., Russo, M., Beraldin, J.A., 2010. *Acquisizione 3D e modellazione poligonale*. McGraw-Hill. Italia.

Juan Vidal, F. & Martínez-Espejo Zaragoza, I., 2011. "Hypotesis for the virtual anastilosis of the main chapel of the Iglesia de Los Desamparados de Les Coves de Vinromá" in *16th International Conference on Cultural Heritage and New Technologies*. Museen der Stadt Wien – Stadtarchäologie. Vienna. ISBN: 978-3-200-02740-4, pp.201-212.

Koska, B. & Křemen, T., 2013. "The combination of laser scanning and structure from motion technology for creation of accurate exterior and interior orthophotos of St. Nicholas Baroque church" in *The International Archives of the Photogrammetry, Remote Sensing and Spatial Information Sciences*, Volume XL-5/W1.

Lerma, J.L., Navarro, S., Cabrelles, M., Villaverde V., 2010. "Terrestrial Laser Scanning and Close Range Photogrammetry for 3D Archaeological Documentation: the Upper Palaeolithic Cave of Parpalló As a Case Study" in *Journal of Archaeological Science* n.37. Elsevier, pp. 499–507.

Lerma, J.L., Cabrelles, M., Navarro, S., Seguí, A.E., 2011. "La documentación patrimonial mediante sensores de imagen o de barrido láser" in *Documentación gráfica del patrimonio, Ministerio de Cultura*,pp. 108-117.

Manferdini, A.M. & Remondino, F., 2010. "Modellazione 3D da immagini. Pipeline fotogrammetrica" in *Modelli Digitali 3D in archeologia: Il caso di Pompei*. Edizioni della Normale. Scuola Normale Superiore Pisa, pp. 162-199.

Marambio, A., Pucci, B., Nuñez, M.A., Buill, F., 2009. "La aplicación del escáner láser terrestre en la catalogación del patrimonio arquitectónico del casco histórico de Castellfollit de la Roca" in *VIII Semana Geomática Internacional*,pp. 1-8.

Martínez-Espejo Zaragoza, I. & Juan Vidal, F., 2012. "Gestión gráfica avanzada de edificaciones antiguas" in *XI Congreso internacional de expresión gráfica aplicada a la edificación*, Universidad Politécnica de Valencia. ISBN: 978-84-8363-964-1.

Merlo, A., Fantini, F., Lavoratti, G., Aliperta, A., López Hernández, J.L., 2013. "Texturing and optimization of reality based digital models: the church of Compañía de Jesús" in *DISEGNARECON*. Vol. 6, n. 12.

Rodríguez Navarro, P., 2012. "La fotogrametría digital automatizada frente a los sistemas basados en sensores 3D activos" in *Revista EGA (Expresión Gráfica Arquitectónica)*. Universidad Politécnica de Valencia. Valencia, pp.100-111.

Wenzel, K., Rothermel, M., Fritsch, D., Haala, N., 2013. "Image acquisition and model selection for multi-view stereo" in *The International Archives of the Photogrammetry, Remote Sensing and Spatial Information Sciences*, Volume XL-5/W1.

This contribution has been peer-reviewed.
doi:10.5194/isprsarchives-XL-5-W4-285-2015

NEW CONTRIBUTIONS ON THE DOME OF THE PANTHEON IN ROME: COMPARISON BETWEEN THE IDEAL MODEL AND THE SURVEY MODEL

L. Aliberti [a], M. Canciani [b], M.A. Alonso Rodriguéz [c]

[a] Departamento de Ideación Gráfica Arquitectónica, Universidad Politécnica de Madrid – licinia.aliberti@gmail.com
[b] Dipartimento di Architettura, Università degli Studi Roma Tre - marco.canciani@uniroma3.it
[c] Departamento de Ideación Gráfica Arquitectónica, Universidad Politécnica de Madrid – miguel.alonso@upm.es

Commission V, WG V/4

KEY WORDS: Pantheon, integrated survey, ideal model, survey model, surface

ABSTRACT:

This work proposes an integrated survey and a study of the intrados of the dome of the Pantheon in Rome. An actualized architectural survey of the interior of the dome can generate useful material for future studies.
The survey has been realized by using in a first stage the digital photogrammetry and in a second stage the three-dimensional laser scan technology. The compared analysis between different methods applied in the same object is useful towards a closer approximation to real dimension.
Among several aspects that arise in dealing with the Pantheon this work focuses mainly on the study of the geometry of the inner surface of the dome. The specific goal of the research is to verify the spherical form of the surface and the coffers' distribution. In this sense it takes an important place the extracting data system. In order to realize the analysis it was applied a critical treatment of selected information contained in the point cloud. The use of plan and section drawings connects to the study of three dimensional models.
The research is based on the construction of an ideal geometrical model that derives from the theoretical model described in the historical documents. The survey points model, which keeps the irregularities of the actual form, determines the creation of an average sphere, that is a regular model defined by clarifying geometrical laws.
The direct comparison between the survey model and the ideal model contributes to the building understanding. It detects irregularities or deformities where they exist, and provides objective and quantifiable data.

1. INTRODUCTION

This work proposes an integrated survey and a study of the *intrados* of the dome of the Pantheon in Rome, based on a PHD research (Aliberti 2014). The intense study of such surprising architecture lasts over the centuries and the monument still offers unsolved questions. An actualized architectural restitution of the interior of the dome can generate useful material for future studies, considering that the last surveys of the building were published by Belardi (2006), Valenti (2009) and Graßhoff (2009).
Among several aspects that arise in dealing with the Pantheon this research focuses mainly on the study of the geometry of the inner surface of the dome. The constructive solutions applied by roman constructors contribute to realize this singular and complex architecture that is still completely conserved. The specific goal of the research is to verify the spherical shape of the surface and the coffers' distribution.
The different analysis developed show the importance of integrating the building metric restitution with the study of its history, geometry and construction. The survey is considered as a comprehensive system of architectural understanding. Bibliographic documentation enriches the research indicating where and when to look in a sharper way.
Nevertheless the metric restitution of the dome may be considered as the first necessary step to develop the research. The integrated survey realized using a 3D laser scan and the digital photogrammetry offers complete and detailed information about the inner surface of the dome.
In order to carry out the geometrical analysis of the dome inside shape, a critical study is applied to data derived from surveys. The research is based on the construction of a geometrical model that approximates the survey points and that is generated with a rigorous average calculation. Survey points, which keeps the irregularities of the constructed object, determines the creation of a regular model defined by clarifying geometrical laws. The direct comparison between the individual positions of the survey points and the geometrical model contributes to the building understanding. By this method we can detect irregularities or deformities where they exist, and provide objective and quantifiable data.
The study of the survey results and their comparison with the ideal model was possible due to a specific data treatment. In order to recognize the specific shape of the object it is necessary to select data from the large information of the point cloud. This process consists in several steps starting from the creation of a continuous surface until the extraction of planar sections. The simplification of the point cloud into a triangulated surface may generate a loss of information. Then in some specific studies, the analysis of part of the dome was realized generating section curves directly from the point cloud. This method can be a further step in the study of the 3D survey in the architectural field (Canciani et al. 2013).
The bi-dimensional drawing allows to study the implicit complexity of architecture by isolating some of its parts and operating in a reduced environment. In this work, the plan and section drawings have been considered as a necessary tool to develop the geometrical analysis.

This contribution has been peer-reviewed.
doi:10.5194/isprsarchives-XL-5-W4-291-2015

At the same time it was realized a comparison study directly working into the three-dimensional work space of vectorial drawing software. These operations have been especially useful in order to realize the study of the continuous surface of the intrados of the dome. The observation of the 3D model permitted the use of calculations of the relative positions of the survey points so that it was possible to determine their distances from the ideal sphere. Furthermore we could detect and indicate the position and the extension of the deformation areas of the intrados of the dome.
The integrated studies realized using plan and section drawings and the three-dimensional models improved the research and generated useful information about the present condition of the Pantheon dome.

2. SURVEY MODEL AND IDEAL MODEL

2.1 Survey model

The complete survey of the *intrados* of the dome was obtained through three-dimensional laser scanner and a digital photogrammetric restitution. Integrated softwares enabled the comparison of results and verified the accuracy of surveys. By analysing the outcome we can state that the results in both methods are coherent and complementary in a certain way. A contrast between different versions of restitution of the same object is useful for a better approximation to the real shape.
The photogrammetric survey was realized using different series of photos and comparing the results (Fig. 1). The inner surface of the dome is completely visible and it can be photographed from different points, except for the hidden area covered by the cornice that separates the drum from the dome. The existence of coffers takes a great importance in the visual perception of the dome and contributes to reconstruct the spherical shape of the surface; meanwhile they are useful in this work of metric restitution as evident reference elements.
The photographic camera used to catch the images was a full-frame digital reflex, with fixed-focus lens that were changed according to need. It is relevant the circumstance that it was used a high definition camera to make the photographs. The software used to realize the photogrammetric restitution is the Photo Modeler Scanner and the point model obtained was analyzed with common vectorial drawing softwares like AutocCAD and Rhinoceros.
Different restitution models were generated applying a selection of photo series with diverse topics. Successively it was selected the model with a better approximation and photo orientation. This model includes all internal and external points of the coffers, several points of the cornice that separates the tambour from the dome and some other points that describe the *oculus*. The points selection was possible due to the coffers' shape and to the marble joints of the cornice (Fig. 2). However some difficulties where found in the restitution of the continuous surface of the *intrados* of the dome. Therefore it was introduced another survey method.
The 3D scanner survey was realized in the summer of 2012 using a Leica HD6200 scanner (Fig. 3). The data were collected in a short interval of time because of the use restrictions of the monument.
The scanner was positioned in several places. The data works include five scanning stations. One of them was realized in high definition from the centre of the space. Other four scanning sessions were developed from different places using a reduced definition. The range registration and alignment was executed using the point cloud data management software Leica Cyclone.

Figure 1. View of the photogrammetric model and the oriented photos used for the restitution.

Figure 2. View of vertices points obtained by photogrammetry.

Figure 3. View of the point cloud.

The central scan of the inner space describes all the intrados of the dome including the coffers' area without any shadow. All the interior levels and the edges are visible from the center of the space, where these elements can be appreciated with a better perception (Licht 1968; Lucchini 1996; Waddell 2008). This fact shows directly the geometrical qualities of Pantheon's coffers. Their design follows complex projective laws.

This contribution has been peer-reviewed.
doi:10.5194/isprsarchives-XL-5-W4-291-2015

In order to carry out the research, a critical study was applied to the outcome of the survey. The necessary information for realizing the geometrical analysis was extracted from the point cloud. The general model obtained from the different scenes alignments was exported to the software 3DReshaper, that offers useful tools to realize a sharp study of the point cloud. Using this software, a continuous mesh surface was generated from the point cloud as a best fit result.

The extraction of a surface from the point cloud is an essential condition to the research development. These procedures, in fact, are aimed to realize the geometrical analysis of the *intrados* of the dome, that requires a work with continuous elements like curves and surfaces (Docci & Maestri 2010).

In the study of the coffers distribution the photogrammetric model has been very useful as a complementary tool. The manual selection of points in the restitution process generates a clear definition of the coffers vertices. On the other hand the triangulation of the mesh surface brings some problems in the identification of the vertices. The use of points from photogrammetric method with those selected from the scanner point cloud allows the study of the coffers' general distribution.

2.2 Ideal model

Historical surveys and documents used to represent the *intrados* of the Pantheon's dome like a perfect hemisphere, with the center situated in the cornice line that divides the drum from the vault. A specific study in the actual spherically-shaped condition of this surface would reconnect to the theoretical model of Pantheon.

To make possible the reconstruction the ideal geometry of the *intrados* of the dome, we have studied the sphere to which the points of the cloud approximate. The ideal sphere gets close to the survey points and is generated with a rigorous average calculation.

In order to calculate the radius and position of the average sphere, the works have been developed in two ways: either by using a selection of the points, or by using the entire points cloud.

The resulting mesh surface has been cut in 20 horizontal sections and 28 vertical sections. 14 of the vertical sections have been produced on the ribs and the other 14 on the axis of the coffers. The intersection points of the horizontal and vertical sections were used for the informatics calculations of the average sphere. These calculations have been realised with the software Rhinoceros under two different conditions. One is considering all the points resulting in a sphere of radius 21.95 meters. The other considers all the points except those of the upper area and results in a radius value of 22.03 meters. The upper part of the dome's model contains a bigger number of horizontal curves. The points in this part show a depression that decreases the value of the resultant radius.

The subsequent studies have been realized by comparing the results with the radius 22.03 m sphere with no deformations.

The diameter of this calculus matches the result of 44.06 meter obtained by the points cloud management program 3DReshaper (best fit). This operation manages a big number of points uniformly distributed.

In this way we obtain a double verification of the accuracy of the average value, by considering the entire cloud in a specific management program, and by considering just some well selected points in the model and proceeding to their analysis with vectorial standard software. The average calculated model could be taken as the ideal sphere.

Most of the publications on the Pantheon describe a value for the diameter of the dome between 43.30 and 43.80 meters, although specific recent studies like Pelletti's (1989), show a value of 22.03 that matches the achieved results. According to a recent survey carried out by the University *"La Sapienza"* in Rome (Valenti 2009) the average radius is close to 21.90 meters in the coffers' part while 21,40 in the oculus part. The laser scanner survey carried out in 2009 by the Bern University (Graßhoff 2009), registers a radius of 21.99m, a gap of 4 cm to the present study.

According to the classic descriptions of the Pantheon, the centre of the dome belongs to the plan of the cornice between the drum and the dome. This fact is collected in the historical surveys from the publications of the Renaissance to the first rigorous study by Desgodetz (1682) and the later representations. Only in the recent surveys another position is set for the centre of the interior sphere.

This work like Licht (1968), Pelletti (1989) or Graßhoff (2009), determines the existence of an offset between the cornice plan and the centre of the inscribed sphere. In order to reconstruct the cornice line it was applied an average calculation to the survey points. The circumference line obtained belongs to a plan forming 0.44 degrees with the horizontal plan. It was fixed a reference height in the horizontal plan passing through the centre of the average circumference of the cornice. The position of the centre of the ideal sphere is calculated at -0,37 m behind this reference system based on the reconstruction of the cornice line that divides the drum from the dome (Fig. 4). This feature suggests a first difference between the ideal geometric model and the theoretical model described in the historical documents.

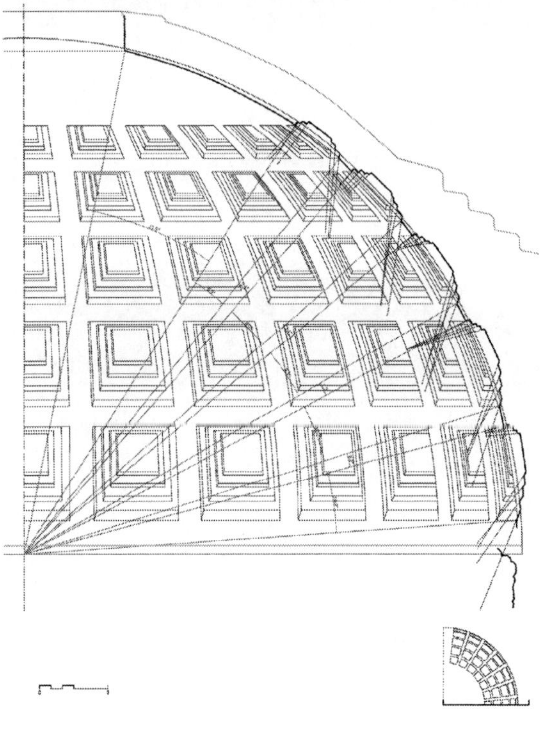

Figure 4. Study of a section of the Nord-East quadrant of the dome showing the offset between the cornice plan and the centre of the inscribed sphere.

This contribution has been peer-reviewed.
doi:10.5194/isprsarchives-XL-5-W4-291-2015

2.3 Comparisons

The point cloud analysis aims to detect and measure the differences between the survey results and the ideal model. Several graphical studies were realized to develop specific topics. The horizontal and meridian sections were overlapped to the curves of the ideal sphere. Some studies were applied directly on the 3D model. Many numerical controls join to the graphical analysis in order to produce quantifiable data.
These architectural analysis tools applied to the point cloud provide important and new data about the geometrical characteristics of the dome.

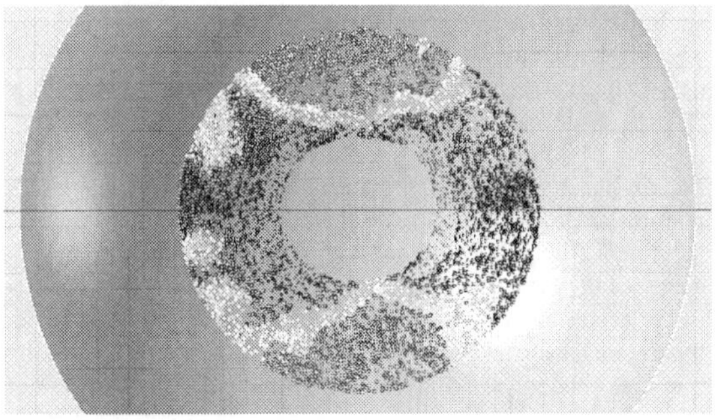

Figure 5. Gradation of distances from the points of the survey-model to the ideal model on top of the dome.

These studies show the almost complete overlapping of the inner surface with the geometrical ideal model, except in a limited deformation area. The good conditions of the spheric shape are surprising, mostly considering its size and age.
In order to calculate the distance of each point to the ideal surface, a reference system with the origin in the centre of the average sphere was established. The data of the outcome show a 10 cm tolerance in 72.5% of the points, while 21% is under 15 cm and only 6.5% is located over this measure.
The results observation indicates that the greater number of points is very close to the spherical surface. The mean distance to the ideal sphere, calculated between all the points, is 7,8 cm, that is the approximated equivalent of a 1/280 of the average radius.
Regarding the separated parts of the dome the work states an average error of 5.5cm in the coffers zone, which means a proportion of 1/400 of the interior radius, while in the upper part, between the end of the coffers and the oculus it is 9.4cm. Data indicate in this part the existence of a special depression where the maximum deformation is held. In order to determine the exact location and dimension of this depression a specific analysis on this part has been properly detailed. As a result of the analysis of the cuts and level curves, data revel a deformation zone along the north-south axis of the dome emphasized in the upper level described before (Figg. 5 and 6).

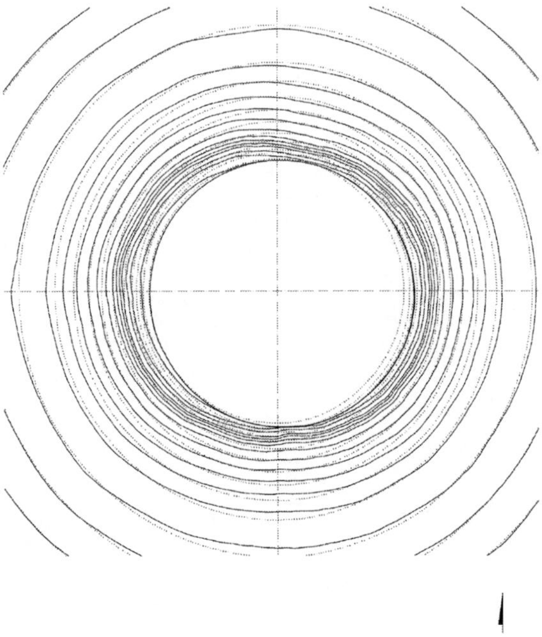

Figure 6. Horizontal sections of the upper area of the *intrados* of the dome: comparison between the survey model (black) and the ideal model (red).

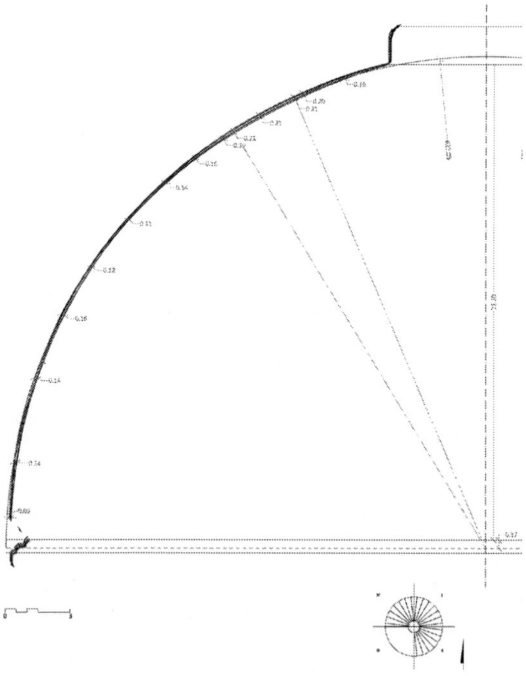

Figure 7. Study of the profile view of the dome by rotating the meridian sections of a quadrant of the dome in the same plan.

This contribution has been peer-reviewed.
doi:10.5194/isprsarchives-XL-5-W4-291-2015

Some recent studies indicate the existence of a deformation on the *intrados* of the dome, but they don't publish detailed information about this argument (Scherer 2009, Valenti 2009). With the purpose of calculating the dimension of that deformation we have analyzed in major quantity the distances of the survey points to the ideal sphere in this area. Part of the point cloud was selected applying a reduction of its density. This method allows determining the position and the extension of the deformed zone.

In order to realise the offset between the different profiles of the dome a rotation of the vertical sections has been applied. The vertical axis for this rotation has been positioned through the centre of the average sphere (Fig. 7).

The dome has been divided in four sectors according to the orthogonal axis of the plant view. Then a rotation has been applied to the curves of sectors I and III, to overlap with the lines of the sectors II and IV. The subsequent contrast of the projection of these lines with the average sphere shows de actual differences.

The maximum value of deviation in quadrant I equals 27cm, and it is found at the base of the dome along a 7 degree arch from the horizontal line. In Quadrant II this value stands at 17 cm, and it is registered in the central-superior part of the dome.

The study of a sector of the scanner model can be of interest in order to provide a gross calculation without any of the common simplifications in data management. In this sense three meridian sections of the north-south sector have been produced. Two of these sections pass through the axis of the ribs and the third cuts the north-south axis through the coffer over the entrance to the building (Fig. 8). The sections of the point cloud in this part register a maximum distance to the ideal sphere of 0.32 m in the south and 0.20 m in the northern part next to the top.

The actual distribution of the coffers can also be the topic of an specific study. There is an evident complexity of design in their position. The partition of the sphere in 28 parts seams geometrically uncomfortable. The division of a quadrant in 7 parts was already a complex geometrical problem at that time.

The tool for the study of this distribution has been again the overlapping of the restitution model and the ideal one. In order to clarify the existence of actual alignments, orders and irregularities, we have used the orthogonal projection of the coffers' vertices. The use of informatics tools helped in the calculations of the average straight lines or arches, considering the results of both 3D scan and photogrammetry methods.

The analysis of the coffers provides additional information about their location. The carried out calculations demonstrate their extraordinarily precise distribution.

Their projection on the ground plan show the deviation of the bisecting lines of each individual sector compared to the axis of the ribs. The error registration stands at 0.6°. The maximum inaccuracy value in the most unfavorable axes equals 1° and it is found in two of the 28 considered ribs. The angles between the axes of the sectors are extremely close to a precise value of 12.85° that means it is close to the result of the subdivision of 360 ° by 28 units.

The coffers are organized in 28 vertical sectors that contain five rows with a decreasing width. The construction lines passing through the external points of the coffers seem to be meridian circles of the maximum sphere inscribed. This study reveals how the vertical side meridians surprisingly do not converge in the centre of the *oculus* (Fig. 9). This indicates a reduced size of these circumferences, and leads to the idea that the construction actual axis were those cutting the coffers vertically through their midpoints in stead of those along the ribs' edges.

The two extensions of the outer lines of each coffer sector intersect in a vertex not coincident with the sphere pole. The distribution of these points does not describe a circle due to evident irregularities. Some of these vertexes approximate the pole while others separate from it significantly. The contrast between these outcomes with the geometrical analysis of the cupola shows specific deformations in the north and south parts of the dome. Indeed, only with a few exceptions, the existing vertexes in a longest distance to the pole are situated in those places.

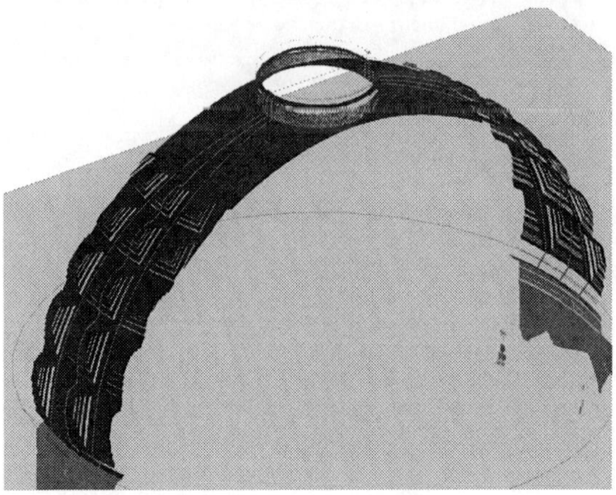

Figure 8. Meridian sections taken directly from the point cloud

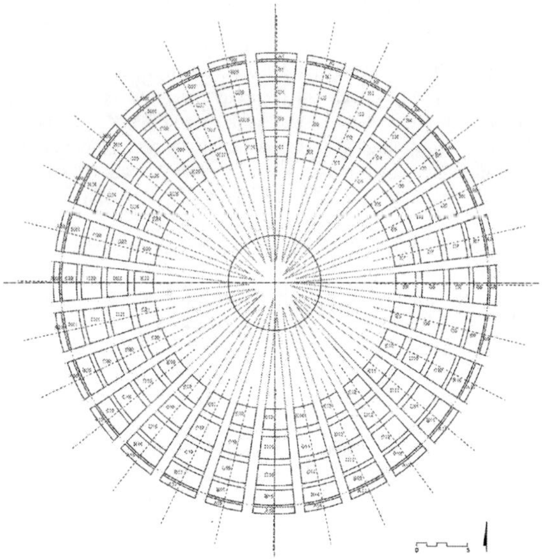

Figure 9. Alignments of the coffers' external edges showing that they don't converge in the center of the *oculus*.

This contribution has been peer-reviewed.
doi:10.5194/isprsarchives-XL-5-W4-291-2015

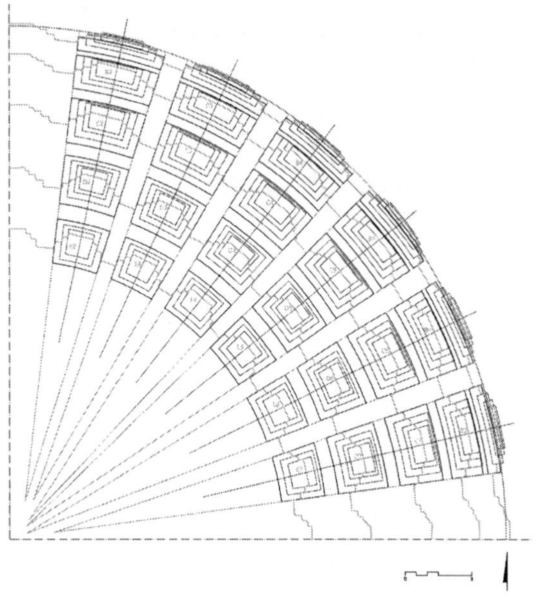

Figure 10. Study of the northeast quadrant of the dome with the convergent lines passing through the external vertices of the coffers.

Figure 8. Comparison between the ideal model and the mesh surface extracted from the point cloud. Top view.

3. CONCLUSIONS

This work shows the efficacy of an integrated survey that includes different methods of metric restitution and a comparative study of the documentation concerning the architectural object.
Some new information about the interior surface of the dome and some about the actual distribution of the coffers have been developed. The restitution of the present configuration of the intrados of the Pantheon can be used to study the current state of the dome and to determine its performance in time.
The system for the geometrical analysis and the data extraction from the point cloud that was applied can be considered as a relevant aspect of this research. The data selection and their treatment belong to a method based on the simplification of the information aimed to carry out specific analysis.
The comparison of the survey model and the ideal model was essential to develop the study. By using average calculations it was possible to determine their shape and relative position. The extraction of surfaces and curves from the point cloud and from the photogrammetric model was a necessary step in the research. This work shows like the bi-dimensional drawing is useful and effective to study the specific topics and to calculate detailed conditions of the object. The parallel analysis realized using three-dimensional models confirms and extends the results. The introduction of the comparison between the survey model and the ideal model can be a useful method for further architectural researches.

ACKNOWLEDGEMENTS

We would like to thank specially the willingness of Arch. Giovanni Belardi (curator of the Pantheon monument in the *Soprintendenza per i Beni Architettonici e Paesaggistici per il Comune di Roma*). Also thanks to the *Leica Geosystems* delegation in Rome and specially to Arch. Valentina Albano for offering the instrumentation used to realize the laser scan survey.

REFERENCES

Aliberti, L., 2014. *Pantheon y cupulas clásicas romanas: geometría y construcción.* PhD. Universidad Politécnica de Madrid.

Aliberti, L., Altozano García, F., 2011. Documentación gráfica mediante fotogrametría digital de la cúpula del Pantheon en Roma. *Actas del VII Congreso Nacional de Historia de la Construcción*, Santiago de Compostela, pp. 33-43.

Belardi, G. 2006. *Il Pantheon. Storia, tecnica e restauro.* BetaGamma, Viterbo.

Canciani, M., Falcolini, C., Saccone, M., & Spadafora, G. 2013. From point clouds to architectural models: algorithms for shape reconstruction. In: *The International Archives of the Photogrammetry, Remote Sensing and Spatial Information Sciences*, Trento, Italy, Vol. XL-5/W1, pp. 27-34.

Canciani, M., Sacccone M. 2011. The use of 3D models in integrated survey: the church of St. Thomas of Villanova in Castel Gandolfo. *The International Archives of the Photogrammetry, Remote Sensing and Spatial Information Sciences*, Trento, Italy, Vol. XXXVIII-5/W16, pp. 591-7.

Desgodetz, A. 1682. *Les edifices antiques de Rome.* 1779th ed. Claude Antoine Jombert fils, Paris.

This contribution has been peer-reviewed.
doi:10.5194/isprsarchives-XL-5-W4-291-2015

Docci, M., Maestri, D. 2010. *Manuale di rilevamento architettonico e urbano*. Laterza, Bari.

Fine Licht, K.d., 1968. *The Rotunda in Rome. A Study on Hadrian's Pantheon*. Nordisk Forlag, Copenhagen.

Geertman, H. 1980. Aedificium caeleberrimum. Studio sulla geometria del Pantheon. *BA Besch,* 55, pp. 203-26.

Graßhoff, G. 2009. *The Bern Digital Pantheon Project. Plates*. Bern Studies, Bern.

Lucchini, F. 1996. *Pantheon*. Carocci, Roma.

Pelletti, M. 1989. Note al rilievo del Pantheon. *Quaderni dell'Istituto di Storia dell'Architettura*, vol. XIII, pp. 10-18.

Scherer, M. 2009. A sustainable geodetic network for documenting and monitoring the Pantheon. In: *The Pantheon in Rome. Contributions to the Conference Bern, November 9-12, 2006*, pp. 215-22.

Valenti, G. M. 2009. A Computing Model for the Pantheon's Cupola: From the Discrete to the Continuous. The Ideal Continuous Model. In: *The Pantheon in Rome. Contributions to the Conference Bern, November 9-12, 2006*, pp. 223-233.

Waddell, G. 2008. *Creating the Pantheon : Design, materials, and construction*. "L'ERMA" di Bretschneider, Roma.

Figure 8. Comparison between the ideal model and the mesh surface extracted from the point cloud. Axonometric view.

This contribution has been peer-reviewed.
doi:10.5194/isprsarchives-XL-5-W4-291-2015

A versatile and low-cost 3D acquisition and processing pipeline for collecting mass of archaeological findings on the field

E. Gattet [a], J. Devogelaere [b], R. Raffin[c], L. Bergerot [a], M. Daniel[c], Ph. Jockey[b], L. De Luca [a]

[a] UMR CNRS/MCC MAP 3495 – (eloi.gattet, laurent.bergerot, livio.deluca)@map.cnrs.fr
[b] Aix-Marseille University, CNRS, UMR 77299 CCJ, LabexMed (devogelaere@mmsh.univ-aix.fr, philippe.jockey@univ-amu.fr)
[c] Aix-Marseille University, CNRS LSIS UMR 7296

Commission V, WG V/4

KEY WORDS: Photogrammetry, 3D acquisition, cultural heritage, low cost solution, web platform

ABSTRACT:
In recent years, advances in the fields of photogrammetry and computer vision have produced several solutions for generating 3D reconstruction starting from simple images. Even if the potentialities of the image-based 3D reconstruction approach are nowadays very well-known in terms of reliability, accuracy and flexibility, there is still a lack of low-cost, open-source and automated solutions for collecting mass of archaeological findings, specially if one consider the real (and non theoretical) contextual aspects of a digitization campaign on the field (number of objects to acquire, available time, lighting conditions, equipment transport, budget, etc...) as well as the accuracy requirements for an in-depth shape analysis and classification purpose. In this paper we present a prototype system (integrating hardware and software) for the 3D acquisition, geometric reconstruction, documentation and archiving of large collections of archaeological findings. All the aspects of our approach are based on high-end image-based modeling techniques and designed basing on an accurate analysis of the typical field conditions of an archaeological campaign, as well as on the specific requirements of archaeological finding documentation and analysis. This paper presents all the aspects integrated into the prototype:
- a hardware development of a transportable photobooth for the automated image acquisition consisting of a turntable and three DSLR controlled by a microcontroller ;
- an automatic image processing pipeline (based on Apero/Micmac) including mask generation, tie-point extraction, bundle adjustment, multi-view stereo correlation, point cloud generation, surface reconstruction ;
- a versatile (off-line/on-line) portable database for associating descriptive attributes (archaeological description) to the 3D digitizations on site ;
- a platform for data-gathering, archiving and sharing collections of 3D digitizations on the Web.
The presentation and the assessment of this prototype is based on an interdisciplinary experience carried out on the study of the Tholos in Delphi within the framework of the CNRS's Eloquentia project and the EFA's archaeological program.

1. INTRODUCTION

The practical aspects of a archaeological findings digitization campaign on the field such as the large number of objects to digitize, the available time, lightning conditions and transportation are today not totally taken into account by the scientific community working on 3D digitization technology design and development. In this paper, we introduce an innovative low cost, open-source and automated solution for a massive 3D acquisition of archaeological fragments on the field. We also introduce a versatile database for the on-site description of the digitized artifacts linked to an on-line platform for gathering, archiving and sharing collections of 3D digitizations on the Web.
This research campaign was conducted in 2014 as part of the program PEPS-CNRS "Eloquenzia", composed of three "Unités Mixtes de Recherche" – the "Centre Camille Jullian" (UMR 7299), the laboratory of "Modèles et simulations pour l'Architecture et le Patrimoine" (UMR 3495), and the "Laboratoire des Sciences de l'Information et des Systèmes" (UMR 7296) – with the help of the Delphi Archaeological Museum, the Ephory of Delphi and the support of the "École Française d'Athènes".
The paper is organized as follows. Section 2 presents the Tholos in Delphi. Section 3 is devoted to related work and our prototype is described is Section 4. We conclude in Section 5.

2. CONTEXT OF THE THOLOS IN DELPHI

2.1 The monument: status, description

Figure 1: The Tholos in the Sanctuary of Athena Pronaia, Marmaria, Delphi, May 2014.

The archaeological site of Delphi, located in Phocis (Central Greece) and excavated by the French School of Archaeology in Athens ("École Française d'Athènes") since the end of the XIX° century (The "Grande Fouille" occurred between 1892 to 1903) (Hellman, 1992a), is composed primarily of two Panhellenic sanctuaries, that of Pythian Apollo at the foot of Phaedriades on Mount Parnassus, and that of Athena Pronaia further below at the place called Marmaria.

This landmark of Greek religion – the *Omphalos* or "navel of the world" – was founded at the beginning between of VIII° century BCE and abandoned during the VI° century CE, with its period of hegemony between the VI° and IV° centuries BCE when the site was a symbol of unity in the Hellenic world.
Our current research campaign focuses on one monument in particular: the Tholos in the sanctuary of Athena Pronaia (Figure 1) . This exploratory project lays the foundation of new collaboration with the "École Française d'Athènes", which holds publication rights to the Tholos ; the publication of its sculptural decoration by Professor Philippe Jockey is forthcoming in the series of the "*Fouilles de Delphes*".
The monument itself has been known since 1836 but was not excavated until 1901, and published for the first time in 1925 (Charbonneaux, 1925). Despite a large number of hypotheses its function is as yet unclear (Hellman, 1992b)(Bommelaer, 1997)(Kyriakidis, 2010).
Built 370-360 BCE, the Tholos is a circular building composed of a Doric colonnade around a central *naos* (main room), constructed in Pentelic marble and dark limestone from Eleusis. The *naos* and its colonnade were both decorated with an architectural frieze of alternating triglyphs and metopes carved in high relief. According to Vitruvius, it architect was likely Theodoros of Phocaea (Vitruvius, 1995). Since Antiquity, the Tholos in Delphi has known great renown by virtue of its round plan, an original and rare feature in Greek architecture (the other Greek *tholoi* of the classical period are those of Athens and Epidaurus), and because of the quality of its decoration. Today it remains one of the most important monuments of the ancient Greek cultural landscape.

2.2 Previous studies and reconstitutions

In 1938 three of its Doric columns and entablature were reconstructed, indicative of its ongoing cultural prominence (Amandry et al., 1942). This anastylosis provides an idea of the monument proportions and of its location in the Delphic landscape, even if, paradoxically, it has only been partially reconstructed (Jockey, 2010). It should be noted, however, that a first partial reconstruction was carried out in 1902 under the aegis of the "École Française d'Athènes". The base of every column was put in place, as well as part of several courses of the *naos*.
Nevertheless difficulties surround our interpretation of the monument, due to the fragmentary state of the material evidence, as the result of the collapse of the building, as well as the reuse, dispersion and destruction (for lime production) of many marble blocks and slabs at the end of Antiquity. It provided this place with its modern name – Marmaria, meaning marble quarry.
This poor state of conservation is even more striking when looking at the remains of the carved metopes in high relief. The corpus comprises approximately 500 fragments of various human figures (torsos, heads, legs, arms, both male and female body parts, nude and draped), horses (which may suggest centaurs in some cases), and marble plaques. The first published study, admirably comprehensive and thorough, was carried out by Jean Marcadé in the 1970s (Marcadé, 1979a, 1979b). Using technical and stylistic criteria, he was able to identify and gather fragments pertaining to the Tholos that had been scattered and reused in later monuments, alongside those rediscovered during excavations. He was further able to reassemble some architectural elements. In this way he revealed the mythological scenes in high relief which decorated the metopes. The forty metopes of the Doric colonnade (65 cm in height) depict scenes of a Gigantomachy and an Amazonomachy, whereas the forty small metopes adorning the *naos* (42 cm in height) depict a group of deities about whom it is difficult to analyse further at present.
Using these early studies, the first trial 3D virtual reconstruction of the monument was carried out successfully by a three-dimensional scanner in 1995 thanks to technological sponsorship (Bommelaer, 1997)(Jockey, 2012). A turntable was used to digitize some architectural fragments with a triangulation based 3D scanner (Bommelaer, 1997). Reconstruction of the metopes was limited by their fragmentary state, the very strong relief decoration and the wear of some archaeological features.

2.3 A new research project: "Eloquenzia"

At a time when experimentations on 3D digitization of heritage artifacts seem having a strong focus on recording and visualization issues, a new look on the cognitive issues of the digital representation seems essential.
Reasoning about shapes, their morphological, structural and semantic similarities is a fairly widespread practice in archaeology and cultural heritage. The analysis of discretised forms is also a research topic for mathematicians and computer scientists studying computer graphics and geometric modeling. To promote an interdisciplinary reflexion about this issue seems particularly interesting today, when observing geometrical features in digitized shapes could make emerge new strategies for analysing and classifying heritage artifacts by reading their morphological signature.
By combining top-down (representing the geometry from high-level semantic descriptors, e.g. vocabulary terms) and a bottom-up (extracting semantics from low-level geometric descriptors - e.g. discrete curvatures) approaches, the CNRS PEPS Eloquentia project looks at the intersection of the two approaches in order to identify some new description models (integrating geometry & semantics) for embedding the richness of 3D digitization into intelligible representations.
These purposes led to the current study, which combines archaeological approaches with 3D digitization and shape analysis, to test our understanding of the sculptural decoration of the Tholos. The contribution of new technologies (photogrammetry, computer vision and 3D modeling) can be seen as a logical development in the digital age to the graphic restitutions of archaeologists and architects of past centuries (Bommelaer, 1997). This close collaboration between archaeologists and computer scientists, based on a continuous and mutual experimentation/assessment process defines the basis of the experimental system presented in this paper. It will also help to better understand the architecture and carved decors of this famous monument that was in Antiquity the Tholos of Marmaria in Delphi.
Last, these new technologies could help to evidence more accurately some material or technical details, such as breaks in the marble, tools traces and final surface treatments. This study, based on both the architecture of the monument and its carved ornaments, will accompany the publication of the sculptural decoration, linked with a relational online database and 3D virtual anastylosis.

3. RELATED WORK

3.1 Artifacts digitizations by photogrammetry

Multi-view photogrammetry applied to dense point cloud generation becomes common today. This is explained by its flexibility and low technological requirements: a digital camera and a laptop are enough. Most software protocols (Agisoft, Bundler/PMVS) recommend between 60 and 120 images all around a simple subject to cover it entirely, but do not provide some exact information about the angles between photographs or their precise relative positions. This means that a typical manual acquisition for a statue or small artifact with one

camera takes around 20 to 30 minutes, without the lightning installation and the high risk to forget some point of views of getting 100% of perfectly stable and sharp images.
That is the reason why recent works are focusing on new solutions for decreasing the acquisition time and for guarantying a better coverage of the artifacts by developing massive acquisition rigs with up to 100 digital reflex cameras. As a DSLR price is at least 400€, a rig cost is extremely high, plus they seem generally difficult to transport and adapt to different subjects. In a parallel way, the Fraunhofer CultLab3D project (cultlab, 2014) , is focusing on the industrialization and automation of the entire 3D digitization process of artifacts by means of state-of-the-art scan and illumination techniques integrated into a digitization laboratory.

3.2 Web solutions for data gathering and sharing 3D

The range of available solutions for visualizing interactive 3D models on the web is growing today (3D PDF, WebGL, Unity, etc.) and many European projects (CARARE, 3D-ICONS, etc.) started international initiatives for making available 3D content into cultural portals (e.g. Europeana) as well as for developing metadata schemas able to enrich the 3D content with information related to the digitization process and to the understanding and interpretation of the digitized artifacts. Even in these projects are forming today the basis for the future 3D libraries of cultural content for the general public, very few works are focusing on a complete integration of the 3D digitization practice within the overall heritage artifacts documentation pipeline (including the on-site observation, classification, description, measurement, annotation, etc.). This underlines the need for versatile databases (working with and/or without internet access) able to assist the documentation process as well as to manage the archiving, the sharing and the publishing of big collections of heterogeneous data (text, photographs, 3D point clouds, 3D models, textures, etc.).

4. PROTOTYPE DEVELOPMENT

4.1 Acquisition protocols

Since 2012, the MAP laboratory, in partnership with ENSG and IGN is developing optimized photogrammetric acquisition protocols for cultural heritage artifacts documentation (archaeological fragments, paintings, sculptures, furniture, architectural elements, buildings, sites, etc.) by basing on the parametrisation of the APERO /MICMAC toolchain (Pierrot-Deseilligny, 2011).
This automated image-based modeling pipeline consists in tie-points extraction (based on SIFT (Lowe, 1999) algorithm), internal and external orientations of each image by bundle adjustment (Apero), and dense image matching (Micmac). The final output is a set of dense and accurate point clouds: the correlation algorithm allows for each pixel of photograph to generate a corresponding 3D point. This means the density is relative to the focal length of the lens and its distance to the subject. A recent paper presents the metrological assessment of this automated processing pipeline by basing on the comparison with 3D laser scanning (Toschi et al., 2014).
By using the recommendations and specifications for an optimal image acquisition protocol (Martin-Beaumont et al., 2013) as starting point, we defined the requirements for a totally automated acquisition process aiming at the development of an experimental acquisition device. The initial requirements are listed below:

- For each desired points cloud take a "master" image and several closed associated images (with low ratio base to distance and important overlapping) (Pierrot-Deseilligny et al., 2011),
- Between each master image take a sufficient number of intermediary images to assure the connection during the orientation step (Figure 2),
- Avoid to move a punctual light source during the photographs acquisition,
- Use a context with enough details spread in all the space photographed as background,
- Fix as many parameters of the cameras as possible.

Figure 2: Spatial configuration of the acquisition of an artifact (Nicolas Martin-Beaumont, 2013)

Nicolas Martin-Beaumont (Martin-Beaumont et al., 2013) then proved that a master image needs at least 4 slave images, all around, taken all around with a angle inferior than 10° for an optimal correlation.
It was then decided to move not the camera but the subject by making it spin on a turntable. Also, 3 DSLR will be used to cover the upper and lower zones of the model. This way, as many photos as wanted can be taken all around. This causes some problems regarding the recommendations: the light should not change between photographs and the context needs details spread around in the background to help the orientation part.

4.2 Hardware development

In accordance with museum curators, it was decided that an artifact measuring less than 60 centimeters in its largest dimension and/or having a weight less than 20 kilograms can be moved and placed on a rotating platform. Those numbers depend of course on each museum rules and policy, but it was a starting point to think about our automated acquisition solution. Also, the whole prototype has to be transportable by plane and used by one person. A large luggage measures 50x80x30 centimeters and the maximum accepted weight is 32 kilograms. Finally, as a challenge, it must be the most cost-effective as possible.
A turn table was salvaged from an old prototype. It could accept a weight up to 100 kilogramd, more than enough keeping in mind this project was meant to stay low cost. It was added a wheel worm drive linked to a stepper motor salvaged from an old printer to move the table. This link has two main advantages: it limits the momentum on the motor and keeps the turntable from moving when the motor is off (Figure 3).

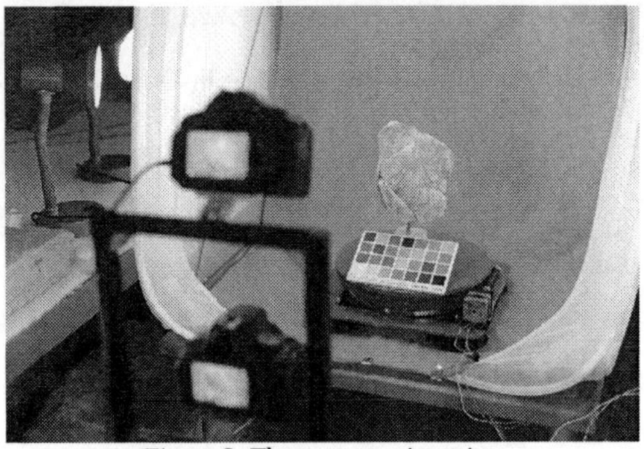

Figure 3: The prototype in action

The 3 digital cameras were already in possession of the laboratory. They are equipped with a 50mm F/1,8 lens. In the case of this first experimentation in Delphi, only the camera trigger was tethered. Each camera has still to be set manually and the memory cards tranfered one by one. The whole system is controlled by an Arduino micro-controller.

Figure 4: Dedicated electronic prototype board

A dedicated electronic prototype board (Figure 4) was developed. Its role is to distribute the power to the turntable, move it and trigged synchronously the camera. A LCD screen and three push buttons (ok/return and left/right) where added to keep the system autonomous. A simple menu allows to set the following parameters:

- Number of acquisitions in a complete rotation
- Time for each acquisition
- Number of photographs per acquisition.

As the cameras are set separately, the waiting time between two displacements has to be manually entered in the controller. Also, the option for bracketed acquisition in order to create high dynamic range images was implemented, so number of photographs per acquisition could be set up. This was yet not used.

To guarantee an optimal image orientation and a good texture quality, the lightning has to be as uniform as possible. We used a folding photo booth of 1x1x1 meter size. The diffusing sides allows to create a uniform illumination. The goal of the Eloquenzia project was to study the geometric aspect of the artifacts, the textures were not the main objective. That's why we used seven cheap standard desk halogen lamps. They were provided with intensity modulators to equalize manually the ambiant light. In a last step, we mounted a green wall on the back of the photo booth. This was used after the acquisition to detach the object and create a uniform white background.

4.3 Software and practice

The acquisition protocol we used during the Eloquenzia mission was the following:

- While the previous object was photographed in the photo booth, the next one was referenced on our database, setting the timestamp, folder name on the camera, number of pictures and other acquisition settings
- The object was then placed on the turntable, stabilized on a weighted platform.
- The cameras were set and adjusted for the object to be seen entirely and meanwhile cover the largest possible portion of the camera's sensor.
- On each camera the image counter was reset and a new folder created. Then the exposure and aperture were set.
- The micro-controller was set if the exposure time has changed, and the number of desired photographs was set before launching the acquisition.

This whole operation lasts an average of 5 minutes for a 3x75 images acquisition.

As the counter is reset for each object, a new folder is created automatically. When emptying the memory cards, each folder name corresponds to one artifact. Then, the images for a same object for the three cameras can be moved automatically, gathered and renamed in a same folder. A simple bash script has been written for this operation. This allows an easier programming for the next steps as we already know that for each number, the image is vertically aligned with the two other camera images.

Then starts the entirely autonomous 3D generation scripts. First the images need to be clipped from the green background (Figure 6, left images). A GIMP script was written for this operation. It automatically picks a median color in 10 pixel radius zone in the upper right corner of each image and selects with a adjustable threshold all the similar colors. This selection is grown to fill the eventually missed pixels and filled with plain white. Then, as said in Section 4.1, a tie-point extraction is performed on all the images. As each image has common points only with the 5 previous and following ones, the SIFT operation is done this way. Also, the 10 last and 10 first images are matched as the acquisition is circular. Also, the images has some strong similarities with the same vertical position in the other lines. This protocol allows an optimized way to compute a maximum of tie-points in a minimum of time. The 3 cameras use the same lens, which was precisely calibrated before.

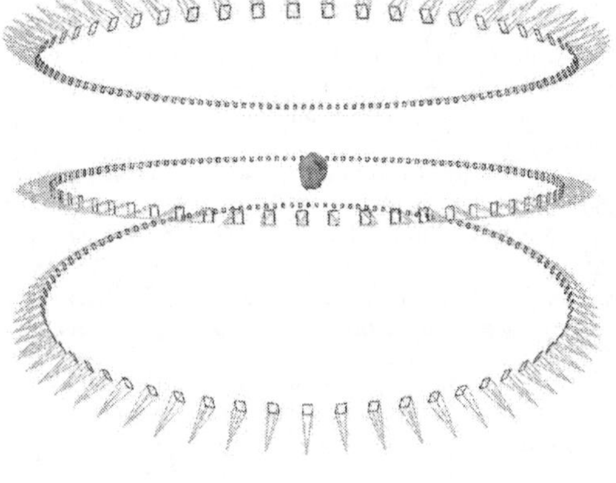

Figure 5: Orientation visualisation

This contribution has been peer-reviewed.
doi:10.5194/isprsarchives-XL-5-W4-299-2015

As the cameras are moved between each acquisition, it was chosen to separately compute the orientation of each object. First the central line orientation is computed, and starting from this result the upper and lower lines are oriented (Figure 5).
Once the orientation computed, a mask is automatically generated via another GIMP script. This accelerate the correlation by deleting the useless white background.
The dense correlation is made every 9 images on each line which gives 8 point cloud for each line. The master images are chosen alternatively to maximize the coverage of the object.

Figure 6: Different steps of the image processing: Raw photograph; clipped image; colored 3d mesh; raw mesh; mesh detail.

The assembled result is composed of an average of five million points, with an important overlap between every separate point clouds. A mesh is then computed using the screened Poisson reconstruction software developed by (Kazdhan and Hoppe, 2013) (Figure 6, right image). The figure 6 subject is an amazon torso N°4312 (Marcadé, 1979b, p161).

4.4 Database development

The whole data produced by the project must be stored. In order to query and manage them we implemented an SQL database containing for each archaeological object: unique id, archiving index, name, type, morphological characteristics, tools marks, photographs and a binary file containing the 3D object mesh. To be easily accessible for scientists we constructed an application framework that will permit to store, modify and share the data. The applications are embedded in a web browser to help diffusion and maintenance, and are available in-line or off-line. We plan two usages: 1) in the laboratory, assuming Internet connection, the web-browser handle a classical interface to query, modify, suppress data in a centralized database that is accessible once the user authenticated (in read-only or read-write mode). 2) while in mission, the user must edit or consult the data in an off-line mode. This is issued by exporting the database into the traveling computer. The application for the user is downloaded on the server. The user carries both data and application, thus he always has the most recent version. After the mission, the data are merged with the server ones and shared again with the community.
A special mode on the server allows an administrator to manage the database scheme and to convert from one scheme to the other one, handling the database evolution (Figure 7).

We based our developments on Javascript language, both on client (embedded in the web-browser) and server (node.js) sides. It makes easier the deployment and the maintenance of the client application and does not bother about computer or operating systems the clients work on. Each time the user opens its browser on-line the current interface is downloaded, and when on departure for a mission the latest client is included. We take advantage of the dissemination of HTML5 standard, in which 3D visualization is made possible without third-party plug-ins. WebGL is specified by the Khronos group (khronos, 2014), as OpenGL or OpenCL, it offers a direct access to the graphic hardware of the client (Joshi, 2012). A 3D canvas can be created in the HTML5 document, and Javascript programming language is used to handle the 3D environment (in the DOM - Document Object Model - of the page). WebGL is an API that brings OpenGL ES (for "Embedded Systems") in a web browser.

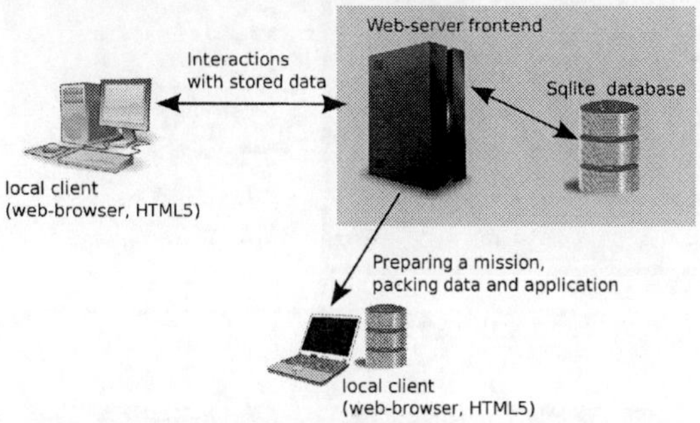

Figure 7: Offline/Online database managment

Working on the database allows the user to insert the 3D digitized mesh model he processed abroad, having a tool to view/rotate/zoom the 3D model. When he will return to the laboratory, the new data will be inserted as other scientific/multimedia ones.
In order to manage these new data when returning from mission, each time a value is added or modified off-line a log is create in a special table in the base, containing author ID, computer ID and field/data modified. These values are used when merging server data with the ones collected by the itinerant researchers. If a coherence problem in the fields or in the values is raised an archaeological scientist must merge manually the data (Figure 8).

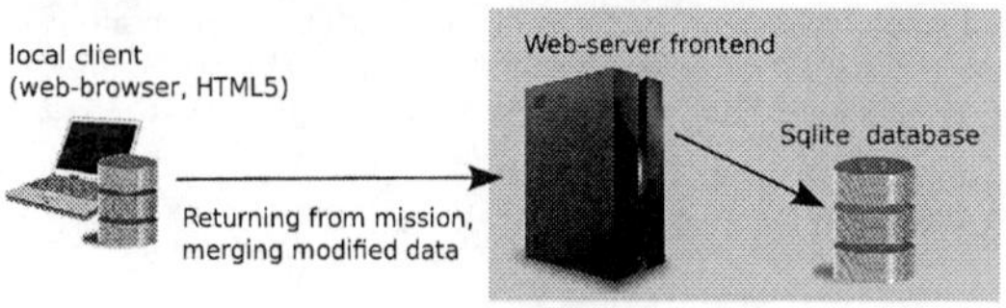

Figure 8: Database merging

4.5 Final web publishing of the gathered data

In order to provide the final web publication of the gathered data (including photographs, 3D point clouds, 3D models and related descriptions) we developed a specific meta-viewer, which can be easily embedded into a common webpage, able to switch the visualisation and interaction properties according to the type of content. Our meta-viewer is based on Web-Gl and Three.js and integrates a set of open source graphics libraries for managing the real time interactive visualization of all the data gathered and elaborated by the presented pipeline: 2D photographs (jpg and png format), dense and colored 3D point clouds (in ply format) by basing on (Potree, 2014), 3D textured meshes (in dae and obj formats). The meta-viewer provides the automatic configuration of the download process as well as of the interaction functions to be added to the selected content (zoom, pan, roll, …). Finally, as it is based on standard W3C certified technologies, it is already compatible with multiple devices (PCs, laptops, tablets and smartphones).

This contribution has been peer-reviewed.
doi:10.5194/isprsarchives-XL-5-W4-299-2015

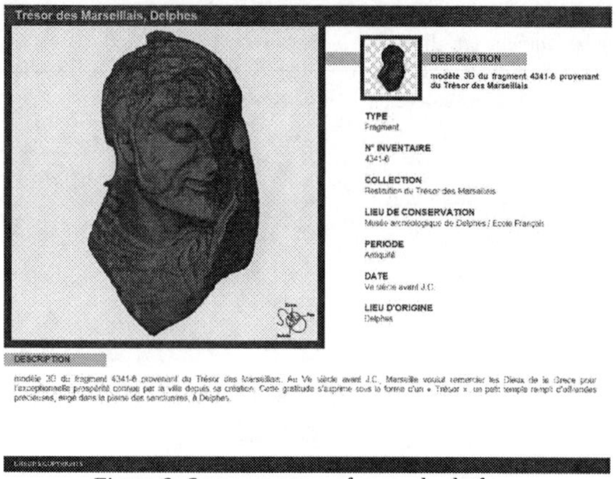

Figure 9: Screen capture of our web platform

5. CONCLUSION

5.1 Assessments

Initially, the PEPS Eloquenzia aimed at capturing a dozain significant models, that were previously carefully selected among the hundreds available. Our approach along the mission proved to be more effective than expected and we acquired in 8 days more than 150 different artifacts. This resulted in 200 gigabytes of JPEG images. In average, the raw 3D point cloud resulting from a automated generation is composed of five million points. As the images used for the point cloud generation have a important overlapping, a sampling is made before meshing. The mesh has an average one million polygons. Also, as the camera global orientation is automatically computed, there is no physical scale and reference. The 3D mesh has to be manually scaled thanks to scale references sticked on the object support before the acquisition.

5.2 Perspectives

This mission was carried out with an exploratory goal, which was outperformed. Thus, a few evolutions are still to implement:
Three cameras were not enough to properly cover the upper and lower part of an artifact. Also the illumination could greatly be improved with dedicated lights. This would result in a better color overall the model. A textured 3D object would greatly enhance the visual experience of the user. It can be combined with compression algorithms to balance quality versus file size because WebGL environment is also limited to "small" objects (65k points for example). This must be done while uploading 3D meshes to the server. It can be possible for instance to create "low-polys" representations which can be associated to normal maps to keep a visually good aspect.
The demand for automatic processes for 3D characterization, semantic recognition, reattachment propositions is increasing, involving that 3D data should be stored in two resolution levels: one simple for visualization, the other at the best level to implement geometry processing algorithms.

ACKNOWLEDGEMENTS

The authors wish to thank the French CNRS-PEPS program and the "École Française d'Athènes" (EFA) for their support, the Delphi Archaeological Museum and the Ephory of Delphi which provided accesses to the archaeological site and reserves.

REFERENCES

Amandry, P. and Bousquet, J., 1992. La colonne dorique de la Tholos de Marmaria. In: *Bulletin de Correspondance Hellénique*, LXIV-LXV, pp. 121-127.

Bommelaer, J.-Fr., 1997. *Marmaria, le sanctuaire d'Athéna à Delphes.* De Boccard, Paris.

Charbonneaux, J., 1925. Fouilles de Delphes, II, Topographie et architecture [4], Le sanctuaire d'Athéna Pronaia. La Tholos. De Boccard, Paris.

Hellmann, M.-Chr. et al., 1992a. *La Redécouverte de Delphes.* De Boccard, Paris.

Hellmann, M.-Chr., 1992b. Delphes, aux sources d'Apollon : un siècle d'archéologie française en Grèce, exposition, EFA, Athènes.

Jockey, Ph., 2010. L'anastylose et la ruine. Brèves réflexions sur la fabrication de mémoire(s) XIXe-XXe siècles. In: M. Crivello et al., *Les échelles de la mémoire en Méditerranée : XIXe-XXIe siècle.* Actes Sud, Arles, pp. 73-98.

Jockey, Ph., 2012. Les enjeux d'une restitution virtuelle. In: M. Garsson (dir.), *Le Trésor des Marseillais : 500 av. J.-C., l'éclat de Marseille à Delphes.* Somogy, Paris, pp. 102-107.

Joshi, P., Bourges-Sévenier, M., Russell, K. and Mo, Z., 2012. Graphics programming for the web. In: *ACM SIGGRAPH 2012 Courses*, SIGGRAPH '12. New York, NY, USA: ACM, pp. 8:1–8:75.

Kazhdan, M., and Hoppe, H., 2013. Screened poisson surface reconstruction. In: *ACM Transactions on Graphics* (TOG), vol. 32, no 3, p. 29.

Kyriakidis, N., 2010. Erreurs à Delphes. La tholos de Marmaria au fil des interprétations (ca 1840-1940). In: *Anabases*, 11, pp. 149-163.

Lowe, D.G., 1999. Object recognition from local scale-invariant features. In: Proceedings of the *International Conference on Computer Vision*, Corfu, Greece (September 1999), pp. 1150-1157.

Marcadé, J., 1979a. Les sculptures décoratives de la tholos de Marmaria à Delphes : état actuel du dossier. In: *Archaische und klassische griechische Plastik*, Mayence, II, pp. 169-173.

Marcadé, J., 1979b. Les métopes mutilées de la Tholos de Marmaria à Delphes. In: *Comptes rendus des séances de l'Académie des Inscriptions et Belles-Lettres*, 123e année, n°2, 1979, pp. 151-170

Martin-Beaumont, N., Nony, N., Deshayes, B., Pierrot-Deseilligny, M. and De Luca, L., 2013. Photographer-friendly work-flows for image-based modeling of heritage artefacts. In: *ISPRS - International Archives of the Photogrammetry, Remote Sensing and Spatial Information Sciences*, XL–5/W2 (July 22, 2013): 421–424.

Pierrot-Deseilligny, M., and Clery, I., 2011. Apero, an open source bundle adjustment software for automatic calibration and orientation of a set of images. In: *Proceedings of the ISPRS Commission V Symposium, Image Engineering and Vision Metrology*, Trento, Italy (2-4 March 2011).

This contribution has been peer-reviewed.
doi:10.5194/isprsarchives-XL-5-W4-299-2015

Toschi, I., Capra, A., De Luca, L., Beraldin, J-A., Cournoyer, L., 2014. On the evaluation of photogrammetric methods for dense 3D surface reconsctruction in a metrological context. *Proceedings of ISPRS Technical Commission V Symposium*, (23-25 June 2014), Riva del Garda, Italy. ISPRS Annals of the Photogrammetry, Remote Sensing and Spatial Information Sciences, Vol. II-5.

Vitruvius, 1995. *De Architectura*, VII, translated by B. Liou et M. Zuinghedau. Paris, Les Belles Lettres.

http://www.cultlab3d.de/

http://www.khronos.org/

http://potree.org/

This contribution has been peer-reviewed.
doi:10.5194/isprsarchives-XL-5-W4-299-2015

MIND YOUR GREY TONES – EXAMINING THE INFLUENCE OF DECOLOURIZATION METHODS ON INTEREST POINT EXTRACTION AND MATCHING FOR ARCHITECTURAL IMAGE-BASED MODELLING

G. Verhoeven [a,c,]*, W. Karel [b,c], S. Štuhec [d], M. Doneus [e,a,c], I. Trinks [a], N. Pfeifer [b]

[a] Ludwig Boltzmann Institute for Archaeological Prospection & Virtual Archaeology (LBI ArchPro), Hohe Warte 38, 1190 Vienna, Austria - (geert.verhoeven, michael.doneus, immo.trinks)@archpro.lbg.ac.at
[b] Department of Geodesy and Geoinformation, Vienna University of Technology, Gusshausstrasse 27-29, 1040 Vienna, Austria – (wilfried.karel, norbert.pfeifer)@geo.tuwien.ac.at
[c] Vienna Institute for Archaeological Science (VIAS), University of Vienna, Franz-Klein-Gasse 1, 1190 Vienna, Austria
[d] University of Ljubljana, Faculty of Arts, Aškerčeva 2, 1000 Ljubljana, Slovenia, seta.stuhec@gmail.com
[e] Department of Prehistoric and Historical Archaeology, University of Vienna, Franz-Klein-Gasse 1, 1190 Vienna, Austria

Commission V, WG V/4

KEY WORDS: Architecture, Colour, Decolourization, Feature, Imagery, Modelling, Point Cloud, Structure-from-Motion

ABSTRACT:

This paper investigates the use of different greyscale conversion algorithms to decolourize colour images as input for two Structure-from-Motion (SfM) software packages. Although SfM software commonly works with a wide variety of frame imagery (old and new, colour and greyscale, airborne and terrestrial, large-and small scale), most programs internally convert the source imagery to single-band, greyscale images. This conversion is often assumed to have little, if any, impact on the final outcome.
To verify this assumption, this article compares the output of an academic and a commercial SfM software package using seven different collections of architectural images. Besides the conventional 8-bit true-colour JPEG images with embedded sRGB colour profiles, for each of those datasets, 57 greyscale variants were computed with different colour-to-greyscale algorithms. The success rate of specific colour conversion approaches can therefore be compared with the commonly implemented colour-to-greyscale algorithms (luma Y'_{601}, luma Y'_{709}, or luminance CIE Y), both in terms of the applied feature extractor as well as of the specific image content (as exemplified by the two different feature descriptors and the various image collections, respectively).
Although the differences can be small, the results clearly indicate that certain colour-to-greyscale conversion algorithms in an SfM-workflow constantly perform better than others. Overall, one of the best performing decolourization algorithms turns out to be a newly developed one.

1. INTRODUCTION

1.1 Image-based modelling

Over the past few years, image-based modelling (IBM) with Structure-from-Motion (SfM) and Multi-View Stereo (MVS) approaches has become omnipresent in all possible fields of research: from medical sciences to a variety of geospatial applications. This success story can largely be attributed to the ease-of-use of such IBM applications, the limited knowledge required to create a geometrical three-dimensional (3D) model and the wide variety of frame imagery that can be used as input: old as well as new, colour and greyscale, airborne and terrestrial, large- and small-scale.

Although many users consider such IBM-pipelines as ideal means to yield visually-pleasing, photo-realistic 3D models in a fast and straightforward way, many applications rely on it to deliver (highly) accurate digital representations of real-world objects and scenes. For those applications, the accuracies of the interior and exterior camera orientations computed during the SfM step are of the utmost importance. To this end, Ground Control Points (GCPs) are generally applied, hereby functioning as constraints in the bundle block adjustment to avoid instability of the bundle solution or to correct for errors such as drift in the recovered camera and sparse point locations.

1.2 Interest points

The camera self-calibration and network orientation heavily depend on the number of image features that can be detected, properly described and reliably matched throughout the entire image collection. Although these features can be edges, ridges or regions of interest, the image features used in most SfM approaches comprise Interest Points (IPs). In past decades, several algorithms have been proposed to compute IPs. Aside from differences in computational complexity, they vary widely in effectiveness.

Ideally, IPs should only depend on the visible scene, but be invariant to perspective transformations and changes in the apparent brightness. As IPs are zero-dimensional, this goal can be achieved to a large extent by consideration of derivatives of the radiometric information with respect to the image and scale spaces. An efficient way to match IPs detected in different

* Corresponding author. Geert.Verhoeven@archpro.lbg.ac.at

This contribution has been peer-reviewed.
doi:10.5194/isprsarchives-XL-5-W4-307-2015

images is to describe their local neighbourhoods in image space at the respective scale and with respect to each neighbourhood's principal direction, and then match these feature vectors by nearest-neighbour searches in the space spanned by them.

However, current, efficient IP description methods are merely invariant to linear changes in apparent brightness and to the geometric transformations of translation, rotation, and scale. Combinations of these basic geometric transformations are valid approximations only to small perspective changes and hence, large perspective changes cannot be handled.

1.3 Greyscale input

Although the input for feature extraction (i.e. IP detection and description) is generally a collection of true-colour frame images, the most common feature detectors and descriptors such as SIFT and SURF have been developed to work on single-band greyscale images, since this – amongst other reasons – greatly reduces the computational complexity of the algorithm compared to the utilization of the common three channels of a full colour image. In practice, this means that the algorithm can be applied onto each colour channel separately, or that a standard greyscale conversion (conventionally the computation of the luma Y'_{601}, luma Y'_{709} or the luminance CIE Y component) provides the necessary single-band input. However, the exact algorithm to convert a colour image into its greyscale variant is often not documented for SfM software and the conversion is commonly assumed to have little, if any, impact on the final SfM outcome.

This paper investigates the effect of different greyscale conversion algorithms used to decolourize colour images as input for an SfM-based architectural IBM-pipeline. To this end, 57 different decolourization methods have been implemented in MATLAB: from very common methods that use a simple weighted sum of the linear R, G, B channels or non-linear, gamma-corrected R', G' and B' components to more complex, perception-based colour-to-greyscale methods that claim to more or less preserve lightness, meaningful colour contrast and other visual features in the greyscale variant.

2. METHODS

2.1 Software and IP extractors

Since none of the decolourization algorithms *a priori* suggest a better performance in terms of feature detection and description, two different feature extractors have been chosen so that their behaviour can be assessed when fed different greyscale versions. One of those IP extractors, SIFT (Lowe, 1999), is implemented in *OrientAL*: a research-based software package developed at TU Vienna aiming to provide a fully automated processing chain from aerial photographs to orthophoto maps.

To this end, high-level command-line scripts and lower level functions enable manual, semi- and fully automatic image orientation, camera calibration and object reconstruction (Karel et al., 2013). OrientAL considers especially the characteristics of archaeological aerial images, including oblique imagery, little overlap, poor approximate georeferencing and historic aerial photographs (Karel et al., 2014). OrientAL (version 20150107) allows the user to apply three different feature extractors: aside from SIFT, SURF (Bay et al., 2006) and Affine-SIFT or ASIFT (Morel and Yu, 2009) are offered as well. The latter two are not investigated in this study.

PhotoScan Pro (1.1.0 build 2004)						
Accuracy	**Pair selection**		**2D IPs limit**	**3D OPs limit**		
High	Disabled (enabled for set 5 & 7)		None	None		
OrientAL (20150107)						
Feature extraction	**Contrast threshold**	**Edge threshold**	**Grid**	**Max IPs per cell**	**Keep**	
SIFT	0.04	10.0	(1, 1)	40 000	Strongest	
Feature matching	**Matching precision**	**Maximum descriptor distance ratio**	**Minimum radius ratio**	**Symmetric matching**	**Minimum matches per pair**	**Maximum matches per pair**
	0.8	0.8	0	True	5	10 000
Reconstruction	**Robust weighting**	**Adjust interior orientation**	**Interior orientation grouping**	**Mimimum PnP ratio of inliers**	**Maximum residual length**	**At least three observations**
	10.0	During	Sequence	0.08	10.0	True

Table 1. Parameters employed for the different SfM solutions.

Next to OrientAL, all datasets have been processed with the well-established commercial package *PhotoScan Professional edition* (PhotoScan Pro 1.1.0 build 2004, 64-bit) from the Russian manufacturer *Agisoft LLC*. The choice for this software was based on its features, cost and completeness. Moreover, recent studies have shown the accuracy and reliability of the results generated by this commercial program (Remondino et al., 2012). The exact algorithms that are programmed in PhotoScan are, however, not publicly known. In the remaining part of the article, the PhotoScan Feature Extractor will therefore be denoted PSFE. All the parameters that were used for the detection and matching of the IPs in both SfM approaches are listed in Table 1.

2.2 Image sets

Besides variation in feature extraction, there should also be variation in the architectural image collections. More specifically, seven different image sets were assembled for this study. They were chosen so as to represent the wide variety of possible historical architectural objects one might wish to geometrically document within an IBM-pipeline. Moreover, the image sets were captured with a mixture of digital still cameras and focal lengths under various illumination conditions (Table 2).

Set	Set name	Images	Format	Profile	Image dimensions	Camera	Focal length
1	Building 5	25	JPEG (8 bit)	sRGB	2880 x 2160 (6 MP)	Kodak EasyShare M590	6.3 mm
2	Fountain	42	NEF (14 bit)	sRGB	4928 x 3264 (16 MP)	Nikon D7000	14 mm
3	Akrotiri	20	NEF (12 bit)	sRGB	4928 x 3264 (16 MP)	Nikon D7000	13 mm
4	Palacol	47	NEF (14 bit)	sRGB	4256 x 2832 (12 MP)	Nikon D700	31-120 mm
5	Heidentor	80	JPEG (8 bit)	sRGB	3000 x 2250 (7 MP)	GoPro HERO3+ Black Edit.	2.8 mm
6	Castel-K19	19	PNG (8 bit)	none	3072 x 2048 (6 MP)	Canon D60	~17.5 mm
7	Piazza Bra	331	JPEG (8 bit)	sRGB	3008 x 2000 (6 MP)	Nikon D50	18 mm

Table 2. Characteristics of the seven different image collections used in this study.

2.2.1 Set 1 – Building 5

The first image set consists of 25 JPEG images which all have been captured with a 6 MegaPixel (MP) Kodak EasyShare M590 compact camera (6.3 mm focal length). As is usual with images from such cameras, they have the sRGB IEC61966-2.1 colour profile embedded. This image set is made publically available by Ceylan and colleagues (http://www.duygu-ceylan.com/duygu-ceylan/symmCalib.html) and accompanies their article which deals with the ambiguity in establishing correspondences encountered by SfM algorithms when dealing with repeated structures in urban facades (Ceylan et al., 2014). Whereas well-established SfM approaches often erroneously reconstruct such scenes, their framework allows for the robust extraction of those repeated façade elements and generates a proper output using a symmetry-based SfM algorithm. The image set used here is their 'Building 5', for which both Bundler (Snavely et al., 2006) and the method of Zach et al. (2010) failed to generate an accurate sparse point cloud and camera orientations. The difficulty of this dataset does not only lie in the repetitive character of the

This contribution has been peer-reviewed.
doi:10.5194/isprsarchives-XL-5-W4-307-2015

architectural elements, but also the amount of indistinctive image features. However, the global illumination of the building is very diffuse. No further information about the building's age or its location is provided.

2.2.2 Set 2 – Viennese fountain (Austria)

The second image collection consists of 42 images captured with a 16 MP Nikon D7000 (14 mm focal length) and stored as losslessly compressed 14-bit NEF (Nikon Electronic Format; i.e. Nikon's RAW format). Subsequently, the images were converted to 8-bit JPEGs using Nikon's Capture NX 2.4.7 software. Except for the conversion of the data into the sRGB colour space, no other parameters were altered so as to simulate an in-camera generated JPEG. The scene comprises a small wall fountain in the middle of the Strudlhofstiege, an outdoor staircase in the city of Vienna. Aside from the running water, the presence of people to the left of the fountain and the similar colour tones of the wall, this scene should not pose any significant difficulties for an SfM algorithm. Moreover, the illumination is very uniform. Given the similarity of the colours in this scene, it is deemed a good example to test different decolourization approaches.

2.2.3 Set 3 – Akrotiri (Greece)

The third set of images depicts some parts of the Bronze Age town of Akrotiri on the Greek island of Thera/Santorini. While most of the prehistoric town is still covered by volcanic ash and pumice, a dozen buildings – sometimes up to four stories high – have been excavated. With the aim to digitally safeguard this earthquake-threatened site, the entire excavated area has been covered with about 850 terrestrial laser scanning positions and several thousand photographs. Twenty of those images constitute the third image collection used in this study. Similar to the second set of images, they have been acquired with a Nikon D7000 (16 MP), but now with a 13 mm focal length lens and stored as 12-bit losslessly compressed NEFs. The conversion to JPEG was identical to the method described for image set 2. Although those twenty images feature quite a lot of structure, they cover a rather small colour gamut. Aside from the very similar colour tones, the scene consists of shadowed and sunlit parts (which also slightly changed over the image set due to the presence of clouds), moving people and a very repetitive set of white pillars supporting the roof over the site.

2.2.4 Set 4 – Palacol palace (Croatia)

Image collection four holds 47 aerial images of the Croatian island of Palacol. A ruined, presumed fortress is located in the middle of the very small island. Although the original image sequence held 165 images, the number of images was heavily reduced in order to have less overlap between neighbouring images and thus to create larger differences regarding their viewpoints. Aside from the very large scale differences, the images feature very repetitive vegetation patterns and difficult to handle water surfaces, often largely affected by sun glint. The original losslessly compressed 14-bit NEF images were captured with a full-frame Nikon D700 (12 MP), while the focal length varied from 31 mm to 120 mm.

2.2.5 Set 5 – Roman Heidentor (Austria)

The fifth dataset comprises a collection of eighty GoPro images (Figure 1). More specifically, the GoPro HERO3+ Black Edition was used and 7 MP (3000 x 2250) JPEG images were saved using the sRGB IEC61966-2.1 colour profile. The images where acquired to test the suitability of the Fotokite, a novel low-altitude aerial platform consisting of a tethered quadcopter. The airborne images depict the Roman monument '*Heidentor*', which is part of the archaeological site of *Carnuntum* (Austria). This antique monument measures approximately 15 m by 15 m and has a height of circa 14 m. The corresponding image dataset is very suitable for the present analysis, given the less than perfect nature of these images. Geometrically, the images are highly distorted. On top of that, they suffer from very serious JPEG compression, whereas the scene itself featured bright skylight (which resulted in lens flare) and moving persons.

Figure 1. A visualisation of five decolourization methods.

2.2.6 Set 6 – Castle K-19 (Switzerland)

The sixth dataset is a long-standing computer vision benchmark image collection. The image set, which is referred to as Castle K19 (Strecha et al., 2008), consists of nineteen 8-bit PNG images that can be downloaded online from: http://cvlabwww.epfl.ch/data/multiview/knownInternalsMVS.html. The 6 MP images were acquired with a Canon EOS-D60 but do not have any colour profile embedded (nor any other metadata). From the camera calibration matrix provided together with the images, a focal length of approximately 17.5 mm can be inferred. All images are underexposed and incorrectly white balanced (they are too blue), making them interesting for the decolourization approaches.

2.2.7 Set 7 – Piazza Bra (Italy)

The last set of images is rather large: 331 JPEG images covering the Piazza Bra in Verona, Italy. The images, which can be found at http://www.diegm.uniud.it/fusiello/demo/samantha/, have been featured in a few publications that introduced a hierarchical SfM scheme (Farenzena et al., 2009; Gherardi et al., 2010). This new SfM pipeline was developed to better cope with drift problems in such large datasets. As such, it is interesting to see whether the greyscale values fed into a normal SfM chain can minimise this possible drift. All images were shot with a 6 MP Nikon D50.

This contribution has been peer-reviewed.
doi:10.5194/isprsarchives-XL-5-W4-307-2015

2.3 Decolourization methods

Over the past decades, several simple and more complex colour-to-greyscale algorithms have been developed to derive the best possible decolourized version of a colour input image. In essence, the decolourization problem is one of dimension reduction since multiple input channels have to be mapped onto a single channel. In doing so, the colour contrast and image structure contained by the colour image should be preserved in the greyscale image.

Although different colour models exist to mathematically code the colours of an image, every pixel of a colour image is usually coded by a triplet of *RGB* numbers (indicating *Red*, *Green* and *Blue*). Perceptually, a colour image can be thought of having three components: an achromatic variable (such as luminance or lightness) and two chromatic channels. Most common decolourization methods use a simple weighted sum of the linear *R*, *G*, *B* channels or non-linear, gamma-corrected *R'*, *G'*, and *B'* components to form a greyscale image that is more or less representative of luminance such as the CIE *Y* channel or the luma channels Y'_{601} and Y'_{709} (Figure 1), the latter two differing by the weighting coefficients used since they are based on the Rec. 601 NTSC primaries and the Rec. 709 sRGB primaries, respectively. The CIE L^* lightness channel is another, very common method for decolourization. Being one of the three channels in the CIE $L^*a^*b^*$ and CIE $L^*u^*v^*$ colour spaces, lightness L^* has almost perceptually uniform values. As such, a numerical increment in the CIE L^* lightness will more or less correspond to an equal increment in tone perception for a human observer. The fifth, rather common greyscale conversion used in this study is average: a very simple method that is completely unrelated to any human sensation. Here, both the linear (using *R*, *G*, *B*) and non-linear (using *R'*, *G'*, *B'*) versions are applied.

Apart from those four common methods, more complex colour-to-greyscale algorithms have been implemented as well. Those methods were developed to retain as much information as possible during the decolourization of the colour image (second row in Figure 1). To this end, they encode the chromaticity contrast rather than the luminance information. Socolinsky and Wolff (2002) were amongst the first to propose an elegant decolourization technique that embedded this principle. During the past years, many new perception-based decolourization approaches have surfaced, all claiming to more or less preserve lightness, meaningful colour contrast and other visual features in the greyscale variant. Very often, these algorithms start from the operating principles of the human visual system, which are then combined with the physics of colour image production.

Although it has been the aim of this study to rather exhaustively implement all of these recently emerged colour-to-greyscale methods, the unresponsiveness or unwillingness of most authors to share their code results in a much more restricted list. Of that list, only those methods are used that can compute a decolourized image in a matter of seconds instead of minutes. Therefore, the popular method of Gooch et al. (2005) is omitted, but the approaches of Grundland and Dodgson (2007), Zhao and Tamimi (2010), Lu et al. (2012a, 2012b) and Wu and Toet (2014) are applied.

Finally, a whole amalgam of new, but computationally simple approaches has been programmed as well (third row in Figure 1). These comprise the standard deviation, variance, norm, minimum and maximum averages (i.e. the minimum/maximum value of the *R'G'B'* triplet averaged with the *R'G'B'* mean), minimum and maximum median, midpoint, PCA, (weighted) NTSC PCA (i.e. a (weighted) PCA after converting the sRGB image to the *Y'IQ* colour space), (weighted) CIE $L^*a^*b^*$ PCA, (weighted) $Y'C_BC_R$ PCA and several variants and combinations of those methods.

In total 57 decolourization methods have been implemented in MATLAB and applied to each image of every dataset described above. To make sure that only the respective colour-to-greyscale method is properly assessed, no other pre-processing of the source imagery is undertaken. This does not necessarily mean that any such pre-processing would be useless. When an image is generated by a digital camera, it can conventionally be stored as a RAW image (i.e. minimally processed sensor data that have a gain, quantisation and some basic correction processes applied), or as a lossy compressed JPEG image. Although RAW files always end up as JPEG or TIFF files, the main difference between both lies in the device that executes the RAW development: the camera itself or the computer. Obviously, any of these development steps, such as white balancing, demosaicking, sharpening, denoising and contrast enhancement, can have a serious impact on the feature extraction performance. Even the bit-depth and output colour space (indicated as 'profile' in Table 2) of an image are very important characteristics, since they define the possible variety of colours contained in an image. As mentioned above, none of these pre-processing choices have been applied for the present study.

All these decolourized, single band images were saved as losslessly compressed 8-bit TIFFs to make sure that no further data loss occurred after the greyscale conversion. Moreover, Phil Harvey's ExifTool (Harvey, 2014) was used to copy all original metadata (except the embedded thumbnail and information on the colour space) back into the newly converted greyscale image.

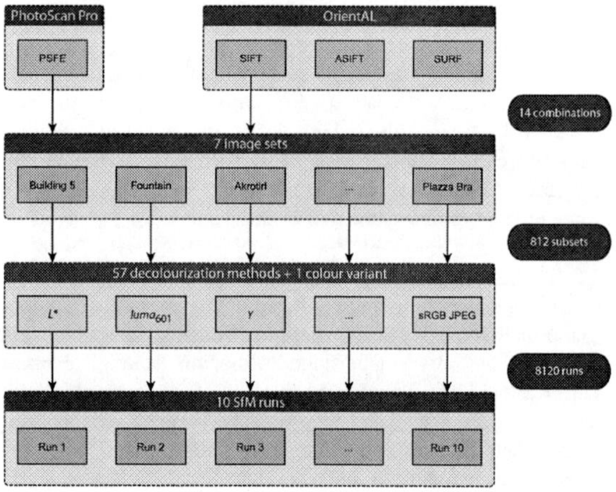

Figure 2. An overview of all possible processing combinations.

2.4 Repeated runs

Given the various decolourization approaches and one colour photo image set, a total of 406 subsets is generated [(57+1)x7]. Since SfM pipelines are non-deterministic, repeated SfM runs can result in (slightly) different results. To take the stochastic nature of these results into account and to make sure that valid information can be obtained from these tests, every SfM software with one of the two specified feature extractors is executed ten times for every subset. This amounts to a total of 8120 executed SfM runs (Figure 2).

2.5 Extracted metrics

Obviously, the first metric that comes to mind when varying the input of an SfM chain is the total amount of IPs that are extracted. This number does, however, not at all express the strength and reliability of those features nor the camera's exterior and interior orientation upon which they are based. As such, over thirty different metrics were computed. Since some of them are ratios and normalized versions of others, Table 3 lists only the most important ones. Furthermore, it is indicated which statistics can be computed by which SfM package.

Metric	PhotoScan Pro	OrientAL
Number of 2D IPs	X	X
Number of 2D IPs matched		X
Number of matches (pairs of 2D IPs)	X	
Number of 2D IPs that passed E-Matrix test		X
Number of 2D Ips with multiple projections removed		X
Number of 3D OP (inliers + outliers)	X	
Number of 3D OPs (inliers)	X	X
Number of 3D OPs (outliers)	X	
Percentage 3D OPs inliers of all 2D IPs	X	
Percentage 3D OPs inliers of all 3D OPs	X	
Percentage 3D OPs outliers of all 2D IPs	X	
Percentage 3D OPs outliers of all 3D OPs	X	
Average number 2D IPs/3D OPs (inliers + outliers)	X	
Average number 2D IPs/3D OPs (inliers)		X
Number of 2D IPs for bundle block adjustment	X	X
Total reprojection error - average	X	X
Total reprojection error - median	X	
Total reprojection error - min	X	
Total reprojection error - max	X	
Number of input images	X	X
Number of oriented images	X	X
Number of unoriented images	X	X

Table 3. Some of the metrics computed by both SfM packages.

Since no extremely accurate reference data are available for most of these seven datasets, the only means of verifying the reconstructed camera positions is through a visual assessment of the SfM result (i.e. the sparse cloud of 3D Object Points – OPs – and the relative camera positions). It goes without saying that this is a hugely time-consuming operation. Since not all SfM outputs could be visually checked so far, the OrientAL results presented in the next section will mainly concentrate on image sets 1, 3 and 6. An overview of those four remaining image collections will be the subject of a forthcoming, more extended paper. The latter will also hold a more rigorous statistical analysis of all the results combined.

3. RESULTS

This section presents some of the first analyses of the data collected so far. Since only the intended PhotoScan runs have been entirely finished at this stage, they are presented first. Afterwards, some observations are made about the image sets that have been completely oriented with OrientAL at this stage (i.e. image collections 1, 3 and 6).

Although it is far from the most relevant metric, many methods that try to optimise the SfM input aim for a high IP count. To this end, Figure 3 shows on overview of all IPs per image set per decolourization method. The results have been normalized to the input of the sRGB JPEG, so that it becomes easy to verify which methods perform better or worse than the standard colour image.

Although many methods yield almost identical IP counts compared to the sRGB JPEG, both the standard deviation and variance methods differ strongly. The variance method – which computes per pixel the variance of its $R'G'B'$ triplet, delivers a very small amount of IPs, whereas the standard deviation delivers on average 20 % more IPs (for the first image set even 98 % more) than the standard input. However, both methods are characterized by flaws: whereas the limited bit depth of the images cannot deal with the various low values of the variance image (hence the low IP count), the standard deviation method often results in waterpainting-like images, as it blurs features in one area and creates artefacts in another. As such, both methods are omitted in the following results.

Upon averaging the results of all image collections, it seems like perception-based methods such as *decol2* and *decol3* (i.e. the decolourize method of Grundland and Dodgson (2007) with different effect parameter values), *copdecol* (i.e. the contrast preserving decolourization of Lu et al. (2012a)) together with some very simple methods such as *band 1* and *band 2* (i.e. the Red and Green image channels, respectively), the minimum decomposition approach (*mindec*, per pixel the minimum of $R'G'B$) and *maxstdv* (i.e. the maximum standard deviation variant of the previously mentioned maximum average method) all generate on average 2-6 % more IPs in PhotoScan Pro than the sRGB input (Figure 3). Most likely, PhotoScan uses the luma Y'_{709} or L^* decolourization methods as their IP counts are very similar to the sRGB input (rounding errors in the conversion might explain the small differences).

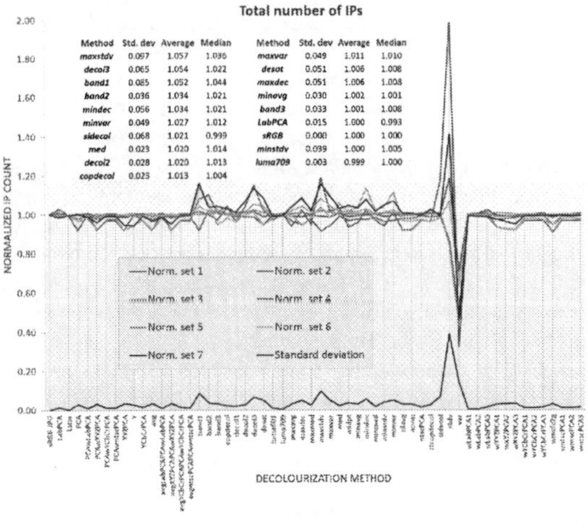

Figure 3. An overview of the number of IPs extracted by PSFE per conversion method and per image set (normalized to sRGB).

The relative importance of total IP count is also illustrated by Figure 4. This graph plots the total Reprojection Error (RE) versus the total IP count. Each of the 56 data points (the *var* and *stdv* methods are omitted) equals the average of all runs on all seven datasets. Looking at the median IP count versus the median RE, it seems that both are unrelated. However, plotting the average IP count versus the RE (both average and median), a positive correlation R^2 can be observed. Despite being rather small, this correlation indicates that the total RE even increases with increasing IP count. For instance, the SfM runs with *decol3* decolourized images – a method which is one of the top performers concerning total IP count – are characterized by average and median REs that are at least 20 % larger than those of standard sRGB SfM runs.

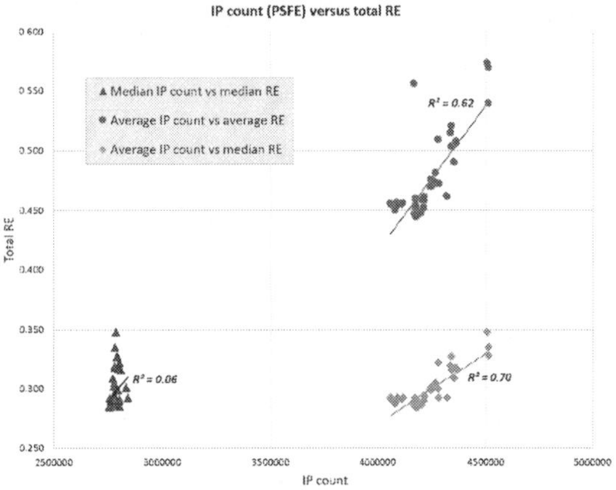

Figure 4. Total IP count (PSFE) versus total RE, averaged over all image collections.

It is often assumed that the total RE is already a very good indication of the global accuracy of the SfM solution. Figure 5 indeed indicates that low REs correspond to results free of gross errors. In this graph, the total RE is plotted against the visual assessment of all PhotoScan-based SfM runs. Both values are averages over all seven datasets.

Despite being necessary, a visual assessment of an SfM output is tricky because one can only assess obvious positional errors of the camera orientations or artefacts in the sparse point cloud. As soon as one of those were detected, the result was given a score '0'. When no obvious inaccuracies in any of those two parameters could be observed, that run got score '1'. Although the runs that seem correct could still incorporate errors, the reverse cannot be said from SfM-runs with score '0'. In the latter category, no difference was made between 'slightly wrong' or 'full with blunders'. The visual assessment score of Figure 5 can thus maximally be 1 (i.e. for every run of every dataset no visual errors could be detected in the output).

First, it is obvious that the correlation coefficient R^2 between both metrics is extremely low, indicating no real relationship between both. Moreover, it seems that the best method for decolourization is the *avgLabPCAPCAwLabPCA* approach. This long and dreadful name indicates that this method starts with a conversion of the non-linear *R*', *G*' and *B*' image bands into the CIE $L^*a^*b^*$ colour space. Afterwards, a Principal Component Analysis (PCA) is executed and the first Principal Component (PC) is extracted. During the PCA, the data are centred and the covariance matrix is used. The output of this process yields an image called *LabPCA*, while the first PC equals a greyscale image. Afterwards, a weighted CIE $L^*a^*b^*$ version is computed from the source sRGB JPEG. This procedure is similar to the one above, but the $L^*a^*b^*$ channels are weighted before being PCA transformed. The algorithm uses three different weighting factors, being 0.25, 0.45 and 0.65. From the three resulting *wLabPCA* images, the first PCs are extracted and concatenated into one new, three channel image. On this new image, a PCA is run again of which the first PC yields an image called *PCAwLabPCA*. The latter is averaged with the initially computed *LabPCA*.

This method, which tries to capture all possible variation in luminance and chrominance data, delivers greyscale images whose SfM output is virtually always perfect (at least visually). However, also the total REs of this method are among the lowest of all decolourization approaches. The performance of this method is very similar to the much faster *rtcopdecol* method developed by Lu et al. (2012b). Their real-time contrast preserving decolourization approach is a faster but lass accurate (in terms of human perception) version of the previously mentioned *copdecol* method. Despite their capability to yield high IP counts, *band 1* and *decol 3* are rather poor performers according to Figure 5.

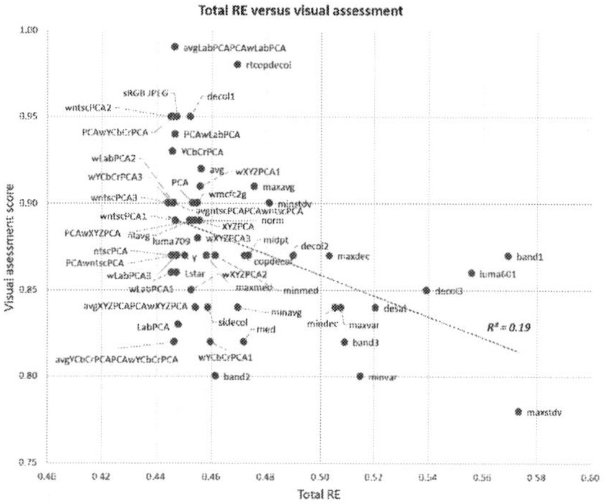

Figure 5. Total RE (PSFE) versus a visual assessment score, averaged over all image collections.

The position of the sRGB JPEG and luma Y'_{601} data points is rather interesting as well. First, the high similarity between the total REs of the luma method and the sRGB JPEGs indicates once more that PhotoScan most likely uses this decolourization method. However, both approaches differ in their visual assessment score. That is, as long as all seven datasets are considered. When only the first six datasets are take into account, both methods also perform virtually the same. This has most likely to do with the limited number of runs executed with image collection seven. Due to the large size of that dataset (331 images), only five runs have been computed. The difference between the average results of both methods could thus be due to the limited sample size.

Although a forthcoming article will be able to quantify all ten runs and more thoroughly assess the performance of the decolourization methods using reference datasets with accurately, pre-computed exterior camera positions, it is still deemed useful to provide Table 4. Here, the five 'best' decolourization methods are enumerated, in which 'best' is determined by the average visual assessment score for both the first six image collections and all seven image collections. From the table (which is based on the PSFE results only), it can be concluded that *avgLabPCAPCAwLabPCA* is constantly one of the best performers, although the standard sRGB JPEG input images can be considered a very viable alternative.

Image collection 1 to 7						
	% 3D OPs inliers of all 3D OPs		RE		Visual assessment	
Method	Average	Std. dev.	Average	Std. dev.	Average	Std. dev.
avgLabPCAPCAwLabPCA	79.02	7.25	0.45	0.19	0.99	0.04
rtcopdecol	79.30	7.83	0.47	0.18	0.98	0.05
sRGB JPEG	79.35	7.44	0.45	0.19	0.95	0.09
PCAwYCbCrPCA	79.70	7.49	0.45	0.19	0.95	0.09
decol1	79.76	7.08	0.45	0.19	0.95	0.09
Image collection 1 to 6						
	% 3D OPs inliers of all 3D OPs		RE		Visual assessment	
Method	Average	Std. dev.	Average	Std. dev.	Average	Std. dev.
PCAwLabPCA	81.07	6.82	0.46	0.20	1.00	0.00
luma601	81.89	6.83	0.46	0.20	1.00	0.00
maxavg	81.13	7.49	0.48	0.20	1.00	0.00
avgLabPCAPCAwLabPCA	80.29	7.04	0.46	0.20	0.98	0.04
sRGB JPEG	80.99	6.63	0.46	0.20	0.98	0.06

Table 4. The five 'best' decolourization approaches, based on the visual assessment of the SfM output and total RE.

Upon checking some of the metrics computed by OrientAL, it becomes clear that the PhotoScan-based results are rather similar to those of the SIFT-approach. In contrast to the commercial application PhotoScan, the default parameters of OrientAL have not been fine-tuned to yield an optimal result in most cases. Despite the – most likely – sub-optimal parameter set, the settings that are reported in Table 1 have been used for every SfM run executed with OrientAL. As such, they still allow for a meaningful comparison between the decolourization approaches. One strength of OrientAL is that it supports plotting the number of OPs that have been observed in at least three photographs. Figure 6 shows this result for each grayscale conversion and the three data sets completely processed so far.

Since there is no reference dataset (with known accuracy) at hand for any of the image collections used in this study, the graphs in Figure 6 are a very useful alternative way of checking the SfM output. It is quite unlikely that OPs that are observed at least three times are gross errors. From this graph, it can be inferred that the decolourization methods *avgLabPCAPCAwLabPCA*, *LabPCA* and *wLabPCA1-3* produce more reliable OPs (that are observed at least thrice) than most other methods. However, those methods only perform slightly better than the sRGB input.

Together with the conclusions drawn from the PSFE-based statistics, the *avgLabPCAPCAwLabPCA* greyscale conversion method seems to perform best from the amalgam of decolourization approaches tested in this study. Although none of the metrics indicates them as the top performer, using the standard sRGB JPEG images in an SfM workflow delivers on average also very good results, often outperforming most of the greyscale conversion methods tested in this paper. Until a more rigorous statistical assessment of all metrics has been performed on these and additional image collections, it remains thus a very valid option to continue the use of standard JPEG images – directly produced by the digital still camera or resulting from a RAW-based workflow – as input for SfM software packages.

4. CONCLUSION AND OUTLOOK

This article compared the output of an academic and a commercial software package to verify the assumption that the result of an SfM algorithm is hardly influenced by the decolourization approach used to convert the input imagery into single band, greyscale images. To this end, seven collections of very dissimilar architectural images were chosen as input for a commercial and academic SfM pipeline, which have an undocumented and SIFT feature extractor implemented, respectively.

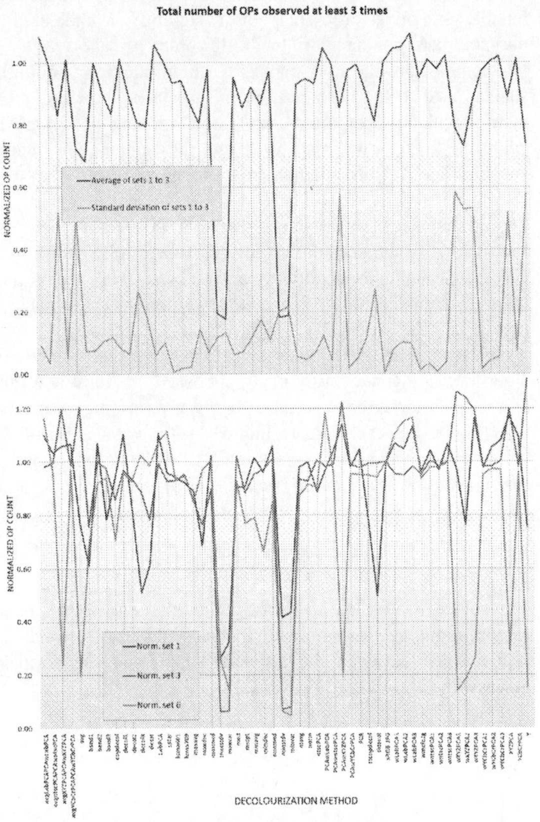

Figure 6. Total number of OPs that have been observed at least three times (SIFT), normalized to sRGB.

Although the computation of all SfM-runs as well as the visual assessment of their outputs is still on-going, the gathered statistics already allow to draw some first conclusions. It was a positive observation that the standard method used in most SfM software performs very well. Using a variety of metrics, it could be confirmed that an sRGB JPEG image generally outperforms a wide variety of specific decolourized input images. However, the *avgLabPCAPCAwLabPCA* method that was newly developed to maximise all chrominance and luminance variation in a single band image proved to be the overall best performing algorithm. This was verified by metrics produced by both software packages. Using the averages of a binary scoring system, *avgLabPCAPCAwLabPCA* was also visually proven to be the most reliable decolourization approach.

In the near future, the presented tests will be seriously expanded. First, several new datasets, which feature images whose exterior orientation is accurately known, will be introduced. This will allow for the unbiased assessment of the influence of the decolourization approaches on the exterior orientation computation of the image collections. Ideally, these new datasets would allow for statistically correlating one (or a few) of the metrics presented here with the accuracies of the final exterior orientation estimates. As such, it may become possible to create a rather limited set of metrics that can express the influence of specific decolourization methods (or other pre-processing steps)

doi:10.5194/isprsarchives-XL-5-W4-307-2015

on the final accuracy of the SfM result. Related to this topic is the development of new metrics that might be more powerful than the ones used so far. Although there is a great need for such reliable statistical measures, most currently available SfM packages do not provide them. In order to allow for the intercomparison between packages, these metrics should be standardized to the maximum extent possible. Therefore, more metrics will be incorporated into future versions of OrientAL, while the authors will also try to get the computation of new statistics into Agisoft's PhotoScan Professional.

Second, more feature extractors should be incorporated in those tests. As was mentioned previously, SURF and ASIFT have been successfully implemented in OrientAL as well. They would thus allow to complete the SIFT-based tests presented here and to investigate the extent by which various IP extractors prefer dissimilar inputs. Because new innovative decolourization algorithms are continuously being published, the third and final aim is to further embed and test the most promising ones. This intention is, however, largely limited by the willingness of the colour-to-greyscale developers to share their code (which was a serious limiting factor for this paper as well).

ACKNOWLEDGEMENTS

The authors would like to thank Alexey Pasumansky (Agisoft) for programming a custom-made Python script. This research has been carried out with the financial support of the Austrian Science Fund (FWF): P24116-N23.

REFERENCES

Bay, H., Tuytelaars, T., Gool, L., 2006. SURF: Speeded Up Robust Features. In: Leonardis, A., Bischof, H., Pinz, A. (Eds.), *Computer Vision – ECCV 2006. 9th European Conference on Computer Vision, Graz, Austria, May 7-13, 2006, Proceedings, Part I*. Springer, Berlin, pp. 404–417.

Ceylan, D., Mitra, N., Zheng, Y., Pauly, M., 2014. Coupled structure-from-motion and 3D symmetry detection for urban facades. *ACM Transactions on Graphics* 33 (1), 2:1–15.

Farenzena, M., Fusiello, A., Gherardi, R., 2009. Structure-and-motion pipeline on a hierarchical cluster tree. *Proceedings of the 2009 IEEE 12th International Conference on Computer Vision (ICCV 2009)*. IEEE, Piscataway, pp. 1489–1496.

Gherardi, R., Farenzena, M., Fusiello, A., 2010. Improving the efficiency of hierarchical structure-and-motion. *Proceedings of IEEE Conference on Computer Vision and Pattern Recognition (CVPR) 2010*. IEEE, pp. 1594–1600.

Gooch, A., Olsen, S., Tumblin, J., Gooch, B., 2005. Color2Gray: Salience-Preserving Color Removal. In: Gross, M. (Ed.), *Proceedings of SIGGRAPH 2005: Special Interest Group on Computer Graphics and Interactive Techniques Conference*. ACM, New York, pp. 634–639.

Grundland, M., Dodgson, N., 2007. Decolorize: Fast, contrast enhancing, color to grayscale conversion. *Pattern Recognition* 40 (11), 2891–2896.

Harvey, P., 2014. ExifTool - Read, Write and Edit Meta Information! http://www.sno.phy.queensu.ca/~phil/exiftool/.

Karel, W., Doneus, M., Briese, C., Verhoeven, G., Pfeifer, N., 2014. Investigation on the Automatic Geo-Referencing of Archaeological UAV Photographs by Correlation with Pre-Existing Ortho-Photos. In: Remondino, F., Menna, F. (Eds.), *Proceedings of the ISPRS Technical Commission V Symposium*, pp. 307–312.

Karel, W., Doneus, M., Verhoeven, G., Briese, C., Ressl, C., Pfeifer, N., 2013. OrientAL – Automatic geo-referencing and ortho-rectification of archaeological aerial photographs. *ISPRS Annals of the Photogrammetry, Remote Sensing and Spatial Information Sciences* II-5/W1, 175–180.

Lowe, D., 1999. Object recognition from local scale-invariant features. In: Werner, B. (Ed.), *The Proceedings of the Seventh IEEE International Conference on Computer Vision. Volume II*. IEEE, Los Alamitos, pp. 1150–1157.

Lu, C., Xu, L., Jia, J., 2012a. Contrast preserving decolorization. Proceedings of the 2012 IEEE International Conference on Computational Photography (ICCP). IEEE, Piscataway, pp. 1–7.

Lu, C., Xu, L., Jia, J., 2012b. Real-time contrast preserving decolorization. *Proceedings of SIGGRAPH Asia 2012 Technical Briefs*. ACM, New York, Article 34.

Morel, J.-M., Yu, G., 2009. ASIFT: A New Framework for Fully Affine Invariant Image Comparison. *SIAM Journal on Imaging Sciences* 2 (2), 438–469.

Remondino, F., Del Pizzo, S., Kersten, T., Troisi, S., 2012. Low-Cost and Open-Source Solutions for Automated Image Orientation – A Critical Overview. In: Ioannides, M., Fritsch, D., Leissner, J., Davies, R., Remondino, F., Caffo, R. (Eds.), *Progress in Cultural Heritage Preservation. Proceedings of the 4th International Conference, EuroMed 2012*. Springer, Berlin, Heidelberg, pp. 40–54.

Snavely, K., Seitz, S., Szeliski, R., 2006. Photo tourism: Exploring photo collections in 3D. *ACM Transactions on Graphics* 25 (3), 835–846.

Socolinsky, D., Wolff, L., 2002. Multispectral image visualization through first-order fusion. *IEEE transactions on image processing* 11 (8), 923–931.

Strecha, C., Hansen, W. von, van Gool, L., Fua, P., Thoennessen, U., 2008. On benchmarking camera calibration and multi-view stereo for high resolution imagery. *Proceedings of the IEEE Conference on Computer Vision and Pattern Recognition, 2008 (CVPR 2008)*. IEEE, Anchorage, AK, pp. 1–8.

Wu, T., Toet, A., 2014. Color-to-grayscale conversion through weighted multiresolution channel fusion. *Journal of Electronic Imaging* 23 (4), 043004-1 - 043004-6.

Zach, C., Klopschitz, M., Pollefeys, M., 2010. Disambiguating visual relations using loop constraints. *Proceedings of the 23rd IEEE Conference on Computer Vision and Pattern Recognition (CVPR 2010)*. IEEE, pp. 1426–1433.

Zhao, Y., Tamimi, Z., 2010. Spectral image decolorization. In: Bebis, G., Boyle, R., Parvin, B., Koracin, D., Chung, R. (Eds.), *Proceedings of the 6th international conference on Advances in visual computing (ISVC'10). Part II, ISVC*. Springer, Berlin, pp. 747–756.

ADVANCES IN IMAGE PRE-PROCESSING TO IMPROVE AUTOMATED 3D RECONSTRUCTION

A. Ballabeni [1], F. I. Apollonio [1], M. Gaiani [1], F. Remondino [2]

[1] Dept. of Architecture – University of Bologna, Italy - (andrea.ballabeni, fabrizio.apollonio, marco.gaiani)@unibo.it
[2] 3D Optical Metrology (3DOM) unit, Bruno Kessler Foundation (FBK), Trento, Italy
remondino@fbk.eu, http://3dom.fbk.eu

Commission V, WG 4

KEY WORDS: Pre-processing, Filtering, Matching, Automation, Denoising, RGB to Gray, Color, 3D reconstruction

ABSTRACT

Tools and algorithms for automated image processing and 3D reconstruction purposes have become more and more available, giving the possibility to process any dataset of unoriented and markerless images. Typically, dense 3D point clouds (or texture 3D polygonal models) are produced at reasonable processing time. In this paper, we evaluate how the radiometric pre-processing of image datasets (particularly in RAW format) can help in improving the performances of state-of-the-art automated image processing tools. Beside a review of common pre-processing methods, an efficient pipeline based on color enhancement, image denoising, RGB to Gray conversion and image content enrichment is presented. The performed tests, partly reported for sake of space, demonstrate how an effective image pre-processing, which considers the entire dataset in analysis, can improve the automated orientation procedure and dense 3D point cloud reconstruction, even in case of poor texture scenarios.

1. INTRODUCTION

It is undoubted that image-based 3D reconstruction methods are definitely back and out of the laser scanning shadow. Indeed the integration of automated computer vision algorithms with reliable and precise photogrammetric methods is nowadays producing successful solutions for automated, detailed and accurate 3D reconstruction from image datasets (Snavely et al., 2008; Barazzetti et al., 2010; Pierrot-Deseilligny et al., 2011). This is particularly true for terrestrial applications where we are witnessing the fact that:

- there is a large variety of optical sensors to acquire images with very different performances and thus quality of the delivered imagery;
- it's constantly growing the impression that few randomly acquired images (or even grab from the Internet) and a black-box software (or mobile app) are enough to produce a decent 3D geometric model;
- automation in image processing is more and more increasing;
- a one-button solution is what non-experts are often seeking to derive 3D models out of inexpensive images.

Due to these reasons, the quality of the acquired and used images is becoming fundamental to allow automated methods of doing their tasks correctly. Motion blur, sensor noise, jpeg artifacts, wrong depth of field are just some of the possible problems that are negatively affecting automated 3D reconstruction methods.

The aim of the present work is to develop an efficient image pre-processing methodology to increase the quality of the results in two central steps of the photogrammetric pipeline:

- the automated alignment of image datasets by:
  (i) increasing the number of correct correspondences, particularly in textureless areas;
  (ii) tracking features along the largest number of possible images to increase the reliability of the extracted correspondences;
  (iii) correctly orienting the largest number of images within a certain dataset;
  (iv) delivering sub-pixel accuracy bundle adjustment results.
- the generation of denser and less noisier 3D point clouds.

In this paper the proposed pre-processing pipeline will be evaluated on the image orientation and dense matching issues. The main idea is (i) to minimize typical failure causes by SIFT-like algorithms (Apollonio et al., 2014) due to changes in the illumination conditions or low contrast blobs areas and (ii) to improve the performances of dense 3D reconstruction methods. The pipeline is grounded on the evaluation of many different state-of-the-art algorithms aiming to give solutions at specific problems and adapt the most promising algorithm (from a theoretical viewpoint), thus creating an ad-hoc methodology and an overall solution to radiometrically improve the image quality.

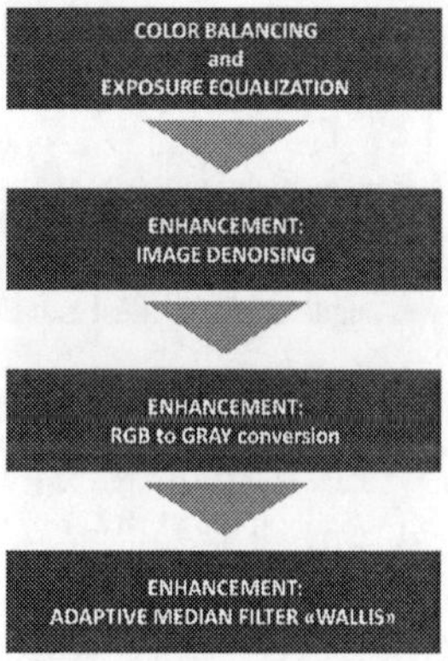

Figure 1: The proposed image pre-processing pipeline.

The developed methodology for image pre-processing consists of color balancing (Section 2), image denoising (Section 3), RGB to grey conversion (Section 4) and image content enhancement (Section 5). All the algorithms are tested and applied to raw images, i.e. images as close as possible to the direct camera output.

2. COLOR BALANCE AND EXPOSURE EQUALIZATION

The aim is essentially to have radiometrically-calibrated images ensuring the consistency of surface colors along all the images (i.e. as much similar as possible RGB values for homologous pixels). Starting from captured RAW images our workflow includes: exposure compensation, optical correction, sharpen and color balance.

This contribution has been peer-reviewed.
doi:10.5194/isprsarchives-XL-5-W4-315-2015

With respect of color characterization, the color targets based technique (Hong et al., 2001) is adopted, using a set of differently colored samples measured with a spectrophotometer. The target GretagMacbeth ColorChecker (McCamy et al. 1976) is employed during the image acquisitions, considering the measurements of each patch as reported in Pascale (2006). The most precise calibration for any given camera requires recording its output for all possible stimuli and comparing it with separately measured values for the same stimuli (Wandell & Farrell, 1993). However, storage of such a quantity of data is impractical, and, therefore, the response of the device is captured for only a limited set of stimuli - normally for the acquisition condition. The responses to these representative stimuli can then be used to calibrate the device for input stimuli that were not measured, finding the transformation between measured CIExyz values and captured RGB values. To find this transformation, several techniques have been developed, including look-up tables (Reinhard et al., 2008).

Summarizing the method for evaluating and expressing color accuracy from this processing includes: a) a physical reference chart acquired under standard conditions; b) a reference chart color space with ideal data values for the chart; c) a means of relating or converting the device color space to the reference chart color space; d) a means of measuring and displaying errors in the device's rendition of the reference chart. The color accuracy is computed in terms of the mean camera chroma relative to the mean ideal chroma in the CIE color metric (ΔE*00) as defined in 2000 by the CIE (Sharma et al., 2005):

$$\Delta E_{00} = \sqrt{\left(\frac{\Delta L'}{k_L S_L}\right)^2 + \left(\frac{\Delta C'}{k_C S_C}\right)^2 + \left(\frac{\Delta H'}{k_H S_H}\right)^2 + R_\tau \frac{\Delta C'}{k_C S_C}\frac{\Delta H'}{k_H S_H}}$$

This formula is a new version of the original one (1976) and is more suitable for our uses. It takes into consideration the problem of non-perceptual uniformity of the colors for which ΔE*00 varies the weight of L* depending on where the brightness range falls. Song & Luo (2000) showed that the perceptible and acceptable color differences in complex images presented on a CRT monitor are approximately 2.2 and 4.5, respectively. In our case, the latter value was used as a strict reference for accuracy, defined from perception tests on the results obtained using this value.

Exposure error in f-stops was also evaluated. The ΔE*00 and the exposure error calculations was made using Imatest Studio software version 3.9.

A captured color image containing the GretagMacbeth ColorChecker is neutralized, balanced and properly exposed for the gamma of the reference data set. Using X-Rite ColorChecker Passport Camera Calibration software is generated an ICC profile that was assigned together with the sRGB color space of the RAW image. Before creating ICC profiles, a standard gamma correction (γ = 2.2) is applied, converting all images to the camera's native linear color space, thus improving the quality of the profiles. A protocol is developed to use the same calibration for groups of images with the same features (i.e., orientation, exposure and framed surfaces) thus to maintain consistency in the process and results. From an operational point of view, the preservation of color fidelity throughout the image processing is ensured by:

- taking pictures in the most possible homogeneous operative conditions (aperture/exposure direction and intensity of light);
- including ColorChecker target inside the photographed scenes in order to correct the image radiometry;
- storing photos in RAW format;
- using an appropriate color space from the beginning of the image processing.

An important and critical issue is the acquisition of the color target. In order to maintain uniform lighting in an external environment, for each image, we need to consider: (i) surfaces illuminated and oriented as the ColorChecker and that presents an angle of incidence with sunlight of approximately 20-45 degrees or (ii) image acquisitions performed with overcast sky. To minimize the light glare, that would give unexpected results in the calibration process, the ColorChecker is normally placed on a tripod with a dark background and orthogonal to the camera optical axis. Finally we verified that a ColorChecker image width of 500 to 1500 pixels is sufficient for ΔE*00 analysis, as suggested also in Imatest user guide.

3. IMAGE DENOISING

Image noise is defined in the ISO 15739 standard as "unwanted variations in the response of an imaging system" (ISO 15739, 2003). It is formed when incoming light is converted from photons to an electrical signal and originates from the camera sensor, its sensitivity and the exposure time as well as by digital processing (or all these factors together). Noise type includes:

- *Fixed pattern noise* ('hot' and 'cold' pixels): it is due to sensor defects or long time exposure, especially with high temperatures. Fixed pattern noise always appears in the same position.
- *Random noise*: it includes intensity and color fluctuations above and below the actual image intensity. They are always random at any exposure and more influenced by ISO speed.
- *Banding noise*: it is caused by unstable voltage power and is characterized by the straight band in frequency on the image. It is highly camera-dependent and more visible at high ISO speed and in dark image. Brightening the image or white balancing can increase the problem.

In real-world photographs, the highest spatial-frequency detail consists mostly of variations in brightness ('luminance detail') rather than variations in hue ('chroma detail'):

- *Luminance noise* is composed of noisy bright pixels that give the image a grainy appearance. High-frequency noise is prevalent in the luminance channel, which can range from fine grain to more distinct speckle noise. This type of noise does not significantly affect the image quality and can be left untreated or only minimally treated if needed.
- *Chrominance noise* appears as clusters of colored pixels, usually green and magenta. It occurs when the luminance is low due to the inability of the sensor to differentiate color in low light levels. As a result, errors in the way color is recorded are visible and hence the appearance of color artifacts in the demosaicked image.

Under all these considerations, the noise model can be approximated with two components:

a) A signal-independent Gaussian noise modeling the fixed pattern noise (FPN);

b) A signal-dependent Poisson noise modeling the temporal (random) noise, called Shot Noise.

Several denoising methods deal directly with the Poisson noise. Wavelet-based denoising methods (Nowak & Baraniuk, 1997; Kolaczyk, 1999) propose adaptation of the transform threshold to the local noise level of the Poisson process. Recent papers on the Anscombe transform by Makitalo & Foi (2011) and Foi (2011), argue that, when combined with suitable forward and inverse variance-stabilizing transformations (VST), algorithms designed for homoscedastic Gaussian noise work just as well as ad-hoc algorithms based on signal-dependent noise models.

This contribution has been peer-reviewed.
doi:10.5194/isprsarchives-XL-5-W4-315-2015

This explains why the noise is assumed to be uniform, white and Gaussian, having previously applied a VST to the noisy image to take into account the Poisson component.
An effective restoration of image signals will require methods that either model the signal a-priori (i.e. Bayesian) or learn the underlying characteristics of the signal from the given data (i.e. learning, nonparametric, or empirical Bayes' methods). Most recently, the latter approach has become very popular, mainly using patch-based methods that exploit both local and nonlocal redundancies or 'self- similarities' in the images (Lebrun et al., 2012). A patch-based algorithm denoises each pixel by using knowledge of (a) the patch surrounding it and (b) the probability density of all existing patches.
Typical noise reduction software reduces the visibility of noise by smoothing the image, while preserving its details. The classic methods estimate white homoscedastic noise only, but they can be adapted easily to estimate signal- and scale-dependent noise.
The main goals of image denoising algorithms are:

- Perceptually flat regions should be as smooth as possible and noise should be completely removed from these regions;
- Image boundaries should be well preserved and not blurred;
- Texture detail should not be lost;
- The global contrast should be preserved (i.e. the low-frequencies of denoised and input images should be equal);
- No artifacts should appear in the denoised image.

Numerous methods have been developed to meet these goals, but they all rely on the same basic method to eliminate noise: *averaging*. The concept of averaging is simple, but determining which pixels to average is not. To meet this challenge, four denoising principles are normally considered:

- transform thresholding (sparsity of patches in a fixed basis),
- sparse coding (sparsity on a learned dictionary),
- pixel averaging and block averaging (image self-similarity)
- Bayesian patch-based methods (Gaussian patch model).

Each principle implies a model for the ideal noiseless image. The current state-of-the-art denoising recipes are in fact a smart combination of all these ingredients.

3.1 Evaluated methods and proposed approach

We investigated different denoise algorithms, mostly commercial, namely:

- *Imagenomic Noiseware* (Imagenomic LLC, 2012; Petrosyan, & Ghazaryan, 2006);
- *Adobe* Camera RAW denoise (Schewe & Fraser, 2010);
- *Non-Local Bayesian* filter (Buades et al. 2005; Kervrann & Boulanger, 2006; Awate & Whitaker, 2006);
- *Noise Clinic* (Lebrun et al. 2014; Colom et al. 2014;
- *Color Block Matching 3D* (*CBM3D*) filter (Dabov et al., 2007b) a color variant of *Block Matching 3D (BM3D)* filter (Dabov et al., 2007a).

Following results of these experiments, an in-house solution was developed starting from the *CBM3D* method. *CBM3D* extends the multi-stage approach of *BM3D* via the YoUoVo color system. *CBM3D* produces a basic estimate of the image, using the luminance data and deliver the denoised image performing a second stage on each of the three color channels separately. *BM3D* is a sliding-window denoising method extending the DCT (Yaroslavsky, 1996) and NL-means algorithms. Instead of adapting locally a basis or choosing from a large dictionary, it uses a fixed basis. The main difference from DCT denoising is that a set of similar patches is used to form a three-dimensional block, which is filtered by using a three-dimensional transform, hence the name collaborative filtering. The algorithm works in two stages: 'basic estimate' of the image and the creation of the final image. A variant of *CBM3D* from the original Matlab code was implemented, with some customizations. Among these, the most significant is the automatic selection of the parameters based on the ISO sensitivity, the type of camera and stored profiles. Stored profiles (camera, ISO) enable presets for each one of those (can be interpolated, potentially), reading initialization data from the EXIF.

4. RGB TO GRAY

Most image-based 3D reconstruction algorithms are conceptually designed to work on grayscale images instead of the RGB triple. Color to grayscale conversion can be seen as a *dimensionality reduction* problem. This operation should not be underestimated, since there are many different properties that need to be preserved. In most of the cases isoluminant color changes are usually not preserved during the RGB to Gray conversion. Many conversion methods have been proposed in recent years. They mainly focus on the reproduction of color images with grayscale mediums, with the goal of a perceptual accuracy in terms of the fidelity of the converted image. These kinds of approaches are not designed to fulfill the needs of image matching algorithms where local contrast preservation is crucial during the matching process. This was observed also in Lowe (2004) where the candidate keypoints with low contrast are rejected in order to decrease the ambiguity of the matching process.
RGB to Gray conversion can be done in:

- *Color Space* (linear or non-linear): the CIE Y method is a widely used conversion, based on the CIE 1931 XYZ color space. It takes the XYZ representation of the image and uses Y as the grayscale value.
- *Image Space* (called functional): following Benedetti et al. (2012), they can be divided in three groups:
  (a) *trivial methods*: they are the most basic and simple ones, as they do not take into account the power distribution of the color channels. They lose a lot of image information as for every pixel they discard two of the three color values, or discard one value averaging the remaining ones, not taking into account any color properties. Despite this loss, they are commonly used for their simplicity. A typical trivial method is the *RGB Channel Filter* that selects a channel between *R*, *G* or *B* and uses this channel as the grayscale value (afterwards called *GREEN2GRAY*).
  (b) *direct methods*: the conversion is a linear function of the pixel's color values. Typically, this class of functions takes into account the spectrum of different colors. The Naive Mean direct method takes the mean of the color channels. The advantage, compared to the other trivial ones, is that it takes information from every channel, though it does not consider the relative spectral power distribution of the RGB channels. The most popular of these methods is *RGBtoGRAY* that uses the NTSC CCIR 601 luma weights, with the formula
  $$Y' = 0.2989R + 0.5870G + 0.1140B.$$
  Another solution embedded inside Adobe Photoshop uses these specific weights for the channels R, G, and B: 0.4, 0.4, 0.2.
  (c) *chrominance direct methods*: they are based on more advanced color spaces, trying to mitigate the problem related to isoluminant colors. Albeit they are local functions of the image pixels, they assign different grayscale values to isoluminant colors, altering the luminance information using the chrominance information. Chrominance direct methods can be performed either locally (Smith et al., 2008; Kim et

al., 2009) or globally (Grundland & Dodgson, 2007). Local methods make pixels in the color image not processed in the same way and usually rely on the local chrominance edges for enhancement. Global methods strive to produce one mapping function for the whole image thus producing same luminance for the same RGB triplets and high-speed conversion.

4.1 Evaluated methods and proposed approach

We investigated different RGB to Gray methods, namely:

- *GREEN2GRAY*;
- Matlab *RGB2GRAY*;
- *GRUNDLAND-DODGSON* (Grundland & Dodgson, 2007);
- *REALTIME* (Lu et al. 2012a,b);
- *Adobe* Photoshop.

Following the tests with the aforementioned methods, a new RGB to Gray conversion technique, named *Bruteforce Isoluminants Decrease* (*BID*), was developed. The aim was to preserve the consistency between the images considering the following matching requirements:

- *Feature discriminability*: the method should preserve the image features discriminability to be matched as much as possible;
- *Chrominance awareness*: the method should distinguish between isoluminant colors;
- *Global mapping*: while the algorithm can use spatial information to determine the mapping, the same color should be mapped to the same grayscale value for every pixel in the image;
- *Color consistency*: the same color should be mapped to the same grayscale value in every image of the dataset;
- *Grayscale preservation*: if a pixel in the color image is already achromatic it should maintain the same gray level in the grayscale image;
- *Unsupervised algorithm*: it should not need user tuning to work properly.

BID computes the statistical properties of the input dataset with the help of a representative collection of image patches. It takes in input and analyses simultaneously the whole set of images that need to be matched. Differently from other similar solutions as the *Multi-Image Decolorize* method (Benedetti et al., 2012), *BID* is a generalization of the Matlab *RGB2Gray* algorithm. . *BID* has its foundation in the statistics of extreme-value distributions of the considered images and presents a more flexible strategy, adapting dynamically channel weights depending on specific input images, in order to find the most appropriate weights for a given color image. *BID* preserves as much as possible the amount of the conveyed information. The algorithm behind *BID* tries to maximize the number of peaks obtained in the image converted and to distribute as evenly as possible the amount of tones present in the converted image by evaluating the goodness of fit of the distribution with respect to a rectangular distribution. To calculate the best rectangular fitting we assumed a 0 slope regression line. If the general equation of the regression line is:

$$\beta = \bar{y} - m\bar{x}$$

β is equivalent to the average of the histogram points. After calculating the average, the minimum error within all the calculated combinations of channel mixings if sought. The error is calculated as least squares error:

$$S = \sum_{i=1}^{n} (y_i - \beta)^2$$

where y_i are the actual points, while β is the best linear fitting of the histogram. *BID* cyclically varies the amount of red, green and blue and for each variation calculates the distribution of the resulting grayscale image and assesses the fitting quality with respect to a rectangular distribution. Finally *BID* chooses the mixing that maximizes the number of tones obtained in the converted image. Similarly to Song et al. (2013), BID uses a measurement criterion to evaluate the decolorization quality, i.e. the newly defined dominant color hypothesis. Main disadvantage of *BID* is the high computational pre-processing cost relieved using sampled patches on each image of the dataset.

5. IMAGE CONTENT ENHANCEMENT

Low-texture surface such as plaster causes difficulties for feature detection such as the Difference-of-Gaussian (DoG) function, which extracts features in pixel level and compares them with adjacent ones and in stereo-matching algorithms.
The Wallis filter (Wallis, 1976), is a digital image processing function that enhances the contrast levels and flattens the different exposure to achieve similar brightness in the images. The filter is normally applied in order to optimize image datasets for subsequent image-matching procedures (Baltsavias, 1991; Baltsavias et al., 1996; Seiz et al. 2002; Ohdake et al. 2005; Remondino et al., 2008; MacDonald et al. 2014).
Wallis uses two parameters to control the enhancement's amount, the *contrast expansion factor A* and the *brightness forcing factor B*. The algorithm is adaptive and adjusts pixel brightness values in local areas only, contrary to a global contrast filter, which applies the same level of contrast throughout an entire image. The resulting image contains greater detail in both low and high-level contrast regions concurrently, ensuring that good local enhancement is achieved throughout the entire image. The Wallis filter requires the user to accurately set a target mean and standard deviation in order to locally adjusts areas and match the user-specified target values. Firstly the filter divides the input image into neighboring square blocks with a user-defined size ('window size') in order to calculate local statistics. Then mean (*M*) and standard deviation (*S*) of the unfiltered image are calculated for each individual block based on the grey values of the pixels and the resulting value is assigned to the central cell of each block. The mean and standard deviation values of all other cells in the block are calculated from this central cell by bilinear interpolation. In this way, each individual pixel gets its own initial local mean and standard deviation based on surrounding pixel values. The user-defined mean and standard deviation values are then used to adjust the brightness and the contrast of the input cells. The resulting Wallis filtered image is thus a weighted combination of the original and user-defined mean and standard deviation of the image. The implementation of the filter of Wallis, given the aforementioned factor *A* and *B*, can be summarized as follows:

- let *S* be the standard deviation for an image
- let *M* be the mean for an image
- for each (x,y) pixel in the image
- calculate local mean *m* and standard deviation *s* using an *NxN* neighborhood
- calculate the enhanced output
  (x,y) = S*(input(x,y) – m)/(s + A)+ M*B + m*(1-B)

5.1 Selection of Wallis parameters

The main problem when using the Wallis filter is the correct selection of its parameters. In fact, although several authors reported parameters for successful projects, the filter is more an 'ad-hoc' recipe than an easily deployable system in an automatic photogrammetry pipelines. To overcome this problem, a Wallis parameters characterization study was carried out using three

This contribution has been peer-reviewed.
doi:10.5194/isprsarchives-XL-5-W4-315-2015

different datasets able to encompass the majority of possible surveying case studies. The three datasets consist of:

1. a cross vault characterized by smooth bright-colored plaster;
2. a building facade with porticos with smooth plaster in both the main and front portico;
3. a Venetian floor, with asphalt and porphyry cubes with a background facade overexposed and an octagonal pillar in the foreground coated with smooth plaster.

For every dataset, three images were matched using a calibrated version (Apollonio et al., 2014) of the SIFT operator available in Vedaldi's implementation (Vedaldi & Fulkerson, 2010).

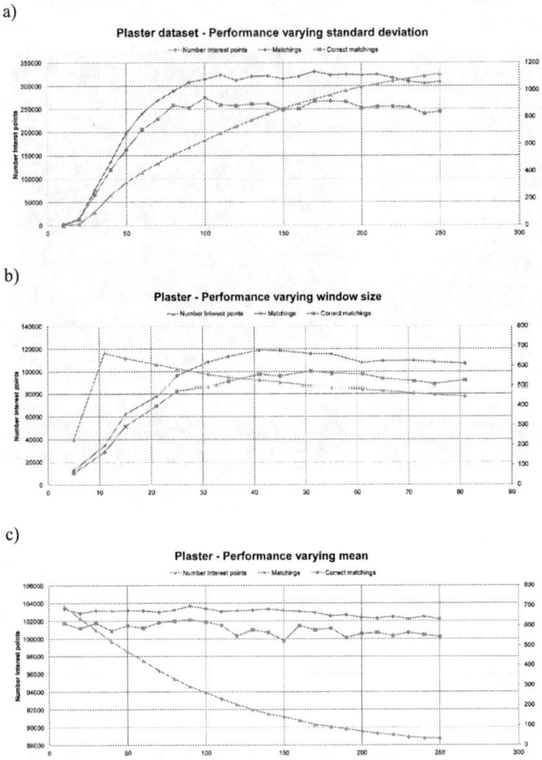

Figure 2: Results of the Wallis evaluation on the plaster dataset (1).

From the characterization procedure (Fig. 2), it can be summarized that:

1. the number of tie points found is inversely proportional to the value of the parameter *A*, but the number of correct matches remains basically stable varying *A*, which can then be set at high values (6-8);
2. varying the user-specified standard deviation, the number of tie points and correct matches increases substantially linearly up to a value of 100 and then remains constant (Fig.2a);
3. sensor resolution and window size are linearly related and the increasing of the window size beyond the optimal value does not involve any improvement in either the number of positive matches and in the number of extracted tie points (Fig. 2b);
4. the mean presents optimal values between 100 and 150 with a decay afterwards (Fig. 2c).

Starting from these observations, a new implementation of the filter was realized to automatically select the filter parameters and achieve the highest possible ratio of corrected matches wrt the number of extracted tie points.

6. EXPERIMENTAL SETUP AND EVALUATION RESULTS

For the evaluation of the implemented methods, different image networks were used with different imaging configurations, textureless areas and repeated pattern/features. The datasets tried to verify the efficiency of the implemented pre-processing pipeline and evaluate its performances. Hereafter, for reasons of space, one peculiar dataset is reported (Building with Porticoes, 35 images – Fig. 3), acquired with a Nikon D3100 (sensor size: 23,1 x 15,4 mm), with a 18 mm nominal focal length. The historical building (19 m height x 10 m width) has three floors characterized by arcades with four aches, pillars/columns, cross vault and plastered wall. The camera was moving along the porticoes, with some closer shots of the columns. The dataset represents an urban test framework and summarizes a typical historical urban scenario. It contains, besides convergent images and some orthogonal camera rolls (ideal for self-calibration), a variety of situations typical of failure cases, i.e. 3D scenes (non-coplanar) with homogeneous regions, distinctive edge boundaries (e.g. buildings, windows/doors, cornices, arcades), repeated patterns (recurrent architectural elements), textureless surfaces and illumination changes.

The performances of the pre-processing strategies previously described are reported using the statistical output of the bundle adjustment (reprojection error), the number of points in the dense point cloud and an accuracy evaluation of the dense matching results. This latter is achieved using a Terrestrial Laser Scanning (TLS) survey as reference (Fig. 4a). Three textureless regions (A1, A2, A3) are identified and, in order to have a reference comparable to the dense matching results, the range data are subsampled to a grid of 5x5 mm. The average image GSD in the three regions of interest is ca 2 mm but the dense matching was carried out using the second-level image pyramid, that is, at a quarter of the original image resolution (i.e. dense point clouds with a sampling resolution of twice the original image GSD).

For every pre-processing method, the Portico dataset is relatively oriented (Fig. 4b) with 4 different tools (Visual SFM, Apero MicMac, Eos Photomodeler and Agisoft Photoscan), trying to keep a uniform number of extracted keypoints and tie points. Then, a dense point cloud is extracted with a unique tool (nFrames SURE) by using (fixing) the same camera parameters for all methods.

6.1 Color balance results

The results of the orientation and dense matching steps are reported in Table 1. The color balancing procedure generally helps in increasing the number of oriented images, except in PS where – as also visible in the next results – the dataset is entirely oriented at every run.

6.2 Image denoising results

After the denoising procedure, the Wallis filtering ia applied to the dataset before running the automated orientation and matching procedures. The achieved adjustment results, according to the different denoising methods, show that more images can be oriented (Table 2) and denser point clouds can be retrieved (Table 3).

This contribution has been peer-reviewed.
doi:10.5194/isprsarchives-XL-5-W4-315-2015

Figure 3: The Portico dataset (35 images) used for the evaluation of the developed image pre-processing pipeline.

	Not enhanced	*Enhanced*
VisualSFM (VSFM)		
Numb. oriented images	31	33
BA quality (px)	0.476	0.481
Apero MicMac (AP)		
Numb. oriented images	33	35
BA reprojection error (px)	0.71	0.85
EOS Photomodeler (PM)		
Numb. oriented images	33	35
BA reprojection error (px)	0.476	0.481
Agisoft Photoscan (PS)		
Numb. oriented images	35	35
BA reprojection error (px)	0.512	0.543
Dense Matching (SURE)		
Numb. 3D points	1.259.795	1.626.267

Table 1: Bundle adjustment (BA) results and dense matching improvement for the enhancement procedure.

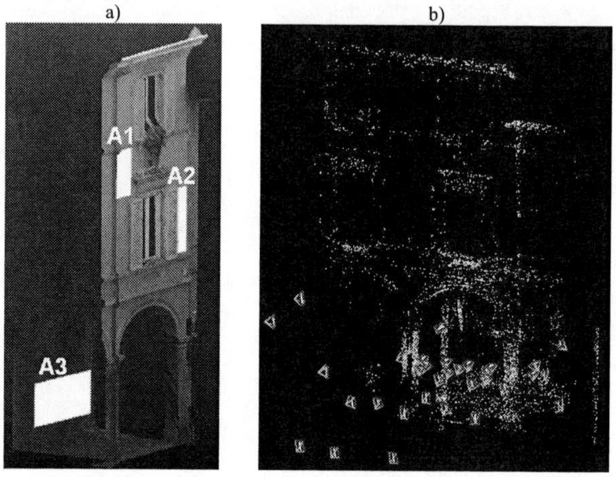

Figure 4: The TLS survey with the 3 textureless regions of interest (a) and the camera network of the Portico dataset (b).

	NO DENOISE	*Adobe*	*CBM3D*	*NL-BAYES*	*NOISE CLINIC*	*IMAGENOMIC*
VisualSFM (VSFM)						
Numb. oriented images	32	35	35	35	35	35
BA quality (px)	0.297	0.476	0.485	0.450	0.694	0.371
Apero MicMac (AP)						
Numb. oriented images	33	35	35	35	35	35
BA reprojection error (px)	0.96	0.88	0.88	0.97	0.87	0.93
EOS PhotoModeler (PM)						
Numb. oriented images	33	33	33	33	33	31
BA reprojection error (px)	0.892	0.838	0.833	0.856	0.872	0.878
Agisoft Photoscan (PS)						
Numb. oriented images	35	35	35	35	35	35
BA reprojection error (px)	0.490	0.493	0.489	0.491	0.527	0.554

Table 2: Bundle adjustment (BA) results of the denoising methods applied to the enhanced dataset.

		NO DENOISE	*Adobe*	*CBM3D*	*NL-BAYES*	*NOISE CLINIC*	*IMAGENOMIC*
Numb. 3D points		998.995	1.308.768	1.456.024	1.456.561	1.428.996	1.346.559

A1 (0,8x1,4m) Ref. pt. cloud:	Numb. 3D points	1.227	27.508	16.257	31.386	23.419	8.194

This contribution has been peer-reviewed.
doi:10.5194/isprsarchives-XL-5-W4-315-2015

36.294 pts	Std Dev (mm)	N/A	12,264	13,613	11,734	12,073	17,219
A2 (0,5x1,9m) Numb. ref. pt.	Numb. 3D points	495	25.327	17.966	29.632	23.541	7.548
cloud: 33.868	Std Dev (mm)	N/A	11,184	11,735	11,897	11,041	19,002
A3 (2,6x1,3m) Numb. ref. pt.	Numb. 3D points	32.014	176.478	179.553	183.166	184.835	160.120
cloud: 120.222	Std Dev (mm)	N/A	11,760	12,354	11,443	10,028	10,991

Table 3: Dense matching results and evaluation of the denoising methods applied to the enhanced dataset. The ground truth is given by a TLS survey resampled to 5x5 mm grid.

	GREEN2GRAY	*Adobe*	*REALTIME*	*GRUNDLAND - DODGSON*	*RGB2GRAY*	*BID*
VisualSFM (VSFM)						
Numb. oriented images	35	33	30	16	33	35
BA quality (px)	0.548	0.424	0.366	0.379	0.353	0.581
Apero MicMac (AP)						
Numb. oriented images	33	34	35	18	35	33
BA reprojection error (px)	0.97	0.91	0.88	0.66	0.90	0.94
EOS PhotoModeler (PM)						
Numb. oriented images	35	33	33	28	32	35
BA reprojection error (px)	0.866	0.890	0.874	0.894	0.876	0.858
Agisoft Photoscan (PS)						
Numb. oriented images	35	35	35	35	35	35
BA reprojection error (px)	0.523	0.530	0.545	0.536	0.518	0.514

Table 4: Bundle adjustment (BA) results of the RGB to GRAY methods applied to the enhanced dataset.

		GREEN2GRAY	*Adobe*	*REALTIME*	*GRUNDLAND - DODGSON*	*RGB2GRAY*	*BID*
Numb. 3D points		1.703.607	1.444.269	1.522.044	1.375.971	1.184.432	1.964.397

A1 – Numb. ref.	Numb. 3D points	27.127	12.776	10.890	N/A	6.863	11.360
pt. cloud: 36.294	Std Dev (mm)	11,279	14,827	12,782	N/A	15,131	18,834
A2 – Numb. ref.	Numb. 3D points	32.619	12.853	15.264	N/A	6.520	9.950
pt. cloud: 33.868	Std Dev (mm)	10,765	13,820	10,768	N/A	12,388	22,287
A3 – Numb. ref.	Numb. 3D points	162.563	153.281	163.246	N/A	152.923	157.837
pt. cloud: 120.222	Std Dev (mm)	11,955	10,818	8,708	N/A	8,892	10,820

Table 5: Dense matching results and evaluation of the RGB to GRAY methods applied to the enhanced dataset. The ground truth is given by a TLS survey resampled to 5x5 mm grid.

6.3 RGB to GRAY results

The grayscale conversion (coupled with the Wallis filtering) shows how algorithms are differently affecting the BA procedure (Table 4) as well as the dense matching results (Table 5). It can be generally noticed a larger number of oriented images, better reprojection errors and denser point clouds.

7. CONCLUSIONS

The paper reported some pre-processing methods to improve the results of the automated photogrammetric pipeline. Among the performed tests, a salient example featuring a complex image network and scenario (lack of texture, repetitive pattern, etc.) - and reflecting also the other outcomes achieved in our experiments - was presented. From the results in Section 6 it is

This contribution has been peer-reviewed.
doi:10.5194/isprsarchives-XL-5-W4-315-2015

clear that the pre-processing procedure, with very limited processing time, is generally positively influencing the performances of the orientation and matching tools. The pre-processing is indeed helping in achieving complete orientation results, sometimes better BA reprojection errors (although it is not a real measure of better quality) and, above all, more dense and complete 3D dense point clouds.

According to our tests, the best procedure implies to apply a color enhancement, a denoise procedure based on the BM3D method, the RGB to Gray conversion using the developed BID method and the Wallis filtering. This latter filtering seems to be fundamental also in the orientation procedure and not only at matching level as so far reported in the literature.

The proposed pre-processing pipeline appears particularly suitable and integrable in a fully automated image processing method as:

- the parameters are strictly related to the image metadata (EXIF header) and could be automatically set;
- the pre-processing / filtering considers the entire dataset and it is not image-dependent;
- it is based on criteria related to algorithms employed for automated 3D reconstruction (e.g. keypoint detection) and not on perceptual criteria typically used in image enhancement algorithms;
- it stems from many experiments and merges state-of-the-art methods;
- it gives advantages also for the texture mapping phase.

Some final concluding remarks derived by our experiments:

- automated orientation is still an open issue;
- the employed tools for automated image orientation do not provide a unique / standardize type of output;
- there is a lack of repeatability, as already demonstrated in Remondino et al. (2012).

In the near future work we plan to publish all the performed results with more detailed analyses.

REFERENCES

Ancuti, C.O., Ancuti C., Bekaert P., 2010. Decolorizing images for robust matching. *ICIP 2010. IEEE International Conference on*, Vol Vol. 1, pp. 149-153.

Apollonio, F., Ballabeni, A., Gaiani, M., Remondino, F., 2014. Evaluation of feature-based methods for automated network orientation. *ISPRS Archives of the Photogrammetry, Remote Sensing and Spatial Information Sciences*, Vol. XL-5, pp. 47-54

Awate, S.P., Whitaker, R.T., 2006. Unsupervised, information-theoretic, adaptive image filtering for image restoration. *IEEE Trans. Pattern Anal. Mach. Intell*, 28(3), pp. 364-376.

Baltsavias, E. P., 1991. *Multiphoto geometrically constrained matching.* Ph.D. dissertation No. 9561, Institute of Geodesy and Photogrammetry, ETH Zurich. 221 pp.

Baltsavias, E. P., Li, H., Mason, S., Stefanidis, A., Sinning, M., 1996. Comparison of two digital photogrammetric systems with emphasis on DTM generation: case study glacier measurement. *Int. Archives of Photogrammetry, Remote Sensing and Spatial Information Sciences*, Vol. XXXI, Part. B4, pp. 104-109.

Barazzetti, L., Scaioni, M., Remondino, F., 2010. Orientation and 3D modeling from markerless terrestrial images: combining accuracy with automation. *The Photogrammetric Record*, 25(132), pp. 356–381.

Buades, A., Coll, B., Morel, J. M., 2005. A review of image denoising algorithms, with a new one. *Multiscale Modeling Simulation (SIAM Interdisciplinary J.)*, Vol. 4(2), pp. 490-530.

Benedetti, L., Corsini, M., Cignoni, P., Callieri, M., Scopigno, R., 2012. Color to gray conversions in the context of stereo matching algorithms: An analysis and comparison of current methods and an ad-hoc theoretically-motivated technique for image matching. *Machine Vision and Application*, Vol. 23(2), pp. 327-348.

Colom, M., Buades, A., Morel, J., 2014. Nonparametric noise estimation method for raw images. *J. Opt. Soc. Am. A*, Vol. 31, pp. 863-871.

Dabov, K., Foi, A., Katkovnik V., Egiazarian, K., 2007a. Image denoising by sparse 3D transform-domain collaborative Filtering. *IEEE Trans. Image Process.*, Vol. 16(8), pp. 2080-2095.

Dabov, K., Foi, A., Katkovnik, V., Egiazarian, K., 2007b. Color image denoising via sparse 3D collaborative filtering with grouping constraint in luminance-chrominance space. *Proceedings of IEEE International Conference on Image Processing (ICIP)*, Vol.1, pp. I-313 - I-316.

Foi, A., 2011. Noise estimation and removal in MR imaging: The variance-stabilization approach. *IEEE International Symposium on Biomedical Imaging: From Nano to Macro*, pp. 1809-1814.

Gooch, A.A., Olsen, S.C., Tumblin, J., Gooch, B., 2005. Color2gray: salience-preserving color removal. *ACM Trans. Graph.*, Vol. 24(3), pp. 634-639.

Grundland, M., Dodgson, N.A., 2007. Decolorize: fast, contrast enhancing, color to grayscale conversion. *Pattern Recognition*, Vol. 40(11), pp. 2891-2896.

Kervrann C., Boulanger J., 2006, Optimal spatial adaptation for patch-based denoising. *IEEE Trans. Image Process.*, Vol. 15(10), pp. 2866-2878.

Kim, Y., Jang, C., Demouth, J., Lee, S., 2009. Robust color-to-gray via nonlinear global mapping, *ACM Transactions on Graphics,* Vol. 28(5).

Kolaczyk E., 1999. Wavelet shrinkage estimation of certain Poisson intensity signals using corrected thresholds. *Statist. Sin.*, Vol. 9, pp. 119-135.

Imagenomic LLC (2012). Noiseware 5 Plug-In User's Guide

Lebrun M., Colom M., Morel J.-M., 2014. The Noise Clinic: a blind image denoising algorithm, *Ipol Journal*, preprint.

Lebrun, M., Colom M., Buades, A., Morel, J.-M., 2012. Secrets of image denoising cuisine, *Acta Numerica*, Vol. 21, pp. 475-576.

Lowe, D., 2004: Distinctive image features from scale-invariant keypoints. *Int. Journal of Computer Vision*, Vol. 60(2), pp. 91-110.

Lu C., Xu L., Jia J., 2012. Contrast preserving decolorization. *ICCP, 2012 IEEE International Conference on*, pp.1-7.

Lu C., Xu L., Jia J., 2014. Contrast preserving decolorization with perception-based quality metrics. *International Journal of Computer Vision*, Vol. 110(2), pp. 222-239.

MacDonald, L., Hindmarch, J., Robson, S., Terras, M., 2014. Modelling the appearance of heritage metallic surfaces. *Int. Archives of Photogrammetry, Remote Sensing and Spatial Information Sciences*, Vol. XL(5), pp. 371-377.

McCamy, C.S., Marcus, H., Davidson, J.G., 1976. A color rendition chart. *Journal of Applied Photographic Engineering*, Vol. 11(3), pp. 95-99.

Makitalo, M., Foi, A., 2011. Optimal inversion of the Anscombe transformation in low-count Poisson image denoising. *IEEE Trans. Image Processing*, Vol. 20, pp. 99-109.

Nowak, R. Baraniuk, R., 1997. Wavelet-domain filtering for photon imaging systems. *IEEE Trans. Image Processing*,Vol. 8, pp. 666-678.

Ohdake, T., Chikatsu, H., 2005. 3D modeling of high relief sculpture using image based integrated measurement system. *Int. Archives of Photogrammetry, Remote Sensing and Spatial Information Sciences*, Vol. XXXVI(5-W17), 6 pp.

Pascale, D., 2006. RGB coordinates of the Macbeth ColorChecker. The BabelColor Company, Montreal, Canada.

Petrosyan, A., Ghazaryan, A., 2006. Method and system for digital image enhancement. *US Patent Application*, Vol. 11(116), pp. 408.

This contribution has been peer-reviewed.
doi:10.5194/isprsarchives-XL-5-W4-315-2015

Pierrot-Deseilligny, M., De Luca, L., Remondino, F., 2011. Automated image-based procedures for accurate artifacts 3D modeling and orthoimage generation. *Geoinformatics FCE CTU Journal*, Vol. 6, pp. 291-299.

Reinhard, E., Arif Khan, E., Oguz Akyüz A., Johnson G., 2008. *Color Imaging Fundamentals and Applications*. A. K. Peters: Wellesley.

Remondino, F., El-Hakim, S., Gruen, A., Zhang, L., 2008. Turning images into 3D models - Development and performance analysis of image matching for detailed surface reconstruction of heritage objects. *IEEE Signal Processing Magazine*, Vol. 25(4), pp. 55-65.

Remondino, F., Del Pizzo, S., Kersten, T., Troisi, S., 2012. Low-cost and open-source solutions for automated image orientation – A critical overview. *Proc. EuroMed 2012 Conference, LNCS*, Vol. 7616, pp. 40-54

Schewe, J., Fraser, B., 2010. *Real World Camera Raw with Adobe Photoshop CS5*. Peachpit Press, Berkeley, CA, USA, 480 pp.

Seiz G., Baltsavias E. P., Grün A., 2002. Cloud mapping from ground: use of photogrammetric methods. *Photogrammetric Engineering and Remote Sensing*, Vol. 68(9), pp. 941-951.

Sharma, G., Wu, W., Dalal, E.N., 2005. The CIEDE2000 Color-difference formula: implementation notes, supplementary test data and mathematical observations. *Color Research and Application*, Vol. 30(1), pp. 21-30.

Smith, K., Landes, P. E., Thollot, J., Myszkowski, K., 2008. Apparent Greyscale: a simple and fast conversion to perceptually accurate images and video. *Computer Graphics Forum*, Vol.27(2), pp. 193-200.

Snavely, N., Seitz, S.M., Szeliski, R., 2008. Modeling the world from internet photo collections. *Int. Journal of Computer Vision*, Vol. 80(2), pp. 189-210.

Song, Y., Bao, L., Xu, X., Yang, Q., 2013. Decolorization: Is rgb2gray() out?, *SIGGRAPH Asia 2013 Technical Briefs (SA '13),* ACM, New York, 15, 4pp.

Song, T., Luo, M.R., 2000. Testing color-difference formulae on complex images using a CRT monitor. *Proceedings of IS&T and SID Eighth Color Imaging Conference*, pp. 44-48.

Vedaldi, A., Fulkerson, B., 2010: VLFeat - An open and portable library of computer vision algorithms. Proc.18th *ACM Intern. Conf. on Multimedia*

Wallis R., 1976, An approach to the space variant restoration and enhancement of images. *Proceedings of Symp. Current Mathematical Problems in Image Science*, Monterey, Naval Postgraduate School, pp. 329-340.

Wandell, B.A., Farrell J.E., 1993. Water into Wine: converting scanner RGB to tristimulus XYZ. *Proceedings of SPIE*, Vol. 1909, Bellingham, pp. 92-101

Yaroslavsky, L.P.. 1996, Local adaptive image restoration and enhancement with the use of DFT and DCT in a running window. *Proceedings of SPIE*, Vol. 2825, Bellingham, pp. 2-13

This contribution has been peer-reviewed.
doi:10.5194/isprsarchives-XL-5-W4-315-2015

HDR IMAGING FOR FEATURE DETECTION ON DETAILED ARCHITECTURAL SCENES

G. Kontogianni, E. K. Stathopoulou*, A. Georgopoulos, A. Doulamis

Laboratory of Photogrammetry, School of Rural and Surveying Engineering, National Technical University of Athens – 15780 Zografou Athens, Greece
georgia.kontog@gmail.com - elliestath@central.ntua.gr - drag@central.ntua.gr - adoulam@cs.ntua.gr

Commission V, WG V/4

KEY WORDS: Feature detection, High dynamic range, Tone mapping, Architectural Scenes

ABSTRACT:

3D reconstruction relies on accurate detection, extraction, description and matching of image features. This is even truer for complex architectural scenes that pose needs for 3D models of high quality, without any loss of detail in geometry or color. Illumination conditions influence the radiometric quality of images, as standard sensors cannot depict properly a wide range of intensities in the same scene. Indeed, overexposed or underexposed pixels cause irreplaceable information loss and degrade digital representation. Images taken under extreme lighting environments may be thus prohibitive for feature detection/extraction and consequently for matching and 3D reconstruction. High Dynamic Range (HDR) images could be helpful for these operators because they broaden the limits of illumination range that Standard or Low Dynamic Range (SDR/LDR) images can capture and increase in this way the amount of details contained in the image. Experimental results of this study prove this assumption as they examine state of the art feature detectors applied both on standard dynamic range and HDR images.

1. INTRODUCTION

3D reconstruction of architectural and cultural heritage objects is commonly used for documentation, visualization, navigation and dissemination purposes. Available Structure from Motion (SfM) algorithms are able to reconstruct large 3D scenes relatively fast, with the use of consequent or randomly taken images of the object. Standard SfM techniques rely on the accurate detection, extraction, description and matching of image features (e.g. keypoints). A huge advantage of such methods is that no prior information about the object scene or the camera path is needed. High quality images are though essential for the performance of such algorithms. Keypoint detection may be complete and more accurate if High Dynamic Range (HDR) images of the scene are used, due to the presence of additional information with respect to standard images (Chermak and Aouf, 2012; Chermak et al. 2014; Jagadish and Sinzinger, 2008). Higher number of detected keypoints will potentially increase the amount of inliers during image matching, resulting consequently to improved SfM results.

Architectural assets are often complex in terms of geometry as well as texture and due to their location are usually exposed to extreme illumination conditions i.e. containing dark shadows or bright sunlight. Under these conditions, important details and colour information of the image may be lost, degrading the result of feature detection and extraction. HDR images could be useful in such cases as they broaden the limits of luminance range that standard images can capture and increase in this way the amount of characteristic features contained in the image. Detail maintenance is crucial towards 3D reconstruction especially in such objects, making them excellent study cases for research in the field of HDR imaging.

This study focuses on the use of HDR images towards optimizing feature detection in images of high detailed architectural scenes such as church altars with complex frescoes, highly decorated arches, columns and other buildings of special architectural style. Tests are performed to investigate the behaviour of common state of the art feature detectors while using HDR images. The rest of the paper is organised as follows: Section 2 is a review of related previous work concerning HDR imaging and feature detection algorithms. Section 3 analyses our approach, followed by experimental results with self-captured images while conclusions are presented in Section 4.

2. RELATED WORK

2.1 Feature Detection

Feature detection is a fundamental research topic in many applications in computer vision and photogrammetry. Camera geometry can be calculated when a sufficient number of correct feature matches between the images is known. Thus, path estimation and 3D reconstruction are strongly influenced by the quality of the correspondences between image pairs and based consequently on reliable key point extraction. A variety of detector operators of distinct points (corners) have been presented in the past (e.g. Moravec, Forstner, Harris). More recent research investigates feature detection and description. We briefly present some important detectors and descriptors below.

SIFT (Scale Invariant Feature Transformation) is an operator introduced by Lowe (Lowe, 1999; Lowe, 2004) which can detect, describe and extract features with the use of Gaussian

* Corresponding author. This is useful to know for communication with the appropriate person in cases with more than one author.

This contribution has been peer-reviewed.
doi:10.5194/isprsarchives-XL-5-W4-325-2015

scale space pyramids. Thus, SIFT features are invariant to image scale and rotation and partially invariant to illumination changes.

Bay et al. (Bay et al 2006) presented an algorithm named SURF (Speed Up Robust Features) that outperformed the existing state of the art detectors and descriptors in terms of repeatability, distinctiveness and robustness as well as speed. The detection part of the algorithm uses integral images through Hessian matrix approximation. The description part, on the other hand, describes the intensity of the neighbourhood around the pixel using Haar wavelets.

FAST (Features from Accelerated Segment Test) comes from the machine learning field (Rosten and Drummond, 2006). It uses a high speed test to distinguish key points from other points, by examining the pixels that lie along a circle around each point. If the pixels are found a lot darker or brighter that the candidate point, it is considered to be a key point. For the sake of speed, instead of checking all pixels, just four of them are tested.

BRIEF (Binary Robust Independent Elementary Features) was introduced by Calonder et al. (Calonder et al., 2010) and is a short binary descriptor using the Hamming distance. The obtained descriptor is not invariant to scale and rotation changes (Alahi et al., 2012).

ORB (Oriented FAST and Rotated BRIEF), introduced by Rublee (Rublee et al., 2011) is a combination of FAST (to detect stable points) and BRIEF (to describe them). Both these operators produce satisfactory results relatively fast, but BRIEF provides also invariance to rotation and robustness to noise (Alahi et al., 2012).

Other efficient detectors and descriptors have also been proposed in the literature (e.g. DAISY, BRISK, FREAK (Tola et al., 2010; Leutenegger et al., 2011; Alahi et al., 2012)). However, few studies regarding feature detection and tracking in HDR images have been published in the past (Chermak and Aouf, 2012; Chermak et al. 2014; Jagadish and Sinzinger, 2008; Cui et al., 2011). Evaluation studies about the behaviour feature detectors also exist. Schmid et al. compared five different detectors of the then state of the art using repeatability rate and information content as evaluation criteria (Schmid et al., 2000). Mikolajczyk et al. introduced in 2005 a reference test set of images and evaluated ten detectors and descriptors (Mikolajczyk et al., 2005). Pribyl et al. compared Harris, Shi-Tomasi, FAST and Fast Hessian detectors on HDR images (Pribyl et al., 2013). Jagadish and Sinzinger evaluated their proposed method for image matching and SIFT on HDR images (Jagadish and Sinzinger, 2008).

2.2 HDR Imaging

High dynamic range imaging (HDR) in image processing and photography is the method of generating images that contain wider dynamic range than one standard digital image can record (Reinhard et al., 2010). This is possible either by using special sensors with extended dynamic range (HDR sensor) or by merging multiple images of the same scene taken with varying exposure settings. Relative literature with techniques for the creation of HDR images exists (Debevec and Malik, 2008; Robertson et al., 1999; Mitsunaga and Nayar, 1999; Rovid et al., 2007).

An HDR image is actually a radiance map whose pixels reach a maximum of 32-bit floating point image representation (Chermak and Aouf, 2012). Common displays can visualise images in a 16- or 8-bit format but are unable to render properly the ones in 32-bit representation. To this end, a procedure called tone mapping was introduced for the compression of HDR image to fit the dynamic range of the display device while preserving the detail. Tone mapping operators can be global or local, empirical or perceptually based, static or dynamic (Cui et al., 2011). Several such operators have been developed in the recent years (Mantiuk et al., 2006; Drago et al., 2003; Durand and Dorsey, 2003; Reinhard et al., 2002; Ashikhmin, 2002; Mertens et al., 2007).

Mantiuk et al. created an algorithm that transforms the luminance values of an image to contrast using gradients on all levels of Gaussian pyramid and consequently transforms contrast values to HVS and scales the response. Finally the image is reconstructed back to luminance values (Mantiuk et al., 2006). On the other hand, Drago et al. proposed a global tone mapping operator that scales the image in the logarithmic domain, using a bias parameter and further radiometrical corrections (Drago et al., 2003). Durand and Dorsey's algorithm decomposes the SDR image into two layers: base layer and detail layer using an edge-preserving bilateral filter. The contrast of the base layer is then compressed, preserving all the details (Durand and Dorsey, 2002). A global, relatively simple tonemapping operator based on photographic practices was introduced by Reinhard et al. in (Reinhard et al., 2002). Another technique of creating HDR images is the so-called exposure fusion according to which the multiple exposures are merged into a high-quality low dynamic range image, ready for the display. Best pixel values from the sequence are selected based on a quality measure and combined into the final result. In this technique the tone mapping procedure is omitted (Mertens et al., 2007).

Although a broadly used technique in the fields of photography and image processing, HDR imaging in architectural heritage is not extended enough according to the existing literature. Ntregka et al. investigate the usage of HDR images in photogrammetric applications in the field of cultural heritage such as calibration and orthoimages (Ntregka et al., 2013). Guidi et al. study the optical pre-processing with HDR imaging that may improve the automated 3D modeling pipeline based on SfM and image matching with special emphasis on optically non-cooperative surfaced of shiny and dark materials (Guidi et al., 2014). Another recent study uses HDR images in the 3D documentation of cultural heritage and investigates how HDR images affect the 3D models, geometrically through point clouds and radiometrically through textured models (Kontogianni and Georgopoulos, 2014).

3. EXPERIMENTAL RESULTS

3.1 Test Datasets

The images used in these tests were taken with a Canon DSL camera EOS 1Ds Mark III. The datasets were produced using exposure bracketing (Figures 1, 2). During the photo shooting the camera was mounted on a tripod. Scene 1 depicts a church altar with highly decorated frescoes and other sculpture details. Scene 2 depicts the columns of an ancient altar, captured from inside. Both scenes are cases where illumination conditions of high contrast occur, i.e. darkness and/or bright sunshine. Sequential images of scene 1 were captured in 5 different exposure layers (1-stop increment, -2EV to +2EV), while for each image of scene 2 we took 7 different ones (1-stop increment, -3EV to +3EV).

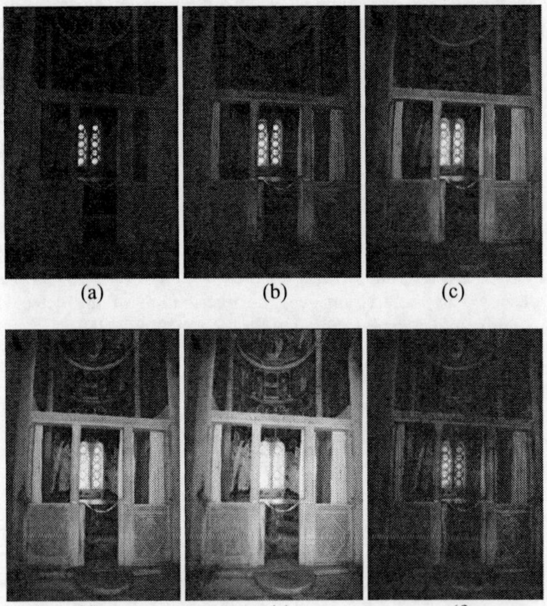

(a) (b) (c)

(d) (e) (f)

Figure 1: Scene1, (a) – (e) 5 frames with exposure values from -2EV to +2EV (f) tone mapped image according to Mantiuk's algorithm (Mantiuk et al., 2006).

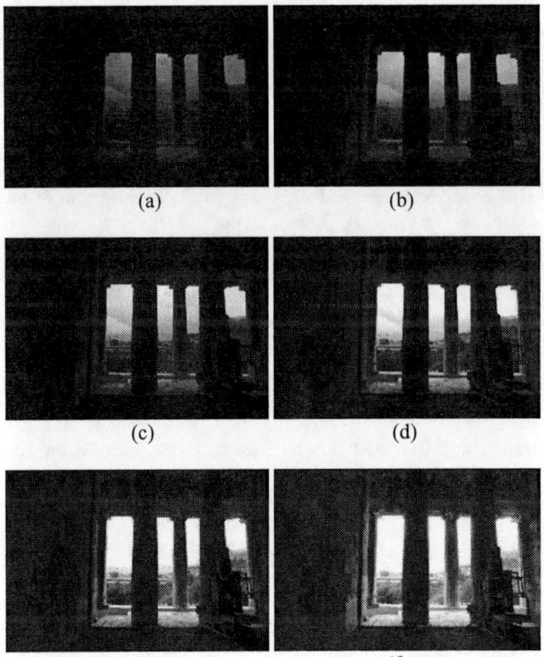

(a) (b)

(c) (d)

(e) (f)

(g) (h)

Figure 2: Scene 2, (a) – (g) 7 frames with exposure values from -3EV to +3EV (h) tone mapped image according to Mantiuk's algorithm (Mantiuk et al., 2006).

3.2 HDR Image Creation

HDR image fusion was implemented according to Debevec and Malik's algorithm (Debecev and Malik, 2008). This approach uses the constraint of sensor reciprocity to linearly recover the camera response function and fuse the multiple images into a high dynamic radiance map. For the tone mapping, Mantiuk's algorithm was used (Mantiuk et al., 2006). They use a gradient domain approach, based on a low pass (Gaussian) pyramid of high contrast values. The main reason of selecting this operator was that it preserves the details needed for feature detection while enhancing the colors of the images.

3.3 Testing Detectors

In our experiments, feature detectors' behaviour is tested on SDR and HDR (their LDR equivalent) images. For this reason, state of the art feature detectors and descriptors are investigated, such as Scale Invariant Feature Transform (SIFT) (Lowe, 2004), Speeded-Up Robust Features (SURF) (Bay et al., 2006), Features from Accelerated Segment Test (FAST) (Rosten and Drummond, 2006) and the relatively new Oriented FAST and Rotated BRIEF (ORB) (Rublee et al., 2011).

Figure 3: Features detected with SIFT (detail) (a) on the SDR image and (b) on the HDR image.

This contribution has been peer-reviewed.
doi:10.5194/isprsarchives-XL-5-W4-325-2015

Figure 4: Features detected with FAST (a) on the SDR image (14073 points) and (b) on the HDR image (68001 points).

(a)

(b)

Figure 5: Features detected with ORB (a) on the SDR image (8144 points) and (b) on the HDR image (20000 points).

(a)

(b)

Figure 6: Features detected with SURF (a) on the SDR image (5047 points) and (b) on the HDR image (16843 points).

Figures 3-6 present some example results of the tested feature detectors on SDR and tone mapped HDR images for both scenes. In terms of objectiveness, the SDR images used in these tests were the ones with 0 EV. As it is shown in the figures, the use of HDR imaging for feature detection certainly increases the number of detected features. In particular, SURF increases its performance by an average of 100% while applied on HDR images. Processing time remains almost stable as it increases slightly by 2%. FAST performs even better, as it detects 350% more features, while in terms of time it is kept on the same level (around half a second). HDR imaging increases the number of detected points by almost 70% using ORB detector while needs around 14% more time, but still less than two seconds. SIFT detects approximately 200% more features on HDR images than on the standard ones, while it needs the same time. Tables 1 and 2 include the average detected feature points for all captured images and the time needed for their detection respectively. As observed in the test images, the distribution of the detected feature points remains the same for both SDR and HDR.

Table 1: Average feature points detected on SDR and HDR images for both scenes.

		SIFT	SURF	FAST	ORB
Scene 1	*SDR*	7062	6292	2730	19106
	HDR	13639	7663	6752	32627
Scene 2	*SDR*	5722	6492	2385	16902
	HDR	19013	15391	9315	23893

Table 2: Average time needed (in sec.) to detect feature points on SDR and HDR images for both scenes.

		SIFT	SURF	FAST	ORB
Scene 1	*SDR*	84.59	27.20	0.31	1.56
	HDR	82.74	27.79	0.32	1.76
Scene 2	*SDR*	75.72	29.34	0.29	1.53
	HDR	79.84	30.09	0.33	1.76

4. CONCLUSIONS

This paper has presented the usage of HDR images in feature detection by testing some state of the art operators. Comparisons are made between SDR and tone mapped HDR images in terms of performance and speed. The results show a marked increase in the number of detected feature points in the same time frame while using HRD imaging. It is believed that a larger number of correctly detected key points will potentially increase the matching performance and consequently the 3D reconstruction of such scenes. Future work will include further investigation towards this direction and evaluate matching results and 3D models while using HDR imaging on such scenes.

ACKNOWLEDGEMENTS

This project has received funding from the European Union's Seventh Framework Programme for research, technological development and demonstration under grant agreement no 608013, titled "ITN-DCH: Initial Training Network for Digital Cultural Heritage: Projecting our Past to the Future." The authors also thank Kostas Makantasis for his helpful advice.

This contribution has been peer-reviewed.
doi:10.5194/isprsarchives-XL-5-W4-325-2015

REFERENCES

Alahi, A., Ortiz, R., & Vandergheynst, P. (2012, June). Freak: Fast retina keypoint. In *Computer Vision and Pattern Recognition (CVPR), 2012 IEEE Conference on* (pp. 510-517). IEEE.

Ashikhmin, M. (2002, July). A tone mapping algorithm for high contrast images. In *Proceedings of the 13th Eurographics workshop on Rendering* (pp. 145-156). Eurographics Association.

Bay, H., Tuytelaars, T., & Van Gool, L. (2006). Surf: Speeded up robust features. In *Computer Vision–ECCV 2006* (pp. 404-417). Springer Berlin Heidelberg.

Calonder, M., Lepetit, V., Strecha, C., & Fua, P. (2010). Brief: Binary robust independent elementary features. In *Computer Vision–ECCV 2010* (pp. 778-792). Springer Berlin Heidelberg.

Chermak, L., & Aouf, N. (2012, August). Enhanced feature detection and matching under extreme illumination conditions with a HDR imaging sensor. In*Cybernetic Intelligent Systems (CIS), 2012 IEEE 11th International Conference on* (pp. 64-69). IEEE.

Chermak, L., Aouf, N., & Richardson, M. (2014). HDR imaging for feature tracking in challenging visibility scenes. *Kybernetes*, *43*(8), 1129-1149.

Cui, Y., Pagani, A., & Stricker, D. (2011). Robust point matching in HDRI through estimation of illumination distribution. In *Pattern Recognition* (pp. 226-235). Springer Berlin Heidelberg.

Debevec, P. E., & Malik, J. (2008, August). Recovering high dynamic range radiance maps from photographs. In *ACM SIGGRAPH 2008 classes* (p. 31). ACM.

Drago, F., Myszkowski, K., Annen, T., & Chiba, N. (2003, September). Adaptive logarithmic mapping for displaying high contrast scenes. In *Computer Graphics Forum* (Vol. 22, No. 3, pp. 419-426). Blackwell Publishing, Inc.

Durand, F., & Dorsey, J. (2002, July). Fast bilateral filtering for the display of high-dynamic-range images. In *ACM Transactions on Graphics (TOG)* (Vol. 21, No. 3, pp. 257-266). ACM.

El Gamal, A., High Dynamic Range Image Sensors, 2002.

Fattal, R., Agrawala, M., & Rusinkiewicz, S. (2007). Multiscale shape and detail enhancement from multi-light image collections. *ACM Trans. Graph.*,*26*(3), 51.

Guidi G., Gonizzi S., Micoli L.L., 2014. Image pre-processing for optimizing automated photogrammetry performances. ISPRS Technical Commission V Symposium, 23-25 June, Riva del Garda, Italy.

Jagadish, K., and Sinzinger, E.D, 2008. ISVC 1, volume 5358 of Lecture Notes in Computer Science, Springer, page 359-369.

Kontogianni, G. and Georgopoulos, A., 2014. Investigating the effect of HDR images for 3D documentation of Cultural Heritage. In: *Proceedings of EuroMed 2014 – International Conference on Cultural Heritage, 3-8 November, Limassol, Cyprus.*

Leutenegger, S., Chli, M., & Siegwart, R. Y. (2011, November). BRISK: Binary robust invariant scalable keypoints. In *Computer Vision (ICCV), 2011 IEEE International Conference on* (pp. 2548-2555). IEEE.

Lowe, D. G. (1999). Object recognition from local scale-invariant features. In*Computer vision, 1999. The proceedings of the seventh IEEE international conference on* (Vol. 2, pp. 1150-1157). IEEE.

Lowe, D. G. (2004). Distinctive image features from scale-invariant keypoints.*International journal of computer vision*, *60*(2), 91-110.

Mantiuk, R., Myszkowski, K., & Seidel, H. P. (2006). A perceptual framework for contrast processing of high dynamic range images. *ACM Transactions on Applied Perception (TAP)*, *3*(3), 286-308.

Mertens, T., Kautz, J., & Van Reeth, F. (2007, October). Exposure fusion. In*Computer Graphics and Applications, 2007. PG'07. 15th Pacific Conference on*(pp. 382-390). IEEE.

Mikolajczyk, K., & Schmid, C. (2005). A performance evaluation of local descriptors. *Pattern Analysis and Machine Intelligence, IEEE Transactions on*,*27*(10), 1615-1630.

Mitsunaga, T., & Nayar, S. K. (1999). Radiometric self calibration. In *Computer Vision and Pattern Recognition, 1999. IEEE Computer Society Conference on.*(Vol. 1). IEEE.

Ntregka, A., Georgopoulosa, A., & Quinterob, M. S. (2013). Photogrammetric Exploitation of HDR Images for Cultural Heritage Documentation. *ISPRS Annals of the Photogrammetry, Remote Sensing and Spatial Information Sciences*, *5*, W1.

Přibyl, B., Chalmers, A., & Zemčík, P. (2013, March). Feature point detection under extreme lighting conditions. In *Proceedings of the 28th Spring Conference on Computer Graphics* (pp. 143-150). ACM.

Reinhard, E., Heidrich, W., Debevec, P., Pattanaik, S., Ward, G., & Myszkowski, K. (2010). *High dynamic range imaging: acquisition, display, and image-based lighting*. Morgan Kaufmann.

Reinhard, E., Stark, M., Shirley, P., & Ferwerda, J. (2002, July). Photographic tone reproduction for digital images. In *ACM Transactions on Graphics (TOG)* (Vol. 21, No. 3, pp. 267-276). ACM.

Robertson, M. A., Borman, S., & Stevenson, R. L. (1999). Dynamic range improvement through multiple exposures. In *Image Processing, 1999. ICIP 99. Proceedings. 1999 International Conference on* (Vol. 3, pp. 159-163). IEEE.

Rosten, E., & Drummond, T. (2006). Machine learning for high-speed corner detection. In *Computer Vision–ECCV 2006* (pp. 430-443). Springer Berlin Heidelberg.

Rovid, A. S., Varkonyi-Koczy, A. R., Hashimoto, T., Balogh, S., & Shimodaira, Y. (2007, May). Gradient based synthesized multiple exposure time HDR image. In *Instrumentation and*

Measurement Technology Conference Proceedings, 2007. IMTC 2007. IEEE (pp. 1-6). IEEE.

Rublee, E., Rabaud, V., Konolige, K., & Bradski, G. (2011, November). ORB: an efficient alternative to SIFT or SURF. In *Computer Vision (ICCV), 2011 IEEE International Conference on* (pp. 2564-2571). IEEE.

Schmid, C., Mohr, R., & Bauckhage, C. (2000). Evaluation of interest point detectors. *International Journal of computer vision*, *37*(2), 151-172.

Tola, E., Lepetit, V., & Fua, P. (2010). Daisy: An efficient dense descriptor applied to wide-baseline stereo. *Pattern Analysis and Machine Intelligence, IEEE Transactions on*, *32*(5), 815-830.

This contribution has been peer-reviewed.
doi:10.5194/isprsarchives-XL-5-W4-325-2015

NON-CONTACT PHOTOGRAMMETRIC METHODOLOGY TO EVALUATE THE STRUCTURAL HEALTH OF HISTORICAL CONSTRUCTIONS

L. J. Sánchez-Aparicio [a, *], A. Villarino [a], J. García-Gago [b], Diego González-Aguilera [a]

[a] Department of Cartographic and Land Engineering; High School of Ávila, University of Salamanca, Hornos Caleros, 50, 05003, Ávila, Spain – (luisj, avillarino, daguilera)@usal.es

[b] Department of Construction and Agronomy, Architectural Area, Polytechnic School of Zamora, University of Salamanca, Avda. Cardenal Cisneros, 34, Zamora 49022, Spain – jesusmgg@usal.es

Commission V, WG V/4

KEY WORDS: Photogrammetry, image-based modelling, digital image correlation, computer-vision, cultural heritage, vernacular heritage, masonry structures, finite element modelling.

ABSTRACT:

Accurate studies of cultural heritage structures usually require the application of combined techniques to understand its structural behaviour. It is in this context, where the article present a set of procedures based on the dual combination of photogrammetric methodologies (image-based modelling and digital image correlation) and the finite element method to understand the structural behaviour of these structures. Through the interpretation of the different obtained products, by the defined approach, it is possible to estimate and evaluate the causes of the structural damage that the analysed infrastructure can suffer and also design subsequent restoration mechanisms, always from a perspective of minimal intrusion to the structure.

This case of study aims to validate the methodology presented, using as an example a case of vernacular architecture (a dome) in Sejas de Aliste (Zamora, Spain).

1. INTRODUCTION

The preservation of architectural heritage is considered a fundamental issue in the cultural life of modern societies. The digitalization and modelling of buildings relevant to the cultural heritage through non-destructive techniques is the first step to form the basis for any kind of decision regarding intervention on cultural heritage. However, their structural analysis is sometimes an aspect that is not considered or only studied separately, being essential for their safeguarding and maintenance.

The conservation of these infrastructures requires understanding their structural behavior and consequently: (i) their boundary conditions,(ii) the mechanical properties of the materials,(iii) the origin of the damage that building suffers and (iv) their vulnerability (Sánchez-Aparicio et al., 2014).

In this process of preserving of cultural heritage structures, procedures based on non-destructive methods are particularly attractive, since they suppose minimum intervention and preservation of the original constructions. It is in this field where geomatic sensors such as terrestrial laser scanner (Armesto-González et al., 2010; Villarino et al., 2014a) and/or digital cameras (Riveiro et al., 2013; Villarino et al., 2014b) have acquired important roles, due to their capacity to acquire accurate geometry without any contact with the structure. However, the terrestrial laser scanner entails some drawbacks such as the difficulty for transport and the lack of flexibility during data acquisition. Wherein this situation photogrammetry can offer several solutions for these drawbacks. Also it is remarkable, in the case of image-based modelling, that this data acquisition system is complemented by the digital image and computer vision approaches, allowing to progress from 2D (images) to 3D (points cloud) in a wide range of applications within the field of cultural heritage structures preservation.

Thanks to the advances in the integration of photogrammetry and computer vision, as well as the numerical methods of finite elements, it is possible to understand the actual behavior of historical constructions with high accuracy. Taking all of this into account, this paper attempts to demonstrate the strong potentialities of image-based modelling, digital image correlation and finite element methods to evaluate the structural behavior and the origin of the damage in these structures.

This paper is organized as follows: Section 1 motivates the importance and complexity of assessing structural health of cultural heritage buildings together with a brief state-of art; Section 2 presents two photogrammetric approaches: image-based modelling and digital image correlation to obtain accurate data for structural analysis; Section 3 is devoted to the pre-processing methodologies to obtain accurate CAD models suitable for finite element simulations, the numerical simulation and also the post-processing strategies to evaluate the robustness of the results; and finally in Section 4, the conclusions are drawn.

2. PHOTOGRAMMETRIC APPROACHES: IMAGE-BASED MODELLING AND DIGITAL IMAGE CORRELATION

2.1 Image-based modelling

Nowadays, the tridimensional reconstruction of objects is generally performed using two types of sensors: in one hand,

* Corresponding author. *E-mail address:* luisj@usal.es (L. J. Sánchez-Aparicio).

This contribution has been peer-reviewed.
doi:10.5194/isprsarchives-XL-5-W4-331-2015

range sensors such as laser scanner or structure light systems; on the other hand, the image-based modelling, namely Structure from Motion (SfM). This approach, the SfM, combines the accuracy and reliability of photogrammetric methods and the automation and flexibility advantages of computer vision (Sánchez-Aparicio et al., 2014).

In the present case of study, an in-house software PW (Photogrammetry Workbench) was developed, which encloses photogrammetric and computer vision algorithms, in a SfM environment.This software applies a multi-stage methodology composed for the following steps (García-Gago et al., 2014): (i) automatic extraction and matching keypoints; (ii) automatic hierarchical orientation of images and (iii) dense surface generation.

2.1.1 Automatic extraction and matching keypoints

The field of architectural documentation, in which often, complex scenarios need to be modeled, classic photogrammetric methods of correspondence as ABM (Area-Based Matching) or LSM (Least Squares Matching), are insufficient, due to the variations in scale, perspective and lightning (García-Gago et al., 2014). It is imperative therefore the use of more sophisticated strategies.

It is in this field where algorithms such as: (i) SUSAN (Smallest Univalue Segment Assimilating Nucleus); (ii) SIFT (Scale-Invariant Feature Transform); (iii) MSER (efficient Maximally Stable Extremal Region) or (iv) SURF (Speeded Up Robust Features), have acquired important roles. However, these new strategies are not unaffected by the perspective variations of different images. In this respect, a variation of SIFT, the ASIFT (Affine Scale-Invariant Transform), has been used and added to the PW tool.

Finally, the matching process is carried out by the employment of the SIFT descriptors, according to the Euclidean distance and secondly filtered by the ORSA (Optimized Random Sampling Algorithm) (Moisa et al., 2012) approach.

2.1.2 Automatic hierarchical orientation

After the extraction and matching of the aforementioned features, the orientation of the images is carried out by a multistage approach. Firstly, a relative orientation of the images is calculated, through the fundamental matrix (Longuet-Higgins, 1987). Secondly, a bundle adjustment of all images is performed by an iterative and least-square process based on the collinearity condition (Tait, 1994).

In cases in which the internal calibration parameters of the camera are unknown, the last step allows the calculation of these parameters, a self-calibration, through the incorporation into the equation as unknowns (Quan, 2010).

2.1.3 Dense surface generation

Given accurate exterior orientation parameters and a sparse point cloud, generated from the keypoints extracted, a robust and reliable automatic dense image matching process is needed.

Nowadays there is a large number of matching algorithm developed by the scientific community, capable of determining the exact 3D coordinate of the object for each pixel. In this field, algorithms like PMVS (Path-based Multi-View Stereo), SGM (Semi-Global Matching), MicMac (Multi-Images Coreespondances, Méthodes Automatiques de Corrélation) or SURE (Surface Reconstruction from imagery) have acquired important roles, and the use of one or other algorithm depends, mainly, on the shooting configuration.

For the present case of study a convergent protocol in the image acquisition has been followed and a MicMac strategy was used. As a result a dense point cloud model was obtained (Figure 1).

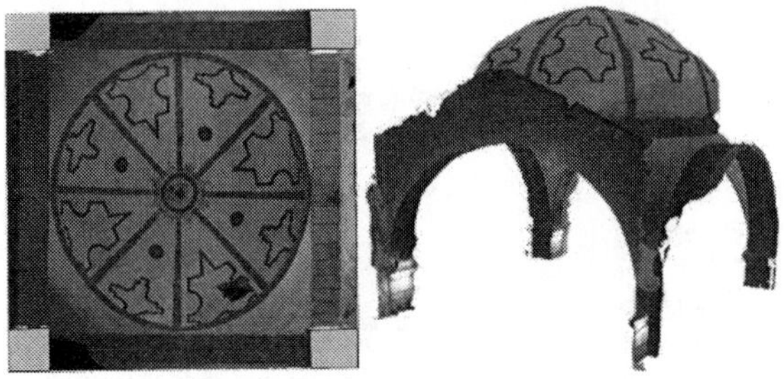

Figure 1. Dense point cloud obtained from the image-based modelling strategy.

2.2 2D Digital image correlation

Although one of the basic requirements to understand the structural behavior of the construction implies the need of an accurate geometric model, this is not the only one. Complementary to a suitable material model is needed in order to represent the mechanical behavior of the materials.

Nowadays, several approaches based on interferometric and white-light methods, have been used successfully in experimental solid mechanics, to evaluate the material behavior at different stress values. These methodologies compared with the traditional measurement systems, like strain gauges or LVDTs (Linear Variable Differential Transformer), provide some important advantages as a full-field data and contact-free. Among these techniques, non-interferometric methods based on image processing, such as digital image correlation (DIC), have been increasingly used (Ghorbani et al., 2014; Pan et al., 2009b).

2.2.1 Pre-processing stage

In contrast to the image-based modelling strategy, DIC requires the preparation of the analyzed specimen. This preparation involves several steps: (i) a suitable test surface pattern; (ii) geometrical camera calibration; and (iii) pose estimation.

The basic principle of DIC is the tracking (or matching) of the same image subsets located in two digital images of the test specimen surface recorded before and after deformation (Figure 2). As an initial approximation a correlation coefficient is used, later, a non-linear strategy is considered, in order to analysis the deformation suffered by the subset in the reference and deformed image. Considering multiple subsets in the images, the analysis of them, can provided a full-field displacement whose resolution depends of the subset size.

This contribution has been peer-reviewed.
doi:10.5194/isprsarchives-XL-5-W4-331-2015

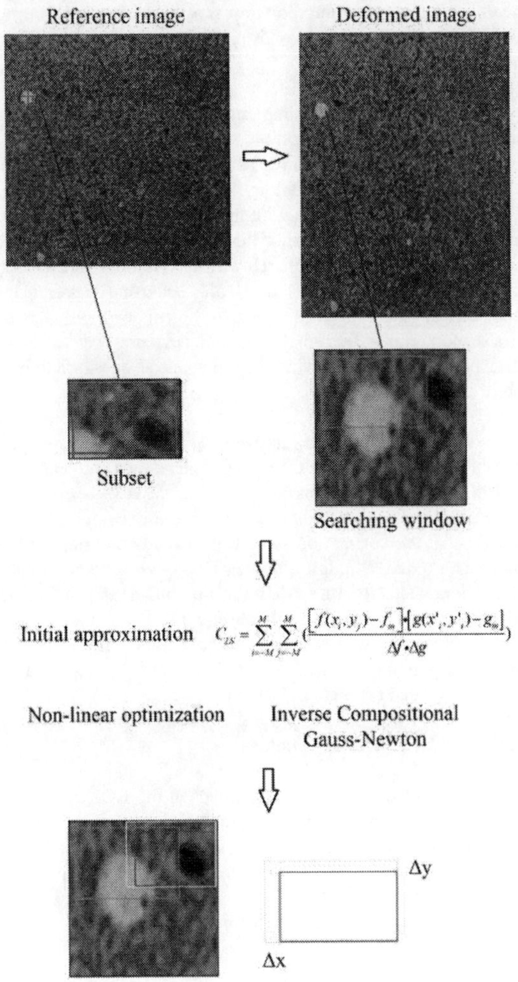

Figure 2. Digital image correlation general outline. In red the reference subset, in blue the initial subset on the deformed image and in yellow the final subset obtained.

It should be noted, that for this purpose, for the case where the natural texture is insufficient, the specimen surface must be covered with a random speckle pattern. This type of patterns provides a surface with random grey levels, whose geometric distribution (size) plays a critical role in the achieved accuracy. In order to minimize this uncertainty, different paint strategies have been tested and evaluated. For these evaluations a global parameter, called MIG (Mean Intensity Gradient) (Pan et al., 2010a) (Equations 1-2), was used.

$$MIG = \sum_{i=1}^{W} \sum_{j=1}^{H} \left| \nabla f(x_{ij}) \right| / (W \times H) \quad (1)$$

$$\nabla f(x_{ij}) = \sqrt{f_x(x_{ij})^2 + f_y(x_{ij})^2} \quad (2)$$

Where: $f_x(x_{ij})$, $f_y(x_{ij})$ are the gradient along x and y direction respectively
W,H are image width and height in pixels

It is remarkable that both mean bias error and standard deviation error of the measured displacements are closely related with the MIG coefficient, as show (Pan et al., 2010a). Large MIG values involves large grey variations and therefore a wide number of singular features. Hence, a good speckle patter will have a large MIG value (Table 1) (Figure 3).

Designation	Print strategy	MIG
Spec_A	Paint-brush (black)	6.7
Spec_B	Paint-brush (black+white)	4.9
Spec_C	Spray (black)	15.8
Spec_D	Spray (black+white)	18.2

Table 1. Comparison, in term of MIG, between the different speckle patterns tested.

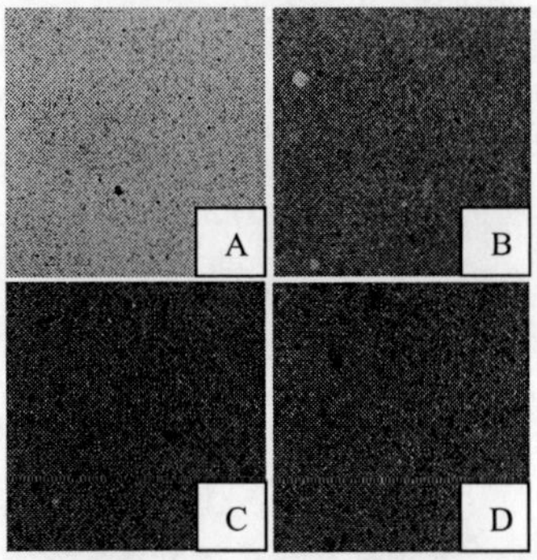

Figure 3. Speckle patterns tested for the DIC test.

Since the captured images are the 2D projection of the specimen surface, the accuracy in the estimation of the displacements is conditioned not only by a proper preparation of the specimen, but also by an accurate geometrical distortion correction and perpendicularity between the sensor and the specimen.

Camera calibration is technically not a part of the DIC technique, but in practice and given the image-based DIC nature, some errors can be introduced due to the geometrical distortions of the image captured. For this reason, a geometrical distortion correction approach, proposed by (Jean-Yves, 2010), is employed.

As discussed above, one of the basic requirements for the successful use of the 2D DIC method involves the use of a flat specimen surface and a parallelism between the specimen and the camera sensor, in order to keep the pixel size as constant as possible. Theoretical analysis done by (Meng et al., 2006) indicates that non-parallel angles less than 5° introduces a displacement error smaller than 0.01 pixel. In practice, this uncertainty is unavoidable due to factors such as (Pan et al., 2009a): (i) deviation of the specimen surface from ideal plane, (ii) non-perfect contact between the loading device and the test specimen; and (iii) the Poisson´s ratio of the material tested. In order to minimize this, whenever the use non-telecentric (i.e standard) lens was used, the camera sensor was placed far from the specimen to approximate this to a telecentric imaging system. Also a robust camera pose estimation approach is considered (Schweighofer and Pinz, 2006), in order to identify and correct the pose of the sensor in relation to the specimen. Since this approach need, at least, four coplanar but non-collinear points to determine the camera pose, a chessboard

This contribution has been peer-reviewed.
doi:10.5194/isprsarchives-XL-5-W4-331-2015

panel was used. Then, the corners are extracted, using for this the enhanced version of the traditional Harris corner detector proposed by (Jean-Yves, 2010).

2.2.2 Processing stage

The correlation process (processing stage) starts with the ROI (Region Of Interest) discretization into clusters called subsets. Once the digital image has been split into subsets, a correlation criterion is used, in order to track the different image subsets between the initial state (reference image) and the currently analyzed image (deformed state). Nowadays, several correlation criterions has been designed and used in literature, among which have acquired special attention, given their robustness (Pan et al., 2009a): (i) Zero-mean Normalized Cross-Correlation (ZNCC); (ii) Zero-mean Normalized Sum of Squared Difference (ZNSSD) and (iii) Parametric Sum of Squared Difference (PSSD). For the present purpose the second one, ZNSSD, is used (Equations 3-7):

$$C_{LS} = \sum_{i=-M}^{M} \sum_{j=-M}^{M} \left(\frac{\left[f(x_i, y_j) - f_m\right] \cdot \left[g(x'_i, y'_i) - g_m\right]}{\Delta f \cdot \Delta g} \right) \quad (3)$$

$$f_m = \frac{1}{(2M+1)^2} \sum_{i=-M}^{M} \sum_{j=-M}^{M} f(x_i, y_i) \quad (4)$$

$$g_m = \frac{1}{(2M+1)^2} \sum_{i=-M}^{M} \sum_{j=-M}^{M} g(x'_i, y'_i) \quad (5)$$

$$\Delta f = \sqrt{\sum_{i=-M}^{M} \sum_{j=-M}^{M} [f(x_i, y_i) - f_m]^2} \quad (6)$$

$$\Delta g = \sqrt{\sum_{i=-M}^{M} \sum_{j=-M}^{M} [g(x'_i, y'_i) - g_m]^2} \quad (7)$$

Where: i,j are the pixel (or subpixel) numbers
x,y are the coordinates of the pixel in the reference image
x',y' are the coordinates of the pixel in the deformed image
f,g are the reference and deformed image functions
M is the number of pixels belonging to the subset
$\Delta g, \Delta f$ are the grey value difference between the point analysed and the subset mean value

Once the different subsets are localized, a non-linear optimization algorithm is required. In contrast to the classical non-linear optimization, based on the forward additive Gauss-Newton method, the Inverse Compositional Gauss-Newton (Pan, 2014) was used. This approach gives equivalent results, in terms of accuracy, but with lower computational costs.

Considering the DIC as an image-based approach, the accuracy of this methodology partially depends of the pixel size. Also, if subpixel interpolation schemes are not employed, a displacement accuracy of up to one pixel can be achieved (Gencturk et al., 2014). In order to improve this accuracy, various interpolation schemes have been proposed (Luu et al., 2011): (i) bicubic; (ii)bicubic spline; (iii) bicubic B-spline; (iv) biquintic B-spline. For the present case a biquintic B-spline had been used, since this interpolation scheme mitigates DIC errors form interpolating signals with high frequency content (Luu et al., 2011).

As described before, the implementation of the DIC analysis includes the use of a ROI on the specimen analyzed. The pixels inside this ROI are considered as points to be computed by the abovementioned methodology. Conventional correlation calculation generally starts with the upper-left point of the ROI, after that the calculation analysis is carried out point by point along each row or column. However, this approach presents important drawbacks as (Pan et al., 2009b): (i) presence of geometrical discontinuities (cracks or holes) or irregular shapes; (ii) discontinuous deformations; and (iii) presence of wrong data points that can provide a wrong initial guess of the next point evaluated.

Considering the stated above, a universally applicable reliability-guided (RG-DIC) method, developed by (Pan et al., 2010b), was used. In this approach, the ZNCC coefficient is used to identify the reliability of the point computed. This means, that the correlation stage begins with an initial seed and then, is guided through the neighbors with highest ZNCC coefficient. This implies, that for the full-field displacement (displacement suffered for all subset evaluated), the calculation path was made along the most reliable direction and possible error propagation of the conventional DIC method can be entirely avoided (Pan et al., 2010b).

2.2.3 Post-processing stage

This full field displacement, obtained on the ROI, is the basis for the subsequent mechanical properties evaluation. For this mechanical evaluation an extensometer is needed. This extensometer gives the displacement values between two subset and consequently the strain values in both axis (x and y). The relation with the stress applied affords the stress-strain curves.

2.2.4 Experimental setup

The main goal of the DIC experimental campaign presented is to obtain the mechanical properties (through the stress-strain curves) of the different material that compose the dome. From these curves the mechanical properties of the material can be extracted (Table 2), namely: (i) Elastic Modulus's; (ii) maximum Strength; and (iii) in plane Poisson ratio are obtained.

Results of brick/mortar characterization	
$E_{m,brick}$ (GPa)	3.10
$f_{c,brick}$ (MPa)	7.80
υ_{brick}	0.17
$E_{m,mortar}$ (GPa)	11.53
$f_{c,mortar}$ (MPa)	10.21
υ_{mortar}	0.15

Table 2. Brick and mortar mechanical properties obtained through DIC.

For the case of study evaluated, and due to the lack of materials to be tested, only compression tests were performed on brick, extracted from a part of the dome and mortar specimens, manufactured with similar mechanical properties, as those expected in the dome, considering the guidelines exposed in the regulation UNE-EN 1052-11 (AENOR, 2007) and UNE-EN 772-1 (AENOR, 2011). The image capture system has been

This contribution has been peer-reviewed.
doi:10.5194/isprsarchives-XL-5-W4-331-2015

composed by a Canon EOS 500D digital camera equipped with a 70-300mm macro lens (fixed in 200mm) and supported by a tripod (Figure 4) (Figure 5) (Table 3).

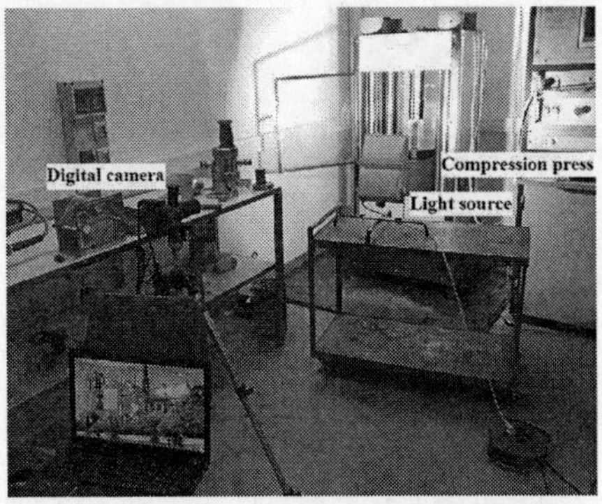

Figure 4. Experimental setup during the DIC test.

Figure 5. Zoomed view of the mortar test specimen.

Canon EOS 500D	
Aperture	7.1
Focal(mm)	200
Working distance(mm)	2700
Pixel size(mm)	0.063
Acquisition frequency(Hz)	0.33

Table 3. Camera configuration during DIC test.

During the processing stage in the DIC workflow a subset size of 40 pixels was considered and a value of 30 pixels for the subset-step (distance between subsets). Once the deformations have been evaluated through DIC, an extensometer is placed between two points in order to obtain the stress-strain curve (Figure 6).

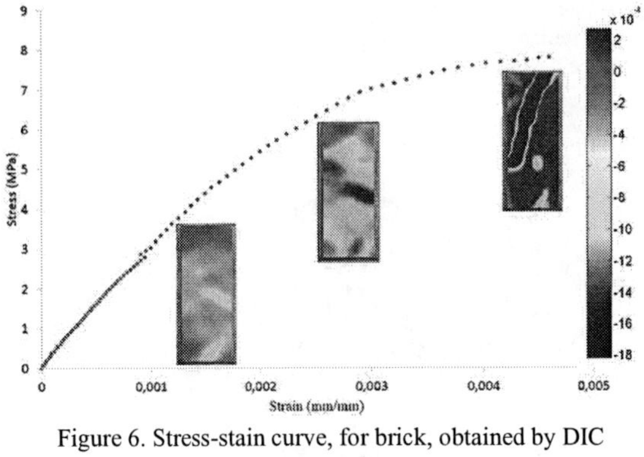

Figure 6. Stress-stain curve, for brick, obtained by DIC approach.

3. CAD AND NUMERICAL MODELLING STRATEGIES FOR STRUCTURAL EVALUATION

3.1 Point cloud pre-processing

The image-based modelling strategy shows great potentialities, in terms of accuracy and dense reconstruction, thanks to the robustness of the presented methodologies. However, the generated point cloud may present some noise. In order to improve the quality of this product, a noise reduction based on a twofold approach is presented: (i) normal estimation proposed by (Grimm and Smart, 2011); and (ii) adaptive Moving Least Square proposed by (Dey and Sun, 2005). Finally a mesh reconstruction is used. As a result a suitable mesh for CAD modelling is generated.

3.2 CAD modelling

For a suitable structural evaluation and thus a comprehensive knowledge about the damage that can suffer the structure, it is imperative to build an accurate CAD model. Thanks to the presence of superabundant and accurate information in the point cloud provided by the image-based modelling this requirement is possible.

Traditionally this step could be made through three different approaches (Sánchez-Aparicio et al., 2014): (i) orthogonal views; (ii) scctions applied along directions and over the mesh; and (iii) Non-Uniform Rational B-Splines (NURBS) generated from the mesh.

For the obtained point cloud, all the strategies defined above, can be applied. Due to the geometrical properties of the structure, which has large deformations, and due to the need of generation a CAD model of the initial state, the second approach out of those mentioned above has been chosen, and complemented with additional modelling strategies defined by (Wang et al., 2012).

3.3 Numerical simulations

Nowadays the numerical simulations through finite element method (FEM) have shown high efficiency in the simulation of historical constructions (Villarino et al., 2014a). This numerical strategy requires different type of input data. Some of this input data can be provided by the presented methodology such as: (i) CAD model (image-based modelling); and (ii) the mechanical properties of the material (digital image correlation).

Through the image-based modelling approach defined above, the deformed shape obtained in combination with the use of orthoimages obtained through the orthographic projection of the dense point cloud can be used to evaluate the agents that caused the structural damage.

For the present case of study, the evaluation of both products (deformed shape and pathological inspection through the orthophoto), clearly show a possible cause of the structural damage (Figure 7) (Figure 8).

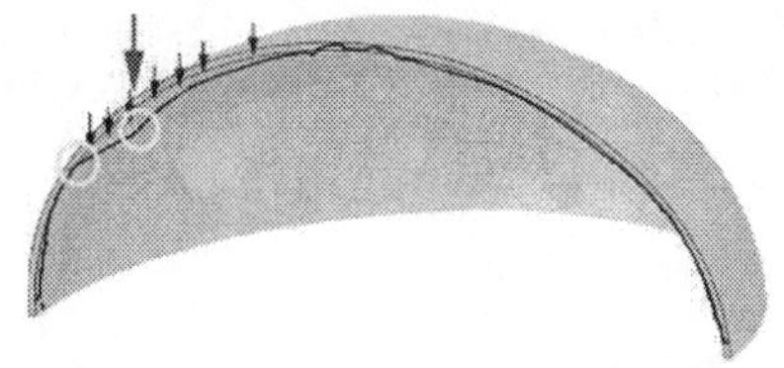

Figure 7. Shape comparison between the initial (in green) and the deformed stage (in blue). In red the load acting on the structure, in black the structural movement and in yellow the hinges observed.

Figure 8. An orthoimage (plan view) of the dome: result of the inspection in search for damage (red: deformed zone, blue: cracks and green: material removal).

The deformation of Aliste´s dome can be attributed to the presence of an asymmetric load. This eccentric load, attributed to the weight of the rooftop, only affects an area equivalent to one of the roof´s divisions. Further research sheds light on the causes, supported by the foregoing observation: a leak on the roof, which causes the deterioration of the timber structure and its deformation until part of its load rests on the dome.

Complementary to what is exposed above, a material model is needed to adequately represent the masonry behavior as well as a modeling strategy. For the material model of the masonry structure, a non-linear constitutive model was considered, based on Total Strain Crack Model (TSCM) (Diana, 2011). This constitutive model shows a good compromise between stability, in the crack opening, and a moderate computational cost. Considering the mechanical properties of the different materials obtained through DIC (Section 2.2.4), homogenization formulas (Tensing, 2013) are used, in order to obtain the masonry mechanical properties of the dome (Table 4).

Name	Value
$E_{masonry}$ (GPa)	3.56
$\delta_{masonry}$ (kg/m^3)	1600
$\upsilon_{masonry}$	0.2
$f_{c,masonry}$ (MPa)	2.38
$f_{t,masonry}$ (MPa)	0.2
$G_{f,t,masonry}$ (N/mm)	0.02

Table 4. Mechanical properties for the masonry.

In the case of the timber structure´s mechanical properties, the content of Table 5 are considered as show (de Oliveira Feio, 2005) into an orthotropic material model.

Name	Value
$E_{longitudinal}$ (GPa)	8.5
$E_{transversal}$$(GPa)$	0.8
δ_{timber} (kg/m^3)	592.12
υ_{timber}	0.3
$f_{c,longitudinal}$$(MPa)$	45
$f_{c,transversal}$ (MPa)	7
$ft_{,longitudinal}$ (MPa)	41.4
$f_{t,transversal}$ (MPa)	1.4

Table 5. Mechanical properties for the timber structure.

Concerning the modelling strategies, several approaches can be use (Lourenço, 2009): (i) detailed micro-modelling; (ii) simplified micro-modelling; and (iii) macro-modelling. For the present case study, a macro-modelling approach is used, in which the tile units and mortars joints are smeared together as one continuum. As a result of the exposed above, a 3D FEM has been performed, considering both structures, the masonry, that constitute the dome and the timber roof structure. The discretization of the model has been carried out considering a 10-node isoparametric tetrahedral element, contact-target elements and pinned supports in their bases. All these result in a total of 58,682 discrete elements (26,903 for the dome and 31,779 for the timber structure) calculated in the numerical model (Figure 9).

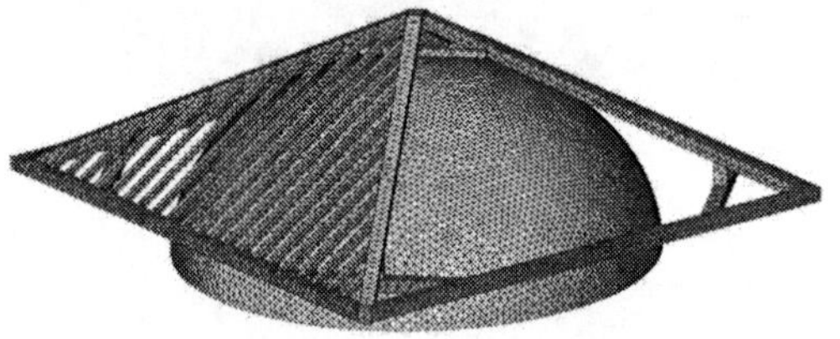

Figure 9. FEM adopted mesh for the dome and timber structure.

One the model has been performed; several hypotheses have been taken into account: (i) the structure´s own-weight; (ii) snow load acting on the roof (considering this as the worst-case estimate); and (iii) a damage model due to the presence of an eccentric load (whose origin is attributed to a timber structure degradation process).

As conclusion derived from the numerical simulations, in a normal case (own-weight/snow-case), the structure seems to be stable and safe. In any case, the maximum tensile strength, with a value of $0.2N/mm^2$, is exceeded. Also the maximum displacements predicted were of 3mm which is insufficient to contact the dome (Figure 10).

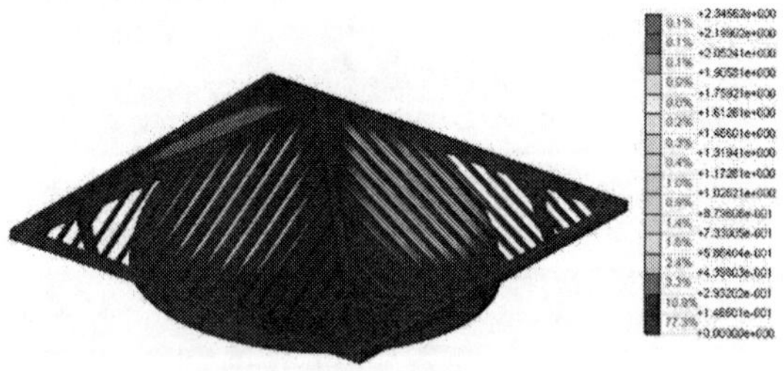

Figure 10. Maximum displacements (snow case), expressed in mm, of the timber structure. The most deformed area is displayed in green-yellow colours.

This contribution has been peer-reviewed.
doi:10.5194/isprsarchives-XL-5-W4-331-2015

As mentioned previously, the origin of the damage that the dome structure suffers is caused by a part of the timber structure that transfer its load onto. For this reason a third model was evaluated, taking into account an eccentric load acting on the dome (which simulate the timber structure degradation process) (Figure 11).

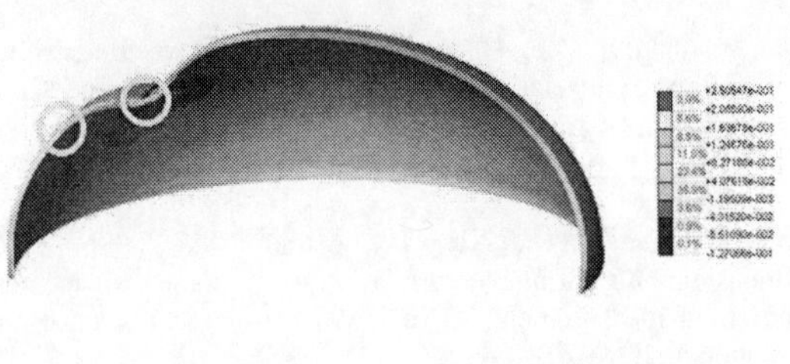

Figure 11. Results of the numerical simulation, in terms of damage (in red colour), considering the degradation process of the timber structure and in yellow the hinges predicted.

3.4 Model-distance comparison approach: a complementary tool to validate the numerical model

At this point three models have been obtained, namely: (i) numerical model (initial and deformed state); and (ii) image-based model (considered as "ground truth"). The distance between these 3D models, provides an overview of the numerical simulation's robustness. This robustness evaluation can be used also to assess additional aspects of the initial model considerations and also evaluate different hypothesis of the damage (Figure 12).

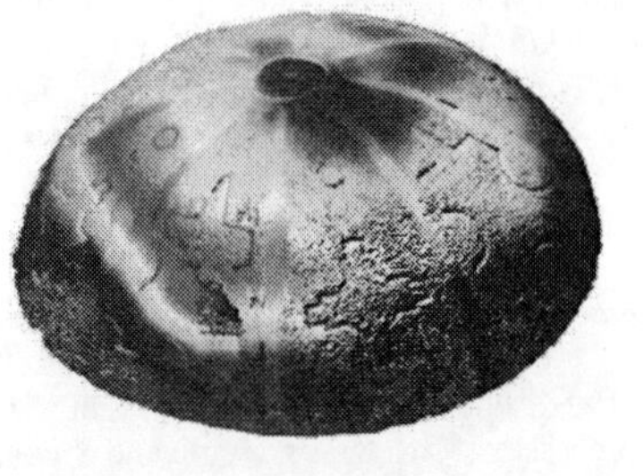

Figure 12. Comparison between the numerical model (initial state) and the image-based model. In red the maximum discrepancies and in green the minimum.

For this purpose, a model comparison has been evaluated through the strategy builds on the Hausdorff distance (Equation 8).

$$d_H = \max\left\{\sup_{x\in X}\inf_{y\in Y} d(x,y), \sup_{y\in Y}\inf_{x\in X} d(x,y)\right\} \quad (8)$$

where d_H distance between models
sup sub-set supremum
inf sub-set infimum
X components (nodes) of the numerical model
Y components (points) of the image-based model

With the considerations exposed in the present paper, the damage has been properly located and the load hypothesis seems to be correct and also, in global terms, the movement of the dome. However the dome displacements, are inaccurate (high rigidity) with discrepancies of several centimetres with respect to the image-based model. Such error can be associated with the presence of elastic supports. Further investigations and sensibility analysis are required in order to improve this part and obtain a better simulation of the dome.

4. CONCLUSIONS

The proposed method shows strong potentialities in the field of built cultural heritage preservation. Through this approach it is possible to solve important drawbacks when the preserving cultural heritage structures. Thanks to the SfM approach it is possible to have a complete tridimensional model of the deformed state, which can be used to estimate the initial state and the global deformation of the structure. In contrast with the traditional damage visual inspections, the use of orthoimages or the dense point cloud can provide the tools needed to evaluate the damages that the structure suffers. Complementary to the exposed above, the use of the DIC technique can provide full-field information of the specimen tested, in contrast to traditional tools like LVDTs or strain gauges. Said information, not only, can be used to estimate the engineering curves (stress-strain) but can be used to understand the behaviour of different construction solutions which are characteristic in the cultural heritage field.

Among the different advantages it should be remarked the following: (i) non-invasive technology; (ii) flexibility; (iii) low associated cost; and (iv) great efficiency, affording the radiometric, geometric and structural characterization of the analysed object with high accuracy and detail even in complex architectural scenarios.

The dual combination of photogrammetric and finite element approaches is a reliable foundation from which to analyse appropriate restoration actions, following the procedure show above. Thanks to the versatility of the methodologies presented and their complementarity, it is possible to estimate the origin of the damage. On one hand, through the analysis of the different photogrammetric products it is possible to establish different hypothesis (which in many cases are unknown) of the initial state or loads acting on the structure and its verification. On the other hand, a comparison between the numerical model and the image-based one, can provided relevant information about its degradation process.

For the case of study that concerns this article (the Aliste's dome), the causes of the damage seem to be clear. Nevertheless, further investigations will be required to approximate the obtained numerical data to the real one. Part of these investigations will be focused on a better understanding of the structural behavior of the dome with elastic supports and also several sensitivity analysis, considering as the variables to analysis the mechanical properties of the material and also elastic supports. The possibility to consider the comparison strategy (once the numerical model can predict better the deformations), defined in Section 3.4 into a deterministic framework for FEM model calibration will also be evaluated.

REFERENCES

AENOR, 2007. Methods of test for masonry units. , Part 11: Determination of flexural and compressive strength of hardened mortar. AENOR, Madrid.

This contribution has been peer-reviewed.
doi:10.5194/isprsarchives-XL-5-W4-331-2015

AENOR, 2011. Methods of test for masonry units., Part 1: Determination of compressive strength. AENOR, Madrid.

Armesto-González, J., Riveiro-Rodríguez, B., González-Aguilera, D., Rivas-Brea, M.T., 2010. Terrestrial laser scanning intensity data applied to damage detection for historical buildings. Journal of Archaeological Science 37, 3037-3047.

de Oliveira Feio, A.J., 2005. Inspection and Diagnosis of Historical Timber Structures: NDT Correlations and Structural Behaviour, Department of Civil Engineering. University of Minho, Guimarães, Portugal.

Dey, T.K., Sun, J., 2005. An Adaptive MLS Surface for Reconstruction with Guarantess. Symposium on Geomtry processing, 43-52.

Diana, T., 2011. Finite Element Analysis User's Manual-Release 9.4. 4. TNO DIANA.

García-Gago, J., González-Aguilera, D., Gómez-Lahoz, J., San José-Alonso, J.I., 2014. A Photogrammetric and Computer Vision-Based Approach for Automated 3D Architectural Modeling and Its Typological Analysis. Remote Sensing 6, 5671-5691.

Gencturk, B., Hossain, K., Kapadia, A., Labib, E., Mo, Y.-L., 2014. Use of digital image correlation technique in full-scale testing of prestressed concrete structures. Measurement 47, 505-515.

Ghorbani, R., Matta, F., Sutton, M., 2014. Full-Field Displacement Measurement and Crack Mapping on Masonry Walls Using Digital Image Correlation, in: Jin, H., Sciammarella, C., Yoshida, S., Lamberti, L. (Eds.), Advancement of Optical Methods in Experimental Mechanics, Volume 3. Springer International Publishing, pp. 187-196.

Grimm, C., Smart, W.D., 2011. Shape classification and normal estimation for non-uniformly sampled, noisy point data. Computers & Graphics 35, 904-915.

Jean-Yves, B., 2010. Camera calibration toolbox for Matlab.

Longuet-Higgins, H.C., 1987. A computer algorithm for reconstructing a scene from two projections. Readings in Computer Vision: Issues, Problems, Principles, and Paradigms, MA Fischler and O. Firschein, eds, 61-62.

Lourenço, P.B., 2009. Recent advances in masonry modelling: micromodelling and homogenisation. Multiscale Modeling in Solid Mechanics: Computational Approaches 3, 251.

Luu, L., Wang, Z., Vo, M., Hoang, T., Ma, J., 2011. Accuracy enhancement of digital image correlation with B-spline interpolation. Optics letters 36, 3070-3072.

Meng, L., Jin, G., Yao, X., 2006. Errors caused by misalignment of the optical camera axis and the object surface in the Dscm Journal-Tsinghua University 46, 1930.

Moisa, L., Moulon, P., Monasse, P., 2012. Automatic homographic registration of a pair of images, with a contrario elimination of outliers. Image Processing On Line 10.

Pan, B., 2014. An Evaluation of Convergence Criteria for Digital Image Correlation Using Inverse Compositional Gauss–Newton Algorithm. Strain 50, 48-56.

Pan, B., Lu, Z., Xie, H., 2010a. Mean intensity gradient: An effective global parameter for quality assessment of the speckle patterns used in digital image correlation. Optics and Lasers in Engineering 48, 469-477.

Pan, B., Qian, K., Xie, H., Asundi, A., 2009a. Two-dimensional digital image correlation for in-plane displacement and strain measurement: a review. Measurement science and technology 20, 062001.

Pan, B., Quian, K., Xie, H., Asundi, A., 2009b. Two-dimensional digital image correlation for in-plane displacement and strain measurement: a review. Measurement science and technology 20(6).

Pan, B., Wang, Z., Lu, Z., 2010b. Genuine full-field deformation measurement of an object with complex shape using reliability-guided digital image correlation. Opt. Express 18, 1011-1023.

Quan, L., 2010. Image-based modeling. Springer.

Riveiro, B., Solla, M., de Arteaga, I., Arias, P., Morer, P., 2013. A novel approach to evaluate masonry arch stability on the basis of limit analysis theory and non-destructive geometric characterization. Automation in Construction 31, 140-148.

Sánchez-Aparicio, L.J., Riveiro, B., González-Aguilera, D., Ramos, L.F., 2014. The combination of geomatic approaches and operational modal analysis to improve calibration of finite element models: A case of study in Saint Torcato Church (Guimarães, Portugal). Construction and Building Materials 70, 118-129.

Schweighofer, G., Pinz, A., 2006. Robust pose estimation from a planar target. Pattern Analysis and Machine Intelligence, IEEE Transactions on 28, 2024-2030.

Tait, D., 1994. Photogrammetry, vol. 1. Fundamentals and standard processes: Karl Kraus. Ferdinand Dümmels Verlag, Bonn, 1993, 402 pp., DM 58.00, ISBN 3-427-78681-1. Elsevier.

Tensing, D., 2013. Experimental study on axial compressive strength and elastic modulus of the clay and fly ash brick masonry. Journal of Civil Engineering and Construction Technology 4, 134-141.

Villarino, A., Riveiro, B., Gonzalez-Aguilera, D., Sánchez-Aparicio, L., 2014a. The Integration of Geotechnologies in the Evaluation of a Wine Cellar Structure through the Finite Element Method. Remote Sensing 6, 11107-11126.

Villarino, A., Riveiro, B., Martínez-Sánchez, J., Gonzalez-Aguilera, D., 2014b. Successful Applications of Geotechnologies for the Evaluation of Road Infrastructures. Remote Sensing 6, 7800-7818.

Wang, J., Gu, D., Yu, Z., Tan, C., Zhou, L., 2012. A framework for 3D model reconstruction in reverse engineering. Computers & Industrial Engineering 63, 1189-1200.

This contribution has been peer-reviewed.
doi:10.5194/isprsarchives-XL-5-W4-331-2015

4D CITY TRANSFORMATIONS BY TIME SERIES OF AERIAL IMAGES

A. Adami

POLITECNICO DI MILANO, Department of Architecture, Built Environment and Construction Engineering, Milano, Italy – andrea.adami@polimi.it

Commission V, WG V/4

KEY WORDS: city transformation, dense image matching, aerial images

ABSTRACT:

Recent photogrammetric applications, based on dense image matching algorithms, allow to use not only images acquired by digital cameras, amateur or not, but also to recover the vast heritage of analogue photographs. This possibility opens up many possibilities in the use and enhancement of existing photos heritage. The research of the original figuration of old buildings, the virtual reconstruction of disappeared architectures and the study of urban development are some of the application areas that exploit the great cultural heritage of photography.

Nevertheless there are some restrictions in the use of historical images for automatic reconstruction of buildings such as image quality, availability of camera parameters and ineffective geometry of image acquisition. These constrains are very hard to solve and it is difficult to discover good dataset in the case of terrestrial close range photogrammetry for the above reasons. Even the photographic archives of museums and superintendence, while retaining a wealth of documentation, have no dataset for a dense image matching approach.

Compared to the vast collection of historical photos, the class of aerial photos meets both criteria stated above. In this paper historical aerial photographs are used with dense image matching algorithms to realize 3d models of a city in different years. The models can be used to study the urban development of the city and its changes through time. The application relates to the city centre of Verona, for which some time series of aerial photographs have been retrieved.

The models obtained in this way allowed, right away, to observe the urban development of the city, the places of expansion and new urban areas. But a more interesting aspect emerged from the analytical comparison between models. The difference, as the Euclidean distance, between two models gives information about new buildings or demolitions. As considering accuracy it is necessary point out that the quality of final observations from model comparison depends on several aspects such as image quality, image scale and marker accuracy from cartography.

1. INTRODUCTION

Nowadays it is widely acknowledged that artistic and cultural heritage of an urban centre is not only determined by its elements of greatest interest, the masterpieces, but also by the urban texture itself, the smaller building and its changes over time. So the study of urban transformations has become an important axis in the investigation of the city (Patias et al, 2011). In literature, there are examples of very different approaches and aims as concerning cultural heritage (El-Hakim et al, 2008), architectural transformations (Stefani et al, 2011), 4D city modelling and urban analysis (Schindler et al, 2012).

In parallel, a topic of great interest is the reconstruction of architectures and lost parts of the city, to provide new views to the visitor and the scholar (Pavlidis et al, 2010). In many cases the virtual reconstruction is based on verbal descriptions, drawings or depictions, however, it involves some uncertainty on the reliability of the result obtained. Another approach involves the reconstruction of the lost parts of the city through the procedural modelling (Ferdani et al, 2011)

Photogrammetry has always played a key role in the description of the city and its heritage and, after a period of relative disappearance, has returned to being considered a fundamental discipline for the survey and the study of urban centre. This revival is due to several aspects, but certainly a key element is the appearance of new software that automate the photogrammetric process, hence making it more affordable also to less specialized people. The meeting with computer vision has encouraged the spread of new digital photogrammetric software that allow to get three-dimensional models mapped with real textures in few steps.

The automation of these software also allows to recover a large part of the existing photographic heritage, from the most recent to the historical one, from terrestrial to aerial ones. Compared to other photogrammetric techniques, dense image matching allows the use of photographs of which some important parameters are unknown and to obtain directly a 3D model. So it is possible to study the development of city centre and also to build virtual reconstructions based on real data.

There are some restrictions in the use of all photographic heritage for automatic reconstruction of buildings. There are problems connected with the quality of photo itself, the limited number of shots and the geometry of acquisition.

These constrains are very hard to solve and it is difficult to discover good dataset in the case of terrestrial close range photogrammetry. Even the photographic archives of museums and superintendence, while retaining a wealth of documentation, have no dataset for a dense image matching approach.

Compared to the vast collection of historical photos, the class of aerial photos meets both criteria stated above. So they can be considered good candidates in the process to discover the urban transformation.

After an introduction about historical photographic heritage and the requirements of new photogrammetric software (based on dense image matching algorythm), this paper will describe

This contribution has been peer-reviewed.
doi:10.5194/isprsarchives-XL-5-W4-339-2015

the use of a time series of aerial images: the orientation based on cartographic coordinates, the calculation of models and their comparison. From the analysis of different model, some changes in the urban context will be evident and described.

2. PHOTOGRAPHIC HERITAGE AND DENSE IMAGE MATCHING

Many scientific papers describe the algorithms of dense image matching and the different approaches used (Haala, 2013). There are also many software which allow, by those algorithms, to build 3D models in an automatic way (Remondino et al, 2012) (Remondino et al, 2014). Although with slight differences, especially during the calculation of the features, orientation and triangulation, these software, both open source and commercial, have a similar behaviour and have the same requirements.

The advantages of these software in the use of historical images are concerning the automation of the process and the knowledge of parameters for orientation. In many cases there are no -or few- information about the images, above all as concerning the camera calibration. This often represents a problem with other techniques of digital photogrammetry. Moreover the use of such images in traditional photogrammetric software, based on DLT algorithm, requires much time to find the homologue points for orientation and then many points have to be extracted to build the 3d model.

On the other side, these software work properly only under certain conditions. A first requirement concerns the quality of the photograph. As the feature recognition within the image is automatic (through algorithms SIFT, SURF, etc), it is necessary that the image is well-focused and that is not damaged (scratches or stains). Also the film grain is problem in this first stage as its points can be considered as feature to match.

Moreover as the orientation takes place through the matching of points of each photo, it is necessary that the shooting conditions are similar: it is not recommended to have shadows that change between images as the software tends to recognize them as different points. The images should be acquired by neighbouring points so as to allow the algorithm to identify the same point in a limited research area between images. Finally, to obtain a good orientation and the subsequent reconstruction, the software requires that images have a good overlap area in which the software can find homologous points.

In order to identify some datasets of images (both terrestrial and aerial) suitable for the reconstruction of missing parts of the city, several archives were visited, but it was very difficult to find a suitable set of photos.

The first problems affect precisely the photos themselves. The further you go back in time, the more the picture quality decreases and the effects of photographic grain are evident. In many cases the state of conservation is very inadequate and there are scratches, creases and stains on the photos themselves.

Another problem concerns the number of photos, especially the terrestrial ones. As it was a rather expensive practise, photographs were carefully chosen and were not taken more photographs than necessary: a photograph was often sufficient to document an architecture. This makes it difficult to find a set of photos large enough to allow a 3D reconstruction

Also with regards to the terrestrial photos, the geometry of acquisition represents a problem too. Perhaps for aesthetic reasons, the images were taken from the same positions, whereby, even in cases where they were available more shots, the geometry of the photographic shot was not suitable for three-dimensional reconstruction.

These observations do not apply with respect to the assets of aerial photographs. Especially aerial photos, captured with photogrammetric purposes, respond sufficiently to the requirements of the software dense image matching.

The flight pattern provides for a sufficient overlap of the images (at least 60% in the longitudinal direction and about 20 in the transverse direction). The pattern of the normal setting also allows a geometric regularity which facilitates the orientation of the frame.

There can be some problems, as presented in (Redecker, 2008), concerning the inaccuracy or total lack of meta-information about calibration (inner orientation and distortion, missing specification about the flight mission, poor radiometric image quality, distortions caused by roll and pitch due to sudden movements of the plane, improper storage of the film and inaccurate processing of original films.

Fortunately, as mentioned before, some information are not necessary in the recent photogrammetric software; instead remain the problems of photographic quality

On the whole Italian territory there are several archives of historical aerial photographs of great informational value. The most important collections include photos planimetric and stereoscopic taken between 1943 and 1945 by the scouts of the

Figure 1: sample images of the time series: 1954, 1981, 1997

doi:10.5194/isprsarchives-XL-5-W4-339-2015

British Royal Air Force (RAF), the United States Army Air Force (USAAF) and the German Luftwaffe. In addition to these also the Italian Air Force flights, Military Geographical Institute and the Property Valuation Office of Florence and numerous collections donated or purchased by private companies detection aerial photography or production of cartography.

3. THE AVAILABLE IMAGES

As for the area of Verona, the subject of this paper, the aerial images have been made available by the Aerial Archive of University IUAV of Venice (Aerial Photo Archive of Iuav). It manages the shots are produced mainly by the Italian Military Geographical Institute (the first flights of 1929 to the flights of the 60s), the Royal Air Force (1944-45), the flight GAI (1954-55 with total coverage of the Veneto) and the Veneto Region (from 1978 to the most recent).

The city centre of Verona appears in 4 flights, but in the first one, made by RAF in 1945, a missing shot does not allow the necessary overlap. So the images used in this reconstruction belong to the following three flights as in Tab. 1

a) GAI Flight (Italian Aeronautics Group) of 1954-55. The flight was commissioned by the US Army Map Service and it is very important because it represents the first stereoscopic coverage of Italy. The acquisition camera was a Fairchild with a focal length of about 153 mm. The altitude of the flight was about 5000 m and the mean scale of the frame is about 1: 30000. Only 4 images have been used in the reconstruction.
b) Flight realised in 1981 by Compagnia Generale Riprese Aree of Parma on commission of Veneto Region. The flight altitude was about 2600 and the mean scale 1:17000. The images were acquired by WILD RC10 with a focal length of about 153mm. Photos used in reconstruction: 9
c) Flight realised in 1997 by Compagnia Generale Riprese Aree of Parma on commission of Veneto Region. The flight altitude was about 2500 and the mean scale 1:16000. The images were acquired with a focal length of about 153 mm. Photos used in reconstruction: 15.

The images of the flight a) and b) are black and white, while the ones of the third flight are in colour. All the images were analogue and acquired in a format 23 x 23 cm and they were digitized at high resolution (600dpi) by the University Iuav of Venice.

Flight	Altitude [m]	Mean scale	Focal length	N. of images
1954	5000	1:30000	153.27	4
1981	2600	1:17000	153.15	9
1997	2500	1:16000	153.04	15

Tab. 1: approximate values

4. PROCESSING OF THE HISTORICAL IMAGES

The image processing in dense image matching software is described exhaustively in (Remondino et al, 2014). So in these paragraphs only some features, strictly connected with aerial historical images, will be highlighted.

The software used to calculate 3d model is Agisoft Photoscan.

After the first automatic alignment of images, in order to georeference the model, the technical regional map (C.T:R.) of Veneto Region at scale 1:5000 has been used for all images.

The problems of this phase have already been described in (Nocerino et al, 2012) and they are related to the difficulty to pinpoint reference points in both the historical and the most recent aerial images. And frequently the identified points are inaccurate and unreliable. All this is due to the changes due to the time shift, but also to the low quality of some images, above all the oldest one.

All points used in the georeferencing are planimetric points (crossroads, property boundaries) as the border of buildings are not represented in their true elevation. However, the morphology of the city itself, with the hills immediately behind, allows to get a good three-dimensional distribution of ground control points.

The analytical results orientation are quite good (Tab. 2). They were calculated by the optimization process in PhotoScan.

Flight	GCP Error [pix]	GCP Error [m]
1954	0.335	0.44
1981	0.485	0.23
1997	0.339	0.27

Tab. 2: results of image orientation

As evident in Figure 2, the model has been created but the geometric configuration is very different. The images of 1954 have a few overlapping and the model is not very stable. The scales of the images caused the difference in the final resolution of the model. In fact the Ground Sample Distance of 1981 and 1997 set of images is very similar and it is about 0.7 m/pix. Instead, for the photos of the flight GAI 1954, the GSD is about 1.4 m/pixel. This resulted in a difference even in the final mesh in terms of the number of polygons and size of the triangles.

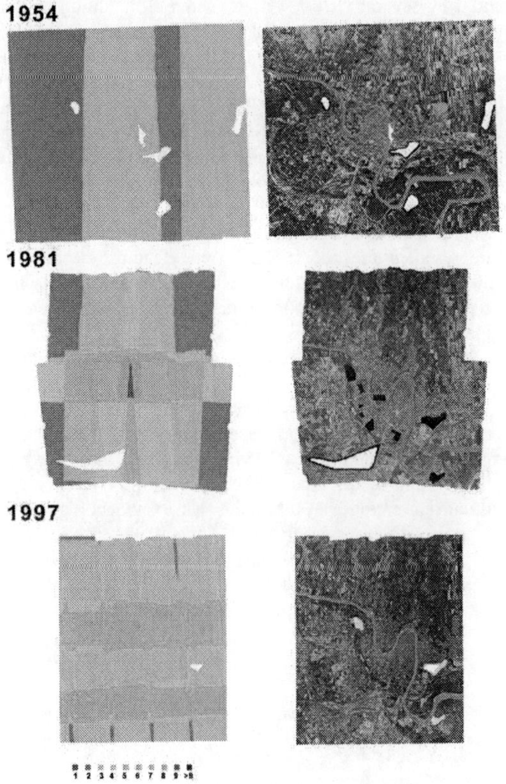

Figure 2: results of image orientation. On the left the representation of overlapping and on the right the orthophoto of the area.

This contribution has been peer-reviewed.
doi:10.5194/isprsarchives-XL-5-W4-339-2015

5. COMPARISONS OF HISTORICAL 3D MODELS

At the end of the image processing, three different models have been obtained, each one for a specific historical phase. The comparison has been split into two time interval: the first from 1954 to 1981, the second from the 1981 to 1997. They allow to observe the urban development of the city, the places of expansion and new urban areas for each period.

But a more interesting aspect emerged from the analytical comparison between models. The comparison between the three phases occurs by the software Cloudcompare (Cloudcompare) that allows to import different mesh and to calculate the difference as the shortest distance between the two corresponding points of the surfaces. If the difference between the surfaces is close to zero, it means that there have been no - or insignificant - changes. In other areas the amount of the different is a clear indicator of changes. The value of change has a sign which helps to understand if the difference represents a new building (addiction) or a demolition (subtraction).

The comparison can be carried out as all the models have been oriented in the same reference system, the cartographic one. However, some errors remain in the process and they are caused by several factors, as seen above. In addition to image quality, the final errors depend on the uncertainty of the points extracted from the cartography and the residuals of orientation.

These errors do not affect, however, the process as it is not aimed exclusively to metrical analysis of the transformations, but especially in locating the sites affected by the biggest changes. For this reason the values of difference between surfaces should not be considered as absolute values, but as indicative parameters.

From the observation of the maps of differences, in Figure 3, represented in false colours, some features become very evident. First of all, the gray holes are due to deficiencies in the initial images as it comes to military areas and thus have been hidden in the aerial photos.

Punctual elements, in red and blue, represent the areas with the greatest changes. The red areas towards the edges of the figure 3 are partly caused by errors on the edges of the image due to the lack of control points, and partly to the fact that those areas were not built-up in 1954 but in 1981 there were new buildings (this is why you are not able to find reference points with current cartography). In general it is evident also that the first map, with changes between 1954 and 1981, is noisier. This may be caused by the different resolution of the first dataset which influences the calculation of the difference, face to face.

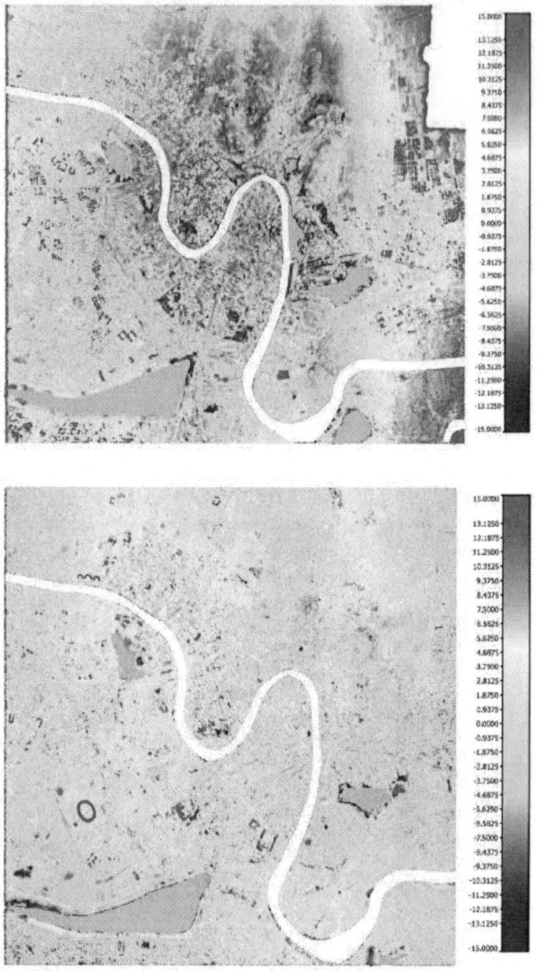

Figure 3: differences between 1981-1954. The maximum values in the scale are + 15 m and -15 m

By analyzing the points highlighted in red or blue, some cases are particularly significant. In the first map the evidences are connected above all with new building and neighbourhoods (Figure 4).

And also the area of the Bentegodi Stadium in the south west of Verona (Figure 5) shows a big transformation. The stadium was built in 1963 on an un-built area and in 1990, in occasion of the Football World Cup in Italy, a further ring was

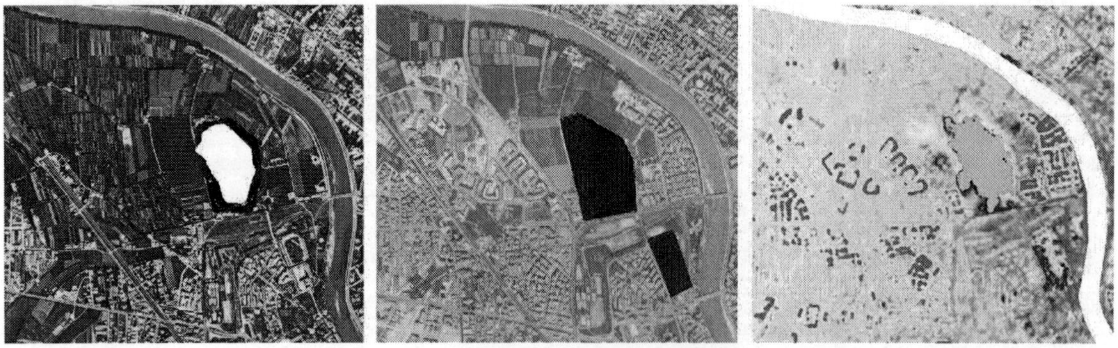

Figure 4: detail of an area: in 1954 it was unbuilt (left) in 1981 there were new buildings. On the right the difference map

This contribution has been peer-reviewed.
doi:10.5194/isprsarchives-XL-5-W4-339-2015

added on the top of the existing building. Because of its elliptical shape, it is easy to recognise in the urban texture its changes. By comparing the flights of 1954 and 1981 it appears that something new was built in a free area (marked by yellow-red colour). Instead in the comparison between 1981 and 1997 there is the evidence of the new ring.

Figure 5: Example of city transformation documented by the comparison of 3d models of different period. On the left the extraction from orthofotos of the area of the Stadio Bentegoni in Verona. On the right two images of the stadio before and after changes. At the bottom the maps of differences

6. CONCLUSIONS

This paper describes the application of the algorithms of dense image matching to the heritage of aerial historical images. This approach allows to enhance the documentary material of the rich Italian archives and also to reconstruct three-dimensional models of cities or parts of cities which now are missing.

The use of historical aerial photographs is favoured by the fact that it is not necessary to have all the information about the calibration of the room, which are often missing or incomplete.

The possibility to apply this approach to multiple time series, when available, allows to relate two or more historical period through a detailed comparison. The heterogeneity of photographic materials leads to results which are not always reliable in terms of metric, but certainly significant to follow the historical development of the city and its architecture.

ACKNOWLEDGEMENTS

The author are really thankful to the architect Marco Tropina who worked on these topic for his bachelor thesis and to the laboratory of cartography of University Iuav of Venice for having made available the aerial images.

REFERENCES

Aerial Photo Archive of Iuav University of Venice, http://mapserver.iuav.it/website/foto_aeree/, last visit on 25 gennuary 2015

Cloudcompare, http://www.danielgm.net/cc/, last visit on 25 gennuary 2015.

El-Hakim, S., Lapointe, J.-F., Whiting, E., 2008: Digital reconstruction and 4D presentation through time. Proc. *SIGGRAPH.*

Ferdani, D., Pescarin, S., 2011. Dal Gis alla ricostruzione 3D del paesaggio urbano antico. http://academia.edu/1884152/Dal_Gis_Alla_ricostruzione_3D_del_paesaggio_urbano_antico.

Haala, N., 2013 : The Landscape of Dense Image Matching Algorithms. In: Fritsch, D. (Ed.): Photogrammetric Week '13, Wichmann, Berlin/Offenbach, 271-284.

Nocerino E., Menna F., Remondino F., 2012: Multi-temporal analysis of landscape and urban areas, International Archives of the Photogrammetry, Remote Sensing and Spatial Information Sciences, Volume XXXIX-B4, 2012, , XXII ISPRS Congress, 25 August – 01 September 2012, Melbourne, Australia, pp.85-90.

Patias, P., Kaimaris, D., Stylianidis, E., 2011: Change detection in historical city centers using multi-source data: the case of historical center of Nicosia – Cyprus. Proc. 23rd Int. CIPA Conference, Prague, Czech Republich.

Pavlidis G., Arnaoutoglou F., Koutsoudis A., Chamzas C., 2010: Virtual Walkthrough in a Lost Town – The Virtual Argostol", in the IASTED Computer Graphics and Imaging CGIM 2010, (Innsbruck, Austria), Feb. 17-19.

Redecker, A.P., 2008. Historical aerial photographs and digital photogrammetry for impact analyses on derelict land sites in human settlement areas. IAPRS&SIS, Beijing, China, Vol. 37(B8), Beijing, China, pp. 5-10.

Remondino F., Del Pizzo S., Kersten T.P., Troisi S. Low-Cost and Open-Source Solutions for Automated Image Orientation ,

2012: A Critical Overview. In: Lecture Progress in Cultural Heritage Preservation. Notes in Computer Science. Proceedings of EuroMed 2012, Limassol, Cyprus, October 29 – November 3, 2012, Ioannides M., et al., Eds., Spinger-Verlag: Berlin Heildelberg, 2012, Vol. 7616, pp. 40-54;

Remondino F., Spera M. G., Nocerino E., Menna F. and Nex, F. 2014: State of the art in high density image matching. The Photogrammetric Record, 29: 144–166.

Schindler G , Dellaert F., 2012: 4D Cities: Analyzing, Visualizing, and Interacting with Historical Urban Photo Collection, "Journal of Multimedia", 2012, 7, 2, pp. 124-131.

Stefani, C., Busayarat, C., Renaudin, N., De Luca, L., Véron, P., Florenzano, M., 2011: An image-based approach for the architectural modeling of past states. IAPRS&SIS, Vol 38(5/W16).

This contribution has been peer-reviewed.
doi:10.5194/isprsarchives-XL-5-W4-339-2015

BIM FROM LASER CLOUDS AND FINITE ELEMENT ANALYSIS: COMBINING STRUCTURAL ANALYSIS AND GEOMETRIC COMPLEXITY

L. Barazzetti, F. Banfi, R. Brumana, G. Gusmeroli, D. Oreni, M. Previtali, F. Roncoroni, G. Schiantarelli

Gicarus Lab, ABC Department, Politecnico di Milano, Piazza Leonardo da Vinci 32, Milan, Italy
(luigi.barazzetti, fabrizio.banfi, raffaella.brumana, daniela.oreni, mattia.previtali, fabio.roncoroni)@polimi.it
(gaia.gusmeroli, giuseppe.schiantarelli)@mail.polimi.it
http://www.gicarus.polimi.it

Commission V

KEY WORDS: BIM, Finite Element Analysis, Laser Scanning, Photogrammetry

ABSTRACT:

This paper describes the use of BIM models derived from point clouds for structural simulation based on Finite Element Analysis (FEA). Although BIM interoperability has reached a significant level of maturity, the density of laser point clouds provides very detailed BIM models that cannot directly be used in FEA software. The rationalization of the BIM towards a new finite element model is not a simple reduction of the number of nodes. The interconnections between the different elements and their materials require a particular attention: BIM technology includes geometrical aspects and structural considerations that allow one to understand and replicate the constructive elements and their mutual interaction. The information must be accurately investigated to obtain a finite element model suitable for a complete and detailed structural analysis. The aim of this paper is to prove that a drastic reduction of the quality of the BIM model is not necessary. Geometric data encapsulated into dense point clouds can be taken into consideration also for finite element analysis.

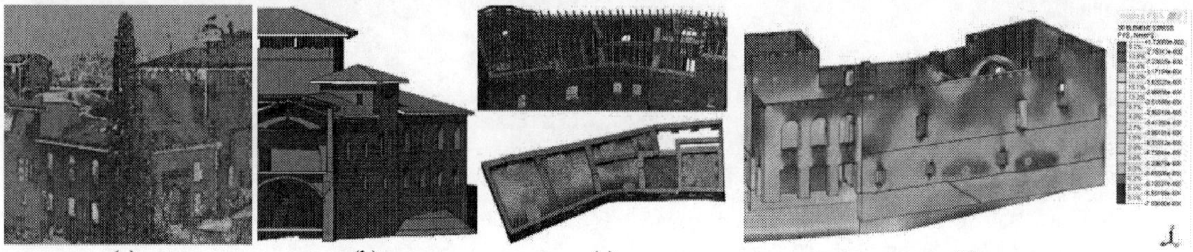

(a) (b) (c) (d)

The phases of the work: (a) point cloud acquisition/registration, (b) creation of the BIM from point clouds by preserving geometric complexity, (c) generation of the finite element model with tetrahedral meshes, (d) structural analysis with the new model. The case study is Castel Masegra, a castle in Sondrio (Italy).

1. INTRODUCTION

This aim of this work was to investigate the possibility to transform an accurate BIM of a historical building (HBIM) into a finite element model (FEM, Zienkiewicz, 1971). The starting point of the work is a set of dense laser scanning point clouds able to capture the geometric complexity of the object. The approach is not the typical structural simulation based on simplified models that fit the requirements of software packages for finite element analysis (FEA).

The recent advances in BIM technology led to an incredible interest towards new efficient methodologies in the construction industry, starting from the survey of the object up to the design and management phases (Osello et al., 2012).

Although specific BIM packages with FEA tools are available on the commercial market, they were developed for new buildings with simple and regular objects. A detailed BIM for architectural purposes (derived from point clouds) is much more complex and must be rigorously discretized through a simplification of its geometry, obtaining a new alveolar mesh that tries to preserve the original complexity (Gusmeroli and Schiantarelli, 2014).

The modification of the original BIM is not intended as a simple reduction of the number of nodes. The interconnection between the different parts requires a particular attention. BIM modelling for FEA includes geometrical aspects along with structural considerations that allow one to replicate the constructive elements and their mutual interaction. Geometric and semantic data contained in the architectural model should be accurately rationalized to obtain a new finite element model for structural analysis.

The aim of this paper is to prove that a drastic reduction of the of the model is not necessary. The geometry encapsulated into dense point clouds can be accurately taken into consideration also for FEA. Complex elements (such as vaults) and structural anomalies (e.g. verticality) are turned into 3D surfaces by an intelligent node-to-node connection, obtaining models that can be managed for static analysis. This is feasible not only for

This contribution has been peer-reviewed.
doi:10.5194/isprsarchives-XL-5-W4-345-2015

modern buildings with predefined mathematical shapes, but also for complex historical objects with several irregularities.
The case study presented in this work is Castel Masegra, a castle located in Sondrio (Italy). A detailed historical BIM was derived from laser scanning point clouds (ca 7 billion points). Building information modelling was carried out by dividing the different structural objects and their constructive logic. Chronological, material, and stratigraphic aspects were also taken into account. This step is not only useful for architectural purposes, but also for further static and dynamic simulations where the temporal evolution of the castle provides additional data about its logic of constrctions.
The historical BIM was then re-adapted for structural purposes obtaining a new sub-model with more details than those usually used for this kind of analysis. A set of tetrahedral meshes was generated to guarantee an adequate representation of the stress/strain distribution. Several problems were taken into consideration, such as the node-to-node correspondence (compatibility of the mesh), very distorted or small elements, and local imprecisions. These effects are extremely important and should be removed to obtain a consistent model for FEA. In fact, small errors and local inconsistencies lead to a failure during the simulation, as the continuity of the model is not guaranteed.
Finally, a linear simulation was carried out in Midas-FEA, a package for finite element analysis. The level of detail of the model allowed one to simultaneously simulate the behaviour of different structural elements (vaults, walls, beams, columns, etc.). The result demonstrates that the creation of a consistent alveolar mesh from the (H)BIM is a crucial step and deserves a special attention.

2. DATA ACQUISITION AND PROCESSING

The complexity and size of the castle required a laser scanning surveying, that is a powerful technique to obtain complex reconstructions (Brumana et a., 2014a; Remondino et al., 2008). The instruments used are a Laser Scanner Faro Focus 3D and a total station Leica TS30. A robust geodetic network made up of 68 stations was measured in about 4 days. Least Squares adjustment provided an average point precision of about ±1.2 mm (Fig. 2).

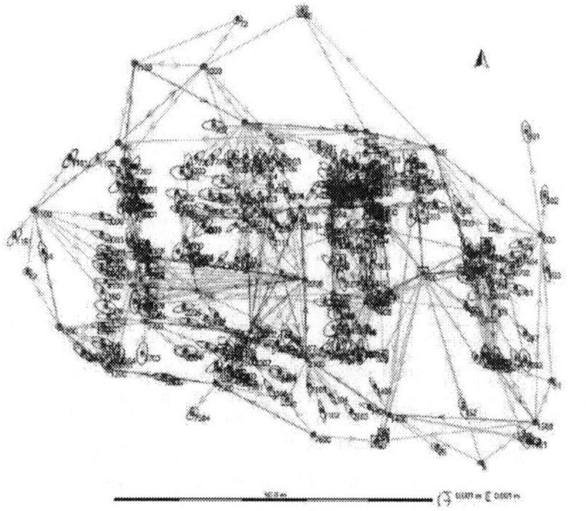

Fig. 2. The geodetic network measured with a total station Leica TS30 (the average precision is ±1.2 mm).

The network provides a reference system for scan registration. 176 scans were acquired to obtain a final point cloud made up of 7.5 billion points, which were registered with an average precision of ±3 mm.
Overall, total station and laser data were acquired in less than a week. Then, small integrations were needed to capture occluded areas in the first surveying phase. This proves the level of maturity reached by these instruments. It is clear that the time needed for data processing (especially BIM generation) is much longer because of the level of detail achievable from such huge dataset.
As the goal is a reconstruction for architectural and structural purposes, the surveying phase cannot be limited to the shape (geometry is just one of the elements to be taken into consideration). As the goal is the creation of an interoperable BIM and its distribution among the different operators that work on the castle (engineers, architects, historians, archaeologists, restorers, etc.) the survey must include historical analysis, materials, construction phases, technological aspects, stratigraphic analysis, and information from other inspections such as infrared thermography or structural tests (flat-jacks, coring, etc.) (Binda and Tiraboschi, 1999; Colla et al., 2008; Gregorczyk and Lourenco, 2000; Rosina and Grinzato, 2001).

Fig. 3. A visualization of the point cloud after scan registration (176 scans, 7.5 billion points).

3. GENERATION OF THE BIM MODEL

3.1 Why use BIM technology?

The creation of a detailed BIM is the second step of the project. The use of a BIM instead of a simple 3D model is motivated by the advantages offered by this technology with advanced parametric modelling tools, functional intelligence, and object attributes.
BIM technology simplifies the work of different operators with new interoperability standards: the different operators involved in the project can update and modify the BIM according to their needs. Then, the other operators can immediately evaluate the effect of the change.
The typical inconsistencies of CAD drawings should be limited. A robust exchange of information can limit the creation of additional documents (drawings, reports, …), towards a better cooperation in the architecture, engineering and construction industry (AEC), where new instruments that allow a consistent exchange of data are required to improve productivity. The interoperability (Drogemuller et al., 2012) has the purpose to connect all the technological aspects involved in construction projects. This is a fundamental requirement not only for a limited number of software packages. Interoperability should be guaranteed for different applications. This approach is extremely important to increase the development of BIM

This contribution has been peer-reviewed.
doi:10.5194/isprsarchives-XL-5-W4-345-2015

oriented projects not only during the design phase, but also for the different phases of the building life cycle.
Although the initial effort for the creation of a BIM is greater than the classical 2D/3D workflows based only on geometry, the advantage of BIM technology becomes effective in the successive phases.
For these reasons, the International Alliance for Interoperability (IAI), today renamed "Building Smart", proposed a new standard open format for the representation of the object in the field of constructions: the Industry Foundation Classes (IFC). Today, IFC2x3 files are able to provide a consistent exchange of information between different software. However, the commonly used IFC format was developed for regular shapes, whereas it does not work with complex surfaces. Probably, the new ICF4 will provide additional tools for advanced geometries.
One of the big limitations of software for finite element analysis concerns the low integration with BIM models. It is clear that there is a growing interest towards BIM technology and some software vendors are updating their application to import BIM projects. This confirms that the construction industry is going in the direction of BIM technology.
As things stand at the present, it is rather clear that BIM represents the future of the constructions in a way similar to CAD technology in the past, when manual drawings were progressively substituted by digital representations.

3.2 BIM from laser clouds: preserving geometric complexity with advanced modelling tools

Starting from laser data a geometric reconstruction with a high level of detail can be carried out. However, detailed building information modelling cannot simply be carried out with the interpolation of the point cloud with mesh-based algorithms often used in photogrammetry and computer vision. Additional information (e.g. materials, construction stages, stratigraphy, …) has to be taken into account to create intelligent parametric objects with attributes (Fig. 4).

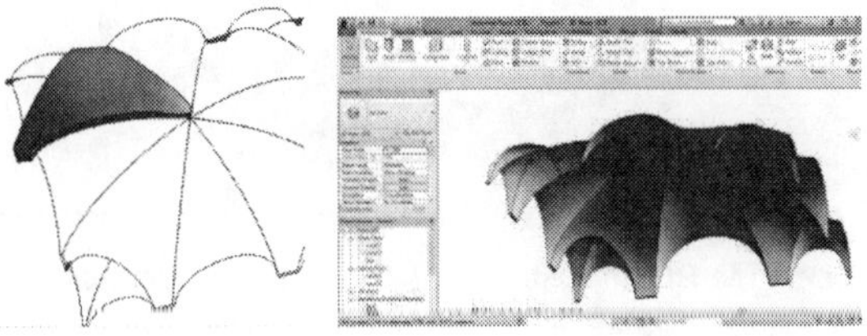

Fig. 4. The vault is a parametric object. The thickness can be set to modify the shape.

An additional problem concerns the lack of parametric software for the management of complex and irregular shapes. BIM software were initially developed to manage new constructions (Eastman et al., 2008; Lee et al., 2006). Building information modelling is usually used for the design, construction, and management phases of new buildings, whereas the use for conservation of historical buildings is still very limited (Fai et al., 2011).
However, they represent an opportunity also for heritage documentation and conservation management, but they still require a methodological discussion and a practical experimentation in order to obtain detailed models of irregular historical objects, that will be really useful for preservation and maintenance activities (Della Torre, 2011; Oreni et al., 2013, 2014a).

The methodology for parametric BIM generation used for the castle is based on a preliminary separation between simple and complex shapes. In the case of simple objects, the tools of most commercial software (Revit or Archicad) can be sufficient. The case of irregular objects (e.g. vaults) is much more complicated. The modelling tools of commercial BIM packages are not adequate to represent the geometric level of detail encapsulated into laser point clouds. For this reason, the procedure described in Oreni et al. (2014b) was used. Vector profiles made up of NURBS curves (Piegl and Tiller, 1997, 1999) provide a curve network for NURBS surfaces, which are then turned into parametric BIM objects.
This methodology was used for the complex shapes found in the castle and it was a very efficient solution to overcome the limitations of BIM software. Some images of the final BIM in Revit are shown in Fig. 5.

Fig. 5. The parametric BIM in Revit with the advanced representation of irregular objects (e.g. the selected wall features variable verticality issues).

4. FROM BIM TO FINITE ELEMENTS

The generated BIM is an interoperable tool that allows different operators to initialize their analyses. Obviously, further data processing could require operations that modify the geometry, not only in terms of different design aspects, but also for specific and advanced simulations requiring simplified models. The aim of this section it to prove that a rationalization can be carried out to transform the BIM model into a tetrahedral mesh for finite element analysis.

4.1 BIM turned into finite elements

The finite element method constitutes a general tool for the numerical solution of partial differential equations. The method has a large acceptance in several engineering applications and it is one of the preferred approaches in structural mechanics.
In the finite element method the region of interest is divided into numerous connected sub-regions (or elements) where the solution is obtained from the analysis of shape functions (Strang and Fix, 1973; Zienkiewicz, 1971). The shape function is the function which interpolates the solution between the discrete values obtained at the mesh nodes.
The traditional FEM approaches used in structural analysis rely on simplifications of the elements of a structure into 1D (beams, trusses) or 2D (plates, shells) elements, which are discretized into segments or 2D finite elements.
The integration of 3D finite element analysis in BIM technology is already a concrete possibility for simple and regular buildings. For instance, the structural analysis tool Autodesk Robot Structure is fully integrated with Revit. Different plugins are available to guarantee interoperability with other structural analysis software, such as Midas FEA. These tools allow one (i)

to transfer the model from Revit and (ii) to obtain a new model for finite element analysis.
The disadvantage of these plugins concern the use of simple load-bearing structures, such as those used for modern buildings with regular columns, beams and walls schematized as 1D or 2D simple elements.
The tri-dimensional finite element analysis has the possibility to deal with 3D objects by using a discretization of the structure by means of 3D elements such as tetrahedra or hexaedra. This last options is less used because the number of elements (and consequently the computational effort) is much larger. Moreover, the approximation given by 1D or 2D objects in modern structures is more than sufficient in several real projects. However, modern computers allow structural engineers to run complex 3D FEM analysis not only for some particular structural elements, but also extended to the whole structure.
The aim of this work is the re-adaptation of the original BIM to include complex shapes that characterize the load-bearing elements of a historical building (mainly vaults and irregular walls) to arrive at a complete and detailed structural analysis.
This approach is still impossible in a full BIM perspective. The aim was the development of a new procedure able to transform the BIM into a finite element model, without redrawing a new model only for structural analysis.
Generally speaking, the conversion of the objects representing the walls (or other load-bearing elements) could be carried out with auto-meshing algorithms that create the finite element model. On the other hand, this is not a trivial task because of several issues:

- *mesh compatibility*: single entities need a perfect node-to-node correspondence to ensure geometric continuity between different objects. This result could be automatically achieved with auto-meshing procedures, but the corresponding faces of two generic elements should be exactly the same (same edges): this is not true in the case of complex BIM composed of multiple connected objects;
- *local distortions*: an ideal mesh must be composed of regular tetrahedral objects. However, the error for the approximation of the solution with an alveolar structure increases with distorted elements. For this reason, very distorted elements should be avoided or at least limited, especially in critical areas;
- *small elements*: small elements (that increase the level of detail of the BIM) without a direct connection with the load-bearing function should be eliminated to avoid the generation of distorted or small finite elements. This reduces the risk of a large number of small components;
- *small imprecisions*: the available Revit tools sometimes give very small imprecisions in the final mesh. A finite element analysis requires the exact correspondence of the nodes in order to avoid the generation of thin faces with very distorted "fissure elements";
- *complex architectural objects*: this is the case of vaults where the parametric modelling capability leads to inconsistent self-intersection between curved objects (e.g. the extrados of a vault modelled from its intrados).

The listed issues are fundamental considerations to obtain an alveolar mesh for FEA. As things stand at the present, there is no software or research algorithm able to transform a complex BIM into a compatible mesh for FEA without approximations.

In other words, the automatic re-adaptation of the historical BIM model towards a consistent mesh required several manual corrections to guarantee a node-to-node continuity that takes into consideration the different elements of the BIM.
The final finite element model (after extensive error removal) is made up 720,393 elements (Fig. 6). Although a simplification of the original shape was still needed, the mesh follows the main irregularities of the structures. In addition, objects with complex geometry (e.g. vaults) did not require excessive simplification.
As mentioned, the aim of this paper is to prove that complex historical BIM created from dense point clouds can be converted into finite element models for structural analysis. As things stand at the present, manual corrections were mandatory but research work can be carried out to develop new data processing algorithms able to consider not only geometrical aspects, but also material properties and logic of construction.

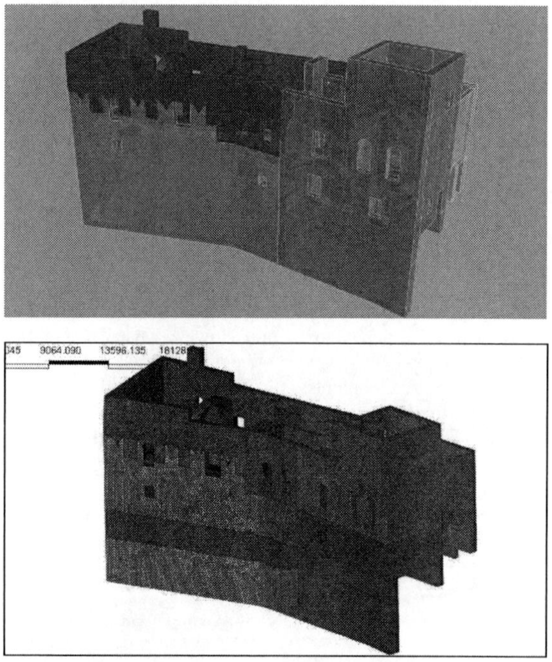

Fig. 6. The BIM turned into a finite element model following construction stages and material properties.

4.2 The finite element analysis

The structural analysis can be run after the creation of the finite element model. Input parameters concern the identification of boundary conditions, material properties, and loading conditions. This information can be partially available from the BIM, which is not limited to geometrical aspects.
The simulation carried out is a linear elastic analysis under self-weight. This is a fundamental step before other analysis for the preliminary check of model reliability and gross error removal. Irregular stress information or deformation patterns can be easily recognized under self-weight. Secondly, it allows one to revise and adjust boundary conditions.
As data processing is relatively fast in the case of linear analysis, several simulations can be run to correct setup errors. Different modifications of boundary conditions can be performed to find the most appropriate for the case under investigation. Moreover, the results in terms of stress can be

This contribution has been peer-reviewed.
doi:10.5194/isprsarchives-XL-5-W4-345-2015

directly compared with single-flat jack measurements. It is possible to have a feedback about model reliability in the different areas. Initial hypotheses are progressively adjusted to reach a better consistence with the experimental tests.
Historical information was extremely important to define input parameters. Material properties were assigned according to the stages of construction. The castle was subdivided into several parts by using the historical evolution and the state of conservation of the materials. Material data were also integrated with other destructive and non-destructive techniques able to distinguish areas with good materials from those with bad materials. This is a fundamental point for historical structures that could reveal a high variability of the same parameter.
The self-weight of the load-bearing elements can be automatically computed from the model (Fig. 7). The dead loads of the elements excluded from the model were applied to the structure as distributed pressure loads. Examples of elements directly converted into dead loads are infilling materials on the vaults and wooden slabs.

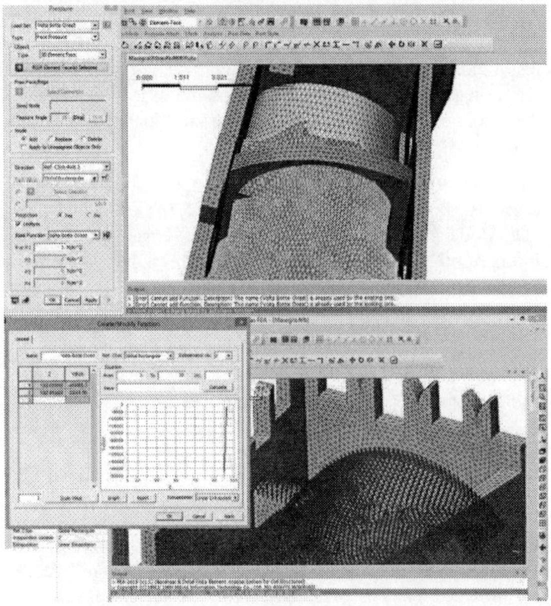

Fig. 7. Assignment of dead load on a barrel vault by using the faces of the mesh

A correct setup of boundary conditions is one of the most important issues. The castle raises on a cliff and its different parts have different foundations. Some walls are directly connected to the rock, whereas others have superficial foundations. The information concerning foundation typologies was obtained from both coring and visual inspection. Additional parameters were available from the work carried out by the Municipality of Sondrio, which performed previous restoration work in the castle. However, in some cases the information was completely missing and some hypotheses were derived by considering the conditions of the surrounding areas.
Shown in Fig. 8 are the general results obtained in terms of stress (for the whole South wing of the castle). The colour ramp of the stress-plot was adjusted to remove the local influence of stress concentration for the foundations. The value of stress distribution is lower than the ultimate strength of masonry estimated by jack tests. Therefore, no particular problem in terms of resistance is expected.

More exhaustive analyses are currently in progress to check the quality of the obtained results. On the other hand, our preliminary solutions have a direct connection to some local damages found in the structure (e.g. cracks).

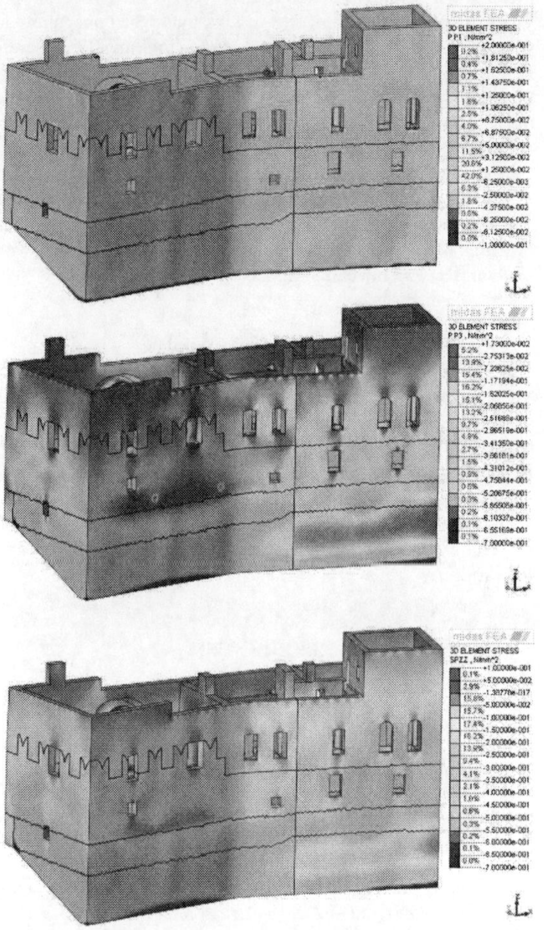

Fig. 8. Stress after simulation via alveolar mesh derived from BIM: (top) first principal stress; (middle) third principal stress; (bottom) vertical stress.

5. CONCLUSION

Point clouds can play a fundamental role in structural simulation of existing constructions, especially in the case complex historical structures. On the other hand, point clouds are just one of the input parameters behind the creation of BIM models with parametric intelligence and attributes. Building information modelling combines geometry with materials, construction phases, technological aspects, stratigraphic analysis, (…), and information from other inspections such as infrared thermography or structural tests.
It is clear that the creation of the BIM cannot be carried by using only the point cloud. The constructive logic must be incorporated into the model in order to understand the behaviour of the different structural elements. This could be intended as a new challenge for operators who work in the fields of photogrammetry, laser scanning, and 3D modelling. The use of the procedure based on curve networks (from

This contribution has been peer-reviewed.
doi:10.5194/isprsarchives-XL-5-W4-345-2015

clouds) and parametrization of NURBS surfaces (Oreni et al., 2014b) was used to obtain a BIM that preserves the complexity offered by laser clouds. Then, a rationalization of the BIM can be carried out to generate a robust and reliable finite element model for structural simulation.

The main aim of this work was to investigate the opportunity to generate such model by exploiting a preliminary accurate building information modelling. The work carried out is only the starting point: manual corrections were mandatory because of the complete lack of "intelligent" algorithms able to perform automated (i) point cloud → (ii) BIM → (iii) FEM conversion. Indeed, the conversion is not only limited to geometrical aspects. Additional architectural and structural considerations must be taken into account. From this point of view, much research is needed to combine geometric considerations with structural and architectural aspects.

ACKNOWLEDGEMENTS

This work was supported by the Interreg project "La Conservazione Programmata nello Spazio Comune Retico" (CPRE). The authors want to thank P. Crespi, N. Giordano and E. Rosina for the technical assistance concerning structural analysis and IR thermography. We are thankful to the staff of Municipality of Sondrio, in particular F. Barri, and Foppoli Moretta e Associati s.r.l. for providing several datasets used in this work.

REFERENCES

Binda, L., Tiraboschi, C., 1999. Flat-Jack Test: a slightly destructive technique for the diagnosis of brick and stone masonry structures. 8th International Conference and Exhibition, Structural Faults and Repair.

Brumana, R., Oreni, D., Cuca, B., Binda, L., Condoleo, P., Triggiani, M., 2014. Strategy for integrated surveing techniques finalized to interpretive models in a byzantine church, Mesopotam, Albania. International Journal of Architectural Heritage, 8, pages 886–924.

Colla, C., Largo, A., Corvaglia, P., Ubertini, F., 2008. Thermography investigations of roman archaeological masonry. In Binda, L., Di Prisco, M., Felicetti, R. editors, On Site Assessment of Concrete, Masonry and Timber Structures, SACoMaTiS 2008: Proceedings of the First International RILEM Symposium, volume 2, pages 923 – 932, Varenna, Italy, 1-2 September. RILEM Publications.

Della Torre, S., 2011. How the Concept of Preventive Conservation implemented on Built Cultural Heritage can work as a Factor for Regional Economic Development. In PRECOAH 2011 International Conference on Preventive Conservation of Architectural Heritage, pages 14 – 22, Nanjing (China), 29-30 October.

Eastman, C., Teicholz, P., Sacks, R., Liston, K., 2008. BIM Handbook - A guide to Building Information Modeling for owners, managers, designers, engineers, and contractors. John Wiley & Sons, Inc..

Fai, S., Graham, K., Duckworth, T., Wood, N. and Attar, R., 2011. Building Information Modeling and Heritage Documentation. CIPA 2011 Conference Proceedings: XXIIIrd International CIPA Symposium, 8 pages.

Gusmeroli, G., Schiantarelli, G., 2014. From Laser Clouds to BIM and Finite Element Analysis: the Case Study of Castel Masegra. Master Thesis in Civil Engineering for Risk Mitigation, Politecnico di Milano, 138 pages.

Lee, G., Sacks, R., Eastman, C. M., 2006. Specifying parametric building object behaviour (BOB) for a Building Information Modeling system. Automation in Construction, 15(6):758–776.

Gregorczyk, P., Lourenço P. B., 2000. A review on Flat-Jack Testing. Engenharia Civil, (9):39 – 50.

Drogemuller, R., Steel, J., Toth, B., 2012. Model interoperability in Building Information Modelling. Software & Systems Modeling, 11(1):99–109.

Oreni, D., Brumana, R., Cuca, B., Georgopoulos, A., 2013. HBIM for conservation and management of built heritage: Towards a library of vaults and wooden bean floors. In CIPA 2013 XXV International Symposium, ISPRS Annals, volume 164, pages 1–6.

Oreni, D., Brumana, R., Della Torre, S., Banfi, F., Barazzetti, L., Previtali, M., 2014a. Survey turned into HBIM: the restoration and the work involved concerning the Basilica di Collemaggio after the earthquake (L'Aquila). ISPRS Annals of the Photogrammetry, Remote Sensing and Spatial Information Sciences, vol.II, pages 267–273.

Oreni, D., Brumana, R., Banfi, F., Bertola, L., Barazzetti, L., Cuca, B., Previtali, M., Roncoroni, F., 2014b. Beyond Crude 3D Models: From Point Clouds to Historical Building Information Modeling via NURBS. In Digital Heritage. Progress in Cultural Heritage: Documentation, Preservation, and Protection, volume 8740, pages 166–175. Springer International Publishing.

Osello, A., 2012. The future of drawing with BIM for Engineers and Architects. Dario Flaccovio Editore s.r.l..

Piegl, L. A., Tiller, W., 1997. The NURBS book. Springer.

Piegl, L. A., Tiller, W., 1999. Computing offsets of NURBS curves and surfaces. Computer-Aided Design, 31(2):147–156.

Remondino, F., El-Hakim, S. F., Grün, A., Zhang, L., 2008. Turning images into 3-D models - Developments and performance analysis of images matching for detailed surface reconstruction of heritage objects. IEEE Signal Processing Magazine, 36(5):55–65.

Rosina, E., Grinzato, E., 2001. Infrared and Thermal Testing for Conservation of Historic Building. Material Evaluation, ASNT Journal, 59/n(8):942 – 954. ASNT Columbus (OH) USA.

Strang, G., Fix, G. J., 1973. An analysis of the finite element method.

Zienkiewicz, O. C., 1971. The finite element method in engineering science. McGraw-Hill.

This contribution has been peer-reviewed.
doi:10.5194/isprsarchives-XL-5-W4-345-2015

Structural Simulations and Conservation Analysis -Historic Building Information Model (HBIM)

C. Dore [a], M. Murphy [a], S. McCarthy [a], F. Brechin [a], C. Casidy [a], E. Dirix [b]

[a] School of Surveying and Construction Management, Dublin Institute of Technology, Maurice.Murphy@dit.ie
[b] KU Leuven - University of Leuven, Raymond Lemaire International Centre for Conservation, Faculty of Engineering, Graduate Student, eveliendirix@outlook.com

Commission V, WG V/4

KEY WORDS: Historic Building Information Modelling, HBIM, Surveying, Historic Structures, Building Conservation, Architectural Conservation, Laser Scanning, Virtual Cultural Heritage

ABSTRACT:

In this paper the current findings to date of the Historic Building Information Model (HBIM) of the Four Courts in Dublin are presented. The Historic Building Information Model (HBIM) forms the basis for both structural and conservation analysis to measure the impact of war damage which still impacts on the building. The laser scan survey was carried out in the summer of 2014 of the internal and external structure. After registration and processing of the laser scan survey, the HBIM was created of the damaged section of the building and is presented as two separate workflows in this paper. The first is the model created from historic data, the second a procedural and segmented model developed from laser scan survey of the war damaged drum and dome. From both models structural damage and decay simulations will be developed for documentation and conservation analysis.

1. INTRODUCTION

1.1 Overview

The specific aim of this paper is to present the research outputs to date of a Historic Building Information Modelling (HBIM) of the Four Courts, a historic classical building in Dublin City. This case study forms part of an overall project to develop a prototype system for national monuments in Ireland for the conservation, maintenance and management of historic properties, based on HBIM.

1.2 Definition

As a multi-disciplinary and evolving system, Historic Building Information Modelling (HBIM) has been developed in the Dublin Institute of Technology. The HBIM process (Murphy, M. et al, 2013) consists of the survey of existing structures using remote sensing followed by the mapping of parametric and information rich objects onto a geometric framework based on the remote survey data. The parametric objects which represent the architectural elements are built using a geometric descriptive language and are based on historic architectural documents (architectural rules and shape grammars). In addition these rules and grammars are exploited to procedurally model parts of the structures to speed up and automate parts of the process (Dore, C. and Murphy, M. 2014).The resultant HBIM can then be used for automatically producing conservation documentation and analysis of historic structures in addition to visualisation.

1.3 Complementary Research

Within the field of Virtual Cultural Heritage, Historic BIM was preceded by researchers (Chevrier, et al 2010; De Luca, L. 2012) developing parametric CAD objects which are then mapped onto laser scan or photogrammetric surveys. The parametric primitives or objects can be modified to fit the range of geometry and texture requirements for virtual representation of historic structures. Parametric modelling was then improved by introducing the benefits of using Building Information Modelling (BIM) for cultural heritage preservation. BIM was developed for the design, build and future management of new buildings and facilities. In BIM the production of virtual models can automatically generate not only standard drawings and schedules but also provides for structural, economic, energy and project management analytical data. Two problems existed for researchers in generating HBIM, the first is the absence of complex historic architectural elements in BIM libraries and secondly a system for mapping the objects onto remotely sensed survey data.

While these problems are been improved upon by the developments of HBIM carried out in the DIT (1), research in HBIM is becoming fairly widespread. This includes for example the work in Carleton University (Fai, S. et al, 2011; 2013; 2014), Polytechnic of Milan (Oreni, D. et al, 2014), KU Leuven (Boeykens, S., 2011; Boeykens, S. et al 2013) and others (Baik, A. et al 2014; Wu, T. C. et al 2013). The later research initiatives are case study based (modelling of historic structures) using various remote sensing hardware and software systems and BIM software platforms. In addition, in the later research the problem of building parametric architectural libraries is mainly overcome graphically as opposed to coding the parametric objects. In contrast, HBIM developed in the DIT; the library objects are built using Geometric Descriptive Language (GDL). Generative Modelling Language (GML) in addition to the existing Geometric Descriptive Language (GDL) is now being examined to allow for the libraries to be used as stand-alone or cross-platform. The systems used for mapping the objects onto point clouds or mesh models are based on manually aligning the objects onto orthographic projects from the point cloud or mesh model. More recent work concentrates on automation for improving the current slow process of converting unstructured point clouds into structured semantic BIM components; this includes DIT research (Dore, C. and

This contribution has been peer-reviewed.
doi:10.5194/isprsarchives-XL-5-W4-351-2015

Murphy, M. 2014) and the University of Bologna (Garagnani, S., 2013). The principal results from this research, which converts laser scan or photogrammetric survey data to HBIM is the creation of intelligent computer models which are used for conservation analysis and solutions for historic structures and their environments.

1.4 Case Study – Four Courts

The Four Courts is a late 18th century classical building in Ireland; the building was partially destroyed in 1922 during the civil war in Ireland which occurred during the early establishment of the Republic (see Figure 1). The effects of the civil war damage to the building are once again a threat to the structural stability of parts of the building. The intention is to use the HBIM to illustrate virtually the current extent of the damage/decay and as a basis for the proposed conservation interventions.

Figure 1: War Damage to the Four Courts

1.5 Summary of Methodologies Used

Two data sets are available for creating the HBIM, the first is the archive of historic surveys carried out since the 1780s through to recent years. The second data set is a recent laser scan survey. The laser scan survey was carried out in the summer of 2014 of the internal and external structure using a Leica HDS C10. For this reason the HBIM work programme is divided into two workflows, the first based on the historic data and the second on the laser scan survey (see Figure 2).

Figure 2: Laser Scan Survey

The historic documents (drawings, documents and iconographic sources of the specific building) are interpreted and identified from an art and architecture historical perspective to accompany the computing research methods. In addition to this, information concerning historic European construction techniques and architectural details can be found in architectural manuscripts, which have evolved from Vitruvius (70 – 35 BC) to the 17th and 18th century architectural pattern books. These architectural pattern books are employed to supplement specific architectural records for creating library objects.

The second model is created as a procedural and segmented model from the laser scan of part of the war damaged drum and dome. Using concepts from procedural modelling, a new set of rules and algorithms have been developed to automatically combine parametric library objects and generate Historic Building Information Models (HBIMs) from survey data. The procedural HBIM workflow is shown in Figure 3. First a laser scan or photogrammetric survey is carried out to acquire data to generate a HBIM. In the second stage survey data is processed to produce segmented point clouds and cut-sections which are imported into BIM software. Next procedural rules are applied to survey data to automatically generate the required BIM geometry. The generated geometry can then be refined by adjusting parameters to accurately match the geometry to specific survey data. Once a HBIM has been generated and precisely mapped to survey data, documentation can be automatically generated from the model. This includes plans, elevation drawings, sections and 3D perspectives. Further analysis can also be carried for structural and conservation analysis of historic buildings.

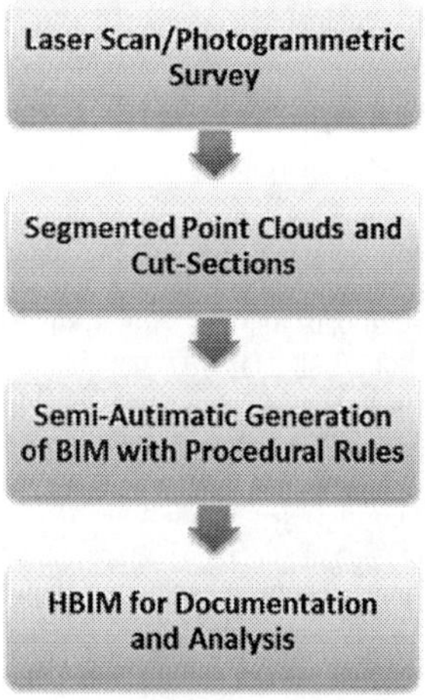

Figure 3: Procedural Workflow

2. WORKFLOW 1 - HISTORIC RECONSTRUCTION

2.1 HBIM

The series of routines for building the first model is based on historic data entirely and no input is introduced from the laser scan survey and is illustrated in the Figure 4 below. Given that the Four Courts is one of the Dublin city's most important historic buildings a large collection of drawings and surveys from a period over 220 years are still available within the state archive. These documents have been scanned and scaled for re-mapping in HBIM. The historic documents are reviewed from an art and architectural history perspective to accompany the computing methodological approaches in order to identify correct uses and interpretation of historic sources.

In Figure 4, (Detail 1) a sample of the historic drawings is shown, the columns, wall and beams are placed on a scaled plan and the section detail which contains imperial measurements allows for vertical positions as shown in detail 2. Other sets of data are used to position the wall (Detail 3), beams and dome (Detail 4). Mouldings and details such as dental block are added (see Detail 5), windows and accompanied openings are inserted, as illustrated in Details 7, 8 and 9. The internal dome is added (see Detail 10) and the supporting buttresses are added as shown in Detail 11.

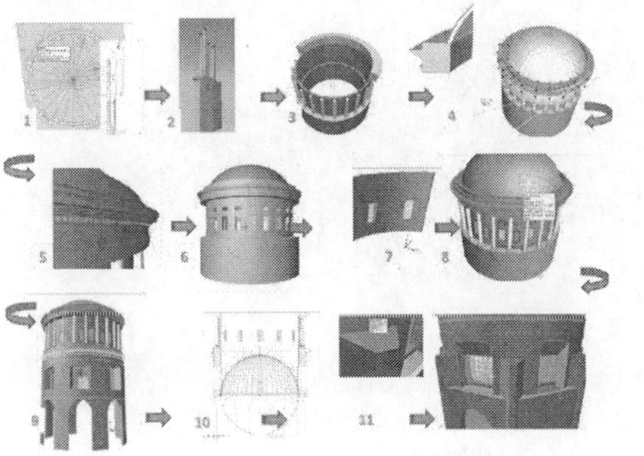

Figure 4: Details 1 to 11 - Historic Model

2.2 Building Morphology

The model illustrated in the figure 4 above, indicates a number of relevant physical characteristics concerning the damaged section of the building. These are the original historic construction techniques used in the 18th century and the interventions to repair to the building after war damage in the 1920s. In summary, almost all of the Corinthian columns surrounding the dome had been shattered. The inner walls of granite appear to have remained intact. The Corinthian columns, which had originally been carved on all sides were removed and rotated so that the damaged sides face inwards. Two of the columns which had been destroyed were replaced with replicas cast in artificial stone. An internal system of steel framing was employed to brace the external walls and reinforced concrete floors replaced the original floors. The dome was replaced using reinforced concrete, which was a very progressive solution for the time, as a thin skin of ferro-concrete was used. This consists of a very slender, complex curved leaves of reinforced concrete used for shell roof construction.

2.3 Creating the Library of Architectural Elements

In the case of this building, an understanding of classical architecture within its correct time-frame is essential. This is required to identify significant rules, which can be applied to developing the library of objects used to plot the model. The historic surveys and drawings (see Figure 5) usually show the spatial positions of architectural elements but do not contain the construction detail, the architectural rules and shape grammars for composing these elements. This information can be found in various historic architectural texts. Therefore a substantial portion of the investigation for creating accurate parametric objects relates to locating and understanding these texts within the correct historic and geographic framework.

Figure 5: 18th Century Section of Four Courts (Architect Gandon)

2.4 Historic Framework for Classical Architecture

The first recorded treatise concerning classical architectural rules was "De Architectura" by the Roman scholar Vitruvius. Later, during the classical revival of the Italian Renaissance the following three tractates on architectural theory are of key importance: "De re Aedificatoria" (1443-1452) by Leon Battista Alberti, "Regole Generali D'Architettura" (1537) by Sebastiano Serlio and "Regola Delli Cinque Ordini D'Architettura" (1562) by Giacomo Barozzi da Vignola. Vitruvius's influence stretched across Europe with the treatises of Claude Perrault who published the only French translation of 'De Architectura' in 1673 and hereby contributed to revive the feeling for the rules and principles in architecture. In 18th century Britain, Andrea Palladio's work "Quattro Libri dell'Architettura" (1570) influenced the production of crucial architectural treatises. 'The Rules for Drawing the Several Parts of Architecture' (1732) by James Gibbs can be considered as a precursor for the production of practical pattern books (see Figure 6 below) throughout the 17th and 18th centuries in Britain (with an extension in 19th century Irish architecture) .

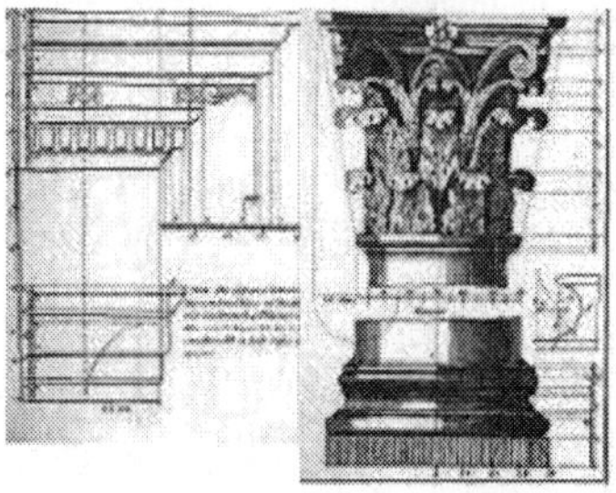

Figure 6: Pattern Book Detail – Corinthian Capital

This contribution has been peer-reviewed.
doi:10.5194/isprsarchives-XL-5-W4-351-2015

The interpretation and understanding of architectural rules can be more easily adapted from these architectural pattern books, which can fill the hiatuses in historic archival documents. The coding or graphic system for developing the libraries follows on from this, below is an example of sash window library parts developed from these rules and pattern book sources (Murphy, M., 2012).

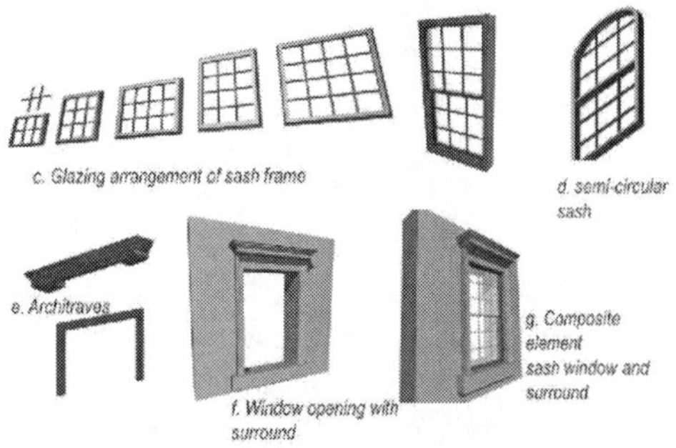

Figure 7: Library Objects

3. WORKFLOW 2 – PROCEEDURAL RECONSTRUCTION

3.1 Procedural Rules for HBIM

Procedural rules have been developed for modelling building façades and complete external building geometry. A prototype plug-in has been developed with these rules for the ArchiCAD BIM software. This prototype plug-in allows for semi-automatic generation of existing buildings from survey data. The design and implementation of the procedural HBIM rules are described in this section.

3.1.1 Design and Conceptual Framework

The design and conceptual framework for the HBIM procedural rules are based on concepts from shape grammars. A shape grammar is a production system used in procedural modelling to automatically generate two or three dimensional geometries from a basic vocabulary of shapes and a set of production or shape rules. A HBIM library of parametric architectural elements (Murphy etal, 2013) are used as a shape vocabulary on which the rules and algorithms are applied.

Architectural rules and proportions are also used to assist the reconstruction process. Architectural rules and proportions outlined in pattern books relating to classical buildings are used to provide an initial estimate for the position and size of building elements on a generated building. This greatly reduces the amount of further editing required when modelling classical buildings from survey data.

3.2 New Procedural Rule for Modelling Deformed Drum Walls

This new procedural rule is designed for accurate and efficient modelling of curved walls that contain deformation such as the walls supporting a dome and drum of the Four Courts. As a result of deformation the walls which support the dome may not be perfectly vertical and may contain different levels of deviation at different heights and locations. A design for accurately modelling this type of irregular wall has been developed using multiple cut sections from a point cloud (see Figure 8). Horizontal cut sections through a circular wall at different heights will show areas where the wall is bulging or warping. Creating a surface from cut sections through the wall at different heights allows the variances of the surface to be accurately represented. This new procedural rule allows users to select a group of horizontal cut sections which are then used to automatically generate the irregular wall surface (Figure 9). The non-vertical wall is created by generating a surface between each horizontal cut section. This rule results in automatically generated geometry for a drum wall which represents the true condition of the wall with deformation based on scan data. The design and implementation of this rule is described more in in this section.

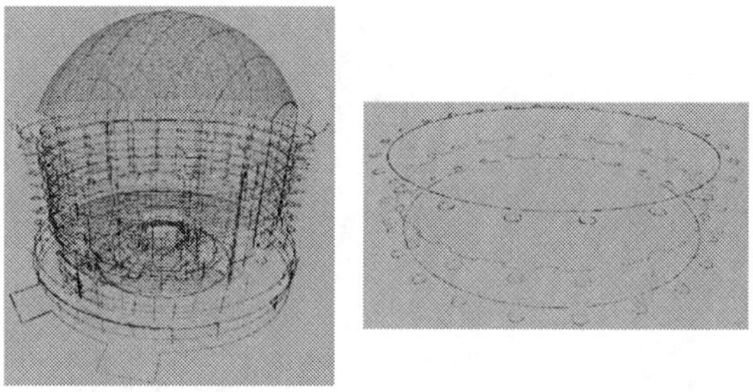

Figure 8: Segmented point cloud and horizontal cut sections imported into BIM software

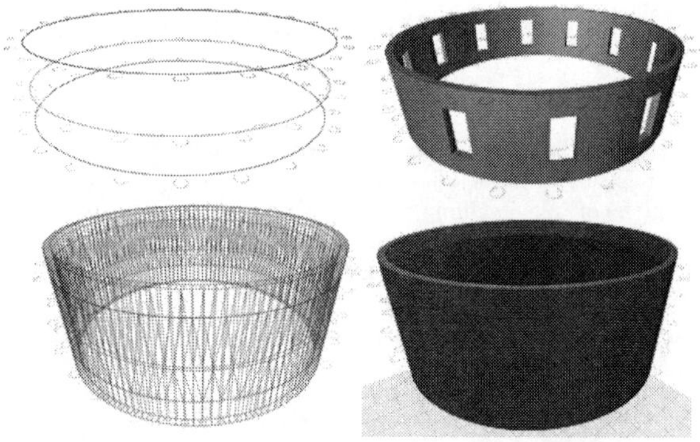

Figure 9: Irregular non-vertical walls containing deformation automatically generated from cut-sections. A split rule is used to automatically add window openings.

3.2.1 Conceptual Design for New Procedural Rule

The input for this rule is a set of horizontal cut sections which are created from the point cloud. For this rule the cut sections need to be represented as closed polygons made up of arcs or lines. Each cut section also must have an associated height parameter for that section. The conceptual design for this rule involves a user selecting any number of cut sections from the point cloud and applying the rule which automatically generates the irregular wall surface.

The main challenge in implementing this rule is in determining how points in one cut section correspond and relate to points in

This contribution has been peer-reviewed.
doi:10.5194/isprsarchives-XL-5-W4-351-2015

the next cut section above and below it. As all cut sections will vary there are no common points between each section. To overcome this problem a common framework is set up to identify relating points between each section (Figure 10). An algorithm is used to calculate coordinates at the intersection of this regular framework and each cut section. The wall surface is then created by connecting these relating points between each section.

3.2.2 Implementation of New Procedural Rule

This new rule has also been implemented as part of the prototype plug-in to the ArchiCAD BIM software. The C++ programming language is first used with the software API to retrieve data relating to the group of cut sections selected by a user. This data is then passed on to a GDL script where the main code for this rule is implemented. An algorithm is used to calculate the corresponding points between cut-sections (Figure 10). The 3D "RULED" function in GDL is then used to create the new surface which connects the cut sections based on corresponding points. Offsets for the inner wall surface are then calculated based on a user parameter for the wall thickness.

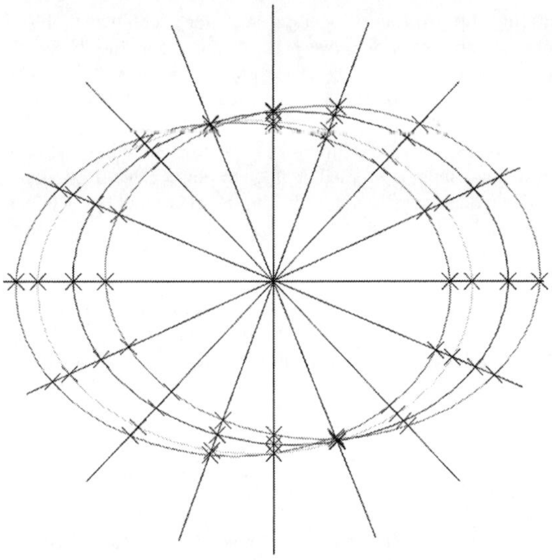

Figure 10. Framework (black lines) for identifying corresponding points between horizontal cut sections (coloured)

Once the walls have been automatically generated with this new procedural rule then other rules can also be applied such as the split rule described in the previous section. This split rule will automatically divide the drum wall into any number of equal tiles with openings (Figure 11). The positions and size of window openings can then be graphically refined to precisely fit the survey data. An additional rule is also used to create a mesh surface for the roof of the dome from a sample of points from the scan data (Figure 11). Continued work will involve developing additional procedural rules for automatically arranging columns and mouldings to complete the HBIM for this section of the Four Courts building.

Existing tools within the ArchiCAD BIM software are very limited for modelling non-vertical circular walls which contain deformation. This new rule allows for more accurate and efficient generation of HBIMs from survey data which can facilitate structural and conservation analysis of existing buildings.

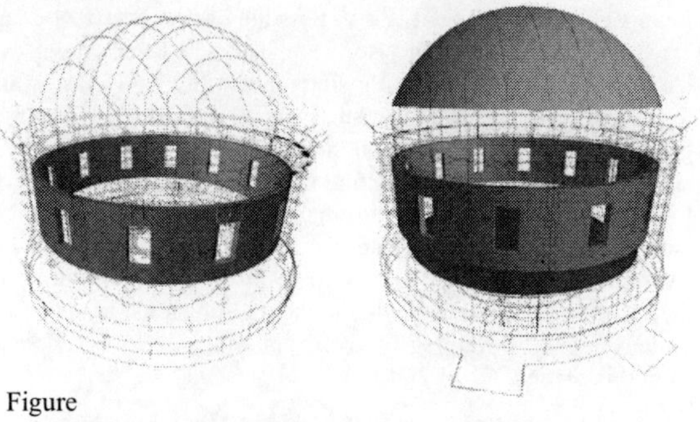

Figure
11. Wall containing deformation automatically generated from cut sections. Additional mesh surfaces used to represent dome roof and floor surfaces (right).

4. NEXT PHASE OF WORK

4.1 Sub-Phases

The project is multidisciplinary; the initial phase for developing the Historic Building Information Model involved surveyors and architectural historians. The next phase encompasses using the model to automate documentation to identify and propose solutions for decay, damage and structural integrity, which will involve material scientists and engineers.

4.1.1 Building Decay and Damage Mapping

Three approaches are recommended: initially, manually mapping of decay onto 2D scaled orthographic projections of the model in plan, elevation and section. The second is an attempt to locate the decay and damage onto the 3D geometry using total station or other instruments, which will locate 3D coordinate positions. Finally the use of digital photogrammetry is proposed, to build the elements where the scan data is not sufficiently dense or cannot access. Below is an example of a digital photogrammetric reconstruction of part of the scroll from the Corinthian capital which broke away from the capital of the column.

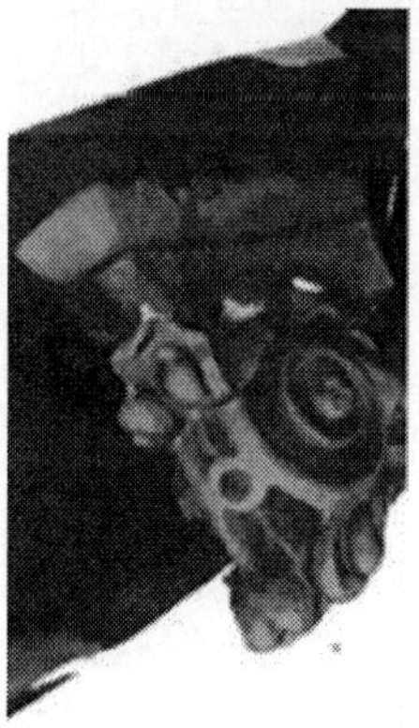

Figure 12: Corinthian capital – Photogrammetric Model

This contribution has been peer-reviewed.
doi:10.5194/isprsarchives-XL-5-W4-351-2015

4.1.2 Structural Analysis

Open BIM is a model-based design approach based on the Industrial Foundation Class or IFC model exchange format. Using standard international formats such as the Industry Foundation Class (IFC) it is also possible to export HBIM data for further analysis in other software. This has been shown by Oreni et al. (2014) where the interoperability of HBIM was tested with other software such as Rhino for performing geometric analysis and simulations and Midas for finite element analysis. An example of structural behaviour is illustrated in the Figure below, a Section through dome and drum walls automatically created from survey data with procedural rules (workflow 2) shows true condition of wall which outlines deformation, which may be due to the traditional materials used or movement. A warping of the drum wall is clearly evident from this section.

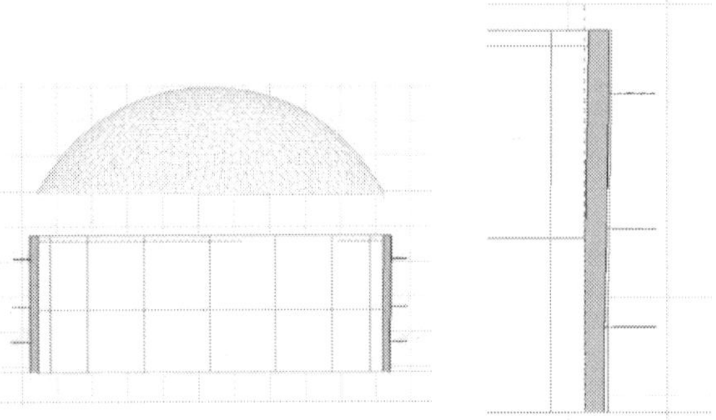

Figure 13. Section through dome and drum walls automatically created from survey data with procedural rules. Section shows true condition of wall which contains deformation.

ACKNOWLEDGEMENTS

The laser scan survey was carried out by **Andrew Moore, Survey Instrument Services Ireland.**

REFERENCES

Baik, A., Alitany, A., Boehm, J., & Robson, S. (2014). Jeddah Historical Building Information Modelling JHBIM – Object Library. ISPRS Ann. Photogramm. Remote Sens. Spatial Inf. Sci., II-5, 41-47. doi: 10.5194/isprsannals-II-5-41-2014

Boeykens, S. (2011, 6-8 July 2011). Using 3D Design software, BIM and game engines for architectural historical reconstruction Paper presented at the CAAD Futures 2011, Liège, Belgium.

Boeykens, S., Himpe, C., & Martens, B. (2012, 12-14 September 2012). A Case Study of Using BIM in Historical Reconstruction - The Vinohrady synagogue in Prague. Paper presented at the The 30th International Conference on Education and research in Computing Aided Architectural Design in Europe, Prague, Czech Republic.

Chevrier, C., Charbonneau, N., Grussenmeyer, P., & Perrin, J.-P. (2010). Parametric Documenting of Built Heritage: 3D Virtual Reconstruction of Architectural Details. International Journal of Architectural Computing, 08(02), 131-145.

De Luca, L. (2012). Methods, Formalisms and Tools for the Semantic-Based Surveying and Representation of Architectural Heritage. Applied Geomatics(1866-9298). doi: 10.1007/s12518-011-0076-7

Fai, S., Graham, K., Duckworth, T., Wood, N., & Attar, R. (2011). Building Information Modelling and Heritage Documentation. Paper presented at the XXIII CIPA International Symposium, Prague, Czech Republic. http://cipa.icomos.org/fileadmin/template/doc/PRAGUE/047.pdf

Fai, S., & Rafeiro, J. (2014). Establishing an Appropriate Level of Detail (LoD) for a Building Information Model (BIM) – West Block, Parliament Hill, Ottawa, Canada. ISPRS Ann. Photogramm. Remote Sens. Spatial Inf. Sci., II-5, 123-130. doi: 10.5194/isprsannals-II-5-123-

Fai, S., & Sydor, M. (2013, Oct. 28 2013-Nov. 1 2013). Building Information Modelling and the documentation of architectural heritage: Between the 'typical' and the 'specific'. Paper presented at the Digital Heritage International Congress (DigitalHeritage), 2013.

Garagnani, S. (2013). Building Information Modeling and real world knowledge: A methodological approach to accurate semantic documentation for the built environment. Paper presented at the Digital Heritage International Congress (DigitalHeritage), 2013.

Dore, C., Murphy, M,. Semi-automatic generation of as-built BIM façade geometry from laser and image data, ITcon Vol. 19, pg. 20-46, http://www.itcon.org/2014/2

Murphy, M., (2012), PhD, Historic Building Information Modelling, Trinity College Dublin, unpublished. https://www.academia.edu/8237173/Historic_Building_Information_Modelling_HBIM_PhD

Murphy, M., McGovern, E., Pavia, S., (2013), Historic Building Information Modelling -Adding intelligence to laser and image based surveys Elsevier, ISPRS Journal of Photogrammetry and Remote Sensing. https://www.academia.edu/5932333/HBIM

Oreni, D., Brumana, R., Della Torre, S., Banfi, F., Barazzetti, L., & Previtali, M. (2014). Survey turned into HBIM: the restoration and the work involved concerning the Basilica di Collemaggio after the earthquake (L'Aquila). ISPRS Ann. Photogramm. Remote Sens. Spatial Inf. Sci., II-5, 267-273. doi: 10.5194/isprsannals-II-5-267-2014

Von Geÿmuller, 1898-1901. Die Baukunst der Renaissance in Frankreich. Stuttgart

Wu, T. C., Lin, Y. C., Hsu, M. F., Zheng, N. W., & Chen, W. L. (2013). Improving Traditional Building Repair Construction Quality Using Historic Building Information Modeling Concept. Int. Arch. Photogramm. Remote Sens. Spatial Inf. Sci., XL-5/W2, 691-694. doi: 10.5194/isprsarchives-XL-5-W2-691-2013.

APPENDIX

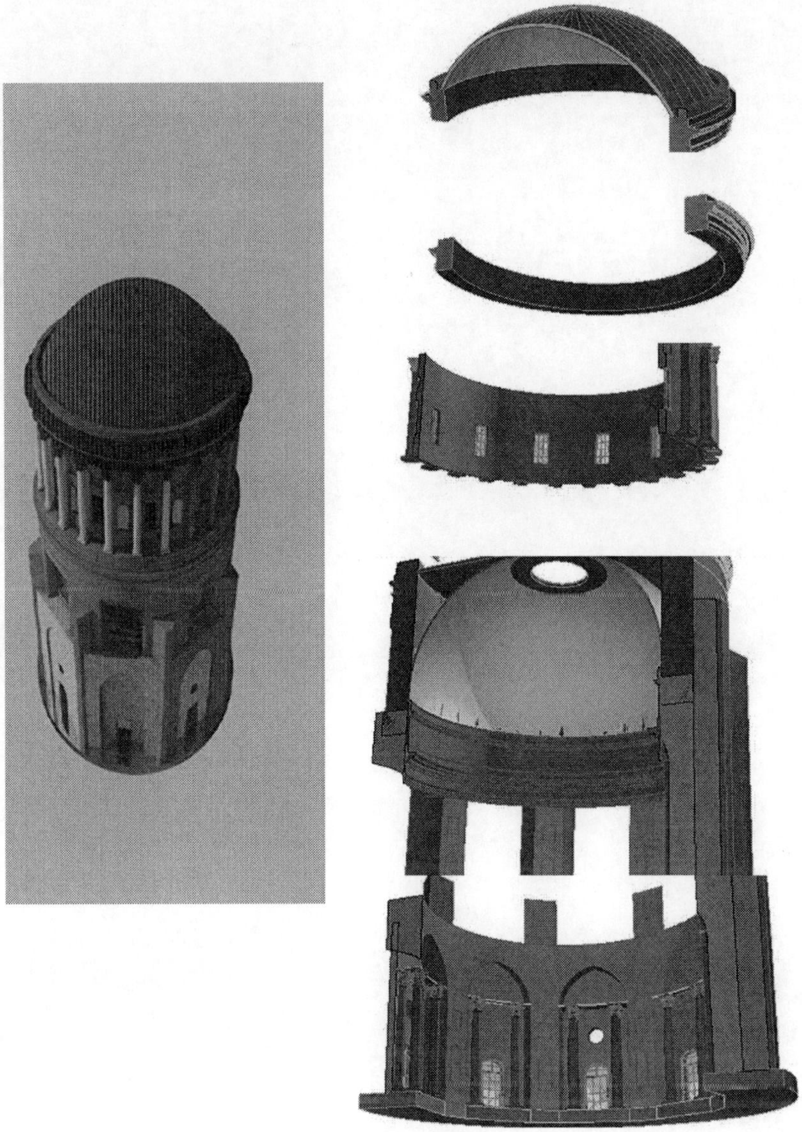

HBIM model of dome and drum

This contribution has been peer-reviewed.
doi:10.5194/isprsarchives-XL-5-W4-351-2015

A NEW IDEA OF BIM SYSTEM FOR VISUALIZATION, WEB SHARING AND USING HUGE COMPLEX 3D MODELS FOR FACILITY MANAGEMENT.

F. Fassi [a] *, C. Achille [a], A. Mandelli [a], F. Rechichi [a], S. Parri [b]

[a] Politecnico di Milano, ABC Department, 20133 Milano, Italy
(francesco.fassi, cristiana.achille, alessandro.mandelli, fabrizio.rechichi@polimi.it)
[b] Newthread srl - parri@newthread.it

Commission V, WG V/4

KEY WORDS: BIM, 3D Visualization, 3D models, web sharing

ABSTRACT:

The work is the final part of a multi-year research project on the Milan Cathedral, which focused on the complete survey and three-dimensional modeling of the Great Spire (Fassi et al., 2011) and the two altars in the transept. The main purpose of the job was to prepare support data for the maintenance operations involving the cathedral since 2009 and still in progress.
The research job had begun addressing our efforts to identify which methods would allow an expeditious but comprehensive measure of complex architectural structure as a whole. (Achille et al., 2012) The following research works were focused mainly to find an efficient method to visualize, use and share the realized 3D model.

1. INTRODUCTION

1.1 General introduction

The presented work will answer the final question asked by the paper presented to 4th ISPRS International Workshop, 3DARCH 2011: "3D Virtual Reconstruction and Visualization of Complex Architectures" Trento, Italy, 2-4 March 2011. (Fassi et al., 2011) The main topic of that job was about the best strategy to survey and model huge and complex architectures. The study case was the research activities conducted on the main spire of the Milan's Cathedral that have led to creation of a huge three-dimensional model of the spire characterized in great detail and high complexity. Initially the aim of the job was to build a 3D complete model of the spire in order to extract all feasible measures useful for the restoration site, but the feeling was that a complex and complete 3D model can give more. For this reason that job was concluded by asking the question: "A huge 3D model: what does it for?" (Fassi et al., 2011). The efforts of these last years were addressed in developing a sort of BIM system working on the web that would allow the use of the three-dimensional models during facility managements, supporting both the filed activities on the yard and the design and data updating in the technical offices (Fassi, 2012).

1.2 The state of art

In the last years we are witnessing a sudden development of. "Building Information Modeling" (BIM) system. It is *"a digital representation of physical and functional characteristics of a facility. A BIM is a shared knowledge resource for information about a facility forming a reliable basis for decisions during its life-cycle"* (NBIMS, 2012). This is only one of the many definitions of BIM but it makes clear in a few words the meaning and the potential of these systems. In fact in the recent years there has been international awareness that BIMs can make the design, construction and the different "building" operations more efficient, optimizing and speeding up building processes (Coates, 2010). Usually and until today, the Architecture, Engineering and Construction Industry is focusing only on new buildings without paying attention to the Cultural Building Heritage. So, the need to refurbish, but also all management and conservation processes of the existing buildings makes BIM for Cultural Heritage an essential research field in the last years (Del Giudice et al., 2013). The use of BIM can help in the understanding and preservation of heritage both for the professional and the general public.
Indeed it is exactly in the field of Cultural Heritage where 3D-BIM systems have big potential, because the "CH objects" are expected to live forever and the maintenance processes, are continuous and endless into the future.
Moreover *"A basic premise of BIM is collaboration by different stakeholders at different phases of the life cycle of a facility to insert, extract, update or modify information in the BIM to support and reflect the roles of that stakeholder"* (NBIMS, 2010) and this is valid more than in other places, right in the world of CH. In fact, all activities require collaborations across disciplines, technologies and sectors, such as history, archaeology, cultural studies, anthropology and other areas of humanities and social science.
The three main tasks are: i) object survey, ii) object representation and iii) visualization/sharing of information. Today the survey phase is relatively fast thanks to laser scanning and automatic image based algorithms. The creation of the 3D model is the real bottleneck of the whole workflow. Nowadays, the research is focusing mainly on this task and in particular developing algorithms able to speed up the modelling phase.
The main topic is normally the simplification of the modelled structures using parametric or semantic modelling procedures.

* Corresponding author. This is useful to know for communication with the appropriate person in cases with more than one author.

This contribution has been peer-reviewed.
doi:10.5194/isprsarchives-XL-5-W4-359-2015

These methodologies have achieved very good results in case of "regular simply study cases" but are inadequate for complex and irregular objects (as in the archaeological case of studies) or for big scale representation where simplification is not possible. Moreover commercial systems are not able to model complex freeform object in a real-based way and also the imported models can not be fully used in the "information system" as smart objects. For these reasons different types of volumetric modelling procedures, typical of Archicad or Revit, or standard parametric libraries (HBIM) make commercial systems inadequate or not sufficient for many CH case of studies.
The third objective faces the difficulty to share and make usable by different people a considerable amount of data. The main issue is the use and sharing of three-dimensional models that are normally of large dimension, both for the high resolution with which are represented and for their effective extension and complessity. Nowadays there are many webservices able to visualize and share 3D models. Normally they are little objects and little 3D models. In the field of Cultural Heritage there are few systems able to visualize complex and large 3D data on the web. One is Cyark viewer able to visualize high resolution pointcloud or mesh surfaces on a dedicated Webgl system directly on the browser (Silla L., 2012).
Another interesting system is Potree (Potree, 2014). It is a free WebGL based point cloud viewer for very large datasets. It is based on Scanopy project with the purpose to develop efficient data structures and algorithm to allow displaying huge point-based models (Wimmer M. et al., 2006). These are very interesting and useful visualization systems. They can be very useful because all survey methods, both image and range based, produce quite huge and dense point clouds, that for many purposes, can be enough to represent the object without any futher elaborations. Another interesting project is 3DHOP (3D Heritage Online Presenter) developed by the Visual Computing Lab, ISTI-CNR. It is a collection of tools and templates for the creation of multimedia interactive Web presentations of digital cultural artifacts. It is a free software able to visualize online a 3D mesh model in the native Nexus format. It is thought for online visualization and museal presentation with the aim to be used by anyone with simple webpage-creation skills, without the need for specific Computer Graphic expertise. Also this application is built using the WebGL API technology and the SpiderGL JavaScript library. An interesting aspect is the possibility to costumize the html pages inserting every kind of textual information.
An important citation is due to the Nubes Project developed by CNRS (De Luca L., 2013); that is an information system at the architectural scale of representation. The main aim of that project is to create a web-system able to allow a multidisciplinary analysis of heritage building creating complex relations between geometric 3D representation of the building with heterogeneus data concerning various domains (technical, historical, documentary). It is the first complete example of information system sharing on the web created ad hoc for Cultural Heritage.

1.3 The three main characteristics of the new system

The system is designed from three main considerations.
The main first idea was to break away from the commercial software that nowadays are unsuitable for the multi-faceted world of Cultural Heritage. They are unable (totally or partially) to handle complex "real based modelled shapes", large scales modelled objects and big dimension areas or buildings. In addition, the main problem is the final use of the model. Commercially, the implemented applications are rarely, and only partially, interesting for a cultural heritage object, because they have been conceived for the design in the maintenance of new buildings. Moreover, they do not include features useful for maintenance operations and restoration activities. This is because "cultural heritage objects" are unique from the artistic, historical, architectural and constructive point of view: each has its own strong identity that is impossible to standardize.
The starting point is the realization of a highly customizable system able to easily adapt itself to the different cases allowing to work with different type of objects, different representation scale and resolutions, also simultaneously showing and sharing different type of information and managing different and unique maintenance operations.
For this reason, the presented system is able to visualize huge dimension and high numbers of different 3D models. They can be built with any external software and can be linear simple blocks (structure) or complex high-resolution multi-scale free form models (decorations) or both integrated together. That mean that:

- there are no constraints on the number of the viewable objects,
- there are no constraints on the level of detail of each model. So the users can visualize each object with different level of detail, depending on their needs,
- the users can place models of any type, regardless of the method or software by which they were created (manual modelling, parametric modelling or semantic modelling or real based modelling),
- the original 3D model format is preserved by the system (nurbs or mesh models) in order to let the modification in the original modelling software, but automatically managed/transformed as mesh (also textured) inside the system viewer.

The second key aspect is a consideration about the operators working in Cultural Heritage field. The modern explosion of 3D models means that even this world will have to create professional roles able to build, modify and update three-dimensional virtual representation of an object. In general, however, the user and the classical operator in the field of cultural heritage will not have the knowledge to use, modify complex three-dimensional models requiring primarily dedicated skills and high-level hardware technologies. The second aim of the system is to solve this problem. The system should be easy to use and easy to learn even for non-experts. It should have a user-friendly interface and should be used with common devices. That is the only way to ensure that the system becomes integral part of the maintenance and restoration practices and very attractive to the variegated world of operators in CH field.
The third consideration is that a digital 3D model is expected to become the representation of the future. This is valid in any contemporary field of study, but can be extremely significant in the field of Cultural Heritage. A 3D model is a complete representation of the "object" that can be analysed and studied in its entirety, without any effort of abstraction by the operators. Digital 3D models add big chances in the process of knowledge, providing new possibilities on the way to study, manage, monitor, visualize, conserve and valorise a cultural heritage object. However, to achieve this purpose it should be dynamic and eternal, that means it should be forever upgradeable and modifiable in the future; it should be "for everybody", i.e. visualized and used easily by different operators and "from everywhere" that means that a digital 3D model should be shared and used in a collaborative way.

This contribution has been peer-reviewed.
doi:10.5194/isprsarchives-XL-5-W4-359-2015

2. THE WEBBIM SYSTEM.

For these reasons, the presented system, named BIM3DSG, is thought divided in three parts:

1. a first part that is thought to be mainly used from professional and 3D specialist. It gives full access to the whole system and all functions inside the modelling software or standalone (work in progress). It is required to add or modify (export process) 3D models into the system; every kind of previously modelled surface are supported.
2. The second part is instead thought for all other users and, therefore, it is specially designed to be easy to use and usable in mobility on different devices, such as PC, notebook, tablet and smartphones. This part requires only a web browser and gives full access to the whole system and all the functions inside the web browser, except the ability to add or modify 3D models. It can also be used on device with few hardware resources because it permits to view objects at different levels of detail, computed automatically by the system during the export process, through approximation of the mesh surface extrapolated from the original model.
3. at last, the system needs a central database that collects both 3D models and all information and additional files related to each object. It is ready-made by system administration but can be easily automatically initialized by the user. The use of a central database allows users to share and synchronize all changes in real-time.

Regardless of whether using the first or the second part, BIM3DSG allows the user to:

- view all or part (selected through some search functions) of the objects previously exported in the system
- view automatically computed BIM information of each object, such as area, volume, position, …
- add/edit/view user BIM information and files (photo, video, documents, dwg, …) associated with one or more objects
- add/edit/view maintenance, restoration and building site activities and also all related information and files (photo, video, documents, dvg, …)
- view past situation, and also view past and present situation at the same time for comparison intent

In both parts powerful caching mechanisms have been implemented, in order to ensure the user-friendliness and the appeal of the system, allowing almost instantaneous loading time, even if the available connection is slow.

2.1 Distinctive feature

What makes our system possible is the innovative idea (patent pending, protocol number MI2014A002016) to split the whole model into each constitutive entities and save all of them separated one from other into a central database; in addition a series of search functions that also allow users to also load only portions of the whole model.

The type of constitutive entities is defined by the user, not only for each project, but also inside a project, to easily adapt to the widest range of needs. For example, if the whole three-dimensional complex model is the representation of a cathedral, the single constitutive entity could be a capital of a column, the column itself, or even an entire spire, but also a constitutive part of the capital of the column.

The only requirement for the user is to use, within the modelling software, different objects for each constitutive entity; during the export process, our system can recognize each object and, automatically and transparently to the user, is responsible to divide each object in a separate file that will be stored into the central database.

When the user needs to view the model, he can search for it and load only the portions of interest through a series of search functions; e.g. the user can make a selection by:

- using one or more filters on identification parameters of the constitutive entities, such as for example the area, the zone, the sector, the type and name of the single element
- including of one or more, or ranges of unique codes, associated automatically by the system for each constitutive entities during the export process
- asking for a cross-section of the entire model (or an area of the entire model), with a height defined by the user and focused on a single element selected by the user
- asking for a selection of a ball around on the whole model (or an area of the entire model), with a radius defined by the user and focused on a single element selected by the user

Other queries can be included by request and can involve maintenance, restoration and building site activities.

The most interesting thing about the ability to load only the portions of interest of the whole model is that it allows users to use so defined model that couldn't be loaded entirely at full quality on the actual hardware, or that could be loaded, but with too poor performance.

In addition, the ability to use different levels of detail (automatically computed) on the web allows you to display always the entire model, even when it should not be possible to do it because of hardware limits, with the only limit of using a lower level of detail.

Thanks to this feature, our system allows the user to:

- insert and manage any number of different objects different from each other, even if they are not parametrizable;
- insert and manage any number of any resolution (chosen by the user) models of objects which geometry is not default;
- use simultaneously simple and complex objects.

2.2 The central database

The central database is rather installed on a remote server or on the cloud, in order to be accessible by a plurality of workstations and devices through the conventional Internet connection. However, if needed, the database can be both local or placed within a local network.

The central database stores not only the internal structure and the textual BIM information inserted by users, but also the models files (both in the format used by modelling software and JSON format, automatically computed and used for web visualization), the BIM files inserted by users and any other necessary data and files.

We chose to use PostgreSQL on Linux, because is Open Source and has a lot of useful features. However, our system can be adapted to use other database, for example Microsoft SQL Server or Oracle Database.

This contribution has been peer-reviewed.
doi:10.5194/isprsarchives-XL-5-W4-359-2015

2.3 The modelling management part

We wrote a complex plug-in for Rhinoceros, that gives users access to all functions and features, except the use of level of detail, that is distinctive for web visualization.
The choice of Rhinoceros derived by the fact that it is one of the few software that support both NURBS and MESH, and it is easy to use and learn.
Our plug-in is written in C# and .NET. For the UI, we used Windows Form in the first version, while in second version we are using WPF embedded inside Windows Form (embedded because Rhinoceros is written in Windows Form), because WPF is more attractive and modern.
Our plug-in also uses ITinnovationsLibrary (Rechichi, 2014) and only in second version Xceed WPF Toolkit and a modded version of AvalonDock.
The export process is written with high multithread parallelism, so that it can use all physical and virtual core available; it can reduce the time of export process to about 7/8 times on consumer processor and at least 11/12 times on even more on high performance workstation processor.
We have planned for the future to release a stand-alone application that can export OBJ files into the system. We can also develop, on request, plug-ins for other modelling software.

2.4 The online platform

The ability to use a common web browser to view 3D models and perform all tasks, except the export process, is a very important feature of our system.
First, it helps to persuade common user towards the use of the system during the daily work. Then it can be also used in mobility and on mobile devices and it is cross-platform. Last but not the least, there is no additional cost for customers because browsers are free, and one organization (for example a public administration) could give read-only access (full access permission and separation, based on UNIX like schema, are supported by BIM3DSG) to models and some data to everyone for free without any cost.
The touch-screen support makes BIM3DSG even more simple to use and let it also run on touch-screen only device, e.g. in a museum where we can have a touch display for visitors.
BIM3DSG, for visualizing 3D models inside a web browser, uses WebGL through SceneJS library. The core system uses also Javascript, JQuery, php, css. The web interface is based on JQuery EasyUI.

2.5 Modules

The main functionality of BIM3DSG can be summarized into a block diagram, as we can see in Figure 1.
The first step is the user authentication: BIM3DSG allows the management of advanced user permissions (users and groups), allowing to prevent or to allow access and/or modification of each single element and each single file inserted, according to a UNIX like access scheme.
Through the export process user can add the constitutive entities and/or modify all the previously inserted entities (only inside modelling management part).
Then we have the import process, by which at any time the user can import and load all or part of the previous inserted entities, in order to view, analyze, modify, or access all the BIM information and data associated with them.
Once the user has imported the entities, it is possible to access a lot of feature for viewing and entering data, information and files on each single constitutive entities and its model, starting from the selection of the same. BIM3DSG let user to:

- insert any type of textual information
- insert multimedia files (photo, video, ...) of any format
- insert other files of any format (documents, reports, pdf, dwg, ...)

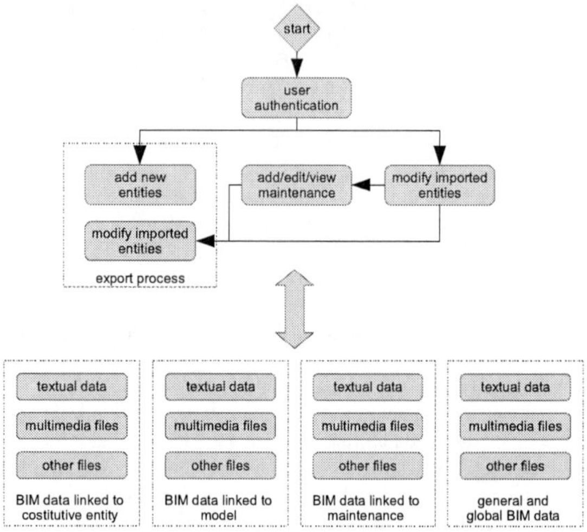

Figure 1. block schema for the main features of BIM3DSG.

Each information can be mapped on both a single constitutive entity, or on its model, or on the maintenance events, or as well on a group of them.
Furthermore, user can view some automatically computed BIM information of each model, such as area, volume, position, ... These data are computed by BIM3DGS during the export process.
Eventually the user can add/edit/view maintenance, restoration and building site events, starting as usual from the selection of constitutive entities involved.

2.6 The export process

The export process is the core of our system, therefore we go on analizing it in more detail.
First, BIM3DSG validates the objects selected for exportation process and write permission; then it displays them to the user, allowing the identification of new objects and modifying objects (identification can be both manual or automatic).
After this first step, BIM3DGS makes a final validation of the previous choice and finally start the actual export process, that is mainly composed from the following step:

1. extraction of identification parameters, such as area, zone, sector, type (from the layer structure) and name
2. saving one temporary file for each constitutive entities to export; each file contains one and only one constitutive entity and it use the specific format of used application (for example 3dm for Rhinoceros); after the system completes the step and the step below, it start to upload the temporary files into the database
3. pre-inizialization of the database for the new entities and cleaning of some records for the modified entities, saving a backup (all user information remains

doi:10.5194/isprsarchives-XL-5-W4-359-2015

untouched); it also locks all items to export to avoid conflicts during web visualization from other user

4. a multithread process, where on threads do some operation on one entity, including:
   - extraction of the mesh, if object is a mesh, or comupute it from NURBS
   - computation of volume, area, bounding box, its center and its radius, ...; volume and area are computed only for high quality LoD (Level of Detal)
   - conversion of the model in JSON format, saving it in one or more temporary file, which eventually include also UV texture map coordinates for applying the texture
   - updating database informations and datas; JSON files are upload in background just were each thread is completed
5. if there are other LoD to create, the system starts a multithread elaboration for decreasing the size of each object: the system uses a own algorithm particularly designed for decimating mesh (described below). When the system completes this multithread process backs to four step
6. when all process and file uploads are completed, the system removes the access lock previously inserted and ends its operations

Our algorithm for mesh decimation permit to approximate it on a grid of increasing size (LoD1 -> 1mm, LoD2 -> 2mm, Lod3 -> 4mm, and so on), by rounding off vertex coordinates on the grid and merging with the same resulting coordinates; our algorithm removes faces collapsed in one line or point and recalculates the normals. It manages UV texture coordinates in the same way of vertexes. Acting in this way, the mesh is reduced in size, but keeps its properties (e.g. if we start from a closed mesh, reduced mesh is also closed).

2.7 BIM3DSG v2 preview

BIM3DSG v2 is currently in development, even if some features are already ready.

For BIM3DSG we rewrite almost all of the code, with a better structure to make it even faster than before; in addittion, also thanks to the decision to adopt WPF for the UI and some framework for WPF, we have focused on giving the software a more visual appeal and easier usability. This could seem a low priority task, but today usability, rapidity and attractive looking makes the difference between a very useful system, but never or rarely used, and a system that spreads out and starts to be used for the daily work.

With regards to the capabilities, we moved mainly toward two directions, both are designed for an use as general as possible of the system: texture support and the ability to change dynamically the information fields.

Texture support is in fact required in some cases of study, where having only the model of the structure is not enough, but also visual information about cover or decoration is required. Texture are now supported both in the modelling management part and the online platform.

Regarding the ability to dynamically change the information fields, it is a very important feature because allows the system to be used in several different situations and fields, adapting it to the user request and need, without the requirement to rewrite the code.

Now we are working on another important feature, the ability to share some information between two or more entities, so that the user can write or modify at the same time all common information to a group of entities (e.g. material information, or survey technique).

3. THE WEBBIM FOR FACILITY MANAGEMENT

All objects belonging to Cultural Heritage have a particular and unique characterization. Also about intervention and maintenance, it is quite impossible to standardize methodologies and techniques because the possible activities on the objects could be countless and also conducted in a very different way according to the different operators and the condition of the object at that particular time of its life. An information system should be able to transform itself in order to fit with the infinite different situations.

The presented system is built with this purpose. An operator can quite easily modify the database from time to time adding ad hoc workings, proper information fields and parameters tables.

In this way many restoration intervention can be easily presented and compiled as textual information using easy dynamic tables linked to the objects as textual list of executed processes with the addition of any associated parameters.

In many cases this is not enough.

The restoration or maintenance activities imply the modification of the structure or part of it. Textual information are inadequate to describe the operations and specially they can not represent the geometric changes of the objects. A Building Information System should be able to support a series of ad hoc created intervention on the object that contemplate the complete replacement or a partial modification of a piece or multiple pieces from a geometric point of view. Moreover, the system is able to connect all the modifications with ad hoc created "father-son" relationship and providing in this way the possibility to estract in real time the modification history of each object, visualizing also its 3D geometric modification.

3.1 The case of the Milan Cathedral.

The case of study of the Milan Cathedral is a very good example of those concepts. Some dedicated features are specially created in the presented Web Information System to manage the ordinary and extraordinary maintenance operated by the Veneranda Fabbrica.

3.1.1 The survey and the model construction.

The survey has been conducted with a multi-scale and multi-data approach. Multi-data means using various measurement sensors, depending on the necessity and used individually or simultaneously. The multi-scale concept, instead, is the integration of objects detected in different scales for different purposes or needs in the model. The use of different technologies is imposed by different factors characterizing the Spire, such as the material, the survey complexity due to the restricted spaces or to the element elevation, to the cable-stayed scaffolding and to the presence of the restoration site.

The most used technique is the photogrammetry, both as regards the construction of simpler and more linear parts of the structure (image modeling), and as the decorated parts (image matching). The choice to use the photogrammetry as the main survey method has been made to direct the research toward a low cost process. Besides the geometric survey, it has also been led a photographic survey describing every single element of the steeple, pointing out the conservation status and the small details hard to model and to be represented graphically.

This contribution has been peer-reviewed.
doi:10.5194/isprsarchives-XL-5-W4-359-2015

The model is built inside Rhinoceros and guarantees a measurements extraction at 1:50 scale. Many details are modelled at bigger scale in order not to lose in the representation important shapes and geometries testimonials of the constructive characterization of the Milan's Cathedral. For this reason many decorative parts are modelled at 1:5 scale (Fassi et al. 2011; Achille et al 2012).

3.1.2 The centuries-old tradition of the cathedral facility management.

It is quite impossible to describe in few words the traditional methods to maintain the cathedral by Veneranda Fabbrica. Synthesizing at most, one can say that the main task of the Fabbrica is to guarantee the structural safety of the cathedral. As said by the Architect of the Veneranda Fabbrica, Benigno Moerlin Visconti Castiglione, the Milan's Cathedral is a living monument and is experienced by people, it has to be safe and functional. It is built entirely in Candoglia marble that undergoes over the years strong structural weakening due to rain washout and especially to air pollution. (Fassi et al., 2011) So all the cathedral is is monitored by careful periodic inspections, from which all maintenance activities are designined. Normally the main interventions operations consist in substitution, in whole or in part, of the degraded marble blocks, anchorages and reinforcement structures. Cleaning and restoration operations are normally scheduled for decorations and statuary. In this way the building site of the Milan Cathedral is an eternal construction site that regularly intervenes to heal those parts that threaten its stability and its security.

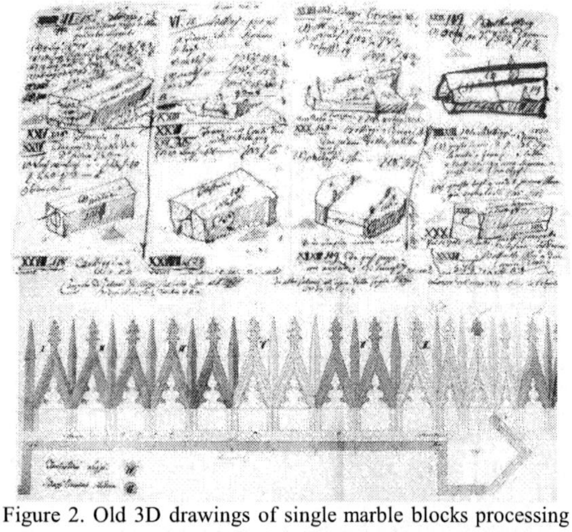

Figure 2. Old 3D drawings of single marble blocks processing with related info and description. 3D representations and colours are used to better represent and describe the management activities. The new BIM will support the activities respecting the old classic way of work

3.1.3 The new facility management system.

In the system, several dedicated features were developed to support the old traditional activities of the Veneranda Fabbrica concerning both ordinary and extraordinary maintenance activities on the Milan's Cathedral and to document all working phases and all restoring activities. The same 3D model was built specifically for this purpose, by modeling each marble block and creating a "big LEGO" of pieces which focus on all the maintenance activities.

Developing this part particular attention was dedicated to create a system able:

1. to manage fast geometric changes in the structures,
2. to increase the real time communication between jard and offices increasing the automatic data updating,
3. to increase the portability of the system in order to encourage its use at the construction site.

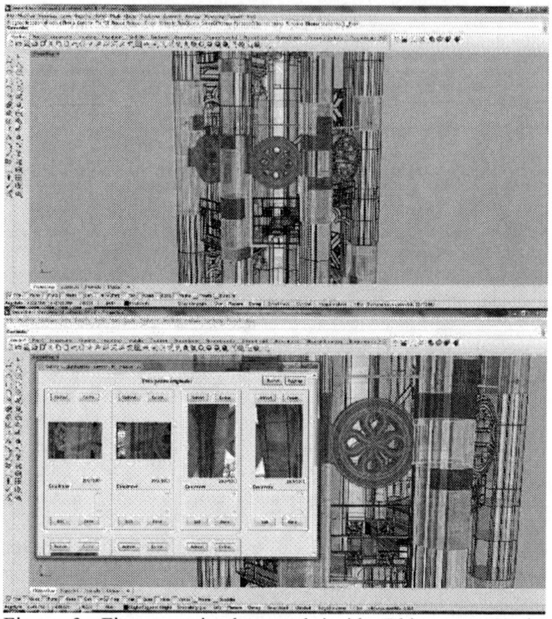

Figure 3. First part implemented inside Rhinoceros Nurbs Modeller. It is possible to see 3D models with different colour in function of different restoration activities. Info and images can be loaded and visualized with ad hoc panel. 3D model and info are loaded from the external DB and can be used without saving them locally

The system lets to:

- enter any maintenance activities with detailed description (text or images) that does not change the structure of the model previously entered,
- insert any kind of maintenance that alter the structure of the model, with either a one to one, or in a one to many, or in a many to one, or in a many to many relationship,
- manage and visualize maintenance objects with different colours in function of the intervention typologies, state of progress, time and history of multi interventions,
- manage and visualize text information and images and other files linked to the intervention,
- manage and visualize text information and images and other files linked to the blocks interested by the intervention,
- apply query or selection in function of intervention typology, time and positions,
- load past situation so users can go back and forth in the history of the building structure, and also load past and actual situations at the same time for comparison intent
- associate any management information to the interventions, such as the assignment of a given job to

an user, planning assistance, recording the hours and costs required or incurred for interventions etc,

- estimate the weight of objects when it is known that the density of the material (weight is automatically computed from the volume); it is also possible to automatically estimate the cost of any replacement if it is known the price of the raw material and its density,
- combine all the information, images and files linked to the single object to an entire area or region or sector or object type,
- connect full resolution models (1:1 scale) for 3D printing and industrial automatic construction.

All the features can be used by both systems (web and desktop), and all changes are always synchronized in real time. All features are usable concurrently from multiple locations and multiple users, both locally and remotely, through the Internet, although, for obvious reasons, some editing operations require a temporary write-lock of the involved objects, in order to prevent concurrency hazards.

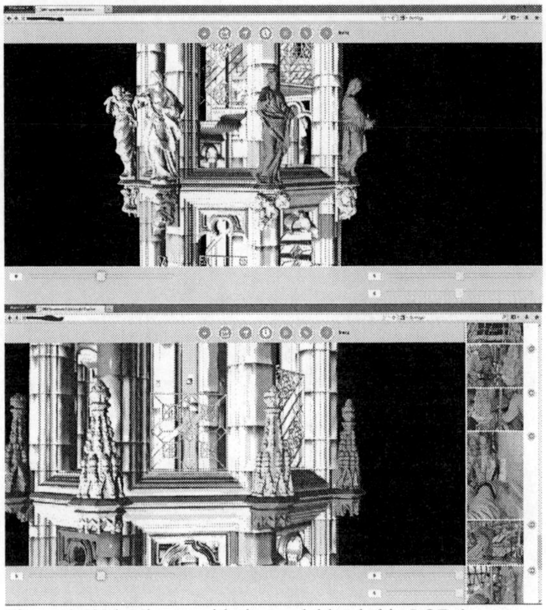

Figure 4. Web Viewer with the model loaded in LOD 1

4. CONCLUSION

The presented system answers the initial question. It has the triple goal to:

1. visualize huge and complex 3D models on the web. It is quite similar to others online viewer, but more flexible in terms of visualizable 3D model types and more powerful in terms of visualized object dimensions,
2. share information and any kind of data (textual, image video, etc.) linked to the model in a simply and immediate way using the web,
3. keep track of the changes of the objects in the years.

Figure 5. Example of daily use of our system for maintenance activities of the Milan Cathedral: (5.1)(5.2) yard operation which requires the insertion of a new maintenance event using the web interface on a tablet; (5.3) a 3D specialistic is updating, in his office, 3d models using photos, information and datas just inserted from operator in fig 5.1 e 5.2

It is important to underline some key aspects that make the system useful and attractive for different stakeholders in the world of Cultural Heritage:

- "the ease-of-use of the system", that can be used by the end user without the aid of any software with fee;
- the data enter and consulting is simple and intuitive;

This contribution has been peer-reviewed.
doi:10.5194/isprsarchives-XL-5-W4-359-2015

- ad hoc choice of the modelling software, even if requires some preparation and exercise, i.e. the end user doesn't need to have a dedicated modeling competence.

It is currently used daily by the Veneranda Fabbrica of Duomo both during the yard activities and in the technical office. The first year of testing gave good results in terms of time saving in designing and scheduling operations. The real time updating of the "drawings" of the cathedral and the real time computations of expenses are particularly successful. The use of the system as method of presentation of the performed works and results was appreciated because immediate and always available on the web. Smaller success but still recognized by the users is the usage of the system during construction activities. In fact, the need of continuous internet connection is the weak point of the system. The study of solutions to this problem is already planned in the future developments.

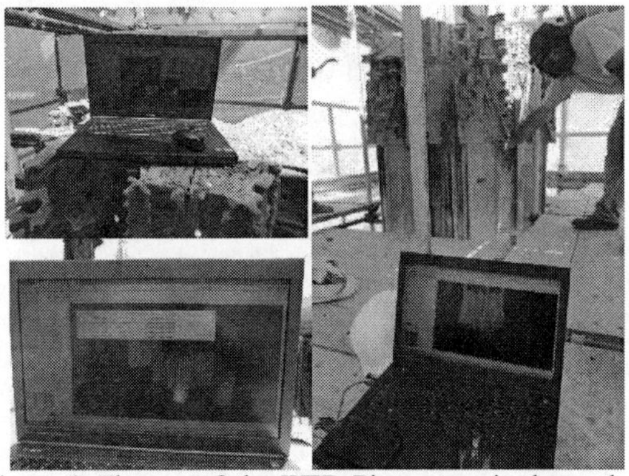

Figure 6. The use of the WEB Bim system in the yard as support for the disassembly of the oldest spire of the Milan Cathedral.

The remarkable interest for this project has propelled the idea of improving and engineering the system and especially adapting it in order to be used by other actors. Tests, further research and developments are in progress to make it even more customizable and suitable for general architectonic and archaeological case studies.

ACKNOWLEDGEMENTS

This work was supported by Veneranda Fabbrica of Milan's Cathedral. Special thanks to the director Eng. Benigno Moerlin Visconti Castiglione for his support and availability.
And to all staff especially Francesco Aquilano and Massimiliano Regis.

REFERENCES

Achille, C., Fassi, F., Fregonese L. 2012, 4 Years History: from 2D to BIM for CH. The main spire on Milan Cathedral. In: *G. Guidi, A. C. Addison. Proceeding of VSMM 2012. Virtual Systems in the Information Society.* Milan, 02/09/2012 - 05/09/2012, p. 377-382, IEEE, ISBN: 9781467325622

Coates, P., Arayici, Y., Koskela, L., Kagioglou, M., Usher, C., O'Reilly, K., 2010, *he key performance indicators of the BIM implementation process*, ICCBE 2010, Jun 30 2010, Nottingham

Cyarc, http://www.cyark.org/

De Luca, L., 2013, 3D Modelling and Semantic Enrichment in Cultural Heritage, 54th Photogrammetric Week in Stuttgart. September 9-13.

Del Giudice, M., Osello, A., 2013, Bim for Cultural Heritage, *Int. Arch. Photogramm. Remote Sens. Spatial Inf. Sci.*, XL-5/W2, 225-229, doi:10.5194/isprsarchives-XL-5-W2-225-2013, 2013.

Fassi, F., Achille, C., Gaudio, F., Fregonese, L., 2011, Integrated strategies for the modelling very large and complex architectures, *Int. Arch. Photogramm. Remote Sens. Spatial Inf. Sci.*, XXXVIII-5/W16, 105-112, doi:10.5194/isprsarchives-XXXVIII-5-W16-105-2011, 2011.

Fassi, F., Achille, C., Fregonese, L., 2011, Surveying and modelling the Main Spire of Milan Cathedral using multiple data sources. *The Photogrammetric Record*, 26: 462–487. doi: 10.1111/j.1477-9730.2011.00658.x

Fassi, F., Parri S., 2012, Complex Architecture in 3D: From Survey to Web. *International Journal of Heritage in Digital Era*, vol. 1, p. 379-398, ISSN: 2047-4970

Potree, 2014, Online at: http://potree.org/wp/ (accessed: 25 January 2015)

Rechichi, F., 2014, ITinnovations Library, online http://www.itinnovations.it

Silla, L., 2012, CyArk Releases New Online 3D Viewer Online at: http://www.cyark.org/news/cyark-releases-new-online-3d-viewer

NBIMS, 2012. National Building Information Model Standard Project Committee, http://www.buildingsmartalliance.org/index.php/nbims/faq/ (accessed: 2 March 2012).

NBIMS, 2010. National Building Information Modeling Standard, online http://www.wbdg.org/pdfs/NBIMSv1_p1.pdf

Wimmer, M., Scheiblauer, C., 2006, Instant Points. In Proceedings Symposium on Point-Based Graphics 2006, pages 129-136. July 2006.

3Dhop, 2014, Online at: http://vcg.isti.cnr.it/3dhop/index.php

WebGL (2013), online https://www.khronos.org/webgl/

SceneJS (2010), online http://scenejs.org

Xceed WPF Toolkit (2010), online http://wpftoolkit.codeplex.com

AvalonDock (2007), online https://avalondock.codeplex.com

JQuery EasyUI (2010), online http://www.jeasyui.com

FROM TLS TO HBIM. HIGH QUALITY SEMANTICALLY-AWARE 3D MODELING OF COMPLEX ARCHITECTURE

R. Quattrini [a, *], E. S. Malinverni [a], P. Clini [a], R. Nespeca [a], E. Orlietti [a]

[a] DICEA, Dept. of Civil and Building Engineering and Architecture, Polytechnic University of Marche, Ancona, Italy
(r.quattrini, e.s.malinverni, p.clini, r.nespeca)@univpm.it - erika.orlietti@tiscali.it

COMMISSION V, WG V/4

KEY WORDS: semantic modeling, 3D accuracy, ontologies, historical architecture, reality based modeling, open source tools

ABSTRACT:

In order to improve the framework for 3D modeling, a great challenge is to obtain the suitability of Building Information Model (BIM) platform for historical architecture. A specific challenge in HBIM is to guarantee appropriateness of geometrical accuracy. The present work demonstrates the feasibility of a whole HBIM approach for complex architectural shapes, starting from TLS point clouds.

A novelty of our method is to work in a 3D environment throughout the process and to develop semantics during the construction phase. This last feature of HBIM was analyzed in the present work verifying the studied ontologies, enabling the data enrichment of the model with non-geometrical information, such as historical notes, decay or deformation evidence, decorative elements etc. The case study is the Church of Santa Maria at Portonovo, an abbey from the Romanesque period. Irregular or complex historical architecture, such as Romanesque, needs the construction of shared libraries starting from the survey of its already existing elements. This is another key aspect in delivering Building Information Modeling standards. In particular, we focus on the quality assessment of the obtained model, using an open-source sw and the point cloud as reference. The proposed work shows how it is possible to develop a high quality 3D model semantic-aware, capable of connecting geometrical-historical survey with descriptive thematic databases. In this way, a centralized HBIM will serve as comprehensive dataset of information about all disciplines, particularly for restoration and conservation. Moreover, the geometric accuracy will ensure also reliable visualization outputs.

1. INTRODUCTION

Building Information Modeling (BIM) is currently used in the architecture & civil engineering (ACE) to integrate a three-dimensional visual model of construction projects and digital data of various fields with properties into a file or a database with a specific format. These platforms allow project designers, project managers, construction units, owners, and clients to view the design through a three-dimensional visual model and to obtain relevant digital data of the project.

In recent times, the HBIM (Historic Building Information Modeling) gives a very interesting research perspective. The main requirement in using BIM for representation and survey of historical architecture is the quality of the model and its reliability regarding the geometry. A second condition involves the addition of a comprehensive database of historical notes regarding each component about materials and changes during the time. If models will be able to meet these requirements, avoiding hard computing processes, in the next years the HBIM will play a key role in restoration, representation and communication of cultural built heritage.

This kind of models offers a lot of advantages in obtaining directly two-dimensional drawings and in the management of a large amount of data concerning each shape.

A specific challenge in 3D reality-based models is to develop easy methods to obtain BIM models for cultural heritage, which guarantee appropriateness of geometrical accuracy. The present work demonstrates the feasibility of a whole BIM approach for complex architectural shapes starting from point clouds.

A novelty of our method is to work in a 3D environment along the process and to develop the semantic of the model during the construction phase. This last feature of HBIM was analyzed in the present work verifying the studied ontologies, enabling the data enrichment of the model with non-geometrical information, such as: historical notes, decay or deformation evidence as well as decorative elements etc.

The semantic organization is central in BIM software. Thanks to parametric models and their attributes, it is able to establish the geometric framework in which multiple data are shared. In the near future, the data collection of all significant aspects in a 3D content will allow the HBIM to become the best way to manage the process from survey to restoration. Another way to facilitate the spread of HBIM would come from developing effective supply chains with commercial and common software.

The case study here presented is the Church of Santa Maria at Portonovo (Ancona, Italy), an abbey from the Romanesque period, recently restored (Fig.1). This building presents characteristics, which are remarkable for the period, i.e. the masonry vaults. We decided to focus mainly on these geometric shapes and on the structural module (pillar, column and vault) in developing ontologies. Indeed, we expected the vaults would represent the weakest point in the modeling process. Moreover, the selected structure is able to describe the whole frame of the architecture, if we obtain the appropriate level of detail.

The survey of the church was performed by TLS (Terrestrial Laser Scanning) and the model was built up in Revit software on the point cloud.

* Corresponding author. This is useful to know for communication with the appropriate person in cases with more than one author.

This contribution has been peer-reviewed.
doi:10.5194/isprsarchives-XL-5-W4-367-2015

The second crucial step was the quality assessment of the 3D BIM model, through the comparison with point cloud. After the check on the model quality, we conceived and created the theoretical semantic organization. This last step opens further development on a reliable model, that is the semantic implementation with open source software.
Following our work structure, here we explain the overview of the paper: after the state of art, two brief paragraphs follow describing the history and features of the church (3.1) and the survey campaign (3.2). The main paragraph (4) is composed by 3 sub-paragraphs about the modeling phase, accuracy tests and assignment of ontologies and attributes, at the end the conclusion and the future perspective of this work.

Figure 1. Photos of outdoor and indoor of Church Santa Maria at Portonovo.

2. STATE OF ART

In data capturing a robust background in laser scanning acquisition and architectural analysis was developed from some authors in past years (Guidi, 2007). Other works carried out 3-D digital virtual models of ancient heritage monuments from on-site data, developing photogrammetry and especially focusing on data acquisition techniques and color processing methods (Remondino, 2011; Barazzetti, 2010). Some authors obtained a very efficient results integrating various "real-based" modeling methodologies (Fassi, 2011).
In modeling, interesting approaches address to produce a digital document that combines metric and photographic data and historical data in a 3D textured model becoming a theoretical model (Lo Buglio, 2012, Quattrini, 2014) or 3D models that recognise the rules of construction becoming access point to architectural criticism (De Luca et al., 2011; Apollonio, 2013). This last perspective opened the discussion about 3D GIS and its semantics: this is a concept encapsulated in HBIM. Using the BIM one can emphasize the use of a semantic construction of the digital model, not only as a means to modeling a building but as a cognitive system (Apollonio, 2013).

Considering a great challenge to standardize BIM procedures for modeling historic buildings, several works create parametric libraries for historical architecture and perform methods for fast and efficient modeling of building façades (Dore, 2014) or entire buildings (Oreni, 2013). Some studies focus on Level of Detail (LOD), (Fai, 2014), others on automatic extraction of features (Garagnani, 2013). A very interesting development is expected also in considering analogy between treatises and BIM (Apollonio, 2012).

In literature, other works show several steps in 3D HBIM modeling, these workflows (Murphy, 2013; Oreni, 2014) use different software with the necessary format conversion and in the majority we observe a 2D simplification with slices of point clouds to build up the 3D model.
The BIM approach is very performing for modern buildings, enabling monitoring all phases of life cycle building nowadays. A complete survey (Volk, 2014) on BIM use in new or existing buildings, demonstrates challenging research opportunities arise from process automation and BIM adaption to existing buildings requirements, despite fast developments and spreading standards.

3. THE ARCHITECTURE AND ITS SURVEY

3.1 Brief description of the church

The Santa Maria at Portonovo church is a very significant heritage from the Romanesque period in the Marche region. The abbey rises up on top of a Conero Mountain cliff, overlooking the sea (Fig.1).
Scholars examined the origins of the Church, finding it difficult to date. More probably, the origins of the building date back to the first half of 11th century, according to a donation to Abbot Paul for the construction of a Benedictine Monastery (document dated 1034). In the next two centuries, the complex was expanded. Perhaps, in 1225 the avant-façade (a sort of a narthex) was built up. In the 1300s, due to a Conero's landslide, the abbey was abandoned. In 16th century, the Turkish invasion destroyed the monastery and the bell tower (Favole, 1993).
Surely, the Church was designed with an overall project, which combined a Greek cross plan with a Latin cross scheme from the Basilica. The Church presents three naves, with seven spans. A rib barrel vault supports the roof of the nave; the crossing vaults cover the lateral naves. Two lateral chapels with tree spans are built up, obtaining the Greek cross scheme. The east end is composed by three circular apses.
Over the crossing of nave and transept, a dome lies on an high drum, with mullioned windows. All the piers are columns, excepting the four pillar (compound rectangular piers) that support the dome. (Piva, 2003).
The outer wall has the typical features of the Romanesque Lombard style with blind arches (niches) along the perimeter.

3.2 The TLS survey campaign

The acquisition phase and the point cloud treatment was performed with standard procedures: with laser scanner Leica C10 we arranged 34 scans and 33 scan stations. The survey operations were conducted in Portonovo, over two days (16 hours of work). The weather conditions were cloudy sky and serene at times, which guaranteed a diffused lighting, ideal sky for the taking phase. The photographic campaign was conducted with the integrated camera. To obtain high-quality photographic images and good textures for the model, the b&w targets were not used as they would have occluded parts of the complex. We

This contribution has been peer-reviewed.
doi:10.5194/isprsarchives-XL-5-W4-367-2015

arranged 6 tilt & turn target, put on beams; they enabled the link between internal and external scans.
A solid background in all-in-one acquisition with this tool (Clini, 2013) demonstrated advantages in speedy and fast acquisition and low-cost methods of representation. However we designed with attention the survey campaign, considering the requested level of detail and the geometry of the plan, i.e. pillar, span or vault shape and height.
While ensuring the visibility of the whole construction, it was not possible to avoid the shadow cone, due to the presence of objects in the path of the photographic camera. However the decision to set, for each station, a large window of taking has led to a good overlap between scans, covering the holes or the gaps. Having performed terrestrial scans only, the main gap in the point cloud model is in the roofs (Fig. 2).
After acquisition phase, the processing and restitution phase were performed in Cyclone 8.1.3 software. Using the images from internal camera, it is possible to apply colors to single scans (Fig. 3).
The alignment of acquired point clouds was carried out cloud to cloud, by identifying manually control points between scans and with the ICP algorithm. We obtained a 3D model of the whole building composed by 416 mln of points with 0,01 m RMS of alignment.
After the alignment, we edited the point cloud model during the phases of decimation, cleaning and noise reduction.
Starting from a point cloud model, which included also the RGB value, we tried to carry out the BIM of Santa Maria at Portonovo.

Figure 2. TLS survey of Church Santa Maria at Portonovo. Aligned Point cloud, exterior (grey) and interior (blue).

4. HBIM AND QUALITY ASSESSMENT

In order to improve the framework for 3D modeling, a great challenge is obtain the suitability of Building Information Model (BIM) platform for historical architecture.
Since the architectural heritage has taken advantages from the development of data capturing technologies, an open issue is the integration of BIM systems and data sources for existing buildings, supporting the creation of 3D knowledge and management systems.

Figure 3. TLS survey of Church Santa Maria at Portonovo. Point cloud mapped with the RGB value of camera acquisition.

4.1 The modeling phase

The 3D point cloud, after cleaning and decimation phases, demonstrated himself as a good model for the case study, without artefacts or significant data voids; this database is able to give all information (geometry, material and colour) for the further development of BIM.
A first challenge is to avoid a model simplification too far from the real shape of the objects in BIM platform. Indeed, when components are generated by existing morphologies, complex abstractions have to be introduced. This problem occurs while remodeling elements, both from acquired point clouds by laser scanning equipment and from SFM (Structure From Motion) capturing. Automatic feature extraction in available software is still not effective. In order to verify reliability and usability of methodology for point cloud processing in a BIM environment with high accuracy, we have developed a procedure for the whole architectural complex at Portonovo in Revit 2014.
Since this version, using the ReCap™ (Reality Capture) tool, Revit allows a very easy management of the 3D point cloud data. From Cyclone we obtained a .pts format file, then we exported a .rcs format file, enabling the importation in Revit software.
Our approach performs the model directly on point clouds, minimizing the number of steps, avoiding to lose accuracy, data quality and details. Revit also shows the point cloud in each views (maps, front & 3D views), due to the encapsulated ReCap tool. In this way we did not need to simplify the building with slices of the point cloud, as well as we did not need a reverse modeling and a polygonalization from the discrete model.
Some objects were made immediately, using the basic commands of the Revit menu. The modeling was performed using parametric elements already in the Revit internal library, some objects were created as families outside, then they were imported into the model. Few libraries are available for free in on-line database. The more complex elements were made in place through B Rep operations. In addition, parametric objects were created specifically for certain items: they will be reused in other similar architecture (Fig. 4).

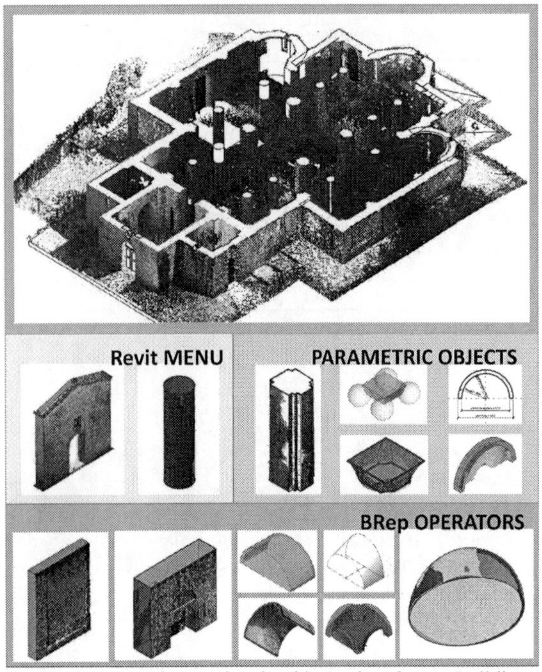

Figure 4. HBIM from point cloud in Revit software. Building methods of architectural shapes

The measures of the contours were extracted from the point cloud by the tool "Measure distance between two references" in different views.
The 3D model was built up in millimeters, according to the coordinate system of the point cloud. A final year student performed the modeling stage, during almost two months.

The walls were created with the tool "Architecture". Along the outer perimeter wall there are several niches (blind arches), typical of Romanesque architecture. The same components were built as local specific. We obtained the niches as empty extrusion.
The modeling of the piers was carried externally by creating families of the "Pillar metric". The piers have been parameterized by the height, the size of the footprint and the number of offsets (with its thickness). This is a very important parametric element for the medieval architecture. Also bases and torus were modeled externally as new families.
The capital was quite simple because it is a block cusion capital. We choose to model it with the "Pillar metric" family. The capital was modeled as a truncated pyramid with a square base. To the latter four balls are subtracted in the corners of the lower base. The profiles were designed according to measures, extracted from the point cloud. Even if the abacus is a rectangular prism, we choose to add parameters in its modeling. This procedure opens a further development: the connection of the proportions between already modeled elements.
The rib vaults (both those in the barrel form and those in cross form) display several problems in the modeling phase; for this reason, we decided to divide vault and rib, like two shapes, then we would have been able to add a link between them, thanks to the Protégé plugin.
The arches and the ribs were modeled by "mass" family, that allowed the change of contour lines. In this way we carried out pointed or segmental or circular arches and vaults. The barrel vaults were obtained by "Special equipment" subtracting each other two full vaults.
A very strong point was to obtain a good geometry that describes the cross vaults: they have a generator non horizontal and they are not regular also in plan (the angles between directrix are not perpendicular).

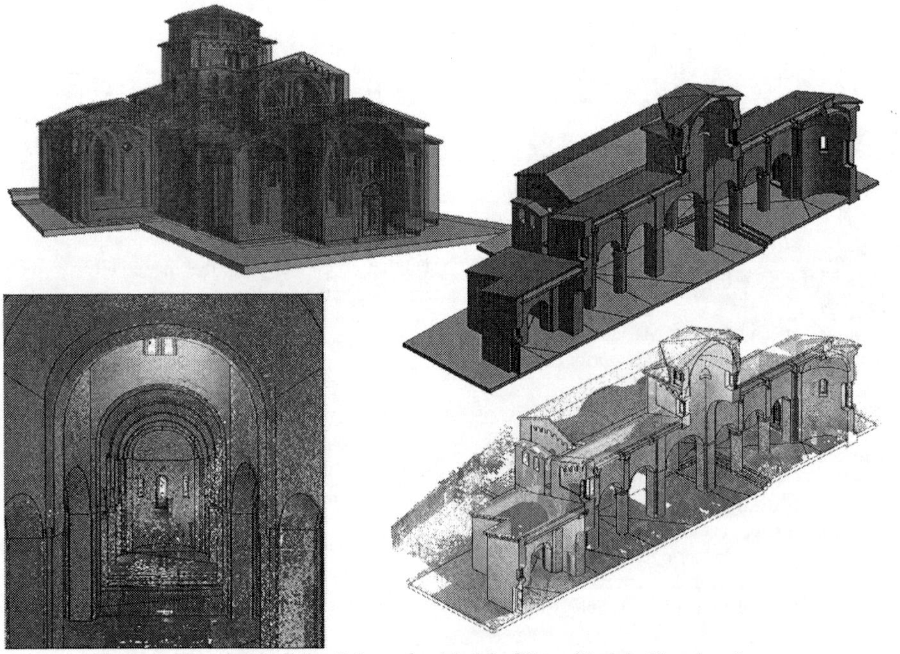

Figure 5. Historical Building Information Model of Santa Maria's Church at Portonovo

This contribution has been peer-reviewed.
doi:10.5194/isprsarchives-XL-5-W4-367-2015

The algorithms that manage intersections in Revit software are not able to perform these irregular shapes. To solve this problem, four full solids and four subtraction solids, as cutting solids, were made, but we observe some artefacts due to errors in Boolean operation. We were not able to add parameter to this kind of vaults, they were tailored to each span. The domes (the principal dome at crossing and the semi-domes over the apses) were modeled like "Special equipment", in composing domes and plans.

From the connection of the several parts, we obtained a BIM model of the church (20 Mb of size) (Fig. 5).

Open points remain in the modeling phase, such as the solid interferences in cross vaults or to complete the parametric management of the order. These issues will be discussed in following works, while the present model is carried out to perform a quality assessment in BIM, regarding a complete building. The main goal of our work was to validate the model, using, as reference database, the point cloud. Only after a positive assessment on the quality of the whole model, we would have proceeded with semantic treatment for the creation of the HBIM.

4.2 Comparison between point cloud and BIM model

The qualitative and quantitative assessment of the accuracy for the 3D BIM of Santa Maria at Portonovo was performed with the open source software CloudCompare, using the point clouds as reference. The software is able to compare the point cloud (format file .pts) and the model exported from Revit software (format file .obj).

The whole model was compared and it presented deviations of less than 9 cm for an amount of 86%, and even reach a level of precision with deviations of less than 3 cm for an amount of 63%.

These results is appropriate, because also no reference plane (as roofs) or noise elements (as plants) are included in this percentage and average. In order to refine the results, we segmented the point cloud and the model. Another way to improve the comparison is to set a maximum distance for computing.

Then the model was separated, according to its semantic, in several categories (e.g.: columns, pillars, cross vaults, barrel vault, walls, roofs etc.) for punctual or more focused assessment. Columns and vaults are adequately modelled (with average distance of 0.05 cm and of 0.01 cm).

The dome has not a perfect geometric shape, therefore the realized model highlights higher deviations (average distance 6.4 cm). This deviation could occur also in CAD modeling, without reverse modeling. The reason of the gap is in deformation of the dome, not represented by an ideal geometry. The analysis here conducted (Fig. 6) is a typical deformation diagram, able to describe damages or carelessness in structural field. The diagram could be attached to the shape as an attribute in BIM and it could become useful for all users of the model.

The greatest differences concern the design of blind arches of the apse due to the lack of curves primitives in B-Rep operation. The numeric results are in the table 7. The overall average distance is 1.5 cm, in other words we get good quality results for a high percentage.

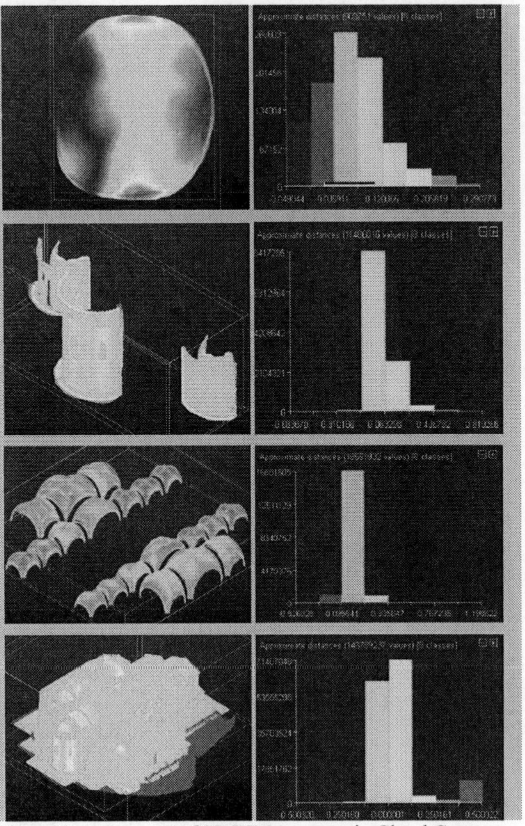

Figure 6. Assessment of 3D BIM accuracy in Cloud Compare software.

The obtained model demonstrated it is possible to build a reality-based 3D model with very good accuracy in BIM environments and without using nurbs operators. This is a crucial step in carrying out standards in order to provide Building Information Modeling new management approaches.

element	D max set	histogram		Gauss's Curve	
		D [cm]	%	D_{media} [cm]	σ [cm]
whole model	-	D<50	93	15,5	66,3
	50	D<12,5	87	4,4	13,7
dome	-	-5<D<12	86	6,4	5,7
apse - niches	-	6<D<25	23	3,1	9,2
cross vaults	-	-9,6<D<12	90	0	6,7
columns	-	-3<D<6,8	83	1,1	7
	6,8	D<3	80	0,05	3
whole model	12	D<9	86	1,5	4,9
		D<3	63		

Table 7. Assessment of 3D BIM accuracy in Cloud Compare software.

This contribution has been peer-reviewed.
doi:10.5194/isprsarchives-XL-5-W4-367-2015

4.3 Semantic modeling

In the 3D model construction, a first semantic partition was made using families and types from Revit: the family refers to models with unique parametric frameworks and can consists in a set of types with different dimension, the single piece are named instances. In the second phase, we studied the relationships among the instances and the types (Fig. 8). The relationships can be horizontal or hierarchical.

The ontology of the church, that is the conceptual system of information related with the geometric model, was carried out and it is a precondition for connection information, such as the construction rules of vaults, the types of ribs, various decorations of capitals, all things that can be the variants. The various data were graphically assigned to each item (Fig. 9). A further development will be the data assignment within Revit through the plug-in Protégé.

Ontologies are increasingly popular among Web developers thanks to the benefits they offer in sharing information. Ontology is intended as a particular conceptual framework or as a specification of a conceptualization (Gruber's definition, 1993).

The most common language in representing ontologies is the OWL. The OWL Web Ontology Language was designed in 2004 for use by applications that needed to process the content of information. OWL facilitates greater machine interpretability of Web content than that supported by XML, RDF, and RDF Schema (RDF-S) by providing additional vocabulary along with a formal semantics (Mc Guinnes, 2004).

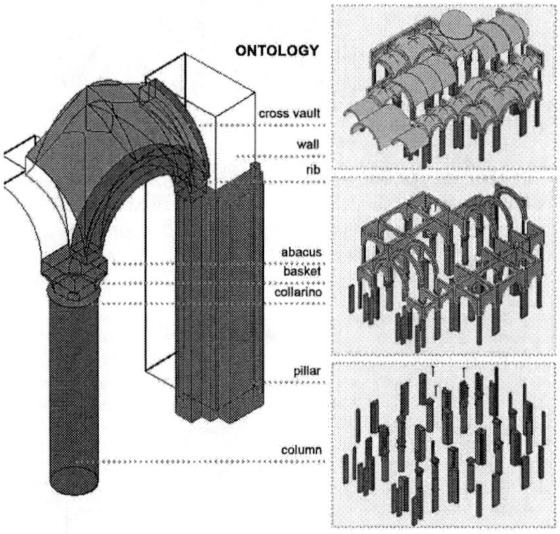

Figure 8. Ontology of whole model and focus on structural module.

OWL Plugins were created in facilitating new practices for building Semantic Web contents, often driven by the needs of and feedback from our users. Among OWL plugins, we choose the open source Protégé plugin. Furthermore, Protégé is flexible open-source platform means that it is easy to integrate custom-tailored components to build real-world applications (Knublauch, 2004).

It enables aggregation, topological and directional relationships. Aggregation (i.e. part of, belong to, etc.), could be modeled with a hierarchical-based tree representation that permits to describe the composition in a local-to-global way.

Starting from historical notes and several collected analysis (about materials, decorative systems, contact-less or invasive analysis etc.) we arranged a data sheet about each shape.

The operational phases linking parametric object with attributes are now on going.

In addition to in-depth study of the modules of vaults, we focused on attributes and description about the following parts: the avant-façade, the internal arch in a wall of the transept (not visible in external wall), the transept door that linked to the destroyed monastery, the chapel on the right that was a funeral chapel (see the cross on the top of the wall), the external bells substituting the destroyed bell tower.

These are the main elements of modification during the time; their representation in our HBIM could enable a further development of historical analysis and could lead to the qualitative increase in studies about the Church

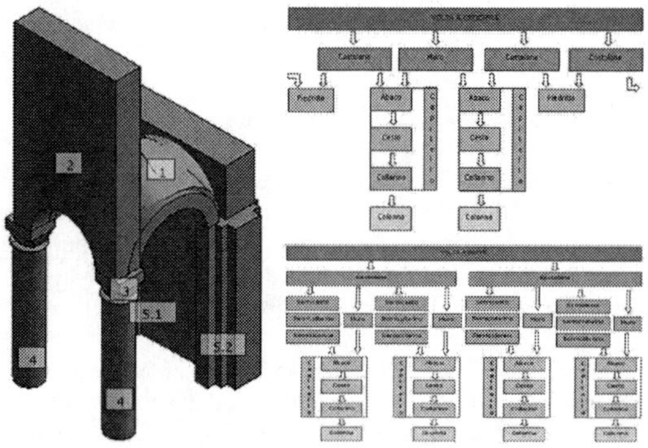

Figure 9. Semantic description of families and types from Revit.

5. CONCLUSIONS AND FUTURE WORKS

The work here presented shows that it is possible to develop a high quality 3D model semantic-aware, able to connect geometrical-historical survey with descriptive thematic databases. In this way, a centralized HBIM will serve as comprehensive dataset of information about all disciplines, particularly for restoration and conservation.

BIM's semantic structure enables multiple data enrichment in different levels of category, family, type and instance, out of different motivations. It gave access to various platforms with the goal of long term management of digital cultural heritage.

The main achieved goal is a very reliable quality of the model, combining the efficient representation procedures of Building Information Modeling with high details, starting from complex surveying data. The geometric accuracy obtained will ensure also consistent visualization outputs.

In our opinion, it is imperative to undertake a detailed documentation and representation strategy in BIM environments for cultural heritage. Consequently, it is possible to develop semantics and management procedures in order to determine the correct LoD of the surveying and of the model.

Our work enables a full exploitation of the TLS accuracy, since objects were mapped on to the point cloud in a 3D environment, without slices.

A specific goal of our approach was to minimize the steps and the format changes along the modeling process, in order to avoid simplification or loss of information. The obtained and validated workflow shows some improvements compared to others in literature (Fig. 10), considering that some procedures involved polygonal model and NURBS operators.
We achieved a good procedure for parametric modelling involving Terrestrial Laser Scanning data, then we verified its suitability and we validated the process thanks to the quality assessment of the model, adopting it in a case of a large monument.

The quality assessment of the BIM model gives very satisfactory results: the whole model presents gaps under 3 cm in 63% of points. The outcomes are good also for single elements. The bi-dimensional drawings, automatically extracted from the Revit model, were also evaluated in the case study. They show distances between 2-3 cm from the compared slices, therefore the model is able to represent the building in a 1:100 scale.

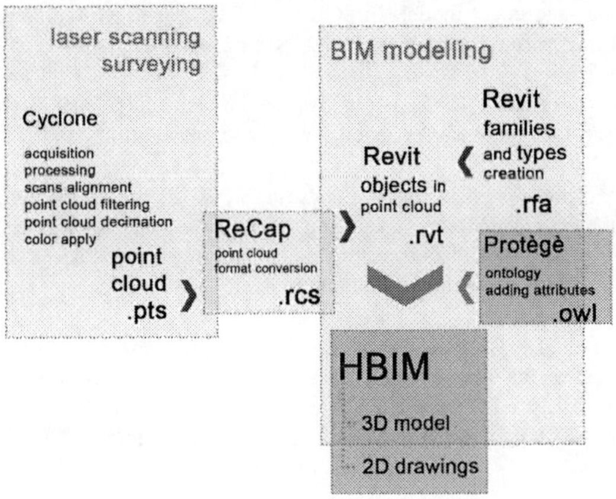

Figure 10. The workflow for HBIM starting from TLS data, minimizing steps and format file

Moreover, another successful aim in our research is to obtain parametric libraries of architectural elements, starting from point clouds or from traditional direct survey.
In HBIM field, several studies faced this issue but a lot of them are referred to classical architecture, particularly they investigated about rules and order in Renaissance age. Irregular or more complex historical architecture, such as Romanesque architecture, needs the construction of shared libraries starting from survey of elements. This is another key aspect in delivering Building Information Modeling standards.
In the construction of parametric libraries of the orders, for example, a further step can involve the study of the proportions between the elements (capital, diameter of column etc.). In several periods of architectural history, the proportions are not fixed or already known, because there are not treatises or theoretical discussions. An interesting point of view in the use of HBIM may occur from collecting parametric libraries of orders and comparing them. That could allow new studies in architectural proportions.

Regarding the description of the elements and its hierarchical organization, we conceived a complete ontology for the case study and we validated it. The prepared ontologies envisage future advances for the classification of the model and the assignment to the individual elements of the metadata collected.
An on going work is the conceiving of a comprehensive standard ontology in Romanesque architecture, to compare with the semantic organization of Santa Maria in Portonovo.
The final HBIM product is the creation of full 3D models including detail behind the surface of the object concerning its methods of construction and material make-up. This set of metadata issues from a robust bibliography or in situ analysis.

BIM model serves as a sustainable dataset comprising comprehensive information, and provides access to life-cycle management and potential simulations of various disciplines.
For the future, we propose several kind of research to develop in BIM field:
a) study of digital stereotomy; this is a research issue that involves analysis of existing architecture (Oreni, 2013) and new forms of design (Rippmann, 2012)
b) development of environments for collaborating BIM and AR (Augmented Reality) applications; nowadays it is only performed in a very simple architectural building (Heidari, 2014; Brundu, 2014).
Recently, some studies are exploring the real-time communication, the integration of BIM to the site and task conditions, and the interaction of BIM. It is envisaged that AR can fulfil this function effectively through visualizing BIM into the physical context of each construction activity. In the near future, we expect to develop a conceptual framework to investigate how BIM can be extended on site via AR.

In conclusion, a new methodology to build a semantic-aware 3D model of complex architecture is here presented. Starting from point clouds by TLS, we obtained a high quality HBIM, we performed its quality assessment and we developed a procedure for its semantic management. This research promotes the HBIM dissemination. Thanks to exploitation of point cloud, the BIM could become easily an environment that produces full engineering drawings for the conservation of historic structure, including 3D documentation, orthographic projections, sections, details and contracts (energy, cost decay, etc.).

ACKNOWLEDGEMENTS

The authors would acknowledge the FAI (Italian Ambiental Found) and the Superintendence for Architectural Heritage and Landscape of the Marche Region for allowing the acquisition of the Church of S. Maria at Portonovo.

This contribution has been peer-reviewed.
doi:10.5194/isprsarchives-XL-5-W4-367-2015

6. REFERENCES

Apollonio, F. I., Gaiani, M., & Sun, Z. (2012). BIM-based modeling and data enrichment of classical architectural buildings SCIRES-IT, 2(2), 41-62.

Apollonio, F. I., Baldissini, S., Clini, P., Gaiani, M., Palestini, C., & Trevisan, C., 2013. The Palladiolibrary geo-models: an open 3D archive to manage and visualize information-communication resources about Palladio. ISPRS-International Archives of the Photogrammetry, Remote Sensing and Spatial Information Sciences, 1(2), 49-54.

Apollonio, F. I., Gaiani, M., & Sun, Z., 2013. 3D Modeling and Data Enrichment in Digital Reconstruction of Architectural Heritage. ISPRS Archives, 5, W2, 43-48

Barazzetti, L., Fangi, G., Remondino, F., and Scaioni, M., 2010. Automation in Multi-Image Spherical Photogrammetry for 3D Architectural Reconstructions. Proceedings VAST 2010. 11th International symposium on virtual reality, archaeology and cultural heritage proceedings. Short and Project Papers, Eurographics Association, Paris, France, 75-81.

Brundu F. G., Patti E., Del Giudice M., Osello A., Macii E., Acquaviva A., 2014. DIMCloud: a distributed framework for district energy simulation and management. In IoTaaS 2014, Lecture Notes of the Institute for Computer Sciences, Social Informatics and Telecommunications Engineering, in press

Clini, P., Quattrini, R., Fiori, F., Nespeca, R., 2013. Integrated Technologies for Surveying Artefacts Damaged by Earthquakes. Application of All-In LIDAR Techniques in the City of L'Aquila. ISPRS Annals of Photogrammetry, Remote Sensing and Spatial Information Sciences, 1(1), 79-84.

De Luca, L., Busayarata, C., Stefania, C., Véron, P., Florenzano, M., 2011. A semantic-based platform for the digital analysis of architectural heritage. Computer & Graphics, 35 (2): 227–241.

Dore, C, Murphy M, 2014. Semi-automatic generation of as-built BIM façade geometry from laser and image data, ITcon, Vol. 19, pg. 20-46, http://www.itcon.org/2014/2

Piva, P., 2003. Marche romaniche, Jaca book, 287 pp.

Fai, S., Rafeiro, J., 2014. Establishing an Appropriate Level of Detail (LoD) for a Building Information Model (BIM)-West Block, Parliament Hill, Ottawa, Canada. ISPRS Annals of Photogrammetry, Remote Sensing and Spatial Information Sciences, 1, 123-130.

Fassi, F., Achille, C., Fregonese, L.,2011. Surveying and modeling the main spire of Milan Cathedral using multiple data sources. The Photogrammetric Record, 26: 462–487.

Favole, P., 1993. Italia Romanica. Le Marche, vol 14, Jaca book, 354 pp.

Garagnani, S., Manferdini, A. M., 2013. Parametric accuracy: Building Information Modeling process applied to the cultural heritage preservation. In International Archive of the Photogrammetry, Remote Sensing and Spatial Information Sciences, XL, 87-92

Guidi, G., Bianchini, C., 2007. TOF laser scanner characterization for low-range applications. In Electronic Imaging 2007 International Society for Optics and Photonics, 649109-649109.

Heidari, M., Allameh, E., de Vries, B., Timmermans, H., Jessurun, J., & Mozaffar, F., 2014 . Smart-BIM virtual prototype implementation. Automation in Construction, 39, 134-144.

Knublauch, H., Fergerson, R. W., Noy, N. F., Musen, M. A., 2004. The Protégé OWL plugin: An open development environment for semantic web applications. In The Semantic Web–ISWC 2004,. 229-243. Springer Berlin Heidelberg.

Lo Buglio, D., & De Luca, L., 2012. Representation of architectural artifacts: definition of an approach combining the complexity of the 3d digital instance with the intelligibility of the theoretical model. In SCIRES-IT SCIentific RESearch and Information Technology, Vol 2, Issue 2, 63-76

McGuinness, D. L., Van Harmelen, F., 2004. OWL web ontology language overview. W3C recommendation, 10(10).

Murphy, M., McGovern, E., Pavia, S., 2013. Historic Building Information Modeling–Adding intelligence to laser and image based surveys of European classical architecture. ISPRS journal of photogrammetry and remote sensing,76, 89-102.

Oreni, D., Brumana, R., Cuca, B., & Georgopoulos, A., 2013. HBIM for conservation and management of built heritage: Towards a library of vaults and wooden bean floors. In CIPA 2013XXV International Symposium, ISPRS Annals, Vol. 164, 1-6

Oreni, D., Brumana, R., Della Torre, S., Banfi, F., Barazzetti, L., & Previtali, M., 2014. Survey turned into HBIM: the restoration and the work involved concerning the Basilica di Collemaggio after the earthquake (L'Aquila). ISPRS Annals of Photogrammetry, Remote Sensing and Spatial Information Sciences,1, 267-273.

Rippmann, M., Block, P., 2012. New design and fabrication methods for freeform stone vaults based on ruled surfaces. In Computational Design Modelling, 181-189. Springer Berlin Heidelberg.

Quattrini, R., & Baleani, E., 2014. Theoretical background and historical analysis for 3D reconstruction model. Villa Thiene at Cicogna. Journal of Cultural Heritage, In Press, Available online 20 February 2014

Remondino, F., 2011. Heritage Recording and 3D Modeling with Photogrammetry and 3D Scanning. Remote Sensing, Vol. 3(6), pp.1104-1138.

Volk, R., Stengel, J., & Schultmann, F., 2014. Building Information Modeling (BIM) for existing buildings—Literature review and future needs. Automation in Construction, 38, 109-127.

This contribution has been peer-reviewed.
doi:10.5194/isprsarchives-XL-5-W4-367-2015

3D TRACKING BASED AUGMENTED REALITY FOR CULTURAL HERITAGE DATA MANAGEMENT

C. Battini [a], G. Landi [b]

[a] DICCA, Department of Civil, Chemical and Environmental Engineering, 16145 Genova, Via Montallegro 1, Italy – carlo.battini@unige.it
[b] ETT s.p.a., 16154 Genova Sestri Ponente, via Sestri 37, Italy - giovanni.landi@ettsolutions.com

Commission V, WG V/4

KEY WORDS: Augmented reality, Laser scanner, Tracking 3D, Reverse modelling, Survey

ABSTRACT:

The development of contactless documentation techniques is allowing researchers to collect high volumes of three-dimensional data in a short time but with high levels of accuracy. The digitalisation of cultural heritage opens up the possibility of using image processing and analysis, and computer graphics techniques, to preserve this heritage for future generations; augmenting it with additional information or with new possibilities for its enjoyment and use. The collection of precise datasets about cultural heritage status is crucial for its interpretation, its conservation and during the restoration processes. The application of digital-imaging solutions for various feature extraction, image data-analysis techniques, and three-dimensional reconstruction of ancient artworks, allows the creation of multidimensional models that can incorporate information coming from heterogeneous data sets, research results and historical sources. Real objects can be scanned and reconstructed virtually, with high levels of data accuracy and resolution. Real-time visualisation software and hardware is rapidly evolving and complex three-dimensional models can be interactively visualised and explored on applications developed for mobile devices.
This paper will show how a 3D reconstruction of an object, with multiple layers of information, can be stored and visualised through a mobile application that will allow interaction with a physical object for its study and analysis, using 3D Tracking based Augmented Reality techniques.

1. INTRODUCTION

Three-dimensional survey techniques are increasingly used today for cultural heritage studies. The collected data may be the basis for morphological and preservation analysis of objects. These survey techniques generate datasets, which are the fundamental basis on which new and general information is overlaid. This process is in a new spatial location and creates relationships between various types of results thereby giving new interpretation tools.
There are numerous 3D collection technologies available today, some of which use laser. These instruments permit the capture of complex geometries with a degree of resolution that varies depending on the technique used. The results must be defined in the design phase of the survey and must be to the scale used for visualization.
Cultural Heritage documentation often includes results rendered in scale 1:50 and, in some cases, 1:20. These scale factors are able to decipher the peculiarities of the piece very accurately, showing not only the form but also the colouring of the materials used. This data is essential for successful restoration planning and design.
By restricting the range of technologies available to scale 1:20, the use of phase-shifting approaches, with a measurement tolerance of 2-3 mm, is a good way recording morphological composition; also for statues, where this type of technology involves information discretisation and highlights the reflectivity of the material exposed to a beam of light.
Dedicated applications process this information and extract dimensional and colouring information. It is unfortunate that the data obtained is only available to those with special training in computer graphics, usually based in distant laboratories.
Other data that need to be collected during the restoration and conservation phases concern conservative and material peculiarities. This information is useful when choosing the most suitable restoration technique and materials. The complexity and the amount of collected data often leads to the creation of complex and not easily transferable maps and reports.
The project described in this article aims at making all this information available on site, simply and immediately. This system permits greater interaction with objects and delivers simple information organisation tools that also store data for easy integration with other software.
In this way, the data will always be available and will facilitate comprehension of the decay conditions.
The aim of the project is also to give basic information to unskilled users, such as tourists.
It will therefore be possible to classify the article and show precise information on the iconographic and material composition, thereby encouraging greater understanding of the design idea behind the piece of art.

2. CASE STUDY

This article deals with the experiment carried out on the statue of Leonardo da Vinci (Figure 1) in Piazza della Scala, Milan, sculptured by Pietro Magni, a local artist. The original planned site of the statue was in Palazzo Brera but this was changed when King Vittorio Emanuele II inaugurated the Galleria Vittorio Emanuele II in 1872, during the Second National Exhibition.
The monument consists of an octagonal base, in Pink Baveno Granite, on which four Carrara marble statues of disciples, by Giovanni Antonio Boltraffio, A Solaino, Marco d'Oggiono and

This contribution has been peer-reviewed.
doi:10.5194/isprsarchives-XL-5-W4-375-2015

Figure 1. The statue of Leonardo da Vinci

Cesare Da Sesto, are mounted. There are also four important bas-relief works. These works show scenes of "Leonardo the painter" while painting The Last Supper, "Leonardo the sculptor" while creating the equestrian statue dedicated to Francesco Sforza, "Leonardo the architect and strategist" in the role of construction manager for the fortification works to the castles owned by the Duca Valentino di Romagna, and

Figure 2. Steps of survey of the monument. 3D scan with elevator

"Leonardo the hydraulic engineer", struggling with the irrigation canals and waterways in Lombardy.

The austere statue of Leonardo da Vinci is mounted on the highest part of the monument.

Laser instruments were used for the geometric characterization survey of this complex monument, and an SLR camera obtained colour information. Laser measurements were made from an elevator, using a laser scanner Z+F Imager 5006h, from 24 positions and at three distinct heights, taking care to capture every detail (Figure 2). At each laser location, 12 megapixel photos were taken with a Nikon D5000 SLR camera using an 8 mm Samyang lens, mounted on a Nodal Ninja 3 with the nodal point linked to the laser's reference system. These images were then processed with PT GUI stitching software to obtain the equirectangular panoramas used to colour the individual laser scans.

During the editing process, a first union of the point cloud within the software Cyclone 8.1 (Figure 3) occurred and, at the same time,

colour information acquired by spherical panoramas was added. This resulted in an almost complete three-dimensional dataset of the statue of Leonardo, useful for the generation of ortho-photo-planes and reusable in other reverse modelling software.

Figure 3. Point cloud of the statue registered in Cyclone

The Geomagic Studio (Figure 4) reverse modelling software was used to process the data. The single point clouds were first filtered to remove isolated points and partially acquired areas. The data were then transformed into meshes. These areas were filtered again to harmonise individual scans. After this, all the portions of the statue were aligned and merged into a single polygonal model.

This contribution has been peer-reviewed.
doi:10.5194/isprsarchives-XL-5-W4-375-2015

Figure 4. Reverse modelling of the monument

This software was also used to test some photographic image re-projection procedures to render the materials of the statue,finding their centres and baking the colour information onto the model's UV Maps.

One problem found was texture non-uniformity, especially in portions of the model where several pictures were taken. It was decided to edit the individual UV Maps using a photo-editing program to balance their RGB curves and make the calculated textures uniform (Figure 5).

Information on the state of conservation of the statue, fundamental for restoration planning, was also added to the geometric and colorimetric data of the monument.

Among the observed defects, it is important to note that:

- the base, in Pink Baveno Granite, is marked by consistent and inconsistent deposit types, with particular reference to traces of pigeon guano deposits;
- the statues of Leonardo da Vinci and the four disciples, in Carrara marble, is marked by inconsistent deposits and black crusts caused by atmospheric pollution;
- in particular, the bas-reliefs placed on the four principal faces of the base are marked by inconsistent deposits and by the beginning of localized disintegration of the stone. There are also noticeable colour changes in many parts of the bas-reliefs, presumably caused by chemical alterations to the limestone surface.

This substantial information is essential for planning and implementation of subsequent phases of restoration.

3. AUGMENTED REALITY USED FOR RESTORATION

This project aims to be an example of the potential that Augmented Reality (AR) can offer. Architecturally, AR systems include mechanisms to track information about user location and the position of real world objects; hardware and software to process information, and devices to show users the digital information integrated with the real environment (Azuma, 1997).

Figure 5. 3D model with applied uv map

Augmented reality is usually employed using special two-dimensional markers or, for special applications, using photographic imaging of the real object. Using these techniques undeniably means that AR cannot be used on rounded objects such as the statue of Leonardo da Vinci.

However, some studies and applications are able to implement 3D tracking, so that AR may then be used for three-dimensional objects.

The project presented in this article uses the ARmedia 3D SDK application. This tracks three-dimensional objects in space. This feature associates coherent three-dimensional information to what is being seen in the camera viewfinder. The application uploads a photoset of investigated case studies, in addition to a three-dimensional CAD model. The SDK algorithm creates relationships and correspondences between the photographs and the geometric characteristics of the model.

This contribution has been peer-reviewed.
doi:10.5194/isprsarchives-XL-5-W4-375-2015

With synergetic use of the edge-based tracking system function (which helps determine the position of the camera by identifying edges and corners on the three-dimensional model) and template-based tracking (a process involving off-line calculations of planar projections of the 3D model) the application creates a 3D target that can be employed to create the mobile application.

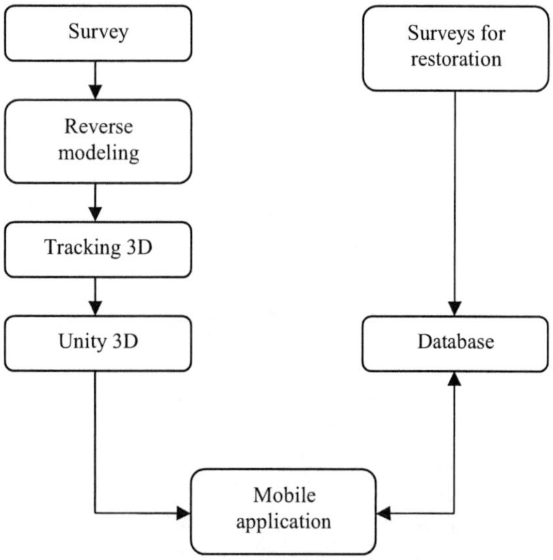

Figure 6. Scheme of the process applied to the creation of the management system

In the Leonardo da Vinci case study, a photoset of the statue was then created as well as an improved three-dimensional model for tracking calculation. This data were then entered into Unity 3D, an integrated multiplatform authoring tool that creates 3D video games and other interactive content (Figure 6).
Having got a prototype AR display of Leonardo da Vinci's monument, the project focused on the preparation of the interaction utility that would be capable of handling the vast quantity of data acquired both during the survey and during the analysis of the statue's state of conservation.
The mobile unit (Figure 7) has an internal application that employs a graphic user interface (GUI), a 3D tracking system and a graphics engine for data visualization and management (Unity 3D). These components operate simultaneously, both to render the image on the device and to aid user interaction.
For the storage and sharing of data, an SQL database is necessary. This will ensure perfect integration with the project App and also with future web-based systems.
Data management script in C# was then designed and written for the Unity platform. System operation buttons were assigned to each function, to be used singly or together with other buttons.
The main functions are for the insertion and editing of up-to-date three-dimensional localised information, or for manual use in user-defined areas. It is therefore possible for a user to select a view of the statue, click on his device and connect to the database in order to combine a database query with spatial coordinates. The system will instantly check for previously saved information to show on the screen, or will create a new query to fill-in.

The points and sensitive areas were identified using ray cast and UV Map functions of the three-dimensional model.
Spatial coordinates of a point were identified using a function that uses ray casting along the perpendicular to the pickup point. This ray intersects the model and gives the spatial coordinates of the meeting point.
In the same way, sensitive areas are identified by multiple ray casting that define a polygonal closed in space. Theme differences in the selected area are defined by the association of distinct colours.
This association was made possible by combining RGB information to the relevant portion of the 3D model UV Map.
Other applied application tools permit point-to-point measurements. This system is not a direct part of the case study but can be very useful in positions difficult to get at from the ground, the statue of Leonardo for example.

Figure 7. Use of the mobile application on the statue of Leonardo da Vinci

4. CONCLUSIONS

This research experimented with the use of applications based on Augmented Reality. This visual system shows great potential both for science and for tourism, and is a first step towards understanding possible innovative future applications in information management.

This contribution has been peer-reviewed.
doi:10.5194/isprsarchives-XL-5-W4-375-2015

Many issues still need to be resolved, mainly related to the customisation and optimisation of the graphical interface, as well as the computing engine performance.
Future research will experiment with solutions proposed in other case studies, comparing solutions and choices with restoration researchers and other possible users.

ACKNOWLEDGEMENTS

The authors would like to acknowledge and thank the developers of AR Media for the opportunity to use their application. System that can lead an important development in the field of cultural heritage.

REFERENCES

Azuma R.T., 1997. A Survey of Augmented reality. In: *Presence: Teleoperators and Virtual Environments*, 6, pp. 355-385.

Heckel W., Yardley J., 2004. *Alexander the great: Historical sources in translation*. Blackwell Publishing, Oxford, UK.

Mohammed-Amin R. K., 2010. *Augmented reality: A narrative layer for historic sites*. MEDes (CMD), University of Calgary.

Di Paola F., Inzerillo L., 2011. Augmented reality. The case of Salinas Museum of Palermo. In: Gambardella C. (a cura di) *Le Vie dei Mercanti*. S.A.V.E. Heritage, La Scuola di Pitagora Editrice, Napoli, Italy, pp. 1-7.

Battini C., 2012. *Rilievo digitale e restituzione, concetti base ed esempi*. Alinea editrice, Firenze, Italy, pp. 114.

Battini C., 2014. *Augmented reality and cultural heritage. New systems of representation*. In: V Congreso Internacional de Expresión Gráfica en Ingeniería, Arquitectura y Carreras Afines y XI Congreso Nacional de Profesores de Expresión Gráfica en Ingeniería, Arquitectura y Carreras Afines, Rosario, Argentina, pp. 452-457.

Bonacini E., 2014. La realtà aumentata e le app culturali in Italia: storie da un matrimonio in mobilità, in: *Il Capitale Culturale. Studies on the Value of Cultural Heritage*, IX, EUM Edizioni Università di Macerata, Italy, pp. 89-121.

Landi G., 2014. *AR for the comprehension of linear perspective in the Renaissance masterpiece The Holy Trinity (Masaccio, 1426)*. In: Mixed and Augmented Reality - Media, Art, Social Science, Humanities and Design (IMSAR-MASH'D), 2014 IEEE International Symposium on, Munich, German, pp. 69-70.

This contribution has been peer-reviewed.
doi:10.5194/isprsarchives-XL-5-W4-375-2015

LASER SCANNER AND CAMERA-EQUIPPED UAV ARCHITECTURAL SURVEYS

N. Maiellaro [a, *], M. Zonno [a], P. Lavalle [b]

[a] ITC-CNR, Construction Technology Institute, via Lembo 38 B, 70125 Bari, Italy - (n.maiellaro, m.zonno)@itc.cnr.it
[b] Pavimental S.p.A., Corso del Mezzogiorno 202, 71122 Foggia, Italy - pierolavalle@hotmail.it

Commission V, WG V/4

KEY WORDS: Laser scanner, UAV, 3D modelling

ABSTRACT:

The paper reports the workflow of 3D modelling and photorealistic texture mapping based on close range imagery acquired through a terrestrial laser scan and a camera mounted on a mini Unmanned Aerial Vehicle (UAV). The 3D models of the same architectural object – a watchtower named "Torre Pelosa" (Bari, Italy) - were compared in order to evaluate strengths and weaknesses of the used instruments and methods. The evaluated parameters where geo-referencing capability, manageability, manoeuvrability, operational constraint, survey completeness and costs. Despite a lower accuracy of the results, the use of a UAV mounted camera is essential to obtain a whole representation of the tower.

1. INTRODUCTION

This paper is based on the results of a postgraduate thesis at the Polytechnic of Bari in partnership with the Construction Technology Institute of the National Research Council. The aim was to compare textured 3D model of the "Pelosa" tower based on two systems: a laser scan with external camera and a camera mounted on a mini Unmanned Aerial Vehicle (UAV).

The use of an UAV complies with the requirement, as a prerequisite for adequate results, to have camera movement between each photo: fixed camera installation on a moveable crane or a tripod might reduce the number of shots and perspectives but influence negatively the result (Kaiser et al., 2014).

The work has been developed on "Torre Pelosa", built in the 15th century in order to defend the coast from pirates and raiders infesting the Adriatic Sea. It is a watchtower that still exists at the centre of the main square of Bari suburb.

Because not all points of the building were visible from the ground, aerial images are fundamental for a full building coverage; moreover a foggy weather is optimal to take pictures without shadows, having all distinguishable. Unfortunately, each image set was acquired on a cloud free and windy day, therefore it was difficult to recognize details in the shadow areas as well as it was not possible to fly on roofing.

Due to insurance-related claims - the tower is located in an urban area (Figure 1) - the UAV was used in a VTOL (Vertical Take-Off and Landing) mode of operation in a limited time: the risks for personal injury had to be kept as low as possible (Eschmann et al. 2012).

During the fieldwork, using both methods, the complete settlement was documented in half-day; the post-processing was done using RiscanPro, Photoscan and Cloudcompare.

Figure 1: The tower location (Google maps).

2. ARCHITECTURAL SURVEYS

2.1 Laser Scanner Architectural Survey

For the image acquisition using a ground station we used the laser scan a Microgeo's RIEGL VZ-400 which has a close range of 500 meters and gives a full wave form analysis of scans, a scanning angle of 360° horizontal and 100° vertical at 122.000 points/second. The station is equipped with an external Nikon D700 camera (10 megapixel; 20 mm lens) and a GPS for an absolute orientation of the images in a global system.

At first, we placed targets all around the tower and acquired data and photos from eight different stations located around the building Targets are very important during the data acquisition because of their retro-reflective material that mirrors the narrow infrared laser beam (Figure 2).

We made a new scan position for each one of those stations in order to make TPoints and targets visible in at least two different images.

* Corresponding author. N. Maiellaro: Introduction and Conclusions; M. Zonno: Results; P. Lavalle: Architectural Surveys

This contribution has been peer-reviewed.
doi:10.5194/isprsarchives-XL-5-W4-381-2015

Figure 2: Laser scanner stations (Google maps).

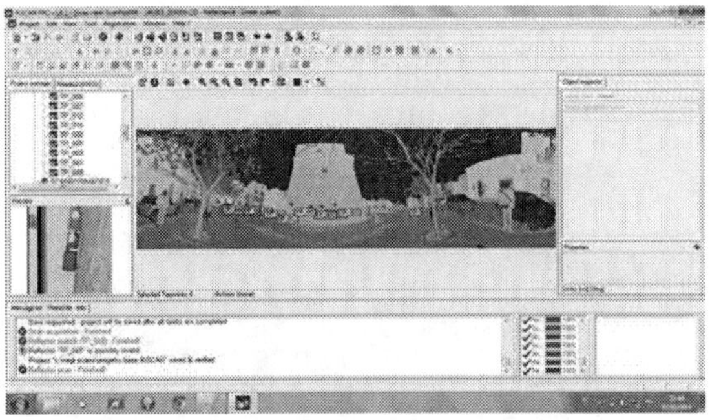

Figure 3: "Find reflectors" function.

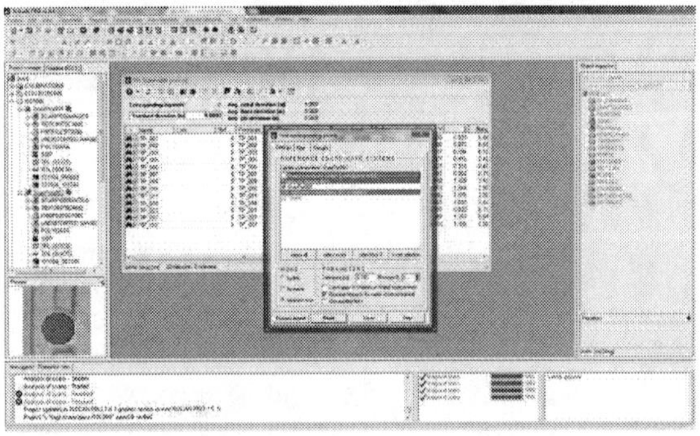

Figure 4: Registration into a PRCS-Project Coordinate System.

Figure 5: 3D Final textured model.

As to the scanner configurations for each scan we had to set up several inputs such as point's range gate, pixel spacing, minimum or maximum reflectance and inclination in order to make a multilevel analysis of the object.

Moreover we had to set up all the camera's settings manually, adjusting aperture and shutter speed for light conditions on site in order to take good quality pictures of the whole Tower for each station.

If this is not possible we could also take pictures with a definite instrument's inclination angle, but in this case we took all the photos with the same inclination angle.

At this point, we started the scan's registration on site. At the end of the first scan registration we hid all the reflection points other than our targets so we created the final scan position (Figure 3). Every scan position was already saved automatically in the Laser's memory.

For the post-processing of data we use RISCANPRO v. 1.6.1, a photogrammetric software which transforms the data acquired during the measurement campaign (such as Scans, Fine scans, digital images, GPS data, coordinates of control points and Tie Points) into a common well-defined coordinate system using transformation matrices.

The registration of each scan position taken on site is based usually on tie point lists. The tie point itself is commonly the centre of the reflective target exposed all over the Tower to the scanner's beam during the image acquisition phase.

In order to register a single scan position into the project coordinate system we choosed one of all the scans and then we registered all the other scan positions onto our reference scan.

The software matches all the corresponding tie points with its algorithm that tries to find corresponding points between the current TP list and the selected coordinate systems of the chosen reference scan. There should be a minimum of 3-5 tie points in common.

If correspondences between points of different scan positions are found, a new tie point in the TPL PRCS (Project Coordinate System) is created.

If the tie point coordinates are available in global coordinates, they should be entered into the global tie point list, TPL GLCS (Global Own Coordinate System) and should be transferred into the TPL PRCS by defining an appropriate transformation matrix (Figure 4).

At the end of the registration into a referenced Global System we have a metrical 2D or 3D model of the entire Tower. RISCANPRO is enable to join the scan information with the high-resolution photos taken. It applies the images as a texture to a meshed surface obtaining a textured model (Figure 5).

In the 2D model's visualisation the angular data (polar and azimuth scan angles) is neglected and the measurements are put in a plane rasterization of the image according to the indices within a 3D data set.

In the 3D visualisation, the complete geometrical information is used. It basically represents a point cloud model in which every measurement is represented as a point in 3D space.

This contribution has been peer-reviewed.
doi:10.5194/isprsarchives-XL-5-W4-381-2015

2.2 Camera-Equipped UAV Architectural Survey

In our approach the process was entirely data-driven, and structure emerges from the data rather than being dictated by a set of potentially incorrect priors (Farenzana et al. 2008).

For the image acquisition using an aerial station we used the low-cost mini UAV Microgeo's AeroMax600. It is a quad-copter equipped with a GPS and an INS-Inertial Navigation System that helps the pilot to control the position and can directly react on wind effects by correcting the position and the flight itself, a barometer that allows the UAV to perform altitude stabilization and position control.

We equipped the UAV with a Canon S100 camera (12 megapixel; 24 mm lens) for aerial images, fixing the focus to infinity (manual mode).

The UAV is capable to fly along a flight path pre-defined by waypoints but in this case, the UAV was controlled by a pilot with a remote controller due to a strong sea wind, thanks to a very good visibility.

For the post-processing data fusion we used Agisoft PhotoScan, a software designed for image-based reconstruction and measurement.

It is an advanced image-based 3D modelling used to create professional quality 3D textured-model from images alignment taken on the UAV's flight providing that the object to be reconstructed is visible on at least two photos (Figure 6).

Agisoft PhotoScan is capable of automatic camera calibration using GCP (Ground Control Points) and the Simultaneous Localization and Mapping (SLAM), largely known as the problem of Structure From Motion (SFM), in which the camera translation and rotation can be calculated by given consecutive views of a scene referring to a project coordinate system.

Figure 6: Pictures taken on the flight.

At first we had to load the photos for the 3D reconstruction. PhotoScan is able to find the camera position and orientation of each photo and build a sparse point cloud model based on the estimated camera positions and pictures themselves that is of almost the same density as LIDAR point clouds.

At this point we built mesh and obtain a Tower's textured model. PhotoScan supports a coordinate system based on either ground control point called Markers taken by ground control stations such as GPS systems that create a .txt list of the Marker's coordinate.

Setting the coordinate system provides a correct scaling of the model allowing for surface area, volume and high measurements and creates a geo-referenced 3D model as a result (Figure 7).

Figure 7: Cloud points and texture using Agisoft Photoscan

3. RESULTS

3.1 Comparison between survey methodologies

The survey carried out on "Torre Pelosa" offered the opportunity to compare the 'range based modelling' survey technique with Riegl VZ400 laser scanner to the 'image based modelling' with Canon S100 12 Megapixel camera (in horizontal attitude) mounted on Microgeo AeroMax600 multi-rotor, availing of a Leica total station for topographic point measurement.

For the survey with laser scanning 8 stations were positioned at a distance of 7 to 10 metres from the object; scans were executed with a 0.040 deg pitch, resulting in a point cloud with a resolution ranging from some millimetres in the lower part to a maximum of about 2 cm on the top.

The drone flight was performed in visual mode moving with regular steps along the building façades, as for a shooting from the ground in manual mode.

The drone was maintained at a distance of 5-6 m for the vertical flight with the camera parallel to façades. In order to cover two façades, 78 frames were recorded.

Contextually, a series of points, corresponding to the targets positioned for laser scanning, were recorded through the total station and subsequently imported in the processing phase. The points that were topographically recorded were used for the model orientation and for the evaluation of the 3D-reconstruction accuracy.

The precision of the methodology was checked by calculating the mean square deviation, i.e. the deviation between the points assigned as markers and the points calculated by the software, indicating that the points were measured with an accuracy of few millimetres.

The shooting of the two façades required more flights, due to environmental conditions – wind in particular - and to the low drone battery life affected both the time committed in survey operations and the quality of acquisitions, that sometimes needed to be replicated.

All flights were performed in manual modality, focusing on the roofing, on the higher parts and on the staircase of the tower where laser scanning had delivered incomplete results.

The roofing also required more flights, due not only to the wind conditions but especially to the pairing of strong lights in some zones and shaded areas, since the bright sunshine made the day not actually suitable (Figure 8).

Despite such environmental conditions, the shooting allowed to reach a much wider coverage of the building compared to the one obtained with laser scanning, resulting in a more complete model (Figure 9).

Light and wind conditions have no significant influence (exclusively in the case of untextured point cloud survey) on the laser scanner equipment, that can be used reliably also in the absence of light.

Both techniques required only few hours of field survey, with a careful organization of work and an adequate number of operators for the simultaneous management of instruments.

The processing of survey data obviously involved more time, both in the case of laser scanning and in the elaboration of images acquired from the drone.

Both software applications, Riscan Pro v.1.6.1 for laser scanning and Agisoft PhotoScan v.1.0.4 for UAV image processing, although with different procedures, delivered satisfying results.

While a laser scanner with integrated GPS allows for automatic geo-referencing of scans, the multirotor-based survey procedure can only rely on a camera, which entails recording coordinates with the help of a GPS receiver and of the total station and importing them in Agisoft to orient the model in a given reference system.

Figure 8: Shot from drone – effect of bright daylight

Figure 9: Point cloud model from drone shooting

The joining of point clouds with Agisoft was obtained also for those façades for which no control point was available, by aligning one model to the other, then scaling, rotating and

This contribution has been peer-reviewed.
doi:10.5194/isprsarchives-XL-5-W4-381-2015

translating the second model on the first one. In this way, all measured points were gathered up and included in a single model.

With Riscan Pro, the joining of scans is automatic and much simpler. At the end of processing with the two applications, it is possible to compare or join point clouds with Riscan Pro or with other dedicated software; starting from the comparison between Image-based and Range-based modelling techniques, our study tried to investigate their combination. This integration is essential to perform the restitution of models with a high level of morphological and chromatic detail.

3.2 Comparison with CloudCompare software

The two point-cloud models obtained with Riscan Pro and Agisoft PhotoScan were analyzed with the help of the OS CloudCompare.

It performs the elaboration of point clouds and meshes; it has been developed advanced algorithms supporting resampling, recording, colour handling, normal calculation, and other functions. In our work, a comparison of the point clouds obtained with the two survey techniques was carried out.

After uploading the ASCII files of the two models with the drag-and-drop function (Figure 10), alignment and overlapping were obtained through the registration of clouds (Figure 11).

Then it is possible to display 3D views of the clouds. CloudCompare associates for each cloud a real scalar value with each point. We can visualize a Scalar Field by means of a colour gradient, applied to the points in the cloud.

Figure 10: CloudCompare, point-cloud models

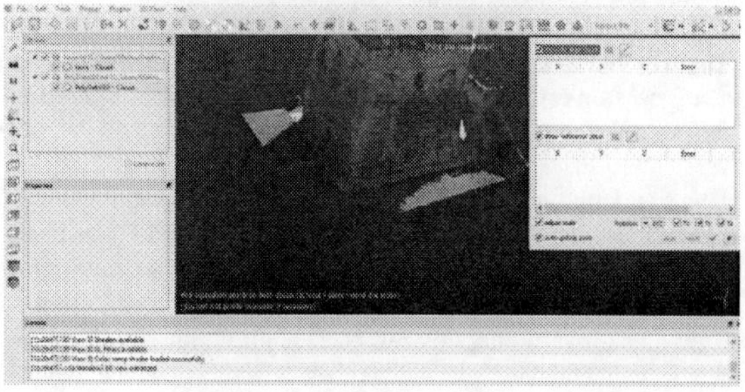

Figure 11: CloudCompare, registration and alignment of points

Each point of the cloud will be coloured as a function of its value, according to the chosen colour scale.

Several toolbars in CloudCompare can be used to register the cloud, align and show the cloud/cloud distances, also obtaining a colour scalar field as a result of the difference analysis.

Figure 12: Cloud points comparison using CloudCompare (blue: minimum deviation; yellow: maximum deviation)

This function calculates the magnitude of the gradient of the active scalar field applying a Gaussian filter to the active scalar field to define the size of the sigma kernel of the Gaussian filter.

The colours in the RGB scale of the 3D model show the deviation between the points of the Agisoft Photo Scan cloud, then from drone survey, and those of the RISCAN Pro cloud, resulting from the laser scanner survey.

Figure 12 shows that maximum deviation occurs on the right part of the tower (yellow-red: maximum deviation = 6 cm); the middle part of the building appears as a homogeneous point cloud, with deviation values less than 1 cm.

The maximum deviation arises from the few images in that area: a staircase on arches were the wind was strong, making hard the fly.

The overlapping of the two models shows that both survey procedures delivered precise and similar results (Tucci et al., 2014) and, by performing also a compensation of missing or insufficient data, it also demonstrates how the integration of techniques was, in this case, the most appropriate solution.

4. CONCLUSIONS

In this work we have shown two data acquisition techniques for the documentation of an architectural site. Both, the terrestrial laser scanner and the camera mounted on an UAV, fulfilled our goals. Both models of the tower where compared using CloudCompare (designed to perform comparison between two 3D point clouds or between a point cloud and a triangular mesh); it returns cloud/mesh signed distances computation in RGB scale as a result of the comparison.

Using CloudCompare, we noticed a point deviation in the 3D models less than a centimetre in the corner of the tower.

The comparison between the two survey systems highlighted strengths and weaknesses; for example, the laser scanner through the integrated GPS gives an automatic geo-referencing from the scan, while the UAV survey needs an additional GPS equipment. On the other hand, the manageability and manoeuvrability of the UAV are better in respect to the laser scanner (which is more expensive respect to the UAV).

The resolution of the laser scanner is higher but blunders occurred for roofing and points in recesses; furthermore, although more viewing directions (e.g. oblique or nadir) are possible from an airborne system, it has a lot of operational and security constraints. Ultimately, only the combination of terrestrial and aerial images gave us a complete, high quality digital representation of the building.

Another approach would be to model the building without laser scanner, combining photos from the same camera mounted on a tripod and on a UAV (Püschel et al., 2008), using post-processing tools able to generate - easily - results comparable to semiautomated equivalents like laser scanning (Spring and Peters, 2014). *Memento,* an Autodesk tool in Alpha stage, could be the right tool: it generates 3D meshes from reality capture input, producing optimized models for Web publishing, virtual and augmented reality, further digital use or 3D printing (Autodesk, 2014).

As a future perspective, the use of images from many uncooperative cameras could be investigated (Cho and Snavely, 2013): despite the formidably difficult problem of social media mining, the availability of many images (at least for main architectural sites) will give the possibility to select those without shades, removing survey costs.

5. ACKNOWLEDGEMENT

We thank the "Soprintendenza per i Beni Architettonici e Paesaggistici per la Provincia di Bari" for the survey permission. Furthermore, we thank MicroGeo for the UAV availability.

REFERENCES

Autodesk, 2014. Project Memento: All About That Space, All About that Space, No Problem http://labs.blogs.com/its_alive_in_the_lab/2014/12/project-memento-all-about-that-space-all-about-that-space-no-problem.html

Cho P., Snavely N., 2013. 3D Exploitation of 2D Imagery, *Lincoln Laboratory Journal,* Vol. 20, n.1, pp.105-137 https://www.ll.mit.edu/publications/journal/pdf/vol20_no1/20_1_8_Cho.pdf

Eschmann C., Boller C., Kuo C.-M., Kuo C.-H., 2012. Unmanned Aircraft Systems for Remote Building Inspection and Monitoring, *proceedings of the 6th European Workshop on Structural Health Monitoring.* Th.2.B.1 http://www.ndt.net/article/ewshm2012/papers/th2b1.pdf

Püschel H., Sauerbier M., Eisenbeiss H., 2008. A 3d Model of Castle Landenberg (CH) from Combined Photogrammetric Processing of Terrestrial and Uav-Based Images in *The International Archives of the Photogrammetry, Remote Sensing and Spatial Information Sciences.* Vol. XXXVII Part B6b. Beijing, pp. 93-98 www.isprs.org/proceedings/XXXVII/congress/6b_pdf/16.pdf

Spring A.P., Peters C., 2014. Developing a low cost 3D imaging solution for inscribed stone surface analysis, *Journal of Archaeological Science* 52 (2014), pp. 97-107 http://sigcis.org/pipermail/members/attachments/20141111/caac6933/attachment-0071.pdf

Tucci G., Bonora V., Korumaz A. G., Salvadori M., Orlandini S., 2014. Fotogrammetria digitale e sistemi UAV: applicazioni al monitoraggio ambientale, *Atti 18a Conferenza Nazionale ASITA,* 14–16 ottobre 2014, Firenze, pp. 1183-1188 http://www.microgeo.it/Droni/pdf/Aeromax%20600%20-2014-UAV-ASITA-Uni%20FI.pdf

Farenzena M., Fusiello A., Gherardi R. 2008. Efficient Visualization of Architectural Models from a Structure and Motion Pipeline, in *EUROGRAPHICS 2008,* www.diegm.uniud.it/fusiello/papers/short2014.pdf

Kaiser A., Neugirg F., Rock G., Müller C., Haas F., Ries J., Schmidt J., 2014. Small-Scale Surface Reconstruction and Volume Calculation of Soil Erosion in Complex Moroccan Gully Morphology Using Structure from Motion, *Remote Sensing* 2014, 6(8), pp. 7050-7080 http://www.mdpi.com/2072-4292/6/8/7050/htm

POLYGONAL SHAPES DETECTION IN 3D MODELS OF COMPLEX ARCHITECTURES

G.B. Benciolini[a], A. Vitti[a,*]

[a] DICAM, University of Trento, 38123 Trento, Italy

Commission V, WG V/4

KEY WORDS: Variational modes, global models, geometric data, corner detection, polygonal shapes

ABSTRACT:

A sequential application of two global models defined on a variational framework is proposed for the detection of polygonal shapes in 3D models of complex architectures.
As a first step, the procedure involves the use of the Mumford and Shah (1989) 1st-order variational model in dimension two (gridded height data are processed). In the Mumford-Shah model an auxiliary function detects the sharp changes, i.e., the discontinuities, of a piecewise smooth approximation of the data. The Mumford-Shah model requires the global minimization of a specific functional to simultaneously produce both the smooth approximation and its discontinuities.
In the proposed procedure, the edges of the smooth approximation derived by a specific processing of the auxiliary function are then processed using the Blake and Zisserman (1987) 2nd-order variational model in dimension one (edges are processed in the plane). This second step permits to describe the edges of an object by means of piecewise almost-linear approximation of the input edges themselves and to detects sharp changes of the first-derivative of the edges so to detect corners. The Mumford-Shah variational model is used in two dimensions accepting the original data as primary input. The Blake-Zisserman variational model is used in one dimension for the refinement of the description of the edges.
The selection among all the boundaries detected by the Mumford-Shah model of those that present a shape close to a polygon is performed by considering only those boundaries for which the Blake-Zisserman model identified discontinuities in their first derivative. The output of the procedure are hence shapes, coming from 3D geometric data, that can be considered as polygons. The application of the procedure is suitable for, but not limited to, the detection of objects such as foot-print of polygonal buildings, building facade boundaries or windows contours.
The procedure is applied to a height model of the building of the Engineering Faculty of the University of Trento and to other data. The obtained results show the effectiveness of the variational methods in handling geometric data describing complex objects.
The most original and innovative contributions of this work are:
- the consecutive and coordinated use of the proper variational models (of first and second order respectively) at first in two dimension and subsequently on the one dimensional sets arising in the first segmentation,
- the application to geometrical data that represent a quite complex architecture.

1. INTRODUCTION

The detection of polygonal shapes as, for example, building footprints, facade boundaries and window contours, can serve for the construction of vector models of architectures, in particular for essential polyhedral volumes with polygonal faces. Polygonal shapes can be selected from a set of shapes of various type detected by the processing of geometric data. A set of shapes can be produced by finding edges in the geometric data.

In this work a procedure for the detection of edges and the subsequent identification of polygonal shapes is presented. Two global models are exploited to accomplish the main goals of the procedure.

The procedure starts with the application of the Mumford-Shah variational model (Mumford and Shah, 1989) to height data for the detection of sharp changes (edges, discontinuities) of a piecewise smooth approximation of the data. The edges of such approximation of height data are then processed using the Blake-Zisserman variational model (Blake and Zisserman, 1987). The output of the Blake-Zisserman model is finally treated to identify those edges that can be approximates as piecewise almost-linear curves connected by corners, i.e., polygonal shapes.

*Corresponding author (alfonso.vitti@unitn.it)

The key idea behind the proposed procedure is deeply related to the characteristics of the same variational models involved. A concise description of the main characteristics of the models is given hereafter so to outline the basic elements that driven the definition of the procedure. Details about the models and their application will be given in the next two sections.

The Mumford-Shah model is an energy functional built up of three terms depending on two unknown quantities of very different kind. One unknown is a piecewise continuous function that must smoothly approximate the data almost everywhere in the data domain. The second unknown is the set of the curves of the data domain where the requirements on the continuity and on the smoothness of the approximating functions are not imposed, i.e. a set of discontinuities. The smoothness of the approximating function is controlled my a specific term of the first order involving the gradient of such function. The other two terms measure the proximity of the unknown approximation to the data and the length of the discontinuity set. The length of the discontinuity set needs to be controlled to avoid over-segmented approximations. The relative strength of the three terms is governed by specific weight parameters. To find the two unknowns one has to minimize the energy functional, globally over the entire data domain.

The Mumford-Shah model is applied to height data to obtain a smooth approximation of the data and a set of homogeneous regions, those bordered by the curves of the discontinuity set repre-

This contribution has been peer-reviewed.
doi:10.5194/isprsarchives-XL-5-W4-387-2015

senting jumps in the height data. Since buildings have, in general, a polygonal footprint, the idea is to select those regions that can be described as polygonal shapes, i.e., a set of linear segments jointed by corners.

The Mumford-Shah model can not be directly used for the recovery of corners or junctions in the curves of the discontinuity set because only the length of the discontinuity set is controlled and because the presence of corners and junctions does not affect the measure of the length of the set. A further process of the Mumford-Shah output is hence needed to explicitly detect corners and edges that can be approximated by a piecewise almost-linear curve, i.e,. a polygonal shapes. A variational model of the second order proposed by Blake and Zisserman (1987) has been involved to accomplish this second step of the procedure. The Blake-Zisserman model is an energy functional depending on a piecewise continuous function that smoothly approximates the data and on two discontinuity sets. In the model not only the approximating function but also its first derivative is requested to be piecewise continuous. Moreover, the smoothness of the function is controlled by a term of the second order involving the second derivative of the approximating function. These two facts leave as unknown the two discontinuity set, one for the discontinuity of the approximating function and one for the discontinuities of its first derivative. Also for the Blake-Zisserman model, the unknowns can be found minimizing the energy functional, globally over the entire data domain.

The Blake-Zisserman model is applied to the boundaries of the homogeneous regions detected by the application of the Mumford-Shah model to height data. The presence of a corner in a boundary will be detectable by observing the discontinuity set of the first derivative of the function that smoothly approximates the boundary.

For technical reasons, the computation of the solution of both variational models requires the discontinuity sets to be described by means of some specific auxiliary functions. The functions that represents the discontinuity sets have been mainly intended as down-weighting functions of the smoothing terms, but in fact in many applications they are a quite useful outcome of the models.

The described procedure has been implemented into a series of original computer programs that have been extensively tested on a variety of both simulated and real data, including optical images and geometric data (Vitti, 2012a,b; Zanetti and Vitti, 2013). After many application with land images and laboratory images this work extends the application range of the Mumford-Shah model and of the Blake-Zisserman model to data of a complex architecture.

The metric quality of the result is obviously related to the resolution and to the overall accuracy of the data. The method is anyway quite claver in filtering out the unavoidable noise.

2. TWO GLOBAL MODELS

The Mumford-Shah model and the Blake-Zisserman model will be described to highlights their main characteristics. Deeper treatments on the subject are available in the literature and will be pointed out at the need.

2.1 The Mumford-Shah model

The Mumford-Shah model in dimension two is:

$$\mathcal{MS}(u,K) = \int_\Omega |u-g|^2\,dx + \lambda \int_{\Omega\setminus K} |\nabla u|^2\,dx + \alpha\mathcal{H}^1(K)\,, \tag{1}$$

where $g : \Omega \to \mathbb{R}$ is the data, $\Omega \subset \mathbb{R}^2$ is the data domain, $u \in C^1(\Omega \setminus K)$ is the piecewise smooth approximating function of the data, K is the discontinuity set and $\mathcal{H}^1(K)$ is the 1-dimensional Hausdorff measure of the set. The first term in (1) is a global measure of the proximity of the solution u to the data, the smoothness of the function u is induced by the second term, the last term is the so called geometric term measuring the length of the unknown set K. The parameters λ and α controls the relative strength of the three terms. The problem here is to find the pair (u, K) that minimizes the Mumford-Shah energy functional.

To solve the problem in practice it is necessary to work on a functional different from (1). Ambrosio and Tortorelli (1992) proposed to solve the Mumford-Shah variational problem by means of the following functional:

$$\mathcal{F}_\epsilon(u,s) = \int_\Omega \left(|u-g|^2 + \lambda s^2|\nabla u|^2\right) dx + \\ + \alpha \int_\Omega \left(\epsilon|\nabla s|^2 + \frac{(1-s)^2}{4\epsilon}\right) dx\,. \tag{2}$$

The explanation of the need for a new functional and of its definition requires mathematical tools and developments that are well outside the scope of this work. We just highlight here the main characteristics of (2) that make it treatable in practice and that distinguish it from (1). All the terms of the functional $\mathcal{F}_\epsilon(u,s)$ are defined over the entire domain Ω, the smooth term is now scaled by a new auxiliary function s and the third term of (1) has been replaced by a functional of s. This last quadratic functional approximates the $\mathcal{H}^1$ measure of the discontinuity set K exploiting the auxiliary function s which is defined on the bi-dimensional domain Ω. The function s has another specific role: to down-weight the gradient of the approximating function u by going to 0 right were the gradient of u takes big values and remaining close to 1 almost everywhere on Ω. The parameter ϵ that appears in (2) is a convergence parameter. In fact it is a sequence of functionals like $\mathcal{F}_\epsilon(u,s)$ that is proved to approximate, according to a particular type of convergence criterion (Braides, 2002), the Mumford-Shah functional (when ϵ goes to zero, the minimizers of (2) converge to the minimizers of (1)).

To better understand the output of the Mumford-Shah model we highlight that the auxiliary function s has values close to 0 either where the approximating function u has very big gradient, i.e., where it is discontinuous, or where the approximating function u is not homogeneous with respect to the smoothness measure expressed by the second term of the functional (scaled by the parameter λ). In this work the function s will be used to identify parts of Ω with homogeneous values of u whose boundaries will be further processed to select those having a polygonal structure.

2.2 The Blake-Zisserman model

The Blake-Zisserman functional in dimension one is:

$$\mathcal{BZ}(u,K_0,K_1) = \int_I |u-g|^2\,dx + \gamma \int_{I\setminus(K_0\cup K_1)} |u''|^2\,dx + \\ + \alpha\#(K_0) + \beta\#(K_1 \setminus K_0)\,, \tag{3}$$

where $g : I \to \mathbb{R}$ is the data, $I \subset \mathbb{R}$ is the data domain, $u \in C^2(I\setminus K_1)$ and piecewise continuous in $I\setminus K_0$ is the approximating function of the data, K_0 and K_1 are the discontinuity sets and $\#(\cdot)$ is the counting measure of a set. The first term in (3) is a global measures of the proximity of the solution u to the data, the smoothness of the function u is induced by the second term, the last two terms measure the size of the unknown sets K_0 and K_1. In dimension one the elements of the sets K_0 and K_1 are points. The parameters λ, α and β control the relative strength of the four terms. The problem here is to find the triplet (u, K_0, K_1) that minimizes the Blake-Zisserman energy functional.

For reasons similar to those requiring a sequence of approximating functionals to find the minimizers of the Mumford-Shah functional, also the Blake-Zisserman functional needs to be properly approximated in order to be used in practice. Ambrosio et al. (2001) proposed to solve the Blake-Zisserman variational problem by means of the following functional:

$$G_\epsilon(u,s,z) = \int_I \left(|u-g|^2 + \gamma z^2|u''|^2 + \mu_\epsilon s^2|u'|^2\right) dx + \\ + (\alpha-\beta)\int_I \left(\epsilon|s'|^2 + \frac{(1-s)^2}{4\epsilon}\right) dx + \\ + \beta \int_I \left(\epsilon|z'|^2 + \frac{(1-z)^2}{4\epsilon}\right) dx\,. \quad (4)$$

In (4) the new auxiliary function z is introduced to approximate the discontinuity set K_1 following the reasoning that lead the introduction of the approximating function s in (2). The function z is close to 0 on the discontinuities of the first derivative of the approximating function u and close to 1 almost everywhere on I.

In this work the function z will be used to identify sharp changes in the coordinates of the points lying on the boundary of the homogeneous regions detected using the Mumford-Shah model and to identify where the variation of point coordinates is almost-linear. Sharp changes in the point coordinates, where z values shrink to 0, are expected to correspond to corners. The function z remains close to 1 on the parts of I corresponding to straight segments.

3. APPLICATIONS

The applications presented hereafter have been realized using original software implementing the Mumford-Shah model in dimension two and the Blake-Zisserman model in dimension one. The source code is available at `www.ing.unitn.it/~vittia/sw` and licensed as Free Software.

3.1 Example 1

A first example illustrates the use of the Blake-Zisserman model for the detection of corners and straight segments on a line. A wall plan of small Church is processed. The plan has been obtained by slicing the 3D point clouds of the Church acquired using a terrestrial laser-scanner.

To apply the 1D BZ model to a plane curve (the boundary of a 2D flat object) we make use of the curvilinear coordinate on the curve. When dealing with discrete data, we work with the set of Cartesian coordinates of points on the plane (assumed to lie on a unknown continuous curve) and with a discrete implementation of the Blake-Zisserman model. The curvilinear coordinate of the points is approximated by the chordal coordinate c computed as the summation of the euclidean distances between successive points. The discrete Blake-Zisserman model is the applied twice: once to the coordinate pair (c, x) and once the the coordinate pair (c, y).

Figures (1) and (2) show a view of the point cloud and the wall plan of the Church respectively. The plot of the auxiliary function z produced by applying the Blake-Zisserman model to the red element of the wall plan of Figure (2) is given in Figure (3). The cyan dots of figure (2) have been placed on the wall plan using the coordinates of the points where the auxiliary function z equals 0 in the plots of Figure (3). The green dot locates the point with coordinate $c = 0$ and the green arrow indicates the verse of increasing of the coordinate. It is possible to note how the auxiliary function z in the plots of Figure (3) remains close to 1 between the points where it shrinks to 0. This fact can be exploited to identify the straight segments that compose the red curve of (2).

Figure 1: View of the point cloud of a small Church

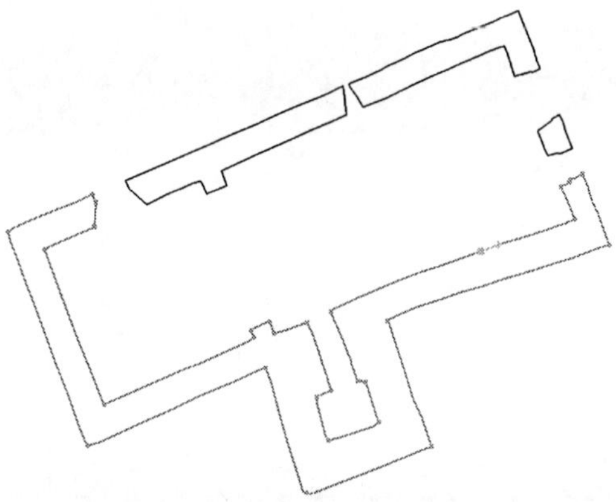

Figure 2: Wall plan of the small Church of Figure (1), with processed element in red, detected corners represented with cyan dots

3.2 Example 2

This example illustrates the plot of the auxiliary function s obtained by applying the Mumford-Shah model to a gridded point cloud of a facade of an industrial building. The boundaries of the homogeneous parts of the facade have been further treated using the Blake-Zisserman model to identify the contours of the facade windows.

Figure (4) shows the geometric data processed using the Mumford-Shah model. Figure (5) shows a detail of the same area of the geometric data (top) and of its smooth approximation (bottom). It is possible to note how the approximating function u is smoother that the data everywhere but along the borders of the window openings. The plot of the auxiliary function s of the facade produced by the Mumford-Shah model is given in Figure (6). Figure (7) shows in green and orange the polygonal shapes identified using the auxiliary function z obtained by the application of Blake-Zisserman model to the boundaries of the homogeneous regions detected after the application of the Mumford-Shah model.

This contribution has been peer-reviewed.
doi:10.5194/isprsarchives-XL-5-W4-387-2015

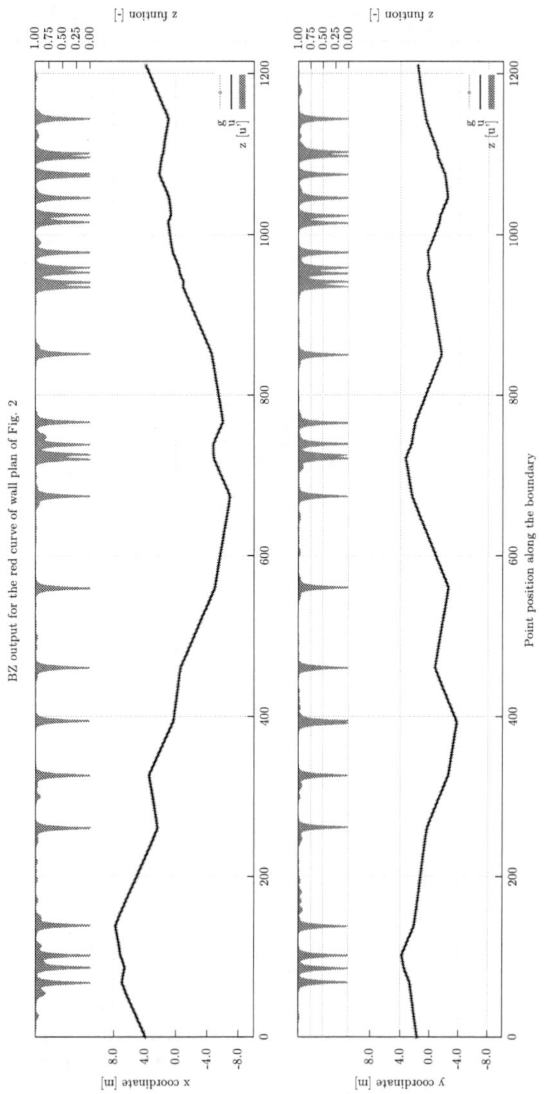

Figure 3: Plot of the auxiliary function z produced by the Blake-Zisserman model applied to the red element of the wall plan of Figure (2)

3.3 Example 3

In this example the Mumford-Shah model is applied to a high-resolution (0.5mx0.5m) Digital Surface Model (DSM) of an area including a polygonal building of complex architecture. Again the boundaries of the homogeneous parts of the DSM detected using the plot of the Mumford-Shah auxiliary function s have been processed using the Blake-Zisserman model to identify the polygonal elements of the building.

Figure (8) shows an aerial view of the building. Figures (9) and (10) give the plot of the Mumford-Shah auxiliary function u for the entire study area and a detail of the close surrounding of the building. It is possible to observe how the auxiliary function describes not only the discontinuities of the approximating function u but also the domain parts where the function u is not homogeneous (in light gray). In figure (10) there are some lines

Figure 4: Geometry model of the facade of an industrial building

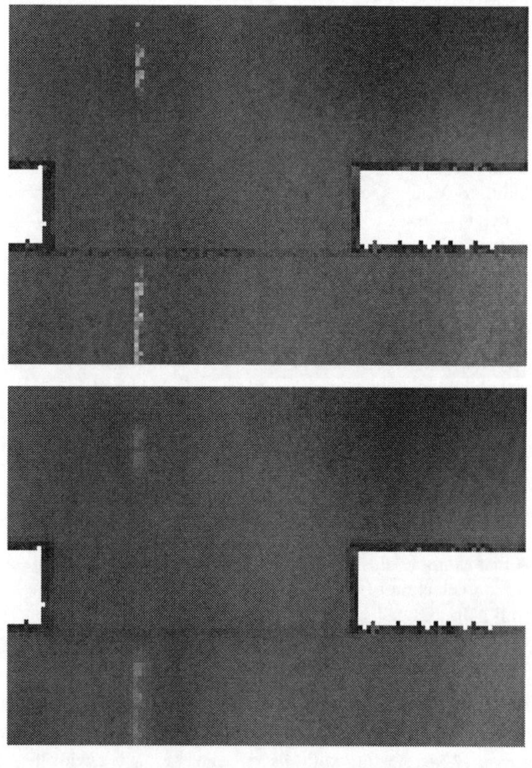

Figure 5: Details of the original model of the facade of Figure (4) (top) and of its smooth approximation (bottom)

with a dead vertex. Those lines describes discontinuities of the function u that gently disappear and that can not be identified as boundaries. This fact is deeply related to the characteristics of the Mumford-Shah model that, right for this reason, is also known as a model for the detection of free discontinuity rather than of free boundaries. The word free refers to the fact that the set K is one of the unknowns involved in the model (De Giorgi, 1991; Ambrosio et al., 2000). Figure (11) shows in color the polygonal shapes identified using the Blake-Zisserman auxiliary function z obtained by the processing of the boundaries of the homoge-

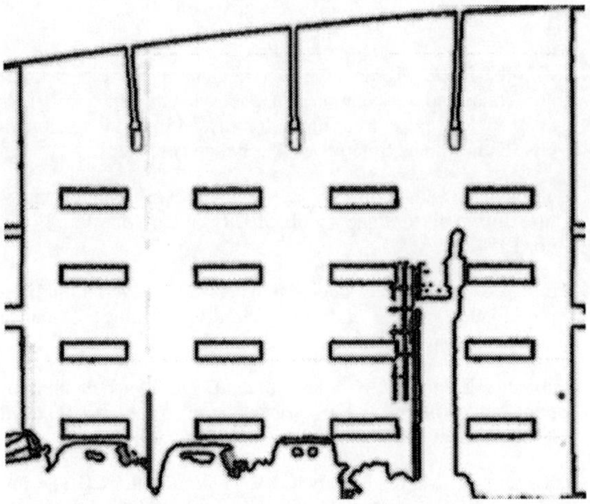

Figure 6: Plot of the auxiliary function s produced by the Mumford-Shah model applied to the facade of Figure (4)

Figure 7: Polygonal shapes (in green and orange) identified from the analysis of the Blake-Zisserman model applied to the boundaries of the homogeneous regions detected after the application of the Mumford-Shah model to the facade of Figure (4)

neous regions detected after the application of the Mumford-Shah model.

4. DISCUSSIONS AND OUTLOOKS

The Mumford-Shah and Blake-Zisserman models produce an optimal segmentation of the data by means of the minimization of specific energy functionals. In practice, the main difference between the two models is in the feature of the smoothing term that is of the first order in (1) and of the second order in (3). Beside this difference the Blake-Zisserman model also permits the detection of discontinuities of the first derivative of the function that approximates the data. In treating geometric data the Mumford-Shah model can manifest an important limitation related to the feature of the smoothing term. In presence of strong variation of the data the smoothing therm, being of the first order, may lead to a over-segmented approximation of the data, i.e., to the appearance of spurious discontinuities. A proper choice of the

Figure 8: Aerial view of a complex building

Figure 9: Plot of the auxiliary function s given by the Mumford-Shah model applied to the DSM

weigh parameters of (2) can however allow to process geometric data without the appearance of this kind of defects. The use of the Blake-Zisserman model with bi-dimensional geometric data is however feasible and may be more appropriate than the use of the Mumford-Shah model (Zanetti and Vitti, 2013).

The variational models used in this work were originally developed in the computer vision field and devoted to the processing of classical images. Nevertheless, the models have been proved to be very well suited for the treatment of geometric data. The 1D Blake-Zisserman the model has also been successfully applied to geodetic data of different kinds (A. Borghi, n.d.; B. Benciolini, 2009).

This contribution has been peer-reviewed.
doi:10.5194/isprsarchives-XL-5-W4-387-2015

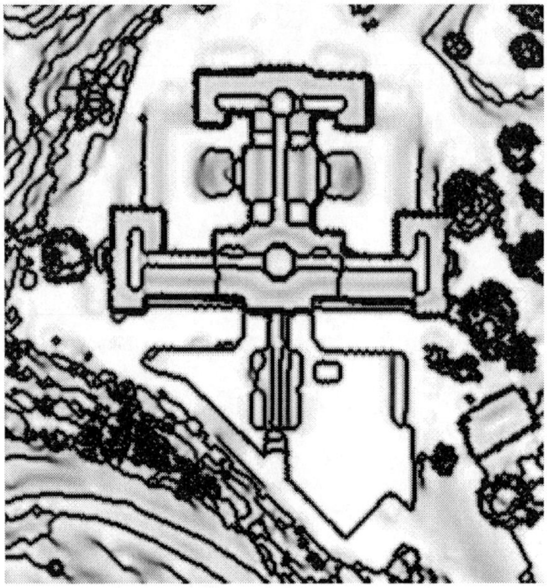

Figure 10: Detail of the plot of the auxiliary function s in the close surrounding of the building of Figure (8)

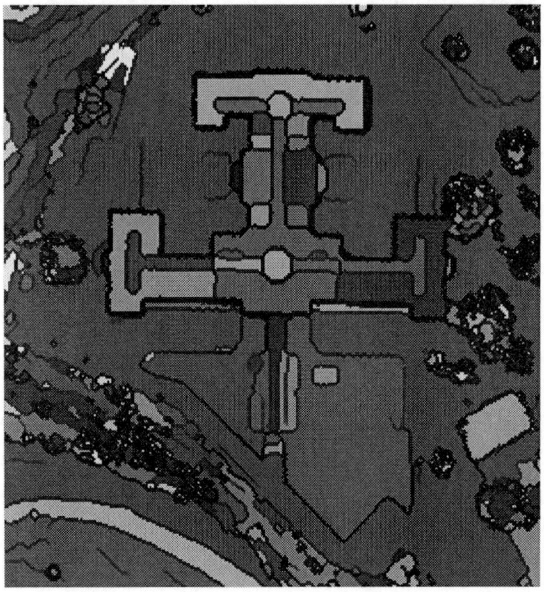

Figure 11: Polygonal shapes (in colors) identified from the analysis of the Blake-Zisserman model applied to the boundaries of the homogeneous regions detected after the application of the Mumford-Shah model to the DSM

In this work the Mumford-Shah model has been applied to gridded data. Changing the numerical scheme would make possible the application of the model to unstructured data (Bourdin and Chambolle, 2000; Zanetti and Bruzzone, n.d.).

An interesting extension of the procedure presented in this work may exploit a variational model proposed by Braides and March (2006) that extends the features of the Mumford-Shah to permit a direct detection of corners along the discontinuity set of the smoothing approximating function.

References

A. Borghi, L. Cannizzaro, A. V., n.d. Geodesy for Planet Earth - International Association of Geodesy Symposia. Vol. 136, chapter Advanced techniques for discontinuity detection in GNSS coordinate time-series. An Italian case study.

Ambrosio, L. and Tortorelli, V. M., 1992. On tha approximation of free discontinuity problems. Boll. Un. Mat. Ital. B (7), pp. 105–123.

Ambrosio, L., Faina, L. and March, R., 2001. Variational approximation of a second order free discontinuity problem in computer vision. Siam J. Math. Anal. 32, pp. 1171–1197.

Ambrosio, L., Fusco, N. and Pallara, D., 2000. Functions of bounded variation and free discontinuity problems (Oxford Mathematical Monographs). Oxford University Press.

B. Benciolini, M. Reguzzoni, G. V. A. V., 2009. VII Hotine-Marussi Symposium on Mathematical Geodesy - International Association of Geodesy Symposia. Vol. 137, Springer, chapter Bayesian and Variational methods for discontinuity detection: theory overview and performance comparison, pp. 147–152.

Blake, A. and Zisserman, A., 1987. Visual Reconstruction. MIT Press.

Bourdin, B. and Chambolle, A., 2000. Implementation of an adaptive finite-element approximation of the mumford-shah functional. Numerische Mathematik 85(4), pp. 609–646.

Braides, A., 2002. Gamma-convergence for Beginners. Oxford University Press.

Braides, A. and March, R., 2006. Approximation by gamma-convergence of a curvature-depending functional in visual reconstruction. Communications on Pure and Applied Mathematics 59(1), pp. 71–121.

De Giorgi, E., 1991. Frontiers in Pure and Applied Mathemathics, a collection of papers dedicated to J.L. Lions on the occasion of his 60th birthday. North-Holland P.C., chapter Free discontinuity problems in calculus of variations, pp. 55–62.

Mumford, D. and Shah, J., 1989. Optimal approximations by piecewise smooth functions and associated variational problems. Comm. Pure Appl. Math. 42, pp. 577–685.

Vitti, A., 2012a. The mumfordshah variational model for image segmentation: An overview of the theory, implementation and use. ISPRS Journal of Photogrammetry and Remote Sensing 69, pp. 50–64.

Vitti, A., 2012b. Sigseg: a tool for the detection of position and velocity discontinuities in geodetic time-series. GPS Solutions 16(3), pp. 405–410.

Zanetti, M. and Bruzzone, L., n.d. Edge-crease detection and surface reconstruction from point clouds using a second-order variational model. In: Proc. SPIE 9244, Image and Signal Processing for Remote Sensing XX.

Zanetti, M. and Vitti, A., 2013. The Blake-Zisserman model for digital surface models segmentation. ISPRS Annals of Photogrammetry, Remote Sensing and Spatial Information Sciences pp. 355–360.

This contribution has been peer-reviewed.
doi:10.5194/isprsarchives-XL-5-W4-387-2015

ATTEMPTS TO AUTOMATE THE PROCESS OF GENERATION OF ORTHOIMAGES OF OBJECTS OF CULTURAL HERITAGE

J.S. Markiewicz[a*], P. Podlasiak[a], D. Zawieska[a]

[a]Faculty of Geodesy and Cartography, Institute of Photogrammetry, Remote Sensing and Spatial Information Systems, Warsaw University of Technology, Warsaw, Poland - (j.markiewicz, p.podlasiak, d.zawieska)@gik.pw.edu.pl

Commission V, WG V/4

KEY WORDS: high resolution orthoimages, Hough algorithm, automatic image orientation, SIFT, orthorectification

ABSTRACT:

At present, digital documentation recorded in the form of raster or vector files is the obligatory way of inventorying historical objects. The orthoimage is a cartometric form of photographic presentation of information in the two-dimensional reference system. The paper will discuss the issue of automation of the orthoimage generation basing on the TLS data and digital images. At present attempts are made to apply modern technologies not only for the needs of surveys, but also during the data processing. This paper will present attempts aiming at utilisation of appropriate algorithms and the author's application for automatic generation of the projection plane, for the needs of acquisition of intensity orthoimages from the TLS data. Such planes are defined manually in the majority of popular TLS data processing applications. A separate issue related to the RGB image generation is the orientation of digital images in relation to scans. It is important, in particular in such cases when scans and photographs are not taken simultaneously. This paper will present experiments concerning the utilisation of the SIFT algorithm for automatic matching of intensity orthoimages of the intensity and digital (RGB) photographs. Satisfactory results of the process of automation, as well as in relation to the quality of resulting orthoimages have been obtained.

1. INTRODUCTION

Within the community of architects a tendency may be observed aiming at substituting the conventional vector drawing with raster products. The architects appreciate the value of orthoimages and they consider these products as the valuable alternative. It is an attractive tool for development of archaeological and architectural documentation, as the combination of the geometric accuracy and the visual quality. In the process of generation of the orthoimages it is important to determined two basic issues: the accuracy and resolution of data. The resolution should be considered as the possibility to discriminate details. It is defined by means of the (object) field pixel size of the resulting orthoimage. The accurate modelling of the surface is the key issue in the process of the orthoimage generation. It directly influences their quality, both, in relation their geometry and their visual correctness. The locally imprecise description of highly complicated surfaces leads to geometric inaccuracies of *„stretched.”* or *„melted.”* type of some parts of the orthoimage. Currently, when the digital object surface model is generated, data from the terrestrial laser scanning is used, which is not always sufficient for the complete description of surfaces of architectural objects. Combination of the point clouds obtained from the laser scanner (as the digital object surface model) and digital images seems to be more useful for the purposes of orthoimage generation. However, the automation of data processing during generation of photogrammetric products is not a separate issue.

2. METHODOLOGY OF ORTHOIMAGES GENERATION

According to the theoretical approach an orthoimage is an image, which projection plane is parallel to the reference plane and all rays are perpendicular to those two planes. In the course of orthoimages generation for the needs of the cultural heritage objects many issues appear, which result from the nature of close-range measurements, and, in particular, from short distances from objects and their complicated geometric characteristics. (Mavromati et al., 2002, 2003). Difficulties related to generation of orthoimages were widely discussed in the context of "true ortho" generation for architectural objects, when orthoimages, as well as other digital projections are the basic tools for documenting the cultural heritage objects (Gianniou et al., 2007; Georgopoulos et al., 2004,2005).

The advantages of TLS is the possibility to support the generation of correct orthoimages. It allows to eliminates many problems, which existed in relation to utilisation of digital images only. The great number of points, the high sampling frequency results in the sufficient density of spatial information concerning the geometry of measured objects. Many professional publications exist which prove the contribution of the TLS data and point clouds from dense image matching to improvements of the accuracy and the increased level of automation for the needs of creation of orthoimages for the cultural heritage objects (Pets et al., 2007; Drap et al., 2003; Georgopoulos et al., 2004, Markiewicz et al., 2014a, Markiewicz et al., 2014b, Remondino et all, 2011). The TLS data allows for acquisition of precise surfaces models of architectural objects, improving, at the same time, automatic generation of orthoimages.

A separate issue concerns integration of the TLS data with "arbitrary" photographs. Procedures are required which would allow for combination of the TLS point clouds with "arbitrary photographs", in order to create a high-resolution orthoimage. Achievement of the sufficient geometric accuracy and the visual quality requires sophisticated abilities at the surface modelling stages and generation of orthoprojections.

At the stage of integration of the TLS data and digital images, as well as during generation of specified photogrammetric products (including digital elevation models or orthoimages) automation of the data processing remains still an open issue. This paper will discuss experimental works concerning automation of generation of the projection plane for the needs of

This contribution has been peer-reviewed.
doi:10.5194/isprsarchives-XL-5-W4-393-2015

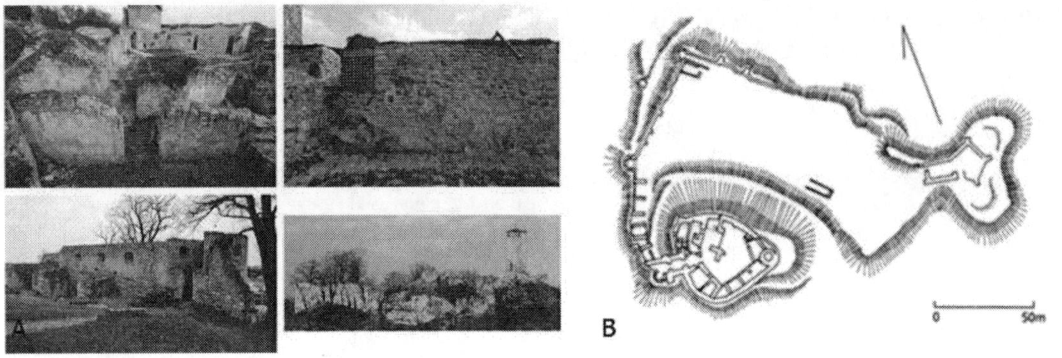

Fig. 1 A sketch of the ruins of the castle in Iłża (source: A) Own source, B) Regional Museum in Iłża)

orthoimages acquisition of the cultural heritage objects and automation of "arbitrary photographs" orientation in relation to the scans.

3. RUINS OF THE CASTLE IN IŁŻA

The objective of performed experiments was to inventory and generate architectural documentation of the ruins of the castle in Iłża. The ruins of the Kraków Bishops' Castle in Iłża is the medieval fortification and, at the same time, the relics of the renaissance residential object. At present, the conservator takes permanent care of this object. It consists of two basic parts: the upper and the lower castle. The following elements, which remained from the past and which are currently exposed on the terrain surface, are: the final defence tower, constructed on the circular plan, a fragment of the upper castle, which has been partially reconstructed, and relics of the lower castle (Fig.2). They are mostly the walls which do not have much architectural adornments. They mostly consist of stones, which create outlines of walls of particular chambers and the cellar.

All above mentioned elements were the subjects of laser measurements (close-range scanner Z+F 5006), photogrammetric measurements (images was taken parallel to walls), photogrammetric documentation and the initial analyses, performed on the basis of acquired data. Due to the existing (conventional) documentation of the object and the current conditions of the ruins, which are permanently accompanied by processes of destruction resulting from the lack of protection against atmospheric conditions or against microorganisms, attempts were made to develop the accurate digital documentation together with the additional analysis of the monuments. These operations aimed at the additional diagnosis of the current conditions of the ruins (Markiewicz et al, 2014a).

4. DESCRIPTION OF PERFORMED EXPERIMENTS

The discussed experimental works were performed according to the diagram presented in Fig.2.

4.1 Automatic plane generation

In many commercial applications the projection plane of point clouds, applied for the needs of generation of orthoimages for the cultural heritage objects the manual intervention of the operator is required. Usually, at least four points must be specified, which allow for its spatial, 3D orientation. In the case of processing orthoimages of architectural objects, this requires intensive works in order to determine many planes. The authors

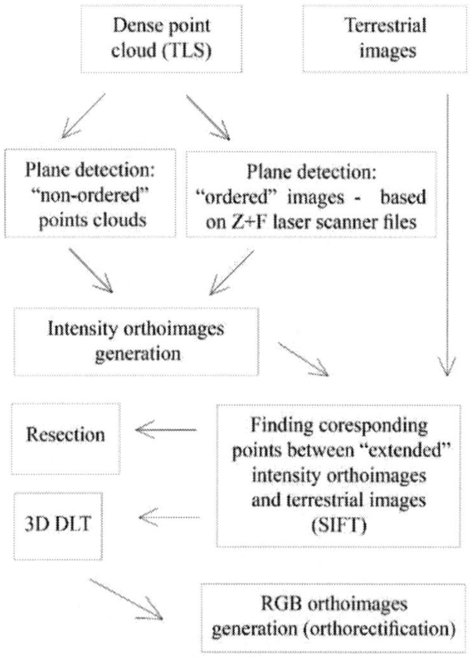

Fig. 2 Diagram of the performed experiment

have developed their own application, which allows for automatic generation of these planes. The first experiments were performed using algorithms from the PCL library (Point Cloud Library). This library utilises RANSAC methods, i.e. it selects three points from the cloud and then it searches through the entire cloud in order to find points, which are located close enough the plane, which is determined by these three points. If the sufficient number of such points (inliners) exists, the plane is accepted; otherwise the selection of points is repeated. For the source data which includes 7 million points, after re-sampling 1:20, 12 areas were obtained, which corresponded to planes (colours in the figure 3), as well as some points, which were not classified into any plane (in grey). The assumed maximum deviation from the plane equalled to 5 cm. The time of computations 37 minutes. Since vertical and horizontal planes are important for documentary purposes, the author's algorithm was developed, which utilises the modified Hough transformation for these purposes.

This contribution has been peer-reviewed.
doi:10.5194/isprsarchives-XL-5-W4-393-2015

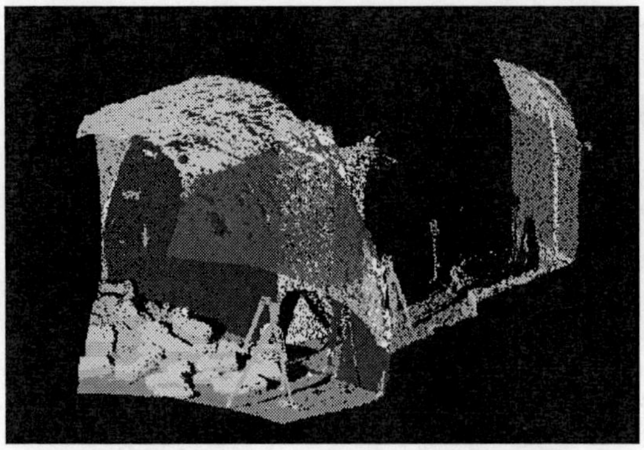

Fig. 3. Planes detected by the PCL algorithm (planes are marked by the same colour)

Since vertical and horizontal planes are important for documentary purposes, the author's algorithm was developed, which utilises the modified Hough transformation for these purposes. In order to limit the time of searching for vertical planes, the projection onto the XY plane is used (Fig. 4a). The histogram is considered as an image according to the brightness scale; it serves as an image in which straight lines are determined using the Hugh algorithm (Fig. 4c). Their equation coefficients allow for the initial determination of the plane equations.

The algorithm has been developed in two versions:

- for "non-ordered" points,
- for "ordered" images, which operates with .zfs files obtained from the Z+F scanner.

The effects of processing are practically identical; they only differ with respect to the time of operations (ordered images require shorter computation time: 2 minutes comparing to 6 minutes).

Possible horizontal planes are detected basing on the analysis of the histogram of points distribution at various altitudes – peaks on the diagram show the possible elevations (Fig.5).

The results of searching are presented in Fig.6 (in order to clarify the figures, two of three determined vertical planes and one of three horizontal planes are presented)

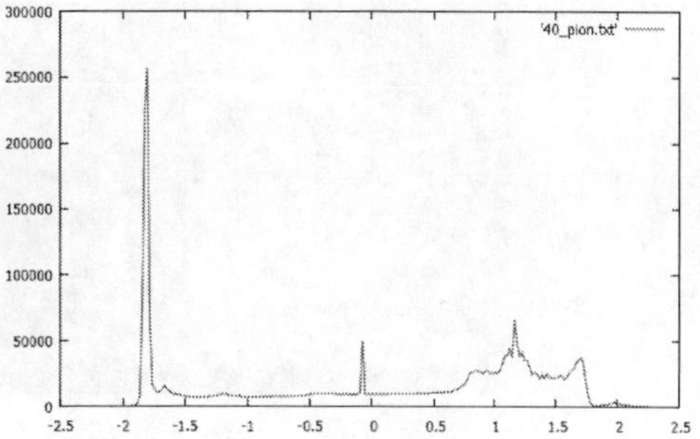

Fig. 5. The histogram of the number of points as the Z co-ordinate function

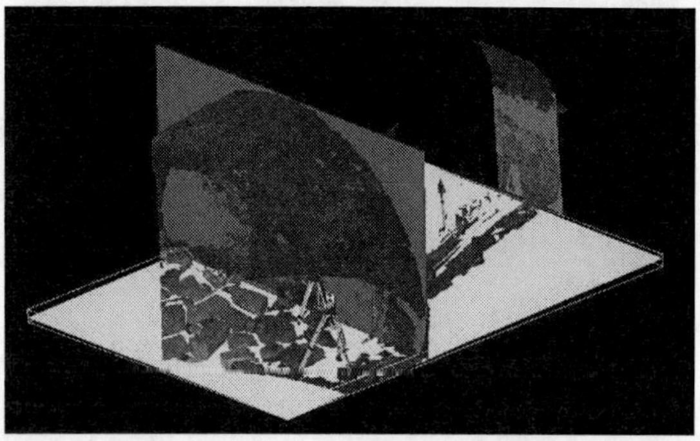

Fig. 6. Planes detected by own algorithm (vertical in red and orange, blue – for the vertical plane

It was assumed during calculations that the maximum distance of points to the theoretical plane equals to 50 mm.

Considering that the wall is constructed of irregular bricks, the distribution of deviations does not correspond to Gaussian distribution (Fig. 7).

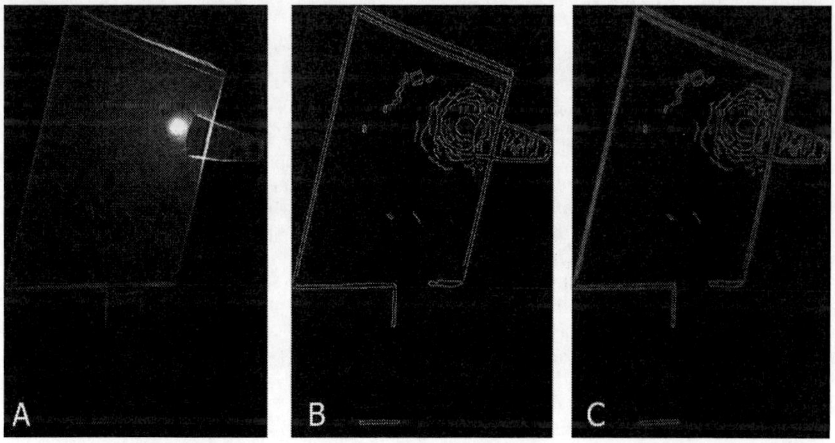

Fig 4. Stages of determination of possible vertical planes: (a) 2D histogram of the number of 3D points, (b) edges detected with the Canny algorithm, (c) approximation of edges by sections of straight lines

This contribution has been peer-reviewed.
doi:10.5194/isprsarchives-XL-5-W4-393-2015

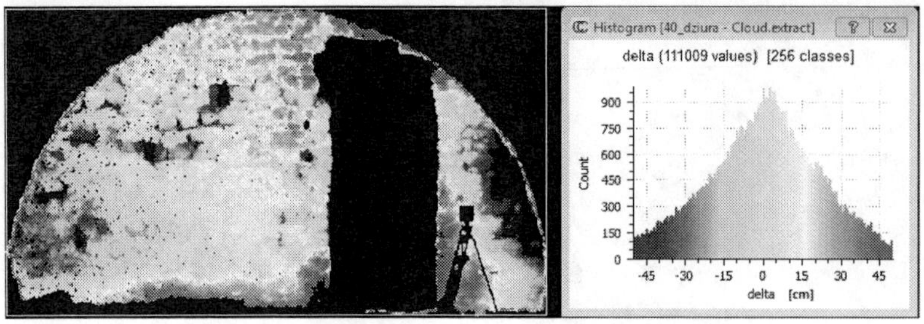

Fig. 7. Distribution of deviations for the plane presented in red in Fig.6

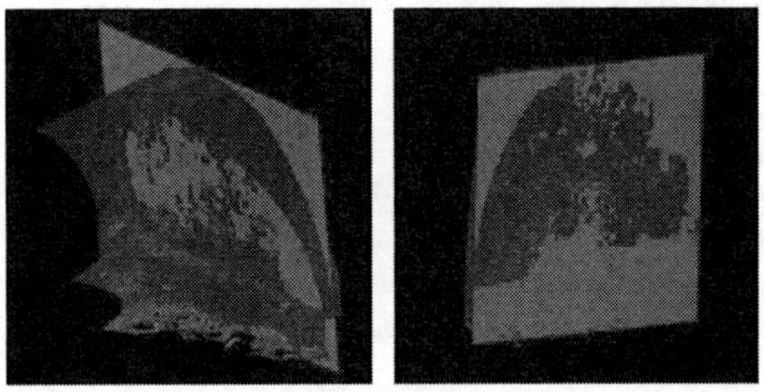

Fig. 8. Another wall of the "test" cellar

Figure 8 presents the results of searching for planes for another part of the cellar, as seen from two different points of view.
In Appendix A was presented results for another architectural object (the building of the Warsaw University of Technology); in this case the wall is not only constructed of smooth bricks, but it also have many parallel planes.

4.2 Automatic generation of orthoimages in the intensity

The process of generation of orthoimages was automatically performed using the LupoScan software. The plane, which was interpolated before and the spatially oriented point cloud were used to perform this process. The TLS data of the ancient cellar of the Castle ruins in Iłża has been utilised; the scanning resolution equalled to 3 mm per 10 m. The following parameters were assumed for the generation of orthoimages:

- NN interpolation
- automatically generated plane with the assumed deviation of 5 cm
- the buffer of projected points +- 3 * 5 cm
- the resolution of orthoimages for the generated object (the ancient cellar) – 2 mm

As a result, orthoimages were obtained which were used for further experiments concerning the automation of orientation of close-range digital photographs (Fig. 9). Besides information concerning the laser beam reflectance intensity, information concerning the depth was also recorded for each pixel.

4.3 Orientation of photographs in relation to orthoimages

The objective of our experiments was to investigate the possibilities to apply "arbitrary" terrestrial photographs, in the process of orthorectification or colouring the point clouds. Although scanners are equipped with embedded digital cameras, which acquire photographs during scanning, they are characterised by the lower resolution and geometric accuracy (Fig. 10).

Fig. 9 Orthoimages in the intensity - GSD 2 mm

Fig. 10 Fragments of RGB orthoimages a) from a camera embedded in the Z+F 5006h scanner b) from the Canon 5D Mark II camera (source: Markiewicz et al., 2014)

The process of orientation of photographs in relation to orthoimages was performed in two stages:

A) Detection of tie points with the use of the SIFT algorithm (the authors' application).

The SIFT (Lowe, 2004) algorithm was applied for searching for and matching the tie points. This algorithm was implemented in the OpenCV function library, with default parameters. In the proposed approach, the plain information concerning images, recorded in raster forms, was utilised. Therefore it was possible to obtain positions of pixels in images. Generated orthoimages were recorded with georeference data and, additionally, they contained information about the depth, what allowed not only to determine co-ordinates of terrain points in the orthoimage plane, but also to add the third dimension.

B) Orientation with the use of two methods - an resection and the DLT 3D

An issue which often occurs in the process of acquisition of complex architectural objects is the need to maintain the constant lens focal length and the limited possibility to frame selected fragments with the use of lenses with constant focal lengths.

Fig. 11 Utilised photographs with points used for orientation (in green) and points deleted from the process of adjustment (in red)

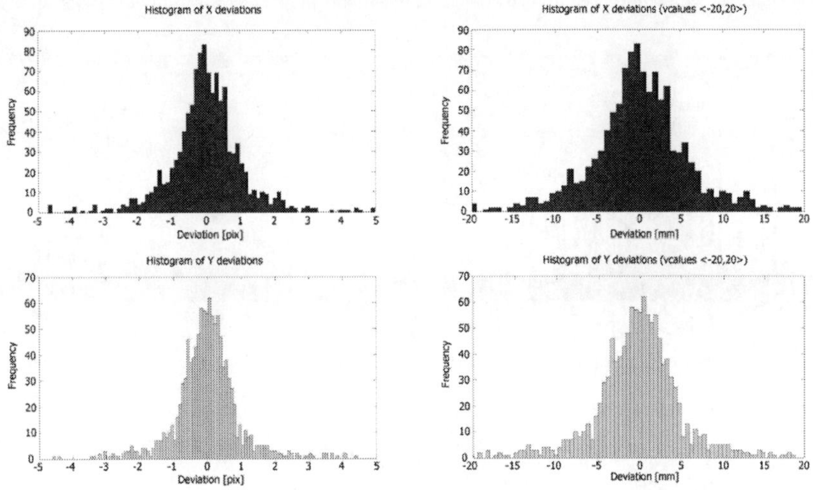

Fig. 12 Histograms of control points deviations (Image 1): A) X direction in pixels (blue), B) Y direction in pixels (green), C) X direction in µm (blue), D) Y direction in µm (green) - resection

Two experiments were performed:

- I: for the first test site (Fig. 11 b) 8 photographs were acquired from different positions, with the use of a zoom lens, with three different focal lengths (each photograph covered the entire orthoimage area).

- II: for the second test site a lens with a constant focal length was applied (4 photographs covered the entire orthoimage area).

It was assumed for the approach, which was based on the adjustment using the resection, that the interior orientation elements, which were determined earlier, are determined in an independent calibration process. In the proposed approach we do not have a network of images or the possibility to perform self-calibration, but we acquire only one image, which covers the entire object; we use the camera, which was earlier calibrated, to acquire this photograph.
In the second approach, which is based on the DLT method, all parameters are determined in the process of adjustment and the impacts of distortion elements is considered in coefficients of the polynomial (Chen et al, 1994, Karara et al, 1971).

Variant I
8 photographs were used for the analysis and testing the assumptions. Results of searching for the number of points are presented in Table 1.

Table 1. Statistic of detected points

Photograph no.	Number of detected points	Number of incorrectly detected points	Percentage of incorrectly detected points
1	1152	4	0.35 %
2	933	3	0.32%
3	1002	35	3.49%
4	477	47	9.58%
5	300	10	3.33%
6	310	30	9.67%
7	530	9	1.70%
8	517	13	2.51%

Figure 11 presents examples of the images with marked points used for orientation. Results of two, independent methods of data orientation were analysed. In order to analyse the accuracy of matching images basing on the network points, the distribution of deviations on matching points were analysed.

Fig. 13 Utilised photographs with points used for orientation (in green) and points deleted from the process of adjustment (in red)

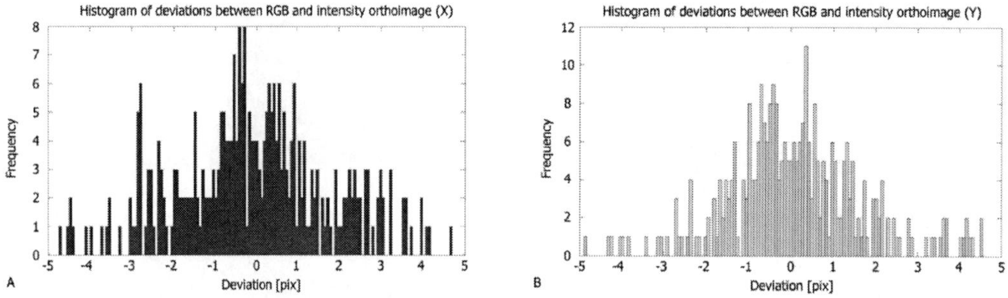

Fig. 14 Histograms of control points deviations (Image 1): A) X direction in pixels (blue), B) Y direction in pixels (green) - 3D DLT

Due to the high number of tie points it was decided to present the results in the form of histograms of deviations, with respect to the x and y axes, respectively. Figure 12 presents examples of histogram for the photograph 1. All analyses were performed in own application made in Matlab software.

After analysing the histograms it could be noticed that their shapes corresponded to Gaussian distribution. This may suggest the correct detection of tie points. It may be noticed that about 95% of points are characterised by the accuracy higher than 2 pixels (12 μm).

Variant II

4 photographs were used for the analysis and testing the assumptions. Figure 13 presents examples of the images with marked points used for orientation . Results of searching for the number of points are presented in Table 2.

Table 2. Statistic of detected points

Photograph no.	Number of detected points	Number of incorrectly detected points	Percentage of incorrectly detected points
1	169	19	11.24 %
2	540	35	19.44%
3	628	43	6.85%
4	324	8	2.47%

Figure 14 presents examples of histogram for the photograph 4. After analysing the histograms it could be noticed that their shapes are similar to Gaussian distribution. This may suggest the correct detection of tie points. It may be noticed that about 95% of points are characterised by the accuracy higher than 3 pixels (18 μm).

4.3 Generation of RGB orthoimages

The process of generation of colour orthoimages was performed automatically, using the LupoScan software, basing on the Digital Model of Object Surfaces, performed earlier and on oriented digital images.

In order to perform the accuracy analysis of the generated RGB orthoimage, it was compared with the intensity orthoimage. The SURF algorithm was used for that purpose, in order to perform independent check (Fig. 15). Obtained results were compared with the DLT photographs and the resection (Table 3).

After analysing the table it could be noticed that that about 95% of points are characterised by the accuracy higher than 2 pixels (4 μm).

Table 3. Average deviation between check points on intensity orthoimages and RGB orthoimages

Id	Avarage		> 2 pix	
	X[pix]	Y[pix]	X[%]	Y%]
Variant_I_DLT	0,82	0,66	1,89	1,21
Variant_I_Resection	0,85	0,33	1,66	1,24
Variant_II_DLT	0,69	1,24	6,35	5,29
Variant_II_Resection	0,71	0,72	5,56	5,29

This contribution has been peer-reviewed.
doi:10.5194/isprsarchives-XL-5-W4-393-2015

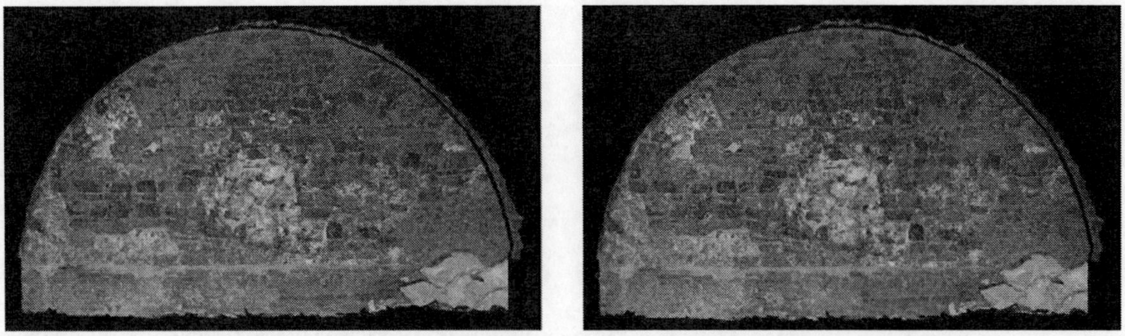

Fig. 15 RGB orthoimages with marked check points (green) used for accuracy analysis (Varian I)

Fig. 16 RGB orthoimages with marked overlaps (Variant II)

Figure 15 presents final result of the single image orthorectification. The final orthomosaic result is shown in figure 16. The red squares show the places where two different images overlap. After analysis of this fragments it can be noticed that there are no parallaxes but only difference in colours.

5. CONCLUSIONS

Performed experiments proved that the process of generation of orthoimages may be automated, from generation of a reference plane to automatic generation of RGB orthoimages.

The proposed, original approach, based on the transformed Hough algorithm, allows for high improvements of the process of searching for planes within entire scans, without the need to divide them into fragments. Thus, the divided scan with automatically generated planes, also specifies the number of orthoimages, and the maximum deviations from the plane define intervals (in front of and behind the plane) of points which are required for generation of orthoimages. This new approach accelerates the process of orthoimage generation and eliminates noises resulting during the stage of data acquisition using close-range scanners.

Utilisation of orthoimages in the form of raster data, amended with the third dimension, allows for the use of CV algorithms to search for homologous points and spatial orientation of any photographs. Such an approach eliminates the necessity to measure the control points.

The proposed approach allows for detecting the great number of evenly distributed points, what allows for detection of self calibration; this is particularly important for zoom lens cameras.

Such an approach is not limited to the use of photographs acquired with cameras integrated with scanners, which are characterised by the lower geometric and radiometric quality. Such solution is particularly recommended for the process of generation of high-resolution photogrammetric documentation, where the resolution and the geometric accuracy of the order of single millimetres is required.

Further works will be focused on testing this approach on more complicated monuments.

This contribution has been peer-reviewed.
doi:10.5194/isprsarchives-XL-5-W4-393-2015

REFERENCES

Abdel-Aziz, Y.I., & Karara, H.M. (1971). Direct linear transformation from comparator coordinates into object space coordinates in close-range photogrammetry. *Proceedings of the Symposium on Close-Range Photogrammetry (pp. 1-18)*. Falls Church, VA: American Society of Photogrammetry

Chen, L., Armstrong, C.W., & Raftopoulos, D.D., 1994. An investigation on the accuracy of three-dimensional space reconstruction using the direct linear transformation technique. *J. Biomech* 27, 493-500

Drap P., Sgrenzaroli M., Canciani M., Cannata G., Seinturier J., 2003. Laser Scanning and Close Range Photogrammetry: Towards a Single Measuring Tool Dedicated to Architecture and Archaeology; CIPA XIXth International Symposium; Antalya, Turkey.

Georgopoulos A., tsakiri M., Ioannidis C., Kakli A., 2004, Large Scale orthophotography Using DTM from Terrestrial Laser Scanning. In: *The International Archives of the Photogrammetry, Remote Sensing and Spatial Information Sciences*, Istambul, Turkey, Vol. XXXV, Part B5, pp. 467-472.

Georgopoulos, A., Makris, G.N., Dermentzopoulos, A., 2005. An alternative method for large scale orthophoto production, Proceedings of CIPA 2005 XX International Symposium, 26 September – 01 October, 2005, Torino, Italy.

Gianniou P., Georgopoulos A., Tsakiri M., della K., 2007 The documentation of the medieval entrance of the Rhodes fortification complex, XXI International CIPA Symposium 1-6 October 2007, Athens Greece

Lowe D. G. Distinctive Image Features from Scale_Invariant Keypoints, 2004. http://www.cs.berkeley.edu/~malik/cs294/lowe-ijcv04.pdf (22.01.2014)

Markiewicz J.S., Zawieska D., 2014. Terrestrial laser scanning or digital images in inventory of monumental objects? - case study. In: *The International Archives of the Photogrammetry, Remote Sensing and Spatial Information Sciences,* Riva del Garda, Italy, Vol. XL-5, pp. 395 - 400.

Markiewicz J.S., Zawieska D.,Kowalczyk M., Zapłata R., 2014. Utilisation of laser scanning for inventory of an architectural object using the example of ruins of the Krakow bishops' castle in Iłża, Poland. In: GeoConference on Informatics, Geoinformatics and Remote Sensing, Albema, Bulgary, Vol. 3, pp. 391-396

Mavromati, D,. Petsa, E., Karras, G., 2002. Theoretical and practical aspects of archaeological orthoimaging *The International Archives of the Photogrammetry, Remote Sensing and Spatial Information Sciences*, 34(B5), pp. 413-418.

Mavromati D., Petsa E., Karras G. E., 2003. Experiences in Photogrammetric Archaeological Recording. XIX International CIPA Symposium, Antalya, Turkey, pp. 666- 669

Petsa E., Grammatikopoulos L., Kalisperakis I., Karras G., Pagounis V., 2007 Laser Scanning And Automatic Multi-Image Texturing Of Surface Projections XXI InternationalCIPA Symposium, 01-06 October 2007, Athens, Greece

Pierrot-Deseilligny M., De Luca L., Remondino F., Automated image-based procedures for accurate artifacts 3d modeling a nd orthoimage generation https://www.conferencepartners.cz/cipa/proceedings/pdfs/B-2/Remondino2.pdf (22.01.2014)

This contribution has been peer-reviewed.
doi:10.5194/isprsarchives-XL-5-W4-393-2015

APENDIX

A) Automatic plane generation

Fig. 1 presents results for another architectural object (the building of the Warsaw University of Technology); in this case the wall is not only constructed of smooth bricks, but it also have many parallel planes

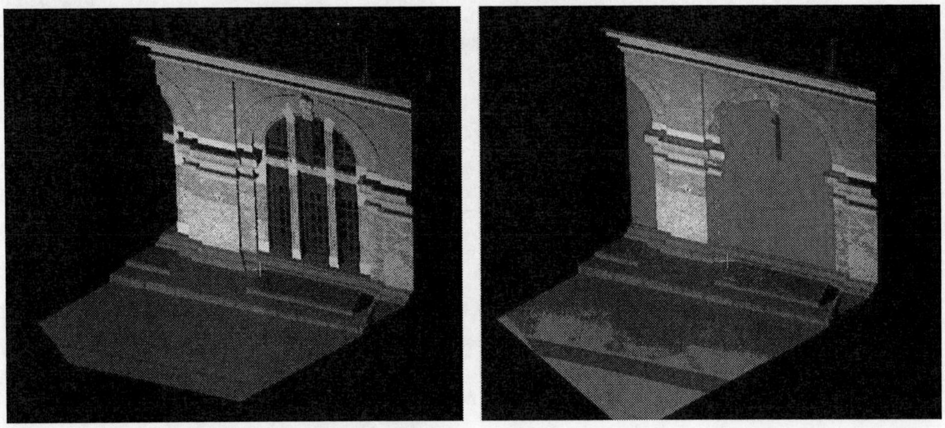

Fig. 1. Results of searching for planes for a building.

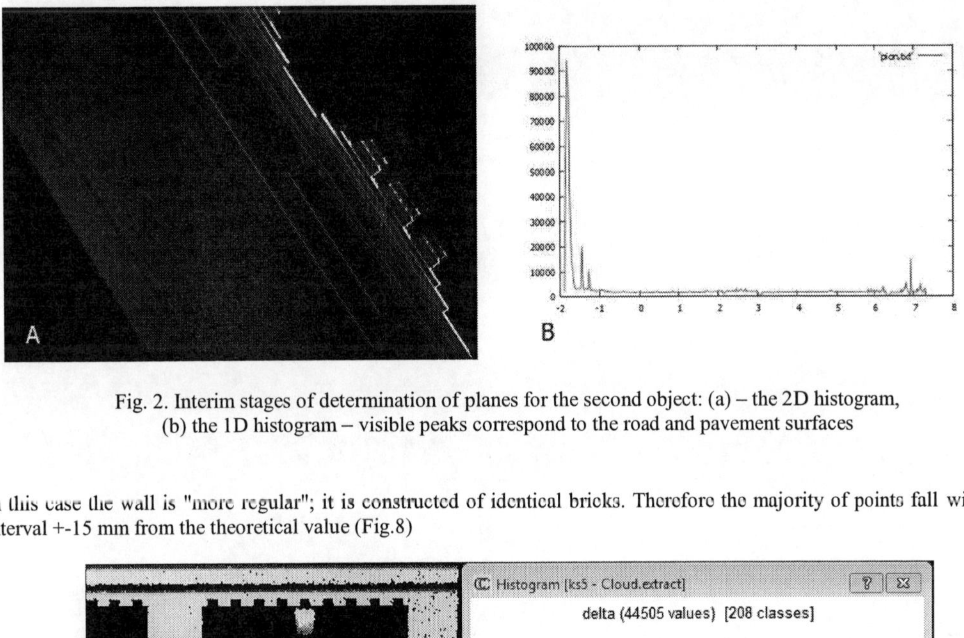

Fig. 2. Interim stages of determination of planes for the second object: (a) – the 2D histogram, (b) the 1D histogram – visible peaks correspond to the road and pavement surfaces

In this case the wall is "more regular"; it is constructed of identical bricks. Therefore the majority of points fall within the interval +-15 mm from the theoretical value (Fig.8)

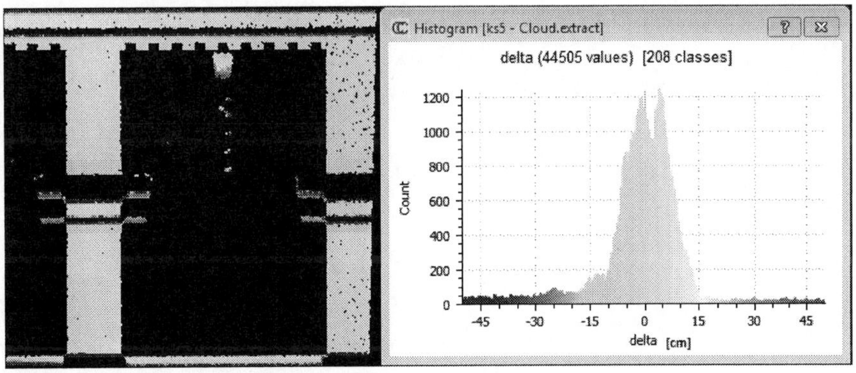

Fig. 3. Deviations from the theoretical plane

This contribution has been peer-reviewed.
doi:10.5194/isprsarchives-XL-5-W4-393-2015

THE INTEGRATED SURVEY FOR EXCAVATED ARCHITECTURES: THE COMPLEX OF CASALNUOVO DISTRICT WITHIN THE WORLD HERITAGE SITE "SASSI" (MATERA, ITALY)

T. Cardinale [a] *, R. Valva [b], M. Lucarelli [c]

[a] PhD Candidate, Dipartimento delle Culture Europee e del Mediterraneo, Università degli Studi della Basilicata, Via Lazzazera Matera, Italy - (tizcardina@hotmail.it)

[b] Msc Student, Dipartimento delle Culture Europee e del Mediterraneo, Università degli Studi della Basilicata, Via Lazzazera Matera, Italy - (roby2110_@hotmail.it)

[c] Msc Student, Dipartimento delle Culture Europee e del Mediterraneo, Università degli Studi della Basilicata, Via Lazzazera Matera, Italy - (marica.lucarelli@libero.it)

Commission V, WG V/4

KEY WORDS: survey, preservation, 3D modelling, historic heritage, non-destructive techniques

ABSTRACT:

Never as within the complex of *Sassi* (Matera, South of Italy), the parties have a volumetric material identity and a special construction condition for which, first of all, you need to know the whole to which they give life, and then the individual components and their connections.

In the course of time, in the Lucan city, there were stable and favorable conditions that allowed the development of an architectural language, of juxtaposition of the materials, interpenetration of space and conformation of the volumes, which generated an exceptional urban phenomenon. The distribution of these building artifacts in symbiotic connection with the connective calcareous texture that hosts them , resulted in a spontaneously harmonious figurative balance that characterizes the constructive expedients employed and the distributive and morphological solutions. This is the reason why the *Sassi*, and the overlooking Park of Rupestrian Churches of Matera Murgia, have been entered in 1993 in the UNESCO World Heritage List.

The complexity of a built space, such as this one, determines the need for a non-traditional approach, so you have to combine last generation tools and canonical ones for survey, drawing and representation, within a dialectic between memory and design, tradition and innovation. For this reason, an appropriate cognitive apparatus has been set up for the entire technical process, making use of different non-destructive and non-contact techniques: digital photogrammetry, total station, laser scanner and thermography, in order to obtain a three-dimensional computer model, useful for the diagnosis and the preservation of the integrity of cultural heritage.

1. INTRODUCTION

1.1 General Instructions

Interpret the signs and put them in conjunction with one or more phenomena is possible if you know the properties of materials and their behaviour in the environment. In this sense, it becomes essential to integrate different methods and tools that can be used not only in the process of understanding and interpretation, but also in those of restitution, in the belief that there is no an "indifferent" survey.

Years of research, survey campaigns and interpretation of the data showed that the *Sassi* of Matera, emblem of the symbiosis between nature, architecture, energy, at the same time, represent the place where tradition and innovation interpenetrate one to each other, and where digital reality cannot replace the real one, but it just goes to run alongside in a perfect union.

The ability to investigate a so complex reality, on the one hand appears as a challenge regarding the usual approach of the restoration and recovery, on the other one offers an important opportunity for the knowledge, the enhancement and the fruition of cultural landscape of Matera, as well as the defence of everyday life in a site where you can break down the constitutive elements of the matter and of the voids keeping in mind that the whole is not simply the sum of the parts.

Traditional instruments are then complemented by a set of data and procedures resulting from new digital technologies capable of giving life to a model that represents in the most effective and interactive possible way any artefact of historical and artistic interest, opening new possibilities of use, classification and study of cultural heritage.

The geometric, material and thermal survey, together with the possible three-dimensional reconstruction of buildings, forcing operators and users to an effort of interpretation, definition of structure and properties of the object's surfaces, and survey of the constructive, technology and energy aspects.

These analyses, carried out by subjects who have different skills and expertise, lead to a semantic interpretation of the data obtained from the survey, which is configured as an operation in which, thanks to the observation of form and architectural composition, you can give back the characters that make up the built through a process that uses the act of "drawing" as a way to "understand" and "communicate" architecture.

So we have tried to prepare an appropriate apparatus to support the entire technical process, making use of different strategies of cognitive and non-destructive diagnostic analyses to investigate the built environment. These methodologies can be aimed at the characterization and qualification of degradation by non-invasive techniques in situ, or at cognitive analysis of the architectural work.

* Corresponding author.

This contribution has been peer-reviewed.
doi:10.5194/isprsarchives-XL-5-W4-403-2015

2. METHODOLOGY

2.1 THE THERMOGRAFIC MAPPING

Among these ones, the thermography can be used for non-destructive investigations to be performed on site in order to check the moisture in the masonry. The use of this "diagnostic" technique is not intended primarily to the quantitative measure of moisture content in a wall, but allows to delineate qualitatively the mapping of the fronts of moisture and to locate the areas of thermal dispersion and/or storage, pathologies not visible with the naked eye.

Thermography is nothing more than the "drawing of the heat", that is the graphic representation of the emissions of radiation of a body in the "thermal infrared range". This representation is carried out by sophisticated electronic devices equipped with particular types of sensors able to convert the signal acquired in the infrared range in a video signal.

The increasingly wide spread of applications in scientific and technical literature shows clearly how in the field of non-destructive investigations, infrared thermography represents, over all, a diagnostic and cognitive technique, extremely versatile and efficient with a very wide horizon of applications.

Reconstruct the history of an artefact, read the constructive characteristics, correlate the state of preservation to the materials that constitute it, are all elements to which the use of infrared thermography is able to offer a significant support.

The use of infrared thermography with well-established and absolutely non-invasive methodologies of acquiring, combined with software tools for image processing can be a valuable aid in the multidisciplinary integrated approach to the knowledge and preservation of architectural artefacts with historical interest, thanks to highlight the discontinuity of physical and thermal properties of materials forming the first few centimetres of the wall.

In this perspective, are then presented the results of several applications within the *Rione Casalnuovo* in the area of the *Sasso Caveoso* (a part of the *Sassi* complex) in the city of Matera (Figs. 1, 2, 3).

Specifically, the analysis began with the collection of archival and documentary data and continued collecting information on the place, mainly geographical, cultural, historical and constructive. Because of the lack of formal-geometric and physical-constructive documentation, we have included a phase of collection of physical data, despite the difficult practicability and accessibility in the area, with most of the spaces that are closed, sometimes with railings, other times because of partial collapses. We have established some criteria for the selection of case studies and then we have made a photo recognition of the traditional core.

The area has an extension of 27000 square meters (three hectares). Its particular morphology, which meant that it was the place of privileged settlement above all religious communities, is characterized by the rock of the church of *Santa Maria dell'Idris* and of *San Giovanni* (387 meters above sea level) and by the large plateau of the monastery of *Santa Lucia alle Malve* on which are the remains of a medieval cemetery, and by a path almost flat, at the foot of *Mont'Errone* at an altitude of approximately 488 mt, which is followed, behind the rocky ridge, by the overhang of the *Gravina*.

Even today it is particularly evident the conformation of the area, made up of settlements mainly in the cave; architectures above ground, in very limited numbers, over time almost completely saturated the very limited building space, taking shape, in most cases, as an extension of the cave and then later of the elementary cell-type of vernacular architecture, the *lamione*, that is the spatial organism basis, then evolved into more complex and original systems: it consists of a very simple rectangular crowned by a barrel vault and is characterized by a jagged gable facade.

The wall that closes underground caves is entirely analogous to that which concludes the front of the elementary living cell leaning against the slope. The tens of hundreds of artefacts of this type are characterized by an entrance door and a fanlight for the circulation of air and light in the indoor environment.

In the specific case of the heritage, object of interest of this work, the most delicate and difficult case is clearly that of extreme situations of decay and neglect. This is the case of this area in the *Casalnuovo District*, which subsequently to the evacuation of the *Sassi* in the Fifties, in a few years, has seen triggering the sequence of a series of events: infiltration of water from the mantle, collapse of roof structures, involvement of the vertical structures. This situation actually unfortunately covers large areas of the national territory and raises dramatic problems about the possibility of permanence of a "minor" heritage that in many cases, first of all that of Matera, presents characteristics of the material culture and remarkable historical and symbolic values that go completely recovered adapting typologies to the contemporary standards of habitability, health and hygiene.

So this is a suggestive urban context of great historical value, constituted precisely not only by monumental buildings, but especially from "minor" buildings that can be treated as decorative elements of a great facade represented by *Sassi* context: any intervention will have as its goal the protection of the *unicum* represented by ancient districts.

So the study of the historical-artistic, archaeological, architectural and environmental heritage, is the prerequisite knowledge to all forms of planning and protection of the area.

For this reason, we made use of the thermal mapping of the site that, through detection of quantitative data about temperature, and qualitative ones about humidity, was used to develop a valuable framework concerning the physical, chemical and biological degradation in the *Sassi* area.

The technique of thermography is an analysis that allows rapid identification of defects. The term *thermal mapping* means a process of monitoring and capture of thermal anomalies of the buildings through the use of digital sensors (infrared thermography), aimed at defining judgments on the quality of the building envelope and related energy losses.

With the thermographic survey campaign, radiations in the infrared range emitted by the buildings are recorded and, on the basis of these ones, temperatures are calculated and represented in a scale of "false colours". The final product of this processing performed by the camera depends on the distance between the object to detect and the camera itself and on the angle of the focal lens of the instrument. In outdoor thermographic survey, in most cases one must necessarily resort to multiple thermographic measures and then to different thermal images. Logically, this also affects the spatial resolution of the image (the shorter the distance, the greater the spatial resolution).

From an instrumental point of view, the images were taken using a Fluke Ti32 thermal camera with detector FPA (Focal Plane Array) micro bolometer operating in the long wave spectral range between 8 and 14 microns. Regarding the spatial resolution of the acquisition system, the FPA is equipped with a matrix of sensitive elements of dimensions 320 (H) x 240 (V) with a field of view of 23.0 °(H) x 17.0 °(V) and is capable of producing a thermal resolution minimum equal to or less to 0.45 °C with a precision of the measured temperature of ± 2 °C within a

This contribution has been peer-reviewed.
doi:10.5194/isprsarchives-XL-5-W4-403-2015

temperature range of 100 °C (the range of measurable temperatures is between -20 °C and 600 °C). This acquisition system offers, in addition to a remarkable handiness, a spatial resolution that allows to investigate an area of dimensions 314 x 309 mm at a shooting distance of 250 mt. From a methodological point of view shots were taken using the solar source as thermal stimulus analyzing qualitatively the transient that passively is created inside the wall to the combined effect of solar radiation and spontaneous thermo-physical processes affecting the border artefacts.

The particular urban structure of the *Sassi*, absolutely stratified, has strongly influenced the outdoor technical survey, forcing the operator, unable to stand in the front position with respect to the complex of buildings, just to fractionate shooting of the area in a high number of frames to obtain a higher resolution. This has led to the need to reconstruct the whole after, in post-processing step, in order to homogenize the color response of every single shot, to recompose the elevations, which have been acquired from *Piazzale Belvedere of the Park of Churches caves*, located frontally to the study area, on the other side of the ravine.

In fact the two-dimensional maps of radiometric measurements processed by the acquisition system in images of the spatial distribution of temperature (thermograms) affecting the outer layers of the architectural surface (the accuracy in the return of textures walls and discontinuities is therefore function of thickness, type, material properties forming the wall and entities of the thermal stress).

The mosaic image shows the surface temperature of the casing therefore measured by the camera in a chromatic scale homogenized in the range of assumed values.

These images were processed with the demo version of the dedicated software Microgeo NRG, one of the best software on the market, able to straighten and create mosaics of the thermograms, optimizing the measurement distance from the object and standardizing the values of emissivity, relative humidity, together with the temperature intervals, in order to subsequently be combined into more complex photo mosaics able to return in a unified manner the entire portion consisting of the facades of the Casalnuovo District facing the ravine.

Important: you cannot change the focus of the machine and then, in the case of blurry photos, you must repeat the analysis.

During post-processing, the obtained image is (geo) - codified and mapped to a reference metric system that allows you to accurately measure the lengths and the surfaces of the represented objects and structures. This superposition of layers is then interrogated within any CAD environment.

Subsequently, the thermal image geometrically exact, is subjected to a procedure of thematic classification of the values with the aim to discretize and account the thermal response for temperature classes. The purpose is to emphasize any heat loss by providing an excellent tool to support the critical analysis of the dispersions.

In this regard it should be emphasized that the thermogram acquired is referred to a very specific moment and represents the thermal response of the building exclusively linked to the time of acquisition.

In order to allow an exhaustive thermographic campaign, we have been complied with recommendations on the boundary conditions: the thermograms were in fact acquired at sunset in a typical day in late June, thus avoiding the risk of misinterpretation, arising from structures directly exposed to solar radiation, with limited variations in wind speed (steady state pressure) and contained differences in internal and external temperature changes (steady state temperature).

Nevertheless, outside, a precise reading is definitely harder than inside since that the data is greatly influenced by the presence of widespread vegetation.

In any case the most evident disease is caused by the internal circulation of moisture and from its evaporation from the surface of the stone, which determines the crystallization of soluble salts contained in it. Another phenomenon of degradation is the washing away of the stone caused by the water of rain, slightly acidic, which, dissolving the calcium carbonate, causes an increase of the surface porosity.

Figure 1. Thermal image of the ravine of an area of the *Casalnuovo District*

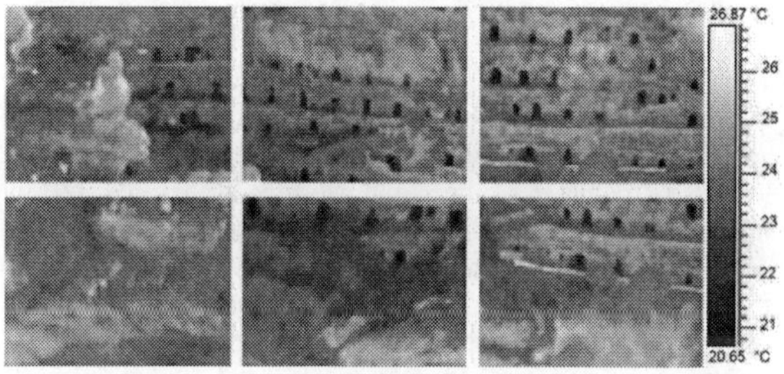

Figure 2. Mosaicking of thermograms processed with the aid of the software NRG © Microgeo

Figure 3. Results of the thermal mosaicking of a big area of the *Casalnuovo District*

This contribution has been peer-reviewed.
doi:10.5194/isprsarchives-XL-5-W4-403-2015

2.2 ACTIVE SENSORS: TIME OF FLIGHT LASER SCANNER

Based on light waves, the active sensors realized, as a direct result of the measurement session, a set of three-dimensional coordinates which are related to a high number of points in a reference system correlated with the instrument. Then, the so measured points are determined by polar spherical coordinates which are then returned in Cartesian coordinates. This is possible because the active optical sensors detect the spatial position of the surface or of the measured point, dividing the framed surface in picture elements similar to the pixels of a digital image. The color information can be recorded by the sensor itself or from a digital external integrated camera. Usually, the light source is integrated in the sensor, for this reason the light that invests the object has a defined structure (Fig. 4).

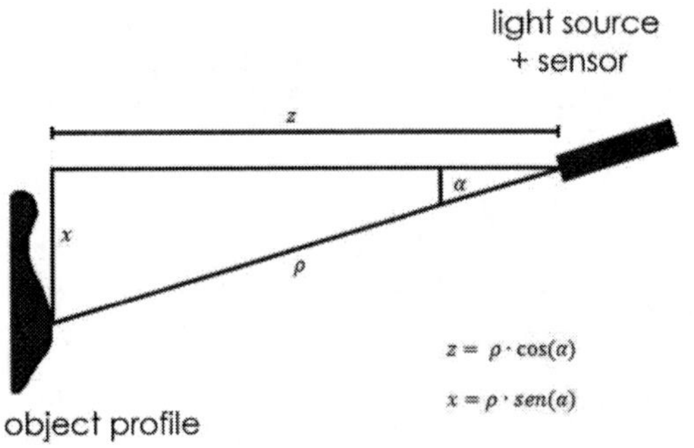

Figure 4. Schematic diagram of a time of flight scanner

The use of this type of controlled light source allows to automate the process of measuring points, which are detected by the sensor itself making repeatable the measuring operation. It is possible to acquire large amounts of three-dimensional data of complex geometries with very high accuracy. You can identify several basic stages for the survey of artifacts: the first phase determines the success of the survey and its metric efficacy and consists of an acquisition project. It is essential the placement of the target, the fixed points that are automatically recognized by the management software of the scans, as each scan is captured with an own reference system. You need at least three targets equal for each pair of scans to be recorded. By paying attention to all the elements that obstruct the recording area you proceed with the acquisition of data. By setting the number of points at a certain distance, you define the scale of the detail to be obtained. Finally there is the processing of data which in turn can be divided into: data filtering, scan registration, extraction of 2D data and/or modeling. Filtering can be done through the manual or automatic elimination of scattered points in the wrong position due to measurement errors; filtering redundant points, or at a distance less than the average of all the mutual distances between the points; elimination of a certain percentage of points to have lower loss of data during the filtering step; reduction of the surface roughness without reducing the number of points. The next step is the registration of two or more clouds of points that is carried out with roto-translational motion of a cloud on the other one performed with the target, automatically recognized, or homologous points selected manually, or with the method of matching surface. After obtaining the complete model, it is possible derive 2D data, such as plans, elevations and sections. By transforming the point cloud into mesh you get a 3D model that can be textured and, by using dedicated software, the display of these models becomes compatible with any media.

Our study on a part of the *Casalnuovo district*, consists in obtaining a three-dimensional geometric model, from which to derive sections, elevations, indirect measurements and, more generally, all the graphic elaborations necessary for the representation and subsequent studies.

The instrument used is a laser scanner Leica HDS 3000 (Fig. 5) which has a field of view up to 360 ° x 270 °, integrated digital camera for photographic overlays calibrated automatically. The emitted laser beam reflected on the surface and acquires each point with coordinates XYZ with respect to a local reference system of the own machine. The figure you get is a very detailed point cloud that follows exactly the geometry of the detected object with an accuracy of 6 mm over a distance of 50 mt from the object.

The first phase was that of the design of digital survey. Due to space constraints, it was necessary to spend more time to this stage in order to ensure that the various scans could be recorded without the obstruction of the affected parts by the walls of the corridors.

To be sure of being able to firmly connect with each other all the scans, we drew some sketches studying what could be the preferred location for each scan and the relative number of target (Fig. 6) sufficient for a proper registration. Another difficulty was to avoid gray areas to have a complete survey. Due to the complexity of the construction site, we had problems of different brightness between the excavated and constructed environments, therefore it was necessary to adjust the brightness of the instrument to find the right balance between overexposed and underexposed areas. The second phase, that of acquisition, lasted 5 days, with a total of 20 scans. Because of the geometry of space and of time at our disposal, the instrument has been set in order to have a distance between the points of 0,03 mt every 10 mt. The third phase, that of registration, was carried out with the roto-translation of a cloud on the other one through the automatic recognition of target HDS and the superficial matching. To minimize the error, the scans were recorded in groups to then obtain the final one with an error of 5 mm. Once you get the overall cloud of points, there is the creation of 2D drawings (plans, sections and elevations), from which it was possible to obtain all the necessary metrics information (Fig. 7). The next step was to make a 3D model that can be queried by a wide audience for educational, technical and touristic purposes.

Figure 5. Laser scanner used for survey

This contribution has been peer-reviewed.
doi:10.5194/isprsarchives-XL-5-W4-403-2015

Figure 6. Identification and positioning of the targets

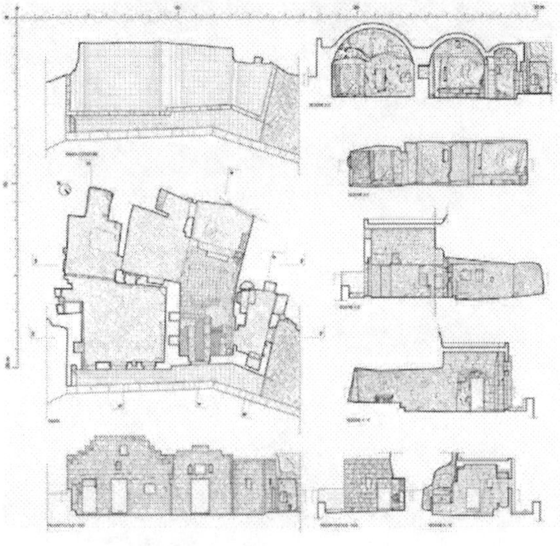

Figure 7. Plans, sections and elevations

2.3 PHOTOGRAMMETRY AND ANALYSIS AIMED AT THE HERITAGE RECOVERY

The techniques based on passive sensors (image-based methods) use the light in the environment to capture images to be processed later to derive 3D information of the scene observed. For cartographic, architectonic, industrial and archaeological surveys, one of the techniques more used and known is the photogrammetry: science that allows accurate measurements and 3D reconstruction from photographs (images).

Photogrammetry is a non-invasive, low-cost and user-friendly interface that allows you to determine metric information on the size, shape and position of an object or scene, starting from the identification of homologous points in the images. This establishes a geometric relationship between the pictures and the actual scene at the time of shooting, using the mathematical model of collinearity (Fig. 8). You can get 3D metric information on the object through the use of at least two images.

To obtain stereoscopic views and derive 3D information from the areas of overlap in the images, in photogrammetry, similarly to what happens in the human vision, you need to incorporate an object with at least two images taken from different viewpoints. For a good photogrammetric restitution and for data processing are required a proper amount of high-resolution images and a right calibration of the machine, knowing the focal length, the offset of the main point and the parameters of lens distortion. It often happens that the object to represented is too complicated as size or geometry to be detected only through stereoscopic shots; in this case a large number of images, from different points of view around the object and with converging axes, are acquired and oriented simultaneously with the method of the compensation with projective bundles (Fig. 9).

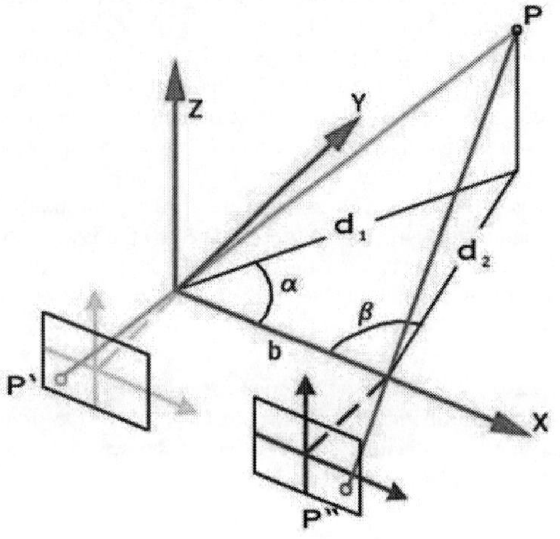

Figure 8. Collinearity

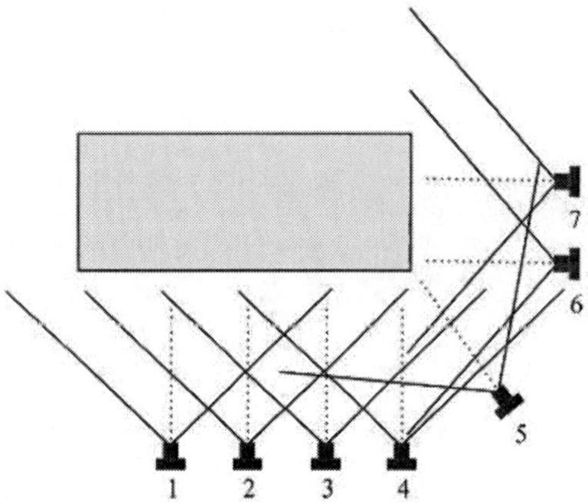

Figure 9. Method of compensating with projective beam

Once captured the images, we proceed with the process of derive metrics and 3D information of a scene:

1. calibration of the camera to determine the internal orientation;
2. triangulation of the images to determine the external orientation;
3. 3D restitution of the scene to derive a unstructured, sparse or dense points cloud;
4. creation of a polygonal geometric structured model (mesh);
5. weaving of the geometric pattern and production of an orthophoto.

This contribution has been peer-reviewed.
doi:10.5194/isprsarchives-XL-5-W4-403-2015

In the case of a detailed geometries, such as low-reliefs or statues, are preferred automatic procedures (image matching), capable of extracting a large number of matches (and thus points in 3D) necessary to return the scene with all its geometric details. These techniques are still developing, but the current results testify to the great potential of the method, comparable to those obtained with active optical sensors. For the orientation of the images can be used control points with known coordinates, geometric constraints of various types, or the resulting model can be scaled by direct distance measurements taken directly on the object. The most common commercial software that use this approach allow to obtain coordinates of points, vector restitution, sections and photo-textured models.
The decision to use photogrammetry in the survey of the environments of Casalnuovo, was dictated by the need to analyze and map the present degradation. The laser scanner at our disposal did not allow us to obtain orthophotos of good quality to work for mapping degradation in a consistent manner with the prospect of recovery project of the environments. Precisely for this reason we had to resort to a photogrammetric survey, this time not to get a complete 3D model, but rather to produce orthophotos (Figs. 10, 11) and proceed in the restitution of a more detailed analysis. The texture obtained by the laser did not respond to our needs because the integrated camera was not with high definition and the overlap of the images was not ideal for the various exposures of light dictated by geometry and setting environments. Furthermore photogrammetry has allowed us to obtain an exact texture point by point because it reports pixel by pixel color information, which does not happen in the project images that show the color information on the detected points, but instead take into account regions of space in the vicinity of the point considered.
On the other hand photogrammetric survey was quite difficult because the computers at our disposal were not in accordance with the image processing. In fact we are still working to get all the interior elevations of the observed environments and to complete the mapping of deterioration.
The integration of the techniques is crucial, especially in relation to the resources and to the desire to achieve goals that with the use of only one of the techniques we would not have got, also according to our knowledge and experience in the use of the aforementioned ones.
The recovery of the environments that we analyzed, as in most cases present in the *Sassi* of Matera, is crucial for the development of a World Heritage Site such this one and it is for this reason that, in our opinion, it is necessary to consider all the possible alternatives analysis in order to achieve consistent considerations.
The degradation factors mainly concern the micro-climatic conditions to which the environments are subjected . The absence of closures makes the internal microclimate particularly influenced by the external one. The mechanisms of alteration of the material are also favored by the porous and tender nature which constitutes the calcareous sandstone. These factors have resulted in the indoor environments degradation phenomena such as:
- stain: localized color variation of the surface, related to the presence of certain natural components of the material (concentration of pyrite in the marbles) and to extraneous material (water, oxidation products of metallic materials, organic substances);
- chromatic alteration: it manifests itself through a variation of one or more parameters that define the color, with different morphologies by referring to large or localized areas;
- biological patina: thin and homogeneous layer caused by the presence of water, mainly consisting of microorganisms, variable to texture, color and adhesion to the substrate;
- exfoliation: formation of one or more laminar portions, of very small thickness and sub-parallel to each other, said layers. It is primarily caused by moisture.

Figure 10. Orthophoto of the interior

Figure 11. Orthophoto of the interior

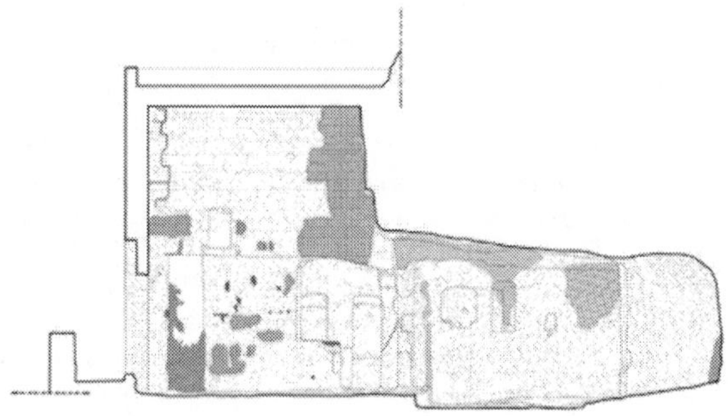

Figure 12. Degradations mapping

This contribution has been peer-reviewed.
doi:10.5194/isprsarchives-XL-5-W4-403-2015

After analyzing the present pathologies (Fig. 12), we moved to the study of the possible recovery interventions. Considering the morphology of the places and respecting the construction techniques, the renovation aims at a cleaning based on the removal of the products of alteration and deposit affecting the surface of the stone. The cleaning will be made with sorghum brush that will not go to affect the surfaces of good consistency. We will proceed with a whitewash, technique of painting or coloring, semi-one coat of lime milk used frequently in the past to protect surfaces. A key aspect to be taken into account are the closures of the environments, the absence of which results in a complete accessibility which obviously creates environmental and human dangers (Fig. 13). In the restoration project were provided suitable closures which also guarantee an improvement of the internal microclimate and are encouraging the protection of the environments. With regard to the coverage of built environments, having been reconstructed in an intervention of the Seventies, does not present big problems of infiltration. The only points where there is the possibility of infiltration are in correspondence of the graft with the masonry realized only with fittings in mortar. Furthermore, the discrete rainfall, the sunshine and the presence of earthy deposits and dust carried by the wind as the result of processes of de-cohesion of the materials, have favored the growth of natural vegetation on the roofs. Indeed terracotta elements of the mantle are hydrophilic in nature and tend to absorb water and encourage the phenomena of delamination of the material. Due to the degradation conditions of the coverage, we are planned interventions aimed at revision of the mantle, reposition and fixation of dispersed elements, removal of vegetation after application of biocide and liberation of the outlets of drain pipes. The interventions take into account the fact that the covering layer rests on a fill in flakes of stone, tiles and earth which ensures breathability but not watertight seal. From the analysis of the elevation we have highlighted the presence of iron elements for the support of the downpipes for the disposal of rainwater coming from the shell. Therefore it is expected the reintroduction of these ones in correspondence to the channels at valley.

Finally we are thinking to work on these areas in order to make them usable, but for this purpose we should increase the size of the project so as not to forget what has been the housing tradition and culture of the *Sassi*.

Figure 13. The elevation and the absence of the closures

3. RESULTS AND DISCUSSION/CONCLUSIONS

The review and reinterpretation of the reading keys of the historic built, thank to traditional and innovative techniques, according to its original and founding energy-environment capacities allows you to return fully to the buildings the quality that has always distinguished them and to rethink, in a sustainable environment, to intervention, rehabilitation and refurbishment procedures.

REFERENCES

Bagavathiappan S. et al. 2013. Infrared thermography for condition monitoring. A review. *Infrared Physics & Technology*, 60, pp.35–55.

Barba S., Mage M. A. 2014. Evaluación ex-ante y ex-post de la precisión de un proyecto fotogramétrico. In: *EGraFIA 2014 Rosario (ARGENTINA)*, CUES Vol.I, pp.548-557.

Bixio A., 2010. Il disegno grafico. Dalla rappresentazione Raster al disegno del Logotipo. Edizioni CUES, Salerno.

Cardinale T., Valva R., Lucarelli M., Advanced representation technologies applied to the temple of neptune, the sphinx and the metope in the archaeological park of Paestum, in: *International Archives of the Photogrammetry, Remote Sensing and Spatial Information Sciences*, Volume XL-5/W1, 2013 3D-ARCH 2013, pp.35-41.

Fiorillo F., Remondino F., Barba S., Santoriello A., De Vita C. B., Casellato A., 2013. 3D digitization and mapping of heritage monuments and comparison with historical drawings. *International Archives of the Photogrammetry, Remote Sensing and Spatial Information Sciences.* Vol. II-5/W1, pp.133-138.

González-Aguilera D., Rodriguez-Gonzalvez P., Armesto J., Lagüela S. 2012. Novel approach to 3D thermography and energy efficiency evaluation. *Energy and Building*, 54, pp. 436-443.

Guidi G., Russo M., Beraldin J., 2010. Acquisizione 3D e modellazione poligonale. McGraw-Hill, Milano.

Kordatos E.Z. et al. 2013. Infrared thermographic inspection of murals and characterization of degradation in historic monuments. *Construction and Building Materials*, 48, pp. 1261-1265.

Kraus K., 2004. Photogrammetry. Geometry from Images and Laser Scan. De Gruyter, Berlin.

Moropoulou A. et al. 2013. Non-destructive techniques as a tool for the protection of built cultural heritage. *Construction and Building Materials*, 48, pp. 1222-1239.

Remondino, F., Del Pizzo, S., Kersten, T., Troisi, S., 2012. Low-cost and open-source solutions for automated image orientation – A critical overview. In: *EuroMed 2012 Conference*, M. Ioannides et al. (Eds.), LNCS 7616, pp. 40-54. Springer, Heidelberg.

Tombari, F., Remondino, F., F., 2013. Feature-based automatic 3D registration for cultural heritage applications. In: *IEEE Conference Digital Heritage 2013*, Vol. 1, pp. 55-62.

3D MODELLING: CROSSING TRADITIONAL BOUNDARIES BETWEEN DIFFERENT RESEARCH AREAS

L. Opgenhaffen [a], *, M.H. Sepers [a]

[a] 4D Research Lab, ACASA, University of Amsterdam, The Netherlands – (l.opgenhaffen, m.h.sepers)@uva.nl

Commission V, WG V/4

KEY WORDS: Interdisciplinarity, 3D modelling, architecture, GIS, dialectic model, common integrative approach

ABSTRACT:

A 3D reconstruction of complex architecture will never be complete nor constructively sound without the combination of knowledge and expertise from different scientific branches. Yet, how should all the different perspectives from these diverse disciplines, leading to different ideas and views on material, be approached, in order to convert the knowledge to a coherent model? It is the goal of this paper to contribute to the development of a common approach to diverse datasets, ideas and interpretations that are involved in the creation of 3D reconstructions. The outline of our suggested approach is demonstrated though two case-studies differing in time and region, but revolving around the same research topics, showing that the approach is universally employable. Therefore, interdisciplinarity - one of the key principles of the Seville Charter - will be the leitmotif of this paper.

1. INTRODUCTION

Reconstructions, be it drawings, descriptions or video's, visualize thoughts and experiences, manage complex data and guide the interpretative process. They are a powerful tool to communicate this otherwise abstract process through images to the outside, material world. The most innovative research tool of today, 3D technology, enable that scientists from different research areas can collaborate and transcend and expand their knowledge through the connection of data, thoughts and ideas in a 3D model. Being enriched with all available data, the 3D models are elevated to dialectic ontologies, creating virtual environments to test and analyse the social dynamics of ancient built environments.

Yet 3D models do not convey the information themselves. It is the process towards the 3D model that does. It is the gathering of data and research to build the model *and* the choices that where made by the modellers and/or scholars (i.e. their agency) that conveys the information that leads to an interpretation (Earl 2012; Forte 2008; Hermon 2008). These interpretations can now be more comprehensive than ever because the extra dimension forces to look for anwers, and frequently these answers are outside the research area and even the field of the specialist. Therefore a 3D reconstruction may be regarded as the *status-quo* of what is known about a specific subject, from the excavated site to the role of the associated material culture with its social implications. It is not the 3D reconstruction of ancient architecture that proves something by itself, but it is how we engage with the model in order to unravel how someone in the past might have engaged with it .

Therefore we cannot stress enough that – and we are certainly not the first - that 3D modelling is primarily a research tool. We stress that 3D techniques are most valuable if used as a method to enhance and enrich *any* ongoing research. It is a transdisciplinary tool that requires to comprise knowledge from all disciplines: it compels specialists to move beyond their specialization and cross their traditional boundaries of research areas.

2. TWO CASE-STUDIES

2.1 Introduction

Through the description of the workflow of two case-studies we try to explain how we collaborate with other disciplines and how we treat the data obtained from these collaborations. The first case describes the pursuit to a famous historical figure who lived in a 'prominent' house in early 16th century Amsterdam. The second case concerns a cluster of houses from the Archaic period of the Etruscan town Acquarossa, Italy.

2.2 The house of Pompeius Occo)

Dr. Madelon Simons, art historian at the University of Amsterdam, does research to the historical figure named Pompeius Occo. Occo was German-Dutch banker who came in 1510 to Amsterdam to establish a branch of the Fugger bankers. He was one of the richest person of the town in his days and a respected mecenas for the arts. Dr. Simons had a collected a lot of historical data and descriptions of Occo and his house and uncovered a local network of the mecenas, artists and other dignitaries. Even the exact adress of Occo's house, called *'t Paradijs*, was found in the city archives: Kalverstraat no. 13 and the adjacent house, Rokin no. 6. These houses do not exist anymore, but on the oldest city-map of Amsterdam, dated to 1544, the houses could be identified, which gave an impression of the features of the houses. Art historians working on Occo's life previous dr. Simons, thought it was the prominent house on the map, the house with the façade with the curlicues. The map was interpreted as it fitted the predetermined ideas of the art historians, and consequently the alley at which the prominent house was located, must have had an angle, which made it the only alley in the city with an angle. Dr. Simons uncovered from literature-, art-, and archival studies much details about specific

* Corresponding author. This is useful to know for communication with the appropriate person in cases with more than one author.

This contribution has been peer-reviewed.
doi:10.5194/isprsarchives-XL-5-W4-411-2015

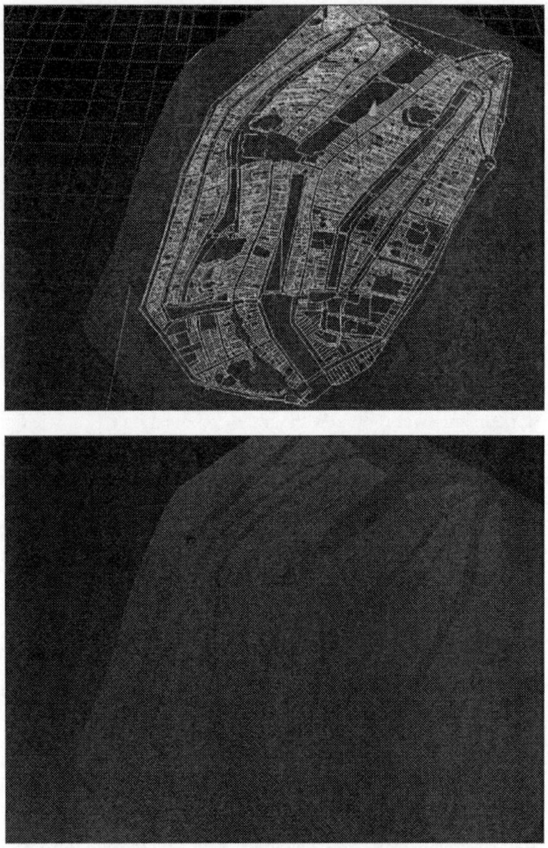

Figure 1. The GIS map imported in Cinema4D and landscape reconstructed (images by L. Opgenhaffen)

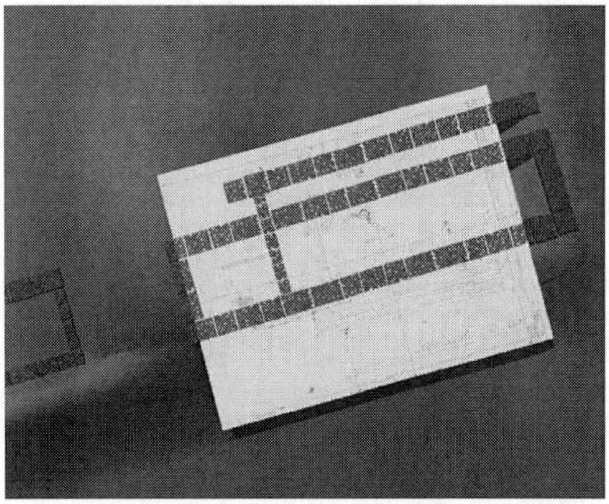

Figure 2. Excavation drawings are imported and build upon (image by L. Opgenhaffen)

spaces in the house, such as a *comptoir*, a library, pieces of art, and a reception hall. But, she also assumed, houses at that time could only have been of modest size: 5 x 6 meters. How then did this erudite person received the Danish king with its entourage and 40 Dutch maidens? Where did these people danced and watched the joust on the canal? Therefore the houses along the alley had to be reconstructed in order to simulate the movement of all these people and how Occo must have lived.

To reach this dr. Simons had to trancend her traditional academic boundaries and visited the 4D Research Lab at the archaeological department (ACASA). At the 4D Lab the archaeologist/modeller asked for actual data, which was a huge surprise for the art historian, who only had art historical descriptions and assumptions. An transdisciplinary project was born, that rapidly expanded to include building historians, the public archaeological service of the city (BMA) and an historical GIS service that provided the digitized map of the oldest register-map of Amsterdam from 1832 (HISGIS).

The GIS map provided the starting point for the reconstruction of at least the area around the houses of Occo, and was imported in the 3D modelling program Cinema4D. First, the GIS map was used in combination with the Dutch national height map (AHN) to reconstruct the landscape with the dykes and canals. This needed additional research to the changing landscape with support from from the city's archaeological service to reconstruct drained canals and ditches. After that, we started with a basic test of placing simple houses in the parcels. That proved directly that the alley was straight and could never have had an angle. This was further confirmed by inserting the archaeological drawings (from the early 1980's, that were now for the first time actually studied, digitized and geo-referenced) into and onto the model, providing the evidence that there was never an angle in the alley and the house of Occo could not have been the 'prominent' house with the curlicues. The house of Occo was on the other side of the alley, perfectly corresponding to the adjacent house at the back of which was known that Occo bought that too. 'Prominent' was therefore in the early sixteenth century understood differently than we perceive it today.

From this point onwards we could start modelling 'for real', from the foundations onwards that were recorded on the archaeological map (which also proved, among a lot of other indications, that the house was not only 5 x 6 m, but 8 x 18 m), and test if these art historical descriptions could literally stand on the foundations, which were, of course, reconstructed first. Immediately another problem occurred when the foundation was laid and the wooden timer frame had to be built on it; how were early sixteenth secular buildings actually constructed? Additional information had to be sought with building historians. The additional research to complement the integrity of the reconstruction of the house of Occo, consisted, for example, of research to timber frame construction, types of late medieval foundation techniques, dendrochronologic analysis, brick building, research to parallells of existing late medieval buildings, fire places and chimneys, windows and doors, and so forth.

When the construction was finished the interior could be reconstructed according to and limited by the construction of the building, preventing random placement of quarters. It is not in the scope of this article to go into further detail, as the main

This contribution has been peer-reviewed.
doi:10.5194/isprsarchives-XL-5-W4-411-2015

Figure 3. Reconstruction of the timber frame (image by L. Opgenhaffen)

goal is to demonstrate the potential of interdisciplinary research and 3D modelling as a coercive reseach tool.

Finally, the computational specialist of the 4D Research Lab developed a database which is able to show 3D models connected with the data and metadata on which the reconstruction is based on. It is possible to rotate and pan the model in the database and click on a random part, which subsequently shows what data is connected with this part of the model, making the model highly transparent and interactive.

2.3 Houses from Acquarossa

The second case-study concerns the reconstruction of a part of the settlement area ("zona B") of Acquarossa. This Etruscan town Acquarossa, near modern Viterbo, Italy, was situated on a tufa plateau, and arose in the Orientalizing period and remained in use until its abandonment in the second half of the 6th century B.C. Several excavation campaigns were undertaken in the late 1960s and 1970s by the Swedish institute at Rome. The excavations yielded a very rich archeological site with clusters of buildings (divided in so-called *zone*) associated with the transitional period from 'huts to houses': the urbanisation process in the Italic peninsula (8th – 6th century B.C.). After the site was abandoned 2500 years ago, the land was mostly used for agricultural purposes thence providing an unique opportunity to study the development an Etruscan settlement.

This general outline of the excavation history of Aqcuarossa indicate that a huge amount of (archaeological) data was uncovered during the excavations resulting in a wide variety of interpretations about the architectural situation in the past. In this case-study special attention is devoted to the area known as 'zona B'. In this area remains of a cluster of buildings was found and attributed to at least two, perhaps even three phases. The buildings were largely constructed from perishable materials such as wood and adobe and have been subject to a variety of post-depositional processes, like the possible re-use of building materials, erosion and most prominent, modern farming activities. These processes ultimately resulted (at least partially) in obscuring the material remains from this period. Notwithstanding these difficult circumstances of the remains, exhaustive research has been carried out to the material culture of Acquarossa that was subsequently published in several journals and books and even resulted in (permanent) museum exhibitions. The interpretation of the architectural remains of

Figure 4. The house of Pompeius Occo, *'t Paradijs*. Left: front side. Right: back side (image by L. Opgenhaffen)

the buildings, and even more so ascribing the remains to different building phases, is not as simple as it may appear. The main enquiry is therefore to how this cluster of houses were constructed in the Archiac period. 3D modelling is applied as a research tool to explore ancient building practices in this particular site. Engineers from the building company Grontmij will check and if necessary revise the constructional part of the recontruction of the houses.

Many reconstructions of the vernacular architecture have been suggested. These reconstructions, however, are still preliminary in the sense that they should be regarded as an illustration of the (temporal) interpretation of how the area might have looked like. The final report on the excavations in 'zona B' is not published yet. But from the research of a specific find categories, the roofing systems proved to be very valuable for reconstructions of the exteriors of the buildings of 'zona B'. The research included recognizing deposition patterns and trying to understand ancient roofing techniques. Based on other, novel insights, earlier visualisations proved to be implausible or even wrong, and with the 'new' data, derived from the research on roofing materials, other reconstructions could be suggested. All known theories and interpretations of this cluster of houses of Acquarossa will be tested in the virtual environment. In the final 3D model, these different hypotheses are visualised as variations of the same model. Alterations and adaptations can be visualised through layers, leaving the actual remains and other data intact and accessible.

The analogue data of the landscape was first digitised in qGIS and subsequently the architectural remains could be georeferenced to this model. The GIS model serves as an intermediary between the 3D reconstruction and the database in which all the collected data is stored. The model of the landscape was then exported to the 3D modelling program Cinema4D for further elaboration and to actual test the constructional methods of the superstructure and roofing system of the houses.

3. CONCLUSION

On the basis of the two case-studies we propose a distinct integrative approach by using - among other digital tools - 3D modelling as a research tool in the interpretative process. For the first time are excavation, geological and (art) historical data, previous analogue research and interpretations on material culture, ideas on construction, ideas concerning visual aspects, knowledge about available resources, combined through digital techniques. These digital techniques include the application of data management systems, GIS and 3D modelling, that make it

This contribution has been peer-reviewed.
doi:10.5194/isprsarchives-XL-5-W4-411-2015

Figure 5. Reconstruction of the alley (image L. Opgenhaffen)

possible to create a visualisation of the current state of knowledge and ideas towards this Etruscan settlement and a late medieval house in Amsterdam. In doing so, through the placement of data in a framework of (visible) relations, a variety of questions occur and choices have to be made. These questions and choices produce unforeseen emerging properties of materials and draw attention to problems in data quality *and* consistency, frequently leading to many different options, constraining the researcher to continuously reconsider, reassess, and most important, demonstrate these choices. As a result large datasets are created that can be directly connected with the visual output. This visualization might be considered as a rendering of our interpretation of the ancient material culture and not as a mere reflective historiographic representation. The availability of the dataset behind the visualization can assist the critical reviewer of the model to determine its validity, add or review data, add interpretations, and ultimately revise the model.

In this light, we developed the concept of a 'dialectic 3D model'. Archaeologists move between information, argumentation, evidence, hypothesis and interpretation, which are moving simultaniously on itself too (after Gooding 2008). This is no lineair progressive path: it moves between the interpretation of new data (added to the database and thence processed into the 3D model), through new hypotheses that give meaning to to this new data (these new hypotheses can be tested in the 3D model/environment), to arguments that may be transformed into evidence for an explanation, the 'final', albeit temporal, 3D model. Moreover, the possibility to move between different dimensions, structures the explanation (Gooding 2008, 17): the 3D models structures the archaeological interpretative process. The dialectic 3D model is always subject to change (to put it in Hegelian terms) and therefore never 'finished', making it an ultimate goal to improve our knowledge of past built environments.

Through these case-studies we try to explain how 3D modelling and interdisciplinary collaborations can be deployed as a common integrative approach to enhance and enrich ongoing research, and let it play an active role in not only the archaeological discours, but the humanities as a whole.

ACKNOWLEDGEMENTS

First and foremost we wish thank dr. Patricia S. Lulof, who founded the 4D Research Lab, and inspired us contribute to the development of a methodology for 3D modelling as a research tool. Further we like to express our gratitude to the research

Figure 6. Reconstruction of the interior of the house of Occo (image L. Opgenhaffen)

project 'CREATE: an e-humanities perspective', headed by prof. dr. Julia Noordegraaf, who financed the reconstruction of the house of Occo. We also are very grateful to Lara and Rafaele Rocchi, who assisted in the onsite exploration of Acquarossa, dr. Madelon Simons for initiating the reconstruction-project of the house of Occo, Jerzy Gawronski, Jorgen Veerkamp and Dick de Roon from BMA, and, finally the engineers of Grontmij. Without you our buildings could not have withstand.

REFERENCES

Earl, G., 2013. Modelling in Archaeology: Computer Graphic and other Digital Pasts. *Perspectives on Science,* 21 (2), pp. 226-244.

Forte, M., 2008. Virtual archaeology: communication in 3D and ecological thinking. In: Frischer, B. & Dakouri-Hild, A. eds. *Beyond Illustration: 2D and 3D Digital Technologies as Tools for Discovery in Archaeology,* Oxford, Archaeopress (BAR International Series 1805), pp. 20-34.

Gooding, D.C., 2008. Envisoning explanation: the art in science. In: Frischer, B. & Dakouri-Hild, A. eds. *Beyond Illustration: 2D and 3D Digital Technologies as Tools for Discovery in Archaeology,* Oxford, Archaeopress (BAR International Series 1805), pp. 1-19.

Huggett, J., 2012. What Lies Beneath: Lifting the Lid on Archaeological Computing. In: Chrysanthi, A. & Murrieta Flores, P. & Papadopoulos, C. eds. *Thinking beyond the Tool. Archaeological computing and the interpretive process.* Oxford, Archaeopress (BAR International Series 2344), pp. 204-214.

Huvila, I., 2012. The Unbearable Complexity of Documenting Intellectual Processes: Paradata and Virtual Cultural Heritage Visualisation. *Human IT,* 12 (1), pp. 97-110.

Lulof, P.S., 2011. The Art of Reconstruction. De wetenschappelijke waarde van 3D-simulaties bij de analyse van pre-Romeinse tempels en daken. *TMA,* 46, pp. 14-24.

Ryan, N., 2001. Documenting and validation virtual archaeology. *Archeologia e Calcolatori,* 12, pp. 254-273.

TECHNIQUES FOR REVEALING 3D HIDDEN ARCHEOLOGICAL FEATURES: MORPHOLOGICAL RESIDUAL MODELS AS VIRTUAL-POLYNOMIAL TEXTURE MAPS

Hugo Pires [a*], José Martínez Rubio [b], Artzai Elorza Arana [c]

[a] CICGE - Research Centre for Geo-Space Sciences, University of Porto (miqhapaqnan@gmail.com)
[b] LFA-DAVAP, University of Valladolid (jmrubio@ega.uva.es)
[c] University of Salamanca (artzai.elorza@gmail.com)

Commission V, WG V/4

KEY WORDS: Morphological Residual Model, Archaeology, Engravings, 3D, Polynomial Texture Mapping

ABSTRACT:

The recent developments in 3D scanning technologies are not been accompanied by visualization interfaces. We are still using the same types of visual codes as when maps and drawings were made by hand. The available information in 3D scanning data sets is not being fully exploited by current visualization techniques.

In this paper we present recent developments regarding the use of 3D scanning data sets for revealing invisible information from archaeological sites. These sites are affected by a common problem, decay processes, such as erosion, that never ceases its action and endangers the persistence of last vestiges of some peoples and cultures.

Rock art engravings, or epigraphical inscriptions, are among the most affected by these processes because they are, due to their one nature, carved at the surface of rocks often exposed to climatic agents. The study and interpretation of these motifs and texts is strongly conditioned by the degree of conservation of the imprints left by our ancestors. Every single detail in the remaining carvings can make a huge difference in the conclusions taken by specialists.

We have selected two case-studies severely affected by erosion to present the results of the on-going work dedicated to explore in new ways the information contained in 3D scanning data sets.

A new method for depicting subtle morphological features in the surface of objects or sites has been developed. It allows to contrast human patterns still present at the surface but invisible to naked eye or by any other archaeological inspection technique. It was called Morphological Residual Model (MRM) because of its ability to contrast the shallowest morphological details, to which we refer as residuals, contained in the wider forms of the backdrop.

Afterwards, we have simulated the process of building Polynomial Texture Maps - a widespread technique that as been contributing to archaeological studies for some years - in a 3D virtual environment using the results of MRM calculations. By this, we wish to benefit from the rendering capabilities of RTI-viewer and from its intuitive graphic interface. At the same time, virtual PTM is a way of applying this to areas barred to conventional PTM, like in the case presented of an entire roman city occupying a plateau of several sq. km.

The results of this research project are presented and discussed using the two case-studies aforementioned, a Latin inscription from a Roman sanctuary in the north of Portugal and a engraved panel with zoomorphic motifs from a rock art site in the north border of Portugal.

1. INTRODUCTION

Erosion is one of the most important factors affecting the conservation of archaeological sites and objects. Every pronounced feature in a rock surface that we know tends to disappear sooner or later affected by this process. Over time, shapes carved by humans become progressively fainted, ultimately leading to its complete loss (Siegesmund et al., 2002) (Pipkin et al. 2014). An increasing public awareness of the damage affecting archaeological sites as lead to major conservation campaigns and scientific studies carried out all over the world that address the origins and consequences of climatic changes in cultural heritage (CH). As a consequence, sites directly exposed to climatic agents are among the most affected but, to a certain degree, every CH tangible asset is endangered.

Our attention is focused in the specificity of open-air sites with engravings or any form of human carving. These sites, often object of study for epigraphy and rock art specialists, are extremely exposed to damage by erosion from natural or human action. One of the principal consequences affecting this types of archaeological evidences is the progressive loss of material from the rock superficial level, seriously affecting theirs state of conservation (Gaddi et al., 2015)

Every incision technique used in the past to carve forms on the stone involved applying a certain amount of energy by impact or friction so to remove material from the surface. The energy used to produce the grooves is the cause for changes to the micro-structural level turning a carving into a weak point in the surface matrix. Cavities are more prone to the accumulation of

* Corresponding author.

This contribution has been peer-reviewed.
doi:10.5194/isprsarchives-XL-5-W4-415-2015

water and debris, fostering the proliferation of biotic colonies and the action of weathering agents. In addition, the edges of grooves are more exposed to erosion and are more fragile than a regular surface. Another important factor in the severity of erosion effects in archaeological sites is the rock origin and composition. The superficial desegregation tends to provoke more serious damages to engravings in rocks with coarse-grained textures than those with smaller crystal structure. Contour definition and groove sharpness are eroded faster in granites or sandstones than in limestones or marbles.

For many years in archaeological research, rock carvings and engravings were inspected using techniques that are nowadays considered unacceptable because of their harmfulness to the rock surface. Those that require mechanical actions in the outcrops, like pressure or friction, or involve chemical interactions with external substances may compromised their conservation are the most detrimental for conservation purposes. Techniques such as scraping with charcoal on paper or pouring melt substances such as plaster or latex to obtain a negative mould have been almost completely abandoned from archaeological field practices.

The pursue for more effective ways of recording, detecting and inspecting ancient engravings is a current commitment for a large number of researchers in the archaeological broad scientific community. 3D Scanning technology as been one of the major contributors for improving recording methods in archaeology. We have now arrived to the time when a simple digital photo-camera with appropriate processing software can be used to 3D scan objects or sites without damaging them, achieving incredible resolutions and at a incredible low cost.

Although a lot of effort is being putted in recording 3D data, we continue to mimic the way we visualized and depicted it before the major event of 3D scanning. As an example, the same graphic codes once used for depicting altitude in flat maps, such as equidistant contours, are still being done, but now from 3D scan data sets, representing a significant cutback on the amount of information available. Building surveys using 3D scanning technology are, in most cases, fate to end in vector drawings like plans, sections or elevations calculated from 3D point clouds. Again, a huge part of the information is lost in translation. Our contribution to this shift concerns the way we use these records after acquisition and processing steps. We believe that, to fully explore the benefits of the increasingly available 3D scanning data, new ways of seeing and interacting must be persecute.

To test our ideas in a challenging environment, we have chosen to focus our efforts in the very delicate remains of ancient engraved archaeological sites, carved in granite outcrops severely damaged by erosion agents.

2. CURRENT TECHNIQUES FOR INSPECTING AND RECORDING ENGRAVED ARCHAEOLOGICAL SITES

The documentation and visualization techniques have evolved along history, following the changes in society attitude towards CH. A vast number of methodologies have been used to record epigraphy and rock art. In recent years, we have witnessed a clear evolution regarding the techniques for visual inspection of engravings. They have matured with a progressive awareness of society towards Cultural Heritage conservation, changing its character from invasive to non-invasive; from non-digital technique to digital technique; and merely documentary to combination of documentary and diffusive.

These can be classified into two groups according to the amount of recorded dimensions:

2D methods:

- Sketching or tracing on translucent or transparent media such as plastic film and vellum paper superimposed to engraved surface.
- Frotagge, a rubbing technique.
- Tracing photographs.
- Polynomial Texture Maps[1]

3D methods:

- Moulding
- Direct discrete tracing. Consist on collection of 3d line strings connecting points measured by touch probes (contact) or total station (non contact)
- Photogrammetric discrete methods. Understood as those that require a human being to perform the digitization of features engraved in the rock by means of a plotter.
- Scanning. Digitizing surfaces by means of laser or structured light scanners that allow a dense collection of points or digital meshes or alternatively by means of recent image-based techniques that use pairs or sets of overlapping photos to produce dense point cloud representation of surfaces.

The first two techniques have been the standard procedure prior to digital era. They are seen today as less scientific but despite being objectionable due to their invasive character, they have honourably served along too many years as a recording technique have been for years the only existing technical methods of graphic documentation, and they owe most of our knowledge of rock art. The so-called frottagge aims to get the imprint of the relief by rubbing a paper placed on the stone with a material that stains such as charcoal, graphite or pastel. This often helps to transfer to the paper some already worn motifs.

Another technique, called bi-chromatic, was thoroughly used by many archaeologists during some decades. Today it is unanimously considered as highly invasive and susceptible of causing harm. It's also intended to enhance the readability. Consists on applying white powder or ink to the engraved stone and, after that, rubbing a dark substance directly on the rock to blacken the areas that are not carved. In this way, the contrast is pushed up and the carvings are turned into a sort of black and white filter. Although cost effective, its aggressiveness isn't acceptable under current standards in CH conservation. The same considerations apply to qualify moulding technique today as an objectionable practice.

For the purposes of this study we have only consider the most frequently used techniques: raking light photography, 3D scanning and PTM. Two different case-studies were selected to assess the suitability of the different methods: a Latin inscription from the Roman sanctuary of Panóias and a panel with engravings from the rock art site of Monte Faro, both in the North of Portugal. Archaeological studies are being

1 For the purposes of this paper, Polynomial Texture Maps and Reflectance Transformation Imaging are considered as a unique solution and are both referred to as PTM.

This contribution has been peer-reviewed.
doi:10.5194/isprsarchives-XL-5-W4-415-2015

conducted in these sites (Correia Santos et al., 2014), (Pires et al., 2015a).

2.1 Raking light technique

Now that we have outlined the different tools and technologies for capture 3D data, lets talk about how to face the problems of reading and interpretation the motifs when they are faded by weathering or other forms of decay. The cinema has shown us many times the image of an archaeologist bringing a torch or a candle and moving the light source to help reading of engraved findings. This gesture of seeking for the best point for a raking light to enhance carved forms to our eyes is certainly the oldest method of inspection, and its usage will surely be contemporary with the creation of the carvings itself. It is still used today in the very same fashion and remains unsurpassed as the primary tool for on-site analysis. Despite of being an active inspection method, it is such a common one that most consider it as unimportant as the simple passive direct inspection.

Figure 1: Top: Raking light photographs of a Latin inscription from the sanctuary of Panóias (Portugal) with two different light directions. Bottom: Raking light photograph from Rock 4 at the rock art site of Monte Faro (Portugal). Photographic credits M. J. Correia Santos (top) and L. Bacelar Alves/M. Reis (bottom).

The recording of this information is usually done in two different ways: photographing the surface illuminated by directional spotlights from different tangential positions and tracing by hand the apparent contours of engravings and main geological features using a transparent film overlaid to the stone surface. Both techniques are often used in conjunction and tracing is done directly on photos taken under raking lighting. Although photographs are far more objective than tracing, they only allow for fragmented records, each corresponding to the unique point of view and shading condition captured in the photography. Usually, several photographs with different light directions are needed to record the entire scene.

One of the major drawbacks this method presents is a fragmented perception of the assemblages because each trace of engraving is better perceived when the light comes from a perpendicular direction to the groove attitude (Figure 1, top). Furthermore, irregular outcrops are very difficult to lighten homogeneously because of shadow effects from spotlights (Figure 1, bottom). Usually, multiple light positions must be used to document the entire scene, therefore a suitable method to combine the feature lines revealed at each lighting condition must be implemented.

Last but not least, this technique is very susceptible to the intrinsic colour texture of rock surfaces that, along with the patina of biotic or chemical origin, envelops and masks the carvings features, preventing archaeologists of having a clear perception of all the existing remains.

2.2 3D Scanning

Digital techniques are changing the way archaeological records are made. Several new solutions have come to light in these last decades. One of the major trends is 3D scanning (Barber et al., 2011), that is increasingly being used in archaeology to produce full records of sites and objects shape, dimensions and position. The improvements achieved by non-contact 3D scanning techniques in archaeological recording are an huge step forward when compared to non-digital techniques. Apart from allowing to record surfaces in three dimensions with astonishing resolution and without any damage to the integrity of the physical materials, they decrease the time and cost needed to accomplish recording tasks in archaeological sites.

These non-invasive techniques allow for detailed and accurate representations of surfaces that can be used to produce conventional drawings (e.g. plans, elevations, sections) or other non-conventional types of visualization like videos or virtual reality. Applied to our case-studies, this technique allowed to calculate dense point clouds with sub-millimetre resolution from rock surfaces previously known for having archaeological evidences. Due to damage inflicted by erosion, both presented a conservation state that prevented any clear reading of the motifs. The interpretation of these ancient testimonies as been therefore conditioned.

Several important advances are achieved by 3D scanning. One is the possibility to separate the colour texture from the 3D model, thus revealing the shape underneath in a homogeneous tonality making motifs easier to detect (Figure2). In these virtual environments, as much as in real situations, light can be

This contribution has been peer-reviewed.
doi:10.5194/isprsarchives-XL-5-W4-415-2015

directed from any point of view, and so, raking light can also be simulated. Despite these clear benefits to recording engravings, it still presents the same drawbacks as raking light.

Figure 2: Snapshot from the 3D scan model of the same panel as in Figure 1, with the simulation of raking light.

2.3 Polynomial Texture Mapping

Another technique that is achieving an important place in archaeological recording is the Polynomial Texture Mapping (PTM) (Malzbender *et a*l., 2011). This technique uses sets of photographs of a scene, all taken from the same point of view, differing only in the light incidence angle used. The results are displayed in a 2D interface, called RTI viewer, allowing the user to interact with the scene by virtually dragging a directional light source and rendering the light changes in real time.

PTM has a promising future as a recording technique for cultural heritage as it is being continuously improved by different researchers in many ways. An innovative approach to PTM visualization is to replicate the process of image capture in a virtual environment. Mostly any 3d modeling programs such as Autodesk 3DS or Blender allow to emulate the behavior of a real camera and can capture snapshots of the 3d scenes as if they were taken in real scenarios (Earl *et al.*, 2010a). The simulation capabilities include also the placement of virtual light sources whose interaction with geometry and physical properties assigned to the virtual objects can faithfully simulate real world scenes. These two features together allow us to emulate the work-flow of the PTM technique.

This process is what is known as Virtual PTM (V-PTM) and presents some interesting advantages. There is no limitation to the size of the object or scene because the construction of a physical assemblage of camera and lights is not required in virtual worlds. Also, the process of lighting is completely under control in a virtual environment, so there are no worries about intruding light sources that could affect the estimation of light source position. The virtual camera is ruled by a mathematical model thus free of chromatic or geometric lens aberrations. Similarly, the virtual snapshot is free of noise and artifacts and is always perfectly sharp. In one word, the simulation is capable of gathering all PTM benefits without any limitations; being the model fidelity the only factor that can negatively affect the rendition of its virtual variant.

Among other major benefits, like the simplification of the capturing process, by using this approach every 3D model can be converted in a multitude of PTM files, one for each possible point of view. Furthermore, V-PTM widens the possibilities of application of PTM to fields where it was not possible before. An example is the 3D model of the Roman city of Clunia, in Spain, that occupies several square kilometers (Figure 3). The 3D data was calculated from photographs taken with digital cameras attached to unmanned aerial vehicles.

V-PTM allows for the effortlessly evaluation of many different light sources distribution by the automation of the spot-light placement and its resulting rendered scene capture. We can even test the ability of RTI builder software to estimate accurately the positions of the light by adding a black sphere to the scene in the way it is used in non-virtual cases. This approach can lead to conclusions about the optimal distribution of light sources, or the choice of the simulated optical parameters and settings to create rendered scenes that introduce less uncertainty.

Tests have enabled us to build an optimized framework to process virtual photos in RTI-builder, obviating the step of estimating the position of the spotlight since that can be stated rather than estimated. We were able to partially automate this process by programming an application that requires the user to enter the number of spotlights, the number of heights where the lights will be distributed, and the maximum dimension of the object. The program returns a complete script that commands the creation and toggling of virtual lights in 3ds Max and executes the corresponding iterated rendering of the scene under the changing light conditions (Figure 4).

Beyond being another iteration of the use of V-PTM technique in which our contribution as been the optimization and automation of the process, the most interesting achievement is the inclusion of a synthetic colour map (the colour coded results of the MRM) as if it was the colours in a photograph.

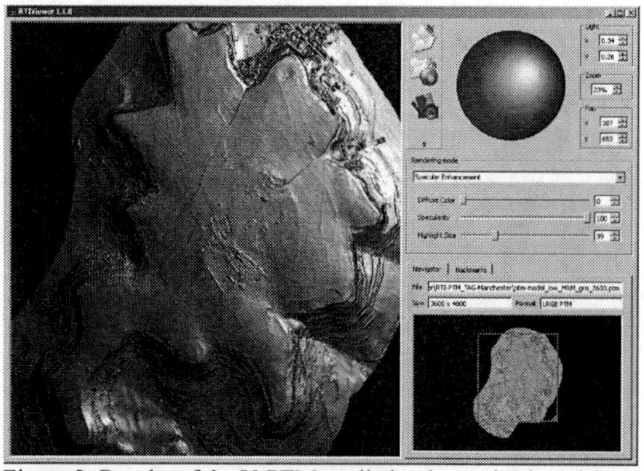

Figure 3: Results of the V-PTM applied to large sized models of earth surface: the Roman archaeological site of Clunia, Spain, (around 4 sq. Km.).

As in the previous techniques, also PTM is light-dependant has an inspection tool. The considerations presented for 3D scanning are also valid to PTM results. When compared to 3D models it lacks of three-dimensionality, but, in a certain way, this same reason can be taken as an advantage. The complexity of operating a 3D data viewer is far more high than in RTI-viewer.

This contribution has been peer-reviewed.
doi:10.5194/isprsarchives-XL-5-W4-415-2015

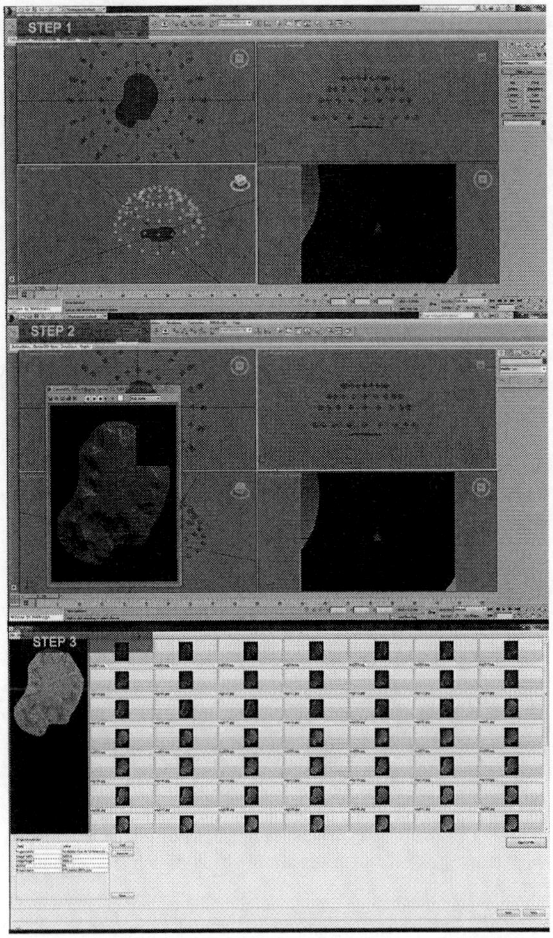

Figure 4: V-PTM processing pipeline (from top): Virtual light dome; image capture using rendering filters; assemblage of PTM using RTI Builder to process the virtual photographs (video available at: http://youtu.be/-QjSMeumSjE).

3. NEW APPROACH FOR ENHANCED 3D MODEL VISUALIZATION: MORPHOLOGICAL RESIDUAL MODELS AS V-PTM

All the aforementioned techniques, digital and non-digital, present the same type of constraints to the visual inspection of fainted motifs carved in stones. Our proposal points in another direction, aiming to contribute to overcome visualization problems related to this conservation conditions, common to many CH sites around the world. It consists in using a combination of two different techniques to achieve a new hybrid output, a V-PTM for inspecting MRM, benefiting from the good features of both worlds. Although apparently pointing in opposite directions, this two techniques can be integrated because of the V-PTM approach.

MRM uses 3d mesh models out of any shape reconstruction technology to calculate contrasted depictions of micro-relief features in 3D models contributing significantly to increase the amount of information available for archaeological researchers. When applied to 3D models of eroded sites or objects it reveals morphological details that are not easily perceived by the human vision, by depicting small shape variation in any 3D surface as a colour code attribute for each in the 3D model. The processing is based in a trend removal technique that allow to calculate residual morphology of engravings. The residual values are coded in a colour scale and assigned to the 3D model as vertex colour attributes. The results are 3D models where small morphological features appear as consistent colours, independent of light orientation (Pires et al., 2015).

This new tool can be considered as an advanced visualization technique, which uses 3D models derived from digital scanning systems. It appears therefore that it is not appropriate to situate it among recording technologies but in another separate category for post-processing 3D models. Perhaps it is so, but if we examine all the listed techniques, we note that scanning techniques, despite being the richest in potential content, are blind in the sense that the recorded is done in an undifferentiated way. The absence of a purpose gives equal importance to areas carved as to those that aren't. In this sense we can think of MRM as a more "intelligent" approach because it contributes to give meaning to underlying the shape captured by scanning. Therefore, more than being just a 3d data treatment, we consider that MRM together with scanning configures a capture technique highly adequate for engravings or any other subtle morphological evidence of human action.

Figure 5: Results of the entire set of techniques discussed in the same engraved panel (from top, left): Raking light photography, 3D scanning, V-PTM, MRM and, at the bottom, MRM in RTI-viewer).

This contribution has been peer-reviewed.
doi:10.5194/isprsarchives-XL-5-W4-415-2015

Once that the surface texture of the 3D model as been remapped with the MRM results, we make use of the V-PTM technique to facilitate the virtually inspection of the engravings, combining the advantages of the new colour coded shape depiction with the effectiveness of the RTI viewer interface. The bi-dimensional intrinsic character of PTM is taken as a way of simplifying the inspection tasks: interacting with 3d models in virtual environments can be an annoying and time consuming operation, specially when dealing with models with large quantity of data, like point cloud based meshes, so maybe a 2D solution can help to improve the receptivity of archaeologists.

The wide range of tests carried out so far in different archaeological contexts confirm that the perception of subtle morphological features is significantly enhanced when using MRM (Correia Santos et al., 2014, Pires et al., 2015a). The traces still present in the stone's surface become depicted in a way that reduces the ambiguity that light-based techniques present.

4. RESULTS

As mentioned above, the proposed methodology was thoroughly applied in two different archaeological sites, an Atlantic rock-art assemblage with a large number of engravings in granite outcrops and a Roman sanctuary where four inscriptions in Latin and Greek still testified the ancient cults. Both are considered to be of very difficult reading due to erosion factors and, before the use of MRM, were not fully "read" by archaeologists. In the case of Panóias, studied for some decades, this lead to controversial interpretations. The contribution of the clear depictions for the archaeological knowledge of these important inscriptions was recently presented to the scientific community (Correia Santos et al., 2014, Pires et al, 2014) and made available online[2].

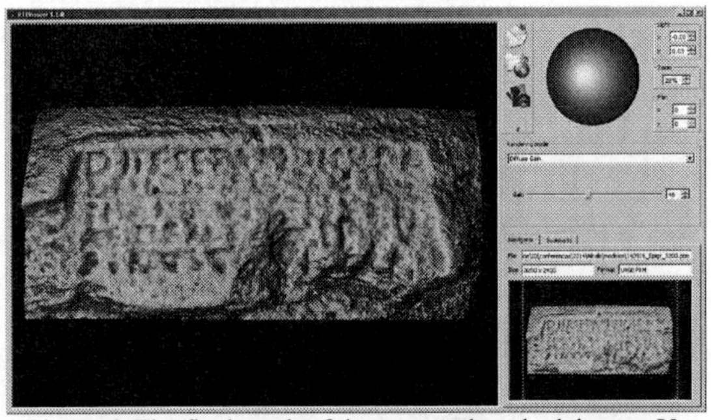

Figure 6: The final result of the proposed methodology: a V-PTM built from a MRM colour-coded 3D model of an inscription from Panóias sanctuary.

Although light is still needed to observe virtual 3D models with MRM codification, the perception of small scale relief is considerably enhanced by the presence of colour as a local deepness factor.

Furthermore, MRM outputs such as digital orthophotos, videos or even PTM, can be considered as scientific depictions *per se*. Indeed, MRM can be considerer a new category of information, as much as non-visible radiation photography is for conventional visible spectrum images. This new category of depictions can in fact reveal non-visible shape information.

From a certain point of view, the use of PTM as an output of MRM, can be considered counter-productive as it corresponds to a significant reduction of the information available in 3D model. Yet, considering the excellent rendering performance of the RTI-viewer and the easiness with which light interaction simulates a 3D environment, it is out opinion that RTI-viewer is an excellent vehicle for delivering information to who needs it.

5. CONCLUSIONS

When dealing with highly eroded surfaces light-dependant techniques will unlikely be able to deliver clear and unambiguous depictions of subtle morphological details and, therefore, new approaches must be pursued. For many CH cases, the time is getting over for apprehending the meanings of ancient carved rock art. Climate changes issues are jeopardizing archaeological sites, worsening conservation problems and so we must urgently respond to the need to retain the most meaningful and comprehensive snapshot of our past.

Having in mind that nowadays a large number of CH 3D records are being accomplished by image-based techniques (like the examples presented in this work), it is rather ironic to realize while looking back, the astonishing amount of images captured for each model processed, from the initial set of photographs taken with digital camera to the virtual processing of PTM. Inspired by this work, we say that digital photographs are an extremely rich source of knowledge, capturing information far beyond what we can imagined each time the shutter release button is pressed. The inspection of MRM using the RTI-viewer interface, is an effective, yet simple, way of delivering valuable and accessible data to archaeologists.

The RTI-viewer is as a promising tool for inspection of 3D models-derived data. It would be very interesting to assist in a near future to the emergence of a digital tracing tool working within RTI-viewer that could contain image orientation parameters for each PTM view, allowing for direct correspondence between the source 3D model and PTM. Multi-view would also be a valuable attribute for profiting from the richness of 3D scanning data.

Other fields of archaeological studies, and other sciences as well, can be potential beneficiaries from the methodological and technological improvements we have presented here. For this purpose, several tests are being carried out at multiple scales, ranging from geomorphological scale of landscape to small artefacts in underwater environments. We hope our work can contribute to foster new developments in 3D data processing and visualization.

2 A digital version can be found at: http://www.eagle-network.eu/eagle-first-international-conference-proceedings-now-online/

This contribution has been peer-reviewed.
doi:10.5194/isprsarchives-XL-5-W4-415-2015

REFERENCES

Barber, D. and Mills J., 2011. 3D Laser Scanning for Heritage (second edition). Advice and guidance to users on laser scanning in archaeology and architecture. English Heritage Publishing.

Bednarik, R.G., 2012. The use of weathering indices in Rock Art Science and Archaeology. Rock Art Research- Volume 29, Number 1, pp. 59-84.

Pipkin, B., Trent, D.D., Hazlett, R, and Bierman, P., 2013, Geology and the Environment, 7th edition, Brookes Cole, Belmont, CA, 624 p.

Correia Santos, M. J., Sousa, O., Pires, H., Fonte, J., Gonçalves Seco, L., 2014. Travelling back in Time to Recapture Old Texts: The use of Morphological Residual Model (M.R.M.) for epigraphical reading. Information Technologies for Epigraphy and Cultural Heritage. Studi Humanisti - Serie Antichista, Sapienza Universitá Editricie, pp. 437-454.

Earl, G., Martinez, K., Malzbender, T., 2010. Archaeological applications of polynomial texture mapping: analysis, conservation and representation. Journal of Archaeological Science, 37, (8), 2040-2050

Earl, G., Beale, G. Martinez, K. and Pagi, H., 2010. Polynomial Texture Mapping and Relate Imaging Technologies for the recording, analysis and presentation of archaeological materials. International Archives of Photogrammetry, Remote Sensing and Spatial Information Sciences, UK, Vol. XXXVIII, Part 5.

Gaddi, R., Cusano, M., Bonanni, P., Cacace, C. and Giovagnoli, A., 2015. Monitoring of cultural heritage decay in Rome. Analysis of soiling and erosion phenomena. Engineering Geology for Society and Territory - Volume 8, pp 535-538.

Happa, J., Mudge, M., Debattista, K., Artusi, A., Gonçalves, A. and Chalmers, A., 2009. Illuminating the Past - State of the Art. The 10th International Symposium on Virtual Reality, Archaeology and Cultural Heritage VAST - State of the Art Reports

Keyser, J.m 2001. Rock Varnishing and Weathering in Handbook of Rock Art Research, ed. D.Whitley, Altamira Press, Lanham

Malzbender T., Gelb D. and Wolters H,. 2001. Polynomial Texture Maps. In SIGGRAPH '01: Proceedings of the 28th annual conference on Computer graphics and interactive techniques

Pires, H.,Gonçalves Seco, L., Fonte, J., Correia Santos, M. J., Sousa, O., 2014 Morphological Residual Model: A Tool For Enhancing Epigraphic Readings Of Highly Erosioned Surfaces. Information Technologies for Epigraphy and Cultural Heritage. Studi Humanisti - Serie Antichista, Sapienza Universitá Editricie, pp. 133-144.

Pires, H., Gonçalves-Seco. L., Fonte,, J., Parcero-Oubiña, P. , Fábrega-Álvarez, P., 2015. Morphological Residuals Model – A mesh decimation filtering tool for detection and contrast of archaeological evidences in point-cloud derived models. Journal of Cultural Heritage. Forthcoming.

Pires, H., Bacelar Alves, L., Martinez Rubio, J.,Elorza Arana, A., Gonçalves Seco, L., Fonte, J. 2015. New depictions for ancient carvings.Results of the MRM pilot project at the rock art site Monte Faro, Valença – Portugal. Proceedings of the XVII World Congress of UISPP. Forthcoming.

Plets, G., Verhoeven, G., Cheremisin, D., Plets, R., Bourgeois, J., Stichelbaut, B., Gheyle, W., and De Reu, J., 2012. The deteriorating preservation of the Altai Rock art: assessing three-dimensional image-based modelling in rock art research and management. Rock Art Research. 29(2). p.139-156.

Purdya, B.A, Jonesb, K., Mecholskyb, J., Bourneb, G., Hulbert Jr.c, R.C., MacFaddenc, B.J., Churchd, K.L., Warrena, M.W., Jorstade, T.M., Stanfordf, D.J., Wachowiakg, M.J., and Speakman, R.J., 2011. Earliest Art in the Americas: incised image of a proboscidean on a mineralized extinct animal bone from Vero Beach, Florida. Journal of Archaeological Science 38 (11): 2908–2913

Siegesmund, S. , Weiss, S., Vollbretch, A. 2002. Natural Stone, Weathering Phenomena, Conservation Strategies and Case Studies. Geological Society Publishing House. Bath,UK

This contribution has been peer-reviewed.
doi:10.5194/isprsarchives-XL-5-W4-415-2015

GOTHIC CHURCHES IN PARIS ST GERVAIS ET ST PROTAIS IMAGE MATCHING 3D RECONSTRUCTION TO UNDERSTAND THE VAULTS SYSTEM GEOMETRY

M.Capone [a], M. Campi [b], R. Catuogno [c]

[a] DiARC Dipartimento di Architettura Università degli Studi di Napoli Federico II, mara.capone@unina.it
[b] DiARC Dipartimento di Architettura Università degli Studi di Napoli Federico II, campi@unina.it
[c] DiARC Dipartimento di Architettura Università degli Studi di Napoli Federico II, raffaele.catuogno@gmail.com

Commission V, WG V/4

KEY WORDS: Structure from Motion, Image Matching, 3D Modeling, Ribbed Vaults, Gothic Flamboyant, 3D reconstruction.

ABSTRACT:

This paper is part of a research about ribbed vaults systems in French Gothic Cathedrals.
Our goal is to compare some different gothic cathedrals to understand the complex geometry of the ribbed vaults. The survey isn't the main objective but it is the way to verify the theoretical hypotheses about geometric configuration of the flamboyant churches in Paris. The survey method's choice generally depends on the goal; in this case we had to study many churches in a short time, so we chose 3D reconstruction method based on image dense stereo matching. This method allowed us to obtain the necessary information to our study without bringing special equipment, such as the laser scanner.
The goal of this paper is to test image matching 3D reconstruction method in relation to some particular study cases and to show the benefits and the troubles. From a methodological point of view this is our workflow:
- theoretical study about geometrical configuration of rib vault systems;
- 3D model based on theoretical hypothesis about geometric definition of the vaults' form;
- 3D model based on image matching 3D reconstruction methods;
- comparison between 3D theoretical model and 3D model based on image matching;

1. THE COMPLEX GEOMETRY OF RIBBED VAULTS IN FRANCE

1.1 The Geometry of Gothic vault

From reading of the *Livre de Portraiture* by Villard de Honnecourt, a fundamental document for the study of Gothic architecture, we understand how the geometric rules used in the Middle Ages constructions are defined.
The vaults system regulates the building of medieval cathedrals, a system in which you can clearly recognize the elements carried by the ribs, whose load-bearing function becomes explicit when the groin vault transforms.
The geometry of medieval vault could be originated by two different shapes: the sphere, the domical vault, from which the vaults with domed trend of Aquitaine, and the cylinder, the groin vault, characterized by perimeter semicircular arches and elliptical diagonal arches, from which derives the rib vault of Ile de France.
The secret of Gothic buildings is the use of geometrical rules, rather than fixed modules, designed to optimize the construction stage, only the knowledge of these rules will help you to understand these architectures where the figurative outcome is closely related to construction methods.
The first Gothic vaults are quadripartite rib vaults on square or rectangular plan. They have curved ridge if they are originated by transformation of the dome or they have horizontal ridge if they are originated by groin vault. The feature of these vaults is to have semicircular diagonal arches and the others are generally pointed arches. The arches can be composed of arcs with the same curvature of the diagonal arch, the keystones can be at the same level (horizontal ridge), or may have different curvatures and keystones placed at different heights (curved ridge). The evolution of quadripartite rib vaults is a more complex rib vault generated by introduction of eight additional ribs (Figure 1) that you can see in the intersection of the nave and transept of the cathedral of Amiens (Figure 2a).

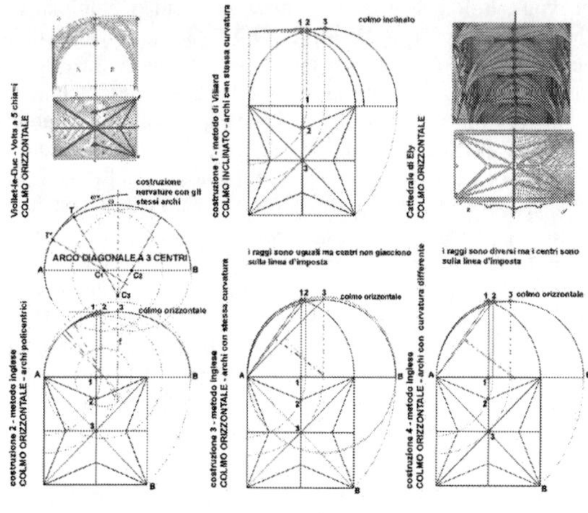

Figure 1 Theoretical model: different way to construct a five keystones vault.

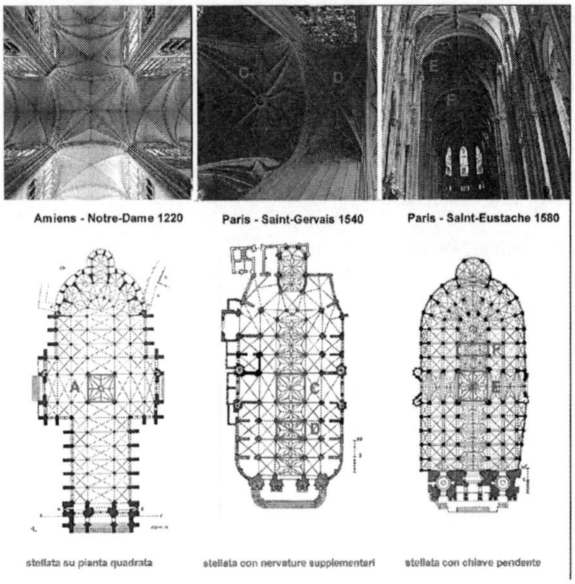

Figure 2 Evolution of ribbed vaults in France.

This kind of vault spreads with the introduction of additional secondary ribs that you can see in St. Gervais et St Protais (Figure 2b), and evolves to generate strange solutions characterized by hanging keys, like those that it is possible to see in Saint-Eustache (Figure 2c).

A Gothic rib vault is composed by a three-dimensional mesh of ribs, which are semicircular arches usually contained in vertical planes, and the webs, that fill the space between the ribs. There are some different ways to construct the rib vaults: the survey has allowed us to compare the theoretical model with the exact configuration of the case studies chosen from among the most meaningful in Paris.

1.2 The evolution of ribbed vaults in France

Gothic architecture was first developed in France, so it is there that you can find the greatest number of transitional buildings.

The great abbey of St. Denis (1140-44) can be considered the earliest gothic church in France. It is characterized by a quadripartite vault system used to make these Gothic churches larger. In which you can see all the main features of gothic cathedral: the high rib vaults system in the central nave and in the transept that allows to open large windows and to generate the clerestory and the triforium below, thanks to a system of external supports and flying buttresses.

From cathedral of St. Denise the buildings became more complex and the vaults systems too: the quadripartite vaults at St. Denise's nave gave way to sexpartite vaults at Notre Dame de Paris (1163), and, in 1220, the introduction of the first rib vault at Amiens Cathedral. We think that the evolution in Gothic churches design was determined by development of the quadripartite vault. From the so called high Gothic Period to late Gothic, known in France as the *flamboyant style*, the interiors underwent a drastic simplification by eliminating the capitals of all the pillars and reducing them to plain masonry supports. In flamboyant churches all architectural ornamentation was concentrated in the vaults, the ribs of which formed an intricate network of even more complicated patterns.

Figure 3 Paris. St. Gervais et St Protais. Nave: rib vault system.

Figure 4 Paris St. Gervais et St Protais transept's rib vault system.

Flamboyant work is characterized by structural simplicity and lavish ornament. Vaulting ribs and pier arch mouldings of many small elements were often carried unbroken down the piers to the floor, or else were allowed to die on the surface of simple cylindrical piers. Mouldings were made to intersect in all sorts of curious ways, and double or triple bases are common. The church of Notre Dame at Alencon and those of St. Maclou at Rouen, St. Gervais et St Protais at Paris (Figures 3 and 4), St. Jacques at Dieppe, are beautiful examples of the developed flamboyant church of the 15th and early 16th centuries. So vital was this flamboyant style, that, in ecclesiastical work, it remained in vogue until long after the Renaissance dominated secular building, and its structural ideas, even longer, as shown in the large church of St. Eustache in Paris, which is structurally late Gothic.

2. CASE STUDY: ST. GERVAIS ET ST. PROTAIS CHURCH

2.1 Church in Paris. Case Study: St. Gervais et St Protais

St. Gervais et St Protais Church is located in the 4th arrondissement of Paris, on Place Saint-Gervais in the Marais district. It is one of the oldest churches in Paris and it is a fine example of the *French flamboyant style*.

The present church was begun in Gothic style in 1494, the chapels of the apse were finished in 1530 and the transept in 1578 (Huisman 1966).

This contribution has been peer-reviewed.
doi:10.5194/isprsarchives-XL-5-W4-423-2015

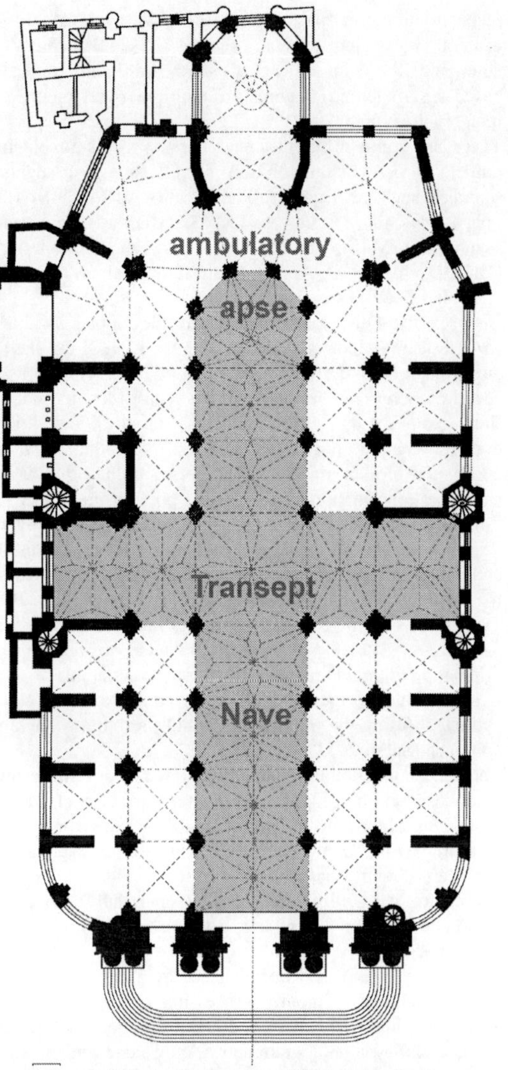

Figure 5 Paris. St. Gervais et St Protais, plan.

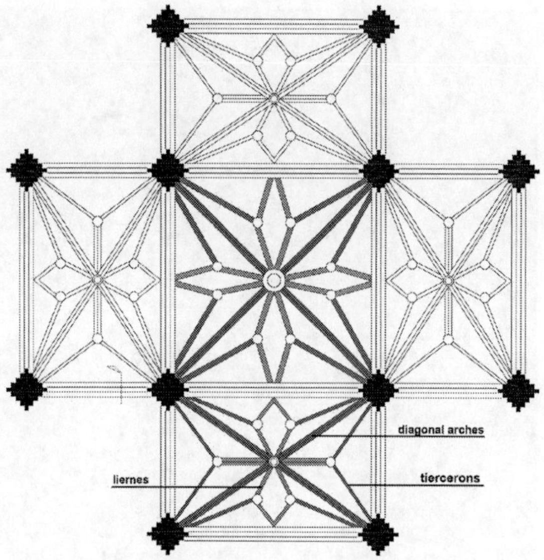

Figure 6 Paris. St. Gervais et St Protais, ribbed vaults: nave and transept.

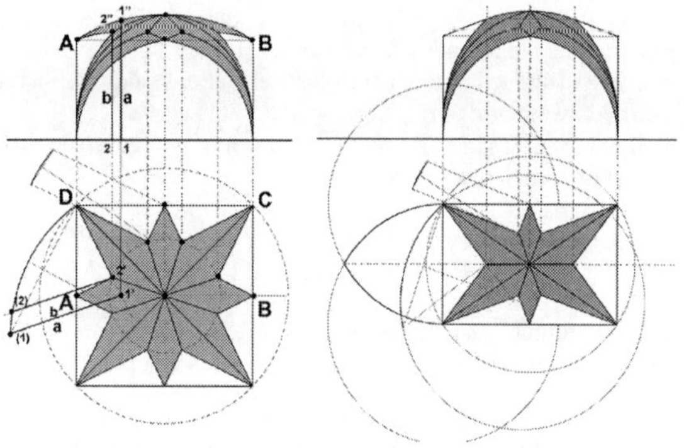

Figure 7 Paris. St. Gervais et St Protais, ribbed vaults: theoretical model

Between 1600 and 1628, a second row of chapels was built on the north side including the golden chapel ornamented with painted woodwork (Huisman 1966). The main façade, in classical style, was built in 1616–1620 by Claude Monnard.

The church is composed of three naves and a choir that ends with a pentagonal apse. Twelve chapels are located longitudinally and five others are located around the ambulatory (Figure 5).

The main nave and the transept are characterized by rectangular ribbed vaults, a square vault at the intersection between the nave and transept and a complex system of ribbed vaults in the pentagonal apse.

Survey and analysis of the ribbed vault system of St. Gervais et St Protais started from the study of the vaults and their geometric construction process. First we have analysed the rectangular ribbed vaults, with six keystones, that are in the nave and in the transept. They are characterised by three ribs, *tiercerons*, that start from each pillar.

Two of them are the diagonal arches that meet in the vault's main keystone and the others are the secondary ribs that meet in the seven minor keystones.

This kind of vault can be considered an evolution of a so called *five keystones vaults* characterised by three main ribs from pillars, *tiercerons*, and four secondary ribs, *liernes*, which link the minor keystones (Figure 1). It is obtained with the introduction of four secondary ribs on longitudinal axis to link the main arches' keystones with the four secondary keystones (Figure 6).

The geometry of the main vault, located on the crossing between nave and transept, is quiet the same, in this vault there are eight secondary keystones connected by sixteen minor ribs, *liernes* (Figure 6).

To construct 3D model based on the theoretical vault construction system we need to recognize which theoretical model we have to consider, because there are some different geometrical constructions for the same type of vault.

We hypothesized that the ridge is curved. If you know the cross section of the vault and the ridge AB, you know the elevations of all the keystones and you can draw all the ribs. For example, to draw the rib D2', you have to extend the D2' line until AB, in point 1'. You have to draw the vertical line from point 1' to determinate point 1''.

This contribution has been peer-reviewed.
doi:10.5194/isprsarchives-XL-5-W4-423-2015

Figure 8 Paris. St. Gervais et St Protais, ribbed vaults: nave and transept. 3D from theoretical model.

Turning over the vertical plane for the rib D2' you can determinate the elevation of keystone 2 (Figure 7).
We have surveyed the keystones' elevations and we have verified that the ridge isn't horizontal but it is curved.
Following the same methodological path we construct the model of rectangular ribbed vault.

3. IMAGE BASED MODELING TO STUDY THE GEOMETRY OF GOTHIC VAULT SYSTEM

3.1 Structure From Motion. 3D model based on image matching 3D reconstruction methods

What is the best way surveying and modeling for this kind of architecture? In recent years it is one of the most common topic of the international meetings, the questions are the comparison between different techniques, Photogrammetry or laserscanning, the estimate of 3D models accuracy and sometimes the advantages from mixed use of more than one survey methods. It's clear that the choice of technique depends on "what" we have to survey and "why".
Generally, when it is required a great metric accuracy you have to use conventional Photogrammetry or laser scanning and you can use methods based on computer vision approach to produce textured models which do not have to be metrically accurate.
Advances in terrestrial laser scanning technology and digital photogrammetry techniques for creation of building documentation allow you to obtain in a short time a point cloud and to create accurate 3D models.
The high costs of data collection and the complex equipment needed often make terrestrial laser scanning not very affordable thus the Close-Range Photogrammetry has been used successfully for documentation of Cultural Heritage for many years. The Photogrammetry traditional workflow consists in Image Acquisition, Camera Calibration and Photogrammetric Modelling Process, the result of this process is a point cloud or a 3D model used for accurate orthophoto production. Traditional softcopy photogrammetric methods require the 3D location and pose of the cameras or the 3D location of ground control points to be known to facilitate scene triangulation and reconstruction, on the contrary the SfM method solves the camera pose and scene geometry simultaneously and automatically, using a highly redundant bundle adjustment based on matching features in multiple overlapping, offset images (Westobya, 2012).
There are some different image-based process to obtain 3D model or point clouds based on IT (Image triangulation) (bundle) such as Bundler, Photosynth, Apero, DIM (Dense image matching) such as PMVS, MicMac, Photomodeler scanner or SfM (Structure from motion) such as Autodesk's 123 Catch3D, Insight3D, Arc3D, PhotoModeler smar, VisualSFM, AgiSoft's PhotoScan.
Between the DIM sofware, MicMac, developed by the IGN MATIS team, which is available within a LINUX OS shell with an end-user GUI, derives the point cloud input with the use of a specific energy minimization function in order to smooth the final geometry and filter unlikely local disorders. Still, the derived camera poses computation and the multi-stereo correlation produces a dense 3D metrics point cloud based on an accurate projection of each scene pixel in the object space. Finally, we can derive a rigorous orthographic view over any desired projection plane; this function is particularly useful in architecture as orthographic metrics are often fundamental for the very expression of the objects characteristics (Saleri 2013).
The first step of the calculation process involves an open-source software - APERO - which is targeted to determine the orientation of a wide range of terrestrial or aerial images.
After estimating the camera poses, a dense point cloud is extracted using a multi-scale, multi-resolution pyramidal approach (Saleri 2013).
The data acquisition has been transformed most significantly by a new generation of high-resolution terrestrial cameras and new processing algorithms of image matching have revolutionised data processing. Using one of the software packages available you can obtain image-based surface models in an almost automatic way, this low-cost, user-friendly image-based technique to obtain 3D models, is the so called 'Structure-from-Motion' (SfM).
SfM diffusion is based on advances in Computer Vision and improvements in computer processing power, this technique allows producing image-based reconstruction and modeling without skills or photogrammetric knowledge and so it's very easy to learn. The increasing number of SfM software solutions free, such as Autodesk's 123d Catch, and commercial solutions such as AgiSoft's PhotoScan or Pix4D's cloud processing software, are fully automated and they are based on bundle adjustment and image matching algorithms.
The main difference between these software are that some work *server side*, such as Autodesk's 123 Catch3D or Microsoft's Photosynth and others work *stand-alone*, such as AgiSoft's PhotoScan, Pix4D and VisualSFM.
For example VisualSFM is a client side open source software, it is a GUI application for 3D reconstruction using structure from motion (SFM). The reconstruction system integrates several previous projects such as SIFT on GPU(SiftGPU), Multicore Bundle Adjustment, and Towards Linear-time Incremental Structure from Motion. VisualSFM runs fast by exploiting multicore parallelism for feature detection, feature matching, and bundle adjustment. For dense reconstruction, this program supports Yasutaka Furukawa's PMVS/CMVS tool chain, and can prepare data for Michal Jancosek's CMP-MVS.
At beginning SfM method was characterize by high quality rendering and it is considered opposed to photogrammetric method for metrically correct outcomes. Recent new developments in high-resolution terrestrial cameras and software applications allow obtaining 3D high-resolution

This contribution has been peer-reviewed.
doi:10.5194/isprsarchives-XL-5-W4-423-2015

models processing photographs in SfM software that can be compared with 3D model from laser scanner or from traditional softcopy photogrammetric methods.
To survey the vault systems of St. Gervais et St Protais Church we used Agisoft PhotoScan. It is a stand-alone software that performs photogrammetric processing of digital images and generates 3D spatial data. It can be used in GIS applications, cultural heritage documentation, and visual effects production as well as for indirect measurements of objects of various scales. Wisely implemented digital photogrammetry technique enforced with computer vision methods results in smart automated processing system that, on the one hand, can be managed by a new-comer in the field of photogrammetry, yet, on the other hand, has a lot to offer to a specialist who can adjust the workflow to numerous specific tasks and different types of data. Processing Images you can obtain: Photogrammetric triangulation, dense point cloud generation and you can generate textured 3D models. It is possible to operate using arbitrary images, photos taken from any position, but you have to provide that the object is visible on at least two photos, the image alignment and 3D model reconstruction are fully automated, it is based on a database of features automatically extracted from a set of multiple overlapping images (Snavely, 2008).

3.2 Structure From Motion - SfM pipeline

Our goal is to describe the photo-based 3d scanning process, "what" and "how" we have done to obtain a high quality 3d models and measurements from photographs.
The main steps of SfM method based on using of PhotoScann are:
- capturing photo
- loading photos into PhotoScan;
- inspecting loaded images, removing unnecessary images;
- aligning photos;
- building dense point cloud;
- editing dense point cloud;
- building mesh (3D polygonal model);
- editing mesh;
- generating texture;
- exporting results.

Our goal is to generate 3D model and a dense point cloud to study the geometry of vault system of St. Gervais et St Protais church. We followed some specific capturing guidelines to make photos for indoor applications.
We have done more photos than it was enough, in this way you can reduce the number of "blind-zones" because PhotoScan is able to reconstruct only geometry visible from at least two cameras. To obtain high resolution textured 3D model we have captured about three hundred photos. The complex geometry of vaults system required a careful survey project and we have done photos from the bottom up, moving along the nave and transept, and around each pillar starting from the basement to the ceiling. Each photo is overlapped by 80%, in this way we have minimized the possibility to have lacks, we were careful each photo should effectively use the frame size and object of interest should take up the maximum area.
We selected photos and separated them in chucks before loading. Once photos are loaded into PhotoScan, they need to be aligned. At this stage PhotoScan finds the camera position and orientation for each photo and builds a sparse point cloud model. When alignment having been completed, computed camera positions and a sparse point cloud will be displayed. In this way, it is possible to inspect alignment results and remove incorrectly positioned photos.

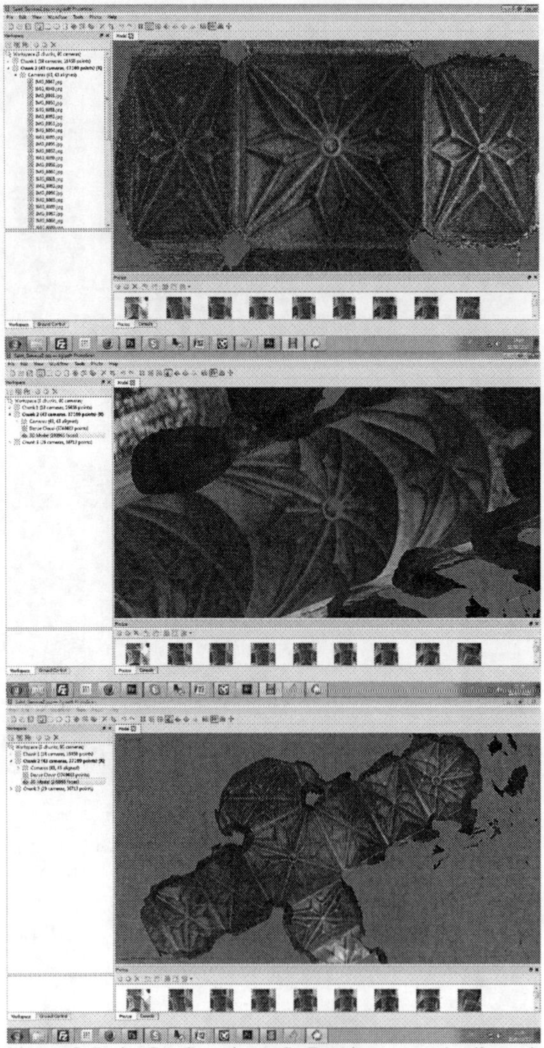

Figure 9 Paris. St. Gervais et St Protais, transept's rib vault system. Image matching 3D reconstruction. Indoor application

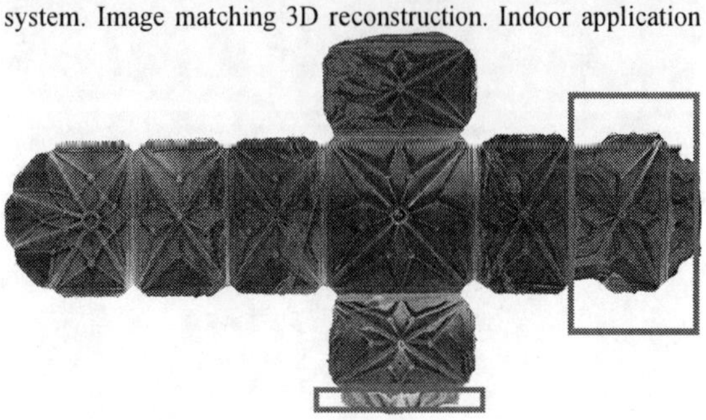

Figure 10 Paris. St. Gervais et St Protais rib vault system. Plan view by 3D textured model from Agisoft PhotoScan.

PhotoScan allows generating and visualising a dense point cloud model. Based on the estimated camera positions, the program calculates depth information for each camera to be combined into a single dense point cloud. A dense point cloud can be edited and classified within PhotoScan environment or exported to an external tool for further analysis.

This contribution has been peer-reviewed.
doi:10.5194/isprsarchives-XL-5-W4-423-2015

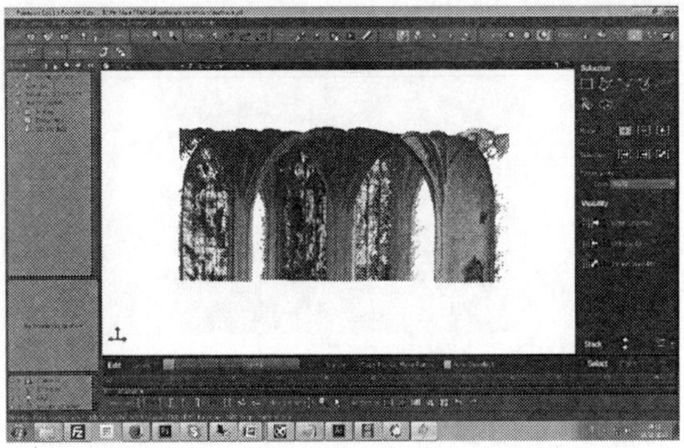

Figure 11 Paris. St. Gervais et St Protais rib vaults system. Section: orthogonal image of dense point cloud obtained using Pointools.

Figure 12 Paris. St. Gervais et St Protais rib vault system. Sections of dense point cloud using Pointools

We edited point cloud in PhotoScan and we construct 3D model using one of several reconstruction methods available within

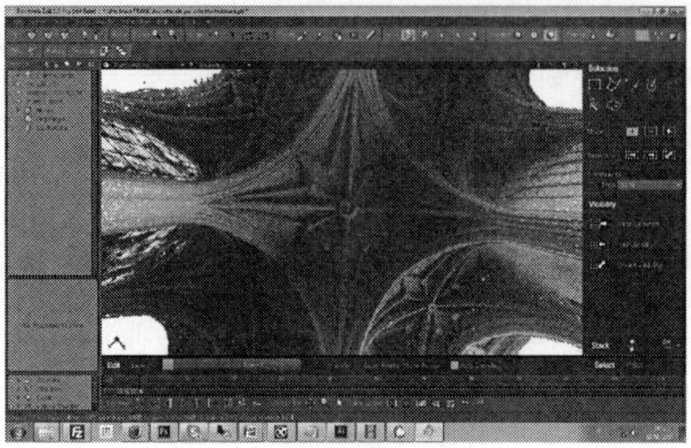

Figure 13 Paris. St. Gervais et St Protais rib vault system. dense point cloud view.

PhotoScan, which helped us to produce very accurate 3D model.

To obtain better visual quality of the final model we tested some of texture mapping mode, Generic and Spherical, and we chose the best outcome.

We used a reflex digital camera APS-C (Advanced Photo System type-C) we did Camera calibration using Agilens software by Agisoft.

We captured photos using two different focal length lenses, and then we had to do two different camera calibrations.

PhotoScan supports export of processing results in various representations: sparse and dense point clouds, camera calibration and camera orientation data, mesh, etc. Orthophotos and digital elevation models (both DSM and DTM) can be generated according to the user requirements. Point clouds and camera calibration data can be exported right after photo alignment is completed. All other export options are available after the 3D model is built.

We exported dense point cloud in Pointools to obtain the sections we needed to verify the geometrical hypothesis that we have done to compare the theoretical model with data from SfM process. Carrying out measurements based on the reconstructed model, we take some measure to scale the orthophotos and 3D model according to the known distances.

We exported orthophoto, to obtain a plan view of the vaults systems, this is high resolution imagery based on the source photos and reconstructed geometry.

We exported 3D model too, as a matter of fact PhotoScan allows to export 3D model in many different formats, such as Wavefront OBJ, 3DS file format, U3D, Autodesk DXF, Adobe PDF and so on Some of these file formats (OBJ, 3DS, VRML, COLLADA, PLY, FBX) save texture image in a separate file then you can rendering 3D model as you like.

3.3 Advantages and problems

SfM method is very convenient when you have a short time collecting data, you have a low budget, don't have sophisticated equipment (you need only a good terrestrial camera) and your aim is to study the geometrical configuration of the objet that you are surveying.

The software used is very easy, you just have to uploads multiple images of an object and the results are produced fully automatically without your interaction.

As all image based methods the result depends on the photos, then the principal disadvantages of the technique are linked to image resolution and some mistakes that you can do capturing photos. Photographs can be taken by any digital high-resolution camera (5 MPix or more), both metric and non-metric, higher is the resolution better is the result. We have to avoid blurred photos, blurriness makes it difficult for the system to process details in a photograph, and one blurry photo can affect the entire reconstruction, than you have to be careful that the aperture value should be high enough to result in sufficient focal depth and shutter speed should not be too fast, otherwise blur can occur due to slight movements. The choice of the correct image focal length is crucial as it will influence the optimal number of shots to perform and the resolution of the final point cloud.

Figure 14 Paris. St. Gervais et St Protais. Point cloud image is a very expressive image: without the weight of the material we can perceive the geometric configuration of the vaults system.

Although not required, we used a tripod to avoid blurriness, loading photos captured used a tripod we got a better result. You have to avoid introducing noise in the image, it is better TIFF files than JPG and ISO should be set to the lowest value because otherwise high ISO values could induce additional noise to images.
To obtain good results good lighting is required, blinks should be avoided and you have to remove sources of light from camera fields of view. For best results, use diffused light to avoid over exposure and try to eliminate any harsh shadows that may conceal details of your subject. Shadows on an object may be confused by the software as a real holes of the object captured, and consequently it will be reflected in the resulting reconstruction. For interior shots, you should avoid the use camera's flash, for exterior shots, it is best to capture scenes during the early morning or later in the afternoon, when sunlight is not so bright. You should avoid under or over exposure because, if photos are too light or too dark, the software will have more trouble locating points of reference.
Sometimes there are some problems linked to object or to scene that you have to reconstruct. You should avoid not textured, shiny, mirror or transparent objects, you have capture the scenery avoiding moving objects within the scene and avoiding flat objects.
SfM technique is image based modeling therefore you can reconstruct only what you see in the photos. To obtain the best result you have to "map" the entire object without lack, you have to take pictures in sequence so that each photo has 80% in common with the following and you have to move the camera. as soon as possible, parallel to the surface to be reconstructed.
In some cases it might be very difficult or even impossible to observe these rules, because there may be obstacles that force you to move randomly and to take pictures in a confused way.
Capturing photos of a little object it's very simple because you can turn around it and "mapping" all surfaces without lack, "mapping" a complex vaults system is more difficult because you can't move in suited way. We had some troubles connected with our case study especially during the step of collecting data. In fact, in this case, to take many overlapping photographs the main problem is the great dimension of ribbed vaults systems that are the feature of French Gothic Cathedrals (Figure 15).
Our goal was to demonstrate the efficiency of the method for this type of objects, we tried to follow all the suggestions above and we obtained a very good textured 3D model and a dense point cloud. We can underline some mistakes in the reconstruction (Figure 10) produced by some noise in the images and a wrong reconstruction of a part that we didn't "mapping".

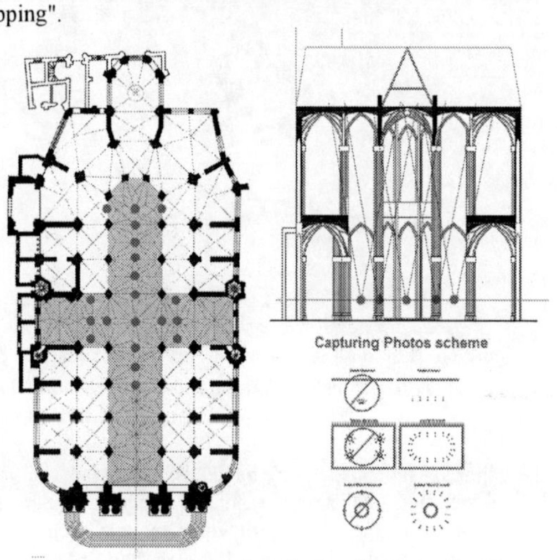

Figure 15 Church of St. Gervais et St Protais. Capturing Photos of complex vaults system

4. COMPARISON BETWEEN 3D THEORETICAL MODEL AND 3D MODEL BASED ON SFM

Analysing the data obtained from image matching 3D model we are able to understand the true geometric configuration of the vaults system. We have analysed the vault system in the nave and in the transept, in particular, we report the results of the study of square vault and rectangular ones.
Our goal is to illustrate the methodological process, to test the SfM image-based modeling technique and to underline the advantages for this kind of study.
Processing dense point cloud in Pointools we obtained the sections, orthogonal image, that we needed to reconstruct the geometry of the vaults.
The square ribbed vault is a kind of vault can be considered an evolution of a so called *five keystones vaults* characterised by three main ribs from pillars, *tiercerons*, and four secondary ribs, *liernes*, that link the minor keystones.
From analysis of the 3D data we can deduced that the square ribbed vault is characterized by two diagonal ribs, AB and CD, that are circular arches and the main arch and the traverse arches, DA, BC and CA, BD, that are pointed arches.
Thanks to 3D data from SfM we were able to verify that the radius of curvature of these arches is the same of diagonal arches AB and CD. The keystone is slightly higher than keystones of the perimeter arches thus the ridge is not
horizontal and, looking at the section, we can see the curvature depends on the webs geometry. The study of the geometric rules allows us to underline that the main ribs, the *liernes*, A2, A3, C1, C8, B7, B6, have the same curvature of diagonal ribs and

This contribution has been peer-reviewed.
doi:10.5194/isprsarchives-XL-5-W4-423-2015

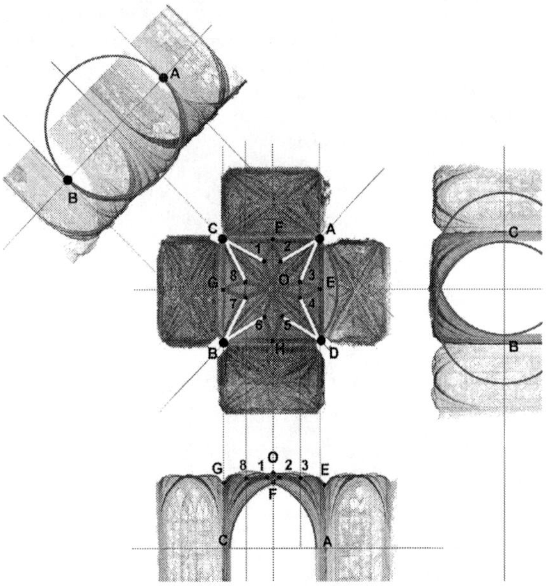

Figure 16 Church of St. Gervais et St Protais, square ribbed vault: study of geometric configuration from SfM 3D data.

the minor keystones 1, 2, 3, 4, 5, 6, 7, 8, which link the *liernes,* are at the same elevation.We studied the rectangular ribbed vaults following the same methodological approach. Looking the data from SfM dense point cloud we can observe that the diagonal arches are AC and BD are circular and the keystones 1 and 2 are almost the same elevation of the main keystone H, then the rib 12 is just a little curved. To verify this geometric rule we have analysed more than one of the rectangular vault and we have obtained the same result.

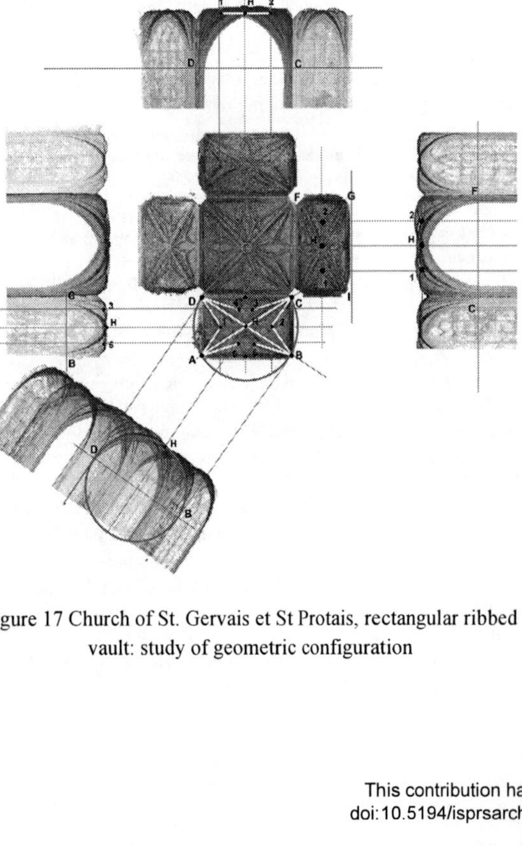

Figure 17 Church of St. Gervais et St Protais, rectangular ribbed vault: study of geometric configuration

In order to compare the two models, theoretical model and 3D model based on SfM process, we can underline that our hypotheses were correct but 3D data from SfM allowed us to understand the geometric configuration of all surfaces.

5. CONCLUSION

Our goal is to test SfM technique to generate a 3D model to study the geometry of vault system of St-Gervais-et-St-Protais church. We'd like to found a no-contact survey methods that is cheap, not needed complex equipment, fast and easy because we are going to use this method to analyse all gothic church in Paris and to extend the study to other Franch Cathedrals.
We can safely conclude that, for our case study , Soft from Motion is the best survey technique that we could choose because it is ideally suited for our low-budget research.

6. REFERENCES

Huisman, G. & Poisson, G.,1966. Les monuments de Paris. Hachette, Paris, pp. 98–99, 169–70

B. Koska, T. Křemen. The combination of laser scanning and structure from motion tecnology for creation of accurate exterior and interior orthophotos of St. Nicolas Baroque Church. In: *International Archives of the Photogrammetry, Remote Sensing and Spatial Information Sciences*, Trento, Italy Volume XL-5/W1, 2013, pp.133-138

D.P. Andrews, J. Bedford, P.G. Bryan, A comparison of laser scanning and Structure from Motion as applied to the great barn Harmondsworth, UK. In: *International Archives of the Photogrammetry, Remote Sensing and Spatial Information Sciences,* Volume XL-5/W2, 2013, pp. 31-36

Westobya M., Brasington J., Glassera N.F., Hambreya M.J., Reynoldsc J.M., 'Structure-from-Motion' photogrammetry: A low-cost, effective tool for geoscience applications. *Geomorphology* Volume 179, 15 December 2012, pp 300–314

Snavely, N., 2008. Scene reconstruction and visualization from Internet photo collections, unpublished PhD thesis, University of Washington, USA.

Saleri R., Cappellini V., Nony N., Pierrot-Deseilligny M., Bardiere E., De Luca L., Campi M., 2013, UAV Photogrammetry for archaeological survey: The Theaters area of Pompeii, in A. C. Addison, L. De Luca, G. Guidi, S.Pescarin, "*Digital Heritage*", Institute of Electrical and Electronics Engineers (IEEE), Danvers, MA (USA)

This contribution has been peer-reviewed.
doi:10.5194/isprsarchives-XL-5-W4-423-2015

INTEGRATION OF POINT CLOUDS ORIGINATED FROM LASER SCANER AND PHOTOGRAMMETRIC IMAGES FOR VISUALIZATION OF COMPLEX DETAILS OF HISTORICAL BUILDINGS

Cihan Altuntas

Selcuk University, Engineering Faculty, Department of Geomatics, Alaeddin Keykubat Kampüsü, 42075 Selcuklu, Turkey - caltuntas@selcuk.edu.tr

Commission V, WG V/4

KEY WORDS: Dense Point Cloud, Image Based Point Cloud, Laser Scanning, Photogrammetry, Integration, Three-Dimensional Model,

ABSTRACT:

Three-dimensional (3D) models of historical buildings are created for documentation and virtual realization of them. Laser scanning and photogrammetry are extensively used to perform for these aims. The selection of the method that will be used in three-dimensional modelling study depends on the scale and shape of the object, and also applicability of the method. Laser scanners are high cost instruments. However, the cameras are low cost instruments. The off-the-shelf cameras are used for taking the photogrammetric images. The camera is imaging the object details by carrying on hand while the laser scanner makes ground based measurement. Laser scanner collect high density spatial data in a short time from the measurement area. On the other hand, image based 3D (IB3D) measurement uses images to create 3D point cloud data. The image matching and the creation of the point cloud can be done automatically.

Historical buildings include more complex details. Thus, all details cannot be measured by terrestrial laser scanner (TLS) due to the blocking the details with each others. Especially, the artefacts which have complex shapes cannot be measured in full details. They cause occlusion on the point cloud model. However it is possible to record photogrammetric images and creation IB3D point cloud for these areas. Thus the occlusion free 3D model is created by the integration of point clouds originated from the TLS and photogrammetric images.

In this study, usability of laser scanning in conjunction with image based modelling for creation occlusion free three-dimensional point cloud model of historical building was evaluated. The IB3D point cloud was created in the areas that could not been measured by TLS. Then laser scanning and IB3D point clouds were integrated in the common coordinate system. The registration point clouds were performed with the iterative closest point (ICP) and georeferencing methods. Accuracy of the registration was evaluated by convergency and its standard deviations for the ICP and residuals on the control points for the georeferencing method.

1. INTRODUCTION

Documentation of historical buildings are very important for rebuilt and reconstruction of them when they are destroyed. Their reconstruction with the keep their original shapes and size is possible with their three-dimensional (3D) models. The creation of their 3D digital model is effective methods on documentation of cultural heritage. It requires measurement 3D spatial data in a particular space from the whole object surface. Photogrammetry is traditional methods on documentation of historical artefacts (Remondino, 2014; Bosch et al. 2005; Toz and Duran, 2004; Lingua et al., 2003). However the measurement to all details wit particular spaces by the photogrammetry require more time and labour. Thus some characteristic points which represent the shape are measured from the images. On the other hand laser scanning has been extensively used to creating 3D model of cultural heritage for two decades (Altuntas et al., 2014; Akca et al., 2006). Terrestrial laser scanner (TLS) collects high density 3D spatial data from the measurement area in a short time. The historical buildings have complex details. All details cannot be measured by TLS since the TLS makes ground based measurement. On the other hand, 3D virtual modelling and realization needs photogrammetric and laser scanning data (Remondino, 2011; Grussenmeyer et al., 2011). Texture data of photogrammetric images are matched with the laser point cloud for creation of 3D virtual model. In addition different measurement techniques have been used for creation 3D model of historical buildings (El-Hakim et al., 2007; Kulur and Sahin, 2008; Chiabrando and Rinaudo, 2013; Altuntas and Yildiz, 2013).

Image based 3D (IB3D) modelling have been performed automatically by the interest points detected from the images. 3D spatial positions and structure from motion are estimated from the matched images in a short time (Chiabrando and Spano, 2013; Grussenmeyer et al., 2012; Barazzetti et al., 2010). The dense IB3D point cloud has been created by matching the image pixels. In this study, IB3D point clouds have been created for the areas that could not be measured by the TLS. Then the point clouds originated from laser scanning and images were integrated by ICP and georeferencing.

2. POINT CLOIUD ACQUISITION METHODS

2.1 Terrestrial laser scanning

TLS measures local coordinates and intensity of the laser light reflected from the measured points. In addition color data can

This contribution has been peer-reviewed.
doi:10.5194/isprsarchives-XL-5-W4-431-2015

be recorded for the measured points. The laser scanner instruments use direct time-of-flight, indirect time-of-flight and triangulation methods to measure distance from instrument to object point. The triangulation laser scanners have small measurement distance because of the inherent characteristics of the instrument. Their measurement distance is about 8 meters. The measurement distance of laser scanner that uses indirect-time-of-flight technique varies between 100 to 300 meters. Finally the measurement distance of laser scanner that uses direct time-of-flight techniques reaches more than three kilometres. On the reverse of the measurement distances, the accuracy of TLS that uses indirect time-of-flight technique is better than that uses direct time-of-flight technique. The specifications and measurement speed of TLS are getting better by the time. Generally, the measurement accuracies of TLSs are a few millimetres in a 100 meter. And they can collect points more than 500000 in per second.

The registration of laser scanning data has been done by control points (Alba and Scaioni, 2007) or point cloud coordinates and photographs (Dold and Brenner, 2006). The many techniques have been executed to registration of point clouds. The iterative closest point (ICP) is extensively used to registration of point clouds (Zhang, 1994). The registration of point clouds by the ICP have been applied iteratively with the conjugate closest points between them. It has high accuracy result. The accuracy of the ICP is a little affected by the space between the measured scan points (Altuntas, 2014).
The main steps on 3D modelling by TLS are below:

-Overlap laser scanning
-Point cloud registration
-Occlusion measurements by other techniques
-Mesh creation
-Texture mapping and 3D virtual model

2.2 Dense image matching

Image based 3D (IB3D) point cloud can be created from the images by dense matching. Hereby high density 3D point positions can be measured with low cost. The procedure needs only camera and software for the processes. Off-the-shelf low cost cameras are used for this aims. Today many, commercial software such as SURE, MicMac, PMVS, Photoscan have been developed for creating IB3D point cloud, dense matching and structure from motion (Remondino et al., 2014). IB3D point cloud generation is attractive method especially for 3D object modelling (Barazzetti et al. 2010, 2011; Remondino et al., 2014) and creation of digital elevation model (Rothermel ve Haala, 2011).

Feature based matching with the interest points has been applied for photogrammetric triangulation of multi-image network. Many interest operators such as Förstner, SUSAN and FAST have been used for matching of images in practices (Jazayeri and Fraser, 2010). In addition, invariant image features (keypoints) have been used to match images. Then many keypoint detection metods such as SURF, ASIFT, BRIEF have been developed after introducing the SIFT. The candidate correspondence for each feature points are find after detecting the feature points from the set of images. The epipolar geometry constraints are used to limit the search area and find good matches (Fathi and Brilakis, 2011). The epipolar geometry is used to detect error matched feature points. It has been applied by RANSAC, least median of squares and maximum posteriori sample consensus (Barazetti et al. 2010). 3D coordinates of the matched features are then calculated. Two sparse point clouds are created from two consecutive stereo pairs, then, camera motion is estimated by registering point clouds using the quaternion motion estimation (Fathi and Brilakis, 2011). After all, bundle adjustment is applied for the block that contains all images matched. So as to creation IB3D point cloud, each pixel is matched with candidate on the other images by the image match techniques. Then 3D point cloud coordinates which represent the map of the image area is calculated by the photogrammetric evaluation. The scale of the point cloud is performed by ratio of distances or coordinates of control points. More information about dense image matching and creation of IB3D point cloud can be found on Brazetti et al., (2010) and Fathi and Brilakis (2011).

3. MEASUREMENT

The studies were performed on front facade of the Sultan Selim mosque, Konya, Turkey (Figure 1). Laser scanning was made from 20m away. The space between the measured points is 2 cm on this measurement distance. Because the facade has more complex details, its fully measurement needs laser scanning from many stations. However, all details should be measured to creating the mesh model. The occluded parts were measured by photogrammetric images and point cloud was created from the images. The images were recorded by Nikon P50 camera (pixel array is 2592x1944, pixel size is 2.14 micron). The record of the images was planned as overlap with the point cloud of TLS. The image pixels (3 micron) corresponds about 1 cm on average imaging distance.

Figure 1. Front facade of the Sultan Selim Mosque.

Two study were performed for creating point clouds from dense matched images. The first study was executed on the left of the front facade. This part includes two window and mihrab among them. Sixty images were recorded to creation IB3D point cloud of this area, and 46130 points were created by Agisoft PhotoScan software (url-1) from dense matched images. The point cloud model was scaled by four control points (Figure 2). The PhotoScan is user friendly and high accuracy software.

The second study was performed on the door that includes more complex shapes. The IB3D point cloud was created from 57 images and 3D positions of 2 million points were calculated (Figure 3). The scale has been executed by four control points.

Figure 2. Dense IB3D point cloud. The first study.

Figure 3. Dense IB3D point clouds of the door. The second study.

4. DATA INTEGRATION

Laser scanning and IB3D point clouds were integrated by the registration with the ICP method. The registration was implemented by PolyWorks software (url-2). The laser scanning data was selected as a reference and the IB3D point cloud was registered relation to reference system.

Figure 4. The TLS point cloud is reference. The IB3D point cloud of the first study was registered into reference system with ICP. The scale bar shows low divergences. [m]

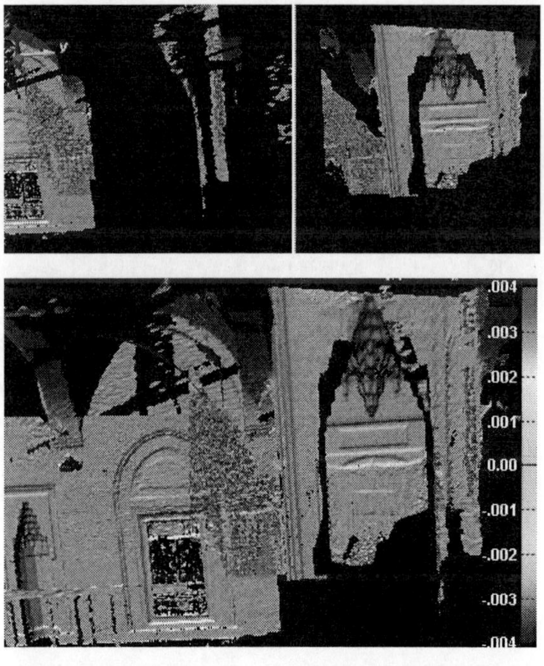

Figure 5. The TLS point cloud is reference. The IB3D point cloud of the second study was registered into reference system with ICP. The scale bar shows low divergences. [m]

This contribution has been peer-reviewed.
doi:10.5194/isprsarchives-XL-5-W4-431-2015

IB3D point cloud of the first study was registered using ICP with 0.003mm mean convergence and 6.4mm standard deviation. The differences between the conjugate points are very low after the registration (Figure 4). Similarly, the IB3D point cloud of the second study was registered to the reference system with mean convergence of 0.059mm and standard deviation of 8.7mm (Figure 5).

Another method for the integration of laser point cloud and IB3D point cloud is registration of each point cloud data into the georeference coordinate system. TLS point cloud and IB3D point cloud has been registered into the geodetic coordinate system by five and four control points respectively. Similarly, IB3D point cloud of the door has been registered by five control points. Thus the TLS point cloud and IB3D point cloud has been integrated by registering into the geodetic coordinate system. On this stage of the processes, we want to compare integration methods of point clouds by ICP and georeferencing. The TLS point cloud was accepted as a reference and ICP was implemented to the georeferenced IB3D point cloud. The changes on the control points were exploited to evaluate the registration methods (Tablo 1).

Table 1. Differences on the control points after applying ICP to the georeferenced point cloud.

	Control points#	Average Differences (m)			Standard Deviations (m)		
		d_x	d_y	d_z	σ_{dx}	σ_{dy}	σ_{dz}
Case1 (left of the front facade)	4	-0.04	-0.05	-0.06	0.02	0.00	0.00
Case2 (Door)	5	-0.07	-0.00	-0.07	0.03	0.01	0.01

5. CONCLUSION

Laser scanner measures 3D spatial data in short time. Nevertheless, all details cannot be measured by TLS. Thus, un-imaged parts of the objects should be measured with the other methods. Dense point cloud creation from the images is the method that can be used as individual or complementary for laser scanning in documentation of cultural heritage. In this study, the surfaces that could not be measured by laser scanner are measured by creating point cloud from dense matched images. Today, many software have been used for dense matching and point cloud creation of the images. PhotoScan is user friendly software. TLS and IB3D point clouds have been integrated by ICP and georeferencing methods. The results of the both methods are satisfactory. According to the results obtained from the comparison, the accuracy of ICP is a little high then the georeferencing. The individual georeferencing of point clouds obtained from TLS and images requires control points that needs more time and labor.

REFERENCES

Akca, D., Remondino, F., Novàk, D., Hanusch, T., Schrotter, G., and Gruen, A., 2006. Recording and modeling of cultural heritage objects with coded structured light projection systems. *2nd International Conference on Remote Sensing in Archaeology*, Rome, Italy, December 4-7, pp.375-382.

Alba, M., Scaioni, M., 2007. Comparison of techniques for terrestrial laser scanning data georeferencing applied to 3-D modelling of cultural heritage, in Proc. ISPRS 3D-ARCH 2007: Virtual Reconstruction and Visualization of Complex Architectures, In: *The International Archives of the Photogrammetry, Remote Sensing and Spatial Information Sciences*, ETH Zurich, Switzerland, Vol. XXXVI-5/W47, p. 8.

Altuntas, C., 2014. The effect of point density on the registration accuracy of a terrestrial laser scanning dataset. *Lasers in Engineering*, 28(3-4), pp. 213-221.

Altuntas, C., Yildiz, F., Baygul, E., 2014. Documentation of historical structures in the courtyard of Mevlana Museum by terrestrial LIDAR and Photogrammetry. *Mediterranean Archaeology and Archaeometry*, 14(2), pp. 249-258.

Altuntas, C., Yildiz, F., 2013. The registration of point cloud data from range imaging camera. *Geodesy and Cartography*, 39(3), pp. 106-112.

Barazzetti, L., Binda, L., Scaioni, M., Taranto, P., 2011. Photogrammetric survey of complex geometries with low-cost software:Application to the G1 temple in Myson, Vietnam. *Journal of Cultural Heritage*, 12(3), pp. 253-262.

Barazzetti, L., Scaioni, M., Remondino, F., 2010. Orientation and 3D modeling from markerless terrestrial images: combining accuracy with automation. *The Photogrammetric Record*, 25(132), pp. 356–381.

Bosch, R., Kulur, S., Gulch, E., 2005. Non-metric camera calibration and documentation of historical buildings. *CIPA XXth International Symposium*, Torino, 25 Sep-1 Oct, p.6.

Chiabrando, F., Rinaudo, F., 2013. TOF Cameras for Architectural Surveys. F. Remondino and D. Stoppa (eds.), *TOF Range-Imaging Cameras*, DOI: 10.1007/978-3-642-27523-4_7, Springer-Verlag Berlin Heidelberg. pp.139-164.

Chiabrando, F., Spano, A., 2013. Points clouds generation using TLS and dense matching techniques. A test on approachable accuracies of different tools. *ISPRS Annals of the Photogrammetry, Remote Sensing and Spatial Information Sciences*, Strasbourg, France, Vol. II-5/W1, pp. 67-72.

Dold, C. and Brenner, C., 2006, Registration of terrestrial laser scanning data using planar patches and image data. In: *The International Archives of the Photogrammetry, Remote Sensing and Spatial Information Sciences*, Dresden, Germany, Vol. XXXVI Part 5, pp. 78-83.

El-Hakim, S., Gonzo, L. Voltolini, F., Girardi, S., Rizzi, A., Remondino, F., Whiting, E., 2007. Detailed 3D modelling of castles. *International Journal of Architectural Computing*, 5(2), pp. 199-220.

Fathi, H., Brilakis, I., 2011. Automated sparse 3D point cloud generation of infrastructure using its distinctive visual features, *Advanced Engineering Informatics*, 25(4), pp. 760-770.

Grussenmeyer, P., Alby, E., Landes, T., Koehl, M., Guillemin, S., Hullo, J.-F., Assali, P., Smigiel, E., 2012. Recording approach of heritage sites based on merging point clouds from high resolution photogrammetry and terrestrial laser scanning.

This contribution has been peer-reviewed.
doi:10.5194/isprsarchives-XL-5-W4-431-2015

In: *The International Archives of the Photogrammetry, Remote Sensing and Spatial Information Sciences*, Melbourne, Australia, Vol. XXXIX-B5, pp. 553-558.

Grussenmeyer, P., Alby, E, Assali, P., Poitevin, V., Hullo, J.F., Smiciel, E., 2011. Accurate documentation in cultural heritage by merging TLS and high-resolution photogrammetric data. *Proc. SPIE 8085, Videometrics, Range Imaging, and Applications XI*, June 21, 808508, doi:10.1117/12.890087, http:/ /dx.doi.org/10.1117/12.890087

Jazayeri, I, Fraser, C.S., 2010. Interest operators for feature-based matching in close-range photogrammetry, *The Photogrammetric Record*, 25(129), pp. 24-41.

Kulur, S., Sahin, H., 2008. 3D cultural heritage documentation using data from different sources. In: *The International Archives of the Photogrammetry, Remote Sensing and Spatial Information Sciences*, Beijing, China, Vol. XXXVII-B5, pp. 353-356.

Lingua, A., Piumatti, P., Rinaudo, F. 2003. Digital photogrammetry: A standard approach to cultural heritage survey. In: *The International Archives of the Photogrammetry, Remote Sensing and Spatial Information Sciences*, Ancona, Italy, Vol. XXXIV-5/W12, pp. 210-215.

Remondino, F., Spera, M. G., Nocerino, E., Menna, F., Nex, F., 2014. State of the art in high density image matching. *The Photogrammetric Record*, 29(146), pp. 144-166.

Remondino, F., 2014. *Photogrammetry - Basic Theory*. In "3D Recording and Modelling in Archaeology and Cultural Heritage - Theory and Best Practices", F. Remondino and S. Campana (Eds), Archaeopress BAR Publication Series 2598, ISBN 9781407312309, pp. 63-72.

Remondino F., 2011. Heritage recording and 3D modeling with photogrammetry and 3D scanning. *Remote Sensing*, 3(6), pp. 1104-1138.

Rothermel, M., Haala, N., 2011. Potential of Dense Matching for the Generation of High Quality Digital Elevation Models, ISPRS Hannover Workshop 2011: High-Resolution Earth Imaging for Geospatial Information, IASPRS, Vol. XXXVIII-4/W19, p.6.

Toz, G., Duran, Z., 2004. Documentation and analysis of cultural heritage by photogrametric methods and GIS: A case study. In: *The International Archives of the Photogrammetry, Remote Sensing and Spatial Information Sciences*, Istanbul, Vol. XXXV, Part B5, pp. 438-441.

url-1: http://www.agisoft.com/

url-2: http://www.innovmetric.com/

Z. Zhang, 1994. Iterative point matching for registration of free-form curves and surfaces. *International Journal of Computer Vision*, 13(2), pp. 119–152.

THE ACCURACY REQUIREMENTS AND SOURCES FOR 3D RECONSTRUCTIONS OF THE PREHISTORIC ARCHAEOLOGICAL SITES: THE CASE OF AGIOS ANTONIOS CHOMATAS (CRETE)

T. Alusik [a], D. Sovarova [b]

[a] Czech Centre for Mediterranean Archaeology/Institute for History of Medicine and Foreign Languages, First Faculty of Medicine, Charles University in Prague, Czech Republic – alusikt@seznam.cz
[b] Faculty of Civil Engineering, Czech Technical University in Prague, Czech Republic – daniela.sovarova@gmail.com

Commission V, WG V/4

KEY WORDS: Accuracy requirements, 3D reconstruction, Crete, Bronze Age, Archaeology

ABSTRACT:

The use of 3D sophisticated visualizations and reconstructions of any preserved architectural remains is becoming more frequent in the modern archaeology. However, it is still not common during the process of reconstruction or recreation of the appearance of any preserved architecture of prehistoric (Bronze Age) Crete. The bad condition of prehistoric sites (the bad preservation of the masonry) or structure is probably the main reason for this state. But let's look at this problem from the other point of view: what is really necessary for the creation of an ideal 3D reconstruction of such sites, which complies with the up-to-date state of the archaeological and architectural knowledge and uses the modern software? In accordance with our experience and findings we think that for the creation of really good and up-to-date 3D reconstruction only the plan of the site (at least the schematic one), enough of general and detailed photos of the site (but not necessary the photogrammetric model) and architectural features, some photos of the site's surroundings incl. the Google Earth image and some photos of similar sites and/or structures/architectural features (type of masonry, entrances etc.) are necessary. In case of accepting our findings and principles by the wider academic audience (especially by the archaeologists and architects) the creation of 3D reconstruction could be – as we hope – more commonly used or even become a standard of archaeological publications.

1. INTRODUCTION

1.1 Virtual reconstructions of archaeological sites

The use of 3D sophisticated visualizations and reconstructions of any preserved architectural remains is becoming more frequent in the modern archaeology. In this way, some ancient cities (e.g. Rome reborn - http://www.romereborn.virginia.edu/; Byzantium1200 - http://www.byzantium1200.com/index.html) or individual structures (e.g. various ancient structures in Greece, Italy, Asia Minor and North Africa - http://www.klassische-archaeologie.uni-hd.de/forschung/3Dmodelle.html) were reconstructed. In some archaeological specializations, for example in Minoan archaeology which is concerned with the studies of prehistoric Crete (especially from the end of the Neolithic till the end of the Bronze Age, i.e. approximately 2nd half of 4th millennium – ca 1100/1050 BC), the 3D models created by computers are still applied relatively rarely. The first example of such reconstructed Minoan structure, as far as the authors know, appeared on the website of the Institute for Classical Archaeology of Friedrich-Alexander Universität Erlangen-Nürnberg in June 1998 (http://www.aeria.phil.uni-erlangen.de/ausstellung_html/lectures_html/ archanes/archanes_1.html).

Referring to the development of more and more sophisticated software for architects and for general 3D modelling of any objects, the quality of few recent years 3D models has certainly been increased. For example in the years of 2009-13 a team managed by K. Glowacki (see e.g. Glowacki – Dafedar 2010) created a virtual reconstruction of the settlement of Kavousi Vronda in Eastern Crete from the end of the Bronze/Minoan Age (ca 1200-1100 B. C.). It is dealt with the first example of computer 3D reconstruction of the whole settlement. In the course of years the series of virtual flyovers of the site (at the same time it is the first example of such initiative), have been placed on the server of Youtube.com (search for "Vronda"). Probably the most elaborated 3D recinstruction was presented by C. Papadopoulos (and G. Earl) in 2009-10 (see e.g. Papadopoulos – Earl 2009; Papadopoulos 2010). These series of visualizations of an important Minoan cemetery of Phourni (3rd and 2nd millennium BC) focus on two burial structures – sc. Tholos C and Burial Building 19 – and possess also interior views, even with very interesting details such as illumination simulation. In 2012, G. Vavouranakis published on the website of "Aegeus-Society for Aegean Prehistory" 3D reconstructions of three sites in Eastern Crete – Gournia-North Cemetery, cemetery on a little islet near the village of Mochlos (both dated to 3rd millennium BC) and the peak sanctuary of Petsofas (2nd millennium BC). Final reconstructions (in more versions) were set in the photographs of the sites in order to achieve the highest possible level of reality (Vavouranakis 2012). All the 3D reconstructions of Minoan sites mentioned hitherto have admittedly been the results of all the available sources exhaustive analysis, but they have often been realized without more considerable involvement of architects.

This contribution has been peer-reviewed.
doi:10.5194/isprsarchives-XL-5-W4-437-2015

The bad condition of the sites (the bad preservation of the masonry) or individual structures is probably the main reason for the low number of ideal 3D reconstructions of Cretan prehistoric sites or structures. The creation of such reconstruction is then very difficult and one has to rely during the reconstruction process on the testimony of the iconographic (representation of the architecture from the area and period in question), archaeological (the similar sites or structures), ethnographical (in case of the preservation of any architectural traditions) and literary sources (if there are any preserved).

This state and our previous experience with the creation of the ideal 3D reconstructions (Alusik et al. 2011; Alusik – Sosnova 2014; Sovarova – Alusik 2013) made the authors (an archaeologist – T.A. – and an architect – D.S.) to look at this problem from the other point of view: what is really necessary for the creation of an ideal 3D reconstruction of such sites, which complies with the up-to-date state of the archaeological and architectural knowledge and uses the modern software? The case study for the discussion of our findings will be the creation of the 3D reconstruction of a small rural Middle to Late Minoan/Bronze Age (2nd mill. BC) site of Agios Antonios Chomatas Site 9 (Haggis 2005, 101-102, fig. 34A), consisting of a single structure and an enclosure wall and located near the village of Kavousi on the north coast of East Crete (Fig. 1-3). Original presumptions for creating the 3D ideal reconstruction of any site or an individual construction is, however, such a preservation of a ground plan that enables – after a relevant hypothetical completion with maximally few unpreserved places –understanding of the basic architectural layout and, at the same time, perfect knowledge of the site based on a personal visit.

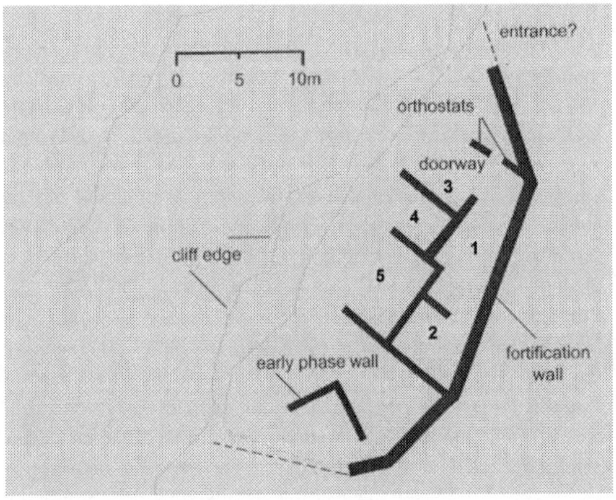

Figure 1. Agios Antonios Chomatas Site 9 groundplan

1.2 Agios Antonios Chomatas Site 9

The site (Fig.) has been discovered within Kavousi-Thrifti Survey project and has not been excavated until the present (Haggis 2005, 101-102, fig. 34A). It is located on northern coast of Eastern Crete, to the north of Kavousi village, on a plateau of the mountain massif of Chomatas, which rises up directly on the coast of Mirabello bay. Its western side is formed by cliffs, eastern side by steep slope which is formed into terraces in its higher parts. Remains of a big building of a rural character have been preserved on one of the plateaus in the size of approximately 30 x 20 m southwards and slightly below the mountain peak (the site altitude is 213.3 m). It was enclosed by a wall along the northern, eastern and southern edges of the plateau. Unfortunately, the basement and walls of the building have been preserved to the height of maximum few dozens of centimetres and were built from rough or only partially worked massive blocks of grey limestone with the length of up to 1 m in the technique of the sc. Cyclopean or oncolithic masonry. The entrance to the walled premises of the site was probably placed in the north; the entrance into the building itself is bordered with two vertical stone blocks (orthostates). The building itself (see Fig.) consists of five main rooms and a short part of a wall to the south of it which is interpreted as remains of an older settlement phase – it is considered as a simple farm-house in our reconstruction.

With respect to its strategic and defensible position – the site has a perfect view of the whole Mirabello bay, of northwardly located islet of Pseira (with a significant Minoan settlement) and of Kambos lowland eastwardly – the enclosure wall is interpreted as a fortification what the both authors agree on and reconstruct it as a high, mostly stone wall without windows with the only entrance.

Figure 2. Agios Antonios Chomatas Site 9: the actual state of the site

Older, poorly documented settlement phase (and probably the architectural one as well) dates back to the beginning of the Middle Bronze/Minoan Age (Middle Minoan I-II phases, late 3rd (?) – early 2nd millennium BC), the main phase(s) – judging by the absolute majority of ceramics preserved on the surface – however, to the beginning of the Late Bronze/Minoan Age (Late Minoan I phase, ca 17th – 16th century BC).

2. RECONSTRUCTION PROCESS

The 3D reconstruction itself has been carried out in the architectural software of ArchiCAD, version 14 (Fig. 7-10). The presented visualisations are rather simple mass studies placed on a terrain created by software with schematically designated basic topographic characteristics. The details of architecture have not been too elaborated in details yet – software pre-set basic textures have been used (rough irregular stone; unburnt bricks; clay) and window and door openings have been left

This contribution has been peer-reviewed.
doi:10.5194/isprsarchives-XL-5-W4-437-2015

without any filler. Their final solution and more suitable specification of façade and roof surface textures will be subject to further project stage.

There were the following stages of the 3D reconstruction process:

1) collecting of archaeological and architectural sources – site layout; site photographs; site surrounding pictures from Google Earth; photographs and plans of similar (architecturally and chronologically) Minoan sites; original (it means from the Bronze Age) iconographical sources (representations of Minoan/Aegean structures and their construction details); architectural-ethnographical parallels (photographs of single farm houses and their remains from 19th to 20th centuries); and, last but not least, also some of already existing 3D reconstructions of Minoan structures
2) architect's acquaintance with the site, sources and basic facts on Minoan archaeology and architecture
3) analysis of sources and creation of the basic conception of the reconstruction (by both authors)
4) creation of 3D reconstruction (by the architect, with the archaeologist's assistance).

Herein presented 3D reconstruction is based on in-depth analysis of all available archaeological, iconographical and architectural sources, supported by functional architectural analysis of the building interior rooms from the view of the access into and movement inside the building and the purpose of individual rooms. Even though it is dealt with a rough mass study without final surface finish and architectural details – and possibly an insertion into a photograph – our final 3D reconstruction corresponds with the most actual state of Minoan archaeology and architecture knowledge.

Individual initial sources will be presented in detail in the following chapters, then basic points of final 3D reconstruction will be characterised, and the role and benefit of the architect will be discussed. Based on these, essential initial presumptions and sources will be consequently defined (including the attempt to define a sort of "minimal standard") for scientifically precise 3D reconstruction that reflects the most up-to-date state of the research.

3. REQUIREMENTS AND SOURCES FOR THE RECONSTRUCTION PROCESS

3.1 Site plan

There exists a sketch plan (i.e. not the stone-by-stone plan) of the site created and published by its discoverer D. Haggis (see above, Fig. 1). One of the authors (T.A.) visited the site in person in 2012 and except for photographs he made there a rough drawing of the preserved architecture.

The existence of a precise plan of the site or the construction is the basic precondition for its ideal 3D reconstruction creation. With respect to frequently difficult circumstances of creating such a plan (administrative procedures, the time-consuming and expensive fieldwork and processing of the results) it is more suitable to choose a site or construction with an already existing plan to create 3D reconstruction. Stone-by-stone plan enables to understand better (and to reconstruct) some of construction elements, but for the creation of the visualization itself even only the sketch plan is sufficient.

3.2 Photographs of the site and its surroundings

The discoverer has published only two photographs: a view of the cliff from the distance where the site is located; and one overall view of the site surface where the state of preserved remains of architecture is visible.

Figure 3. Agios Antonios Chomatas Site 9: the actual state of the site

Therefore several dozens of photographs of the site and its closest surroundings have been taken (both detail and overall views; see Fig. 2-3). To understand some topographic details and general local context the site have been localised in Google Earth application and several pictures of its surroundings up to the distance of approximately 3 km from it have been taken (Fig. 4). According to authors' opinion, it is not possible to create an authentic 3D reconstruction without a thorough photo-documentation and knowledge of the site and its surroundings, because different specific elements or details indeed reflect a specificity of a certain environment. The particular architectural solution could have been influenced not only by the place, where the site or building was precisely located, but also by the position towards the surrounding landscape as well.

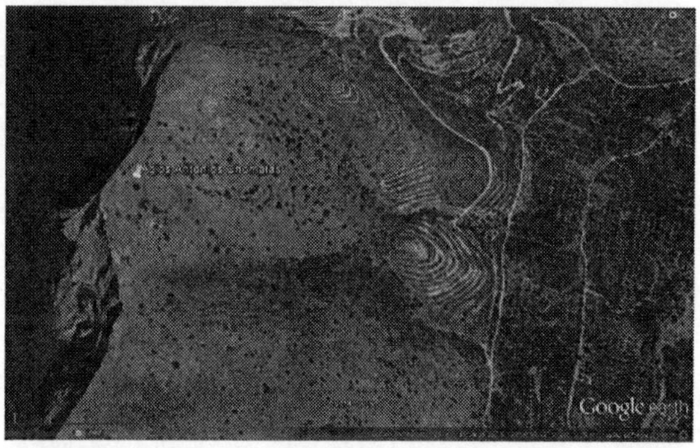

Figure 4. Agios Antonios Chomatas Site 9: Google Earth image of the site's vicinity

This contribution has been peer-reviewed.
doi:10.5194/isprsarchives-XL-5-W4-437-2015

3.3 Archaeological sources

As for the archaeological sources, it is inevitable to mention especially an extensive group of similar rural constructions (or sites) built up especially from rough or partially worked massive blocks of local stone (limestone, basalt or conglomerate; this construction technique is being labelled as "Cyclopean" or "oncolithic" masonry). But mudbricks were also used (most likely for higher storeys and parts of buildings and walls) as indicated by the findings of these bricks fragments on the surface of many sites. They are most frequently labelled as rural villas, megalithic farmsteads (an overview and typology of these constructions are presented by Schlager 2006) or guard houses where agricultural and/or military function is assumed (basic information on this construction type and their typology is mentioned by Tzedakis et al. 1989, Chryssoulaki 1999 and Alusik 2007, 124-35; see Fig. 5). Unfortunately, most of them have not been documented in detail yet (there is at least a sketch plan of a few dozens of them and only a few of them were at least partially excavated). These sites come from the Middle and Late Bronze/Minoan Age, thus from 2nd millennium BC. Some sections of masonry have been preserved in several sites up to the height of 2 metres. Houses preserved in the site of Akrotiri on Thera belong among other important architectural sources (see e.g. Marinatos 1968, 1969, 1970, 1971, 1972, 1974, 1976; see Fig. 6). A town with very fine houses (which are labelled with a Greek term "xestai") built from half-timbered masonry, which façades have been preserved up to the height of several storeys, were covered by the volcanic sediments during a volcano eruption at the beginning of the Late Bronze/Minoan Age (Late Minoan I phase, 2nd half of 17th century BC). Of course, stone for ground floor and mudbricks, or the sc. rubble masonry for higher storeys were used as construction materials. All the walls were reinforced with wooden beams to get the better fixity and stability. "Xestai" present us lots of architectural details such as windows and doors, staircases and other interior details.

Figure 5. "Guard house" of Choiromandres, East Crete

In the last years, the first author has visited several dozens of sites with above-mentioned rural buildings and has taken hundreds of photographs there. With respect to the relatively bad state of preservation of most Agios Antonios Chomatas Site 9 walls, the photographs of several best preserved rural sites have served as an initial point for our reconstruction and from this we have derived a type of masonry, way of construction and some another knowledge concerning the interior rooms. The first author has also visited Akrotiri. The photographs of local xestai provide an evidence of design and construction of more quality and advanced architecture than is in our site in question. Therefore they will be of bigger importance only in the next stage of our project where the design of some surfaces and constructional elements (windows, doors, etc.) will be finally defined.

Figure 6. Akrotiri on Thera, view of the site

3.4 Iconographical sources

Quite a number of architectural representations have been preserved in Minoan art that mostly date back to the end of Middle and the beginning of Late Bronze/Minoan Age (2nd millennium BC) and they can become a valuable source when creating ideal 3D reconstructions of sites or structures. They show us a range of important architectural details (such as shapes of windows and doors, design of columns), they indicate materials used (stone, bricks, wood) and most of all they capture the shape of superstructure of the building including the number of floors. With respect to the prevailing religious aspect of Minoan iconography the sacral architecture or whole towns were depicted most often. The towns represented served sometimes as the background of religious scenes. In many cases the architecture is also depicted in a simplified way, or the whole construction is represented only by a column or a pillar. The best depictions of architecture have naturally been preserved in Minoan frescoes (including the Minoanizing frescoes from Aktoriri on Thera), but some architectural models, relief plates and seals or their imprints are also important. But it is neither necessary to mention an overview of at least the most important depictions of architecture herein, nor to discuss their exactness and benefit to the studies of Minoan architecture; this was a subject to lots of studies of other authors (e.g. Hallager 1985; Kontorli-Papadopoulou 1996; Nörling 1995).

But in our reconstruction – with respect to the specifics and rural character of the site – we have used the iconographic sources in a minimal way. They will be of a bit bigger value in a further stage of the project where we are going to focus on the final surface finish and filling elements.

This contribution has been peer-reviewed.
doi:10.5194/isprsarchives-XL-5-W4-437-2015

3.5 Architectural-etnographical sources

The first author has seen and taken photographs of many examples of traditional Cretan rural buildings which were built in 19th and 20th centuries. They were placed in the villages or they were built separately in the countryside. Lots of them have been in ruins but some of them have been still used.

Such buildings are a good example of local architectural traditions and reactions to specific conditions concerning topography, weather and available material. They represent long-lasting experience of Cretans with the life and conditions on the island and reflect a pursuit of the simplest and most effective possible way of solving the problems that can substantially influence a house construction, such as subsurface, terrain slope, prevailing direction and strength of the winds, sunshine etc. In our case the photographs of them have been used only to verify some of our initial presumptions and final hypotheses (i.e. overall building orientation or the shape of the roof), they were of a minimum meaning (rather a supportive one).

3.6 The already existing 3D reconstructions of Minoan structures or sites

Before beginning the reconstruction process itself several previously made 3D reconstructions of other Minoan buildings had been collected. But they were only used to find out the way of other specialists' work – which software they used (for architects or just for general 3D modelling), which way the terrain was indicated, if the 3D reconstruction was set into the site photograph etc. – and also to be inspired for the overall final finish of our visualisation.

4. THE FINAL 3D RECONSTRUCTION

The basic concept of the reconstruction has been created after a thorough analysis of all above-mentioned sources and assessment of their contribution. The basic scope of the building is certainly given by the size of a plateau, preserved ground plan and enclosure wall. After a detailed study of all architectural and topographic characteristics of the site we have decided to reconstruct the main building as a partially two-storey, the second floor has been built above the western part of the building towards the sea. A long space in the eastern part of the site adjacent to the bounding wall has been reconstructed as a walkable roof or a "terrace" where the rooms of the second floor open to. The ground floor of all the buildings in our reconstruction has been carried out in stone as massive irregular blocks, the upper floor from mudbricks reinforced with wooden beams (but these are not visible in the final images). A short section of the wall southward from the building has been interpreted by the authors as a simple production extension – workshops, kitchens or any similar room(s) which were better to situate outside the main rural building due to spatial and practical reasons (i.e. noise, smell).

The final ideal 3D reconstruction (Fig. 7-10) of the site has been based on the above-mentioned basic concept supported by a functional architectural analysis of interior rooms of the building from the viewpoint of the access into and movement in the building and the purpose of individual rooms. Therefore, according to our opinion, there were mostly magazines on the ground floor (eastern parts; rooms 1-2) and workshops/kitchens (rooms 3-4) or the room (no. 5) were the inhabitants stayed during the day. The dwelling rooms themselves were situated on the first floor, over the western part of the ground floor (rooms 3-5). The open space in the eastern part of the first floor (above the stocks no. 1-2) served as a balcony and, at the same time, as a defensive platform from the inside of the fortification.

Figure 7. Agios Antonios Chomatas Site 9: ideal 3D reconstruction of the site

Based on the sources analysis and summary of the archaeological facts the basic points of the final 3D reconstruction have been defined by both the authors. With respect to the fact the site has not been excavated, it is not possible to deduce the purpose and equipment of individual rooms based on the analysis of the findings. Therefore the only possibility was to carry out a functional architectural analysis that was nearly exclusively in the architect's management. The final alternations of rough mass 3D reconstruction – such as the placement of th ewindows, doors and staircase(s) and setting of the interior communication pattern – have been carried out by the architect as well. Her fundamental benefit consists in the application of static, construction, functional and technological knowledge and rules. Herewith, an application of purely "construction-technological " aspect, it means verification, clarification and completion of imaginations of functional, static and engineering-constructional functioning of the building, will be used in the process of creation of the ideal 3D reconstructions of Minoan tructures, together with "artistic-historical" and "archaeological" viewpoints.

5. CONCLUSIONS

The authors assume that it is necessary to apply the latest computer technologies and sophisticated software in modern archaeology. According to their opinion, it should become a standard in the near future that in case of publishing the sites with preserved architecture, an ideal 3D reconstruction should be created, even though sometimes objective archaeological reasons prevent these aims to be carried out. As was mentioned above, the initial presumptions for creation of such a reconstruction must be a preservation of the ground plan in such a condition which enables to understand the architectural dispositions and also thorough knowledge of the site based on a personal visit.

After the actual reconstruction process was finished, the authors were able to finally evaluate the meaning, potential and value of the individual categories of the sources for the reconstruction. It was absolutely clear that some (categories of the) sources had been important and absolutely necessary, but the other ones had had only the character of an „added value", i.e. had helped to find an easier architectural element or expression. On the basis of their experience and findings the authors believe that for the creation of really good and up-to-date 3D reconstruction of a rather simple structure used primarily for habitation (that means e.g. an usual town or rural house, farmstead etc.) only the following three categories of sources are necessary: 1) the plan of the site (at least the schematic one; Fig 1); 2) enough of general and detailed photos of the site (but not necessary the photogrammetric model), its architectural features and some photos of the site's surroundings incl. the Google Earth image (Fig. 2-4); and 3) some photos of similar sites and/or structures/architectural features (type of masonry, entrances etc.; Fig. 5-6). These three categories of sources, the satisfactory state of preservation of the site and good personal knowledge of the site and its close vicinity can be defined as the minimal standards for the creation of an ideal 3D reconstruction (Fig. 7-10). The actual reconstruction in the special software should be done by the architect – and not by just a specialist in 3D software modelling – in close collaboration with an archaeologist. Only the architect is able to fully cope with the functional and structural rules and practical aspects and details of the reconstructed structures, i.e. statics, inner partitions, air-flow, and structural details such as windows, doors, staircases etc. So, in authors' opinion, there are not so many accurate sources necessary for the creation of the up-to-date 3D reconstructions.

The authors hope that this paper contribute to the discussion what are really the prerequisites, sources and accuracy requirements for the creation of the scientifically reliable 3D reconstruction of the prehistoric archaeological sites or structures. In case of accepting our findings and principles by the wider academic audience (especially by the archaeologists and architects), the creation of software 3D reconstruction could be – as we hope – more commonly used or even become a standard of archaeological publications.

ACKNOWLEDGEMENTS

The authors would like to thank Prof. Donald Haggis for the permission to create the 3D reconstructions of the site in question.

REFERENCES

Alusik, T., 2007. *Defensive Architecture of Prehistoric Crete*. Archaeopress, Oxford.

Alusik, T. – Sosnova, A. B., 2014. Möglichkeiten einer 3D-Rekonstruktion der Architektur und der Fundorte auf dem minoischen Kreta. In: E. Trinkl, Hrsg. *Akten des 14. Österreichischen Archäologentages am Institut für Archäologie der Universität Graz vom 19. bis 21. April 2012,* Phoibos Verlag, Wien 2014, pp. 21-31.

Alusik, T. – Chalupka, M. – Kopernicky, D., 2011. 3D Virtual Reconstructions of Minoan Rural Sites: The Case of Livari Cheromylia (East Crete). In: F. Remondino – S. El-Hakim, eds. *Proceedings of the 4th ISPRS International Workshop 3D-ARCH 2011: "3D Virtual Reconstruction and Visualization of Complex Architectures", Trento, Italy, 2-4 March 2011.* International Archives of Photogrammetry, Remote Sensing and Spatial Information Sciences, Volume XXXVIII-5/W16, http://www.int-arch-photogramm-remote-sens-spatial-inf-sci.net/XXXVIII-5-W16/ (25 Jan. 2015).

Chryssoulaki, S., 1999. Minoan Roads and Guard Houses – War Regained. In: R. Laffineur, ed. *Polemos. Le contexte guerrier en Egée a l'age du bronze. Actes de la 7e Rencontre égéenne internationale, Université de Liége, 14 – 17 avril 1998.* Université de Liége, Liége – Austin, pp. 75-84.

Glowacki, K. – Dafedar, S. K., 2010. Modeling Domestic Architecture at Late Minoan IIIC Vronda, Kavousi, Crete. In: *Archaeological Institute of America 111th Annual Meeting Abstracts,* Archaeological Institute of America, Boston, pp. 42-43.

Haggis, D. (with contributions by Michael W. Morris, John T. Ammons, Michael E. Timpson, John E. Foss, Margaret S. Mook, Peter M. Day, Louise Joyner, Evangelia Kiriatzi, and Maria Relaki), 2005. *Kavousi I: The Archaeological Survey of the Kavousi Region*. INSTAP Academic Press, Philadelphia.

Hallager, E., 1985. *The Master Impression.* Paul Åströms Förlag, Göteborg.

Kontorli-Papadopoulou, L., 1996. *Aegean Frescoes of Religious Character.* Paul Åströms Förlag, Göteborg.

Nörling, T., 1995. *Altägäische Architekturbilder.* Philipp von Zabern, Mainz.

Papadopoulos, C. – Earl, G., 2009. Structural and Lighting Models for the Minoan Cemetery at Phourni, Crete. In: Debattista, K. – Perlingieri, C. – Pitzalis, D. – Spina, S. eds. *Proceedings of the 10th VAST International Symposium on Virtual Reality, Archaeology and Cultural Heritage (Malta 22-25 September 2009),* Faculty of ICT, University of Malta, Valletta.

Papadopoulos, C., 2010. *Death management and virtual pursuits: A virtual reconstruction of the Minoan cemetery at Phourni, Archanes, examining the use of Tholos Tomb C and Burial Building 19 and the role of illumination, in relation to mortuary practices and the perception of life and death by the living.* Archaeopress, Oxford.

Marinatos, S., 1968. *Excavations at Thera. First Preliminary Report (1967 Season).* Archaeological Society at Athens, Athens.

Marinatos, S., 1969. *Excavations at Thera II (1968 Season).* Archaeological Society at Athens, Athens.

This contribution has been peer-reviewed.
doi:10.5194/isprsarchives-XL-5-W4-437-2015

Marinatos, S., 1970. *Excavations at Thera III (1969 Season).* Archaeological Society at Athens, Athens.

Marinatos, S., 1971. *Excavations at Thera IV (1970 Season).* Archaeological Society at Athens, Athens.

Marinatos, S., 1972. *Excavations at Thera V (1971 Season).* Archaeological Society at Athens, Athens.

Marinatos, S., 1974. *Excavations at Thera VI (1972 Season).* Archaeological Society at Athens, Athens.

Marinatos, S., 1976. *Excavations at Thera VII.* Archaeological Society at Athens, Athens.

Schlager, N. 2006. "Cyclopean" or "Megalithic" Buildings in East Crete: Distribution, Form, Date, and Function. In: E. Tampakaki and A. Kaloutsakis, eds. *Acts of the 9th Cretological Congress (Elounda 2001), Vol. A3*, Etairia Kritikon Istorikon Meleton, Herakleion, pp. 365-78.

Sovarova, D. – Alusik, T., 2013. Rekonstrukce prehistorických archeologických lokalit na Krétě: Agios Antonios Chomatas Site 9 (Reconstruction of Prehistoric Archaeological Sites in Crete: The Case of Agios Antonios Site 9). In: P. Holubec, ed. *Člověk, stavba a územní plánování 7 (Man, Structure and Urban Planning 7)*, Faculty of Civil Engineering, Czech Technical University in Prague, Prague, pp. 217-25.

Tzedakis, Y., Chryssoulaki, S., Voutsaki, S. and Venieri, Y., 1989. Les Routes Minoennes: Rapport Préliminaire – Défense de la Circulation ou Circulation de la Défense? *BCH,* 113, 43-75.

Vavouranakis, G., 2012. Topography, Architecture and Socio-historical Structure in East Crete, during the Bronze Age. http://www.aegeussociety.org/en/index.php/excavations-and-research/vavouranakis-topography-in-east-crete/ (25 Jan. 2015).

Figure 8. Agios Antonios Chomatas Site 9: ideal 3D reconstruction of the site

doi:10.5194/isprsarchives-XL-5-W4-437-2015

Figure 9. Agios Antonios Chomatas Site 9: ideal 3D reconstruction of the site

Figure 10. Agios Antonios Chomatas Site 9: ideal 3D reconstruction of the site

This contribution has been peer-reviewed.
doi:10.5194/isprsarchives-XL-5-W4-437-2015

ASSESSING PATHOLOGIES ON VILLAMAYOR STONE (SALAMANCA, SPAIN) BY TERRESTRIAL LASER SCANNER INTENSITY DATA

J. García-Talegón[a,*], S. Calabrés[b], J. Fernández-Lozano[a], A.C. Iñigo[c]; H. Herrero-Fernández[c], B. Arias-Pérez[b], D. González-Aguilera[b]

[a] Department of Geology; Faculty of Sciences, University of Salamanca, Salamanca, Spain – (talegon@usal.es, jfl@usal.es)
[b] Department of Cartographic and Land Engineering; High School of Ávila, University of Salamanca, Ávila, Spain – (sergio.calabres@gmail.com, daguilera@usal.es, benja@usal.es)
[c] Spanish National Research Council, Instituto de Recursos Naturales y Agrobiología de Salamanca, Salamanca (CSIC) (adolfo.inigo@irnasa.csic.es, quinto_cicerón@yahoo.es)

[*] Corresponding author

Commission V, WG V/4

KEY WORDS: Pathologies, Cultural Heritage, Villamayor Stone, Terrestrial laser scanner, Photogrammetry, Single-lens reflex camera, Remote sensing, Supervised classification.

ABSTRACT:

This paper deals with the assessing of pathologies in façades using a variety of intensity data provided by different terrestrial laser scanner. In particular, a complex building built in the Villamayor Stone that is to be candidate as a Global Heritage Stone Resource has been chosen as study case. The Villamayor Stone were quarrying for the construction and ornamentation of monuments in Salamanca, declared World Heritage City by UNESCO in 1988. The objective of this paper is to assess the pathologies of Villamayor Stone and compare the results obtained through the laser techniques with the classical techniques of mapped pathologies (i.e. visual inspection). For that intensity data coming from laser scanners will be used as non-destructive techniques applied to the façades and several retired plaques (after of building restoration) of Villamayor Stone with pathologies (fissures, scales, loss of matter, humidity/biological colonization) carried to the laboratory. Subsequently it will perform different comparisons between the accuracy reached with the different sensors and a high precision model setup on laboratory which performs as "ground truth". In particular, the following objectives will be pursued: i) accuracy assessment of the results obtained in *in situ* and laboratory; ii) an automation or semi-automation of the detection of pathologies in Villamayor Stone; iii) discriminate the different types of Villamayor Stone used in the façades in function of the radiometric response; iv) establish a methodology for detection and assessing of pathologies based on laser scanner intensity data applied to monuments and modern buildings built in Villamayor Stone.

1. INTRODUCTION

The conservation of cultural heritage is one of the fundamental concerns and potential of modern society and particularly, historic infrastructures largely characterize modern Europe. Historical constructions present a cultural legacy that needs to be preserved. This generic term includes a wide range of constructions such as residential buildings, religious buildings, or even civil structures such as bridges. Many of these constructions are still in service, which is a proof of their historic and engineering value, so their preservation must be a vital concern of modern and advanced reflective societies. However, when these constructions need to be evaluated from the point of view of engineering, many restrictions make it difficult to obtain accurate and realistic analyses. Among others, the main aspects that characterize the engineering evaluation of historical constructions are:

- Complex and often unique geometry.
- The uncertainty and diversity of the mechanical properties of the materials, and construction technologies.
- Unknown techniques and processes during the staged construction.
- Vulnerability and monumental status of the construction.

Close-range remote sensing has evolved importantly in the last decades in terrestrial environments: new technologies such as laser scanning have significantly contributed to this growth. In particular, terrestrial laser scanners have gained popularity in the last decade related to the tasks involving geometric characterization. The terrestrial laser scanner provides a great volume of information quickly and with high accuracy. The information is formed, at least, by a point cloud with spatial coordinates of the points and metric capabilities. In addition, the main advantage of this technology is their non-destructive nature (Genovese, 2005), together with the crosscutting applications of the technologies to many different fields: archaeology (Lamberts et al., 2007), civil engineering (Lubowiecka, et al., 2009; González-Aguilera et al., 2008), geology (Buckley et al., 2008) and geomorphological analysis (Armesto et al., 2009), among others. Thus, this technology represents a very attractive and powerful tool for the study of historical constructions, but they still need to be explored further. In fact, radiometric information, provided by the intensity measurements in the case of laser technology, is used less frequently, so that its potential for classification tasks and recognition of different materials should be investigated. Recently radiometric information has started to use in the detection of different pathologies in granites (González-Jorge et al., 2012) and detection of humidity and lichens in granite (Rodriguez-Gonzálvez et al., 2013). Although, nowadays several works have been performed in granites and other materials, never an exhaustive study has been realized over the Villamayor Stone.

This contribution has been peer-reviewed.
doi:10.5194/isprsarchives-XL-5-W4-445-2015

The objective of this paper is to assess the pathologies of Villamayor Stone and compare the results obtained with the intensity data coming from laser scanners with the classical techniques of mapped pathologies (i.e. visual inspection). For that it will be used several kind of laser scanners applied to the façades of a complex building "El Tormes" located in Salamanca (Spain) and to several retired plaques (after of building restoration) of Villamayor Stone with pathologies.

The paper has been structured as follows: after this introduction, Section 2 presents the Villamayor Stone as a Global Heritage Stone and its pathologies. Section 3 outlines in details the methods developed; and the final two sections summarize the experimental results and present conclusions drawn from the study.

2. MATERIALS

2.1 Villamayor Stone: Heritage Stone

In Salamanca (Spain), which was declared a World Heritage City by UNESCO in 1988, the construction of the principal religious and civil monuments maintains a fixed pattern in the use and position of the Villamayor Stone that is used. In the lower parts of monuments and columns (base, shaft and capital) other types of Heritage Stones have been used such as silicified conglomerates, Sandstone Salamanca Formation (Tosca Stone); monzogranite, Granite of Los Santos; and Leucogranite with tourmaline of Martinamor (Stone of Vaugnerite) (García-Talegón et al., 2014; López-Plaza *et al.*, 2007a, b).

Villamayor Stone is an arkosic stone of the Middle Eocene age and belongs to the Cabrerizos Sandstone Formation that comprises braided fluvial systems and paleosoils at the top of each stratigraphic sequence. The sandstone is known by several names: i) It is known locally and in historical documents as Franca Stone; ii) Villamayor Stone because the quarries are located in Villamayor de Armuña; iii) the Golden Stone due to its patina that produces a ochreous/golden colour on the façades of monuments of Salamanca which are built with this natural stone. Villamayor Stone (Golden Stone) is presented as a candidate to be designated Global Heritage Stone Resource (García-Talegón et al, 2014). Currently, Villamayor Stone is using for cladding of the façades of the new buildings. Villamayor stone (Golden Stone) has several varieties (García-Talegón et al, 2014): i) Carbonated Variety: the fine-grained stone, microporous, is partially cemented by dolomite, 27% (bulk porosity), ii) Clay and fine Variety: ochre colour, microporous, with smectite with iron oxides, 30% (bulk porosity), is very compact with abundant clay matrix, iii) Medium grained Variety: without cement, low clay matrix and with iron oxides present, 38% (bulk porosity) and macroporous. Main components for all three varieties: Quartz (up to 60%), feldspars, 2:1 layered silicates (smectites), palygorskite-type fibrous silicates, and small amounts of micaceous minerals (illite/mica).

Golden Stone (García-Talegón et al., 2014) has a particular physic characteristic like its absorption index (about 13.55%). It does that this stone be highly porously and therefore very sensible of the humidity.
Furthermore, the façades of the monuments demonstrate a surprising uniformity in which it can be observed seen that the ashlars have different intrinsic characteristics given the wide variety of types of Villamayor stone.

2.2 Villamayor Stone: Pathologies

Looking at the Villamayor stone monuments it can be deduced that on the façades exposed to the sun and the joint action of wind and rain (Rives and Vicente, 1993), but without the constant humidity to produce their arenization, the patina is more intense. Two types of surface modification of the Villamayor Stone in these monumental microenvironments have been defined: i) on vertical surfaces which have good drainage and are exposed to the sun, good conservation of the stone can be observed and it acquires a beautiful golden patina and ii) if the stone is located in zones of frequent humidity, water with or without salts (nitrates, chlorides , sulfates, phosphates, etc.) which is located in the lower parts of the monument rises from their bases by capillary action. The patina is a symptom of the condition of the stone, and contributes to its conservation (Vicente and Brufau, 1986).

During the years 2007/2008, detailed mapping pathologies and different varieties of Villamayor Stone are preceded, due to the detachment of small portions of the façade cladding close to the copings of the cornices and the terraces of the building (Figure 1). The decay of the Villamayor Stone was consequence of the bad drainage system in the copings of the building, together join to the humidity sensibility of this stone and abrupt changes of temperature (frost weathering "ice wedge effect" by increasing water volume when its physical state changes) were producing a deterioration in the copings in a short time (less than 7 years). The mean decay of the Villamayor Stone is leading to microfissures, scales, flakes and loss of matter (Figure 1) following ICOMOS glossary (2011).

From the climatic data of Salamanca city, it is concluded that the climate of Salamanca can be defined as a Mediterranean with a continental trend, with an annual mean temperature of 12.2 ºC. The range of daily temperature fluctuations may be as large as 35 ºC. The absolute maximum temperature measured over the last 30 years is 39.4 ºC, while the absolute minimum is -12.6 ºC. Yearly average precipitation is about 372.5 mm, mainly in spring and autumn, July and August being characterized by strong droughts.

Figure 1.Villamayor Stone decay. A.- Southwest cladding façade with significant decay of the Villamayor Stone. B.- Material loss and scaling in the copings of the cornices. C.- Fissures in the grained Villamayor Stone. D.- Fissures and material loss in the clay-fine Villamayor Stone and its anchoring. E.- Fissures and cracking in the clay-fine Villamayor Stone below the coping of the cornice.

This contribution has been peer-reviewed.
doi:10.5194/isprsarchives-XL-5-W4-445-2015

3. METHODS

3.1 Equipments

For the documentation of the façades and specific plaques on the laboratory two different terrestrial laser scanners were used with different characteristics and data acquisition principles:

The FARO Focus 3D laser scanner is based on the principle of continuous wave, with a wavelength of 905 nm. This device measures distances in a range of 0.60–120 m with a point measurement rate of 976,000 points per second. It has an accuracy of 0.015° in normal lighting and reflectivity conditions and a beam divergence of 0.19 mrad, equivalent to 19 mm per 100 m range. The field of view covers 320° vertically and 360° horizontally with a 0.009° of angular resolution and the returning intensity is recorded in 11 bits.

The Trimble GX is a laser scanner bases in the principle of time of fly, with a wavelength of 534 nm. The range measure is between 2-350m with a point measurement rate of 5000 points per second. It has an accuracy of 2,5mm per 100m. The field of view covers 60º vertically and 360º horizontally.

The main characteristics and technical specifications of the lasers equipments used in this study are resumed in Table 1.

	TRIMBLE GX	Faro Focus 3D
Principle of measure	Time of Flight	Continuous Wave
Wavelength	534 nm (green)	905 nm (NIR)
Deflection System	Oscillating mirror	Rotary mirror
Field of vision	360º H x 60º V	360º H x 305º V
Standard deviation	1.4mm in D<50m	2mm in D=25m
Measure rank	2-350 m	0.6-20m
Angular resolution	Horizontal: 12" Vertical: 14"	Horizontal: 33" Vertical: 33"
Laser divergence	3mm in 50 m	0.16mrad
Scan Velocity	5000 points/s	122.000 points/s

Table 1: Specifications of the different terrestrial laser scanners used in this study, features referred to vendors documents.

3.2 Data acquisition

The study was divided in two parts. First part allowed getting information with terrestrial scanner laser after previously inspection of the zone with potential interest in pathological analysis. Second part was focused in the analysis and treatment of the obtained point cloud with different software.

The data acquisition had two phases that where carry out in 2007 and 2014. The first phase had the objective the 3D documentation of the complex building "El Tormes" located in Santa Marta de Tormes (Salamanca, Spain). The equipment used was Trimble GX. It was realized 5 stations with a base distance between 30-50 m covering three of the four façades and obtaining a point cloud with a resolution of 10 mm in the mains parts of the façades and 5mm in the copings where the main pathologies were presented.

The second part of the study was realized in December of 2014 in the photogrammetric laboratory of the Polytechnic High School of Ávila (Spain). The terrestrial laser scanner used in this part, was the Faro Focus 3D. In this occasion many plaques retired of the façades in the 2007 restoration were scanned and analysed. A resolution of 1.5 mm with an object distance between 3-7 m was used.

In both campaigns the laser surveying was supported by a photographic documentation realized with a single-lens reflex (SLR) digital camera, Panasonic Lumix DMC-FZ38. The photographs had the objective to grant additional RGB information to the intensity values provided by laser scanners.

3.3 Data processing

The raw laser point clouds were filtered and segmented in order to remove those points that were not part of the Villamayor Stone. In the case of the complex building "El Tormes", the point cloud alignment was done by a solid rigid transformation using common surfaces and supported by the iterative closes point algorithm (Besl and McKay, 1992). In addition, the different images acquired with the SLR digital camera were registered through a camera resection process based on collinearity condition (Kraus, 1993). As a result, RGB information can be used as additional radiometric information in the assessment of pathologies. Finally, orthoimages with the different intensity values, including RGB information, were generated in order to establish a common cartographic product for analysing the Villamayor Stone.

Two types of orthoimages were created. The first type has a ground sample distance (GSD) of 10 mm and covers the complete façade surveyed in 2007. The second type has a GSD of 5 mm and only covers the copings and those plaques analysed on laboratory in 2014. Both types were created using the RGB and intensity values. Orthoimages were generated based on the anchor points method (Kraus, 1993). This method consists of applying an affine transformation to each one of the planes formed by the optimised triangular mesh, which was obtained from the point cloud determined by the laser. Through the collinearity condition (Albertz and Kreiling, 1989), the pixel coordinates of the vertices of the mesh are calculated, and the mathematical model of the affine transformation directly relates the pixel coordinates of the registered image and of the orthoimage.

On the other hand, an expert analysis of the pathologies was performed *in situ*. As a result, thematic cartography was obtained using the orthoimages and drawing manually the pathologies.

3.4 Recognition of Villamayor Stone varieties and humidity/biological colonization: Non-supervised classification

With the purpose to compare the results obtained in 2007 with pathological cartography realized in the classic way by the expert, a non-supervised classification using the intensity values of the orthoimages of the copings was carried out.

This classification has been realized with the Maximum Likelihood (ML) method using Isodata clustering algorithm to determine the characteristics of the natural groupings of cells in multidimensional attribute space and stores. The ML classifier quantitatively evaluates both the variance and covariance of the category spectral response patterns when classifying an unknown pixel. The resulting bell-shaped surfaces are called probability functions (Equation (1)), and there is one such function for each spectral category (Lillesand et al., 2004).

$$\Pr(k/g) = -\ln\left|\sum k\right| - (g-\mu_k)^T \sum k^{-1} (g-\mu_k) \quad (1)$$

where $\sum k$ = covariance matrix
μ_k = mean vector

3.5 Villamayor Stone pathologies: fissures and scaling

3.5.1 High resolution Digital Model: The analysis of fissures and scales was performed by mean of the digital model obtained after processing the laser point cloud. The mathematical approach used for the calculation of the surface grid was completed by applying the nearest neighbour interpolation method with a 0.2x0.2 mm grid spacing. Subsequent analysis of the interpolated, surface was completed by applying a low-pass Gaussian filter based in the convolution matrix of the signal (Z values) and following the Gaussian functions proposed by Nixon (2008).

Once the grid was obtained, a colour scale with normal shading was applied in order to quantify the volume lost. The volume reconstruction was finally calculated by subtracting the non-altered rock surface from the study samples. The visual inspection of the grid and the orthoimages allowed the digitalization and mapping of pathologies (fissures and scaling) in the Villamayor Stone.

3.5.2 Fourier Analysis: The digital model was mathematically processes under a two-dimensional Fourier analysis in order to assess the contribution of fissures and scaling to the final surface geometry observed. Mathematically, the Fourier transform can be expressed in the space domain as follows (Bracewell and Bracewell, 1986) (Equation (2)):

$$f(\omega) = \int_{-\infty}^{\infty} f(x) e^{i\omega x} dx \quad (2)$$

Since the space domain function f(x) has a single and unique wavenumber domain function, and viceversa, it can be assumed that, by a given wavenumber, the function can be transformed into the space domain by using the inverse Fourier Transformation (Arfken and Weber, 1985). The reciprocal relation is represented by (Equation (3)):

$$f(x) = \frac{1}{2\pi} \int_{-\infty}^{\infty} f(\omega) e^{-i\omega x} d\omega \quad (3)$$

where ω is an angular wavenumber in radians per millimetre (for a given x in millimetres). The wavenumber in cycles per millimetre is $\omega /2p$ (p accounts for period).

The inverse Fourier transform aimed to convert from the wavenumber domain back into the original space domain using the Oasis Montaj MAGMAP module and applying a raised cosine filter (Kekre, et al., 1982) (Equation (4)):

$$L(k) = \cos^n \left[\frac{\pi}{2} \left(\frac{k-k_0}{k_1-k_0} \right) \right] \quad \text{for } k_0 \le k \le k_1 \quad (3)$$

$$L(k)=1 \quad \text{for } k < k_0$$
$$L(k)=0 \quad \text{for } k > k_0$$

where K_0 = initial wavenumber
K_1 = final wavenumber
n = degree of function

4. EXPERIMENTAL RESULTS

4.1 Results obtained from Trimble GX

The cartographic results obtained during the campaign executed in 2007 are outlined in Figure 2. The field work was carried out around the different façades of the complex building "El Tormes", the most interesting façades from a pathological point of view were those with orientation north and south-west. A total of 5 stations were required to cover the façades at a distance between 30-50 m. The resolution of the data capture was 10 mm for the façades and 5 mm in the copings where the main pathologies were presented. Moreover, the façades were photographed at the same distance with a SLR digital camera, Panasonic Lumix DMC-FZ38. The number of points per side is 3millions in the laterals and 7 millions in the main façades.

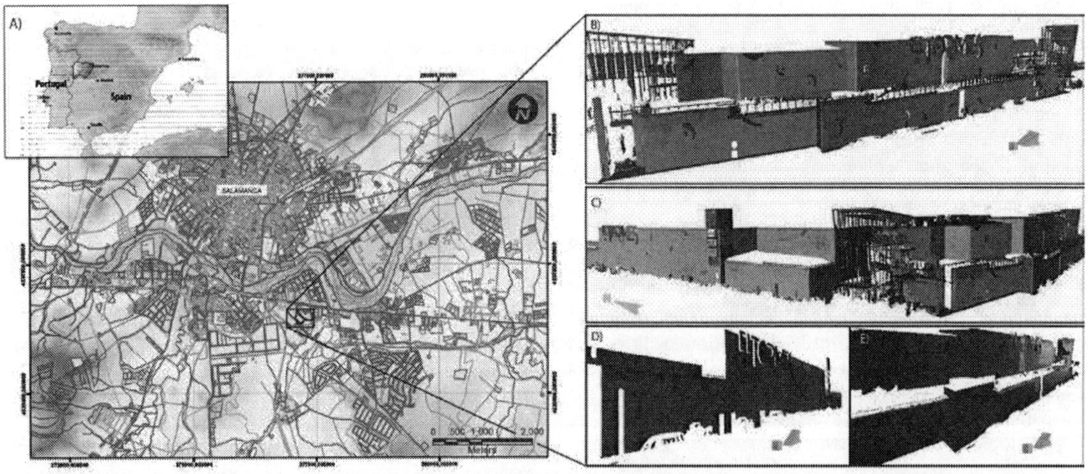

Figure 2. A,- Localization of the complex building studied ("El Tormes", Salamanca, Spain). B and C.- Intensity 3D models acquired with the Trimble GX laser scanner. D.- True colour 3D model. E.- Height Map.

This contribution has been peer-reviewed.
doi:10.5194/isprsarchives-XL-5-W4-445-2015

4.2 Results obtained from the non-supervised classification.

The non-supervised classification executed over the orthoimages using the different intensity data has shown different grades of humidity affection. These differences respond to a combination of factors controlled by the microenvironments of the façades (orientation, the copings of the terraces, etc.) and the diversity of intrinsic properties of Villamayor Stone. For instance, façades facing north (Figure 3 A) are prone to maintain high rates of humidity aimed by an important decrease in daily sun insulation and hence low evaporation rates.

The Villamayor Stone porosity (>30%) favors humid conditions, increasing the development of lichen growth (Figure 2 D). The non-supervised classification of rock pathologies shows the impact of humidity in the cladding, affecting the first four top-lines, and clearly marked by different degrees of intensity along the north façade (Figure 1 and Figure 3 A); whereas the western façade only presents humidity affection in the coping and first top-line of the cladding (Figure 2 C and Figure 3 B). In addition, the classification allowed us to differentiate between two different varieties of Villamayor Stone represented by changes in rock reflectivity. These differences were based on composition, internal structure and degree of surface weathering (Figure 2 C and 2 B).

In the eastern part of the building there was no cladding façade and the southern one did not show any sights of humidity nor the presence of lichens.

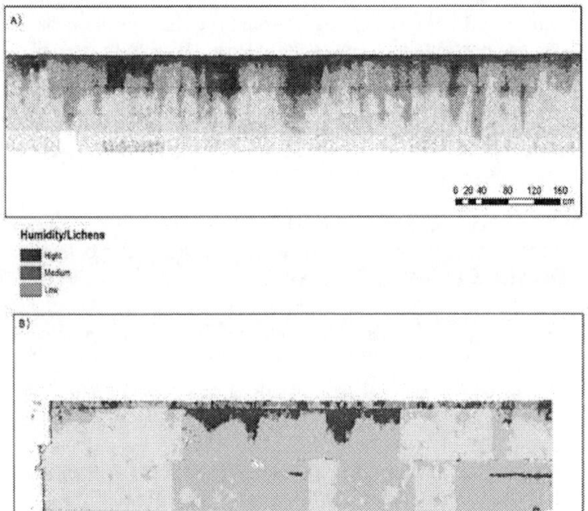

Figure 3. A and B.- Results of the non-supervised classification zoned North and Southwest of the studied building, respectively. Variety 1: clay-fine Villamayor Stone; Variety 2: carbonated Villamayor Stone

4.3 Results obtained from the cracking and detachment analysis

Digital Model obtained from TLS point cloud of an altered Villamayor Stone plaques removed from the façade of the studied building were analyzed by adding different hypsometric shading. The color patterns aimed to identify different areas of damage characterized by scaling and fissures (Figure 4). These pathologies were digitized in the obtained orthoimages (Figure 4 B) and several profiles were performed (Figure 4 C). Fissures were clearly identified in the profiles, marked by a sharp pattern in elevation (Figure 4 C profile 1-4). However, this method strongly depends on the visual interpretation and experience of the observer and the resolution of the method of data acquisition.

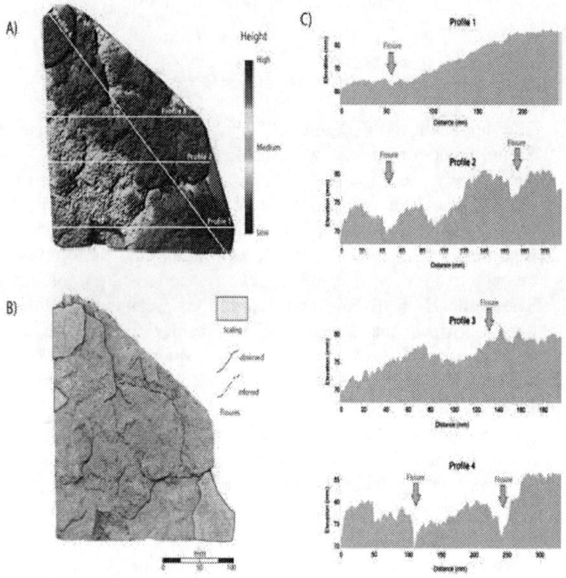

Figure 4. A. Hypsometric shading of the D M. B.- Orthoimage showing the fissures and scaling affecting the Villamayor Stone plaque. C.-Longitudinal and transversal profiles of the plaque marking the position of the fissures (arrows).

In addition to the analysis of the Digital Model (D M) and the visual interpretation of the different pathologies observed, it carried out an inverse 2D-Fourier Analysis over the obtained DM grid. The results aimed to identify two characteristic wavelengths at 25mm and 50mm. The subsequent application of a low pass raise cosine filter at 25mm shows the imprinted pattern related to fissuring, characterized by short-spaced highs and lows (Figure 5 A). In addition, the 50mm wavelength marked the position of scaling pathologies represented by the presence of clearly defines highs. The Fourier analysis offers a good quality control of this type of pathologies affecting the Villamayor Stone and could be use for the study and identification of patterns of fissuring and scaling in building façades. It is also a fast method for mapping of rock pathologies which needs to be combined with other methods of analysis such as orthoimages or true-color laser-derived images based on point clouds obtained from TLS.

This contribution has been peer-reviewed.
doi:10.5194/isprsarchives-XL-5-W4-445-2015

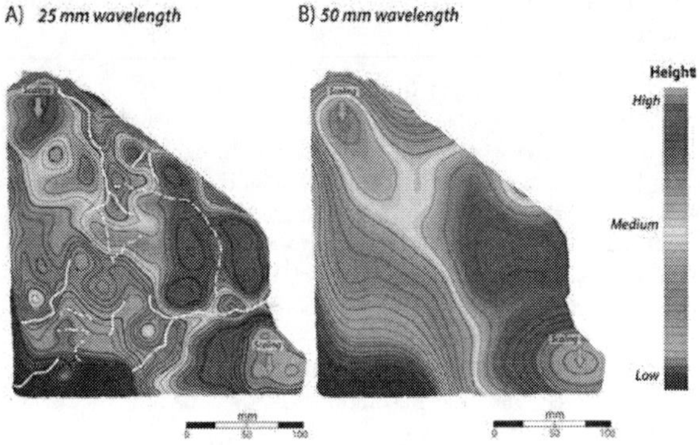

Figure 5. Results of the Fourier transformation. A.- Results from 25mm grid. B.-Results from 50mm grid.

5. CONCLUSIONS

The application of geomatic technology in the Villamayor Stone, recognized as a Global Heritage Resource, has successfully contributed to improve the analysis of rock pathologies. The integration of terrestrial laser scanner and photogrammetry was used for the first time in Villamayor Stone in order to complement visual observations and petrophysical characterization carried out by experienced rock-pathologists. The comparison of Faro and Trimble GX scanner technology has confirmed the capacity of the former to improve data resolution and lowering the data acquisition time. Non-supervised classification of these data enhances to distinguish between different patterns of spectral intensity aiming to:

- Differentiate among pathologies such as humidity and biological colonization.
- Assess the total surface affected by rock pathologies.
- Measure the surface involved in rock scaling and fissuring.

Finally, the new approach based on the treatment of the point cloud allowed the generation of high resolution digital models, which were used for volume lost quantification in the past 10 years. In addition, the Fourier analysis has demonstrated the capability to provide a fast and accurate response in the interpretation of rock pathologies. This methodology could help, in addition to visual exploration, for the identification of fissures and scaling in façades.

ACKNOWLEDGEMENTS

The authors are grateful for funding from an agreement with the ING INMUEBLES S.A.U. to perform this work. We would like also to recognize Mrs. Cuca Gonzalo, C.C.Tormes´s manager and her team, who gave us all of the information and facilities that we needed.

REFERENCES

Albertz, J., Kreiling, W., 1989 *Photogrammetrisches* Taschenbuch. Wichmann.

Arfken, G., Weber, H. J., 1985. *Mathematical Methods for Physicists Academic.*New York, 1985, 309.

Armesto J, Lubowiecka I, Ordóñez C, Rial F. FEM modeling of structures based on close range digital photogrammetry. *Automation in Construction*, 2009; 18: 59 69.

Besl, P., McKay, N., 1992. A Method for Registration of 3-D Shapes. *IEEE Transactions on pattern analysis and machine intelligence*, 14(2), pp. 239-256.

Bracewell, R.N., Bracewell, R., 1986. *The Fourier transform and its applications*, McGraw-Hill New York.

Buckley, S.J., Howell, J.A., Enge, H.D., Kurz, T.H., 2008. Terrestrial laser scanning in geology: data acquisition, processing and accuracy considerations. *Journal of the Geological Society,* 165, pp. 625-638.

García-Talegón, J., Iñigo, A. C., Alonso-Gavilán, G., Vicente-Tavera, S., 2014. Villamayor Stone (Golden Stone) as a global heritage stone resource from Salamanca (NW of Spain). *Journal of Geological Society of London (Special Publication)*, DOI:10.1144/SP407.19.

Genovese, R., 2005. Architectural, archaeological and environmental restoration planning methodology: historic researches and techniques of survey aiming to conservation, In: *CIPA 2005 XX International Symposium*, Torino, Italy.

Gonzalez-Aguilera, D., Gomez-Lahoz, J., Munoz-Nieto, A., Herrero-Pascual, J., 2008. Monitoring the health of an emblematic monument from terrestrial laser scanner. *Nondestructive Testing and Evaluation* 23 (4), pp. 301-315.

González-Jorge, H., Gonzalez-Aguilera, D., Rodriguez-Gonzalvez, P., Arias, P., 2012. Monitoring biological crusts in civil engineering structures using intensity data from terrestrial laser scanners. *Construction and Building Materials,* 31, pp. 119-128.

ICOMOS, 2011. *Illustrated Glossary on Stone Deterioration Patters*. Paris: Monuments and Sites XV. ICOMOS.

Kekre, H.B., Sahasrabudhe, S.C., Goyal, N.C., 1982. Raised cosine function for image data interpolation, *Computers & Electrical Engineering,* 9, 131-152.

Kraus, K., 1993. *Photogrammetry .Fundamentals and Standard Processes Vol 1. 4th Edition*. Dümmler. Bonn, Germany.

Lambers, K., Eisenbeiss, H., Sauerbier, M., Kupferschmidt, D., Gaisecker, T., Sotoodeh, S., et al., 2007. Combining photogrammetry and laser scanning for the recording and modelling of the Late Intermediate Period site of Pinchango Alto, Palpa, Peru. *Journal of Archaeological Science*, 34 (10), pp. 1702-1712.

Lillesand, T. M., Kiefer, R. W., Chipman, J. W., 2004. *Remote sensing and image interpretation* (No. Ed. 5). John Wiley & Sons Ltd.

López-Plaza, M., González-Sánchez, M., Iñigo, A. C., 2007a. La utilización del leucogranito turmalinífero de Martinamor en los monumentos de Salamanca y Alba de Tormes. *Studia Geologica Salmanticensia*, 43, pp. 247-257.

López-Plaza, M., González-Sánchez, M., García de los Ríos-Cobo, J. I., Cortazar-Estibaliz, J., Iñigo, A. C., Vicente-Tavera, S., López-Moro, J. J., 2007b. La utilización de rocas vaugneríticas en los monumentos de Salamanca. *Studia Geologica Salmanticensia*, 43, pp. 115-125.

Lubowiecka, I., Armesto, J., Arias, P., Lorenzo, H., 2009. Historic bridge modelling using laser scanning, ground penetrating radar and finite element methods in the context of structural dynamics. *Engineering Structures,* 31 (11), pp. 2667-2676.

Nixon, M., 2008. *Feature extraction & image processing.* Academic Press. 399pp

Rodríguez-Gonzálvez, P., Pascual, J.H., Aguilera, D.G., Nieto, Á.L.M., Mancera-Taboada, J., Martín, N.S., 2013. Aplicación de técnicas geomáticas al análisis y diagnóstico de patologías en el Patrimonio Arquitectónico. *Mapping*, 161, pp. 4-19.

Rives, V., Vicente, M. A., 1993. Formas de alteración de la arenisca de Villamayor en distintos microambientes de edificios salmantinos. In: *Alteración de granitos y rocas afines, empleados como materiales de construcción*, Madrid, Spain, pp. 75-82.

Vicente, M. A., Brufau, A., 1986. Weathering of the Villamayor arkosic sandstone used in buildings, under a continental semi-arid climate. *Applied Clay Science*, 1(3), pp. 265-272.

This contribution has been peer-reviewed.
doi:10.5194/isprsarchives-XL-5-W4-445-2015

EXTENDING A MOBILE DEVICE WITH LOW-COST 3D MODELING AND BUILDING-SCALE MAPPING CAPABILITIES, FOR APPLICATION IN ARCHITECTURE AND ARCHAEOLOGY

M. Ancona[a, *], P. Clini[b], A. Dellacasa[a], P. Falzone[c], A. La Camera[a], R. Quattrini[b], E. Sommariva[d], J. Stephens[e]

[a] DIBRIS, University of Genoa, Via Dodecaneso 35, Genoa, Italy
[b] DICEA, Polytechnic University of Marche, Via Brecce Bianche 12, Ancona, Italy
[c] DSA, University of Genoa, Stradone S. Agostino 37, Genoa, Italy
[d] IES, Department of Urban Design And Planning, Leibniz University Hannover
[e] MJC2, 33 Wellington Business Park, Crowthorne, Berkshire, Rg45 6Ls, UK

Commission V WG V/4

KEY WORDS: Cultural Heritage, Architecture, LIDAR, GIS, Multisensor, Mobile, Three-dimensional, Reconstruction

ABSTRACT:

One of the most challenging problem in architecture is the automated construction of 3D (and 4D) digital models of cultural objects with the aim of implementing open data repositories, scientifically authenticated and responding to well accepted standards of validation, evaluation, preservation, publication, updating and dissemination. The realization of such an ambitious objective requires the adoption of special technological instruments.
In this paper we plan to use portable devices (i.e. smartphones, tablets or PDAs eventually extended to wearable ones), extended with a small plug-in, for automatically extracting 3D models of single objects and building-scale mapping of the surrounding environment. At the same time, the device will provide the capability of inserting notes and observations.
Where the instrument cannot be directly applied, for example for exploring the top of a complex building, we consider mounting our device, or using equivalent existing equipment, on a drone, in a modular approach for obtaining data de-facto interchangeable. The approach based on the expansion packs has the advantage of anticipating (or even promoting) future extensions of new mobile devices, when the spectrum of possible applications justify the corresponding increased costs.
In order to experiment and verify this approach we plan to test it in two specific scenarios of the cultural heritage domain in which such devices seem particularly promising: *Strada Nuova* in Genoa and *Palazzo Ducale* in Urbino, both located in Italy.

1. INTRODUCTION

Three/four-dimensional (3D/4D) model is a digital virtual content geometric-aware and reality based. In the field of architectural or archaeological heritage the 3D model must guarantee accuracy and quality of representation, regarding geometry, colour and building rules. The 4D concept improves the digital copy of cultural heritage objects with awareness of modification along the time and semantic shape, becoming in a knowledge based system.

Archiving and linking our 3D cultural heritage objects ensures the preservation of Archaeological and Architectural Heritage (AAH). The preservation has two different aspects. The first one is the preservation of the real AAH object owing the capability to perform advanced analysis over linked data (e.g. change detection). The second aspect is the preservation of a digital copy of an AAH object; this is particularly suitable for virtual museums where a user can interactively interact with the AAH object.

Cultural heritage sites and artefacts get a significant added value from high-resolution 3D models. A large number of models is increasingly available due to evolutions in technology and higher integration of survey techniques such as laser scanning and computer vision. Several attempts have been made to address the problem of automation the workflow in 3D model construction, reducing integration of formats and procedures.

In the field of survey for cultural heritage, it is mandatory to cite London Charter conceived to enhance the rigour with which computer-based visualisation methods and outcomes are used and evaluated in heritage contexts. The London Charter for the Computer based Visualisation of Cultural Heritage (Denard, 2009) was expanded to encompass computer-based visualisation embracing 2D, 3D, 4D and even hard-copy printouts or computer generated physical objects such as replicas of museum artefacts (Denard, 2012). The London Charter states that an evaluation of the goals to achieve should prove first of all if 3D visualisation is an appropriate method, and if so, which 3D visualisation method is the most adequate to reach the goals. The Charter has also given birth to a new charter, the Seville Charter, which will propose specific implementation guidelines, in the field of Virtual Archaeology (VA).

Progresses are obtained in data management of 1 and 2 dimensions and they are already shared thanks to *Europeana* and its linked project. The more complex issues of 3D and 4D data management are still open, but remarkable tools and operational procedures have been developed to facilitate the cultural institutions in the creation of integrated 3D digital collections and to make easy use of these resources in their daily work (Arnold, 2014). Some recent works for digital repositories bring to important goals: sustainable and centralized safeguarding and archiving of 2D/3D data; free access to metadata; secure access to data for the different actors involved in scientific projects and the support and advice for these actors in the 3D data production and exploration through the latest digital technologies, modeling tools and virtual reality systems.

Today ICT can guarantee a new audience for cultural places and a sustainable management for public institutions. The current need

*Corresponding author. E-mail: massimo.ancona@unige.it

This contribution has been peer-reviewed.
doi:10.5194/isprsarchives-XL-5-W4-453-2015

for museums to augment their own revenue and improve their performance coupled with the transformation of cultural models towards more experiential services has led many to adopt a consumer orientation in an effort to make museums and their collections increasingly accessible to visitors.

The digital process of visual improvement of the artefacts is currently carried out based on photographic or TLS documentation of some archetypes. The commercial framework for mobile devices enables development of Augmented Reality applications. Using a pattern that is recognized by the device, the virtual model is shown as it is in the real world. Some works have the major novelty in that the experts in the cultural heritage field as well as museum staff, could benefit of this work and implement such a system.

2. STATE OF THE ART

2.1 Architectural and Cultural Heritage framework description

Some authors developed in the past years a robust background in laser scanning acquisition and architectural analysis (Guidi and Bianchini, 2007, Clini et al., 2014), other works for 3D digital virtual models of ancient heritage monuments from on-site data, especially focusing on three-dimensional data acquisition techniques and color processing methods (Remondino, 2011) or focusing on simulation of large architectural heritage complex, restoration or colour information starting from real world data (Apollonio et al., 2014).

Clini et al. highlighted the need of validation of easy and low cost methods for architectural survey (Clini et al., 2013) while other authors investigated urban surveys and their use in the interpretation and design about cities and the methods for their modeling.

From the point of view of the automatic survey of AAH outdoor objects, the adoption of unmanned ground and aerial robotic platform (RPAS/UGV) represents an increasing opportunity. Actually the unmanned/manned robotic platforms are suitable to host complex survey payload as high-resolution camera or LIDAR (EU project ROVINA). Robot aided monitoring or surveying is involved in some projects such as After the Dark, winner of 2014 IK prize, and the MuseBot (Cigola et al., 2014). The currently available tools for AAH are shown in Table 1.

Tools and Devices for surveying and analysis systems
Laser scanner tof, ps, triangulation (LIDAR)
Total Station (TS)
Structured light scanning
Ground Penetration Radar (GPR)
RGB-D sensor
Infrared Thermography (IRT) and multispectral cameras
CamerasCMOS o CCD sensor
Drone
Inertial Measurement Unit (IMU)
Micro Processor Unit (MPU)
Global Positioning System (GPS)

Table 1: Available tools in AAH survey field

Although techniques for the acquisition of 3D building geometries via TLS have constantly been improved, a fully automated procedure for constructing automatically reliable 3D building models is not yet in sight. This is due essentially to the difficulties of exploring directly and automatically valuable spatial information from the huge amount of 3D data. Thus many post-processing operations must be performed before accessing to reconstruction of a reliable 3D model. Many studies investigated algorithms for automatic 3D modelling of building façades; however we consider more focused on our study the approaches addressed to produce a digital document that combines metric and photographic data and historical data in a 3D textured model becoming a theoretical model (Lo Buglio and De Luca, 2012, Quattrini and Baleani, 2014).

The use of 3D model is developed in several levels of detail and/or representation scales from GIS to 3D Cognitive-Information System through semantic modeling and Decision Making System (DMS) or information system dedicated to building the project proposal, fully directing its communicative complexity to the synthetic and interactive support of 3D modeling. The list of 3D acquisition systems is shown in Table 2.

3D acquisition and modelling	
range-based	Mobile Laser Scanning System Terrestrial Laser Scanning
image-based	Close-range photogrammetry Spherical photogrammetry
systems range or image-based	Remotely Piloted Aircraft Systems (RPAS) Unmanned ground vehicle (UGV) system

Table 2: 3D acquisition systems

2.2 Information Technology background

Fallon et al. (Fallon et al., 2012) describe a system enabling rapid multi-floor indoor map building using a body-worn sensor system fusing information from RGB-D cameras, LIDAR, inertial, and barometric sensors. While map building in real-time is fundamental for our proposal, several differences exist from our environment and the original applicative context adopted in (Fallon et al., 2012). Fallon's context is centered on the need of emergency responders, security and military personal to quickly develop knowledge about a given environments. In such an environment, the user main task is not map building: he takes advantage of the constructed maps but has little time to dedicate to their construction.

In our case the user is completely dedicated to the map building process and the linking to the constructed maps with references to digital objects reporting a 3D representation of all relevant entities located in the area represented in the constructed maps. The user has no other mandatory task to perform during the map construction and the use of a lightweight hand-held device seems more convenient giving to its user more flexibility and interaction capabilities.

Today, the problem of indoor map building is considered of primary importance, see (Apple Indoor Mapping Project and iBeacon, 2014) for example. Moreover, low-cost full-color holographic 3D displays for smartphones are expected in the very next future. Their coming will be certainly an event that will give a wide relevance to 3D reconstructions. The companies are moving in this direction and, in particular, HP has already developed a wide-angle, glass-free display prototype (Lewin, 2014). Fattal et al. introduce diffractive backlight solutions that seem well suited to realizing a multi-view 3D display for mobile devices (Fattal et al., 2013).

Several authors have used mobile phones for 3D map reconstructions (Tanskanen et al., 2013, Pirchheim and Reitmayr, 2011, Pan et al., 2011, Martin et al., 2014) obtaining interesting results despite the limitations of the device used.

Martin et al. prove that, with the today smartphones, it becomes possible to foresee Augmented Reality (AR) applications on such

doi:10.5194/isprsarchives-XL-5-W4-453-2015

devices (Martin et al., 2014). Nevertheless most AR Apps only use sensors such as IMU, compass and GPS and barely consider image-based localization approaches. Apart from the computational cost of such methods, one of the main reasons is that they usually rely on the use of a 3D model or assumption on either the structure of the scene (supposed to be planar) or the camera motion (supposed to be rotational). Nevertheless, since the introduction of vision-based AR on mobile devices (by using ARToolkit Markers) impressive progresses have been made. The authors' approach is based on the decoupling of the mapping process from the tracking step in order to save computational power for the end-user application, an approach successfully considered for vehicle localization and augmented reality. Their system has been tested on both Android and iOS (Samsung Galaxy S2, Samsung Galaxy S3 and iPhone 4S). On both the Galaxy S3 and the iPhone 4S the tracking and image-based re-localization are done in real-time for an image of size of 320x240 and 480x360 pixels, respectively.

In (Pirchheim and Reitmayr, 2011) the authors present a real-time camera pose tracking and mapping system which uses the assumption of a planar scene to implement a highly efficient mapping algorithm. Their light-weight mapping approach is based on keyframes and plane-induced homographies between them. They solve the planar reconstruction problem of estimating the keyframe poses with an efficient image rectification algorithm.

Mapping of a-priori unknown scenes in full 3D and tracking with a monocular camera still remains a hard task, or, at least, on mobile phones. Classical EKF-based approaches known from the robotics community seem to be computationally infeasible. Only the well-known PTAM system (Pan et al., 2011) has been shown to run as a prototype on the iPhone 3G. The authors tested their systems scalability on the Nokia N900, a phone equipped with a 600 MHz single-core ARMv7 and 256 MB memory. They processed camera images with a resolution of 320x240 pixels. In order to achieve the goal of rapid 3D reconstruction of environments, authors of (Martin et al., 2014) propose a method using several wide field-of-view images instead of a large collection of images, by adapting a cheap online space carving approach based on Delaunay triangulation to obtain a dense, polygonal representation of the environment.

Finally, in (Tanskanen et al., 2013) the authors propose a complete on-device 3D reconstruction pipeline for mobile monocular hand-held devices, generating dense 3D models with absolute scale on-site and simultaneously supplying the user with realtime interactive feedback. The method fills a gap in current cloud-based mobile reconstruction services as it ensures at capture time that the acquired image set fulfills desired quality and completeness criteria. In contrast to existing systems, the developed framework offers explore innovative solutions such as the usability of the on-device inertial sensors to make the tracking and mapping process more resilient to rapid motions and to estimate the metric scale of the captured scene.

3. INSERTING RECONSTRUCTED 3D DIGITAL OBJECT INTO THE BUILD MAPS

For the exhaustive description of an AAH object, we need 3D-4D model, additional data and metadata. The object can be adapted in order to extract new models with different purposes and for different users (a sort of custom view), in order to produced different views on request.

Remote sensing techniques are evolving and improving at an incredible speed. The software and hardware required to process the acquired data is generally less available. Furthermore, there are very few packages able to simultaneously deal with different sensors (and data). The development and use of new sensors requires the study and test of innovative sensor models and the investigation of the related network structures and accuracy performance. Also UAV platforms need some improvements for more reliable and fast acquisition procedures.

The recently introduced technology of human portable devices seems particularly suitable for extracting 3D models from the surrounding environment during an inspection of a museum or of an historical building (Fallon et al., 2012, Ancona et al., 2012). Their use is not intrusive nor dangerous for goods such as paintings, ancient drawings, potteries etc. and renders their adoption acceptable also in museums and in buildings containing precious and prestigious artefacts. Moreover, the wearer of the device has the capability of inserting photos, notes and observations with the underlining smartphone or tablet by adding customized information to the automatically collected data, for pointing out the most relevant aspects. In the external part of buildings, that a human with an hand-held device cannot reach, the use of drones could be applied. In this case is important the unification of the two approaches in order to produce indistinguishable digital objects. Also the data model adopted should be unified for making acquired data completely integrable.

There is an increasing use of hybrid sensors and platforms for the collection of different types of heritage information. The combination of different data sources allows the creation of different geometric levels of detail (LoD) and the exploitation of the intrinsic advantages of each sensor. The integration so far is mainly done at model-level (i.e., at the end of the modelling pipeline) while it should be exploited also at data-level to overcome the weakness of each data source. We need an all-in-one instrument that allow the acquisition of each important characteristic of object: shape, color, material, state of conservation.

3.1 Device extension

The main objectives of this approach are:

- quality/precision of the obtained model,
- low costs,
- conformity to standards and well accepted procedures,
- simplicity and rapidity of use.

An objective of our research is the application of a new methodology supporting the digital reconstruction of a wide spectrum of cultural heritage ranging from buildings and museums to single paintings, sculptures or drawings through a set of procedures supporting automatic digital 3D reconstruction of real architectural artefacts. Also works of art designed, but never built, can be considered for a virtual reconstruction based on the remaining documentation.

The paradigm we are proposing will be open, i.e. integrable with other mechanisms and any format of information, essentially for:

1. Supporting the use of 3D models of an object as the reconstruction of the reality and as a metaphor of the object itself.

This contribution has been peer-reviewed.
doi:10.5194/isprsarchives-XL-5-W4-453-2015

2. Adopting models (i.e., virtual objects) as a systematic architectural replacement of the reality with an approach that uses descriptive replacements as a collection of structured models, with a well defined naming mechanism, based on a specific Architectural dictionary, in order to naturally extend (maybe merge) the concept of 2D GIS into that of 3D GIS.

3. Implementing a scalable system supporting a common data set usable from each kind of users (Superintendents of Artistic and Cultural Heritage, scientific researchers, tourists etc.)

Thus, we propose a systematic interactive and collaborative mechanism for the acquisition of data to be integrated in standard 3D models in order to construct an information framework for the scientific visualisation, complete integration and web-based presentation of sets of spatial and temporal data naturally heterogeneous in order to facilitate the interpretation exploration and analysis of large quantities of data for pointing out their Geo-spatial, temporal and semantic properties.

Automation in 3D data processing is one of the most important issues when it comes to efficiency or costs. Now, different research solution and commercial packages have turned towards semi-automated approaches, where the human capacity in data interpretation is paired with the speed and precision of computer algorithms. Indeed the success of fully automation in image understanding or point clouds processing depends on many factors and is still a hot topic of research. The progress is promising but the acceptance of fully automated procedures, judged in terms of handled datasets and accuracy of the final 3D results, depends on the quality specifications of the user and final use of the produced 3D model.

In some applications, fast processing is essential and thus requires new algorithmic implementation, sequential estimation, GPU and multi-core processing. The internet is also helping a great deal in this sector and web-based processing tools for image analysis and 3D model generation are available although limited to specific tasks and not ideal for collecting CAD data, metric information and accurate 3D models.

The extraction and the derivation of geometric structures and semantic information are fundamental tasks in modelling. The existing procedures require some improvements to speed up the entire 3D restitution pipeline from images or range data. Images theoretically allow an easier interpretation and extraction with respect to range data. An ideal solution is able to produce truly 3D geometries with attributes and topologies and it should have an internal quality control procedure. A true sensor and data integration at an early point in the processing chain makes the information extraction more reliable, precise and effective.

Although heritage 3D recording and documentation should be an interdisciplinary task, clear protocols and user-friendly packages should be released to facilitate the use of data processing techniques to non-technical users. This will result in an increased adoption of standards for data storage and data exchange as well as sensor characterization and semantic organization of architectural models.

Embracing the Linked Open Data Philosophy will open the use and re-use of the data for different purposes with different level of details. The user (expert or not) must be able to perform queries connecting different concepts for different purposes: visualisation, predictive maintenance of cultural heritage, change-detection, etc.

Therefore we will extend smartphones or tablets PCs in order to implement an automatic construction of 3D digital models of cultural objects ranging from statues and fronts/façades to scale mapping of buildings. We are developing a prototype device by using a smartphone extended with a plug-in which contains on its back two color cameras, a Micro-LIDAR (Pulsed Light, Inc), MPUs and a supplementary battery in order to power supply these sensors. A scheme of our device is shown in Fig. 3.

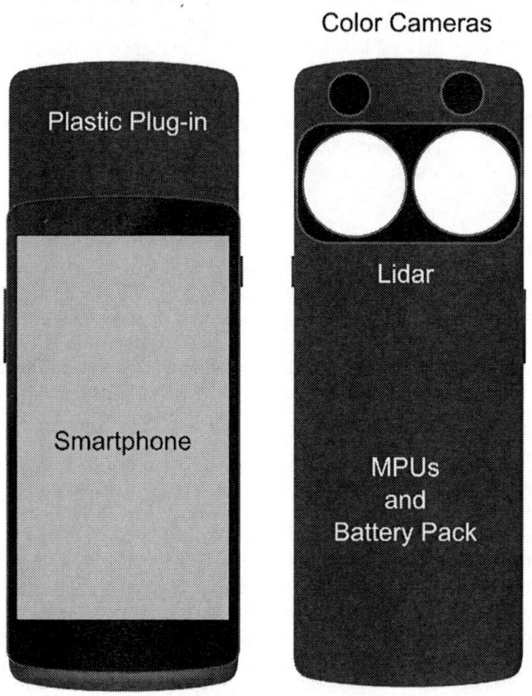

Figure 3: Device prototype scheme in which we show the smartphone with its extension: the two color cameras, the Micro-LIDAR, the MPUs and the supplementary battery.

3.2 National gallery of Marche (indoor demonstrator)

The National Gallery of Marche in *Palazzo Ducale* (Ducal Palace) at Urbino is the indoor demonstrator in our research. Our research already demonstrated how useful is to carry out high resolution acquisitions, to store those and finally to perform mobile or portable application for tourism or insiders (Clini, 2014). We developed a very cost-effective workflow in National Gallery of Marche, based on the *Città Ideale* painting (Fig. 4).

Now it is possible to involve the whole building of *Palazzo Ducale*, thanks to the new proposed device. This kind of test-bench provides solutions both for management and preservation of cultural heritage objects in the museum both for amazing architectural and historical value of the *Palazzo Ducale*. For the first task, the indoor demonstrator leads to obtain and validate operational practices through:

- acquisition/survey of exhibition space to solve the onerous and difficult survey of complex indoor spaces;
- acquisition/survey of museum pieces at different LoD, strictly connected with the intrinsic characteristics, and validation of their accuracy in team of experts;

This contribution has been peer-reviewed.
doi:10.5194/isprsarchives-XL-5-W4-453-2015

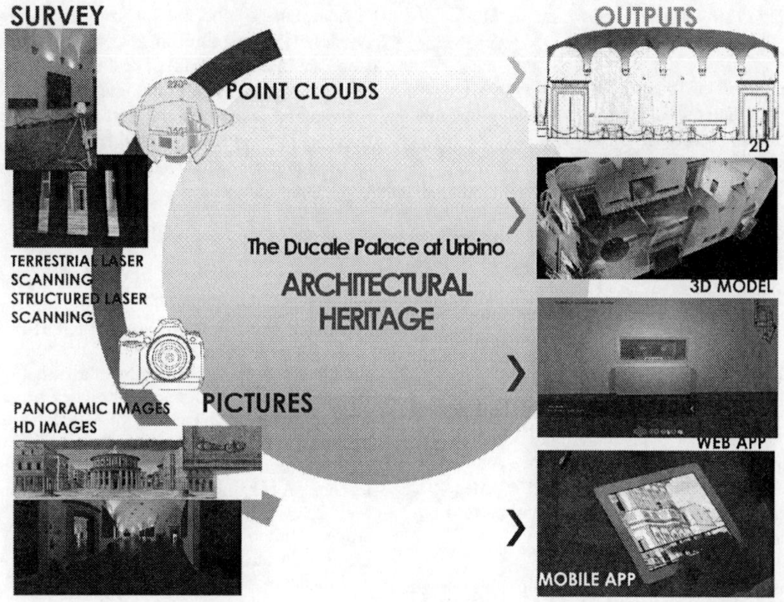

Figure 4: Workflow from acquisition to outputs in AAH fields: the *Palazzo Ducale* at Urbino demonstrator.

- real time (or quasi real time) monitoring of the evidences / artefacts conserved in the museum, acquiring in different times. The development of this section is mainly devoted to verify results in automation processing and feature extraction challenges;
- 3D modelling of same pieces and semantic segmentation based on an ontological representation;
- 3D tool for non-expert users addressing the lack of digital tools in documentation and management of Cultural heritage, overcoming non digital (on paper) cataloguing, already diffused.

For the second task, the indoor demonstrator develops the following issue:

- Virtual exhibition or AR application using architectural models carried out in the same museum demonstrator and related to acquisition about paintings, statues or mobile heritage.

3.3 Strada Nuova in Genoa (outdoor demonstrator)

The example of the Renaissance "Strada Nuova" (nowadays Via Garibaldi, Fig. 5) in Genoa is so relevant that Mario Labò in 1956 (Labò, M., Roma, 1956) wrote:

> *The interest of artists and historians, for Genoese historical city center is timeless, and even more perceived by the international literature than the Italian one.*

A heritage that is not yet appropriately and comprehensively documented, especially with the most recent ICT technologies, suitable for widespread diffusion, concerning the different urban-architectural and artistic aspects as well as the acquisition and processing of the existing documentary materials (Falzone, 1985).

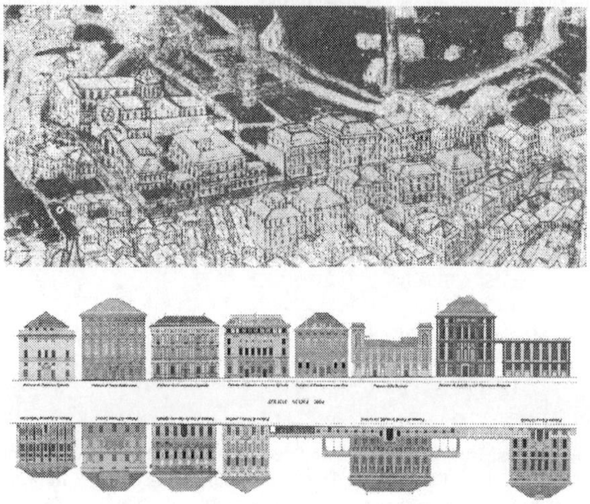

Figure 5: A synthesis image of *Strada Nuova* compared to an historical illustration.

The necessity to define a complete communication plan to underline the current state of conservation of this enormous patrimony, inscribed since 2006 in the list of UNESCO World Heritage Sites, should consider:

1. the whole environmental landscape values of this organic urban space, from its origin to the late transformation period into a residential district for noble families (a sort of small court of the Renaissance Genoese Republic);
2. the morphological aspect of the site facing north toward the hills of Montalbano and limited by the hanging gardens of the palaces Brignole and Durazzo Grimaldi, with its unique

This contribution has been peer-reviewed.
doi:10.5194/isprsarchives-XL-5-W4-453-2015

access from *Piazza Fonte Maroso* and *Piazza della Meridiana*;

3. the stereometric rhythm of 15th to 17th century noble palaces, know especially for their architectural decorations and painted façades (*Palazzo Bianco*, *Palazzo Rosso* and *Palazzo Tursi*) as well as museum heritage.

The Genoese painted façades represent an important characteristic of *Palazzi dei Rolli* system, which has no equal in Europe (Vagnetti, 1970, Poleggi, 1972). An issue worthy of being documented and that, for its exposure and its conservation requires a specific depth of documentation, to be acquired with 3D digital models that are particularly precise to visualize the chromatic fidelity. This requires designing specific software, which can be implemented by a GIS standard tool, able to confront all the existing drawings, surveys, literary documentation, archival documents and maps with the new acquired data in time. A software that is characterized by: simplicity and rapidity of the procedure, cost containment quality, accuracy of the image and pattern obtained.

In particular the most useful technical survey equipment known today are LIDAR (TLS, Laser scanner), GPS and total station (TS), GPR (Ground Penetration Radar - GPR), and multispectral imagers. Moreover, the new mobile acquisition technology (used for smartphone) here proposed, could replace or supplement largely these tools. The key elements on which the work is essentially structured are:

- 3D digital models as a metaphor of the detected object for navigation through the data;
- survey as a cognitive system, described as a collection of structured objects, identified by a precise architectural vocabulary that allows easily extension of the concept of 2D GIS to 3D GIS;
- scalable applications with a common database for every kind of end users (asset management, scientific study, tourism, etc.) by using filtered data according to the requirements.

The research will work on the definition of best practices for the acquisition of the data and the realization of 3D models, in order to build a real information system by providing an uniform framework for a web-based scientific visualisation of data. The goal is to facilitate interpretation, exploration and analysis of large volumes of data with significant geo-spatial, temporal and semantic issues.

4. CONCLUSIONS

A relevant and time consuming activity in applied architecture is the 3D/4D indoor map building of historic buildings and of all internal 3D cultural objects like statues, potteries, gables decorated in relief etc.

In order to speed-up the data acquisition needed for supporting such an activity we plan to use a hand-held sensor system composed by a smartphone (or tablet computer) extended with few optical sensors for precise distance evaluation. In order to verify our device we choose two specific demonstrators of the cultural heritage domain: *Palazzo Ducale* in Urbino and *Strada Nuova* in Genoa, both located in Italy.

A final remark concerns 3D reconstruction performed on smartphones. It is a promising activity in view of the next advent (forecast in 2 or 3 years) of real 3D holographic displays. Three-dimensional display-technologies offer a wide applicative potential ranging from GIS and cartography to 4D object visualization (Schratt and Riedl, 2005). Holographic 3D visualisation will revolution most of today applications centered on 2D displays.

REFERENCES

Ancona, M., Briola, D., Locoro, A., Mascardi, V., Tacchella, A. and Traverso, A., 2012. Exploiting human portable devices for 3d models extraction. In: Proc. of RICH-2012, in print.

Apollonio, F. I., Gaiani, M., Fallavollita, F., Ballabeni, M. and Zheng, S., 2014. Bologna porticoes project: 3d reality-based models for the management of a wide-spread architectural heritage site. In: Digital Heritage. Progress in Cultural Heritage: Documentation, Preservation, and Protection, Euromed14 Conference proceedings, Vol. 8740, Springer, pp. 499–506.

Apple Indoor Mapping Project and iBeacon, 2014. http://www.businessinsider.com/apple-indoor-mapping-project-and-ibeacon-2014-6.

Arnold, D., 2014. Computer graphics and cultural heritage: From one-way inspiration to symbiosis, part 1. Computer Graphics and Applications, IEEE 34(3), pp. 76–86.

Cigola, M., Gallozzi, A., Ceccarelli, M., Carbone, G., De Stefano, C. and di Freca, A. S., 2014. Strategie robotiche ed informatiche per la fruizione museale. SCIRES-IT 4(1), pp. 59–68.

Clini, P., Frontoni, E., Quattrini, R. and Pierdicca, R., 2014. Augmented reality experience: From high-resolution acquisition to real time augmented contents. Advances in Multimedia.

Clini, P., Quattrini, R., Fiori, F. and Nespeca, R., 2013. Integrated technologies for surveying artefacts damaged by earthquakes. application of all-in-one lidar techniques in the city of laquila. PRS ANN. PHOTOGRAMM. REMOTE SENS. SPATIAL INF. SCI II-5/W1, pp. 79–84.

Denard, H., 2009. The london charter for the computer-based visualisation of cultural heritage. no. February pp. 1–13.

Denard, H., 2012. A new introduction to the london charter. Paradata and Transparency in Virtual Heritage Digital Research in the Arts and Humanities Series(Ashgate, 2012) pp. 57–71.

Fallon, M., Johannsson, H., Brookshire, J., Teller, S. and Leonard, J., 2012. Sensor fusion for flexible human-portable building-scale mapping. In: Intelligent Robots and Systems (IROS), 2012 IEEE/RSJ International Conference on, pp. 4405–4412.

Falzone, P., 1985. Le facciate dipinte. In: AA.VV. (ed.), Le ville del genovesato, vol. I, Sagep, Genova.

Fattal, D., Peng, Z., Tran, T., Vo, S., Fiorentino, M., Brug, J. and Beausoleil, R. G., 2013. A multi-directional backlight for a wide-angle, glasses-free three-dimensional display. Nature 495(7441), pp. 348–351.

Guidi, G. and Bianchini, C., 2007. Tof laser scanner characterization for low-range applications. In: Electronic Imaging 2007, International Society for Optics and Photonics, pp. 649109–649109.

Labò, M., Roma, 1956. Strada nuova (più che una strada un quartiere). scritti in onore di Lionello Venturi.

This contribution has been peer-reviewed.
doi:10.5194/isprsarchives-XL-5-W4-453-2015

Lewin, S., 2014. Holographic displays coming to smartphones [news]. Spectrum, IEEE 51(8), pp. 13–14.

Lo Buglio, D. and De Luca, L., 2012. Representation of architectural artifacts: definition of an approach combining the complexity of the 3d digital instance with the intelligibility of the theoretical model. SCIRES-IT 2(2), pp. 63.

Martin, P., Marchand, E., Houlier, P. and Marchal, I., 2014. Mapping and re-localization for mobile augmented reality. In: IEEE Int. Conf. on Image Processing, ICIP'14, Paris, France, pp. 3352–3356.

Pan, Q., Arth, C., Reitmayr, G., Rosten, E. and Drummond, T., 2011. Rapid scene reconstruction on mobile phones from panoramic images. In: Mixed and Augmented Reality (ISMAR), 2011 10th IEEE International Symposium on, pp. 55–64.

Pirchheim, C. and Reitmayr, G., 2011. Homography-based planar mapping and tracking for mobile phones. In: Mixed and Augmented Reality (ISMAR), 2011 10th IEEE International Symposium on, pp. 27–36.

Poleggi, E., 1972. Strada nuova: una lottizzazione del Cinquecento a Genova. Sagep, Genova.

Pulsed Light, Inc. http://pulsedlight3d.com.

Quattrini, R. and Baleani, E., 2014. Theoretical background and historical analysis for 3D reconstruction model. Villa Thiene at Cicogna. Journal of Cultural Heritage, In Press, Corrected Proof, Available online 20 February 2014.

Remondino, F., 2011. Heritage recording and 3d modeling with photogrammetry and 3d scanning. Remote Sensing.

Schratt, A. and Riedl, A., 2005. The potential of three-dimensional display technologies for the visualization of geo-virtual environments. In: Proceedings of the 22. ICA Cartographic Conference, A Coruna, Spain.

Tanskanen, P., Kolev, K., Meier, L., Camposeco, F., Saurer, O. and Pollefeys, M., 2013. Live metric 3d reconstruction on mobile phones. In: Computer Vision (ICCV), 2013 IEEE International Conference on, pp. 65–72.

Vagnetti, L., 1970. Genova, Strada nuova. Genova.

This contribution has been peer-reviewed.
doi:10.5194/isprsarchives-XL-5-W4-453-2015

AERIAL MAPPING AND MULTI-SENSORS APPROACHES FROM REMOTE SENSING APPLIED TO THE ROMAN ARCHAEOLOGICAL HERITAGE

P. Uribe [a, *], J. Angás [b], F. Pérez-Cabello [c], J. de la Riva [c], M. Bea [d], A. Serreta [e], Mª. A. Magallón [f], C. Sáenz [f], M. Martín-Bueno [f]

[a] Tecnitop, Postdoctoral Research Torres Quevedo (MINECO), Zaragoza, Spain - uribe@unizar.es
[b] 3D Scanner Patrimonio e Industria, *Spin-off* University of Zaragoza, Zaragoza, Spain - j.angas@3dscanner.es
[c] Department of Geography and Spatial Management, University of Zaragoza, Zaragoza, Spain - <fcabello, delariva> @unizar.es
[d] 3D Scanner Patrimonio e Industria, *Spin-off* University of Zaragoza, Postdoctoral Research Torres Quevedo (MINECO), Zaragoza, Spain - manubea@unizar.es
[e] Department of Design and Manufacturing Engineering, University of Zaragoza, Zaragoza, Spain - serreta@unizar.es
[f] Department of Antiquity Studies, University of Zaragoza, Zaragoza, Spain- <amagallo, casaenz, mmartin> @unizar.es

Commission V, WG V/4

KEY WORDS: Aerial Mapping, UAV, Remote Sensing, multispectral image, NDVI.

ABSTRACT:

This report details the preliminary results of the research focused on Roman archaeological heritage in the Middle Ebro Valley (Spain). The principal objective of this project was to obtain several different readings by means of a UAV equipped with different sensors. Firstly, it has been possible to obtain accurate maps, 3D models and digital elevation models of the site. Secondly, it has been possible to investigate and define archaeological remains still underground, via a new methodology which utilises visible and near-infrared wavelengths.

1. INTRODUCTION

The aim of this study was to mitigate various problems currently encountered in the study of the archaeological site by applying remote sensing.

Currently, there is much that is unknown regarding urban and the architectural form of many of the buildings in these pre-Roman and Roman settlements. Given the economic inability to carry out archaeological excavations, low-altitude multispectral images were employed by UAVs (Unmanned Aerial Vehicles) or RPAS (Remotely Piloted Aircraft Systems) with the aim of investigating and defining the still-buried remains.

In recent years, various new technologies have emerged making it possible to obtain accurate data quickly at a lower cost in the field of archaeology.

Firstly, the reliability of UAVs and RPAS has rapidly improved, allowing archaeologists obtain accurate images in variable weather conditions. The possibilities offered to the field of archaeology by this emergent, cutting-edge technology is now generating great interest. Reflecting this interest are new scientific publications concerning the topic: Verhoeven *et al.* (2013), Hill (2013) o Eisenbeiss (2009) y Casana *et al.* (2014), to cite some examples.

Secondly, in parallel to UAV market, various types of software as Agisoft PhotoScan, Pix 4D and APS processes have improved mosaicing, orthorectification and image georeferencing, allowing the archaeologist to obtain a wide range of results such as accurate maps, 3D models or digital terrain and elevation models.

Likewise, the evolution of trade sensors (cameras) adapted for UAVs, above all those related to the field of precision agriculture, today allow the use of devices with a high spatial and a wide spectral resolution. Thus, it is possible to place multispectral cameras on these platforms, covering visible, near-infrared, SWIR and thermal (Casana *et al.* 2014) wavelengths, similar to those provided by satellite imagery.

All these technological advances have been utilised in this project. Consequently, we present preliminary results in which the capabilities of this technique can be seen by their application in different types of deposits. Our aim was to discriminate between the possibilities of identification offered by remains buried beneath flat cultivated fields and steep orography with little vegetation, and the possibilities of identifying remains in unexcavated settlements.

2. CASE STUDY

The choice of the Middle Ebro Valley as geographical framework is due to the detection there of some of the oldest Roman structures in Hispania. This is the result of the crucial role played by this region in the Roman conquest of the Iberian Peninsula. Similarly, the Middle Ebro basin is a natural region with relatively well-defined geographical boundaries: the Catalan Coastal Range, the narrowing of the Riojan Valley, the Pyrenees and the Iberian mountain range.

According to Beltrán (2006, 218.), Rome would have had a recognized perception of this territory as an entity in Imperial times when Augusto created a judicial district - the juridical convent *Caesauragustanus* - which coincided almost exactly with this natural region, as is apparent from the Plinian description (Plin., Nat., 3.4.24.).

* Corresponding author. This is useful to know for communication with the appropriate person in cases with more than one author.

This contribution has been peer-reviewed.
doi:10.5194/isprsarchives-XL-5-W4-461-2015

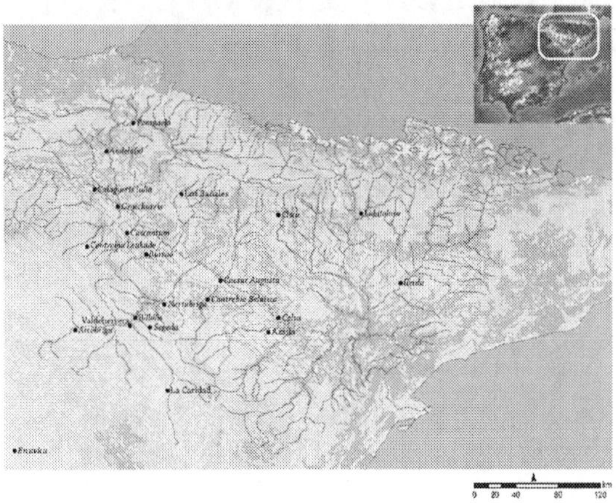

Fig. 1. Roman urban settlements in the Middle Ebro Valley (in-house cartography).

Consequently, the object of our study covers the archaeological remains found in the current territory of Aragon. Among the many aspects of study offered by this territory, we can identify determined areas of interest in relation to the morphology, the chronology and current knowledge about the ancient settlements. The choosing of these fields has given us the opportunity to evaluate the method with the objective of examining its role in the assessment of the orography of sloping terrain, terrain with little vegetation and unknown sites.

The city of *Valdeherrera*, whose ancient name is unknown, was selected. This city is in the vicinity of Calatayud, to the southwest of Saragossa (Zaragoza), the ancient *Caesar Augusta* and capital of *Conventus.* In the northern zone we carried out a study in the pre-Pyrenean city of *Labitolosa* (La Puebla de Castro, Huesca). The city of *Valdeherrera* (Calatayud, Zaragoza) was located in one of the neuralgic points of Celtiberia between the confluence of the rivers Jiloca and Jalon. These rivers were two of the most important ancient natural routes, since they permitted rapid access to both the plateau and the coast (Sáenz and Martín-Bueno 2014, 206). Using terrestrial archaeological prospections, it has been estimated that the enclave would have covered around 40 hectares, of which 30 have survived. Although the name of the city is unknown, some authors have conjectured it was the location of the pre-Roman *Bilbilis* or the *Platea* mentioned by the famous Hispanic poet Martial (4.55.13; 18.12.11.). The oldest preserved remains date from the 4th and 5th centuries BC and had an important role in the Celtiberian Wars in the 2nd and 3rd centuries BC. The city would later be destroyed during the Sertorian conflict, surviving as a minor settlement in Imperial times. The ancient city was partially surrounded by a wide ditch to the East and South. It also had houses with central courtyards, decorated with *opera signina* pavements and walls in the First Style, as recent excavations have shown. These homes have been dated by their excavators as coming from the second half of the 2nd century BC.

The Hispano-Roman city of *Labitolosa* is located in *Cerro del Calvario*, the Ribagorzanan town located at the confluence of the rivers Cinca-Ésera some 2 km east of *La Puebla de Castro,* in the vicinity of Huesca. The original of the name *Labitolosa* was deduced from the Gentilic *labitolosani* that appeared in an inscription found in the 16th century, in a dedication made by the *cives Labitolosani et incolae* (CIL, II, 3008=5837). The recognition of *Labitolosa* as a Latin municipality was confirmed in 1994 by the discovery of an epigraph in the building of the municipal *genius*, located in the forum, which made an explicit reference to the municipality of *Labitolosa* (Navarro and Magallón 1999, 30-56). Thus the city, which originated in the late 1st century BC, developed into a stipendiary community and become a municipality during the Flavian period. In over twenty years of uninterrupted[1] archaeological work in the city of *Labitolosa,* two complete thermal installations have been discovered, along with the remnants of several houses and a significant section of the forum, in which the Curia is particularly notable because of its epigraphic set.

3. AERIAL MAPPING AND MULTISPECTRAL IMAGE

The term UAV refers to an unmanned aerial vehicle that can be controlled remotely, its operation being either semi-autonomous or entirely autonomous. The use of this technology consists of the creation of a photogrammetric measurement platform that permits digitalization to obtain a 3D model of each one of the selected archaeological sites from aerial images. The necessary flight paths are pre-programmed, thus this particular platform operates autonomously. After takeoff, this UAV system is able to follow a programmed flight path over each site using pre-programmed GPS coordinates. The incorporation of high-resolution calibrated RGB and NIR cameras allows the obtaining of high-resolution graphic models.

Although there are diverse types available today, an 8-engined UAV with a folding carbon structure - manufactured in-house - was used in our project, with a Mikrokopter v. 2.02 ME flight control board. This system integration has been designed by Tecnitop SA together with the University of Zaragoza Department of Design and Manufacturing, which was enabled by the SME (Small and Medium Enterprise) Technology Transfer Program of the Aragonese Foundation for Research and Development. The project arose from the necessity of a platform capable of integrating particular sensors, in this case photogrammetric cameras already available in the research group but which could not be utilised with commercial systems. Two different sensors, that have allowed us to capture visible and near-infrared wavelength, have been used.

To capture visible wavelengths, an OLYMPUS E-PM2[2] - a micro four thirds compact camera with a 16-megapixel Live MOS sensor (max. resolution 17.2 megapixels) and TruePic VI processor - was used. This camera has a primary color (RGB) filter known as a *Bayer filter*. To capture near-infrared data, the Olympus E-PM1[3] (similar to the above-mentioned model*)*, with a 4/3" sensor, was used. The micro 4/3 was equipped with an 1-inch diameter IR Long Pass optical filter produced by *Edmund Optics*. The sensitivity of the CMOS of the E-PM1 camera is between 300 and 1000 nm, and the longitude range that the filter can capture is between 700 and 2250 nm. This permitting us to obtain a radiation between 700 to 1000 nm., with the

[1] Regarding the bibliography provided during these years of investigations, we refer to the recent monographic volume of the city: Magallón & Sillières 2013.

[2] For more information, see: http://www.olympus.es/site/es/archived_products/cameras_1/pen_1/e_pm_2/index.pdf

[3] For more information, see: http://www.olympus.es/site/es/archived_products/cameras_1/pen_1/e_pm1/index.pdf

radiation being absent or very low, between 300 and 700 nm. This type of sensor has already been applied in aircraft by Verhoeven (2008; 2012).

Fig. 2. *Example of a generic UAV prototype (Unmanned Aerial Vehicle) used in this project.*

4. DATA ACQUISITION

The acquisition of field data requires the determination of several control points on the ground, known as GCPs (Ground Control Points). This acquisition required a good geometric distribution within the defined flight area with the objective of scaling the final model to a metric system, and inserting it into the selected coordinate reference system. In both tasks, the geodetic reference system used was the official system of Spain - UTM ETRS89 (Royal decree 1071/2007 of 27 July) - in UTM zone 30. The GCPs were acquired using a Leica GS14 GNSS system. In Valdeherrera, the GNSS system was connected via radio link to the geodesic IGN vertex 43756 (Valdegalindo), located in the municipality of Paracuellos de Jiloca. In Labitolosa, the connection between the mobile equipment to a fixed network was carried out via GPRS connection to the ARAGEA active geodesic network of the Aragonese Government (http://gnss.aragon.es/). Subsequently between 15 and 20 points, distributed within the defined area, were recorded. Landmarks ranged from easily-identified points in the field, such as roads or stones, along with targets and survey stakes in areas with vegetation. Consequently, a file with the format "x, y, z, cod" was obtained, to permit later photogrammetric support. The flights in both cases, Valdeherrera and Labitolosa, were planned using the *MKTools* application. The purpose of this planning was to define a route that would allow a longitudinal and transverse image overlap of 80%. The average speed for both flights was 10 m/s at an altitude of approximately 100 metres. The images were recorded in a *.raw* format (the proprietary *Olympus* format being *.orf*) with the same fixed Zuiko Digital ED 12 mm f/2.0 lens with an ISO of 200 and with a manual focus to infinity. In Valdeherrera the images were obtained on the 7th July 2013 between 12 noon and 2:00 pm. In *Labitolosa* the same schedule was used, but the images were acquired on October 5th, 2014.

5. IMAGE PROCESSING

The orthorectification and mosaicing processed were performed by *Agisoft PhotoScan* (2014) software, for all wavelengths. The data processing is summarised in the following steps:

1. Importation of selected images.
2. Synchronization of pictures, coordinates and angles (yaw, pitch, roll).
3. Alignment of images using bundle adjustment or common points.
4. Importation of the *.txt* file with the acquired GCPs.
5. Recognition of the images of the GCPs and x, y, z error control.
6. Adjustment of the layout, selection of the coordinate system used and triangulation of the point cloud.
7. Application of the texture of each of the images that make up the 3D model using a radiometric adjustment by *Adaptive Orthophoto* [4].
8. Exportation of errors in the model, point cloud, 3D model, orthophotos and elevation map.

After obtaining the orthophotos and 3D models, the data was adapted to work with *Erdas Imagine (2013).* This adaption consisted of the creation of a layer stack with the VIS and the NIR images in *.tiff* format, generating multilayer files with 6 bands of multispectral information in each case. (VIS = 123; NIR = 456).
For the identification of buried archaeological remains the following techniques of digital image processing have been applied in combination: (a) the generation of new bands of information (neo-channel and PCAs) from the original bands; and (b) the application of radiometric and spatial enhancements and the creation of compositions in false color (RGB), applied both to the original bands as well as the neo-channels.

a) The generation of neo-channels allows us to extract continuous information and enhance significant aspects of the images from the original bands. For this study the metric relationship between bands has been employed to estimate biophysical parameters such as plant vigor, in the case of obtaining vegetation indices, and the statistical relationship between bands to create the Principal Component Analysis (PCA).
The vegetation indices consist of the pixel-by-pixel building of ratios or quotients, between the digital levels of two or more image bands (Chuvieco 2010, 337). Specifically, for this project have been used: NDVI (*Normalized Difference Vegetation Index*, Rouse *et al.*, 1974), Green NDVI (*Green Normalized Difference Vegetation Index*, Gitelson *et al.*, 1996) y *Soil Adjusted Vegetation Index* (SAVI) (Huete, 1988). The potential of this technique in the field of archaeology and was raised by Verhoeven and Doneus (2011), Verhoeven (2012) y Bennett *et al.* (2012).

The PCA (Principal Components Analysis) is a statistical procedure that reduces redundant information by transforming a number of quantitative variables (common underlying dimensions) into a number of new uncorrelated variables. This reduction is performed using a variance-covariance matrix

---

[4] To create the texture of the 3D model, the software performs a radiometric adjustment based on the average of each pixel in each image taken.

This contribution has been peer-reviewed.
doi:10.5194/isprsarchives-XL-5-W4-461-2015

without losing significant portions of the original information. This capacity of synthesis is of great interest in the field of remote sensing as the acquisition of images in adjacent spectrum bands often involves the detection of redundant information, since canopy cover often exhibit similar behavior in close wavelengths. In the field of archeology, it has been used in particular when working with hyperspectral sensors (Traviglia 2006). In this work, it has been applied to the VIS-NIR image (containing all six bands) for the creation of neo-channels.

b) With respect to radiometric and spatial enhancements, radiometric histogram adjustments, compositions false color (RGB) and spatial filters have been applied. According to Jensen (1996, 145), the linear contrast tends to expand the original digital levels to the minimum and maximum (0-255) to exploit the sensitivity of the device. In this case, statistical (standard deviation) thresholds were used, which were defined statistically and interactively using breakpoints. These settings have been applied to the following images and neo-channels: VIS, NIR, and VIS-NIR images, NDVI neo-channels, Green NDVI, SAVI and ACP. False color compositions are three-band combinations, to each of which were assigned additive primary colors (Red, Green and Blue -RGB-). Here, in place of the original bands, the three NDVIs (those created with the three bands of NIR, *vid.* infra), or the six ACP bands, have been used.

Filters or spatial enhancements are based on the concept of spatial frequency - that is, the number of changes in the brightness per unit of distance (Jensen 1996, 153). These seek to strengthen or weaken spatial contrasts between adjacent pixels via the use of a mobile matrix of filtering coefficients, or kernel. Two types of filters can be distinguished by their filtering coefficients. Firstly, low-pass filters, that have a tendency to highlight the homogeneity component of the image, thereby softening it. Secondly, high pass filters, which sharpen the contrasts between the digital levels of adjacent pixels, leading us to believe they could act as effective tools for the detection of structures. Thus they have been applied to the following images: VIS, NIR, VIS-NIR, as well as individual images without orthorectification or mosaicing. Lasaponara and Masini (2012, 51) describe this type of image processing for archaeological remote sensing using satellite imagery.

6. RESULTS

While the timing of data acquisition was not optimal, due to the phenological time of the vegetation (the images of Valdeherrera were taken in July), a series of structures in the Roman city of Valdeherrera have been identified (figs. 3 and 4) (Fig. 3 y Fig. 4)[5].

[5] The archaeological findings of Labitolosa (Huesca) have not been included as they are still in the processing phase (the field data was collected in October 2014). However, we note that, applying the same methodology used in the case of Valdeherrera, initial results have been obtained.

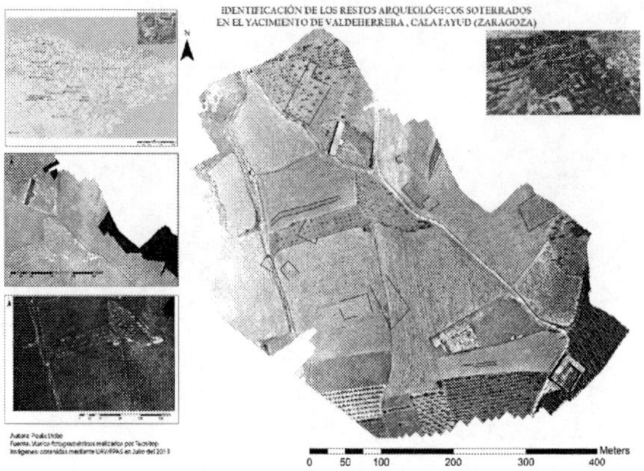

Fig. 3. Mapping with the results obtained for the Valdeherrera site (VIS image).

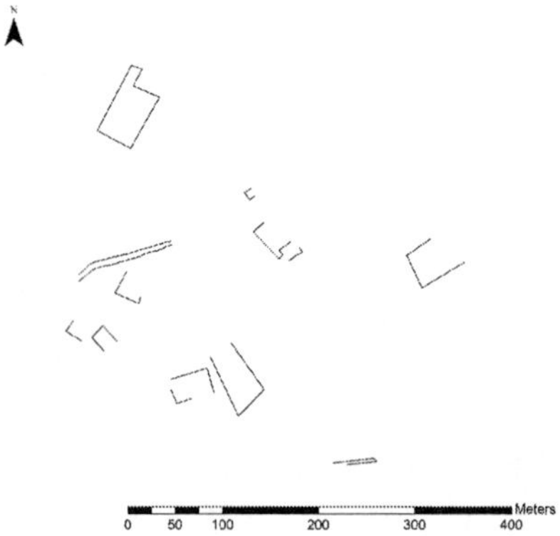

Fig. 4. Documented archeological remains at the site of Valdeherrera (VIS image).

Of the structures identified, we focus on the structure known as A (figs. 5 and 6). We believe it delineates an important building in the city due to its elevated position (as shown in the Digital Elevation Model), its dimensions (25 x 26.67 m) and the discovery of large rectangular ashlars on the ground.

Fig. 5. Documentation of rectangular ashlars on the ground that confirm the existence of a structure.

This contribution has been peer-reviewed.
doi:10.5194/isprsarchives-XL-5-W4-461-2015

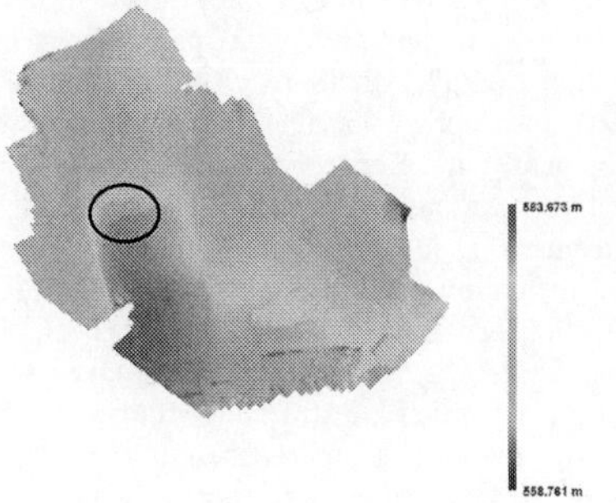

#

Fig. 6. Digital Elevation Modelo (DEM) with the location of structure A.

Similarly, with the structure we identify as B (fig. 7), we posit that it delineates a large structure well-located in the river basin. From its form and dimensions (40 x 20 m.) we believe it was one of the entrances to the city.

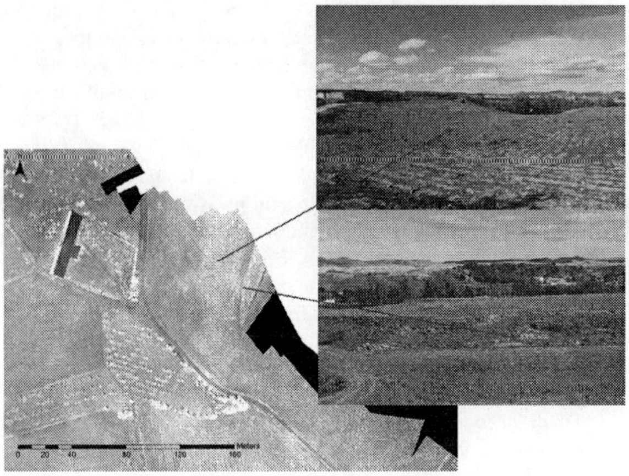

Fig. 7. Archaeological remains on the ground that confirm the existence of a structure B.#

Figure 8 shows the results of a linear contrast enhancement using breakpoints. Its application to vegetation indices has been that which has provided the best results - the clear visual definition of underground structures -. As can be observed in the results obtained by linear contrast enhancement (fig. 8), the digital levels corresponding to the remains are distributed at the extremes of the histogram, being either extremely dark or extremely light.

Fig. 8. Detail of the result of the linear contrast enhancement of the histogram with breakpoints of images NDVI_5 and NDVI_6 of Valdeherrera.

As discussed in the section on methodology, it was initially thought that high-pass filters could be effective tools for the detection of structures. However, as shown below in Figure 9, the results have not been optimal for the purposes of this work.

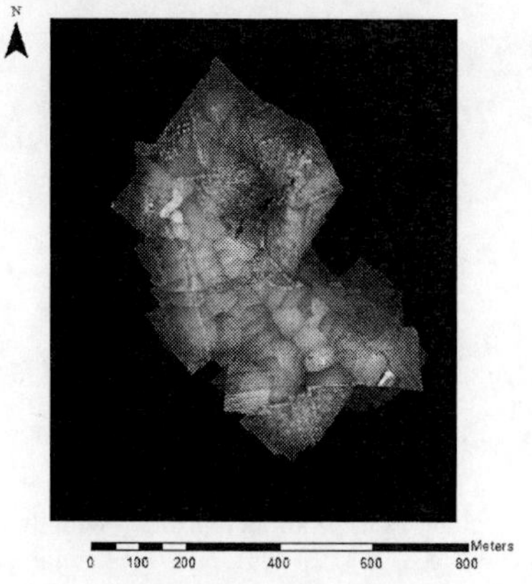

Fig. 9. Results of the digital processing of the VIS-NIR image of Valdeherrera. The filter used was high-pass, with a 5x5 kernel (composition 643) with a standard deviation linear contrast enhancement of 2.5.

As shown in the above image (fig. 9) the results obtained with these tools were not as hoped. The *PhotoScan* mosaicing process is reflected in the image after applying a 5x5 high-pass filter. We have observed that by reducing the kernel to 3x3, this does not occur. In this way, the 5x5 high-pass filter digital processing has been used as a tool for discriminating the buried residues in the overlapping of the *PhotoScan* images, becoming a necessary tool for the processing of the results given by the mosaicing.

The process of determining the Principal Components has not been particularly revealing, identifying practically the same archaeological remains as with vegetation indices, as shown in the detail image above (fig. 10). However, the process vastly improves the identification of archaeological remains when using the PCAs displayed using false color compositions.
Likewise, the use of the vegetation indices (figs.11 and 12) to display underground structures shows itself to be a suitable tool with high potential [6].

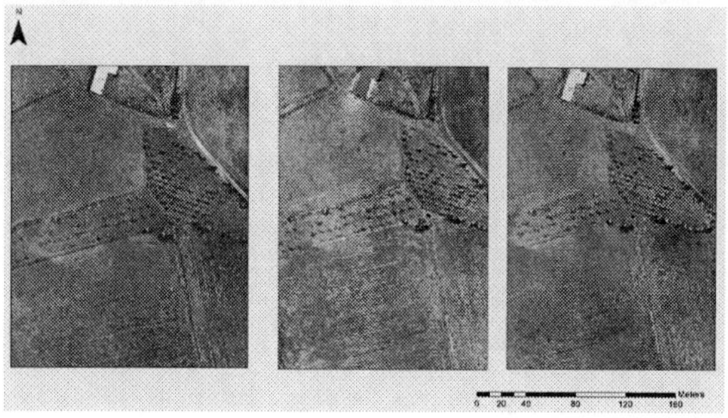

[6] Acording to the NDVI application to identify archaeological remains, vid: Lasaponara y Masini (2006); Verhoeven and Doneus (2011); Lasaponara y Masini (2012); Verhoeven (2012); Bennett *et. al* (2012).

This contribution has been peer-reviewed.
doi:10.5194/isprsarchives-XL-5-W4-461-2015

Fig. 10. Display using false color compositions of the results obtained from the PCA. The first image combines the PCA 123; the second 234, and the third 323. The three images were processed with standard deviation enhancement of 2.5.

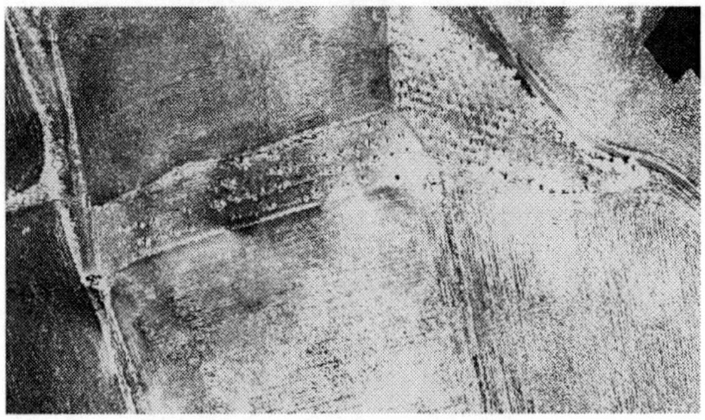

Fig. 11. Detail of false color composition with the three NDVIs (R: NDVI_6; G: NDVI_5; B: NDVI_6) and linear contrast enhancement of the histogram with breakpoints in Valdeherrera.

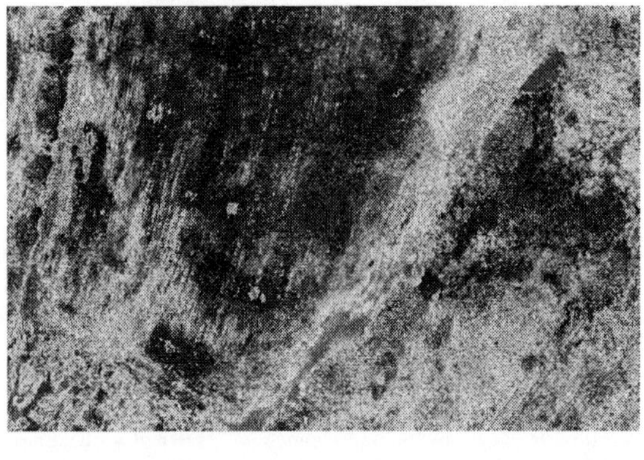

Fig. 12. Detail of false color composition with the three NDVIs (R: NDVI_6; G: NDVI_5; B: NDVI_6) and linear contrast enhancement of the histogram with breakpoints in Labitolosa.

7. CONCLUSIONS AND PROSPECT

In conclusion, it can be seen that the use of UAVs in the field of archeology, primarily, reduces costs. The acquisition of a very high-resolution commercial satellite image or the carrying out of a manned flight greatly increases the cost of fieldwork. Secondly, it is possible to obtain custom data, where the researcher defines the pixel resolution required for the project, and also obtains mapping results: drawings, 3D models, and digital terrain and elevation models. In the same way, the incorporation of multispectral sensors permits the visualization of elements beyond visible wavelengths, that the human eye is incapable of detecting. This, however, introduces a new set of problems such as orthorectification and the processing and obtaining of images.

Despite these problems, and based on the results, it has been proved that this methodology is optimal for identification of buried archaeological remains. After reaching this conclusion, there are many issues concerning future work:

- Without doubt, it has been demonstrated that UAVs, in this day and age, are highly effective in remote sensing studies requiring a maximum spatial resolution tool. However, particular aspects remain to be developed sufficiently, such as: flight time, compatibility with particular sensors, and the integration of RTK GNSS receivers to eliminate the need of support topography. We feel that this area, which is under development, will offer in next years to come more stable and innovative solutions such as the possibility of the use of 3D laser scanners on such aerial platforms.
- Experimentation with different sensors is still required. A new sensor, the *Tetracam Mini-MCA6* has been available since October 2014. The various approaches taken with this camera (the results of which are still under study) have shown that the management of such approaches is not simple, as many problems surface, such as orthorectification and mosaicing. Consequently, we feel, although the precise ranges of wavelengths is unknown, the use of the *Olympus* E-PM2 and E-PM1 cameras is currently the most effective and inexpensive method for reaching our archaeological objectives.
- With respect to image treatment, normalized vegetation indices and their subsequent treatment have proved to be the most effective for our purposes. The use of filters has allowed us to discriminate the mosaicing effects of *Agisoft PhotoScan.* However, other techniques, such as ACP, have produced relevant results when visualised in false color compositions. Thus several lines of research remain open, such as the deepening of the discrimination of soil marks or zonification of the vegetation index pixels.

In conclusion, we believe that the results have been satisfactory. From an archaeological perspective, despite subsequent checks in the field, only the archaeological excavation of documented structures can validate our method and hypothesis.

8. ACKNOWLEDGEMENTS

This research is included within the project led by Dr. M. Martín-Bueno: (HAR2013-48456-C3-1-P) URBS: *Public and private ornamental repertoires in NE Hispania,* funded by the Ministry of Economy and Competitiveness and Dra. Mª. A. Magallon (HAR2011- 24390) - *Roman Hydraulics in the Middle Ebro Valley II. The Roman dams of Muel (Zaragoza), Pared de los Moros (Muniesa. Teruel) and Virgen del Pilar (Moyuela, Teruel)* Ministry of Economy and Finance.

We would like to thank *Tecnitop* and the Spin Off project *Scanner Patrimonio e Industria* for all its support and help during the post-processing of the images.

9. REFERENCES

Angás, J., Uribe, P. 2012. Social spreading of geometric, recorded data from a range of types of 3D scanners via a web data server, *Archaeology in the Digital Era Volume II, 40th Conference on Computer Applications and Quantitative Methods in Archaeology*, Southampton, pp. 363-375.

Barrientos, A., del Cerro, J., Gutiérrex, P., San Martín, R. Martínez, a. y Rossi C. 2007. "Vehículos aéreos no tripulados para uso civil. Tecnología y aplicaciones" http://webdiis.unizar.es/~neira/docs/ABarrientos-CEDI2007.pdf

Braasch, O. 2007. "Gallipoli Ahead-air survey between the Baltic and Mediterranean" In: *Publikácia vznikla v rámci Centra excelentnosti SAV Výskumné centrum Najstarsích dejín Podunajska pri Archeologickom ústave SAV,* Nitra, pp. 84-96.

This contribution has been peer-reviewed.
doi:10.5194/isprsarchives-XL-5-W4-461-2015

Bennett, R., Welham, K., Hill, R. A. y Ford, L. J. 2010. The application of Vegetation Indices for the prospection of Archaeological features in grass-dominated Environments, *Archaeological Prospection,* 19, pp. 209-218.

Beltrán, F. 2006. El valle medio del Ebro durante el periodo republicano: de *limes* a *conventus,* In: *La invención de una geografía de la Península Ibérica,* Málaga-Madrid, pp. 217-240.

Campana, S. 2011. *Remote Sensing, GIS, GPS e tecniche tradizionali. Percorsi integrati per lo studio dei paesaggi archeologici: Murlo-Montalcino e bassa Val di Cornia. Tesi di Dottorato. https://cambridge.academia.edu/StefanoCampana*

Casana, J., Kantner, J., Wiewel, A. and Cothren, J. 2015. Archaeological aerial thermography: a case study at the Chaco-era Blue J community, New Mexico, *Journal os Archaeological Science,* 45, pp. 207-219.

Chuvieco, E. 2010. *Teledetección ambiental. La observación de la tierra desde el espacio,* Ariel Ciencia, Barcelona.

Doneus, M., Verhoeven, G., Atzberger, C., Wess, M. y Rus, M. 2014. New ways to extract archaeological information from hyperspectral pixels, *Journal of Archaeological Science*, 52, pp. 84-96.

Eisenbeiss, H. 2009. UAV Photogrammetry, Zurich, Phd. http://www.igp-data.ethz.ch/berichte/blaue_berichte_pdf/105.pdf

Eisenbeiss, H. y Zang, L. 2006. Comparison of DSMs generated from mini UAV imagenery and terrestrial laser scanner in a cultural heritage application, *International Archives of the Photogrammetry and Remote Sensing,* 36 (5), pp. 90-96.

Erdas Imagine 2013. Erdas Imagine 2013. Huntsville, AL: Intergraph Geospatial

GMES Initial Operations emergency mapping validation service. Technical Specifications JRC/IPR/2012/G02/007/OC, JRC - European Commission.

Gitelson A. A., Kaufman Y.J., Merzlyak, M. N. 1996. Use of green channel in remote sensing of global vegetation from EOS-MODIS, *Remote Sensing Environment* 58, pp. 289-298.

Hill, A. C. 2013. UAVs at Marj Rabba, In: *Israel: low-cost high-tech tools for aerial photography and photogrammetry. Society for American Archaeological Record* 13 (1), pp. 25-29.

Huete, A. R. 1988. A soil-adjusted vegetation index (SAVI), *Remote Sensing Environment*, 25, pp. 295-309.

Jensen, J. 1996. *Introductory digital Image processing*, Prentice Hall Series in Geographic Information Science, Prentice-Hall Inc., Upper Saddle River, New Jersey.

Lasaponara, R. y Masini, N. 2012: *Satellite Remote Sensing A New Tool for Archaeology,* Springer, New York.

Magallón, M. A. y Sillières. P. 2013. *Labitolosa* une cité romaine de l'Hispanie Citérieure, Insitut Ausonius Col. Mémoires nº 33, Burdeos.

Martín-Bueno, M., Sáenz, C., Krausz, S. y Mathé, V. 2009. La ciudad celtibérica de Valdeherrera (Calatayud, Zaragoza). Prospecciones geofísicas, *Salduie*, 9, pp. 419-439.

Montero Ruiz, I.; Rodríguez Alcalde, A. L.; Vicent García, J. M., y Cruz Berrocal, M. 2000. Técnicas analíticas basadas en el proceso de imágenes digitales multiespectrales, *Arkeos,* 7. pp. 13-34.

Nagy, P., Unold, W. and Vogt, S. 2001. *Flug in die Vergangenheit. Die Luftbildpropektion im Dienste der Arhaeologie, Helvetia Archaeologica,* 125/126, Zürich.

Navarro, M and Mª. A. Magallón, 1999. Las ciudades del prepirineo occidental y central en época alto-imperial: sus habitantes y su *status*" In: *Simposio Internacional de Epigrafía E. I. E. G. L. Ciudades privilegiadas en el Occidente Romano,* Sevilla, pp. 30-56.

Rejas, J. G. *et al.* (forthcoming). Anomaly Detection using remote sensing for the Archaeological Heritage registration, *Digital Heritage Congress,* Marsella.

Rouse, J. W., Hass, R. H., Schell, J. A., Deering, D. W. 1974. Monitoring vegetation systems in the Grat Plains with ERTS. In : Proceedings Third Earth Resources Technology Satellite-1 Symposium, Greenbelt, NASA SP-351, pp. 309-317.

Sáenz, C. and Martín-Bueno, M. 2012. La ciudad celtibérica de Valdeherrera (Calatayud. España), *Aquitania*, 28, pp. 7 - 32.

Sáenz, C. and Martín-Bueno, M. 2014. Valdeherrera: la ocupación del territorio en época celtibérica en el valle medio del Jalón, In : *La guerre et ses traces. Conflits et sociétés en Hispanie à l'époque de la conquête romaine (IIIe-Ier s. a.C.),* Burdeaux, pp. 213-229.

Traviglia, A. 2006. Archaeological usability of hyperspectral images: successes and failures of images processing techniques, In: *From Space to place. Proceedings of the 2nd International Conference on Remote Sensing in Archaeology,* Roma, pp. 123-130.

Verhoeven, G. 2008. Imaging the invisible using modified digital still cameras for straightforward and low-cost archaeological near-infrared photography, *Journal of Archaeological Science, pp.* 3087-3100.

Verhoeven, G. 2012. Near_infrared aerial crop marks archaeology: from its historical use to current digital implementations, *Journal of Archaeological Method and Theory*, 19 (1), pp.132-160.

Verhoeven, G., Doneus, M., Atzberger, C., Wess, M., Rus, M., Pregesbauer, M. y Briese, C. 2013. New approaches for archaeological feature extraction of airborne imaging spectroscopy data, In: *Archaeological prospection. proceedings of the 10th International Conference on Archaeological Prospection,* pp. 13-15.

This contribution has been peer-reviewed.
doi:10.5194/isprsarchives-XL-5-W4-461-2015

BUILT BY ALGORITHMS – STATE OF THE ART REPORT ON PROCEDURAL MODELING –

C. Schinko[a,b]*, U. Krispel[a,b], T. Ullrich[a,b], D. Fellner[b,c]

[a]Fraunhofer Austria Research GmbH, Visual Computing
[b]Institute of ComputerGraphics and KnowledgeVisualization (CGV), TU Graz, Austria
[c]GRIS, TU Darmstadt & Fraunhofer IGD, Darmstadt, Germany

Commission V, WG V/4

KEY WORDS: Generative Modeling, Procedural Modeling, Inverse Modeling, Modeling Applications, Shape Description, Language Design

ABSTRACT:

The idea of generative modeling is to allow the generation of highly complex objects based on a set of formal construction rules. Using these construction rules, a shape is described by a sequence of processing steps, rather than just by the result of all applied operations: Shape design becomes rule design. Due to its very general nature, this approach can be applied to any domain and to any shape representation that provides a set of generating functions. The aim of this report is to give an overview of the concepts and techniques of procedural and generative modeling as well as their applications with a special focus on Archaeology and Architecture.

1. INTRODUCTION

The task of generating highly complex objects based on a set of formal construction rules is called generative modeling (Krispel et al., 2014). In contrast to classical modeling, where the object is just the end result of applied operations, this modeling paradigm describes a shape by a sequence of processing steps. The result is a paradigm shift from shape design to rule design. This general approach can be applied to many domains.

1.1 Ruler and Compass

Geometry from the days of the ancient Greeks placed great emphasis on problems of constructing various geometric figures using only a ruler without markings (to draw lines) and a compass (to draw circles). Ruler-and-compass constructions are based on EUCLID's axioms (Heiberg, 2007) using points, lines and circles that have already been constructed. The resulting geometric primitives together with the ruler-and-compass constructions are the first algorithmic descriptions of generative models.

When a line is considered constructed when its two endpoints are located, all constructions possible with a compass and straightedge can be done with a compass alone. The reverse is also true, since JACOB STEINER showed that all constructions possible with straightedge and compass can be done using only a straightedge, as long as a fixed circle and its center have been drawn beforehand. Such a construction is known as a Steiner construction.

The long history of geometric constructions (Martin, 1998) is also reflected in the history of civil engineering and architecture (Mitchell, 1990). Gothic architecture, especially window tracery, exhibits a good example of these constructions. Their complexity is achieved by combining only a few basic geometric patterns. SVEN HAVEMANN and DIETER W. FELLNER show how constructions of prototypic Gothic windows can be formalized using generative modeling techniques (Havemann and Fellner, 2004). By combining modular construction rules it is possible that complex configurations can be obtained from elementary constructions. The different combinations of specific parametric features can be grouped together, leading to the concept of styles. A differentiation between basic shape and appearance allows, for example, the creation of ornamental decoration (Thaller et al., 2013a). This leads to an extremely compact representation for a whole class of shapes (Berndt et al., 2005a).

*Corresponding author.

1.2 Civil Engineering

Because the generative modeling approach is very general, it can be applied to any domain and is not restricted to shape representations (Chakrabarti et al., 2011), (Compton and Mateas, 2006). The discipline of civil engineering focuses on computer-aided design, shape design, and 3D modeling.

Each design process which involves repetitive tasks is perfectly suited for a generative approach. Engineering processes can be classified in repetitive and creative processes. In contrast to creative processes, repetitive ones consist of nearly identical tasks and are therefore independent of creative decisions. This is a precondition for modeling them in a system of rules as demonstrated by GERALD FRANK (2012): Liebherr manufactures and sells an extensive range of products including different kinds of cranes. Each crane has to be partially or fully engineered tailored to the needs of the customer. Nevertheless, the design process of ascent assemblies is based on repetitive tasks that are described by a set of invariant rules. These rules have been modeled and stored by Liebherr. The integration into the existing CAD pipeline now allows a construction engineer to create ascent assemblies only by determine the defining parameters and filling out the corresponding input fields in a user interface. Using the procedural approach, the efforts of engineering ascent assemblies have been reduced to 10%.

1.3 Natural Patterns

In today's procedural modeling systems, scripting languages and grammars are often used to create a set of rules to achieve a description of an object or pattern. Early systems based on grammars were Lindenmayer systems, short L-systems, named after ARISTID LINDENMAYER (Prusinkiewicz and Lindenmayer,

1990). They were successfully used for modeling plants (Deussen and Lintermann, 2005) or fractal structures (Mandelbrot, 1982). Given a set of string rewriting rules, complex strings are created by applying these rules to simpler strings. Starting with an initial string the predefined set of rules form a new, possibly larger string. In order to use L-systems to model geometry an interpretation of the generated strings is necessary.

The modeling power of L-systems was limited to creating fractals and plant-like branching structures. This limitation lead to the introduction of parametric L-systems. The idea is to associate numerical parameters with L-system symbols to address continuous phenomena which were not covered satisfactorily by L-systems alone.

In combination with additional 3D modeling techniques, Lindenmayer systems can be used to generate complex geometry. ROBERT F. TOBLER et al. (2002) introduce a combination of subdivision surfaces, fractal surfaces, and parametrized L-systems to create models of natural phenomena. Different combinations can be used at each level of resolution. Since the whole description of such multi-resolution models is procedural, their representation is very compact and can be exploited by level-of-detail renderers.

This trade-off between data storage and computation time can be found in various fields of computer graphics, e.g. the tessellation of curved surfaces specified by a few control points directly on the GPU. The result is low storage costs allowing the generation of complex models only when needed, while also reducing memory transfer overheads. Although L-systems are parallel rewriting systems, derivation through rewriting leads to very uneven workloads. Since the interpretation of an L-system is an inherently serial process, they are not straightforwardly applicable to parallel processing. In 2010, MARKUS LIPP et al. (2010) presented a solution to this algorithmic challenge.

2. LANGUAGES & GRAMMARS

Scripting languages have been designed for a special purpose, e.g., for client-side scripting in a web browser. Nowadays, scripting languages are used for many different applications. JavaScript, for example, is used to animate 2D and 3D graphics in VRML (Brutzman, 1998) and X3D (Behr et al., 2007) files. It checks user forms in PDF files (Breuel et al., 2011), controls game engines (Di Benedetto et al., 2010), configures applications, defines 3D shapes (Schinko et al., 2011a), and performs many more tasks. According to JOHN K. OUSTERHOUT (1998) scripting languages use a higher level of abstraction compared to system programming languages as they are often typeless and interpreted to emphasize the rapid application development purpose. System programming languages, on the other hand, are designed for creating algorithms and data structures based on low-level data types and memory operations. Consequently, graphics libraries (OpenGL Architecture, 1993), shaders (NVidia, n.d.) and scene graph systems (Reiners et al., 2002), (Voß et al., 2002) are usually written in C/C++ dialects (Eckel, 2003), whereas procedural modeling frameworks incorporate scripting languages such as Lua, JavaScript, etc.

2.1 Language Processing & Compiler Construction

For the evaluation of procedural descriptions typically techniques used for description of formal languages and compiler construction are used (Parr, 2010). There is a wide range of different concepts of languages to describe a shape including all kinds of linguistic concepts (Chomsky, 1956). The main categories to describe a shape are

- *rule-based:* using substitutions and substitution rules to generate complex structures out of simple starting structures (Özkar and Kotsopoulos, 2008), (Krecklau et al., 2010), (Müller et al., 2006c), (Snyder and Kajiya, 1992).
- *imperative and scripting-based:* using a scripting engine and techniques from predominant programming languages (Havemann, 2005), (Schinko et al., 2011a), (Krecklau and Kobbelt, 2011), or
- *GUI and dataflow-based:* using new graphical user interfaces (GUI) and intelligent GUIs to detect structures in modeling tasks, which can be mapped onto formal descriptions (Lipp et al., 2008), (Thaller et al., 2012).

The general principles of formal descriptions and compiler construction are the same in all cases – independent of ahead-of-time compilation, just-in-time compilation or interpretation (Schinko et al., 2012). In the first stage of the compilation process, the input source code is passed to lexer and parser. A first step here is to convert a sequence of characters into a sequence of tokens, which is done by special grammar rules forming the lexical analysis. Typically only a limited number of characters is allowed for an identifier: all characters `A-Z`, `a-z`, digits `0-9` and the underscore `_` are allowed with the condition that an identifier must not begin with a digit or an underscore. The lexer rules are embedded in another set of rules – the parser rules. They are evaluating the resulting sequence of tokens to determine their grammatical structure. The complete grammar is of hierarchical structure and consists of rules for analyzing all possible statements and expressions that can be formed in the language, thus forming the syntactic analysis.

For each available language construct a set of rules ensures syntactic correctness and incorporates mechanisms to report possible syntactic errors and warnings. These rules are also used to create the intermediate AST structure that is a representation of the input source code to be used for the next stage: semantic analysis. Once all statements and expressions of the input source code are collected in the AST, a tree walker checks their semantic relationships for errors and warnings. After performing all compile-time checks, a translator uses the AST to generate platform-specific files possibly involving other intermediate structures.

As mentioned in the Introduction, the first procedural modeling systems were L-systems. Later on, L-systems were used in combination with shape grammars to model cities. YOGI PARISH and PASCAL MÜLLER (2001) presented a system that generates a street map enriched with geometry for buildings using a number of image maps as input. The resulting framework called *CityEngine* is a modeling environment for the shape grammar *CGA Shape*. MARKUS LIPP et al. (2008) presented another modeling approach based on *CGA Shape* following the notation of PASCAL MÜLLER (2006). It enables more direct local control of the underlying grammar by introducing visual editing. Principles of semantic and geometric selection are combined as well as functionality to store local changes persistently over global modifications.

SVEN HAVEMANN (2005) takes a different approach to generative modeling. He proposes a stack based language called *Generative Modeling Language* (GML). The postfix notation of the language is very similar to that of *Adobe Postscript*. High-level shape operations are created by using low-level shape functionality. A number of applications are based on the GML platform because it is easily extensible and offers an integrated visualization engine. Current efforts in the context of the GML are devoted to directly creating interactive generative visualizations for the web.

This contribution has been peer-reviewed.
doi:10.5194/isprsarchives-XL-5-W4-469-2015

Generative modeling inherits methodologies of 3D modeling and programming (Ullrich et al., 2008a), which leads to drawbacks in usability and productivity. The need to learn and use a programming language is a significant inhibition threshold especially for non-computer scientists. The choice of the scripting language has a huge influence on usability and effectiveness of procedural modeling. *Processing* is a good example of how an interactive, easy to use, yet powerful, development environment can open up new user groups. It has been initially created to serve as a software sketchbook and to teach students fundamentals of computer programming. It quickly developed into a tool that is used for creating visual arts (Reas et al., 2007).

Processing is a Java-like interpreter offering new graphics and utility functions together with some usability simplifications. The large community behind the tool produced libraries to facilitate computer vision, data visualization, music, networking, and electronics. The success of *Processing* is based on two factors: the simplicity of the programming language and the interactive experience. Instant feedback of the scripting environments allow the user to program via "trial and error".

2.2 Scripting Languages for Generative Modeling

There are many different programming paradigms in software development that are also used in the field of generative modeling, where some paradigms emerged to be useful for specific domains.

imperative: Many generative models are described using classical programming paradigms: A programming language is used to generate a specific object possibly using a library that utilizes some sort of geometry representation and operations to perform changes. Any modeling software that is scriptable by an imperative language or provides some sort of API falls into this category.

dataflow based: A generative description can be represented by a directed graph of the data flowing between operations. This graph representation also allows for a graphical representation; Visual Programming Languages (VPL) allow to create a program by linking and modifying visual elements. Many VPL's are based on the dataflow paradigm. Examples in the domain of generative modeling are the Grasshopper3D[1] plug-in for the Rhinoceros3D[2] modeling suite, or the work of GUSTOVA PATOW et al. (2012) built on top of the procedural modeler Houdini[3].

rule based systems: Another different representation for generative modeling are rule-based systems. These systems provide a declarative description of the construction behavior of a model by a set of rules. An example are L-Systems, as described in the Introduction. Furthermore, the seminal work of GEORGE STINY and JAMES GIPS (1971) introduced shape grammars, as a formal description of capturing the design of paintings and sculptures. Similar to formal grammars, shape grammars are based on rule replacement.

Shape Grammars

In the classical definition of GEORGE STINY and JAMES GIPS, a shape grammar is the 4-tuple $SG = (V_T, V_M, R, I)$, where V_T a set of shapes, $V_T{}^*$ denotes the set of the shapes of V_T with any scale or rotation. V_M is a finite set of *non-terminal* shapes (markers) such that $V_T{}^* \cap V_M = \emptyset$. R denotes the set of rules, which consists of pairs (u, v), such that $u = (s, m)$ consists of a shape $s \in V_T{}^*$ combined with a marker of $m \in V_M$, and v is a shape consisting of either

- $v = s$
- $v = (s, \tilde{m})$ with $\tilde{m} \in V_M$
- $v = (s \cup \tilde{s}, \tilde{m}$ with $\tilde{s} \in V_T{}^*$ and $\tilde{m} \in V_M$

Elements of the set $V_T{}^*$ that appear in rules of R are called *terminal shapes*. I is called the *initial shape*, and typically contains an $u \in (u, v) \in R$. The final shape is generated from the shape grammar by starting with the initial shape and applying matching rules from R: for an input shape and a rule (u, v) whose u matches a subset of the input, the resulting shape is another shape that consists of the input shape with the right side of the rule substituted in the matching subset of the input. The matching identifies a geometric transformation (scale, translation, rotation, mirror) such that u matches the subset of the input shape and applies it to the right side of the rule. The *language* defined by a shape grammar SG is the set of shapes that will be generated by SG that do not contain any elements of V_M.

Split Grammars

The work of PETER WONKA et al. (2003) applied the concepts of shape grammars to derive a system for generative modeling of architectural models. This system uses a combination of a spatial grammar system (split grammar) to control the spatial design and a control grammar, which distributes the design ideas spatially (e.g. set different attributes for the first floor of a building). Both of these grammars consist of rules with attributes that steer the derivation process. The grammar consists of two types of rules: *split* and *convert*. The *split* rule is a partition operation which replaces a shape by an arrangement of smaller shapes that fit in the boundary of the original shape. The *convert* rule replaces a shape by a different shape that also fits in the boundary of the original shape.

This system has further been extended by the work of PASCAL MÜLLER et al. (2006), which introduced a *component split* to extend the split paradigm to arbitrary 3d meshes, as well as occlusion queries and snap lines to model non-local influences of rules. For example, two wall segments that intersect each other should not produce windows such that the window of one wall coincides with the other wall, therefore occlusion queries are used to decide if a window should be placed or not.

The derivation of a split grammar, starting from an initial shape, yields a tree structure, which suggests that the derivation can be speed up by a parallel implementation, which has been shown by JEAN-EUDES MARVIE et al. (2012). Parallel generation is especially useful in an urban context, with scenes with high complexity and detail. The work of LARS KRECKLAU et al. (2013) used gpu accelerated generation in the context of generating and rendering high detailed building façades; the work of ZHENGZHENG KUANG et al. (2013) proposes a memory-efficient procedural representation of urban buildings for real-time visualization.

With more advanced shape grammar systems, the non-local influences are a problem because they introduce dependencies between arbitrary nodes of the derivation tree. Recent work by MARKUS STEINBERGER et al. (2014) shows how to overcome this problem in an GPU implementation. Furthermore, the same authors presented methods to interactively generate and render only the visible part of a procedural scene using procedural occlusion culling and level of detail (Steinberger et al., 2014b)

[1] http://www.grasshopper3d.com
[2] http://www.rhino3d.com
[3] http://www.sidefx.com

This contribution has been peer-reviewed.
doi:10.5194/isprsarchives-XL-5-W4-469-2015

3. MODELING BY PROGRAMMING

3D objects consisting of organized structures and repetitive forms are well suited for procedural descriptions, e.g. by the combination of building blocks or by using shape grammars.

3.1 Building Blocks & Elementary Data Structures

Creating shapes with elementary data structures requires the definition of modeling operations. Depending on the underlying representation, certain modeling operations are difficult or impossible to implement. The selection of operations for these data structures are manifold and can be grouped as follows:

- *Instantiations* are operations for creating new shapes.
- *Binary Creations* are operations involving two shapes such as constructive solid geometry (CSG) operations.
- *Deformations* and *Manipulations* stand for all deforming and modifying operations like morphing or displacing.

Building blocks can also be regarded as modeling operations. When creating an algorithmic description of a shape, an important task is to identify inherent properties and repetitive forms. These properties must be accounted for in the structure of the description. Identified subparts or repetitive forms are best mapped to functions in order to be reusable. However, the true power of an algorithmic description becomes obvious when parameters are introduced for these functions. Even if only used to position a subpart at a different location. From that point on, the algorithmic description no longer stands for a single object, but for a whole object family.

3.2 Architectural Modeling with Procedural Extrusions

This method utilizes the paradigm of footprint extrusion to automatically derive geometry from a coarse description. Input to this system are polygons whose segments can be associated with an extrusion profile polygon. The system utilizes the weighted straight skeleton method (Aurenhammer, 2008) to calculate the resulting geometry. Examples can be seen in Figure 1.

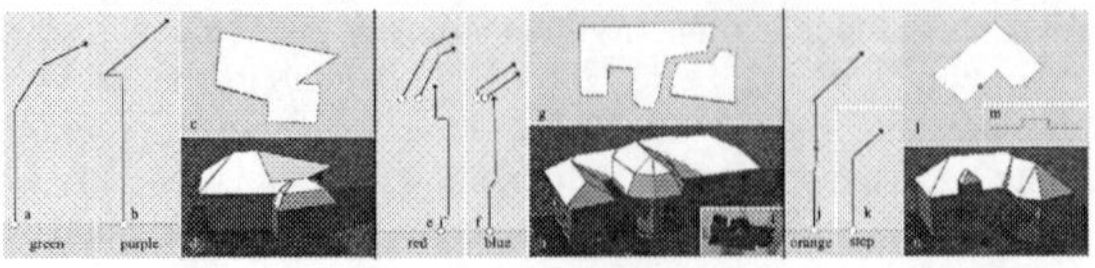

Figure 1: The work of TOM KELLY and PETER WONKA (2011) offers a framework to specify the geometry of a building by extrusion profiles. The segments of footprint polygons (*c*) are associated with extrusion profiles - the green segments are associated to the profile *a*, the purple segments to the profile *b*. The resulting geometry can be seen in *d*.

The growing demand for new building models for virtual worlds, games, and movies, makes the easy and fast creation of modifiable models more and more important (Watson and Wonka, 2008). Nevertheless, 3D modeling of buildings can be a tedious task due to their sometimes complex geometry (Whiting et al., 2009). For historic buildings, especially the roofs can be challenging. JOHANNES EDELSBRUNNER et al. (2014) present a new method of combining simple building solids to form more complex buildings, and give an emphasis on the blending of roof faces. Their method can be integrated in common pipelines for procedural modeling of buildings and extends their expressiveness compared to existing methods.

3.3 Deformation Aware Shape Grammars

Generative models based on shape and split grammar systems often exhibit planar structures. This is the case because these systems are based on planar primitives and planar splits. There are many geometric tools available in modeling software to transform planar objects into curved ones, e.g. *free-form deformation* (Sederberg and Parry, 1986). Applying such a transformation as a post-processing step might yield undesirable results. For example, if a planar facade of a building is bent into a curved shape, the windows inside the façade will have a curved surface as well. Another possibly unwanted property arises when an object is deformed by scaling: the windows on a façade would have different appearances.

RENÉ ZMUGG et al. (2013) introduced *deformation aware shape grammars*, which integrate deformation information into grammar rules. The system still uses established methods utilizing planar primitives and splits, however, measurements that determine the available space for rules are performed in deformed space. In this way, deformed splits can be carried out, the deformation can be baked at any point to allow for straight splits in deformed geometry. An example is shown in Figure 2.

3.4 Procedural Shape Modeling

The effectiveness of procedural shape modeling can be demonstrated with mass customization of consumer products (Berndt et al., 2012). A generative description composed of a few well-defined procedures can generate a large variety of shapes. Furthermore, it covers most of the design space defined by an existing collection of designs – in this case wedding rings.

The basic shape of most rings can be defined using a profile polygon, the angular step size defined by the number of supporting profiles to be placed around the ring's center, the radius, and a vertex transformation function. A ring's design variations are decomposed into a set of transformation functions. Each function transforms selected parts of the profile in a certain way. Effects can be combined by calling a sequence of different transformations. The creation of the basic shape is separated from optional steps to create engravings, change materials, or add gems. Engravings are implemented as per-vertex displacements (to maintain the option for 3D-printing) and can be applied on quadrilateral parts of the ring's mesh using half-edges to specify position and spatial extend.

Materials like gold, silver, and platinum are used for wedding rings. Their surfaces can be treated with various finishing tech-

Figure 2: Deformation aware shape grammars allow the integration of free-form deformation into a grammar-based system based on planar primitives and splits. An undeformed building with rooms (top image) is deformed using two different deformations (middle, bottom).

This contribution has been peer-reviewed.
doi:10.5194/isprsarchives-XL-5-W4-469-2015

niques like polishing, brushing, or hammering. In order to account for these effects, a per-pixel shading model is used featuring anisotropic highlights. By using a cube map, visually appealing reflections are created and predefined surface finishes can be applied using normal mapping techniques. Procedural gem instances can also be placed on the ring.

The presented approch is used in a hardware accelerated server-side rendering framework (Schinko et al., 2014), which has been included in an online system called *REx* by *JohannKaiser*. It offers intuitive web interface for configuring and visualizing wedding rings.

This work demonstrates the efficiency of procedural shape modeling for the mass customization of wedding rings. The presented generative description is able to produce a large variety of wedding rings. Figure 3 shows a few results of the parametric toolkit.

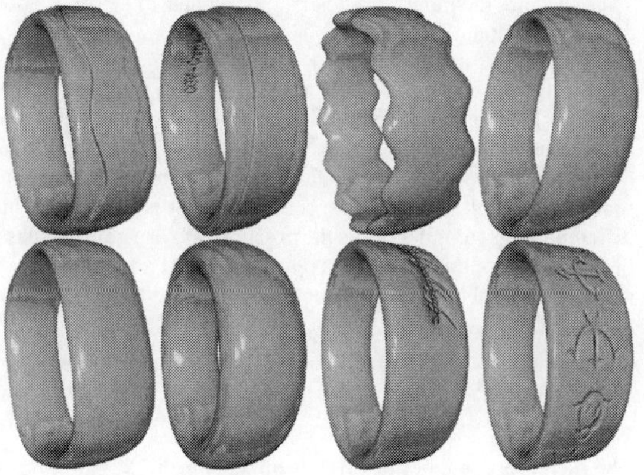

Figure 3: The presented generative description is able to produce a large variety of wedding rings. Features like engravings, recesses, different materials, unusal forms and gems can be created and customized.

3.5 Variance Analysis

The analysis and the visualization of differences of similar objects is important in many research areas: scan alignment, nominal/actual value comparison, and surface reconstruction to name a few. In computer graphics, for example, differences of surfaces are used to validate reconstruction and fitting results of laser scanned surfaces. Scanned representations are used for documentation as well as analysis of ancient objects revealing smallest changes and damages. Analyzing and documentation tasks are also important in the context of engineering and manufacturing to check the quality of productions.

CHRISTOPH SCHINKO et al. (2011) contribute a comparison of a reference / nominal surface with an actual, laser-scanned data set. The reference surface is a procedural model whose accuracy and systematics describe the semantic properties of an object, whereas the laser-scanned object is a real-world data set without any additional semantic information. The first step of the process is to register a generative model (including its free parameters) to a laser scan. Then, the difference between the generative model and the laser scan is stored in a texture, which can be applied to all instances of the same shape family.

A generative models represent an ideal object rather than a real one. The combination of noisy 3D data with an ideal description enhances the range of potential applications. This bridge between both the generative and the explicit geometry description is very important: it combines the accuracy and systematics of generative models with the realism and the irregularity of real-world data as pointed out by DAVID ARNOLD (2006). Once the procedural description is registered to a real-world artifact, we can use the fitted procedural model to modify a 3D shape. In this way we can design both low-level details and high-level shape parameters at the same time.

3.6 Semantic Modeling

In the context of digital libraries, semantic meta data plays an important role. It provides semantic information that is vital for digital library services: indexing, archival, and retrieval. Depending on the field of application, meta data can be classified according to the following criteria (Ullrich et al., 2010b):

Data Type The data type of the object can be of any elementary data structure (e.g. Polygons, NURBS, Subdivision Surfaces, ...).

Scale of Semantic Information This property describes, whether meta data is added for the entire data set or only for a sub part of the object.

Type of Semantic Information The type of meta data can be descriptive (describing the content), administrative (providing information regarding creation, storing, provenance, etc.) or structural (describing the hierarchical structure).

Type of creation The creation of the semantic information for an object can be done manually (by a domain expert) or automatically (e.g. using a generative description).

Data organization The two basic concepts of storing meta data are storing the information within the original object (e.g. EXIF data for images), or storing it separately (e.g. using a database).

Information comprehensiveness The comprehensiveness of the semantic information can be declared varying from low to high in any gradation.

Many concepts for encoding semantic information can be applied to 3D data, unfortunately only a few 3D data formats support semantic markup (Settgast, 2013):

Collada The XML-based Collada format allows storing meta data like title, author, revision etc. not only on a global scale but also for parts of the scene. This file format can be found in Google Warehouse where meta data is, for example, used for geo-referencing objects.

PDF 3D PDF 3D allows to store annotations separated from the 3D data even allowing annotating the annotations. An advantage is that the viewer application is widely spread and PDF documents are the quasi standard for textual documents.

Due to the persistent naming problem, a modification of the 3D model can break the integrity of the semantic information. Any change of the geometry can cause the referenced part of the model to no longer exist or being changed. There are a lot of examples for semantic modeling in various contexts (Boulch et al., 2013), (Haegeler et al., 2009), (Mendez et al., 2008), (Thaller et al., 2013b), (Van Gool et al., 2013), (Yong et al., 2012).

This contribution has been peer-reviewed.
doi:10.5194/isprsarchives-XL-5-W4-469-2015

4. INVERSE MODELING

The full potential of generative techniques is revealed when the inverse problem is solved; i.e. what is the best generative description of one or several given instances of an object class? This problem can be interpreted in different ways. The simplest way is to create a generative model out of a given 3D object and to store it in a geometry definition file format. Obviously, this is not the desired result as the generative model can only represent a single object, not a family of objects.

4.1 Parsing shape grammars

Shape grammars can be used to describe the design space of a class of buildings / façades . An interesting question in this context is: given a set of rules and measurements of a building, typically photographs or range scans, which application of rules yields the measurements? Here, the applied rules can also be seen as *parse tree* of a given input.

The work of HAYKO RIEMENSCHNEIDER et al. (2012) utilizes shape grammars to enhance the results of a machine learning classifier that is pre-trained to classify pixels of an orthophoto of a façade into categories like windows, walls, doors and sky. The system applies techniques from formal language parsing to parse a two-dimensional split grammar consisting of horizontal and vertical splits, as well as repetition and symmetry operations. For the reduction of the search space, an irregular grid is derived from the classifications, and the parsing algorithm is applied to yield the most probable application of rules that yields a classification label per grid cell. These parse trees can easily be converted into procedural models.

FUZHANG WU et al. (2014) also address the problem of how to generate a meaningful split grammar explaining a given facade layout. Given a segmented facade image, the system uses an approximate dynamic programming framework to evaluate if a grammar is a meaningful description. However, the work does not contribute to the problem of facade image segmentation.

4.2 Model synthesis

PAUL MERELL and DINESH MANOCHA (2008) present an approach that given an object (i.e. a mesh) and constraints, derives a locally similar object. This method is related to texture synthesis. It computes a set of acceptable states, according to several types of constraints and constructs parallel planes that correspond to faces orientations of the input model. The intersections of these planes yield possible vertex positions in the output model. Acceptable states are assigned to a vertex while incompatible states are removed in its neighbourhood. The system terminates, if every vertex has been assigned a state.

4.3 Inverse procedural modeling of trees

The method proposed by ONDREJ STAVA et al. (2014) estimates the parameters of a stochastic tree model, given polygonal input tree models. This is done in such a way that the stochastic model produces trees similar to the input. The parameters are estimated using Markov Chain Monte Carlo (MCMC) optimization techniques. A statistical growth model consisting of 24 geometrical and environmental parameters is used. The authors propose a similarity measure between the statistical model and a given input mesh that consists of three parts: *shape distance*, measuring the overall shape discrepancy, *geometric distance*, reflecting the statistics of geometry of its branches, and *structural distance*, encoding the cost of transforming a graph representation of the statistical tree model into a graph representation of the input tree model. The MCMC method has also been applied by other methods to find parameters of a statistical generative model: (Talton et al., 2011), (Vanegas et al., 2012), (Yu et al., 2011).

4.4 Parameter Fitting and Shape Recognition

TORSTEN ULLRICH and DIETER W. FELLNER (2011) presented an approach that uses generative modeling techniques to describe a class of objects and to identify objects in real-world data e.g. laser scans. A point cloud P and a generative model M are the input data sets of the algorithm. It answers the questions

1. whether the point cloud can be described by the generative model and if so,
2. what are the input parameters x_0 such that $M(x_0)$ is a good description of P.

A hierarchical optimization routine based on fuzzy geometry and a differentiating compiler is used. The complete generative model description $M(x_1, \dots, x_k)$ (including all possibly called subroutines) is differentiated with respect to the input parameters. This differentiating compiler offers the possibility to use gradient-based optimization routines in the first place. Without partial derivatives many numerical optimization routines cannot be used at all or in a limited way.

5. SEMANTIC ENRICHMENT

The increasing number of (3D) documents makes digital library services become more and more important. A digital library provides markup, indexing, and retrieval services based on available metadata. In a simple case, metadata is of the Dublin Core (1995) type: title, creator/author, time of creation, etc. This is insufficient for large collections of 3D objects, because of their versatility and rich structure.

Scanned models are used in raw data collections, for documentation archival, virtual reconstruction, historical data analysis, and for high-quality visualization for dissemination purposes (Settgast et al., 2007). Navigation and browsing through the geometric models should be possible on the semantic level - this requires higher-level semantic information. The need for semantic information becomes immediately clear in the context of electronic data exchange, storage and retrieval (Fellner, 2001), (Fellner et al., 2007). The problem of 3D semantic enrichment is closely related to the shape description problem (Maybury, 2012):

How to describe a shape and its structure on a higher, more abstract level?

The traditional way of classifying objects, pursued both in mathematics and, in a less formal manner, in dictionaries, is to define a class of objects by listing their distinctive properties. This approach is hardly realizable because of the fact that definitions cannot be self-contained. They depend on other definitions, which leads to circular dependencies that cannot be resolved automatically by strict reasoning, but rely on intuitive understanding at some point.

An alternative, non-recursive approach for describing shape uses examples. Each entry in a picture dictionary is illustrated with

This contribution has been peer-reviewed.
doi:10.5194/isprsarchives-XL-5-W4-469-2015

a photo or a drawing. This approach is widely used, for example in biology for plant taxonomy. It avoids listing an exhaustive list of required properties for each entry. However, it requires some notion of similarity, simply because the decision whether object x belongs to class A or B requires measuring the closeness of x to the exemplars $a \in A$ resp. $b \in B$. This decision can be reached by a classifier using statistics and machine learning (Bishop, 2007), (Ulusoy and Bishop, 2005). A survey on content-based 3D object retrieval is provided by BENJAMIN BUSTOS et al. (2007). Statistical approaches clearly have their strength in discriminating object classes. However, feature-based object detection, e.g., of rectangular shapes, does not yield object parameters: width and height of a detected rectangle must typically be computed separately.

To describe a shape and its construction process, its inner structure must be known. Structural decomposition is well in line with human perception. In general, shapes are recognized and coded mentally in terms of relevant parts and their spatial configuration or structure (King and Wertheimer, 2005). One idea to operationalize this concept was proposed, among others, by MASAKI HILAGA et al. (2001), who introduce the Multiresolution Reeb Graph, to represent the skeletal and topological structure of a 3D shape at various levels of resolution. Structure recognition is a very active branch in the field of geometry processing. The detection of shape regularities (Pauly et al., 2008), self-similarities (Bokeloh et al., 2010) and symmetries (Mitra et al., 2006), (Mitra et al., 2007) is important to understand a 3D shape. To summarize, structural decomposition proceeds by postulating that a certain type of general regularity or structure exists in a class of shapes. This approach clearly comes to its limits when very specific structures are to be detected, i.e., complicated constructions with many parameter interdependencies.

A possibility to describe a shape is realized by the generative modeling paradigm (Özkar and Kotsopoulos, 2008), (Ullrich et al., 2010a). The key idea is to encode a shape with a sequence of shape-generating operations, and not just with a list of low-level geometric primitives. In its practical consequence, every shape needs to be represented by a program, i.e., encoded in some form of programming language, shape grammar (Müller et al., 2006c), modeling language (Havemann, 2005) or modeling script (Autodesk, 2007).

The implementation of the "definition by algorithm" approach is based on a scripting language (Ullrich and Fellner, 2011): Each class of objects is represented by one algorithm M. Furthermore, each described object is a set of high-level parameters x, which reproduces the object, if an interpreter evaluates $M(x)$. As this kind of modeling resembles programming rather than "designing", it is obvious to use software engineering techniques such as versioning and annotations. In this way, model M may contain a human-readable description of the object class it represents.

In contrast to other related techniques using fitting algorithms, such as "Creating Generative Models from Range Images" by RAVI RAMAMOORTHI and JAMES ARVO (1998), the approach by TORSTEN ULLRICH (2011) can classify data semantically. Although RAVI RAMAMOORTHI and JAMES ARVO also use generative models to fit point clouds, they modify the generative description during the fitting process. As a consequence the optimization can be performed locally with a computational complexity, which is significantly reduced. But starting with the same generative description to fit a spoon as well as a banana does not allow to generate or preserve semantic data.

An example illustrates this process. The generative model to describe a vase takes 13 parameters: $R(r_x, r_y, r_z)$ is the base reference point of the vase in 3D and $T(t_x, t_y, t_z)$ is its top-most point. The points R and T define an axis of rotational symmetry. The remaining seven parameters define the distances $d_0, \ldots, d_6$ of equally distributed Bézier vertices to the axis of rotation (see Figure 4). The resulting 2D Bézier curve defines a surface of revolution – the generative vase.

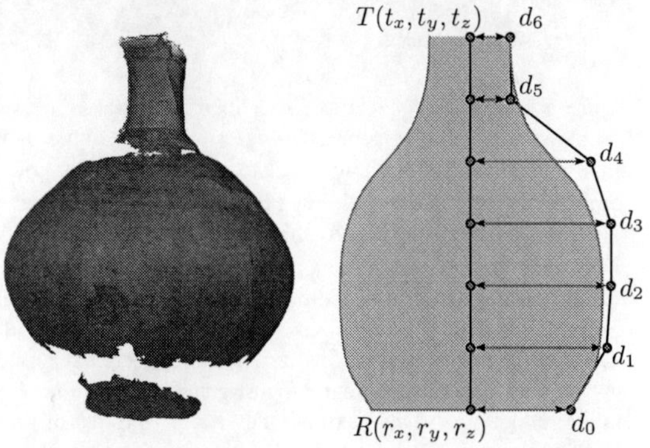

Figure 4: The vase on the left hand side is a digitized artifact of the "Museum Eggenberg" collection. It consists of 364 774 vertices and 727 898 triangles. The example of a procedural shape on the right hand side takes two points R and T in 3D and distance values, which define the control vertices of a Bézier curve.

6. ARCHAEOLOGY & ARCHITECTURE

The huge volume of cultural objects is a challenge even for the most ambitious plans for digitization campaigns (Arnold, 2014b). The fact that probably 90 percent of museum collections are in storage and not accessible to the public is almost demanding for digitization and public accessibility. However, the digitization alone is only part of a larger process that begins at a field excavation and does not end with the presentation in museum exhibitions. Secondary exploitation, database access and sustainable long-time archival of digitized artifacts is also part of the process (Havemann et al., 2006). A very important aspect is the choice of the 3D format used during this process (Niccolucci, 2002), (Niccolucci and D'Andrea, 2006). However, the availability of large quantities of cultural heritage data will enable new methods for analysis and new applications (Arnold, 2014a).

The presented modeling system by CHRISTOPH SCHINKO et al. (2010) is restricted to techniques to meet sustainability conditions. By using JavaScript, the inhibition threshold to use a programming language is reduced resulting in a beginner-friendly tool with a high degree of usability. RENÉ BERNDT et al. (2005) present a system for the production of three-dimensional interactive illustrations in the domain of medieval castles. A special focus is on creating generic modeling tools that increase the usability with a unified 3D user interface.

One of the advantages of procedural modeling techniques is the included expert knowledge within an object description (Ullrich and Fellner, 2011). Classification schemes used in architecture, archaeology and other domains can be mapped to procedures (Ullrich et al., 2008b). When a procedural object description is available, only type and instantiation parameters have to be identified in order to create an object (Ullrich et al., 2013) (see Figure 5).

This contribution has been peer-reviewed.
doi:10.5194/isprsarchives-XL-5-W4-469-2015

Figure 5: Gothic architecture is defined by strict rules with its characteristics. The generative description of Gothic cathedrals encodes these building blocks and the rules on how to combine them. These building blocks have been created by MICHAEL CURRY, http://www.thingiverse.com/thing:2030.

The usage of generative modeling techniques in architecture is not limited to buildings of the past (Müller et al., 2006a), (Müller et al., 2006b). Over the last few decades, architects have used a new class of design tools that support generative design. Generative modeling software extends the design abilities of architects and may even help to reduce costs by harnessing computing power in new ways. Computers, of course, have long been used to capture and implement the design ideas of architects by means of CAD and 3D modeling. Generative design actually helps architects design by using computers to extend human abilities (Hohmann et al., 2009).

In the context of urban modeling, procedural systems can be used to cover different levels of detail (Musialski et al., 2012b). On a coarse scale, the procedural paradigm is applicable to the generation of terrain using methods based on hydrology (Génevaux et al., 2013), as well as the generation of roads (Galin et al., 2010), entire city layouts (Lipp et al., 2011) and urban spaces (Vanegas et al., 2010). Within the scale of a building, layouts can be generated (Merrell et al., 2010), (Bao et al., 2013b) exhibiting different façades (Musialski et al., 2012a), (Bao et al., 2013a). Exterior Lighting can be designed even for buildings with complex constraints (Schwarz and Wonka, 2014). When it comes down to the interior of a building, furniture can be placed following interior design guidelines (Merrell et al., 2011).

7. OPEN RESEARCH QUESTIONS

According to DIETER W. FELLNER (2001, 2005) and SVEN HAVEMANN (2011) several research challenges have to be met: from the classification of shape representations via generic, stable, and detailed 3D markup to 3D query operations (Havemann and Fellner, 2007).

A particularly important problem occurs in the context of internal structure organization and interfaces. Within a composition of modeling functions, where each function is attached via its parameters to topological entities defined in previous states of the model, referenced entities must be named in a persistent way in order to be able to reevaluate the model in a consistent manner. In particular, when a reevaluation leads to topological modifications, references between entities used during the design process are frequently reevaluated in an erroneous way, giving results different from those expected. This problem is known as "persistent naming problem" (Marcheix and Pierra, 2002).

REFERENCES ON "THEORY OF MODELING"

Aurenhammer, F., 2008. Weighted skeletons and fixed-share decomposition. Computational Geometry 40(2), pp. 93 – 101.

Deussen, O. and Lintermann, B., 2005. Digital Design of Nature: Computer Generated Plants and Organics. Springer.

Galin, E., Peytavie, A., Marechal, N. and Guerin, E., 2010. Procedural Generation of Roads. Computer Graphics Forum 29, pp. 429–438.

Havemann, S., 2005. Generative Mesh Modeling. PhD-Thesis, Technische Universität Braunschweig, Germany 1, pp. 1–303.

Heiberg, J. (ed.), 2007. Euclid's Elements of Geometry. Fitzpatrick, Richard.

Krecklau, L., Pavic, D. and Kobbelt, L., 2010. Generalized Use of Non-Terminal Symbols for Procedural Modeling. Computer Graphics Forum 29, pp. 2291–2303.

Krispel, U., Schinko, C. and Ullrich, T., 2014. The Rules Behind – Tutorial on Generative Modeling. Proceedings of Symposium on Geometry Processing / Graduate School 12, pp. 2:1–2:49.

Lipp, M., Wonka, P. and Wimmer, M., 2010. Parallel Generation of Multiple L-Systems. Computers & Graphics 34, pp. 585–593.

Mandelbrot, B. B., 1982. The Fractal Geometry of Nature. W. H. Freeman and Co.

Marvie, J.-E., Buron, C., Gautron, P., Hirtzlin, P. and Sourimant, G., 2012. GPU Shape Grammars. Computer Graphics Forum 31, pp. 2087–2095.

Maybury, M. T. (ed.), 2012. Multimedia Information Extraction. John Wiley & Sons.

Merrell, P. and Manocha, D., 2008. Continuous Model Synthesis. ACM Transactions on Graphics 27, pp. 158:1–9.

Özkar, M. and Kotsopoulos, S., 2008. Introduction to shape grammars. International Conference on Computer Graphics and Interactive Techniques ACM SIGGRAPH 2008 (course notes) 36, pp. 1–175.

Prusinkiewicz, P. and Lindenmayer, A., 1990. The Algorithmic Beauty of Plants. Springer-Verlag.

Schinko, C., Ullrich, T. and Fellner, D. W., 2012. Minimally Invasive Interpreter Construction – How to reuse a compiler to build an interpreter. Proceedings of the International Conference on Computational Logics, Algebras, Programming, Tools, and Benchmarking (Computation Tools) 3, pp. 38–44.

Schinko, C., Ullrich, T., Schiffer, T. and Fellner, D. W., 2011. Variance Analysis and Comparison in Computer-Aided Design. Proceedings of the International Workshop on 3D Virtual Reconstruction and Visualization of Complex Architectures XXXVIII-5/W16, pp. 3B21–25.

Schwarz, M. and Wonka, P., 2014. Procedural Design of Exterior Lighting for Buildings with Complex Constraints. ACM Transactions on Graphics 33, pp. 166:1–166:16.

Sederberg, T. W. and Parry, S. R., 1986. Free-form Deformation of Solid Geometric Models. Proceedings of the Conference on Computer Graphics and Interactive Techniques 13, pp. 151–160.

Settgast, V., 2013. Processing Semantically Enriched Content for Interactive 3D Visualizations. PhD-Thesis, Technische Universität Graz, Austria 1, pp. 1–233.

This contribution has been peer-reviewed.
doi:10.5194/isprsarchives-XL-5-W4-469-2015

Snyder, J. M. and Kajiya, J. T., 1992. Generative modeling: a symbolic system for geometric modeling. Proceedings of 1992 ACM Siggraph 1, pp. 369–378.

Stava, O., Pirk, S., Kratt, J., Chen, B., Měch, R., Deussen, O. and Benes, B., 2014. Inverse Procedural Modelling of Trees. Computer Graphics Forum p. to appear.

Stiny, G. and Gips, J., 1971. Shape Grammars and the Generative Specification of Painting and Sculpture. Best computer papers of 1971 1, pp. 125–135.

Talton, J. O., Lou, Y., Lesser, S., Duke, J., Mech, R. and Koltun, V., 2011. Metropolis Procedural Modeling. ACM Transactions on Graphics 30, pp. 11:1–14.

Thaller, W., Krispel, U., Havemann, S. and Fellner, D., 2012. Implicit Nested Repetition in Dataflow for Procedural Modeling. Proceedings of the International Conference on Computational Logics, Algebras, Programming, Tools, and Benchmarking (Computation Tools) 3, pp. 45–50.

Thaller, W., Krispel, U., Zmugg, R., Havemann, S. and Fellner, D. W., 2013. A Graph-Based Language for Direct Manipulation of Procedural Models. International Journal on Advances in Software 6, pp. 225–236.

Ullrich, T., 2011. Reconstructive Geometry. PhD-Thesis, Technische Universität Graz, Austria 1, pp. 1–322.

Ullrich, T. and Fellner, D. W., 2011. Generative Object Definition and Semantic Recognition. Proceedings of the Eurographics Workshop on 3D Object Retrieval 4, pp. 1–8.

Ullrich, T., Krispel, U. and Fellner, D. W., 2008a. Compilation of Procedural Models. Proceeding of the 13th International Conference on 3D Web Technology 13, pp. 75–81.

Ullrich, T., Schinko, C. and Fellner, D. W., 2010a. Procedural Modeling in Theory and Practice. Poster Proceedings of the 18th WSCG International Conference on Computer Graphics, Visualization and Computer Vision 18, pp. 5–8.

Ullrich, T., Schinko, C., Schiffer, T. and Fellner, D. W., 2013. Procedural Descriptions for Analyzing Digitized Artifacts. Applied Geomatics 5(3), pp. 185–192.

Ullrich, T., Settgast, V. and Berndt, R., 2010b. Semantic Enrichment for 3D Documents: Techniques and Open Problems. Publishing in the Networked World: Transforming the Nature of Communication, Proceedings of the International Conference on Electronic Publishing 14, pp. 374–384.

Ullrich, T., Settgast, V. and Fellner, D. W., 2008b. Semantic Fitting and Reconstruction. Journal on Computing and Cultural Heritage 1(2), pp. 1201–1220.

Ulusoy, I. and Bishop, C. W., 2005. Generative versus Discriminative Methods for Object Recognition. Proceedings of the 2005 IEEE Computer Society Conference on Computer Vision and Pattern Recognition 2, pp. 258 – 265.

Voß, G., Behr, J., Reiners, D. and Roth, M., 2002. A multithread safe foundation for scene graphs and its extension to clusters. Proceedings of the Fourth Eurographics Workshop on Parallel Graphics and Visualization 4, pp. 33–37.

Watson, B. and Wonka, P., 2008. Procedural Methods for Urban Modeling. IEEE Computer Graphics and Applications 28(3), pp. 16–17.

REFERENCES ON "ARCHITECTURE"

Bao, F., Schwarz, M. and Wonka, P., 2013a. Procedural Facade Variations from a Single Layout. ACM Transactions on Graphics 32, pp. 8:1–8:13.

Bao, F., Yan, D.-M., Mitra, N. J. and Wonka, P., 2013b. Generating and Exploring Good Building Layouts. ACM Transactions on Graphics 32, pp. 122:1–122:10.

Berndt, R., Fellner, D. W. and Havemann, S., 2005. Generative 3D Models: a Key to More Information within less Bandwidth at Higher Quality. Proceeding of the 10th International Conference on 3D Web Technology 1, pp. 111–121.

Edelsbrunner, J., Krispel, U., Havemann, S., Sourin, A. and Fellner, D. W., 2014. Constructive Roof Geometry. Proceedings of the International Conference on Cyberworlds 12, pp. 63–70.

Génevaux, J.-D., Galin, E., Guérin, E., Peytavie, A. and Beneš, B., 2013. Terrain Generation Using Procedural Models Based on Hydrology. ACM Transactions on Graphics 32, pp. 143:1–143:13.

Havemann, S. and Fellner, D. W., 2004. Generative Parametric Design of Gothic Window Tracery. Proceedings of the 5th International Symposium on Virtual Reality, Archeology, and Cultural Heritage 1, pp. 193–201.

Hohmann, B., Krispel, U., Havemann, S. and Fellner, D. W., 2009. Cityfit: High-Quality Urban Reconstructions by Fitting Shape Grammars to Images and Derived Textured Point Clouds. Proceedings of the ISPRS International Workshop 3D-ARCH 3, pp. 61–68.

Kelly, T. and Wonka, P., 2011. Interactive Architectural Modeling with Procedural Extrusions. ACM Transactions on Graphics 30, pp. 14:1–15.

Krecklau, L., Born, J. and Kobbelt, L., 2013. View-Dependent Realtime Rendering of Procedural Facades with High Geometric Detail. Comput. Graph. Forum 32(2), pp. 479–488.

Lipp, M., Scherzer, D., Wonka, P. and Wimmer, M., 2011. Interactive Modeling of City Layouts using Layers of Procedural Content. Computer Graphics Forum 30, pp. 345–354.

Lipp, M., Wonka, P. and Wimmer, M., 2008. Interactive Visual Editing of Grammars for Procedural Architecture. ACM Transactions on Graphics 27(3), pp. 1–10.

Merrell, P., Schkufza, E. and Koltun, V., 2010. Computer-generated residential building layouts. ACM Transactions on Graphics 29, pp. 181:1–11.

Merrell, P., Schkufza, E., Li, Z., Agrawala, M. and Koltun, V., 2011. Interactive Furniture Layout Using Interior Design Guidelines. ACM Transactions on Graphics 30, pp. 87:1–10.

Müller, P., Vereenooghe, T., Ulmer, A. and Van Gool, L., 2006a. Automatic Reconstruction of Roman Housing Architecture. Recording, Modeling and Visualization of Cultural Heritage 1, pp. 287–298.

Müller, P., Wonka, P., Haegler, S., Andreas, U. and Van Gool, L., 2006b. Procedural Modeling of Buildings. Proceedings of 2006 ACM Siggraph 25(3), pp. 614–623.

Musialski, P., Wimmer, M. and Wonka, P., 2012a. Interactive Coherence-Based Facade Modeling. Computer Graphics Forum 31, pp. 661–670.

Musialski, P., Wonka, P., Aliaga, D. G., Wimmer, M., van Gool, L. and Purgathofer, W., 2012b. A Survey of Urban Reconstruction. Proceedings of EUROGRAPHICS, State of the Art Report (STAR) 31, pp. 1–28.

This contribution has been peer-reviewed.
doi:10.5194/isprsarchives-XL-5-W4-469-2015

Parish, Y. and Müller, P., 2001. Procedural Modeling of Cities. Proceedings of the 28th annual conference on Computer graphics and interactive techniques 28, pp. 301–308.

Patow, G., 2012. User-Friendly Graph Editing for Procedural Modeling of Buildings. IEEE Computer Graphics and Applications 32, pp. 66–75.

Riemenschneider, H., Krispel, U., Thaller, W., Donoser, M., Havemann, S., Fellner, D. W. and Bischof, H., 2012. Irregular lattices for complex shape grammar facade parsing. Proceedings of the IEEE Conference on Computer Vision and Pattern Recognition (CVPR) 25, pp. 1640–1647.

Steinberger, M., Kenzel, M., Kainz, B., Müller, J., Peter, W. and Schmalstieg, D., 2014a. Parallel Generation of Architecture on the GPU. Computer Graphics Forum 33, pp. 73–82.

Steinberger, M., Kenzel, M., Kainz, B., Wonka, P. and Schmalstieg, D., 2014b. On-the-fly generation and rendering of infinite cities on the GPU. Comput. Graph. Forum 33(2), pp. 105–114.

Thaller, W., Zmugg, R., Krispel, U., Posch, M., Havemann, S. and Fellner Dieter, W., 2013. Creating Procedural Windowbuilding Blocks using the Generative Fact Labeling Method. Proceedings of the ISPRS International Workshop 3D-ARCH 5, pp. 235–242.

Tobler, R. F., Maierhofer, S. and Wilkie, A., 2002a. A Multiresolution Mesh Generation Approach for Procedural Definition of Complex Geometry. Proceedings of the Shape Modeling International 6, pp. 35 – 44.

Tobler, R. F., Maierhofer, S. and Wilkie, A., 2002b. Mesh-Based Parametrized L-Systems and Generalized Subdivision for Generating Complex Geometry. International Journal of Shape Modeling 8, pp. 173–191.

Van Gool, L., Martinovic, A. and Mathias, M., 2013. Towards Semantic City Models. Proceedings of Photogrammetric Week 1, pp. 217–232.

Vanegas, C. A., Aliaga, D. G., Wonka, P., Müller, P., Waddell, P. and Watson, B., 2010. Modelling the Appearance and Behaviour of Urban Spaces. Computer Graphics Forum 29, pp. 25–42.

Vanegas, C. A., Garcia-Dorado, I., Aliaga, D. G., Benes, B. and Waddell, P., 2012. Inverse Design of Urban Procedural Models. ACM Transactions on Graphics 31, pp. 168:1–.

Whiting, E., Ochsendorf, J. and Durand, F., 2009. Procedural Modeling of Structurally-Sound Masonry Buildings. ACM Transactions on Graphics 28, pp. 112:1–9.

Wonka, P., Wimmer, M., Sillion, F. and Ribarsky, W., 2003. Instant Architecture. International Conference on Computer Graphics and Interactive Techniques, ACM SIGGRAPH 2003 22(3), pp. 669 – 677.

Wu, F., Yan, D.-M., Dong, W., Zhang, X. and Wonka, P., 2014. Inverse Procedural Modeling of Facade Layouts. ACM Transactions on Graphics 33, pp. 121:1–121:10.

Yong, L., Mingmin, Z., Yunliang, J. and Haiying, Z., 2012. Improving procedural modeling with semantics in digital architectural heritage. Computers & Graphics 36, pp. 178–184.

Zmugg, R., Thaller, W., Krispel, U., Edelsbrunner, J., Havemann, S. and Fellner, D. W., 2013. Procedural Architecture using Deformation-Aware Split Grammars. The Visual Computer 12, pp. 1–11.

REFERENCES ON "CULTURAL HERITAGE"

Arnold, D., 2014a. Computer Graphics and Cultural Heritage: Continuing Inspiration for Future Tools. Computer Graphics and Applications 34, pp. 70–79.

Arnold, D., 2014b. Computer Graphics and Cultural Heritage: From One-Way Inspiration to Symbiosis. Computer Graphics and Applications 34, pp. 76–86.

Berndt, R., Gerth, B., Havemann, S. and Fellner, D. W., 2005. 3D Modeling for Non-Expert Users with the Castle Construction Kit. Proceedings of the 6th International Symposium on Virtual Reality, Archaeology and Cultural Heritage (VAST) 6, pp. 1–9.

Haegeler, S., Müller, P. and Van Gool, L., 2009. ProceduralModeling for Digital Cultural Heritage. Journal on Image and Video Processing 9, pp. 1–11.

Havemann, S., Settgast, V., Krottmaier, H. and Fellner, D. W., 2006. On the Integration of 3D Models into Digital Cultural Heritage Libraries. Proceedings of the 7th International Symposium on Virtual Reality, Archaeology and Cultural Heritage (VAST) 1, pp. 161–169.

Müller, P., Vereenooghe, T., Wonka, P., Paap, I. and Van Gool, L., 2006. Procedural 3D Reconstruction of Puuc Buildings in Xkipche. Proceedings of Eurographics Symposium on Virtual Reality, Archaeology and Cultural Heritage (VAST) 1, pp. 139–146.

Schinko, C., Strobl, M., Ullrich, T. and Fellner, D. W., 2010. Modeling Procedural Knowledge – a generative modeler for cultural heritage. Proceedings of EUROMED 2010 - Lecture Notes on Computer Science 6436, pp. 153–165.

Settgast, V., Ullrich, T. and Fellner, D. W., 2007. Information Technology for Cultural Heritage. IEEE Potentials 26(4), pp. 38–43.

REFERENCES ON "CAD / ENGINEERING"

Berndt, R., Schinko, C., Krispel, U., Settgast, V., Havemann, S., Eggeling, E. and Fellner, D. W., 2012. Ring's Anatomy – Parametric Design of Wedding Rings. Proceedings International Conference on Creative Content Technologies 4, pp. 72–78.

Bokeloh, M., Wand, M. and Seidel, H.-P., 2010. A Connection between Partial Symmetry and Inverse Procedural Modeling. Proceedings of ACM SIGGRAPH 2010 29, pp. 104:1–104:10.

Boulch, A., Houllier, S., Marlet, R. and Tournaire, O., 2013. Semantizing Complex 3D Scenes using Constrained Attribute Grammars. Proceedings of Eurographics Symposium on Geometry Processing 32, pp. 33–42.

Frank, G. and Hillbrand, C., 2012. Automatic support of standardization processes in design models. Proceedings of the International Conference on Intelligent Engineering Systems (INES) 16, pp. 393–398.

Krecklau, L. and Kobbelt, L., 2011. Procedural Modeling of Interconnected Structures. Computer Graphics Forum 30, pp. 335–344.

Mendez, E., Schall, G., Havemann, S., Fellner, D. W., Schmalstieg, D. and Junghanns, S., 2008. Generating Semantic 3D Models of Underground Infrastructure. IEEE Computer Graphics and Applications 28, pp. 48–57.

This contribution has been peer-reviewed.
doi:10.5194/isprsarchives-XL-5-W4-469-2015

Ramamoorthi, R. and Arvo, J., 1999. Creating Generative Models from Range Images. Proceedings of ACM Siggraph 1, pp. 195–204.

Schinko, C., Berndt, R., Eggeling, E. and Fellner, D., 2014. A Scalable Rendering Framework for Generative 3D Content. Proceedings of the International ACM Conference on 3D Web Technologies 19, pp. 81–87.

Yu, L.-F., Yeung, S.-K., Tang, C.-K., Terzopoulos, D., Chan, T. F. and Osher, S., 2011. Make it Home: Automatic Optimization of Furniture Arrangement. ACM Transactions on Graphics 30, pp. 86:1–11.

FURTHER REFERENCES

Arnold, D., 2006. Procedural methods for 3D reconstruction. Recording, Modeling and Visualization of Cultural Heritage 1, pp. 355–359.

Autodesk, 2007. Autodesk Maya API. White Paper 1, pp. 1–30.

Behr, J., Dähne, P., Jung, Y. and Webel, S., 2007. Beyond the Web Browser – X3D and Immersive VR. IEEE Virtual Reality Tutorial and Workshop Proceedings 28, pp. 5–9.

Bishop, C. M., 2007. Pattern Recognition and Machine Learning. Springer.

Breuel, F., Bernd, R., Ullrich, T., Eggeling, E. and Fellner, D. W., 2011. Mate in 3D – Publishing Interactive Content in PDF3D. Publishing in the Networked World: Transforming the Nature of Communication, Proceedings of the International Conference on Electronic Publishing 15, pp. 110–119.

Brutzman, D., 1998. The virtual reality modeling language and Java. Communications of the ACM 41(6), pp. 57 – 64.

Bustos, B., Keim, D., Saupe, D. and Schreck, T., 2007. Content-based 3D Object Retrieval. IEEE Computer Graphics and Applications 27(4), pp. 22–27.

Chakrabarti, A., Shea, K., Stone, R., Cagan, J., Campbell, M., Vargas-Hernandez, N. and Wood, K. L., 2011. Computer-Based Design Synthesis Research: An Overview. Journal of Computing and Information Science in Engineering 11, pp. 021003:1–10.

Chomsky, N., 1956. Three models for the description of language. IRE Transactions on Information Theory 2, pp. 113–124.

Compton, K. and Mateas, M., 2006. Procedural Level Design for Platform Games. Proceedings of the Artificial Intelligence and Interactive Digital Entertainment Conference 2, pp. 109–111.

Di Benedetto, M., Ponchio, F., Ganovelli, F. and Scopigno, R., 2010. SpiderGL: a JavaScript 3D graphics library for next-generation WWW. Proceedings of the 15th International Conference on Web 3D Technology 15, pp. 165–174.

Eckel, B., 2003. Thinking in C++: Introduction to Standard C++, Practical Programming. Prentice Hall.

Fellner, D. W., 2001. Graphics Content in Digital Libraries: Old Problems, Recent Solutions, Future Demands. Journal of Universal Computer Science 7, pp. 400–409.

Fellner, D. W. and Havemann, S., 2005. Striving for an adequate vocabulary: Next generation metadata. Proceedings of the 29th Annual Conference of the German Classification Society 29, pp. 13 – 20.

Fellner, D. W., Saupe, D. and Krottmaier, H., 2007. 3D Documents. IEEE Computer Graphics and Applications 27(4), pp. 20–21.

Havemann, S. and Fellner, D. W., 2007. Seven Research Challenges of Generalized 3d Documents. IEEE Computer Graphics and Applications 3, pp. 70–76.

Havemann, S., Ullrich, T. and Fellner, D. W., 2012. The Meaning of Shape and some Techniques to Extract It. Multimedia Information Extraction 1, pp. 81–98.

Hilaga, M., Shinagawa, Y., Kohmura, T. and Kunii, T. L., 2001. Topology Matching for Fully Automatic Similarity Estimation of 3D Shapes. Proceedings of the 28th annual conference on Computer graphics and interactive techniques 28, pp. 203–212.

Initiative, D. C. M., 1995. Dublin Core Metadata Initiative. http://dublincore.org/.

King, B. D. and Wertheimer, M., 2005. Max Wertheimer & Gestalt Theory. Transaction Publishers. ISBN 0-7658-0258-9.

Kuang, Z., Chan, B., Yu, Y. and Wang, W., 2013. A Compact Random-access Representation for Urban Modeling and Rendering. ACM Trans. Graph. 32(6), pp. 172:1–172:12.

Marcheix, D. and Pierra, G., 2002. A Survey of the Persistent Naming Problem. Proceedings of the ACM Symposium on Solid Modeling and Applications 7, pp. 13–22.

Martin, G. E., 1998. Geometric Constructions. Springer.

Mitchell, W. J., 1990. The Logic of Architecture: Design, Computation, and Cognition. MIT Press.

Mitra, N. J., Guibas, L. J. and Pauly, M., 2006. Partial and approximate symmetry detection for 3D geometry. ACM Transactions on Graphics 25, pp. 560 – 568.

Mitra, N. J., Guibas, L. J. and Pauly, M., 2007. Symmetrization. International Conference on Computer Graphics and Interactive Techniques 26, pp. 1–8.

Niccolucci, F., 2002. XML and the future of humanities computing. SPECIAL ISSUE: First European workshop on XML and knowledge management 10, pp. 43–47.

Niccolucci, F. and D'Andrea, A., 2006. An Ontology for 3D Cultural Objects. Proceedings of the 7th International Symposium on Virtual Reality, Archaeology and Cultural Heritage (VAST) 7, pp. 203–210.

NVidia, n.d. NVIDIA CUDA C Programming Guide.

OpenGL Architecture, R. B., 1993. OpenGL Reference Manual. Addison-Wesley Publishing Company.

Ousterhout, J. K., 1998. Scripting: Higher Level Programming for the 21st Century. IEEE Computer Magazine 31(3), pp. 23–30.

Parr, T., 2010. Language Implementation Patterns: Create Your Own Domain-Specific and General Programming Languages. Pragmatic Bookshelf.

Pauly, M., Mitra, N. J., Wallner, J., Pottmann, H. and Guibas, L. J., 2008. Discovering structural regularity in 3D geometry. ACM Transactions on Graphics 27, pp. #43, 1–11.

Reas, C., Fry, B. and Maeda, J., 2007. Processing: A Programming Handbook for Visual Designers and Artists. The MIT Press.

Reiners, D., Voss, G. and Behr, J., 2002. OpenSG: Basic concepts. Proceedings of OpenSG Symposium 2002 1, pp. 1–7.

Schinko, C., Strobl, M., Ullrich, T. and Fellner, D. W., 2011. Scripting Technology for Generative Modeling. International Journal On Advances in Software 4, pp. 308–326.

This contribution has been peer-reviewed.
doi:10.5194/isprsarchives-XL-5-W4-469-2015

AUTHOR INDEX

AUTHOR INDEX